Applied Analytical Mathematics for Physical Scientists

Applied Analytical Mathematics for Physical Scientists

JAMES T. CUSHING

Associate Professor of Physics
University of Notre Dame

John Wiley & Sons, Inc.
New York / London / Sydney / Toronto

Library of Congress Cataloging in Publication Data:

Cushing, James T. 1937–
Applied analytical mathematics for physical scientists.

Bibliography: p.
1. Mathematical analysis. I. Title.

QA300.C87 515′.02′45 75–9611
ISBN 0–471–18997–9

Printed in the United States of America

10 9 8 7 6 5 4 3 2 1

For Rose, Chris, and Pat

Preface

The selection of material and the form of presentation for this book grew out of several courses in applied mathematics that I have taught to graduate and undergraduate students of physics, chemistry, and engineering over the past 10 years. Two main problems face an instructor of applied mathematics courses that are intended for advanced students in the physical sciences. Since there are more topics of importance than can be covered in a typical one-year course, a choice of subjects must be made, based on the needs of the students and the prejudices of the instructor. Second, the style of presentation of the topics can range from the cookbook method of developing many results and tricks by studying physical applications to the completely general and rigorous abstract approach.

It is for these two reasons that this textbook on applied mathematics now competes for the attention of physical science students. I have been unable to find a single textbook that covered all of the necessary topics and treated them with sufficient rigor and generality while still remaining intelligible to most students in the class. This is a value judgment that cannot be defended in absolute terms. The fact that many other texts exist in this general area shows that several competent authors have reached a conclusion very different from mine. This in itself is not criticism of their work nor, I hope, of mine. Readers will choose the approach that best fits their needs.

The purpose of this text is to cover the topics that are necessary to give the student a sufficiently broad background for his advanced studies in the physical sciences and to present this material with enough rigor and generality to provide him with a unified view and a solid understanding of this material. The philosophy underlying the development of this text is that it is as important for a student to know *when* a given theorem or result applies and to be aware of the subtlety of certain questions as it for him to be familiar with the mechanics of the applications of these principles. One of the most important goals to be achieved in studying mathematics is the development of a certain method of thinking and a style of approach to a problem. Nevertheless, technical skills must be developed and there are many problems included to accomplish this, as well as to extend the material of the text.

I assume that students have a calculus background through the differentiation and integration of functions of several real variables. On this assumption I have usually been able to cover Chapters 1 through 6 the first semester and Chapters

7 through 9 the second. It is generally true that physical science students have strengths in certain areas of mathematics and weaknesses in others as a result of the use they have made of mathematics in their physical sciences courses. The text attempts to build on these strengths and not to cover old material unnecessarily. It does not abound with "typical" hackneyed examples from physics and engineering, since students taking this course invariably have had or are taking courses in mechanics, electricity and magnetism, quantum mechanics, or other such courses that apply many of the techniques developed here. No attempt is made to provide direct motivation for each topic covered by illustrating these techniques with specific applications. The student should have enough sophistication to study the mathematics itself. It has been my experience that students do not enjoy or greatly profit from going over the same applications in several different courses. Such specific problems receive much more detailed consideration in other advanced courses. The most general and abstract discussions and proofs are not always given but, instead, those that the student's background will allow him to understand well. Nevertheless, the assumptions made in proving important theorems are kept as weak as is consistent with this goal. Many details of proofs and some advanced topics have been relegated to chapter appendices to facilitate the development of central ideas in each chapter. Useful results are listed in tables at the end of various chapters for ease of reference.

Since most physical science students for whom this text is designed have a working familiarity with, or little difficulty in grasping, three-dimensional vector analysis and even the elements of tensor analysis, no time is spent on these topics, although frequent reference is made to simple analogies from these fields. No knowledge of the Lebesgue theory of integration is assumed, even though it is only with this that concepts such as completeness and the theory of linear operators on a Hilbert space can be fully appreciated. At several points in the first chapter, and later in the text, the symbol $\mathscr{L}_2$ (standing for Lebesgue square-integrable) appears. The reader can usually grasp the essence of the point being made by thinking in terms of the more familiar Riemann definition of an integral. A brief discussion of measure and of the Lebesgue definition of an integral is given in Appendix II. Symbolic notation has been used often in the text since it sometimes makes an argument sufficiently compact that is more easily understood. Furthermore, the student will find reference to more advanced works easier once he has become accustomed to it.

Clearly the most important unifying concepts in the text are those of a linear vector space and of the theory of linear operators on such spaces. Chapter 2 sets the style for most of the further considerations of the book. The debt owed for this central chapter, and for much of the text, goes to Bernard Friedman's *Principles and Techniques of Applied Mathematics*. In fact it is my belief that, had the late Professor Friedman chosen to write another book covering a few more topics and with more background material, but with the same beautiful style of clarity and logic as his *Principles and Techniques*, then many of the applied mathematics texts published in the last 15 years would have been unnecessary. Chapter 2 also lays the foundation for the work on integral equations in the discussion of completely continuous linear operators. Chapter 3 develops many important properties of linear operators, mainly in finite-dimensional Hilbert spaces.

Chapter 4 deals with the very important topic of complete sets of functions in terms of the Weierstrass approximation theorem and the basic theorem on Fourier series. Other complete sets of functions, such as Legendre polynomials, are also introduced. Separation of variables and the subsequent applications to boundary value problems are not given any great space either here or later in the text. Most students already know how to apply the method but have little understanding of the concept of completeness, which is discussed extensively here. Rigorous treatments of Fourier series and of Fourier integrals are given in as general a form as is possible without using results from Lebesgue integration theory.

The principal result on the existence and uniqueness of the solution to Volterra integral equations, established early in Chapter 5, serves as a basis for a later discussion of ordinary differential equations. The main business of Chapter 5 is a development of Fredholm theory culminating in a proof of the Hilbert-Schmidt theorem. The basic approach taken for several topics in this chapter is that of the limit as $n \to \infty$ of a set of linear algebraic equations in a representation-dependent framework. Although this is somewhat cumbersome in detail for a few of the proofs, it is very easily understood conceptually. The more elegant and powerful technique of defining completely continuous operators in terms of sequences and convergent subsequences of vectors is also introduced in the text and developed in an appendix.

Chapter 6 is a brief introduction to the calculus of variations and follows the basic approach of the classic small volume by Gilbert A. Bliss. There is also a discussion of Noether's theorem that plays such an important role in modern formulations of mechanics and field theories.

The long Chapter 7 on complex variables stresses methods of analytic continuation, especially as applied to the classical functions of mathematics. Very little is done with conformal mapping, since this is a rather specialized tool presenting no conceptual difficulty but best learned in some specific area of application. Complex variables is a very useful subject for physical scientists, who seem to have relatively little difficulty with it, unlike discussions of sets, completeness, and Hilbert spaces, for example. Perhaps this is easily understood since Riemann formulated many of his great theorems in complex variable theory by considering the idealized flow of electric charges along thin conducting surfaces. Illustrations prove very useful here and are used abundantly.

Chapter 8, on linear differential equations and Green's functions, begins with a brief introduction to distributions or ideal functions. A fairly complete and rigorous treatment is then given of Green's functions for second-order linear differential equations in a single variable. A discussion of the important topic of the continuous spectrum for differential operators in an infinite domain is given. Finally Green's functions are developed for some common differential operators in three dimensions, but with considerably less rigor than in the one-dimensional case.

Chapter 9 on group theory has been included because this is a subject of considerable importance to physical scientists today, and it is one that they are very often unfamiliar with. After a general introduction most of the emphasis is placed on continuuus Lie groups and in particular on the three-dimensional rotation group and its representations. Eugene P. Wigner's classic text, first published in 1931, serves as the basic reference, as well as Giulio Racah's famous lectures on continuous groups. The latter half of this chapter is biased in favor of those students of chemistry or physics who will study quantum mechanics.

Appendix I is a convenient reference for some useful results from that area of real analysis often referred to as advanced calculus. Appendix II contains an extremely sketchy discussion of the concept of measure and of the Lebesgue definition of integration, as well as a collection of several important theorems from functional analysis for the reader who is interested in seeing how some of the results given in the text, especially those of Chapters 4 and 5, can be generalized. However these advanced theorems are never used in the text itself.

Naturally, very little new will be found in a text of this type covering these topics in applied mathematics for the beginning graduate or advanced undergraduate student. Each subject covered is treated more extensively in works devoted to that subject alone. This book provides the student with a single, unified, understandable presentation of the material.

Unfortunately, when a text is developed from so many different sources over a period of several years, it is not always possible to recall exactly the sources used and to quote them. I have tried to indicate my major references at the end of each chapter and in the bibliography. Nearly four hundred years ago Sir Francis Bacon in his *Nuovum Organum* charged: "... let a man look carefully into all that variety of books with which the arts and sciences abound, he will find everywhere endless repetitions of the same thing, varying in the method of treatment, but not new in substance ...." Indeed it is often so in the mathematical literature that the original proofs and discussions given by the great classical mathematicians are so clear and beautiful and that subsequent works have essentially copied them. This is not a criticism of lesser authors since it would serve no purpose to concoct an original proof at the expense of preciseness or clarity. The present text is not free from such charges of nonoriginality in several key discussions.

I have tried to keep my personal prejudices as a theoretical physicist from coming through too blatantly in the selection and presentation of subject matter, although it is clear that I have yielded to temptation on several occasions, especially in the last sections of Chapter 9.

Finally I thank Brian Cheng-jean Chen, Gerald L. Jones, and William D. McGlinn for discussions of various points during the writing of this text and my students, who have endured the preliminary versions of the notes used for this course, for their questions and suggestions. I also thank Sharon Duram and Eleanor Klingbeil, who typed the bulk of the lengthy first draft, and Jo Robertson, who typed and retyped the extensive revisions. Naturally, those errors that remain are mine, and I welcome any comments on, corrections to, or criticisms of this text.

JUNE, 1974 JAMES T. CUSHING

Contents

3 Spectral Analysis of Linear Operators 101

8 Second-Order Linear Ordinary Differential Equations and Green's Functions 431

1 Linear Vector Spaces

1.0 INTRODUCTION

As we will repeatedly stress throughout the text, many of the concepts we will use are simply generalizations of those already encountered in ordinary three-dimensional vector analysis. Consider a three-vector $\mathbf{x}=(x_1, x_2, x_3)$, where x_1, x_2, x_3 are the components of $\mathbf{x}$ along three mutually orthogonal Cartesian axes. We say that $\mathbf{x}$ is a vector in E_3 (i.e., three-dimensional Euclidean space). Suppose we wish to find an $\mathbf{x}$ such that

$$\begin{aligned} a_{11}x_1+a_{12}x_2+a_{13}x_3&=b_1\\ a_{21}x_1+a_{22}x_2+a_{23}x_3&=b_2\\ a_{31}x_1+a_{32}x_2+a_{33}x_3&=b_3 \end{aligned} \tag{1.1}$$

Such systems of inhomogeneous linear equations are studied in elementary algebra.

Let us state some results for these systems by way of review. The reader who is unfamiliar with the following results will find proofs outlined in Problem 1.1 at the end of this chapter. If we define a quantity Δ as

$$\Delta\equiv\begin{vmatrix} a_{11} & a_{12} & a_{13}\\ a_{21} & a_{22} & a_{23}\\ a_{31} & a_{32} & a_{33} \end{vmatrix}$$

$$\equiv a_{11}(a_{22}a_{33}-a_{32}a_{23})-a_{12}(a_{21}a_{33}-a_{31}a_{23})+a_{13}(a_{21}a_{32}-a_{31}a_{22}) \tag{1.2}$$

then the set of homogeneous equations

$$
\begin{aligned}
a_{11}x_1+a_{12}x_2+a_{13}x_3&=0\\
a_{21}x_1+a_{22}x_2+a_{23}x_3&=0\\
a_{31}x_1+a_{32}x_2+a_{33}x_3&=0
\end{aligned}
\tag{1.3}
$$

has only the trivial solution $\mathbf{x}=0$ if and only if $\Delta\neq 0$; that is,

$$\Delta\neq 0 \Leftrightarrow \exists \text{ only trivial solution (i.e., } \mathbf{x}=0\text{) to Eq. 1.3} \tag{1.4}$$

(See p. 640 for a list of the symbols used in this text.) Also $\Delta=0$ is the necessary and sufficient condition that, in addition to the trivial solution that *always* satisfies Eq. 1.3, there also exist nontrivial solutions $\mathbf{x}\neq 0$ to Eq. 1.3; that is,

$$\Delta=0 \Leftrightarrow \exists \mathbf{x}\neq 0 \text{ solution to Eq. 1.3} \tag{1.5}$$

Equation 1.4 guarantees the existence and uniqueness of the solution to Eq. 1.3 when $\Delta\neq 0$.

If we return to the original inhomogeneous set, Eq. 1.1, and in analogy with Eq. 1.2 define the quantities

$$
\Delta_1=\begin{vmatrix} b_1 & a_{12} & a_{13}\\ b_2 & a_{22} & a_{23}\\ b_3 & a_{32} & a_{33}\end{vmatrix}
\qquad
\Delta_2=\begin{vmatrix} a_{11} & b_1 & a_{13}\\ a_{21} & b_2 & a_{23}\\ a_{31} & b_3 & a_{33}\end{vmatrix}
$$

$$
\Delta_3=\begin{vmatrix} a_{11} & a_{12} & b_1\\ a_{21} & a_{22} & b_2\\ a_{31} & a_{32} & b_3\end{vmatrix}
\tag{1.6}
$$

then by direct algebraic manipulation we can deduce that Eq. 1.1 is completely equivalent to the set of equations

$$a_{ij}\,\Delta x_k=a_{ij}\,\Delta_k \qquad i,j=1,2,3 \qquad j\neq k \tag{1.7}$$

That is, Eq. 1.1 implies Eq. 1.7 and vice versa. Since we may assume that *at least* one of the a_{ij} in each column is not zero (for otherwise we would have at most two simultaneous linear algebraic equations and that case is trivially dispatched), we conclude that

$$\Delta x_j=\Delta_j \qquad j=1,2,3 \tag{1.8}$$

Therefore if $\Delta\neq 0$ a solution to Eq. 1.1 is

$$x_j=\frac{\Delta_j}{\Delta} \qquad j=1,2,3 \tag{1.9}$$

which is simply *Cramer's rule*. Furthermore, this solution is unique since the corresponding set of homogeneous equations has only the trivial solution $\mathbf{x}=0$. Clearly, if $\mathbf{b}=(b_1, b_2, b_3)\neq 0$, then this unique solution, Eq. 1.9, to the inhomogeneous equations is not zero since *all* Δ_j's can vanish if and only if $\mathbf{b}=0$, once $\Delta \neq 0$. This is easily seen since the conditions $\Delta_j = 0$, for *all* $j = 1, 2, 3$ simultaneously, can be regarded as a set of homogeneous equations for the quantities b_1, b_2, and b_3. If $\Delta \neq 0$, then necessarily $b_1 = b_2 = b_3 = 0$. Also if the solution to the set of inhomogeneous equations, Eq. 1.1, is unique, then $\Delta \neq 0$ so that

$$\Delta \neq 0 \Leftrightarrow \exists \text{ unique solution to Eq. 1.1} \tag{1.10}$$

If $\Delta = 0$ for the set of inhomogeneous equations, Eq. 1.1, then there can be a solution to the inhomogeneous problem only if

$$\Delta_j = 0 \qquad \forall j = 1, 2, 3$$

This follows from Eq. 1.8. This solution is not unique.

It is important to note that even if $\Delta_j = 0$, $j = 1, 2, 3$, when $\Delta = 0$ and $\mathbf{b} \neq 0$, there is no guarantee that the inhomogeneous set has a solution. A simple counter-example will suffice to prove this point.

$$x_1 + x_2 + x_3 = 1$$

$$x_1 + x_2 + x_3 = 2$$

$$x_1 + x_2 + x_3 = 3$$

Direct calculation shows that

$$\Delta = 0 = \Delta_1 = \Delta_2 = \Delta_3$$

However this set of equations is inconsistent and therefore possesses no solution.

Let us now take stock of our results. In order to simplify the notation, we may define a *matrix* A as

$$\mathrm{A} = \begin{pmatrix} a_{11} & a_{12} & a_{13} \\ a_{21} & a_{22} & a_{23} \\ a_{31} & a_{32} & a_{33} \end{pmatrix}$$

and write the vectors $\mathbf{x}$ and $\mathbf{b}$ as

$$\mathbf{x} = \begin{pmatrix} x_1 \\ x_2 \\ x_3 \end{pmatrix} \qquad \mathbf{b} = \begin{pmatrix} b_1 \\ b_2 \\ b_3 \end{pmatrix}$$

The set of simultaneous linear equations, Eq. 1.1, can then be written as

$$\mathrm{A}\mathbf{x} = \mathbf{b} \tag{1.11}$$

with an obvious definition of matrix multiplication (i.e., Eq. 1.11 and Eq. 1.1 are to be identical). (We will discuss matrices in more detail in Section 2.1.)

For this linear problem, the question of the uniqueness of the solution to the inhomogeneous problem,

$$\mathrm{A}\mathbf{x}_1 = \mathbf{b}$$

$$\mathrm{A}\mathbf{x}_2 = \mathbf{b}$$

reduces to the existence of nontrivial solutions to the homogeneous problem

$$A(\mathbf{x}_1-\mathbf{x}_2)=0$$

If the only solution to the homogeneous problem is $\mathbf{x}=0$, then $\mathbf{x}_1=\mathbf{x}_2$ and the solution to the inhomogeneous problem is unique.

We summarize these results in the following table.

TABLE 1.1

	$A\mathbf{x}=\mathbf{b}$	$A\mathbf{x}=0$ (always trivial solution $\mathbf{x}=0$)
$\Delta=0$	solution (nonunique) *only* if $\Delta_j=0 \quad j=1, 2, 3$ (i.e., necessary but not sufficient condition for existence of solutions)	$\Leftrightarrow \exists\mathbf{x}\neq 0$ (as well as $\mathbf{x}=0$)
$\Delta\neq 0$	$\Leftrightarrow \exists\mathbf{x}$ (unique) $\{x_j=\Delta_j/\Delta$—Cramer's rule$\}$	$\Leftrightarrow \exists\mathbf{x}=0$ *only*

To illustrate these statements let us consider the following simple example of a linear set.

Example 1.1

$$-x_1+x_2+\sqrt{2}\,x_3=-2$$
$$x_1-x_2-\sqrt{2}\,x_3=2$$
$$\sqrt{2}\,x_1-\sqrt{2}\,x_2-2x_3=2\sqrt{2}$$

$$A=\begin{pmatrix}-1 & 1 & \sqrt{2}\\ 1 & -1 & -\sqrt{2}\\ \sqrt{2} & -\sqrt{2} & -2\end{pmatrix} \qquad \mathbf{b}=\begin{pmatrix}-2\\ 2\\ 2\sqrt{2}\end{pmatrix}$$

Direct calculation shows that

$$\Delta=0$$

Therefore from the discussion above we *know* that this set cannot have a unique solution, but may still have (many) solutions, as well as (nontrivial) solutions to the homogeneous set

$$A\mathbf{x}=0$$

In fact, solutions of

$$A\mathbf{y}=\mathbf{b}$$

are, for example,

$$\mathbf{y}_1 = \begin{pmatrix} 1/2 \\ -1/2 \\ -\sqrt{2}/2 \end{pmatrix} \qquad \mathbf{y}_2 = \begin{pmatrix} -1 \\ 1 \\ -2\sqrt{2} \end{pmatrix}$$

Hence nonunique solutions do exist. Furthermore solutions of

$$A\mathbf{x} = 0$$

are, for example,

$$\mathbf{x}_1 = \begin{pmatrix} 1 \\ 1 \\ 0 \end{pmatrix} \qquad \mathbf{x}_2 = \begin{pmatrix} 1 \\ -1 \\ \sqrt{2} \end{pmatrix}$$

We also see that

$$\mathbf{x}_1 \perp \mathbf{x}_2 \qquad \mathbf{x}_1 \perp \mathbf{b} \qquad \mathbf{x}_2 \perp \mathbf{b}$$

Geometrically we see that the vector $\mathbf{b}$ is orthogonal to the *plane* containing those vectors that are solutions to the homogeneous problem. In fact, if we use $\mathbf{x}_1$, $\mathbf{x}_2$, and $\mathbf{b}$ to set up a new coordinate system, we have the following picture in

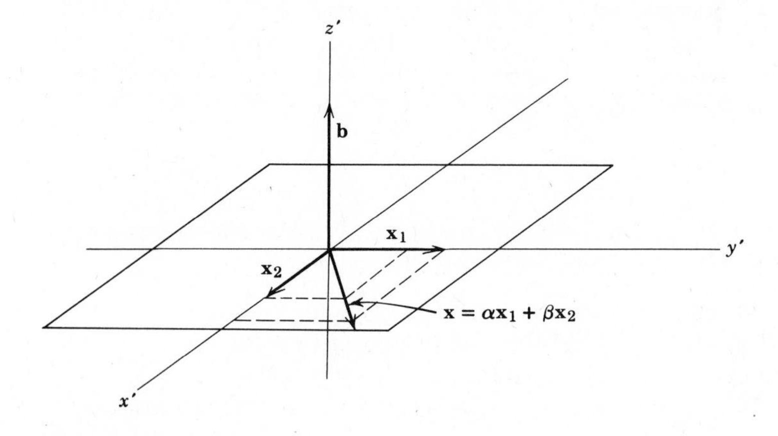

Figure 1.1

E_3. (Here $\mathbf{x}_1 \times \mathbf{x}_2 = -1/\sqrt{2}\,\mathbf{b}$.) Actually we could have seen this immediately since the three homogeneous equations are identical and all reduce to

$$\sqrt{2}\, x_3 = x_1 - x_2$$

the equation of a plane in E_3. In our new set of coordinates this corresponds to

$$z' = 0$$

We have chosen to discuss this simple problem of a set of three linear algebraic equations since it is quite rich in interesting questions (i.e., existence of solutions, uniqueness, etc.) and in a geometrical interpretation that gives us insight into more general problems. The case of two linear equations is too trivial to be of much interest, while the algebra of four or more equations becomes too cumbersome for the direct pedestrian approach taken above, the purpose of which was to give an elementary discussion of these results for those not already familiar with them. The length of the discussion above should point up the fact that the most obvious and direct approach can sometimes become extremely involved. We will develop elegant and powerful techniques that will allow us to handle the general case of n linear algebraic equations, as well as those of differential and integral equations. However as we progress we will see that many of the features of this simple problem are quite general.

1.1 DEFINITION OF A LINEAR VECTOR SPACE

The previous example considered three vectors, which exist independent of any coordinate system. The concept of a linear vector space is a generalization of an abstract vector, independent of any reference frame.

We begin by defining the *field of scalars.* These quantities obey the same axioms as do the scalars. The definition of this field becomes much more compact if we first define a *group.*

a. Groups

Consider a set of elements $G=\{g_1, g_2, g_3, \ldots, g_n, \ldots\}$ and an operation, $\otimes$. Then these elements form a group provided ($\forall g_l \in G$):

i. $\{g_i \in G, g_j \in G\} \Rightarrow \exists g_k = g_i \otimes g_j \in G, \forall g_l \in G$; *closure*
ii. $g_i \otimes (g_j \otimes g_k) = (g_i \otimes g_j) \otimes g_k$; *associative* operation w.r.t. $\otimes$
iii. $\exists e \in G \ni g_i \otimes e = g_i = e \otimes g_i, \forall g_i \in G$; *identity* element
iv. $g_i \in G \Rightarrow \exists g_i^{-1} \in G \ni g_i^{-1} \otimes g_i = e = g_i \otimes g_i^{-1}$; *inverse* for every element $g_i \in G$

The identity of the left and right inverses can be proven from the preceding axioms and need not be assumed. If, furthermore, the operation is *commutative,*

v. $g_i \otimes g_j = g_j \otimes g_i \qquad \forall g_i, g_j \in G$

then the group is a commutative or *abelian* group.

As some simple examples of groups, we list the following.

1. *Integers*
   a. Addition (positive and negative integers including zero).
   b. Multiplication (zero excluded)—*NO,* no inverse; however rational numbers—*YES.*

2. *Rotations*
   a. In the plane (commutative group).
   b. In three dimensions (a noncommutative group).

In Chapter 9 we will return to a more complete study of groups.

b. Fields

We now define a *field* as a collection of elements $F=\{f_1, f_2, f_3, \ldots\}$ having two operations, "+" and "·," defined for these elements and satisfying the following properties.

i. With respect to the operation + the elements F form an abelian group with the identity element denoted by 0.
ii. With respect to the operation · the elements of F *with the exclusion of 0* form an abelian group with the identity element denoted by e.
iii. If f_i, f_j, $f_k \in F$, then $f_i \cdot (f_j+f_k)=f_i \cdot f_j+f_i \cdot f_k$; that is the operation · is *distributive* with respect to the operation +.

Common examples of fields are the set of all rational numbers, the set of all real numbers, and the set of all complex numbers with the usual definitions of addition for the operation + and of multiplication for the operation · above. As a matter of fact, in this text we use either the real scalar field or the complex scalar field. Therefore the reader may simply take the expression *scalar field* henceforth to mean the set of all real (or complex) numbers with the ordinary definitions of addition and multiplication.

c. Linear Vector Spaces*

We now define a *linear vector space* as follows. The operation, "+," on a set of elements $S=\{x, y, z, \ldots\}$ is to have the following properties:

i. $\{x \in S, y \in S\} \Rightarrow z \equiv x+y \in S \qquad \forall x, y \in S$
ii. The operation + is commutative and associative.

$$x+y=y+x$$

$$(x+y)+z=x+(y+z)$$

iii. $\exists$ a unique element $0 \ni x+0=x \qquad \forall x \in S$
iv. $\forall x \in S \, \exists -x \ni x+(-x)=0$

* The order of topics presented in Sections 1.1c, 1.2, 1.4, 1.7, and 1.8 of the present chapter follows closely the order of the corresponding sections of Chapter 1 of Bernard Friedman's excellent *Principles and Techniques of Applied Mathematics* (cf. references at the end of this chapter and the bibliography at the end of the text). Here and elsewhere we have attempted to supply sufficient background material, illustrative examples, and discussion to make our presentation of these topics more readily accessible to a wider audience of physical science students than Professor Friedman's may have been.

That is, the elements of S form an abelian group with respect to addition. Furthermore, if the quantities α, β, $\gamma, \ldots$ are the elements of some field F, that we simply call *scalars*, then we require that multiplication of an element of S by a scalar satisfy $(\forall x, y \in S, \forall \alpha, \beta \in F)$:

v. $\alpha(\beta x) = (\alpha\beta)x$
vi. $(\alpha + \beta)x = \alpha x + \beta x$
vii. $\alpha(x + y) = \alpha x + \alpha y$
viii. $e \cdot x = x$

No claim is made that all of these axioms are logically independent.

Henceforth we call any space S that is closed under the operations of addition and of multiplication by scalars a *linear vector space*. Its elements will be termed *vectors*.

The following are examples of linear vector spaces.

1. Vectors in an n-dimensional Euclidean space, E_n.

$$x = (\xi_1, \xi_2, \xi_3, \ldots, \xi_n)$$
$$y = (\eta_1, \eta_2, \eta_3, \ldots, \eta_n)$$
$$x + y = (\xi_1 + \eta_1, \xi_2 + \eta_2, \xi_3 + \eta_3, \ldots, \xi_n + \eta_n)$$
$$\alpha x = (\alpha\xi_1, \alpha\xi_2, \alpha\xi_3, \ldots, \alpha\xi_n)$$

The vector space is real or complex according to whether the $\{\xi_j\}$ (i.e., the *components* of the vectors) are real or complex. We denote both the real and the complex n-dimensional vector space by E_n. That is, we hereafter take our field of scalars to be the field of all real or of all complex numbers (not, for instance, just the rational numbers). We will usually deal with complex spaces and, for our purposes, simply treat real spaces as special cases of these.

2. An infinite-dimensional vector space E_∞ where a vector in this space is represented as

$$x = (\xi_1, \xi_2, \xi_3, \ldots, \xi_n, \ldots)$$

with a *countable* infinity of components, ξ_j, $j = 1, 2, 3, \ldots$.

3. The set of all complex numbers. This is a complex vector space where $x + y$ and αx are ordinary complex addition and multiplication, respectively.

4. The set of all continuous functions, $f(x)$, $x \in [0, 1]$. Here each vector has an *uncountable* infinity of components.

5. The set of all complex-valued functions, $f(x)$, $x \in [0, 1]$, such that $|f(x)|^2$ is (Lebesgue) square-integrable over $[0, 1]$. This space is usually denoted by $\mathcal{L}_2[0, 1]$.

1.2 INNER PRODUCT

Above we defined the operation of addition, +, for elements of our linear vector space S and of multiplication by scalars. We now define a multiplication of two of these elements, whose product we denote by $\langle x, y\rangle$ or by $\langle x \mid y\rangle$. For the spaces E_n and E_∞ we simply extend the definition of the inner (or scalar) product from E_3. For E_n we write

$$\langle x, y\rangle \equiv \sum_{j=1}^{n} \xi_j^* \eta_j \tag{1.12}$$

while for E_∞

$$\langle x, y\rangle \equiv \sum_{j=1}^{\infty} \xi_j^* \eta_j \tag{1.13}$$

provided this infinite series converges.

We restrict E_∞ to contain only those vectors such that $\langle x, x\rangle$ is finite, $\forall x \in E_\infty$. Here ξ_j^* is the complex conjugate of ξ_j. Again, in complete analogy with E_3, we define the length or *norm* of a vector as

$$\|x\| \equiv \langle x, x\rangle^{1/2} = \left(\sum_{j=1}^{\infty} |\xi_j|^2\right)^{1/2} \geq 0 \tag{1.14}$$

With this definition we have that

$$\|x\| = 0 \Leftrightarrow x = 0$$

Two vectors are said to be *orthogonal* provided

$$\langle x, y\rangle = 0$$

We can also give an abstract definition of the scalar product by requiring that the operation $\langle x, y\rangle$ satisfy the following properties for the elements of an *abstract space S* defined previously.

i. $\langle x, y\rangle = \langle y, x\rangle^*$
ii. $\langle \alpha_1 x_1 + \alpha_2 x_2, y\rangle = \alpha_1^* \langle x_1, y\rangle + \alpha_2^* \langle x_2, y\rangle$
iii. $\langle x, x\rangle > 0,\ x = 0$

Notice that these three axioms are sufficient to establish that the norm of a vector is zero if and only if $x = 0$. If $x = 0$, then $\alpha x = 0$, $\forall \alpha$, so that

$$\|x\|^2 = \|\alpha x\|^2 = |\alpha|^2 \, \|x\|^2 \qquad \forall \alpha$$

or

$$\|x\| = 0$$

so that

$$x = 0 \Rightarrow \|x\| = 0$$

Now either $\langle x, x\rangle > 0$ or $\langle x, x\rangle < 0$ for *all* $x \in S$ (when $\langle x, x\rangle \neq 0$). Otherwise, if we choose a pair such that $\langle x, y\rangle = 0$, then

$$\langle \alpha x + \beta y, \alpha x + \beta y\rangle = |\alpha|^2 \langle x, x\rangle + |\beta|^2 \langle y, y\rangle = 0$$

for some $\alpha \neq 0$ and $\beta \neq 0$ unless both $\langle x, x\rangle$ and $\langle y, y\rangle$ are of the same sign. Therefore from (iii) we conclude that $\langle x, x\rangle \geq 0$. Since $x \neq 0 \Rightarrow \langle x, x\rangle > 0$, we see that $\langle x, x\rangle = 0 \Rightarrow x = 0$.

In the space $\mathscr{L}_2$ consisting of all (in general) complex-valued functions such that

$$\int_a^b |f(x)|^2\, dx < \infty$$

we can define the scalar product as

$$\langle f, g\rangle \equiv \int_a^b f^*(x) g(x)\, dx$$

Now $\|f\|=0$ implies that $f(x)=0$ *almost everywhere* (i.e., except on a set of measure zero).

1.3 CONVERGENCE AND COMPLETE SPACES

a. Continuity and Uniform Continuity of a Function

In what follows we require the concepts of continuity of a function and of the convergence of an infinite series. Therefore we state the definitions here.

i. Definition of continuity A function $f(x)$ is continuous on $x\in[a, b]$ if and only if $f(x)$ is defined for every $x\in[a, b]$ and if corresponding to any $\varepsilon>0$ and $x_0\in[a, b]$ a number $\delta>0$ can be found such that $|f(x)-f(x_0)|<\varepsilon$ whenever $x\in[a, b]$ and $\|x-x_0\|<\delta$.

Notice that here δ may depend on the choice of x_0 (for fixed ε).

Example 1.2 $f(x)=1/x$, $0<x<\infty$. $|f(x)-f(x_0)|=|x-x_0|/|xx_0|$

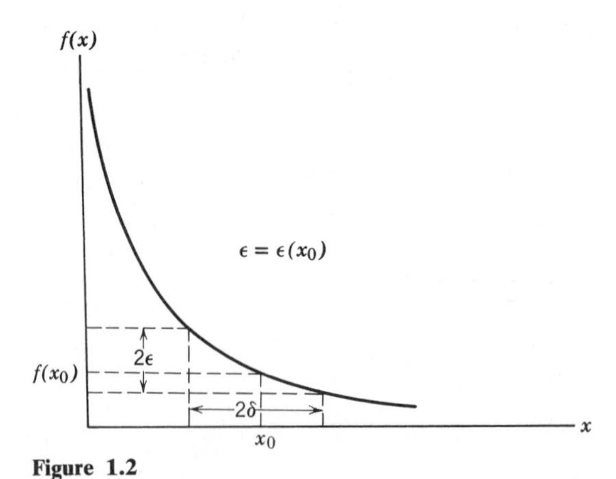

Figure 1.2

---

ii. Definition of uniform continuity A function $f(x)$ is uniformly continuous on $x \in [a, b]$ if and only if $f(x)$ is defined for all $x \in [a, b]$ and if, for any $\varepsilon > 0$, it is possible to find a $\delta > 0$, independent of x_0, such that $|f(x) - f(x_0)| < \varepsilon$ when $|x - x_0| < \delta$, for all x and x_0 on $[a, b]$.

Notice that here δ may *not* depend on the choice of x_0.

Example 1.3 $f(x) = 1/x$, $1 \leq x < \infty$.

---

b. Convergence of a Series of Functions

iii. Definition of convergence of a series of functions Let each of the functions $\{u_n(x)\}$ be defined for the points $x \in [a, b]$. Then the series $\sum_n^\infty u_n$ is said to converge pointwise on $[a, b]$ if and only if $\sum_n^\infty u_n(x)$ converges for each $x \in [a, b]$.

Note that tests such as the ratio test are simply means of testing for the convergence of a series. They are *not* definitions of convergence.

iv. Definition of uniform convergence of a series of functions Let

$$f(x) \equiv \sum_{j=1}^{\infty} u_j(x) \equiv \sum_{j=1}^{n} u_j(x) + R_n(x) \equiv S_n(x) + R_n(x)$$

The given infinite series is said to be uniformly convergent on $x \in [a, b]$ if a number N, which is independent of the value of $x \in [a, b]$, can be found for any $\varepsilon > 0$ such that $|f(x) - S_n(x)| < \varepsilon$ for all $n \geq N$.

v. Definition of uniform convergence of a sequence of functions A sequence of functions $\{f_n(x)\}$ is said to converge uniformly to the function $f(x)$ if and only if for any $\varepsilon > 0$ $\exists$ an $N \ni$

$$|f(x) - f_n(x)| < \varepsilon \qquad n \geq N$$

for all $x \in [a, b]$ independent of the choice of x.

Example 1.4

$$f_n(x) = \frac{n}{1 + nx^2} \qquad x \in [1, \infty)$$

$$f(x) \equiv \lim_{n \to \infty} f_n(x) = \frac{1}{x^2}$$

$$|f(x) - f_n(x)| = \left| \frac{1}{x^2} - \frac{n}{1 + nx^2} \right|$$

$$= \left| \frac{1}{x^2(1 + nx^2)} \right| \leq \left| \frac{1}{1 + n} \right|$$

This last bound is *independent* of the choice of $x \in [1, \infty)$. Therefore for $n \geq N$, N sufficiently large, $|f(x) - f_n(x)|$ can be made as small as we please, independent of the choice of x, so that this sequence is uniformly convergent for $x \in [1, \infty)$.

---

c. Cauchy Convergence

A sequence of vectors $\{x_n\}$ is said to converge to a vector x if, given any $\varepsilon > 0$, $\exists N = N(\varepsilon) \ni \|x - x_n\| < \varepsilon$, $\forall n > N$. Similarly an infinite sum of vectors, $\sum_{m=1}^{\infty} y_m$, is said to converge to a sum x if the sequence of partial sums

$$x_n \equiv \sum_{m=1}^{n} y_m$$

converges to the limit x.

If we consider a sequence of vectors in E_n, we write

$$x_j = (\xi_1^j, \xi_2^j, \xi_3^j, \ldots, \xi_n^j)$$

where n is the dimensionality of S. Therefore, since the equality of two vectors in E_n, $v = w$, implies the equality of their components, $v_j = w_j$, $\forall j$, the convergence of a sequence of partial sums of vectors in E_n implies the corresponding convergence of the sequences of partial sums for each component.

By the *Cauchy convergence* of a sequence of numbers $\{s_n\}$ we mean $\exists N \ni$, $\forall m$, $n > N$

$$|s_n - s_m| < \varepsilon$$

for a given $\varepsilon > 0$. This is, in fact, simply a criterion for the convergence of the sequence. We can see this as follows. If the sequence is convergent, then

$$|s_n - s_m| \equiv |s - s_m + s_n - s| \leq |s - s_n| + |s - s_m| < \varepsilon' + \varepsilon'' \equiv \varepsilon$$

To prove the converse we begin by assuming the sequence $\{s_n\}$ consists of real numbers. We shall base our proof on the *Bolzano-Weierstrass theorem* (cf. Appendix I, Th. I.1) which states that every bounded infinite sequence $\{s_n\}$ has at least one *limit point* (i.e., a sequence $\{s_n\}$ has a limit point S if for every positive ε, no matter how small, there exists an unlimited number of terms in the sequence such that $S - \varepsilon < s_n < S + \varepsilon$). We are given the Cauchy criterion

$$|s_n - s_m| < \varepsilon \qquad \forall m, n > N \tag{1.15}$$

Let us put $\varepsilon = 1$ and $m = n + p$ so that

$$|s_{n+p} - s_n| < 1 \qquad n > N \qquad \forall p = 0, 1, 2, \ldots$$

This inequality implies that the sequence $\{s_n\}$ is bounded so that by the Bolzano-Weierstrass theorem there exists at least one limit point. Furthermore this limit point S is unique. For suppose there were two limit points S and S'.

Then by the definition of a limit point there would have to exist positive numbers q and r such that

$$|S - s_{n+q}| < \varepsilon \qquad |S' - s_{n+r}| < \varepsilon$$

However these inequalities plus Eq. 1.15, imply that

$$|S - S'| = |S - s_{n+q} + s_{n+q} - S' - s_{n+r} + s_{n+r} + s_n - s_n|$$
$$\leq |S - s_{n+q}| + |s_{n+q} - s_n| + |-S' + s_{n+r}| + |-s_{n+r} + s_n| < 4\varepsilon$$

Since ε can be made arbitrarily small, we see that the limit point is unique.

Now let $\{s_n\}$ be a sequence of complex numbers. If we let

$$s_n = \text{Re } s_n + i \text{ Im } s_n$$

then since

$$|\text{Re } s_{n+p} - \text{Re } s_n| \leq |s_{n+p} - s_n|$$
$$|\text{Im } s_{n+p} - \text{Im } s_n| \leq |s_{n+p} - s_n|$$

the Cauchy criterion for convergence, Eq. 1.15, for a sequence of complex numbers implies the existence of unique limit points for the sequences $\{\text{Re } s_n\}$ and $\{\text{Im } s_n\}$ separately and hence for $\{s_n\}$.

Similarly a sequence of vectors $\{x_n\}$ is said to be Cauchy convergent provided that for every $\varepsilon > 0 \exists N \ni \forall m, n > N$

$$\|x_n - x_m\| < \varepsilon \tag{1.16}$$

If the sequence $\{x_n\}$ converges to a limit vector x, then this sequence is Cauchy convergent since

$$\|x_n - x_m\| = \|x - x_m + x_n - x\| \leq \|x - x_n\| + \|x - x_m\| < \varepsilon' + \varepsilon'' \equiv \varepsilon$$

This follows from the triangle inequality for vectors,

$$\|x + y\| \leq \|x\| + \|y\|$$

the proof of which is left to a problem.

That the Cauchy convergence of a sequence of vectors in E_n implies the existence of a limit vector is evident. Since the Cauchy convergence of the sequence of vectors, $\{x_m\}$, implies the Cauchy convergence of the corresponding sequences of components, $\{\xi_j^m\}$, which are just complex numbers, one sees that each component tends to a definite limit so that $x_n \xrightarrow[n\to\infty]{} x$, a definite limit vector.

d. Proof of Completeness of E_∞

We now consider whether or not the Cauchy convergence criterion, Eq. 1.16, for a sequence of vectors $\{x_n\}$ in a space S such that $\|x_n\| < \infty$ necessarily implies the existence of a limit vector x also in S (i.e., such that $\|x\| < \infty$). This is the question of the *completeness* of the space S.

The completeness of E_n is evident because convergence in the Cauchy sense implies the existence of a limit vector each of whose n components is finite. But

$$\|x\|^2 = \langle x, x\rangle = \sum_{j=1}^{n} |\xi_j|^2$$

so that the norm is finite since the sum contains only a finite number of finite terms.

However for E_∞ we must also prove the convergence of the infinite sum defining the norm of the limit vector. We are given that

$$\|x_m - x_n\| < \varepsilon \qquad m, n > N$$

and that for all finite m and n

$$\|x_n\| < \infty \qquad \|x_m\| < \infty$$

while a vector in the sequence

$$x_n \equiv (\xi_1^n, \xi_2^n, \ldots, \xi_j^n, \ldots)$$

has countably many components. The following inequality is obvious.

$$|\xi_j^m - \xi_j^n|^2 \leq \sum_{j=1}^{\infty} |\xi_j^m - \xi_j^n|^2 \equiv \|x_m - x_n\|^2 < \varepsilon$$

Therefore the jth components also form a Cauchy sequence. Hence, as previously, we have $\xi_j^n \underset{n\to\infty}{\longrightarrow} \xi_j$, some complex numbers each of which has finite norm. We must now show that

$$\langle x, x\rangle \equiv \sum_{j=1}^{\infty} |\xi_j|^2 < \infty$$

But it is evident that

$$\sum_{j=1}^{M} |\xi_j^m - \xi_j^n|^2 \leq \sum_{j=1}^{\infty} |\xi_j^n - \xi_j^m|^2 \equiv \|x_n - x_m\|^2 < \varepsilon \qquad n, m > N$$

Letting $m \to \infty$ we obtain

$$\sum_{j=1}^{M} |\xi_j - \xi_j^n|^2 < \varepsilon \qquad n > N$$

Now let $M \to \infty$ to obtain

$$\sum_{j=1}^{\infty} |\xi_j - \xi_j^n|^2 < \varepsilon \qquad n > N$$

We may write

$$\langle x, x\rangle \equiv \sum_{j=1}^{\infty} |\xi_j|^2 = \sum_{j=1}^{\infty} |\xi_j - \xi_j^m + \xi_j^m|^2$$

However for any complex numbers α and β we have

$$(|\alpha|-|\beta|)^2 \equiv |\alpha|^2+|\beta|^2-2\,|\alpha|\,|\beta| \geq 0,$$

so that

$$2\,|\alpha|\,|\beta| \leq |\alpha|^2+|\beta|^2$$

or

$$|\alpha+\beta|^2 \leq \big|\,|\alpha|+|\beta|\,\big|^2 = |\alpha|^2+|\beta|^2+2\,|\alpha|\,|\beta| \leq 2(|\alpha|^2+|\beta|^2)$$

which finally yields

$$\langle x, x\rangle \leq 2\left(\sum_{j=1}^{\infty} |\xi_j-\xi_j^n|^2+\sum_{j=1}^{\infty} |\xi_j^n|^2\right)=2(\|x-x_n\|^2+\|x_n\|^2)<\infty$$

There are many examples of spaces that are not complete. If we consider the linear vector space of all functions $f(x)$ continuous on a closed interval $x \in [a, b]$ with the scalar product

$$\langle f, g\rangle \equiv \int_a^b f^*(x)g(x)\,dx$$

then this space is not complete.

Example 1.5 An example of this can be seen from the sequence of functions defined on $-1 \leq x \leq +1$ as

$$f_n(x)=\begin{cases} -1 & x \leq -\dfrac{1}{n} \\ nx & -\dfrac{1}{n} \leq x \leq \dfrac{1}{n} \\ +1 & x \geq \dfrac{1}{n} \end{cases}$$

The sequence function and its limit are shown graphically in Fig. 1.3. While each

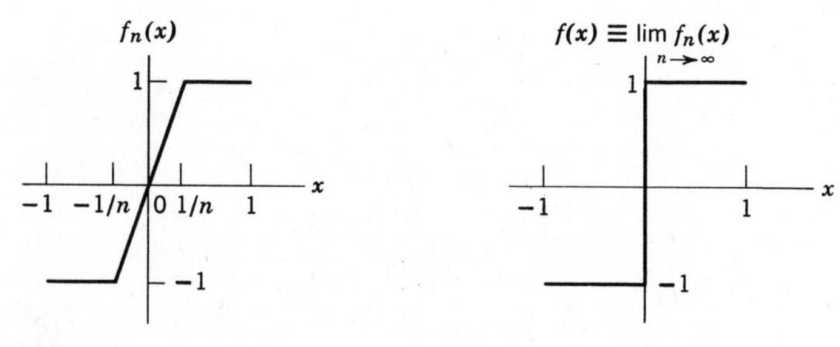

Figure 1.3

function $f_n(x)$ in the sequence is continuous, the limit function $f(x)$ is discontinuous. It is easy to see that the sequence is Cauchy convergent since

$$\|f_n - f_m\|^2 \equiv \langle (f_n - f_m), (f_n - f_m) \rangle = \int_{-1}^{1} [f_n(x) - f_m(x)]^2 \, dx$$

$$= 2\left[\int_0^{1/n} (nx - mx)^2 \, dx + \int_{1/n}^{1/m} (1 - mx)^2 \, dx\right]$$

$$= \frac{2}{3} \frac{1}{m} \left(1 - \frac{m}{n}\right)^2$$

where we have taken $n > m$. For m and n that are sufficiently large this can be made as small as we please. Therefore the sequence is convergent in the Cauchy sense, but the actual limit, $f(x)$, while it exists (except at $x = 0$), does not belong to the original vector space of *continuous* functions. This space is not complete.

We see then that Cauchy convergence does not necessarily imply completeness. We can remove the difficulty encountered in the example above by so defining the space S that the limit of the sequence is included in S. For example, take all $f_n(x) \in \mathscr{L}_2$ and consider sequences that converge in the Cauchy sense. The limit functions are also in $\mathscr{L}_2$ so that this space is complete. This is the *Riesz-Fischer theorem*, which we will not prove.

If we have a sequence of vectors $\{x_n\}$ defined in a space S, and if the Cauchy convergence of this sequence implies that there exists a limit vector also in S, then S is said to be *complete.* A complete linear vector space with a complex-type scalar product is a *Hilbert space.* Hereafter we will limit our discussion to linear vector spaces that are complete.

1.4 LINEAR MANIFOLDS AND SUBSPACES

a. Linear Independence

We simply generalize the concept of the linear independence of vectors in E_3. Let

$$\alpha_1 x_1 + \alpha_2 x_2 + \cdots + \alpha_m x_m \equiv \sum_{j=1}^{m} \alpha_j x_j = 0 \qquad \textbf{(1.17)}$$

i. If this equation can be satisfied with not all of the $\{\alpha_j\}$ zero, then the set of vectors $\{x_j\}$ is *linearly dependent.*
ii. If this equation can only be satisfied when all of the $\{\alpha_j\}$ vanish, then the set of vectors $\{x_j\}$ is *linearly independent.*

With this definition, as with many of the definitions and theorems in the next few chapters, the reader may find it helpful to visualize it in terms of vectors in E_3.

b. Linear Manifolds

Let $\mathcal{M}$ stand for some set of vectors, x, y, z, . . . , in a space S. We say that $\mathcal{M}$ is a *linear manifold* provided that

$$\{x \in \mathcal{M},\ y \in \mathcal{M},\ \forall x, y \in \mathcal{M}\} \Rightarrow (\alpha x + \beta y) \in \mathcal{M},\ \forall \alpha, \beta$$

A set of vectors $\{x_n\}$ *spans* $\mathcal{M}$ if, for *all* $y \in \mathcal{M}$, we can write y as

$$y = \sum_{j=1}^{k} \eta_j x_j$$

or as the limit of such linear combinations. If the expansion of any $y \in \mathcal{M}$ in terms of the vectors $\{x_j\}$ is *unique*, then this set forms a *basis* for $\mathcal{M}$. The *dimension* of a linear manifold is the number of vectors in the basis.

It follows from these definitions that any set of $n+1$ vectors in an n-dimensional linear manifold is linearly dependent. For if this were not the case, then there would exist at least one set of $n+1$ vectors that would be linearly independent and this set of $n+1$ vectors could be used as a basis. This would define a linear manifold of dimension $n+1$, contrary to the assumption that the linear manifold is of dimension n.

As a technical point that will not concern us in our applications, we mention that in an infinite-dimensional manifold $\mathcal{M}_\infty$ an infinite set of linearly independent vectors that spans $\mathcal{M}_\infty$ need not form a basis.

A linear manifold $\mathcal{M}$ is *closed* if, whenever a sequence of vectors in $\mathcal{M}$ converges to a limit, this limit is in $\mathcal{M}$. A closed linear manifold is called a *linear subspace*. In E_n (defined over the field of real or complex scalars; cf. Sec. 1.1c) every linear manifold is a subspace.

Recall that at the end of Section 1.3d we stated that we will consider only *complete* linear vector spaces so that we avoid any logical inconsistency in our definition of a subspace (i.e., the possibility that a *subspace* would be defined as closed while a space need not be).

We write

$$\mathcal{N} = \mathcal{M}_1 + \mathcal{M}_2$$

if for $\forall x \in \mathcal{N}$ we can always find an $x_1 \in \mathcal{M}_1$ and an $x_2 \in \mathcal{M}_2$ such that

$$x = x_1 + x_2$$

We say that the space $\mathcal{N}$ is the *sum* of subspaces $\mathcal{M}_1$ and $\mathcal{M}_2$. If this decomposition is *unique*, we have a direct sum and write

$$\mathcal{N} = \mathcal{M}_1 \oplus \mathcal{M}_2$$

Example 1.6 As an example of a sum of subspaces which is not a direct sum, consider two orthogonal planes in E_3. As shown in Fig. 1.4 these form a sum, but not a direct sum, since the decomposition of an $x \in E_3$ is not necessarily unique.

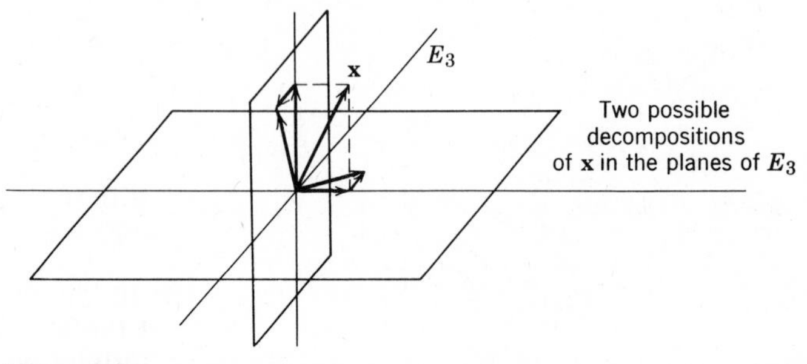

Figure 1.4

1.5 BASIS FOR E_n AND E_∞

So far we have discussed the properties of an abstract vector space S. Furthermore we have stated that we will work with complete linear vector spaces and in particular with Hilbert spaces that we now denote by $\mathcal{H}$. An especially important type of Hilbert space is a *separable* one. A space is separable provided it has a countable (i.e., finite or denumerably infinite) basis. We will restrict our work in this text almost exclusively to separable Hilbert spaces. In fact we now show that any separable Hilbert space $\mathcal{H}$ can be represented by E_∞ (or by E_n in some special cases). In Chapter 4 we will prove a similar result for piecewise continuous functions on a finite range $[a, b]$ when we develop the expansion of a function in terms of Fourier series.

For any separable Hilbert space $\mathcal{H}$ we may always construct a basis $\{x_j\}$ by finding an appropriate set of linearly independent vectors that span $\mathcal{H}$. Therefore for all $x, y \in \mathcal{H}$ we have

$$x = \sum_j \xi_j x_j$$

$$y = \sum_j \eta_j x_j$$

so that $\mathcal{H}$ is identical to E_∞ (or E_n) w.r.t. addition and multiplication by scalars. However we must also consider the form of the scalar product in $\mathcal{H}$.

$$\langle x, y \rangle = \left\langle \sum_j \xi_j x_j, \sum_l \eta_l x_l \right\rangle = \sum_{j,l} \xi_j^* \eta_l \langle x_j, x_l \rangle$$

This will reduce to the definition of $\langle x, y\rangle$ given for E_∞, namely,

$$\langle x, y\rangle = \sum_j \xi_j^* \eta_j$$

if and only if

$$\langle x_j, x_l\rangle = \delta_{jl} = \begin{cases} 1 & j = l \\ 0 & j \neq l \end{cases}$$

where δ_{jl} is the *Kronecker delta symbol.* Therefore we must have an orthonormal basis for $\mathcal{H}$ if $\mathcal{H}$ is to be represented by E_∞ (or by E_n). We now show how to obtain such an orthonormal basis for $\mathcal{H}$.

1.6 SCHMIDT ORTHOGONALIZATION PROCESS

The *Schmidt orthogonalization process* allows us to construct an orthogonal set of vectors from any linearly independent set. Given any linearly independent set, $x_1, x_2, \ldots, x_n$, we proceed as follows. Let

$$y_1 \equiv x_1$$

$$y_2 \equiv x_2 - \frac{\langle y_1 \mid x_2\rangle}{\langle y_1, y_1\rangle} y_1 \qquad \ni \langle y_2, y_1\rangle = 0$$

$$y_3 \equiv x_3 - \frac{\langle y_1, x_3\rangle}{\langle y_1, y_1\rangle} y_1 - \frac{\langle y_2 \mid x_3\rangle}{\langle y_2 \mid y_2\rangle} y_2 \qquad \ni \langle y_3 \mid y_1\rangle = 0 = \langle y_3, y_2\rangle$$

$$\vdots$$

$$y_n \equiv x_n - \sum_{j=1}^{n-1} \frac{\langle y_j \mid x_n\rangle}{\langle y_j \mid y_j\rangle} y_j \qquad \ni \langle y_n \mid y_j\rangle = 0 \qquad j = 1, 2, \ldots, n-1$$

We can easily interpret this construction graphically in a real space as shown in Fig. 1.5. That is in general we take $y_n \to x_n - \{$the projection of x_n onto $y_1, y_2, \ldots, y_{n-1}\}$ and properly normalize each term with $\langle y_j, y_j\rangle^{-1}$. We can continue the process until we have $y_1, y_2, \ldots, y_n$ all orthogonal. Notice that none of the $\{y_j\}$ yielded by this construction vanish since, if one did, this would imply that the $\{x_j\}$ are linearly dependent, contrary to our starting assumption.

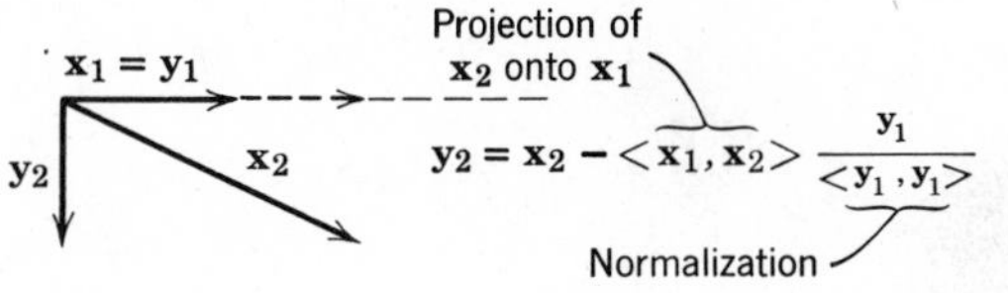

Figure 1.5

Therefore we may always choose this set to be orthonormal since any orthogonal set of vectors can always each be normalized to unit length.

1.7 PROJECTION THEOREM

We now state and prove a theorem that will be of central importance for the sections that follow.

THEOREM 1.1 (Projection theorem)

If $\mathcal{M}$ is a subspace of a complete space S, then every vector $y \in S$ can be expressed uniquely as

$$y = w + z$$

where $w \in \mathcal{M}$ and $\langle z \mid x \rangle = 0$ for all $x \in \mathcal{M}$.

Symbolically we may express this theorem as follows.

$$\{\exists \text{ a subspace } \mathcal{M} \& y \notin \mathcal{M}\} \Rightarrow \{\exists w \in \mathcal{M} \ni \langle y - w, x \rangle = 0, \forall x \in \mathcal{M}\}$$

Before we give a proof for the theorem, let us see what it means in E_3 (Fig. 1.6) and that it is trivially true for E_n.

Notice that if we take $x \in \mathcal{M}$ and, for a *fixed* $y \notin \mathcal{M}$, form the norm

$$\|y - x\|^2 \equiv \langle (y - x), (y - x) \rangle > 0$$

and vary x until we minimize this norm, then $x = x_{\min} = w$ is such that

$$\langle y - x_{\min}, x \rangle = 0 \qquad \forall x \in \mathcal{M}$$

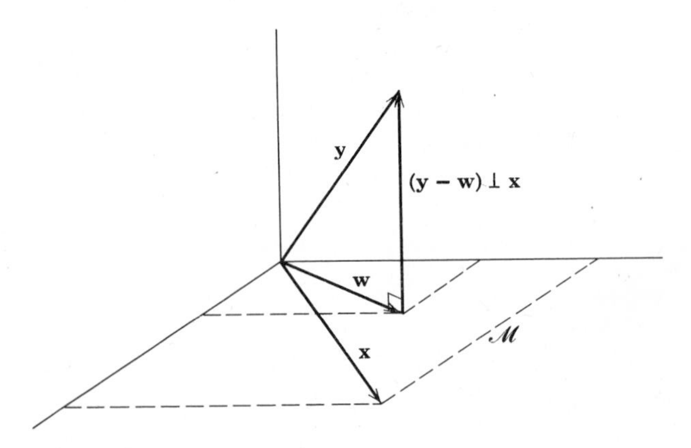

Figure 1.6

It is precisely this observation that provides a standard proof of this theorem. If our space is E_n and $\mathcal{M}=E_m$, $m<n$, the proof is trivial. In a suitable basis we may write

$$y=(\eta_1, \eta_2, \ldots, \eta_n)$$

$$x=(\xi_1, \xi_2, \ldots, \xi_n)=(\xi_1, \xi_2, \ldots, \xi_m, 0, \ldots, 0)$$

$$\|y-x\|^2=\sum_{j=1}^{n} |\eta_j-\xi_j|^2=\sum_{j=1}^{m} |\eta_j-\xi_j|^2+\sum_{j=m+1}^{n} |\eta_j|^2$$

Since every term in these sums is nonnegative, we minimize $\|y-x\|$ by adjusting ξ_j for fixed η_j so that

$$\xi_j=\eta_j \qquad j=1, 2, \ldots, m$$

or

$$x_{\min}\equiv w=(\eta_1, \eta_2, \ldots, \eta_m, 0, 0, \ldots, 0)$$

By direct calculation we see that

$$\langle y-w, x\rangle=\sum_{j=1}^{m} (\eta_j-\eta_j)^*\xi_j+\sum_{j=m+1}^{n} \eta_j^*(0)=0$$

In fact this proof will hold for any separable Hilbert space as long as $\mathcal{M}$ is a finite-dimensional subspace.

PROOF

We now give a general proof of this theorem. As in our discussion above we begin by seeking that vector $w\in\mathcal{M}$ that minimizes the norm $\|y-x\|$ where $x\in\mathcal{M}$ and y is *fixed.* We define the minimum value of this norm as

$$\|y-w\|\equiv\delta \tag{1.18}$$

We must first establish the *existence* of such a $w\in\mathcal{M}$. For all $x\in\mathcal{M}$, since $y\notin\mathcal{M}$,

$$\|y-x\|>0$$

which implies the existence of a greatest lower bound δ for this norm. Let us define a sequence of vectors $\{x_n\}\in\mathcal{M}$ such that

$$\lim_{n\to\infty} \|y-x_n\|=\delta$$

We are guaranteed that this sequence of norms converges to a limit since any bounded monotone sequence is convergent (as follows from the Bolzano-Weierstrass theorem). We can now show that these $\{x_n\}$ form a Cauchy sequence of vectors

$$\delta\leq\|y-\tfrac{1}{2}(x_n+x_m)\|\leq\tfrac{1}{2}\|y-x_n\|+\tfrac{1}{2}\|y-x_m\| \xrightarrow[n,m\to\infty]{} \delta$$

But since

$$\|2y-x_n-x_m\|^2+\|x_n-x_m\|^2\equiv 2\,\|y-x_n\|^2+2\,\|y-x_m\|^2$$

we see that

$$\|x_n-x_m\| \xrightarrow[n,m\to\infty]{} 0$$

so that $\{x_n\}$ indeed forms a Cauchy sequence.

Now by assumption $\mathcal{M}$ is a *subspace* of a *complete* space S. By the latter property we know that every Cauchy sequence has a strong limit in S (i.e., $\lim_{n\to\infty} x_n=x\in S$). Moreover since $\mathcal{M}$ is a subspace this limit vector must be contained in $\mathcal{M}$ itself.

By the definition of w in Eq. 1.18 we see that for *all* $x\in\mathcal{M}$

$$\|y-x\|\geq\delta$$

Therefore if we let $\hat{x}=x/\|x\|$ we can write

$$\begin{aligned}\|y-w\|^2 &\leq\|y-w-\langle\hat{x},(y-w)\rangle\hat{x}\|^2\\ &\equiv\|y-w\|^2-|\langle\hat{x},(y-w)\rangle|^2\end{aligned}$$

so that

$$0\leq-|\langle\hat{x},(y-w)\rangle|^2$$

or

$$\langle(y-w),x\rangle=0 \qquad x\in\mathcal{M}$$

Also w is unique since if it were not there would exist two vectors, w_1 and w_2, such that

$$\|y-w_1\|=\delta=\|y-w_2\|$$

which would imply that

$$\delta\leq\|y-\tfrac{1}{2}(w_1+w_2)\|\equiv\|\tfrac{1}{2}(y-w_1)+\tfrac{1}{2}(y-w_2)\|\leq\tfrac{1}{2}\|y-w_1\|+\tfrac{1}{2}\|y-w_2\|\equiv\delta$$

or that

$$\|y-\tfrac{1}{2}(w_1+w_2)\|=\delta=\tfrac{1}{2}\,\|y-w_1\|+\tfrac{1}{2}\,\|y-w_2\|$$

We leave it to the reader to show that since $\|x+y\|\leq\|x\|+\|y\|$, then when $\|x+y\|=\|x\|+\|y\|$ it follows that $y=\alpha x$, where α is a real nonnegative constant. This allows us to write $(y-w_1)=\alpha(y-w_2)$. If $\alpha=1$, then $w_1=w_2$ as claimed; if $\alpha\neq 1$, then

$$y=\frac{w_1-\alpha w_2}{(1-\alpha)}$$

which places $y\in\mathcal{M}$, contrary to assumption. Q.E.D.

1.8 LINEAR FUNCTIONALS

We define a linear functional as a mapping of the linear vector space S onto the space of scalars. We assign to every vector $x \in S$ a scalar $f(x)$ which is a *functional* defined on S. Below we list the various properties we will consider.

i. *Functional.* To every $x \in S$ we associate a unique scalar, $f(x)$.
ii. $f(\alpha x + \beta y) = \alpha f(x) + \beta f(y)$, $\forall x, y \in S \Rightarrow f(x)$ is a *linear* functional.
iii. $f(x)$ is *continuous* if

$$\lim_{n\to\infty} x_n = x \Rightarrow \lim_{n\to\infty} f(x_n) = f(x)$$

iv. $f(x)$ is *bounded* if $|f(x)| \leq \mu \, \|x\|$ where μ is some constant (independent of x), $\forall x \in S$.

Notice that for a *linear* functional $|f(x)| < C$, $\forall \, \|x\| = 1$ implies that $|f(x)| < C \, \|x\|$, $\forall x \ni x \neq 0$, and conversely.

THEOREM 1.2

For any continuous linear functional $f(x)$ there exists a unique $z \in S$ for which

$$f(x) = \langle z, x \rangle \qquad \forall x \in S$$

PROOF

We define $\mathcal{M}$ to be all of those vectors y such that for a given continuous lin
functional f

$$f(y) = 0$$

$$\left.\begin{array}{l} y_1 \in \mathcal{M} \\ y_2 \in \mathcal{M} \end{array}\right\} \Rightarrow f(\alpha y_1 + \beta y_2) = 0 \Rightarrow \mathcal{M} \text{ is a } \textit{manifold}$$

Since $f(x)$ is given as being *continuous* the existence of the limit
sequence $\{x_n\}$, $x_n \in \mathcal{M}$, implies that

$$\lim_{n\to\infty} f(x_n) = 0 = f(x)$$

Hence $x \in \mathcal{M}$ so that $\mathcal{M}$ is a *subspace*, which means t
Therefore either

i. $\mathcal{M} \equiv S$, the whole space $\Rightarrow z = 0$, and proo

or

ii. $\mathcal{M} \subset S$, $\mathcal{M} \not\equiv S \Rightarrow z \perp \mathcal{M}$

In case (ii) we must demonstrate the *existence* of such a z. By assumption now there exists a vector $v \notin \mathcal{M}$ such that $f(v) \neq 0$. Since Th. 1.1 applies we know there exists a w such that $v - w \equiv u$ where $w \in \mathcal{M}$ and $u \perp \mathcal{M}$, $y \neq 0$ so that

$$f(u) = f(v) \neq 0$$

Since z, if it exists in case (ii), must be orthogonal to $\mathcal{M}$, we set $z = \alpha u$ and determine the normalization so that

$$f(z) = \langle z, z \rangle$$

which is always possible since $y \neq 0$ and $f(y) \neq 0$. Since

$$\alpha f(u) = |\alpha|^2 \langle u, u \rangle$$

we have

$$\alpha^* = \frac{f(u)}{\langle u, u \rangle} \neq 0$$

However since for all $x \in S$

$$f\left[x - \frac{f(x)}{f(z)} z\right] = f(x) - f(x) = 0$$

we see that

$$x - \frac{f(x)}{f(z)} z \in \mathcal{M}$$

so that

$$\left\langle \left[x - \frac{f(x)}{f(z)} z\right], z \right\rangle = 0 = \langle x, z \rangle - f^*(x)$$

or

$$f(x) = \langle z, x \rangle$$

Q.E.D.

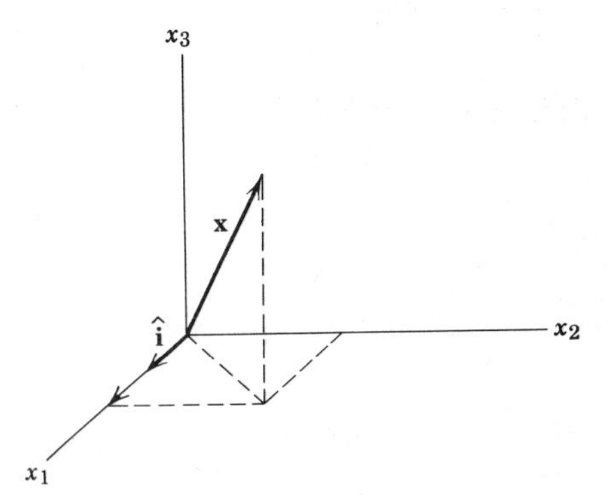

Figure 1.7

Example 1.7 Let $S=E_3$ and f be the length of the x_1 component of x.

$$z=\hat{\mathbf{i}}$$
$$f(\mathbf{x})=\langle z, \mathbf{x}\rangle=\langle \hat{\mathbf{i}}, \mathbf{x}\rangle=x_1$$
$$f(\hat{\mathbf{i}})=\langle \hat{\mathbf{i}}, \hat{\mathbf{i}}\rangle=1$$

The result of problem 1.18 shows that for *linear* functionals

$$f(x) \text{ bounded} \Leftrightarrow f(x) \text{ continuous}$$

With this result we can now outline a method for expressing any $x \in E_n$ in terms of an arbitrary nonorthogonal basis. If the set $\{x_j\}$ forms a basis for E_n, then any $x \in E_n$ can be written as

$$x=\sum_{j=1}^{n} \alpha_j x_j \tag{1.19}$$

Clearly the $\{\alpha_j\}$ are linear functionals of x. We now show that they are *bounded* linear functionals of x. Let $\{e_j\}$ be an orthonormal basis for E_n. We have seen in Section 1.6 that such an orthonormal basis can always be constructed from a nonorthonormal basis $\{x_j\}$. Therefore we may express the $\{e_j\}$ in terms of the $\{x_j\}$ as

$$e_j=\sum_{l=1}^{n} \gamma_{jl} x_l \qquad j=1, 2, \ldots, n$$

where the γ_{jl} are simply constants (independent of x). Since

$$x=\sum_j \xi_j e_j \qquad \ni \langle x \mid x\rangle=\sum_j |\xi_j|^2 \tag{1.20}$$

we see that

$$|\xi_j| \leq \|x\| \qquad j=1, 2, \ldots, n$$

However from Eqs. 1.19 and 1.20 we see that

$$\alpha_l=\sum_j \xi_j \gamma_{jl} \tag{1.21}$$

since the $\{x_j\}$ being a basis is a linearly independent set. Therefore we may use Eq. 1.21 and the Schwarz inequality to obtain.

$$|\alpha_l|=\left|\sum_j \xi_j \gamma_{jl}\right| \leq \left(\sum_j |\xi_j|^2\right)^{1/2}\left(\sum_j |\gamma_{jl}|^2\right)^{1/2}=\left(\sum_j |\gamma_{jl}|^2\right)^{1/2} \|x\|$$
$$\equiv C\,\|x\| \qquad \forall l=1, 2, \ldots, n$$

where $\sum_{j=1}^{n} |\gamma_{jl}|^2$ is finite since we are in a finite-dimensional vector space. This proves that the $\{\alpha_j\}$ are bounded and therefore continuous linear functionals of the vector, x. By Th. 1.2 we know there exist vectors $\{z_j\}$ such that

$$\alpha_j = \langle z_j, x \rangle \qquad j = 1, 2, \ldots, n$$

or that

$$x_i = \sum_{j=1}^{n} \alpha_j^{(i)} x_j = \sum_{j=1}^{n} \langle z_j, x_i \rangle x_j$$

From the linear independence of the $\{x_j\}$, we deduce that

$$\langle z_j, x_i \rangle = \delta_{ji}$$

Therefore for every basis $\{x_j\}$ in a finite-dimensional vector space there always exists a *reciprocal basis* $\{z_k\}$ such that

$$\langle z_k, x_j \rangle = \delta_{jk} \qquad \textbf{(1.22)}$$

$$x = \sum_{j=1}^{n} \langle z_j, x \rangle x_j \qquad \textbf{(1.23)}$$

SUGGESTED REFERENCES*

B. Friedman, *Principles and Techniques of Applied Mathematics.* This text is a model of clear and rigorous presentation of many topics of great value in applied mathematics. The outline of much of the material in Chapters 1, 2, and 3 of the present text follows the first two chapters of Friedman.

P. R. Halmos, *Finite-Dimensional Vector Spaces.* This is the standard introductory text for the study of finite-dimensional vector spaces that we have denoted as E_n. There are many examples and problems that illustrate the material well.

W. Kaplan, *Advanced Calculus.* This is an excellent reference text for the theorems and techniques of advanced calculus. The presentation is rigorous but illustrated with many applications.

A. E. Taylor, *Introduction to Functional Analysis.* This is a rather advanced text on functional analysis that the mathematically sophisticated reader may find useful.

The reader wishing to review the elements of three-dimensional vector analysis might consult either of the following works.

H. Lass, *Vector and Tensor Analysis.*

H. B. Phillips, *Vector Analysis.*

* Full bibliographical description of the references listed at the end of each chapter will be found in the bibliography at the end of the book.

PROBLEMS

1.1 In order to obtain the results stated in Eqs. 1.4 and 1.5 for the homogeneous system of Eq. 1.3, begin by considering the special case

$$a_{11}x_1 + a_{12}x_2 + a_{13}x_3 = 0$$
$$0 + a_{22}x_2 + a_{23}x_3 = 0$$
$$0 + \quad 0 + a_{33}x_3 = 0$$

and assume $a_{11} \neq 0$. You might proceed systematically as follows.

a. Take $\Delta \neq 0$, assume the existence of a nontrivial solution, and argue that this implies a contradiction. Hence when $\Delta \neq 0$, $\mathbf{x} = 0$ is the *only* solution.

b. Next assume that $\Delta = 0$, which implies that at least one of a_{22} or a_{33} is zero, and exhaust the three distinct possibilities

$$\text{i. } a_{33} = 0$$
$$\text{ii. } a_{33} \neq 0 \qquad a_{22} = 0$$
$$\text{iii. } a_{33} \neq 0 \qquad a_{22} \neq 0$$

to exhibit explicitly a nontrivial solution. Hence $\Delta = 0$ implies the existence of a (nonunique) nontrivial solution.

c. Now negate both sides of the logical implication established in (b) to show that

$$\nexists \mathbf{x} \neq 0 \Rightarrow \Delta \neq 0$$

The results of (a) and (c) taken together yield the statement of Eq. 1.4.

d. Assume that a nontrivial solution does exist and deduce that either $a_{22} = 0$ and/or $a_{33} = 0$ so that $\Delta = 0$. This and the conclusion of (b) establish Eq. 1.5.

e. Finally argue that the system of Eq. 1.3 can always be reduced to the simpler form considered in this problem.

1.2 Prove the following properties of the inner product for E_n and E_∞ using the definition given in Eqs. 1.12 and 1.13.

a. $\langle \alpha x, y \rangle = \alpha^* \langle x, y \rangle = \alpha^* \langle y, x \rangle^*$

b. $\langle (x_1 + x_2), y \rangle = \langle x_1, y \rangle + \langle x_2, y \rangle$

c. $\|\alpha x + \beta y\|^2 \equiv \langle (\alpha x + \beta y), (\alpha x + \beta y) \rangle$
$\equiv |\alpha|^2 \langle x, x \rangle + |\beta|^2 \langle y, y \rangle + 2 \operatorname{Re} (\alpha^* \beta \langle x, y \rangle)$

Here* denotes the complex conjugate of a complex number,

$$z \equiv x + iy \equiv \operatorname{Re} z + i \operatorname{Im} z \qquad i = \sqrt{-1} \qquad z^* = x - iy$$

1.3 Prove that E_∞ is a linear vector space.

1.4 Prove the results of Problem 1.2 using the abstract definition of the scalar product given in Section 1.2. Furthermore show that this definition implies the *Schwarz inequality*,

$$|\langle x, y \rangle| \leq \|x\| \, \|y\|$$

1.5 Prove the triangle inequality

$$\|x + y\| \leq \|x\| + \|y\|$$

1.6 Consider a linear vector space of once-differentiable continuous complex-valued functions on the interval $x \in [a, b]$. Which—if any—of the following defines a scalar product?

a. $\langle f, g \rangle = f^*(a)g(a)$
b. $\langle f, g \rangle = \int_a^b f^*(x)g(x)\,dx$
c. $\langle f, g \rangle = f'^*(a)g'(a)$
d. $\langle f, g \rangle = \int_a^b f'^*(x)g'(x)\,dx$
e. $\langle f, g \rangle = \int_a^b f'^*(x)g'(x)\,dx + f^*(a)g(a)$

1.7 Use the definition for continuity of a real function of two variables $f(x, y)$ for (x, y) in a region R (cf. Appendix I) to demonstrate that if $f(x, y)$ is defined in R and positive and continuous at $(x_0, y_0) \in R$, then there is a neighborhood about (x_0, y_0) such that $f(x, y) > \frac{1}{2}f(x_0, y_0)$ [i.e., $\sqrt{(x-x_0)^2+(y-y_0)^2} < \delta$].

1.8 Discuss the continuity or uniform continuity if applicable of the following functions.

a. $f(x) = x^2,\ 0 < x < \infty$
b. $f(x) = \begin{cases} \dfrac{1}{1-e^{1/x}}, & x \in (-\infty, \infty),\ x \neq 0 \\ 0 & x = 0 \end{cases}$
c. $f(x) = \tan^{-1}(\sinh x),\ x \in (-\infty, \infty)$

1.9 Discuss the convergence or uniform convergence of the following sequences of functions and find their limits

a. $f_n(x) = \dfrac{n}{1+nx^2},\ x \in [1, \infty)$
b. $f_n(x) = \tanh(nx),\ x \in (-\infty, \infty)$
c. $f_n(x) = \dfrac{1}{1+nx},\ x \in (0, 1]$

1.10 If $\sum_{n=1}^{\infty} a_n$ is an absolutely convergent series, prove that $\sum_{n=1}^{\infty} a_n \sin(nx)$ is uniformly convergent for $0 \le x \le 2\pi$.

1.11 Prove that the series

$$x^2 \sum_{n=0}^{\infty} \frac{1}{(1+x^2)^n}$$

is not uniformly convergent in any interval that includes $x=0$.

1.12 Show that the series

$$\sum_{n=1}^{\infty} \frac{(-1)^{n+1}}{\sqrt{n}}$$

converges conditionally but not absolutely.

1.13 The projection theorem (Th. 1.1) proves the existence of vectors orthogonal to every subspace. Prove the "converse" statement that all vectors orthogonal to a fixed vector form a subspace.

1.14 Consider the space $\mathscr{L}_2$, $x \in [-1, +1]$, defined at the end of Section 1.1c. Let $\mathscr{M}$ be that subspace of $\mathscr{L}_2$ spanned by the linearly independent functions $P_0 = 1/\sqrt{2}$, $P_1 = \sqrt{(3/2)}\,x$, $P_2 = \sqrt{(5/2)}\,[(3x^2-1)/2]$ and $\mathscr{N}$ be that subspace of functions $g(x)$ of $\mathscr{L}_2$

such that

$$\langle g, P_j \rangle = 0 \qquad j = 0, 1, 2$$

Prove that $\mathscr{L}_2 = \mathscr{M} \oplus \mathscr{N}$. Show that if $f(x) \in \mathscr{L}_2$ then $f(x) - \sum_{j=0}^{2} \alpha_j P_j(x)$ is orthogonal to $\mathscr{M}$, where $\alpha_j \equiv \langle P_j \mid f \rangle$.

1.15 Prove that the necessary and sufficient condition that a subspace, which is the sum of two subspaces, be the direct sum of these subspaces is that the zero vector be the only vector that is contained in *both* of the smaller subspaces.

1,16 Let $\{e_j\}$, $j = 1, 2, \ldots, n$, be an orthonormal set of vectors. Find those values of $\{\alpha_j\}$ such that

$$\sum_{j=1}^{n} \alpha_j e_j$$

minimizes the norm

$$\left\| x - \sum_{j=1}^{n} \alpha_j e_j \right\|$$

for a fixed vector x. From this result deduce Bessel's inequality,

$$\langle x, x \rangle \geq \sum_{j=1}^{n} |\langle e_j, x \rangle|^2$$

You may find it useful to realize that if $f(x_1, x_2, \ldots, x_n)$ is a real function of n independent real variables, and if for *arbitrary* infinitesimal $\{\delta x_j\}$,

$$f(\bar{x}_1 + \delta x_1, \bar{x}_2 + \delta x_2, \ldots, \bar{x}_n + \delta x_n) - f(\bar{x}_1, \bar{x}_2, \ldots, \bar{x}_n) > 0$$

then $f(x_1, x_2, \ldots, x_n)$ has a *minimum* at $(\bar{x}_1, \bar{x}_2, \ldots, \bar{x}_n)$. Furthermore we have

$$f(\bar{x}_1 + \delta x_1, \bar{x}_2 + \delta x_2, \ldots, \bar{x}_n + \delta x_n) - f(\bar{x}_1, \bar{x}_2, \ldots, \bar{x}_n) \simeq \sum_{j=1}^{n} \frac{\partial f}{\partial x_j}\bigg|_{\bar{x}} \delta x_j + \sum_{j,k=1}^{n} \frac{\partial^2 f}{\partial x_j \, \partial x_k}\bigg|_{\bar{x}} \delta x_j \, \delta x_k$$

while in the present problem

$$\frac{\partial f}{\partial \alpha_j}\bigg|_{\bar{\alpha}} = 0 \; \forall j \qquad \text{and} \qquad \frac{\partial^2 f}{\partial \alpha_j \, \partial \alpha_k} \text{ is diagonal}$$

1.17 a. Prove that it is impossible to have $|f(x)| < M$ (where M is independent of x) for all $x \neq 0$ for a linear functional unless $f(x) \equiv 0 \; \forall x$.

b. Prove that for a linear functional

$$|f(x)| < C \qquad \forall x \ni \|x\| = 1 \Leftrightarrow |f(x)| < C \, \|x\| \qquad \forall x \ni 0 < \|x\| < \infty$$

Here C is a constant independent of x.

1.18 Prove that a linear functional is continuous if and only if it is bounded. The proof that continuity implies boundedness is easily done by contradiction using the result of Problem 1.17 and considering a sequence of vectors $\{x_n\}$ such that $\|x_n\| < 1$.

2 Operators on Linear Spaces

2.0 INTRODUCTION

In this chapter we discuss linear operators and their representations on linear vector spaces. We begin with a consideration of a very well-known and useful class of linear operators—*matrices*—in a finite-dimensional vector space. One's intuition from a real three-dimensional space is of great value in understanding the theorems and techniques that we will develop in the first few sections.

In the last three sections of this chapter we extend our discussion to general linear vector spaces, which include function spaces, and lay the foundations for our study of integral equations in Chapter 5.

2.1 MATRICES AND DETERMINANTS

a. Definitions and Basic Operations with Matrices

A *matrix* is simply an $m \times n$ array of elements that we will conventionally exhibit as follows.

$$A = m\overset{\longleftarrow n \longrightarrow}{\begin{pmatrix} a_{11} & a_{12} & a_{13} & \cdots & a_{1n} \\ a_{21} & a_{22} & a_{23} & \cdots & a_{2n} \\ \cdot & \cdot & \cdot & \cdot & \cdot \\ \cdot & \cdot & \cdot & \cdot & \cdot \\ \cdot & \cdot & \cdot & \cdot & \cdot \\ a_{m1} & a_{m2} & a_{m3} & \cdots & a_{mn} \end{pmatrix}} \tag{2.1}$$

Here A is an "$m \times n$" matrix having m *rows* and n *columns* and is referred to as a matrix of *order* $m \times n$. A matrix has no numerical value associated with it (unlike determinants that we will define in Section 2.1b). We will often use a compact notation to write Eq. 2.1 as

$$A = \{a_{ij}\} \tag{2.2}$$

where the index i labels the rows and j the columns. A matrix with only one row (or column) is called a row (or column) vector.

$$(\xi_1 \quad \xi_2 \quad \xi_3 \quad \cdots \quad \xi_n) \qquad \textit{row} \text{ vector}$$

$$\begin{pmatrix} \xi_1 \\ \xi_2 \\ \xi_3 \\ \cdot \\ \cdot \\ \cdot \\ \xi_n \end{pmatrix} \qquad \textit{column} \text{ vector}$$

For matrices we now *define* certain operations that will prove useful later. We first agree that two matrices are equal if and only if they are equal element by element. It is then evident that two matrices can be equal only if they are of the same order.

Given two matrices $A = \{a_{ij}\}$ and $B = \{b_{ij}\}$ of the same order we define the *sum* $A+B$ to be that matrix $C = \{c_{ij}\}$ such that $c_{ij} = a_{ij} + b_{ij}$, $i = 1, 2, \ldots, m$, $j = 1, 2, \ldots, n$. That is, we add the matrices element by element. This sum is defined only for matrices of the same order.

Example 2.1

$$A = \begin{pmatrix} 1 & i \\ 2 & 1 \end{pmatrix} \qquad B = \begin{pmatrix} 0 & 3 \\ 1 & 0 \end{pmatrix} \qquad C = A + B = \begin{pmatrix} 1 & 3+i \\ 3 & 1 \end{pmatrix}$$

A matrix is zero if and only if all of its elements are zero.

Next we define an operation of *multiplying* one matrix by another. Let A be a matrix of order $r \times s$ and B a matrix of order $m \times n$.

$$A = r \overbrace{\begin{pmatrix} a_{11} & a_{12} & \cdots & a_{1s} \\ a_{21} & a_{22} & \cdots & a_{2s} \\ \cdot & \cdot & \cdot & \cdot \\ \cdot & \cdot & \cdot & \cdot \\ \cdot & \cdot & \cdot & \cdot \\ a_{r1} & a_{r2} & \cdots & a_{rs} \end{pmatrix}}^{s} \qquad B = m \overbrace{\begin{pmatrix} b_{11} & b_{12} & \cdots & b_{1n} \\ b_{21} & b_{22} & \cdots & b_{2n} \\ \cdot & \cdot & \cdot & \cdot \\ \cdot & \cdot & \cdot & \cdot \\ \cdot & \cdot & \cdot & \cdot \\ b_{m1} & b_{m2} & \cdots & b_{mn} \end{pmatrix}}^{n}$$

The product matrix $C \equiv AB$ exists if and only if $s = m$ (i.e., the number of columns in A equals the number of rows in B) and is defined as $C = \{c_{ij}\}$ where

$$c_{ij} = \sum_{l=1}^{m} a_{il} b_{lj} \qquad \forall i = 1, 2, \ldots, r \qquad j = 1, 2, \ldots, n \tag{2.3}$$

That is, loosely speaking, we multiply the rows of A by the columns of B to obtain the columns of C. The result is a matrix C of order $r \times n$.

Example 2.2

$$A = \begin{pmatrix} 1 & i \\ 2 & 1 \end{pmatrix} \qquad B = \begin{pmatrix} 0 & 3 \\ 1 & 0 \end{pmatrix}$$

$$C \equiv AB = \begin{pmatrix} 1 & i \\ 2 & 1 \end{pmatrix} \begin{pmatrix} 0 & 3 \\ 1 & 0 \end{pmatrix} = \begin{pmatrix} i & 3 \\ 1 & 6 \end{pmatrix}$$

Multiplication of a matrix A by a scalar α is defined as a matrix $C = \alpha A = \{c_{ij}\}$ where

$$c_{ij} = \alpha a_{ij}$$

That is, every element of the matrix A is multiplied by the scalar α.

Example 2.3

$$A = \begin{pmatrix} 1 & i \\ 2 & 1 \end{pmatrix} \qquad \alpha = 2 \qquad C \equiv \alpha A = 2 \begin{pmatrix} 1 & i \\ 2 & 1 \end{pmatrix} = \begin{pmatrix} 2 & 2i \\ 4 & 2 \end{pmatrix}$$

The *transpose* of a matrix $A = \{a_{ij}\}$ is denoted by $A^{T} = \{a_{ij}^{T}\}$ where

$$a_{ij}^{T} \equiv a_{ji}$$

That is, A^{T} is obtained from A by taking the columns of A to be the rows of A^{T}.

Example 2.4

$$A = \begin{pmatrix} 1 & i \\ 2 & 1 \end{pmatrix} \qquad A^{T} = \begin{pmatrix} 1 & 2 \\ i & 1 \end{pmatrix}$$

The *hermitian adjoint* (or simply *adjoint*) of a matrix $A = \{a_{ij}\}$ is denoted by $A^{\dagger} = (a_{ij}^{\dagger})$ where

$$a_{ij}^{\dagger} = a_{ji}^{*}$$

That is, the adjoint is the complex conjugate of the transpose [i.e., $A^{\dagger}=(A^{T})^{*}=(A^{*})^{T}$].

Example 2.5

$$A=\begin{pmatrix}1 & i\\ 2 & 1\end{pmatrix} \qquad A^{\dagger}=\begin{pmatrix}1 & 2\\ -i & 1\end{pmatrix}$$

A very important class of matrices are *square matrices* of order $n\times n$. In particular the *unit matrix* is a square matrix all of whose elements are zero, except for the diagonal elements, all of which are unity.

$$I=\begin{pmatrix}1 & 0 & 0 & \cdots & 0\\ 0 & 1 & 0 & \cdots & 0\\ 0 & 0 & 1 & \cdots & 0\\ \cdot & \cdot & \cdot & \cdot & \cdot\\ \cdot & \cdot & \cdot & \cdot & \cdot\\ \cdot & \cdot & \cdot & \cdot & \cdot\\ 0 & 0 & 0 & \cdots & 1\end{pmatrix}$$

We now summarize the properties of the matrix operations.

i. $AO=OA=O$
ii. $IA=AI=A$
iii. $A(B+C)=AB+AC$ multiplication distributive w.r.t. addition
iv. $(A+B)C=AC+BC$ addition distributive w.r.t. multiplication
v. $A(BC)=(AB)C$ associative w.r.t. multiplication
vi. $A+B=B+A$ commutative w.r.t. addition

b. Determinants

Corresponding to every square matrix $A=\{a_{ij}\}$, $i, j=1, 2, \ldots, n$, we define a *determinant* of A, denoted by det A, as follows,

$$\det A\equiv \varepsilon_{ijk\cdots}a_{1i}a_{2j}a_{3k}\cdots\equiv\begin{vmatrix}a_{11} & a_{12} & \cdots & a_{1n}\\ a_{21} & a_{22} & \cdots & a_{2n}\\ \cdot & \cdot & \cdot & \cdot\\ \cdot & \cdot & \cdot & \cdot\\ \cdot & \cdot & \cdot & \cdot\\ a_{n1} & a_{n2} & \cdots & a_{nn}\end{vmatrix} \tag{2.4}$$

where summation over repeated indices is implied and where the completely antisymmetric n-index *Levi-Civita symbol* $\varepsilon_{ijk\cdots}$ is defined as

$$\varepsilon_{ijk\cdots} = \begin{cases} 0, & \text{if any two of the } n \text{ indices are equal} \\ +1, & \text{if the indices } i, j, k, \ldots, \text{ are an } \textit{even} \text{ permutation of } 1, 2, 3, \ldots, n \\ -1, & \text{if the indices } i, j, k, \ldots, \text{ are an } \textit{odd} \text{ permutation of } 1, 2, 3, \ldots, n \end{cases} \tag{2.5}$$

A *permutation* is even or odd according to whether it takes an even or odd number of *transpositions* to bring the permutation $ijk \cdots$ into the standard permutation $1, 2, 3, \cdots n$. A transposition is the interchange of the positions of two elements in a permutation. It is evident from Eqs. 2.4 and 2.5 that there are $n!$ nonvanishing terms in Eq. 2.4 each term being a product of n elements (one from every row and column in the matrix A). In other words det A is simply the sum of all possible permutations of products of one element from each row and column, each term weighted with a factor $(-1)^P$ where P is the *parity* of the permutation (i.e., $P=+1$ for an even permutation and $P=-1$ for an odd permutation).

A direct term-by-term examination of the sum indicated in Eq. 2.4 will show that we could have begun by ordering the column indices as $1, 2, 3, \ldots, n$ and summing over the row indices so that a completely equivalent definition of the determinant of A is

$$\det A = \varepsilon_{ijk\cdots} a_{i1} a_{j2} a_{k3} \cdots \tag{2.6}$$

That is, we have

$$\det A = \det A^{\mathrm{T}}$$

These determinants (which *do* have a numerical value associated with them, unlike matrices) have several important properties that we now list, the obvious ones without proof.

i. If every element in a row (or column) has a common factor, then this constant may be factored out of the determinant.
ii. The determinant of the unit matrix is unity.
iii. If all of the elements of a given row (or column) are zero, then the determinant vanishes.
iv. The effect of interchanging any two rows (or columns) of a determinant is to change the sign of the determinant.

PROOF

This follows directly from the definition in Eq. 2.4. Denote by $\bar{A}$ that matrix obtained from A by interchanging the first two rows. Then we have

$$\begin{aligned} \bar{A} &= \varepsilon_{ijk\cdots} a_{2i} a_{1j} a_{3k\cdots} = \varepsilon_{ijk\cdots} a_{1j} a_{2i} a_{3k\cdots} \\ &= \varepsilon_{jik\cdots} a_{1i} a_{2j} a_{3k\cdots} = -\varepsilon_{ijk\cdots} a_{1i} a_{2j} a_{3k\cdots} \\ &= -\det A \end{aligned}$$

This shows that property (iv) holds for the interchange of any two neighboring rows. However the reader may easily convince himself that any two rows may be interchanged by an *odd* number of interchanges of neighboring rows. That is, if rows i and f have n other rows between them, then by means of $(n+1)$ transpositions we may move f up and arrange to have rows i and f next to each other in the order f, i and then by n more transpositions we can move i down to the initial position of f, so that a total of $(2n+1)$ transpositions is required to interchange i and f. Q.E.D.

v. Any determinant having two rows (or columns) identical vanishes.
vi. The determinant of the product of two matrices, $\det(AB)$, has the value

$$\det(AB) = (\det A)(\det B)$$

PROOF

We begin by writing the definition of the determinant, Eq. 2.4, in a slightly more general form,

$$\varepsilon_{\alpha\beta\gamma\cdots} \det A = \varepsilon_{ijk\cdots} a_{\alpha i} a_{\beta j} a_{\gamma k} \cdots$$

If $\alpha\beta\gamma \cdots = 123 \cdots n$, this reduces to the original definition, Eq. 2.4. If any two of the indices $\alpha\beta\gamma \cdots$ are identical, both sides vanish since the right side is then a determinant with two identical rows. If $\alpha\beta\gamma \cdots$ is some permutation of parity P of the standard permutation $123 \cdots n$, then both sides pick up a factor $(-1)^P$ in the process of normal ordering. Similarly we may write

$$\varepsilon_{\alpha\beta\gamma\cdots} \det A = \varepsilon_{ijk\cdots} a_{i\alpha} a_{j\beta} a_{k\gamma} \cdots$$

Therefore we have

$$\begin{aligned}(\det A)(\det B) &= (\det A)\varepsilon_{ijk\cdots}(b_{i1} b_{j2} b_{k3} \cdots) \\ &= \varepsilon_{\alpha\beta\gamma\cdots} a_{\alpha i} a_{\beta j} a_{\gamma k} \cdots b_{i1} b_{j2} b_{k3} \cdots \\ &= \varepsilon_{\alpha\beta\gamma\cdots}(a_{\alpha i} b_{i1})(a_{\beta j} b_{j2})(a_{\gamma k} b_{k3}) \cdots = \det(AB) \qquad \text{Q.E.D.}\end{aligned}$$

vii. A determinant is left unchanged when a scalar multiple of a row (or column) is added to another row (or column).

PROOF

For instance, if A is a matrix with elements a_{j2} in the second column, let $\bar{A}$ be that matrix with elements $(\alpha a_{j1} + a_{j2})$ for its second column. Then we have

$$\begin{aligned}\det \bar{A} &\equiv \varepsilon_{ijk\cdots} a_{i1}(\alpha a_{j1} + a_{j2}) a_{k3} \cdots \\ &= \alpha \varepsilon_{ijk\cdots} a_{i1} a_{j1} a_{k3} \cdots + \varepsilon_{ijk\cdots} a_{i1} a_{j2} a_{k3} \cdots = \det A\end{aligned}$$

by use of property (v). This argument holds for any two rows (or columns).

Q.E.D.

viii. The necessary and sufficient condition that the row (or column) vectors of a matrix A be linearly dependent is that $\det A = 0$.

PROOF

a. *Necessary.* If we assume the linear dependence of the row (or column) vectors, then the result follows from properties (vii) and (iii).

b. *Sufficient.* By property (vii) we can always arrange, for example, to have $a'_{j1} = 0\ \forall j$, except $j = 1$. Now if $a'_{11} = 0$ also, then these vectors are linearly dependent since we then have n vectors in a manifold of dimension $(n-1)$ (cf. Section 1.4b), while $\det A = 0$ by property (iii). If $a'_{11} \neq 0$, we proceed to make all of the elements in the second column zero except the top two. If we do this n times, we have, pictorially,

$$\begin{vmatrix} a_{11} & a_{12} & a_{13} & \cdots & a_{1n} \\ a_{21} & a_{22} & a_{23} & \cdots & a_{2n} \\ a_{31} & a_{32} & a_{33} & \cdots & a_{3n} \\ \cdot & \cdot & \cdot & \cdot & \cdot \\ \cdot & \cdot & \cdot & \cdot & \cdot \\ \cdot & \cdot & \cdot & \cdot & \cdot \\ a_{n1} & a_{n2} & a_{n3} & \cdots & a_{nn} \end{vmatrix} = \begin{vmatrix} a'_{11} & a'_{12} & a'_{13} & \cdots & a'_{1n} \\ 0 & a'_{22} & a'_{23} & \cdots & a'_{2n} \\ 0 & a'_{32} & a'_{33} & \cdots & a'_{3n} \\ \cdot & \cdot & \cdot & \cdot & \cdot \\ \cdot & \cdot & \cdot & \cdot & \cdot \\ \cdot & \cdot & \cdot & \cdot & \cdot \\ 0 & a'_{n2} & a'_{n3} & \cdots & a'_{nn} \end{vmatrix}$$

$$= \begin{vmatrix} a''_{11} & a''_{12} & a''_{13} & \cdots & a''_{1n} \\ 0 & a''_{22} & a''_{23} & \cdots & a''_{2n} \\ 0 & 0 & a''_{33} & \cdots & a''_{3n} \\ 0 & 0 & a''_{43} & \cdots & a''_{4n} \\ \cdot & \cdot & \cdot & \cdot & \cdot \\ \cdot & \cdot & \cdot & \cdot & \cdot \\ \cdot & \cdot & \cdot & \cdot & \cdot \\ 0 & 0 & a''_{n3} & \cdots & a''_{nn} \end{vmatrix} = \cdots = \begin{vmatrix} \bar{a}_{11} & \bar{a}_{12} & \bar{a}_{13} & \cdots & \bar{a}_{1n} \\ 0 & \bar{a}_{22} & \bar{a}_{23} & \cdots & \bar{a}_{2n} \\ 0 & 0 & \bar{a}_{33} & \cdots & \bar{a}_{3n} \\ \cdot & \cdot & \cdot & \cdot & \cdot \\ \cdot & \cdot & \cdot & \cdot & \cdot \\ \cdot & \cdot & \cdot & \cdot & \cdot \\ 0 & 0 & 0 & \cdots & \bar{a}_{nn} \end{vmatrix}$$

where this last determinant is in triangular form. From Eq. 2.6 we see that

$$\det A = \bar{a}_{11}\bar{a}_{22}\bar{a}_{33} \cdots \bar{a}_{nn} \tag{2.7}$$

If we now assume that $\det A = 0$, then at least one of the $\bar{a}_{jj} = 0$ and the vectors are linearly dependent since if $\bar{a}_{kk} = 0$ then there will be $(n-k+1)$ vectors in a manifold of dimension $(n-k)$, $k = 1, 2, \ldots, (n-1)$. Clearly, if $\bar{a}_{nn} = 0$, this implies that the last two row vectors in the previous step were linearly dependent. Q.E.D.

We will now prove some theorems that will be very useful for evaluating determinants and for constructing inverses of matrices.

THEOREM 2.1

Any determinant can be expressed as

$$\det A=\sum_{j=1}^{n} a_{Nj} \operatorname{cof}(a_{Nj}) \tag{2.8}$$

where the cofactor of a_{Nj}, denoted by $\operatorname{cof}(a_{Nj})$, is defined as $(-1)^{N+j}$ times the determinant of order $(n-1)\times(n-1)$ obtained from the determinant of A by striking the Nth row and jth column.

Here N is a fixed-row index (i.e., we are expanding across a given row by cofactors). Once this formula has been proved for an expansion by a row, it will obviously hold for an expansion by columns since $\det A=\det A^{T}$.

PROOF

$$\begin{aligned}\det A &= \varepsilon_{ijk\cdots\alpha\beta\gamma\cdots} a_{1i}a_{2j}a_{3k} \cdots a_{N-1,\alpha}a_{N,\beta}a_{N+1,\gamma} \cdots \\ &= \sum_{\beta=1}^{n} a_{N\beta}[\varepsilon_{ijk\cdots\alpha\beta\gamma\cdots} a_{1i}a_{2j}a_{3k} \cdots a_{N-1,\alpha}a_{N+1,\gamma} \cdots] \\ &= \sum_{\beta=1}^{N} a_{N\beta}(-1)^{N-1}[\varepsilon_{\beta ijk\cdots\alpha\gamma\cdots} a_{1i}a_{2j}a_{3k} \cdots a_{N-1,\alpha}a_{N+1,\gamma} \cdots]\end{aligned} \tag{2.9}$$

From the definition of the Levi-Civita symbol, Eq. 2.5, we see that for a *fixed* value of β, and once the indices $ijk\cdots\alpha\gamma\cdots$ are normal ordered, $\varepsilon_{\beta ijk\cdots\alpha\gamma\cdots}$, when it does not vanish, is $+1$ when β is odd and -1 when β is even. That is, $\varepsilon_{\beta ijk\cdots}$ is equal to $(-1)^{\beta-1}$ times a Levi-Civita symbol with $(n-1)$ indices where the normal order of the indices is, for $\beta=m\leq n$, $1, 2, 3\cdots(m-1), (m+1),\ldots, n$. Therefore in Eq. 2.9 we have

$$(-1)^{N-1}[\varepsilon_{\beta ijk\cdots\alpha\gamma\cdots} a_{1i}a_{2j}a_{3k} \cdots a_{N-1,\alpha}a_{N+1,\gamma} \cdots]=\operatorname{cof}(a_{N\beta})$$

so that

$$\det A=\sum_{\beta=1}^{n} a_{N\beta} \operatorname{cof}(a_{N\beta}) \qquad \text{Q.E.D.}$$

Notice that there is *no* summation implied over the fixed index N in Eq. 2.8.

THEOREM 2.2

$$\delta_{NM} \det A=\sum_{j=1}^{n} a_{Nj} \operatorname{cof}(a_{Mj}) \tag{2.10}$$

PROOF

a. The case $N=M$ is simply Th. 2.1.
b. If $N\neq M$, then the l.h.s. of Eq. 2.10 vanishes identically. If we now use Eq.

2.9 to write out Eq. 2.10 when $M \neq N$, we find

$$\sum_{j=1}^{n} a_{Nj} \operatorname{cof}(a_{Mj})$$

$$= \sum_{\beta=1}^{n} a_{N\beta}(-1)^{M-1}[\varepsilon_{\beta i \cdots jkl \cdots \alpha\gamma \cdots} a_{1i} \cdots a_{N-1,j} a_{N,k} a_{N+1,l} \cdots a_{M-1,\alpha} a_{M+1,\gamma} \cdots]$$

$$= \varepsilon_{i \cdots jkl \cdots \alpha\beta\gamma \cdots} a_{1i} \cdots a_{N-1,j} a_{N,k} a_{N+1,l} \cdots a_{M-1,\alpha} a_{N,\beta} a_{M+1,\gamma} \cdots$$

which is a determinant with two identical rows (i.e., the Nth row appears twice) and is zero by property (v) of Section 2.1b. Q.E.D.

c. Inverse of a Matrix

THEOREM 2.3

A matrix $A = \{a_{ij}\}$ has an inverse matrix $A^{-1} \equiv \{(a^{-1})_{ij}\}$, where

$$(a^{-1})_{ij} \equiv \frac{1}{\det A} \operatorname{cof}(a_{ji}) \tag{2.11}$$

and such that $AA^{-1} = I = A^{-1}A$ if and only if $\det A \neq 0$.

PROOF

a. If $\det A \neq 0$, define a matrix $B = \{b_{ij}\}$ with

$$b_{ij} \equiv \frac{1}{\det A} \operatorname{cof}(a_{ji})$$

Then let $C = AB$ so that

$$c_{rs} = a_{rj} b_{js} = \sum_j a_{rj} \frac{\operatorname{cof}(a_{sj})}{\det A} = \delta_{rs} \frac{\det A}{\det A} = \delta_{rs}$$

by Th. 2.2, which implies that $C = I$ so that $B = A^{-1}$. Similarly we see that $A^{-1}A = I$.

b. If we are given the existence of an A and an A^{-1} such that $AA^{-1} = I = A^{-1}A$, then by property (vi) of Section 2.1b we have

$$\det(AA^{-1}) = (\det A)(\det A^{-1}) = \det I = 1$$

Since we are dealing with finite-dimensional matrices, $\det A^{-1}$ is finite so that $\det A \neq 0$. Q.E.D.

Notice that in the process of proving this theorem we have also given a formula, Eq. 2.11, for constructing A^{-1} given A and the fact that $\det A \neq 0$. If and only if $\det A \neq 0$, the matrix A is termed *nonsingular*. This inverse is *unique* and has the property

$$(AB)^{-1} = B^{-1}A^{-1} \tag{2.12}$$

Example 2.6 To construct the cofactor of the element a_{ij} of A delete the ith row and jth column from A, form the determinant of the remaining matrix, and weight this with a factor $(-1)^{i+j}$.

$$A=\begin{pmatrix} a_{11} & a_{12} & a_{13} \\ a_{21} & a_{22} & a_{23} \\ a_{31} & a_{32} & a_{33} \end{pmatrix}$$

$$\operatorname{cof}(a_{21})=(-1)^{2+1}\begin{vmatrix} a_{12} & a_{13} \\ a_{32} & a_{33} \end{vmatrix}=-(a_{12}a_{33}-a_{32}a_{13})$$

$$A=\begin{pmatrix} 1 & 0 & 2 \\ 0 & 3 & 1 \\ 0 & 1 & 2 \end{pmatrix}$$

$$\det A=5$$

$$A^{-1}=\frac{1}{5}\begin{pmatrix} 5 & 2 & -6 \\ 0 & 2 & -1 \\ 0 & -1 & 3 \end{pmatrix}$$

We see from this theorem and the properties of matrix operations listed at the end of Section 2.1a that square matrices of order n form an abelian group with respect to matrix addition with identity element O and that all such nonsingular matrices form a group (in general noncommutative) with respect to matrix multiplication with identity element I.

d. Direct Product of Matrices

In Section 2.1a, Eq. 2.3, we defined one type of matrix multiplication. We shall now define the *direct*, or *Kronecker*, *product* of a matrix A of order $r\times s$ by a matrix B of order $m\times n$ as

$$C=A\otimes B \tag{2.13}$$

where

$$c_{ij;kl}\equiv(A\otimes B)_{ij;kl}\equiv a_{ik}b_{jl} \qquad \begin{matrix} i=1,2,\ldots,r & k=1,2,\ldots,s \\ j=1,2,\ldots,m & l=1,2,\ldots,n \end{matrix} \tag{2.14}$$

Obviously C is a matrix of order $(rm)\times(sn)$. We will use this direct product mainly for square matrices and for column vectors. In this latter case if we have a set of basis vectors $\{\phi_i\}$, $i=1,2,\ldots,m$, in a space of dimension m and $\{\psi_j\}$, $j=1,2,\ldots,n$, in a space of dimension n, then we obtain a set of basis vectors

$\{\chi_{ij}\}$ for a space of dimension mn as

$$\chi_{ij}=\phi_i\otimes\psi_j \qquad \begin{array}{l} i=1,2,\ldots,m \\ j=1,2,\ldots,n \end{array} \tag{2.15}$$

We see that this is a way to build higher dimensional vector spaces from two lower dimensional ones. This technique will prove to be important in our study of group representations in Chapter 9.

For an $m\times m$ matrix A and an $n\times n$ matrix B, C is an $(mn)\times(mn)$ matrix. To see what this operation looks like explicitly in terms of matrices we must establish a correspondence for the elements of C labeled as c_{rs}, $r, s,= 1, 2, \ldots, mn$, in terms of the $c_{ij,kl}$. We do this by letting r (and s too) take on the values

$$r=11, 12, 13, \ldots, 1n, 21, 22, 23, \ldots, 2n, \ldots, m1, m2, m3, \ldots, mn$$

Therefore $C=\{c_{rs}\}=\{c_{ij,kl}\}$ has the form

$$\begin{pmatrix} c_{11;11} & c_{11;12} & c_{11;13} & \cdots & c_{11;1n} & \cdots & c_{11;mn} \\ c_{12;11} & c_{12;12} & c_{12;13} & \cdots & c_{12;1n} & \cdots & c_{12;mn} \\ \cdot & \cdot & \cdot & \cdot & \cdot & \cdot & \cdot \\ \cdot & \cdot & \cdot & \cdot & \cdot & \cdot & \cdot \\ \cdot & \cdot & \cdot & \cdot & \cdot & \cdot & \cdot \\ c_{mn;11} & c_{mn;12} & c_{mn;13} & \cdots & c_{mn;1n} & \cdots & c_{mn;mn} \end{pmatrix}$$

$$=\begin{pmatrix} a_{11}b_{11} & a_{11}b_{12} & a_{11}b_{13} & \cdots & a_{11}b_{1n} & \cdots & a_{1m}b_{1n} \\ a_{11}b_{21} & a_{11}b_{21} & a_{11}b_{23} & \cdots & a_{11}b_{2n} & \cdots & a_{1m}b_{2n} \\ \cdot & \cdot & \cdot & \cdot & \cdot & \cdot & \cdot \\ \cdot & \cdot & \cdot & \cdot & \cdot & \cdot & \cdot \\ \cdot & \cdot & \cdot & \cdot & \cdot & \cdot & \cdot \\ a_{m1}b_{n1} & a_{m1}b_{n2} & a_{m1}b_{n3} & \cdots & a_{m1}b_{nm} & \cdots & a_{mm}b_{nn} \end{pmatrix} \tag{2.16}$$

Symbolically we may write Eq. 2.16 in the compact form

$$A\otimes B=\begin{pmatrix} a_{11}B & a_{12}B & \cdots & a_{1n}B \\ a_{21}B & a_{22}B & \cdots & a_{2n}B \\ \cdot & \cdot & \cdot & \cdot \\ \cdot & \cdot & \cdot & \cdot \\ \cdot & \cdot & \cdot & \cdot \\ a_{m1}B & a_{m2}B & \cdots & a_{mm}B \end{pmatrix} \tag{2.17}$$

We now prove two useful properties of direct products.

i. $(A\otimes B)(C\otimes D)=(AC)\otimes(BD)$ **(2.18)**

Here A and C are taken to be square matrices of the same order, for example, $m\times m$, and B and D square matrices of order $n\times n$.

PROOF

$$[(AC)\otimes(BD)]_{ij;kl}=(AC)_{ik}(BD)_{jl}=a_{ir}c_{rk}b_{js}d_{sl}$$
$$=a_{ir}b_{js}c_{rk}d_{sl}=(A\otimes B)_{ij;rs}(C\otimes D)_{rs;kl}=[(A\otimes B)(C\otimes D)]_{ij;kl}\quad \text{Q.E.D.}$$

ii. If we define the trace of a matrix A, Tr A, as

$$\text{Tr A}\equiv\sum_{j=1}^{n} a_{jj} \tag{2.19}$$

then

$$\text{Tr}(\text{A}\otimes B)=\text{Tr A Tr }B \tag{2.20}$$

PROOF

$$\text{Tr}(\text{A}\otimes B)=\sum_{i,j}(\text{A}\otimes B)_{ij;ij}=\sum_{i,j}a_{ii}b_{jj}=\text{Tr A Tr }B \qquad \text{Q.E.D.}$$

The following two properties are easily proved.

iii. $(\text{A}+B)\otimes C=\text{A}\otimes C+B\otimes C$
iv. $(\text{A}\otimes B)^{-1}=\text{A}^{-1}\otimes B^{-1}$, *provided* det $\text{A}\neq 0$ and det $B\neq 0$

2.2 SYSTEMS OF LINEAR ALGEBRAIC EQUATIONS

a. Cramer's Rule

We are now in a position to generalize the results of Section 1.0 to a system of n linear algebraic equations in n unknowns,

$$\begin{aligned} a_{11}x_1+a_{12}x_2+\cdots+a_{1n}x_n&=b_1\\ a_{21}x_1+a_{22}x_2+\cdots+a_{2n}x_n&=b_2\\ \vdots\qquad\qquad&\quad\vdots\\ a_{n1}x_1+a_{n2}x_2+\cdots+a_{nn}x_n&=b_n \end{aligned} \tag{2.21}$$

We simplify the notation considerably by defining a matrix $\text{A}=\{a_{ij}\}$ and column vectors

$$\mathbf{x}=\begin{pmatrix}x_1\\x_2\\ \cdot\\ \cdot\\ \cdot\\ x_n\end{pmatrix}\qquad \mathbf{b}=\begin{pmatrix}b_1\\b_2\\ \cdot\\ \cdot\\ \cdot\\ b_n\end{pmatrix}$$

and writing our system of equations as

$$\mathbf{A}\mathbf{x}=\mathbf{b} \tag{2.22}$$

As long as the system is nonsingular, that is $\Delta \equiv \det A \neq 0$, we can trivially exhibit the unique solution to Eq. 2.22 since A^{-1} exists so that

$$\mathbf{x}=\mathrm{A}^{-1}\mathbf{b} \tag{2.23}$$

If we express Eq. 2.23 in component form and use Eq. 2.11, we obtain

$$x_i=(a^{-1})_{ij}b_j=\frac{1}{\Delta}\sum_{j=1}^{n} b_j \operatorname{cof}(a_{ji})\equiv\frac{\Delta_i}{\Delta} \qquad i=1,2,\ldots,n \tag{2.24}$$

where, from Eq. 2.8, we see that Δ_i is just that $n\times n$ determinant obtained from $\det A$ by replacing the ith column elements, a_{ji}, $j=1,2,\ldots,n$, by the components of $\mathbf{b}$, b_j, $j=1,2,\ldots,n$. Therefore we have proved *Cramer's* rule.

THEOREM 2.4 (Cramer's rule)

If $\Delta\equiv\det A\neq 0$ for the system of equations

$$\mathbf{A}\mathbf{x}=\mathbf{b}$$

then the unique solution $\mathbf{x}=(x_1, x_2, \ldots, x_n)$ is given by

$$x_j=\frac{\Delta_j}{\Delta} \qquad j=1,2,\ldots,n$$

where Δ_j is that determinant of order n obtained from $\det A$ by replacing the elements of the jth column by the corresponding components of $\mathbf{b}$.

b. Singular Homogeneous Systems

We are now given

$$\mathbf{A}\mathbf{x}=0 \tag{2.25}$$

subject to the condition

$$\Delta\equiv\det A=0 \tag{2.26}$$

From property (viii) of Section 2.1b we know that the row vectors of A are linearly dependent since $\det A=0$. Let us denote a set of vectors as $\mathbf{a}_i=(a_{i1}^*, a_{i2}^*, a_{i3}^*, \ldots, a_{in}^*)$, $i=1,2,3,\ldots,n$. Then Eq. 2.25 can be written

$$\langle \mathbf{a}_i, \mathbf{x}\rangle=0 \qquad i=1,2,\ldots,n \tag{2.27}$$

Since the $\{\mathbf{a}_i\}$ are linearly dependent [i.e., *at most* $(n-1)$ of them can be linearly independent], they are contained in a subspace of dimension (at most) $(n-1)$. Since $\mathbf{x}$ can be *any* vector in the space of dimension n, we only need choose any $\mathbf{x}$

contained in the one-dimensional subspace orthogonal to the subspace of dimension $(n-1)$ containing the vectors $\{\mathbf{a}_i\}$. Obviously of course the vectors $\{\mathbf{a}_i\}$ may be contained in an even lower dimension subspace, say $m \leq (n-1)$. In that case there will be $(n-m)$ linearly independent nontrivial solutions to Eq. 2.25, $\{\mathbf{x}_j\}$, $j=1, 2, \ldots, (n-m)$.

From this point of view it is also clear why $\mathbf{x}=0$ is the only solution to Eq. 2.25 when $\Delta \neq 0$ since then we are seeking a vector $\mathbf{x}$ in an n-dimensional space that is orthogonal to n linearly independent vectors $\{\mathbf{a}_i\}$ (c.f. Eq. 2.27).

Similarly, if there exists a nontrivial solution $\mathbf{x}$ to Eq. 2.25, then take this $\mathbf{x}$ to be one of the basis vectors in this n-dimensional space. Since Eq. 2.27 states that all n of the $\{\mathbf{a}_i\}$ are orthogonal to this vector $\mathbf{x}$, all the $\{\mathbf{a}_i\}$ must be contained in a subspace of dimension no greater than $(n-1)$ from which it follows that $\Delta=0$.

This completes the proof of the following theorem.

THEOREM 2.5

The necessary and sufficient condition that the equation

$$A\mathbf{x}=0$$

have a nontrivial solution $\mathbf{x}$ is that $\Delta \equiv \det A=0$.

c. Singular Inhomogeneous Systems

We now consider the case

$$A\mathbf{x}=\mathbf{b} \tag{2.28}$$

$$\Delta \equiv \det A=0 \tag{2.29}$$

If there is a solution $\mathbf{x}$ for Eqs. 2.28 and 2.29, then let the matrix $B \equiv \{b_{ij}\}$

$$b_{ij} \equiv \operatorname{cof}(a_{ji})$$

operate on both sides of Eq. 2.28 to yield

$$BA\mathbf{x} \equiv \Delta I\mathbf{x}=0=B\mathbf{b}$$

or in component form,

$$\operatorname{cof}(a_{ji})b_j=0 \qquad i=1, 2, 3, \ldots, n$$

However, from Eq. 2.24, we see that this is simply the condition

$$\Delta_j=0 \qquad j=1, 2, \ldots, n \tag{2.30}$$

THEOREM 2.6

The equation

$$A\mathbf{x}=\mathbf{b}$$

with $\Delta=\det A=0$ can have a solution only if

$$\Delta_j=0 \qquad j=1,2,\ldots,n$$

We have now proven all of the statements contained in Table 1.1 for the general system of linear equations, Eq. 2.21 or 2.22. In Section 2.9 we shall derive a necessary and sufficient condition for the system (2.28)–(2.29) to have a solution.

2.3 GRAM DETERMINANT

a. Test for Linear Independence

We now obtain a simple criterion for the linear independence of a set of vectors. Let

$$\mathbf{x}_i=(\xi_1^i,\xi_2^i,\ldots,\xi_n^i) \qquad i=1,2,\ldots,m \tag{2.31}$$

which is a set of m vectors in a space of dimension n. The *Gram determinant* is defined as that determinant whose elements are the inner products $\langle\mathbf{x}_i,\mathbf{x}_j\rangle^*$ (or equivalently the $\langle\mathbf{x}_i,\mathbf{x}_j\rangle$), $i,j=1,2,\ldots,m$,

$$G(\mathbf{x}_1,\mathbf{x}_2,\ldots,\mathbf{x}_m)\equiv\det\{\langle\mathbf{x}_i,\mathbf{x}_j\rangle^*\}=\det\{\langle\mathbf{x}_j,\mathbf{x}_i\rangle^*\}=\det\{\langle\mathbf{x}_i,\mathbf{x}_j\rangle\}\equiv\det\Gamma \tag{2.32}$$

where $\Gamma=\{\langle\mathbf{x}_i,\mathbf{x}_j\rangle^*\}$ is an $m\times m$ matrix. We can express Γ as

$$\Gamma=\begin{pmatrix}\xi_1^1 & \xi_2^1 & \cdots & \xi_n^1\\ \xi_1^2 & \xi_2^2 & \cdots & \xi_n^2\\ \cdot & \cdot & \cdot & \cdot\\ \cdot & \cdot & \cdot & \cdot\\ \cdot & \cdot & \cdot & \cdot\\ \xi_1^m & \xi_2^m & \cdots & \xi_n^m\end{pmatrix}\begin{pmatrix}\xi_1^{1*} & \xi_1^{2*} & \cdots & \xi_1^{m*}\\ \xi_2^{1*} & \xi_2^{2*} & \cdots & \xi_2^{m*}\\ \cdot & \cdot & \cdot & \cdot\\ \cdot & \cdot & \cdot & \cdot\\ \cdot & \cdot & \cdot & \cdot\\ \xi_n^{1*} & \xi_n^{2*} & \cdots & \xi_n^{m*}\end{pmatrix}\equiv AA^\dagger \tag{2.33}$$

where $A=\{\xi_j^i\}$ is, in general, an $m\times n$ matrix.

If $m=n$, Eq. 2.32 reduces to

$$G(\mathbf{x}_1,\ldots,\mathbf{x}_n)=\det A\,\det A^\dagger=|\det A|^2$$

But from property (viii) of Section 2.1b $\det A=0$ is the necessary and sufficient condition for the linear dependence of the vectors $\{\mathbf{x}_i\}$. Clearly, if the $\{\mathbf{x}_i\}$ are linearly independent, $G(\mathbf{x}_1,\ldots,\mathbf{x}_m)>0$.

If $m>n$, the vectors $\{\mathbf{x}_i\}$ are obviously linearly dependent. In this case we may define a new $m \times m$ matrix $\bar{A}$ as

$$\bar{A} \equiv \begin{pmatrix} \xi_1^1 & \xi_2^1 & \cdots & \xi_n^1 & 0 & \cdots & 0 \\ \xi_1^2 & \xi_2^2 & \cdots & \xi_n^2 & 0 & \cdots & 0 \\ \cdot & \cdot & \cdot & \cdot & \cdot & \cdot & \cdot \\ \cdot & \cdot & \cdot & \cdot & \cdot & \cdot & \cdot \\ \cdot & \cdot & \cdot & \cdot & \cdot & \cdot & \cdot \\ \xi_1^m & \xi_2^m & \cdots & \xi_n^m & 0 & \cdots & 0 \end{pmatrix}$$

such that

$$\Gamma = \bar{A}\bar{A}^\dagger$$

so that again

$$G(\mathbf{x}_1, \ldots, \mathbf{x}_n) = |\det \bar{A}|^2 = 0$$

since $\det \bar{A} = 0$ because it has a column of zeros.

Finally for $m < n$ we can argue as follows. If all the $\{\mathbf{x}_i\}$, $i = 1, 2, \ldots, m$, are linearly independent, then we may use the Schmidt orthogonalization process (Section 1.6) to construct an orthonormal basis, $\{\mathbf{e}_j\}$, $j = 1, 2, \ldots, m$. Let $\bar{\xi}_j^i$ be the components of the $\{\mathbf{x}_i\}$ in this basis,

$$\mathbf{x}_i = (\bar{\xi}_1^i, \bar{\xi}_2^i, \ldots, \bar{\xi}_m^i) \qquad i = 1, 2, \ldots, m$$

We then have m linearly independent vectors in a subspace of dimension m and may use the same argument as for the case $m = n$ above to conclude that $G(\mathbf{x}_1, \ldots, \mathbf{x}_m) > 0$. On the other hand, if the vectors $\{\mathbf{x}_i\}$ are linearly dependent and only $M < m$ of them are linearly independent, we can construct only M orthonormal basis vectors so that

$$\mathbf{x}_i = (\bar{\xi}_1^i, \bar{\xi}_2^i, \ldots, \bar{\xi}_M^i) \qquad i = 1, 2, \ldots, m$$

and we may argue as in the case $m > n$ that $G(\mathbf{x}_1, \ldots, \mathbf{x}_m) = 0$.

We summarize the various possibilities considered in this argument in Table 2.1.

TABLE 2.1

	Linearly Dependent	Linearly Independent
$m = n$	$G = 0$	$G > 0$
$m > n$	$G = 0$	
$m < n$	$G = 0$	$G > 0$

Since by its definition $G(\mathbf{x}_1, \ldots, \mathbf{x}_m)$ cannot be negative, we may complete the table above by negation as

$$\left.\begin{matrix}\text{l. dep.} \Rightarrow G=0 \\ \text{l. indep.} \Rightarrow G>0\end{matrix}\right\} \Rightarrow (\text{via negation}) \left\{\begin{matrix} G>0 \Rightarrow \text{l. indep.} \\ G=0 \Rightarrow \text{l. dep.}\end{matrix}\right.$$

THEOREM 2.7

If the set of vectors $\{\mathbf{x}_i\}$, $i=1, 2, \ldots, m$, is linearly independent then the Gram determinant, $G(\mathbf{x}_1, \ldots, \mathbf{x}_m) \equiv \det\{\langle \mathbf{x}_i \mid \mathbf{x}_j \rangle\}$, is positive and if the set is linearly dependent, then $G(\mathbf{x}_1, \ldots, \mathbf{x}_m)$ vanishes.

Of course since any Gram determinant can be only zero or positive (i.e., never negative) and since a set of vectors is either linearly dependent or linearly independent, Th. 2.7 is equivalent to stating that the necessary and sufficient condition for the linear dependence of a set of vectors is the vanishing of the Gram determinant or that the necessary and sufficient condition for the linear independence of a set of vectors is that the Gram determinant be positive (cf. Table 2.1).

Let $\mathbf{y}$ be a vector contained in the subspace spanned by the vectors $\{\mathbf{x}_i\}$, $i=1, 2, \ldots, m$, let $\mathbf{z}$ be a vector orthogonal to this subspace, $\langle \mathbf{x}_i \mid \mathbf{z} \rangle = 0$, $i=1, 2, \ldots, m$, and let $\mathbf{x}=\mathbf{y}+\mathbf{z}$. Then,

$$G(\mathbf{x}_1, \mathbf{x}_2, \ldots, \mathbf{x}_m, \mathbf{x}) = \begin{vmatrix} \langle \mathbf{x}_1, \mathbf{x}_1 \rangle & \cdots & \langle \mathbf{x}_1, \mathbf{x}_m \rangle & \langle \mathbf{x}_1, \mathbf{y} \rangle \\ \langle \mathbf{x}_2, \mathbf{x}_1 \rangle & \cdots & \langle \mathbf{x}_2, \mathbf{x}_m \rangle & \langle \mathbf{x}_2, \mathbf{y} \rangle \\ \cdot & \cdot & \cdot & \cdot \\ \cdot & \cdot & \cdot & \cdot \\ \cdot & \cdot & \cdot & \cdot \\ \langle \mathbf{x}_m, \mathbf{x}_1 \rangle & \cdots & \langle \mathbf{x}_m, \mathbf{x}_m \rangle & \langle \mathbf{x}_m, \mathbf{y} \rangle \\ \langle \mathbf{y}, \mathbf{x}_1 \rangle & \cdots & \langle \mathbf{y}, \mathbf{x}_m \rangle & \langle \mathbf{y}, \mathbf{y} \rangle + \langle \mathbf{z}, \mathbf{z} \rangle \end{vmatrix}$$

$$= G(\mathbf{x}_1, \ldots, \mathbf{x}_m, \mathbf{y}) + \|\mathbf{z}\|^2 G(\mathbf{x}_1, \ldots, \mathbf{x}_m) = \|\mathbf{z}\|^2 G(\mathbf{x}_1, \ldots, \mathbf{x}_m) \quad \textbf{(2.34)}$$

since the first Gram determinant vanishes by Th. 2.7. Therefore, since $\|\mathbf{x}\|^2 = \|\mathbf{y}\|^2 + \|\mathbf{z}\|^2 \geq \|\mathbf{z}\|^2$, we see that

$$G(\mathbf{x}_1, \ldots, \mathbf{x}_m, \mathbf{x}) \leq \|\mathbf{x}\|^2 G(\mathbf{x}_1, \ldots, \mathbf{x}_m)$$

which, if applied repeatedly, yields

$$G(\mathbf{x}_1, \ldots, \mathbf{x}_m) \leq \|\mathbf{x}_1\|^2 \|\mathbf{x}_2\|^2 \cdots \|\mathbf{x}_m\|^2 \quad \textbf{(2.35)}$$

Clearly the equality sign holds in Eq. 2.35 if and only if

$$\langle \mathbf{x}_i, \mathbf{x}_j \rangle = \delta_{ij} \|\mathbf{x}_i\|^2 \qquad \forall i, j = 1, 2, \ldots, m$$

b. Hadamard's Inequality

We will now derive an inequality for the value of a determinant that we will need in Section 5.6 for our study of integral equations. We wish to obtain a bound on the magnitude of the value of the determinant $\Delta \equiv \det A$ for the matrix $A = \{a_{ij}\}$. Let us consider the elements of the ith row as the components of a vector $\mathbf{a}_i = \{a_{ij}\}$. Then we have

$$G(\mathbf{a}_1, \mathbf{a}_2, \ldots, \mathbf{a}_n) = \{\langle \mathbf{a}_i, \mathbf{a}_j \rangle\} = \det A^* \det A^T = |\Delta|^2$$
$$\leq \prod_{i=1}^{n} \|\mathbf{a}_i\|^2 = \sum_{j=1}^{n} |a_{ij}|^2 \sum_{j=1}^{n} |a_{2j}|^2 \cdots \sum_{j=1}^{n} |a_{nj}|^2 \tag{2.36}$$

This is *Hadamard's inequality*. Finally, if every matrix element a_{ij} is bounded as

$$|a_{ij}| \leq M \qquad \forall i, j \tag{2.37}$$

then

$$|\Delta| \leq n^{n/2} M^n \tag{2.38}$$

2.4 DEFINITION OF A LINEAR OPERATOR†

By an operator acting on the elements of a space we mean a mapping of these elements onto those of some space. We will be particularly concerned with mappings of a space onto itself in the theory of linear operators. An operator L is *linear* provided that

$$\left.\begin{array}{l} y_1 = Lx_1 \\ y_2 = Lx_2 \end{array}\right\} \Rightarrow L(\alpha x_1 + \beta x_2) = \alpha L x_1 + \beta L x_2 = \alpha y_1 + \beta y_2 \tag{2.39}$$

where this must be true for all x_1, x_2 in the space and all scalars α and β.

By no means all operators are linear. Consider the operation of taking the norm of a vector. Let $y = Sx \equiv \langle x, x \rangle$.

$$\left.\begin{array}{l} y_1 = Sx_1 \equiv \langle x_1, x_1 \rangle \\ y_2 = Sx_2 \equiv \langle x_2, x_2 \rangle \end{array}\right\} \Rightarrow S(\alpha x_1 + \beta x_2) = \langle \alpha x_1 + \beta x_2, \alpha x_1 + \beta x_2 \rangle$$
$$= |\alpha|^2 \langle x_1, x_1 \rangle + |\beta|^2 \langle x_2, x_2 \rangle + \alpha^* \beta \langle x_1, x_2 \rangle + \beta^* \alpha \langle x_2, x_1 \rangle$$
$$\neq \alpha y_1 + \beta y_2$$

An important example of a nonlinear operator is that of taking the complex conjugate. Let $x = (\xi_1 \cdots \xi_n)$, a vector in E_n, and define $y = Kx \equiv x^* = (\xi_1^* \cdots \xi_n^*)$. Then we have

$$\left.\begin{array}{l} y_1 = Kx_1 \\ y_2 = Kx_2 \end{array}\right\} \Rightarrow K(\alpha x_1 + \beta x_2) = \alpha^* x_1^* + \beta^* x_2^*$$
$$= \alpha^* y_1 + \beta^* y_2 \neq \alpha y_1 + \beta y_2$$

† Our Sections 2.5, 2.7, 2.9, 2.10, and A2.3 are, in part, expanded versions of similar sections in Chapter 1 of Friedman's *Principles and Techniques of Applied Mathematics.*

The matrices that we have studied earlier in this chapter are examples of linear operators. Two other types of linear operators that will be of great interest to us later are integral and differential operators on function spaces. If $K(x, y)$ is continuous in both variables and if we consider those $f(x) \in \mathcal{L}_2[0, 1]$, then the operation

$$g(x) \equiv Kf(x) = \int_0^1 K(x, y) f(y)\, dy$$

is linear. Actually in many practical applications the *kernel* $K(x, y)$ will not be continuous but only $\mathcal{L}_2$ integrable. The case of linear differential operators, for example,

$$g(x) = Df(x) = \frac{d^2 f(x)}{dx^2}$$

poses new problems of existence since not all $f(x) \in \mathcal{L}_2[0, 1]$ necessarily have a derivative, let alone an $f''(x) \in \mathcal{L}_2[0, 1]$. Clearly, these are *unbounded* operators. We will devote a good deal of effort to extending our theory of linear operators to differential operators in Chapter 8.

The set of vectors $x \in S$ for which the mapping

$$y = Lx$$

onto some $y \in S$ is defined is the *domain* of L. The set of vectors y produced by Lx, x in the domain of L, is the *range* of L. A linear operator is *bounded* if its domain is the entire space S and there exists a constant C (independent of x) such that

$$\|Lx\| < C\,\|x\| \qquad \forall x \in S \tag{2.40}$$

A linear operator is *continuous* if

$$x = \lim_{n\to\infty} x_n \Rightarrow Lx = \lim_{n\to\infty} L(x_n) \tag{2.41}$$

As will be proved in Problem 2.15, for a linear operator continuity implies boundedness and vice versa.

2.5 REPRESENTATION OF A LINEAR OPERATOR BY A MATRIX

We saw in Section 1.5 that any separable space $\mathcal{H}$ can be represented by E_n (or E_∞). We now prove something similar for linear bounded operators.

THEOREM 2.8

Every linear bounded operator in a separable space $\mathcal{H}$ can be represented by a matrix.

PROOF

We denote the orthonormal basis vectors in $\mathcal{H}$ by $\{e_j\}$. Any vector in $\mathcal{H}$ may be represented as

$$x = \sum_j \xi_j e_j$$

and since L is assumed to be bounded we know that Lx is in $\mathcal{H}$ whenever x is so that

$$y = Lx = \sum_j \eta_j e_j \tag{2.42}$$

Since L is a linear operator we may write

$$Lx = L \sum_j \xi_j e_j = \sum_j \xi_j L e_j \tag{2.43}$$

Actually, if this is an infinite sum we must justify taking the operator L inside the summation. The boundedness of L allows us to do this (cf. Problem 2.15). If we now apply Eq. 2.42 to a basis vector e_j, we have

$$Le_j = \sum_l \alpha_{lj} e_l \tag{2.44}$$

From Eqs. 2.42, 2.43, and 2.44 we have

$$\sum_j \eta_j e_j = \sum_{j,l} \xi_j \alpha_{lj} e_l = \sum_{j,l} \alpha_{jl} \xi_l e_j$$

so that

$$\eta_j = \sum_l \alpha_{jl} \xi_l \tag{2.45}$$

Therefore if we write

$$y = \begin{pmatrix} \eta_1 \\ \eta_2 \\ \cdot \\ \cdot \\ \cdot \\ \eta_n \end{pmatrix} \qquad x = \begin{pmatrix} \xi_1 \\ \xi_2 \\ \cdot \\ \cdot \\ \cdot \\ \xi_n \end{pmatrix} \qquad L = \{\alpha_{jl}\} \tag{2.46}$$

then the $\{\alpha_{jl}\}$ are the matrix elements of the representation of L in E_n (or in E_∞). In fact explicitly we see that

$$\langle e_k, Le_j \rangle = \sum_l \alpha_{lj} \langle e_k, e_l \rangle = \sum_l \alpha_{lj}\, \delta_{kl} = \alpha_{kj}$$

or that

$$\alpha_{jl} = \langle e_j, Le_l \rangle \qquad \text{Q.E.D.} \tag{2.47}$$

Relation 2.47 is true in an *orthonormal* basis only.

We now introduce and use frequently a notation due to Dirac for row vectors, column vectors, and inner products. We begin with vectors in E_n. A *ket* vector (e.g., a column vector) is written as $|x\rangle$ or as $x\rangle$,

$$|x\rangle = \begin{pmatrix} \xi_1 \\ \xi_2 \\ \cdot \\ \cdot \\ \cdot \\ \xi_n \end{pmatrix} \tag{2.48}$$

A *bra* vector (e.g., a row vector) is written as $\langle x|$ or as $\langle x$,

$$\langle x| = (\xi_1^* \xi_2^* \cdots \xi_n^*) \tag{2.49}$$

Notice that in this orthonormal basis representation Eq. 2.49 is just the adjoint (cf. Section 2.1a) of Eq. 2.48,

$$\langle x| = |x\rangle^\dagger \tag{2.50}$$

In this notation the inner product of a bra and a ket vector (a "bracket") is written as $\langle x \mid y\rangle$ or as $\langle x, y\rangle$ so that in an orthonormal basis of E_n we have

$$\langle x \mid y\rangle = \langle x, y\rangle = \sum_j \xi_j^* \eta_j \tag{2.51}$$

In this same notation we write

$$|x\rangle\langle y| = x\rangle\langle y = \begin{pmatrix} \xi_1 \\ \xi_2 \\ \cdot \\ \cdot \\ \cdot \\ \xi_n \end{pmatrix} (\eta_1^* \eta_2^* \cdots \eta_n^*) = \begin{pmatrix} \xi_1\eta_1^* & \xi_1\eta_2^* & \cdots & \xi_1\eta_n^* \\ \xi_2\eta_1^* & \xi_2\eta_2^* & \cdots & \xi_2\eta_n^* \\ \cdot & \cdot & \cdot & \cdot \\ \cdot & \cdot & \cdot & \cdot \\ \cdot & \cdot & \cdot & \cdot \\ \xi_n\eta_1^* & \xi_n\eta_2^* & \cdots & \xi_n\eta_n^* \end{pmatrix} \tag{2.52}$$

which is an $n \times n$ matrix in E_n.

If an operator can be written as a sum of outer products of vectors,

$$L = \sum_j |a_j\rangle\langle b_j| \tag{2.53}$$

it is called a *separable operator*.

In terms of a complete orthonormal basis, $\{e_j\}$, we can give a useful representation of the identity operator I as

$$I = \sum_j |e_j\rangle\langle e_j| \tag{2.54}$$

since

$$|x\rangle \equiv I\,|x\rangle = \sum_j \langle e_j \mid x\rangle |e_j\rangle \tag{2.55}$$

From Eq. 2.55 we see that Eq. 2.54 is a statement that the vectors $\{e_j\}$ form a basis [i.e., that they are a *complete* set in the sense that *any* vector x in E_n can be written as a linear combination of them as in Eq. 2.55 (cf. Section 3.3)]. If we have a nonorthogonal basis $\{x_j\}$ and its reciprocal basis $\{z_j\}$ (cf. end of Section 1.8) then, since (cf. Eq. 1.23),

$$x = \sum_j |x_j\rangle\langle z_j \mid x\rangle \equiv Ix$$

we can resolve the identity operator as

$$I = \sum_j |x_j\rangle\langle z_j| \tag{2.56}$$

In such a nonorthogonal basis the inner product becomes

$$\langle x \mid y\rangle \equiv \langle x|\, I\,|y\rangle = \sum_j \langle x \mid x_j\rangle\langle z_j \mid y\rangle \tag{2.57}$$

For later use let us now generalize this notation slightly. We begin in E_n. Suppose we agree to let $|x\rangle$ represent the abstract vector independent of any particular basis. If we want a representation of $|x\rangle$ in a particular basis (e.g., as in Eq. 2.48), we must first specify a basis, $\{e_j\}$, here assumed to be orthonormal. We then obtain the components, ξ_j, of $|x\rangle$ in this particular basis as

$$\langle e_j \mid x\rangle = \xi_j \qquad j = 1, 2, \ldots, n \tag{2.58}$$

Similarly the components of $\langle x|$ would be

$$\langle x \mid e_j\rangle = \xi_j^* \qquad j = 1, 2, \ldots, n$$

Of course we could have chosen another basis, for example, $\{e_j'\}$, in terms of which the components of $|x\rangle$, ξ_j', would be

$$\langle e_j' \mid x\rangle = \xi_j'$$

Similarly the resolution of the identity would be

$$I = \sum_j |e_j'\rangle\langle e_j'|$$

Suppose we now consider a function space [say $\mathscr{L}_2(-\infty, \infty)$]. There are many different bases in which we may represent the abstract function (or vector) $|f\rangle$. The most familiar is the coordinate representation, $f(x)$, which we write as

$$\langle x \mid f\rangle = f(x) \tag{2.59}$$

in analogy with Eq. 2.58. As an example of another basis for $|f\rangle$, let us mention the *Fourier transform* (cf. Section 4.5), $f(k)$, with which the reader may already be familiar,

$$f(k) \equiv \langle k \mid f \rangle = \frac{1}{\sqrt{2\pi}} \int_{-\infty}^{\infty} e^{+ikx} f(x)\, dx \tag{2.60}$$

with the inversion formula

$$f(x) \equiv \langle x \mid f \rangle = \frac{1}{\sqrt{2\pi}} \int_{-\infty}^{\infty} e^{-ikx} f(k)\, dk \tag{2.61}$$

From Eqs. 2.60 and 2.61, which we will prove later, we see that $\langle x \mid f \rangle$ and $\langle k \mid f \rangle$ each serve equally well (up to some possible subtleties of sets of measure zero) to specify the abstract function $|f\rangle$. We can make the analogy with the finite-dimensional case even stronger if we write

$$\langle x \mid k \rangle = \frac{1}{\sqrt{2\pi}} e^{-ikx} \tag{2.62}$$

so that Eqs. 2.60 and 2.61 become

$$\langle k \mid f \rangle = \int_{-\infty}^{\infty} dx\, \langle k \mid x \rangle \langle x \mid f \rangle$$

$$\langle x \mid f \rangle = \int_{-\infty}^{\infty} dk\, \langle x \mid k \rangle \langle k \mid f \rangle$$

from which we might expect to be able to write the identity as

$$I = \int dx\, |x\rangle\langle x| = \int dk\, |k\rangle\langle k| \tag{2.63}$$

This is all consistent with our definition of inner product for functions in $\mathscr{L}_2$ (cf. Section 1.2),

$$\langle f \mid g \rangle = \int f^*(x) g(x)\, dx = \int dx\, \langle x \mid f \rangle^* \langle x \mid g \rangle$$

$$= \int dx\, \langle f \mid x \rangle \langle x \mid g \rangle = \langle f | \, I \, | g \rangle \tag{2.64}$$

Of course we must consider the manipulations implied in Eqs. 2.63 and 2.64 as purely formal ones until we prove in what sense Eq. 2.63 is true; that is, until we prove the completeness relation,

$$\int_{-\infty}^{\infty} |f(x)|^2\, dx = \int_{-\infty}^{\infty} |f(k)|^2\, dk$$

in Section 4.5b.

2.6 EFFECT OF A LINEAR TRANSFORMATION

We now consider the geometrical interpretations of the effect of a linear transformation that can be represented by a matrix.

a. Rotation of Coordinate Axes and of Vectors

We begin with a specific example of a rotation in a real two-dimensional space. Suppose we have a vector **r** and an orthogonal set of axes. We may consider the coordinate axes (i.e., the basis vectors) as rotated through a *counterclockwise* angle φ or the vector as rotated through a *clockwise* angle φ. The components of the rotated vector $\mathbf{r}'$ w.r.t. the original basis will be numerically the same as the components of the original vector **r** w.r.t. the new basis. The first interpretation (i.e., the transformation applied to the basis) is often referred to as the *passive* interpretation, while the second (i.e., transformation applied to the vector itself) is termed the *active* interpretation. Figures 2.1*a* and *b* shows the passive and active interpretations, respectively, of this transformation.

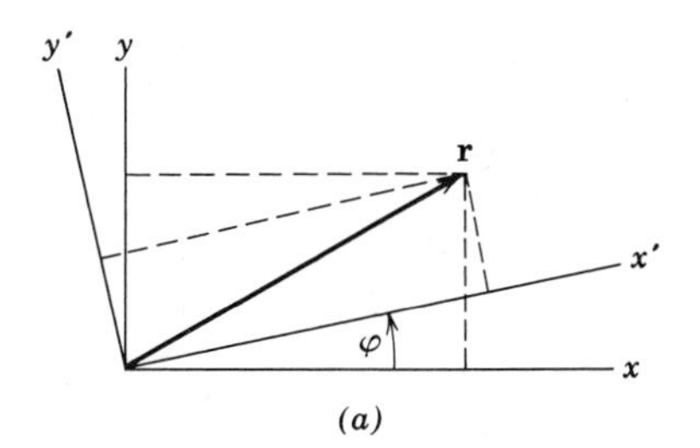

(*a*)

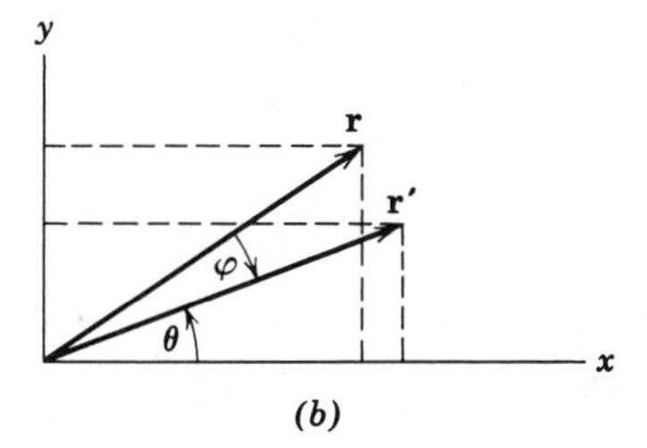

(*b*)

Figure 2.1

In case (*a*) we have

$$x' = x \cos\varphi + y \sin\varphi$$
$$y' = -x \sin\varphi + y \cos\varphi$$

or

$$\mathbf{r}' = \mathrm{A}\mathbf{r} \tag{2.65}$$

where

$$\mathbf{r}' = \begin{pmatrix} x' \\ y' \end{pmatrix} \qquad \mathbf{r} = \begin{pmatrix} x \\ y \end{pmatrix} \qquad \mathrm{A} = \begin{pmatrix} \cos\varphi & \sin\varphi \\ -\sin\varphi & \cos\varphi \end{pmatrix}$$

This gives us the components of the *same* vector in two different coordinate systems. In case (*b*) we have

$$x = r\cos(\varphi+\theta)$$
$$y = r\sin(\varphi+\theta)$$
$$x' = r\cos\theta = r\cos(\varphi+\theta-\varphi) = x\cos\varphi + y\sin\varphi$$
$$y' = r\sin\theta = r\sin(\varphi+\theta-\varphi) = -x\sin\varphi + y\cos\varphi$$

so that again

$$\mathbf{r}' = A\mathbf{r}$$

This gives the components of a vector, once before and once after rotation, w.r.t. the same coordinate system.

Therefore, as claimed, the numerical values of the components of the vectors are the same for both cases, and the same matrix A effects the transformation. We customarily think in terms of changing from one coordinate system to another [i.e., case (*a*) here].

The form of a general transformation of coordinates in terms of the components, $\{\xi_j\}$, of the vectors and the matrix elements, $\{\alpha_{ij}\}$, is (cf., e.g., Eq. 2.65)

$$\xi_j' = \sum_l \alpha_{jl}\xi_l \tag{2.66}$$

In a (real or complex) Euclidean space we often make the physical requirement that the norm of a vector x be independent of the coordinate system used so that

$$\sum_j |\xi_j'|^2 = \sum_j |\xi_j|^2 \tag{2.67}$$

for *all* vectors in the space. This is a restriction on the class of transformations we are considering. It implies that

$$\sum_{j,k,l} \alpha_{jl}^*\alpha_{jk}\xi_l^*\xi_k = \sum_j |\xi_j|^2 = \sum_{l,k} \delta_{lk}\xi_l^*\xi_k$$

or that

$$\sum_{l,k}\left[\sum_j \alpha_{jl}^*\alpha_{jk} - \delta_{lk}\right]\xi_l^*\xi_k = 0 \tag{2.68}$$

We can express Eq. 2.68 in matrix form as

$$\langle x|\,(A^\dagger A - I)\,|x\rangle = \langle x|\,C\,|x\rangle = 0 \qquad \forall x \tag{2.69}$$

where

$$C \equiv A^\dagger A - I \qquad C^\dagger = C = \{c_{ij}\} \tag{2.70}$$

Since Eq. 2.69 must hold for *all* vectors in E_n, we have

$$\langle (x+y)|\,C\,|(x+y)\rangle = 0 = \langle x|\,C\,|y\rangle + \langle y|\,C\,|x\rangle = \langle x|\,C\,|y\rangle + \langle x|\,C\,|y\rangle^* \tag{2.71}$$

now for all x and y in E_n. Let

$$x = \{\xi_j\} = \{\delta_{jl}\}$$
$$y = \{\eta_j\} = \{\delta_{jk}\}$$

where l and k are fixed (but arbitrary) indices. Equation 2.71 then implies that

$$c_{ij} + c_{ij}^* = 0$$

so that

$$\text{Re } c_{ij} = 0 \qquad \forall i, j \tag{2.72}$$

Now let

$$x = \{\xi_j\} = \{\delta_{jl}\}$$
$$y = \{\eta_i\} = \{i\delta_{jk}\}$$

so that Eq. 2.71 becomes

$$ic_{ij} - ic_{ij}^* = 0$$

or

$$\text{Im } c_{ij} = 0 \qquad \forall i, j \tag{2.73}$$

Equations 2.72 and 2.73 are equivalent to

$$c_{ij} = 0 \qquad \forall i, j$$

so that

$$C = 0 = A^\dagger A - I$$

Therefore Eq. 2.68 is true if and only if

$$A^\dagger A = I \tag{2.74}$$

or in index form,

$$\sum_{j=1}^{n} \alpha_{jk}^* \alpha_{jl} = \delta_{kl} \qquad \forall k, l \tag{2.75}$$

If we take the adjoint of Eq. 2.75 we obtain

$$AA^\dagger = I$$

or

$$\sum_{j=1}^{n} \alpha_{kj} \alpha_{lj}^* = \delta_{kl} \qquad \forall k, l \tag{2.76}$$

Any matrix A satisfying Eq. 2.74 is called *unitary*. Clearly if A is a real matrix, then Eq. 2.74 becomes

$$A^T A = I = AA^T \tag{2.77}$$

which defines an *orthogonal* matrix. It is evident from Eqs. 2.74 and 2.77 that unitary and orthogonal matrices have the important property that the inverse is simply given as, respectively,

$$A^\dagger A = I = AA^\dagger \Rightarrow A^{-1} = A^\dagger \tag{2.78}$$
$$A^T A = I = AA^T \Rightarrow A^{-1} = A^T \tag{2.79}$$

Notice that this is obviously satisfied for the rotation matrix given in Eq. 2.65 (i.e., $A \leftrightarrow A^{-1} \Leftrightarrow \varphi \leftrightarrow -\varphi$). It is also evident that unitary (or, in a real space, orthogonal) transformations take one orthonormal set of basis vectors into another orthonormal set of basis vectors.

For the passive interpretation let us consider now in general the effect that a transformation applied to the basis vectors (i.e., a change of coordinate system) has on the components of a vector. That is, we can ask what matrix $B=\{\beta_{ij}\}$ when applied to the basis vectors $\{\hat{x}_j\}$ (not necessarily orthonormal) as

$$\hat{x}_j' = \sum_k \beta_{jk}\hat{x}_k \tag{2.80}$$

will produce the same effect on the components of a vector $x=\{\xi_j\}$ as

$$\xi_j' = \sum_l \alpha_{jl}\xi_l \tag{2.81}$$

so that

$$x = \sum_j \xi_j'\hat{x}_j' = \sum_j \xi_j\hat{x}_j \tag{2.82}$$

If we substitute Eqs. 2.80 and 2.81 into Eq. 2.82 and use the linear independence of the $\{\hat{x}_j\}$, we find

$$\sum_{j,l} \alpha_{jl}\beta_{jk}\xi_l = \sum_l \delta_{lk}\xi_l$$

Since this must hold for all $\{\xi_l\}$, we obtain

$$\sum_j \alpha_{jl}\beta_{jk} \equiv \sum_j (\alpha^{\mathrm{T}})_{lj}\beta_{jk} = \delta_{lk}$$

which as a matrix equation is just

$$A^{\mathrm{T}}B = I \tag{2.83}$$

As long as A is nonsingular, B is related to A as

$$B = (A^{\mathrm{T}})^{-1} = (A^{-1})^{\mathrm{T}} \tag{2.84}$$

Therefore, for the passive interpretation, applying $(A^{-1})^{\mathrm{T}}$ to the basis vectors in passing from one coordinate system to another is equivalent to relating the components of the one vector as seen in those two frames as $\mathbf{r}' = A\mathbf{r}$ (i.e., Eq. 2.80). If A is orthogonal then $B=A$.

Similarly, suppose that for the active interpretation we want the general relation between a transformation that changes the coordinate system (i.e., the basis vectors as in Eq. 2.80) and one that transforms the vector $\mathbf{r}$ into the vector $\mathbf{r}'$.

Let $\{\xi_j\}$ be the components of the vector in the original basis. If we transform the vector, then with $\bar{A}=\{\bar{\alpha}_{ij}\}$ we have for the components of the two vectors referred to the original basis

$$\xi_j' = \sum_l \bar{\alpha}_{jl}\xi_l \tag{2.85}$$

If instead we transform vector x along with the basis vectors $\{\hat{x}_j\}$, then the numerical value of the vector components will still be $\{\xi_j\}$ w.r.t. the new basis (i.e., the $\{\hat{x}_j'\}$). Therefore, according to Eqs. 2.85 and 2.80 we have

$$x = \sum_j \xi_j \hat{x}_j' = \sum_j \xi_j' \hat{x}_j$$

so that

$$\sum_{j,k} \xi_j \beta_{jk} \hat{x}_k = \sum_{j,k} \bar{\alpha}_{kj} \xi_j \hat{x}_k$$

or

$$\beta_{jk} = \bar{\alpha}_{kj} \qquad \bar{A} = B^{T} \tag{2.86}$$

Again for orthogonal transformations $\bar{A} = B^{-1}$.

It may be instructive to demonstrate the results stated in Eqs. 2.84 and 2.86 with an example in two dimensions. We have seen that the matrix of Eq. 2.65 effects a rotation of the coordinates through a counterclockwise angle φ. Suppose, instead, that we rotate the vector $\mathbf{r}$ through this same angle as shown in Fig. 2.2.

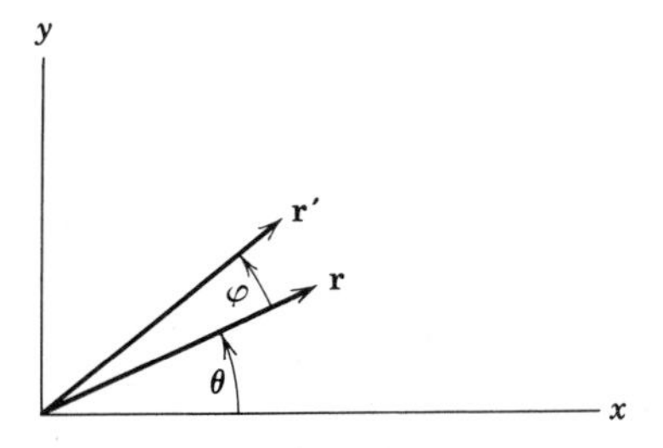

Figure 2.2

Since

$$x = r \cos\theta$$
$$y = r \sin\theta$$
$$x' = r \cos(\theta + \varphi) = x \cos\varphi - y \sin\varphi$$
$$y' = r \sin(\theta + \varphi) = x \sin\varphi + y \cos\varphi$$

we have

$$\mathbf{r}' = \mathrm{A}^{T}\mathbf{r} = \mathrm{A}^{-1}\mathbf{r}$$

which is the result expected from Eq. 2.86 for the orthogonal matrix

$$\mathrm{A} = \begin{pmatrix} \cos\varphi & \sin\varphi \\ -\sin\varphi & \cos\varphi \end{pmatrix} \tag{2.87}$$

Again, in a two-dimensional space, let us find a new set of basis vectors in terms of the original ones so indicated in Fig. 2.3.

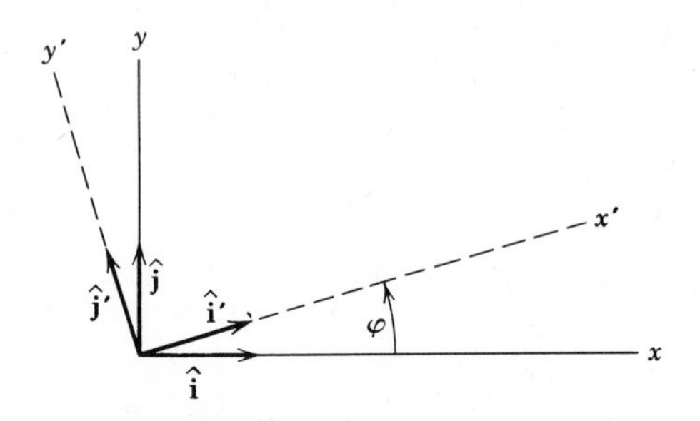

Figure 2.3

According to Eqs. 2.87, 2.80, and 2.84, we have

$$\begin{aligned} \hat{\mathbf{i}}' &= \cos\varphi\,\hat{\mathbf{i}} + \sin\varphi\,\hat{\mathbf{j}} \\ \hat{\mathbf{j}}' &= -\sin\varphi\,\hat{\mathbf{i}} + \cos\varphi\,\hat{\mathbf{j}} \end{aligned} \tag{2.88}$$

However we can also obtain this relation by considering the *components* of $\hat{\mathbf{i}}'$ and $\hat{\mathbf{j}}'$ in terms of the original basis (i.e., rotate $\hat{\mathbf{i}}$ and $\hat{\mathbf{j}}$ into $\hat{\mathbf{i}}'$ and $\hat{\mathbf{j}}'$ and use Eq. 2.86).

$$(\hat{\mathbf{i}}')_x = \cos\varphi \qquad (\hat{\mathbf{j}}')_x = -\sin\varphi$$
$$(\hat{\mathbf{i}}')_y = \sin\varphi \qquad (\hat{\mathbf{j}}')_y = \cos\varphi$$

Then, by definition, we have

$$\hat{\mathbf{i}}' \equiv (\hat{\mathbf{i}}')_x\hat{\mathbf{i}} + (\hat{\mathbf{i}}')_y\hat{\mathbf{j}} = \cos\varphi\,\hat{\mathbf{i}} + \sin\varphi\,\hat{\mathbf{j}}$$
$$\hat{\mathbf{j}}' \equiv (\hat{\mathbf{j}}')_x\hat{\mathbf{i}} + (\hat{\mathbf{j}}')_y\hat{\mathbf{j}} = -\sin\varphi\,\hat{\mathbf{i}} + \cos\varphi\,\hat{\mathbf{j}}$$

which is again Eq. 2.88.

b. Elements of a Matrix in Different Bases

We have seen how to represent bounded linear operators in spaces of denumerable dimension in terms of matrices. The particular representation we obtain (i.e., the numerical values of the matrix elements) depends on the basis we use. If we have a matrix given in one basis, we can ask how we obtain the representation of that same operator in another basis. That is, suppose we are given a matrix A, which carries a vector $\mathbf{x}$ into a vector $\mathbf{x}'$, and another matrix S, which takes us from one basis to another. Given the elements of A w.r.t. the first basis (which we denote by $A_{(\mathrm{I})}$) and S w.r.t. the first basis (i.e., $S_{(\mathrm{I})}$), we must find A w.r.t. a second basis (i.e., $A_{(\mathrm{II})}$). The situation is depicted graphically in Fig.

2.4, although the argument we give is perfectly general (provided only that S is not singular).

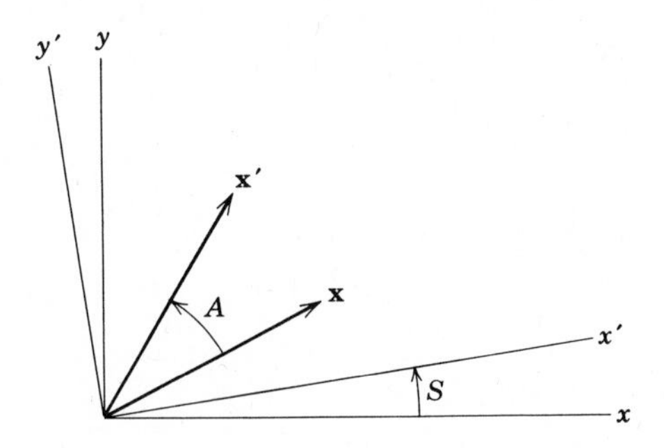

Figure 2.4

Since A carries $\mathbf{x}$ into $\mathbf{x}'$, we have w.r.t. frame I,

$$\mathbf{x}'_{(\mathrm{I})} = A_{(\mathrm{I})}\mathbf{x}_{(\mathrm{I})} \tag{2.89}$$

If we take S to be the matrix that takes us from one set of basis vectors to another then, according to Eq. 2.84, the vectors $\mathbf{x}$ and $\mathbf{x}'$ as seen from either frame are related as

$$\begin{aligned} \mathbf{x}_{(\mathrm{II})} &= (S^{\mathrm{T}}_{(\mathrm{I})})^{-1}\mathbf{x}_{(\mathrm{I})} \\ \mathbf{x}'_{(\mathrm{II})} &= (S^{\mathrm{T}}_{(\mathrm{I})})^{-1}\mathbf{x}'_{(\mathrm{I})} \end{aligned} \tag{2.90}$$

Finally we can express the relation between $\mathbf{x}$ and $\mathbf{x}'$ in the second frame as

$$\mathbf{x}'_{(\mathrm{II})} = A_{(\mathrm{II})}\mathbf{x}_{(\mathrm{II})} \tag{2.91}$$

If we use all of these equations, we obtain

$$\begin{aligned} \mathbf{x}'_{(\mathrm{I})} &= S^{\mathrm{T}}_{(\mathrm{I})}\mathbf{x}'_{(\mathrm{II})} = S^{\mathrm{T}}_{(\mathrm{I})}A_{(\mathrm{II})}\mathbf{x}_{(\mathrm{II})} \\ &= S^{\mathrm{T}}_{(\mathrm{I})}A_{(\mathrm{II})}(S^{\mathrm{T}}_{(\mathrm{I})})^{-1}\mathbf{x}_{(\mathrm{I})} = A_{(\mathrm{I})}\mathbf{x}_{(\mathrm{I})} \end{aligned} \tag{2.92}$$

Since Eq. 2.92 must hold for *all* vectors $\mathbf{x}$, the desired relation is

$$A_{(\mathrm{II})} = (S^{\mathrm{T}}_{(\mathrm{I})})^{-1}A_{(\mathrm{I})}S^{\mathrm{T}}_{(\mathrm{I})} \tag{2.93}$$

If S is an orthogonal transformation, then this reduces to

$$A_{(\mathrm{II})} = S_{(\mathrm{I})}A_{(\mathrm{I})}S^{-1}_{(\mathrm{I})} \tag{2.94}$$

In either case S is a *similarity transformation.* In general two matrices A and B are termed *similar* provided there exists a nonsingular matrix C such that

$$B = CAC^{-1} \tag{2.95}$$

c. Covariant Vectors, Contravariant Vectors and the Reciprocal Basis

Let us begin with the example of a nonorthogonal coordinate transformation shown in Fig. 2.5.

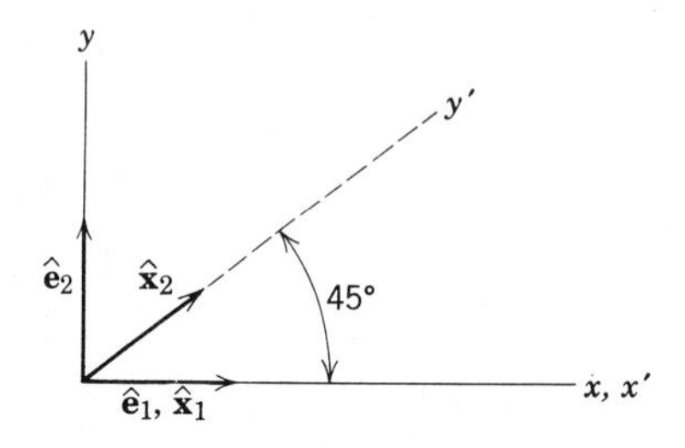

Figure 2.5

If we let $\hat{\mathbf{e}}_1$ and $\hat{\mathbf{e}}_2$ denote orthonormal basis vectors along the x and y axes, respectively, then we can express the nonorthogonal basis vectors $\hat{\mathbf{x}}_j$, $j=1, 2$, as

$$\hat{\mathbf{x}}_1 = \hat{\mathbf{e}}_1$$

$$\hat{\mathbf{x}}_2 = \frac{1}{\sqrt{2}}(\hat{\mathbf{e}}_1 + \hat{\mathbf{e}}_2)$$

and the reciprocal basis as

$$\hat{\mathbf{z}}_1 = \hat{\mathbf{e}}_1 - \hat{\mathbf{e}}_2$$

$$\hat{\mathbf{z}}_2 = \sqrt{2}\,\hat{\mathbf{e}}_2$$

where

$$\langle \hat{\mathbf{z}}_j \mid \hat{\mathbf{x}}_k \rangle = \delta_{jk} \qquad j, k = 1, 2 \tag{2.96}$$

We can take an arbitrary vector $\mathbf{x}$ and expand it in terms of any of the basis sets $\{\hat{\mathbf{e}}_j\}$, $\{\hat{\mathbf{x}}_j\}$, or $\{\hat{\mathbf{z}}_j\}$, the components being denoted by $\{x_j\}$, $\{\xi^j\}$, and $\{\xi_j\}$, respectively, in each of these bases. If $A = \{\alpha_{ij}\}$ is the matrix that takes us from the original orthogonal basis to the nonorthogonal one as

$$\xi^j = \sum_k \alpha^{jk} x_k \tag{2.97}$$

where for our example

$$A = \{\alpha_{ij}\} = \{\alpha^{ij}\} = \begin{pmatrix} 1 & -1 \\ 0 & \sqrt{2} \end{pmatrix}$$

then from Eq. 2.84 we have

$$\hat{\mathbf{x}}_j = \sum_k [(A^{\mathrm{T}})^{-1}]_{jk} \hat{\mathbf{e}}_k \tag{2.98}$$

as we can also easily verify directly from Eq. 2.97 and the requirement that $\|x\|^2$ must be left invariant under this transformation. From Eqs. 2.96 and 2.98 we see that

$$\hat{\mathbf{z}}_j = \sum_k [A^*]_{jk} \hat{\mathbf{e}}_k \tag{2.99}$$

Notice that Eqs. 2.96–2.99 are perfectly general and hold in any (real or complex) E_n. They do not depend on our specific illustrative example depicted in Fig. 2.5. Also we denote the matrix elements of A as either $\{\alpha^{ij}\}$ or $\{\alpha_{ij}\}$ without distinction. Since we have

$$\mathbf{x} = \sum_j x_j \hat{\mathbf{e}}_j = \sum_j \xi^j \hat{\mathbf{x}}_j = \sum_j \xi_j \hat{\mathbf{z}}_j \tag{2.100}$$

we see from Eq. 2.99 that

$$\xi_l = \sum_j [(AA^\dagger)^{-1}]_{lj} \xi^j = \sum_j [(A^\dagger)^{-1}]_{lj} x_j \tag{2.101}$$

so that

$$\|x\|^2 = \sum_{j,k} \xi_j^* \xi^k \langle \hat{\mathbf{z}}_j \mid \hat{\mathbf{x}}_k \rangle = \sum_j \xi^j \xi_j^* = \sum_{j,k} \xi^j [(AA^\dagger)^{-1}]_{jk}^* \xi^{k*}$$

$$= \sum_{j,k} \xi^{k*} [(AA^\dagger)^{-1}]_{kj} \xi_j \equiv \sum_{i,j} \xi^{i*} g_{ij} \xi_j \tag{2.102}$$

where g_{ij} is defined as the *metric tensor*. Of course for unitary or real orthogonal transformations $AA^\dagger = I$ (i.e., $g_{ij} = \delta_{ij}$) and Eq. 2.102 reduces to the usual definition of the norm in an orthonormal basis in E_n. In our example we have

$$G = \{g_{ij}\} = (AA^T)^{-1} = \begin{pmatrix} 1 & \frac{1}{\sqrt{2}} \\ \frac{1}{\sqrt{2}} & 1 \end{pmatrix}$$

In tensor analysis in a real space a vector whose components transform as

$$\xi^{i'} = \sum_j \alpha^{ij} \xi^j \tag{2.103}$$

is called a *contravariant vector* (cf. Eq. 2.97). We will consider only those transformation matrices $A = \{\alpha^{ij}\}$ whose elements are constants (i.e., independent of the coordinates ξ^i). If $\varphi(x)$ is a *scalar* function w.r.t. the transformation [i.e., $\varphi(x') \equiv \varphi(x)$], then the quantities

$$\xi_i \equiv \frac{\partial \varphi(x)}{\partial \xi^i}$$

transform as

$$\xi_i' \equiv \frac{\partial \varphi(x')}{\partial \xi^{i'}} = \sum_j \frac{\partial \varphi(x)}{\partial \xi^j} \frac{\partial \xi^j}{\partial \xi^{i'}} = \sum_j (A^{-1})_{ji} \xi_j = \sum_j [(A^T)^{-1}]_{ij} \xi_j \tag{2.104}$$

Such quantities are termed *covariant vectors* (cf. Eq. 2.101, remembering that $A^\dagger = A^T$ for a real transformation). The product, $\sum_j \xi^i \eta_i$, is an invariant w.r.t. the transformation A for *any* contravariant and *any* covariant vector since

$$\sum_i \xi^{i'} \eta_i' = \sum_{i,k,l} [A]^{ik} \xi^k [(A^T)^{-1}]_{il} \eta_l = \sum_{i,k,l} \xi^k [A^{-1}]_{li} [A]^{ik} \eta_l$$

$$= \sum_{k,l} \xi^k \eta_l \, \delta_{kl} = \sum_i \xi^i \eta_i \tag{2.105}$$

If we define

$$g_{ij} \equiv [(AA^T)^{-1}]_{ij} = \sum_{k,l} [(A^T)^{-1}]_{ik}[(A^T)^{-1}]_{jl}\, \delta_{kl} \tag{2.106}$$

then we see that this is a *covariant* (second rank) tensor since in an orthogonal basis the metric tensor is just δ_{ij}. Furthermore the contravariant quantity

$$g^{ij} = [AA^T]^{ij} \tag{2.107}$$

is such that

$$\sum_k g_{ik} g^{kj} = \sum_k [(AA^T)^{-1}]_{ik}[AA^T]^{kj} = \delta_{ij} \tag{2.108}$$

The basic definitions given in Eqs. 2.103 and 2.104 allow us to raise and lower indices as

$$\xi^i = \sum_j g^{ij} \xi_j \tag{2.109}$$

$$\xi_i = \sum_j g_{ij} \xi^j \tag{2.110}$$

since

$$\xi_i' \equiv \sum_j g_{ij}' \xi^{j\prime} = \sum_{\substack{j,k,\\ l,m}} [(A^T)^{-1}]_{ik}[(A^T)^{-1}]_{jl} g_{kl} [A]^{jm} \xi^m$$

$$= \sum_{k,l} [(A^T)^{-1}]_{ik} g_{kl} \xi^l = \sum_j [(A^T)^{-1}]_{ij} \xi_j$$

and, similarly, for Eq. 2.110.

Finally, if we are in a complex space, then in addition to the quantities defined in Eqs. 2.103 and 2.104, we must define other contravariant and covariant quantities, $\xi^{\bar{i}}$ and $\xi_{\bar{i}}$, as (cf. Eq. 2.101)

$$\xi^{\bar{i}\prime} = \sum_j [A^*]^{ij} \xi^{\bar{j}} \tag{2.111}$$

$$\xi_{\bar{i}}' = \sum_j [(A^\dagger)^{-1}]_{ij} \xi_{\bar{j}} \tag{2.112}$$

That is, we must distinguish between barred and unbarred indices and write the metric tensor as

$$g_{\bar{i}j} = [(AA^\dagger)^{-1}]_{ij} \tag{2.113}$$

$$g^{i\bar{j}} = [AA^\dagger]_{ij} \tag{2.114}$$

with

$$\sum_k g_{\bar{i}k} g^{k\bar{j}} = \delta_{ij} \tag{2.115}$$

so that

$$\xi_{\bar{i}} = \sum_j g_{\bar{i}j} \xi^j \tag{2.116}$$

$$\xi^i = \sum_j g^{i\bar{j}} \xi_{\bar{j}} \tag{2.117}$$

Again the quantity $\sum_i \xi_{\bar{i}}\xi^{\bar{i}}$ will be a scalar. In fact we can always generate a scalar by contracting (i.e., summing) upper and lower unbarred indices or upper and lower barred indices (but *not* by contracting upper unbarred with lower barred or vice versa). In such a space the norm of a vector can be expressed as

$$\|x\|^2 = \sum_{i,j} \xi^{\bar{i}} g_{\bar{i}j} \xi^j$$

In most of our future work in this text we will not find it necessary to distinguish between covariant and contravariant quantities.

2.7 INVERSION OF OPERATORS

We now begin to consider the type problems that we will meet in practice. Given a linear operator L and a vector a such that

$$Lx = a \tag{2.118}$$

we must find out when such an x satisfying (2.118) exists, when it is unique and, specifically, what it is. We have already seen in Section 2.1c how to construct the inverse L^{-1} when L is represented by a matrix. We now treat some other simple special cases that will prove very useful later, especially in our study of Volterra and Fredholm integral equations in Chapter 5.

a. Separable Operators

In Section 2.5, Eq. 2.53, we defined L to be separable provided it could be expressed as a finite sum of outer products,

$$L = \sum_j |a_j\rangle\langle b_j| \tag{2.53}$$

The simplest example of such a separable operator would be a single outer product,

$$L = |a_1\rangle\langle b_1| \tag{2.119}$$

so that

$$L\,|x\rangle = |a\rangle \tag{2.120}$$

or

$$|a_1\rangle\langle b_1 \mid x\rangle = |a\rangle \tag{2.121}$$

Since we have been given the vectors a_1, b_1, and a, Eq. 2.121 tells us that the two vectors, a_1 and a, must be proportional. *A priori* there is no reason that a_1 and a should be proportional. Therefore a solution exists only if *in fact*

$$|a\rangle = \alpha\,|a_1\rangle \tag{2.122}$$

This turns out to be a necessary and sufficient condition for the existence of a solution as we will see. If this condition is fulfilled, then we know α. Hence a particular solution is

$$x_p = \frac{\alpha}{\|b_1\|^2} b_1 \tag{2.123}$$

Furthermore if c is *any* vector orthogonal to $|b_1\rangle$,

$$\langle c \mid b_1 \rangle = 0 \tag{2.124}$$

then the most general solution is

$$x = \frac{\alpha}{\|b_1\|^2} b_1 + c \tag{2.125}$$

We prove that *every* solution must be of this form as follows (if S has a denumerable basis). Choose for an orthogonal basis in S b_1 plus an orthogonal set, $\{\sigma_j\}$. Then any vector in S may be written as

$$x = \rho b_1 + \sum_j \alpha_j \sigma_j$$

In particular assume x is a solution to

$$|a_1\rangle\langle b_1 \mid x\rangle = |a\rangle$$

Then we have

$$Lx = |a_1\rangle\langle b_1 \mid x\rangle = |a_1\rangle\langle b_1 \mid \rho b_1\rangle + |a_1\rangle\Big\langle b_1 \Big| \sum_j \alpha_j \sigma_j \Big\rangle$$
$$= \rho \, \|b_1\|^2 \, |a_1\rangle = |a\rangle = \alpha \, |a_1\rangle$$

so that

$$\rho = \frac{\alpha}{\|b_1\|^2}$$

provided a solution does exist. From this we also see that Eq. 2.122 is the necessary and sufficient condition for Eqs. 2.119 and 2.120 to have a (generally nonunique) solution. Therefore a solution, if it exists, is not in general unique.

We next consider the case in which L is of the form (2.53) with n terms.

$$L = \sum_{j=1}^{n} |a_j\rangle\langle b_j| = |a_1\rangle\langle b_1| + |a_2\rangle\langle b_2| + \cdots + |a_n\rangle\langle b_n|$$

In this case

$$L\,|x\rangle = |a\rangle$$

implies

$$\sum_{j=1}^{n} \langle b_j \mid x\rangle \, |a_j\rangle \equiv \sum_j \alpha_j \, |a_j\rangle = a \tag{2.126}$$

This condition must be fulfilled or there will exist no solution. Therefore it is a necessary condition. If this condition is fulfilled, then we know the values of the $\{\alpha_j\}$. Seek a particular solution $x_p \in \mathcal{M}_B$ (i.e., the manifold spanned by the $\{b_j\}$) such that

$$|x_p\rangle = \sum_{l=1}^{n} \xi_l \, |b_l\rangle \tag{2.127}$$

This requires that

$$L\,|x_p\rangle = \sum_{j,l=1}^{n} \xi_l \langle b_j \mid b_l\rangle \, |a_j\rangle = |a\rangle = \sum_{j=1}^{n} \alpha_j \, |a_j\rangle$$

or

$$\alpha_j = \sum_{l=1}^{n} \xi_l \langle b_j \mid b_l\rangle \tag{2.128}$$

We have assumed here that the $\{|a_j\rangle\}$ are linearly independent. If they were not, then we should have reduced the set initially. If such $\{\xi_l\}$ exist, then the most general solution is

$$|x\rangle = \sum_{l=1}^{n} \xi_l \, |b_l\rangle + |c\rangle$$

where

$$\langle c \mid b_l\rangle = 0 \qquad l = 1, 2, \ldots, n \tag{2.129}$$

If the linear algebraic equations for the $\{\xi_l\}$, Eq. 2.128, have no solution, then no such $x_p \in \mathcal{M}_B$ exists. The only possible solution would be one orthogonal to $\mathcal{M}_B$. But this can be so only if $\langle b_j \mid x\rangle \equiv \alpha_j = 0, \forall j$, which implies $a = 0$. Therefore, if $a \neq 0$, no solution exists when Eq. 2.128 has no solution. If $a = 0$ then $x = c$. Notice that Eq. 2.126 would also be a sufficient condition for the existence of a solution *provided* we knew that Eq. 2.128 always had a solution. However Eq. 2.128 is simply a system of n linear algebraic equations for the n unknowns $\{\xi_l\}$. It will have a unique solution provided $\det\{\langle b_j \mid b_l\rangle\} \neq 0$. But this is just the Gram determinant, $G(b_1, b_2, \ldots, b_n)$. If all the $\{b_j\}$ are linearly independent, then Eq. 2.128 will have a unique solution for the $\{\xi_l\}$. This allows us to state the following theorem.

THEOREM 2.9

If the vectors $\{a_j\}$, $j = 1, 2, \ldots, n$, are linearly independent among themselves, and if vectors $\{b_j\}$, $j = 1, 2, \ldots, n$, are also linearly independent among themselves, and if L is a separable operator of the form

$$L = \sum_{j=1}^{n} |a_j\rangle\langle b_j|$$

then for a given vector $|a\rangle$ the equation

$$L\,|x\rangle=|a\rangle \tag{2.130}$$

has a solution if and only if $|a\rangle$ is some linear combination of the $\{|a_j\rangle\}$,

$$|a\rangle=\sum_{j=1}^{n}\alpha_j\,|a_j\rangle \tag{2.131}$$

If condition (2.131) is satisfied, then a (generally nonunique) solution to (2.130) is

$$|x\rangle=\sum_{j=1}^{n}\xi_j\,|b_j\rangle+|c\rangle \tag{2.132}$$

where the $\{\xi_j\}$ is the (unique) solution to

$$\alpha_j=\sum_{l=1}^{n}\xi_l\langle b_j\mid b_l\rangle$$

and $|c\rangle$ is any vector orthogonal to all the $\{|b_j\rangle\}$,

$$\langle b_j\mid c\rangle=0 \qquad j=1,2,\ldots,n$$

It is clear that under the conditions of Th. 2.9 the solution (2.132) becomes unique (i.e., $|c\rangle\equiv 0$) if and only if the $\{b_j\}$ form a basis for the space containing x. Furthermore, since both the $\{a_j\}$ and $\{b_j\}$ are each contained in some manifolds of dimension n (which may or may not overlap), we can often use a new set of $\{a_j\}$ and $\{b_j\}$ for which

$$\langle b_j\mid b_k\rangle=\delta_{jk}=\langle a_j\mid a_k\rangle \tag{2.133}$$

$$\langle a_j\mid b_k\rangle=\langle a_j\mid b_j\rangle\,\delta_{jk} \tag{2.134}$$

In this case Eq. 2.128 is trivial to solve.

Finally, if there exists a complete set of orthonormal basis vectors $\{e_j\}$ in terms of which L takes on the particularly simple diagonal form

$$L=\sum_{j=1}^{n}\lambda_j\,|e_j\rangle\langle e_j| \tag{2.135}$$

where the $\{\lambda_j\}$ are some complex numbers, then we can write

$$|x\rangle=\sum_j\xi_j\,|e_j\rangle \tag{2.136}$$

$$|a\rangle=\sum_j\alpha_j\,|e_j\rangle \tag{2.137}$$

and the equation

$$Lx=a$$

implies

$$\lambda_j\xi_j=\alpha_j \tag{2.138}$$

so that if none of the $\{\lambda_j\}$ vanish, then the unique solution is

$$x = \sum_j \left(\frac{\alpha_j}{\lambda_j}\right) e_j \tag{2.139}$$

Another simple case that will prove useful later is that in which L is the identity I plus a separable operator,

$$L = I + \sum_{j=1}^{n} |a_j\rangle\langle b_j| \tag{2.140}$$

Again the simplest case here is

$$L = I + |a_1\rangle\langle b_1| \tag{2.141}$$

Then the equation

$$Lx = a \tag{2.142}$$

becomes

$$|x\rangle + \langle b_1 \mid x\rangle \, |a_1\rangle = |a\rangle \tag{2.143}$$

If we knew the *number* $\langle b_1 \mid x\rangle$, then the solution would be given as

$$|x\rangle = |a\rangle - \langle b_1 \mid x\rangle \, |a_1\rangle \tag{2.144}$$

However we can obtain an equation for $\langle b_1 \mid x\rangle$ from Eq. 2.143 as

$$[1 + \langle b_1 \mid a_1\rangle]\langle b_1 \mid x\rangle = \langle b_1 \mid a\rangle \tag{2.145}$$

If

$$1 + \langle b_1 \mid a_1\rangle \neq 0 \tag{2.146}$$

Eq. 2.145 has the unique solution

$$\langle b_1 \mid x\rangle = \frac{\langle b_1 \mid a\rangle}{1 + \langle b_1 \mid a_1\rangle} \tag{2.147}$$

The solution (2.144) to Eq. 2.142 then becomes

$$|x\rangle = |a\rangle - \frac{\langle b_1 \mid a\rangle}{1 + \langle b_1 \mid a_1\rangle} |a_1\rangle \tag{2.148}$$

We see by substitution that (2.148) is a solution to (2.142). Therefore Eq. 2.146 is the necessary and sufficient condition that Eqs. 2.141 and 2.142 have a *unique* solution. If

$$1 + \langle b_1 \mid a_1\rangle = 0 \tag{2.149}$$

then Eq. 2.145 can have a solution only if

$$\langle b_1 \mid a\rangle = 0 \tag{2.150}$$

in which case the now nonunique solution is given as

$$|x\rangle = |a\rangle + \alpha\,|a_1\rangle \tag{2.151}$$

where α is an arbitrary scalar.

Example 2.7 A particularly important example of such an operator is that of an integral equation for $\phi(x)$

$$\phi(x) - \lambda\int_a^b K(x, y)\phi(y)\,dy = f(x) \tag{2.152}$$

where $f(x)$ is a given function, when the kernel $K(x, y)$ separates as

$$K(x, y) = a(x)b^*(y)$$

Then Eq. 2.152 becomes

$$\phi(x) - \lambda a(x)\int_a^b b^*(y)\phi(y)\,dy = f(x)$$

so that

$$\langle b \mid \phi\rangle[1 - \lambda\langle b \mid a\rangle] = \langle b \mid f\rangle \tag{2.153}$$

Again if

$$\lambda\int_a^b b^*(x)a(x)\,dx \neq 1$$

then Eq. 2.152 has the unique solution

$$\phi(x) = f(x) + \frac{\lambda a(x)\displaystyle\int_a^b b^*(y)f(y)\,dy}{1 - \lambda\displaystyle\int_a^b b^*(y)a(y)\,dy} \tag{2.154}$$

Finally we can treat the case for L given by Eq. 2.140,

$$L = I + \sum_{j=1}^{n} |a_j\rangle\langle b_j| \tag{2.140}$$

In this case

$$L\,|x\rangle = |a\rangle$$

requires that

$$|x\rangle + \sum_{j=1}^{n} \langle b_j \mid x\rangle\,|a_j\rangle = |a\rangle$$

or

$$\langle b_l \mid x\rangle + \sum_{j=1}^{n} \langle b_j \mid x\rangle\langle b_l \mid a_j\rangle = \langle b_l \mid a\rangle \qquad l = 1, 2, \ldots, n \tag{2.155}$$

We then solve this set of n simultaneous linear equations for the n unknowns, $\{\langle b_j \mid x\rangle\}$, and the solution to Eq. 2.142 is just

$$|x\rangle = |a\rangle - \sum_{j=1}^{n} \langle b_j \mid x\rangle \, |a_j\rangle \tag{2.156}$$

Of course this set of n simultaneous linear equations for the n unknowns $\langle b_j \mid x\rangle$ may be singular.

This problem becomes especially easy to solve if L is in a form such that

$$\langle b_j \mid a_l\rangle = \delta_{jl}\langle b_j \mid a_j\rangle \tag{2.157}$$

Then we have

$$Lx = a$$

so that

$$\sum_{j=1}^{n} \langle b_j \mid x\rangle \, |a_j\rangle + |x\rangle = |a\rangle$$

or

$$\langle b_l \mid a\rangle = \langle b_l \mid x\rangle + \sum_{j=1}^{n} \langle b_j \mid x\rangle\langle b_l \mid a_j\rangle = \langle b_l \mid x\rangle + \langle b_l \mid x\rangle\langle b_l \mid a_l\rangle$$

Therefore, unless $1 + \langle b_l \mid a_l\rangle = 0$ for some l,

$$\langle b_l \mid x\rangle = \frac{\langle b_l \mid a\rangle}{1 + \langle b_l \mid a_l\rangle}$$

which yields

$$|x\rangle = |a\rangle - \sum_{j=1}^{n} \frac{\langle b_j \mid a\rangle}{1 + \langle b_j \mid a_j\rangle} |a_j\rangle \tag{2.158}$$

As previously, if $1 + \langle b_l \mid a_l\rangle = 0$ for some values of l, then no solution exists *unless* $\langle b_l \mid a\rangle = 0$ for these same values of l, in which case there are arbitrary terms $\alpha_l \, |a_l\rangle$ in the solution.

THEOREM 2.10

If the operator L has the form

$$L = I + \sum_{j=1}^{n} |a_j\rangle\langle b_j|$$

such that

$$\langle b_j \mid a_k\rangle = \delta_{jk}\langle b_j \mid a_j\rangle \qquad j = 1, 2, \ldots, n$$

then the necessary and sufficient condition that there exists a unique solution x to the equation

$$Lx = a$$

is that

$$1 + \langle b_j \mid a_j\rangle \neq 0 \qquad \forall j = 1, 2, \ldots, n$$

This unique solution is given by Eq. 2.158. If

$$1+\langle b_j \mid a_j\rangle=0$$

for some values of $j=j_1, j_2, \ldots, j_m$, then there is no solution unless

$$\langle b_j \mid a\rangle=0 \qquad \forall j=j_1, j_2, \ldots, j_m$$

in which case the (nonunique) solution is given as

$$|x\rangle=|a\rangle-\sum_{\substack{j=1\\ j\neq\{j_m\}}}^{n} \frac{\langle b_j \mid a\rangle}{1+\langle b_j \mid a_j\rangle}|a_j\rangle+\sum_{j=\{j_m\}} \alpha_j\,|a_j\rangle$$

where the $\{\alpha_j\}$ are arbitrary constants.

The main results of this section are summarized in Table 2.2 at the end of this chapter.

b. Identity Plus an Infinitesimal Operator

We now construct the inverse of an operator of the form $(I+A)$ provided A is an operator satisfying a stringent enough bound.

THEOREM 2.11

If the operator A satisfies the bound

$$\|Ax\|\leq\alpha\,\|x\| \qquad 0<\alpha<1 \tag{2.159}$$

for all vectors x in S, where α is a constant independent of x, then the inverse of $(I+A)$ is given by the *Neumann series*

$$(I+A)^{-1}=I+\sum_{j=1}^{\infty}(-1)^j A^j \tag{2.160}$$

PROOF

In order to prove that (2.160) is the desired inverse we must show that the operator defined by the infinite series exists and then that

$$(I+A)\left[I+\sum_j (-1)^j A^j\right]=I$$

First we obtain a bound on $\|A^j x\|$ as

$$\|A^j x\|\equiv\|A(A^{j-1}x)\|\leq\alpha\,\|A^{j-1}x\|\equiv\alpha\,\|A(A^{j-2}x)\|\leq\alpha^2\,\|A^{j-2}x\|\leq\cdots\leq\alpha^j\,\|x\| \tag{2.161}$$

Therefore from the triangle inequality for norms (Problem 1.5) we have

$$\left\|\sum_{j=1}^{\infty}(-1)^j A^j x\right\|\leq\sum_{j=1}^{\infty}\|A^j x\|\leq\left(\sum_{j=1}^{\infty}\alpha^j\right)\|x\|=\frac{1}{1-\alpha}\|x\|$$

The operator defined in (2.160) is, in fact, a bounded operator also. Finally, since the product of two bounded operators is a bounded operator, the product of (2.160) and $(I+A)$ will exist and we obtain

$$\left[I+\sum_{j=1}^{n}(-1)^j A^j\right][I+A]x=\left[I+A-A+\sum_{j=2}^{n}(-1)^j A^j+(-1)^n A^n-\sum_{j=1}^{n-1}(-1)^{j+1}A^{j+1}\right]x$$
$$=[I+(-1)^n A^{n+1}]x$$

Since

$$\|A^n x\|\leq \alpha^n \|x\|$$

we see that

$$\lim_{n\to\infty} A^n x=0$$

for all $x\in S$. Therefore,

$$\|[I+(-1)^n A^{n+1}]x\|^2=\|x\|^2+\langle x \mid (-1)^n A^{n+1}x\rangle$$
$$+\langle (-1)^n A^{n+1}x \mid x\rangle+\|A^{n+1}x\|^2 \xrightarrow[n\to\infty]{} \|x\|^2$$

which implies that

$$\left[I+\sum_{j=1}^{\infty}(-1)^j A^j\right][I+A]=I$$

By the same type argument we see that (2.160) is also the right inverse. Q.E.D.

An alternative approach often used in physical applications is that of a *perturbation series.* Let a be a given vector, A an operator, and ε a continuous "small" parameter and seek a solution to

$$x=a-\varepsilon Ax \tag{2.162}$$

We *assume* there exists a solution of the form

$$x=\sum_{n=0}^{\infty}\varepsilon^n a_n \tag{2.163}$$

where the $\{a_n\}$ are vectors independent of ε. Substituting (2.163) into (2.162) we obtain

$$\sum_{n=0}^{\infty}\varepsilon^n a_n=a-\sum_{n=0}^{\infty}\varepsilon^{n+1}Aa_n \tag{2.164}$$

If we equate like powers of ε^n, we find

$$\begin{aligned} a_0&=a \qquad n=0 \\ a_{n+1}&=-Aa_n \qquad n\geq 1 \end{aligned} \tag{2.165}$$

We can solve (2.165) recursively as

$$a_n = -Aa_{n-1} = -A(-Aa_{n-2}) = A^2 a_{n-2} = \cdots = (-1)^n A^n a$$

so that

$$x = \sum_{n=0}^{\infty} (-1)^n A^n \varepsilon^n a \tag{2.166}$$

If we set $\varepsilon = 1$, then we have the result

$$x = \sum_{n=0}^{\infty} (-1)^n A^n a \equiv (I+A)^{-1} a$$

Of course this perturbation argument as given here is *not* a proof since we have not proven the convergence of the perturbation series (2.163). We would still have to go through the proof of Th. 2.11 to be rigorous.

As an example of the application of the Neumann or perturbation series, let us reexamine Example 2.7 given in (2.152),

$$\phi(x) - \lambda a(x) \int_a^b b^*(y)\phi(y)\, dy = f(x)$$

$$\equiv (I + \lambda K)\phi(x) \tag{2.152}$$

Assume that

$$\phi(x) = \sum_{n=0}^{\infty} \lambda^n \phi_n(x)$$

Equation 2.152 then becomes

$$\sum_{n=0}^{\infty} \lambda^n \phi_n(x) - a(x) \sum_{n=0}^{\infty} \lambda^{n+1} \langle b \mid \phi \rangle = f(x)$$

$$= \phi_0(x) + \sum_{n=0}^{\infty} \lambda^{n+1} [\phi_{n+1}(x) - a(x)\langle b \mid \phi_n \rangle]$$

from which we see that

$$\phi_0(x) = f(x)$$

$$\phi_{n+1}(x) = a(x)\langle b \mid \phi_n \rangle, \; n \geq 1$$

$$\phi_n(x) = a(x)\langle b \mid \phi_{n-1} \rangle = a(x)\langle b \mid a \rangle \langle b \mid \phi_{n-2} \rangle$$

$$= a(x)\langle b \mid a \rangle \langle b \mid a \rangle \langle b \mid \phi_{n-3} \rangle = \cdots = a(x)\langle b \mid a \rangle^{n-1} \langle b \mid f \rangle, \; n \geq 1$$

or

$$\phi(x) = f(x) + \left(\sum_{n=1}^{\infty} \lambda^n \langle b \mid a \rangle^{n-1} \right) \langle b \mid f \rangle a(x)$$

$$= f(x) + \lambda \left(\sum_{n=0}^{\infty} \lambda^n \langle b \mid a \rangle^n \right) \langle b \mid f \rangle a(x)$$

$$= f(x) + \frac{\lambda a(x) \langle b \mid f \rangle}{1 - \lambda \langle b \mid a \rangle} \tag{2.167}$$

This is just the solution given in (2.154). However the series in (2.167) converges only for $|\lambda\langle b \mid a\rangle|<1$, whereas the solution (2.154) is valid except when $\lambda\langle b \mid a\rangle=1$. If we had wanted to satisfy the conditions of Th. 2.11 in order to be guaranteed of the convergence of the Neumann series, we could have used the Schwarz inequality to estimate the norm of $\|K\phi\|$ as

$$\|K\phi\|=\|\lambda a\langle b \mid \phi\rangle\|=|\lambda|\,|\langle b \mid \phi\rangle|\,\|a\|\leq|\lambda|\,\|a\|\,\|b\|\,\|\phi\|$$

so that we would require for convergence

$$|\lambda|<[\|a\|\,\|b\|]^{-1}$$

which is a more stringent requirement than

$$|\lambda|<|\langle b \mid a\rangle|^{-1}$$

This is simply an example of a case in which the Neumann series may actually converge to a solution even through the conditions of Th. 2.11 are not fulfilled.

Let us consider one more instructive example of an infinitesimal operator. Begin with

$$R(\theta)\equiv\begin{pmatrix}\cos\theta & \sin\theta\\ -\sin\theta & \cos\theta\end{pmatrix}=I\cos\theta+\begin{pmatrix}0 & 1\\ -1 & 0\end{pmatrix}\sin\theta$$

$$\equiv I\cos\theta+C\sin\theta \tag{2.168}$$

If we let $\theta\rightarrow d\theta$, we obtain the infinitesimal transformation

$$\begin{aligned}R(d\theta)&=\begin{pmatrix}1 & d\theta\\ -d\theta & 1\end{pmatrix}=I+C\,d\theta\\ R(-d\theta)&=I-C\,d\theta\end{aligned} \tag{2.169}$$

Infinitesimal operators of the form $A=I+\varepsilon B$, where B is a bounded operator and ε an infinitesimal parameter, have the important property that

$$A^{-1}=I-\varepsilon B$$

since

$$(I+\varepsilon B)(I-\varepsilon B)=I-\varepsilon^2B^2\simeq I$$

to first order in ε. In many applications it is often relatively easy to write down the infinitesimal form of a transformation operator. A question of great importance then is whether or not the infinitesimal form of an operator will imply its general form for finite transformations. In Section A2.2 at the end of this chapter we show that $R(\theta)$, Eq. 2.168, can be recovered from $R(d\theta)$, Eq.

2.169. We might *guess* this result since

$$R^n(d\theta) \simeq I + n\, d\theta\, C = \exp\left[(n\, d\theta)C\right]$$

where the last equality can be established by direct expansion. If we let $n \to \infty$, $d\theta \to 0$ in such a way that $n\, d\theta \to \theta$, a finite limit, then we would have

$$R(\theta) = \exp(\theta C)$$

Of course this argument as presented here is not a proof.

These matrices $R(d\theta)$ are the elements of a *Lie group* of continuous transformations (i.e., one in which we have been able to build up the finite form of the transformation from the infinitesimal one by simple exponentiation). One crucial property here is that $R(d\theta)$ is continuously connected to the identity; that is,

$$\lim_{d\theta \to 0} R(d\theta) = I$$

We shall study these Lie groups in some detail in Section 9.5.

2.8 ADJOINT OF AN OPERATOR

An operator $L^\dagger$ is said to be the *adjoint* of a linear operator of L if for all x and y in S,

$$\langle y, Lx \rangle = \langle L^\dagger y, x \rangle \tag{2.170}$$

The adjoint is unique for a bounded operator. If

$$L^\dagger = L \tag{2.171}$$

then L is *self-adjoint* or *hermitian.*

THEOREM 2.12

There exists an adjoint operator for every bounded linear operator.

PROOF

We are given that L is a bounded linear operator

$$\|Lx\| \le C\,\|x\|$$

so that

$$|\langle y, Lx \rangle| \le \|y\|\,\|Lx\| \le (\|y\|\, C)\,\|x\|$$

Therefore $\langle y, Lx \rangle$ is a bounded linear functional of x. By Th. 1.2 we know there exists a vector z such that

$$f(x) \equiv \langle y, Lx \rangle = \langle z, x \rangle \tag{2.172}$$

Let

$$L^{\dagger} y \equiv z \tag{2.173}$$

This defines a linear operator (or linear correspondence between y and z).

Q.E.D.

We now consider finding the adjoints of some operators.

i. Let $L = A = \{\alpha_{ij}\}$, an $n \times n$ matrix.

$$\langle y, Lx \rangle = \sum_{j,l} \eta_j^* \alpha_{jl} \xi_l = \sum_{j,l} \alpha_{jl} \eta_j^* \xi_l = \sum_l \left(\sum_j \alpha_{lj}^{\dagger} \eta_j \right)^* \xi_l$$

$$= \langle A^{\dagger} y, x \rangle \tag{2.174}$$

Hence the adjoint of a matrix operator is precisely the hermitian adjoint defined previously (cf. Section 2.1a). (For a real matrix it is simply the transpose, A^{T}.)

ii. Consider an integral operator defined as

$$Lf(x) \equiv \int_a^b K(x, y) f(y)\, dy$$

$$\langle g, Lf \rangle = \int_a^b dx g^*(x) \int_a^b K(x, y) f(y)\, dy$$

$$= \int_a^b dy \left[\int_a^b dx K^*(x, y) g(x) \right]^* f(y)$$

$$\equiv \int_a^b dx \left[\int_a^b dy K^{\dagger}(x, y) g(y) \right]^* f(x) = \langle L^{\dagger} g, f \rangle$$

Therefore the adjoint kernel is

$$K^{\dagger}(x, y) = K^*(y, x) \tag{2.175}$$

so that

$$L^{\dagger} g(y) \equiv \int_a^b K^{\dagger}(y, x) g(x)\, dx = \int_a^b K^*(x, y) g(x)\, dx$$

iii. Let L be a differential operator

$$Lf(x) = \frac{d^2 f(x)}{dx^2} + g(x) f(x) \qquad x \in [a, b] \tag{2.176}$$

defined on the linear manifold of twice-differentiable functions specified as

$$\begin{aligned} f(a) &= f(b) = 0 \\ h(a) &= h(b) = 0 \end{aligned} \tag{2.177}$$

We will assume that g(x) is a real function [i.e., $g^*(x)=g(x)$].

$$\begin{aligned}
\langle h \mid Lf\rangle &\equiv \int_a^b h^*(x)\left[\frac{d^2f(x)}{dx^2}+g(x)f(x)\right]dx \\
&= h^*(x)\frac{df(x)}{dx}\Big|_a^b - \int_a^b \frac{dh^*(x)}{dx}\frac{df(x)}{dx}\,dx + \int_a^b g(x)f(x)\,dx \\
&= \frac{dh^*(x)}{dx}f(x)\Big|_a^b + \int_a^b\left[\frac{d^2h^*(x)}{dx^2}+g(x)h^*(x)\right]f(x)\,dx \\
&= \langle Lh, f\rangle \\
&\equiv \langle L^\dagger h, f\rangle
\end{aligned} \tag{2.178}$$

Therefore in this particular case $L^\dagger = L$ so that L is self adjoint. Notice that the boundary conditions, Eq. 2.177, satisfied by the functions on which L acts have entered in an essential manner. We will return to adjoints of differential operators in Section 8.4b. Remember also that differential operators are not bounded operators so that Th. 2.12 as just proven does not necessarily apply.

2.9 EXISTENCE AND UNIQUENESS OF THE SOLUTION OF $Lx = a$

As we have stated previously, given the equation

$$Lx = a$$

we must consider the uniqueness and the existence of the solution. These are distinct questions. We extend some of the results of Section 2.2 on linear algebraic systems to more general linear operators.

THEOREM 2.13

If and only if the homogeneous equation

$$Lx = 0$$

possesses a nontrivial solution, then the inhomogeneous equation

$$Lx = a$$

does not have a unique solution.

The proof of this theorem is immediate and is left to the reader.

A linear operator L is said to have *closed range* if, whenever we take a sequence of vectors $\{x_n\}$ in the domain of L and define a sequence $\{y_n\}$ in the range of L as

$$y_n \equiv Lx_n \tag{2.179}$$

such that

$$y_n \xrightarrow[n\to\infty]{} y$$

where y is a vector in S, then there always exists a vector x in S such that

$$y = Lx \tag{2.180}$$

Notice that the sequence $\{x_n\}$ need not tend to a limit and that, in particular, x need not be such that $x = \lim_{n\to\infty} x_n$. A bounded operator need not have closed range as is shown by the example in Problem 2.18. Also let the reader beware that an operator with closed range as defined here (and by Friedman) is *not* the same as a closed operator as commonly defined in functional analysis (cf. Riesz and Nagy, Schmeidler, or Stone). The purpose of defining a closed range as done here is to allow us to apply the projection theorem (Th. 1.1) to those operators having closed range. This definition of closed range is in some sense the inverse of the definition of the continuity of an operator given in Eq. 2.41. It is clear that the range of a linear operator with closed range is a subspace since if $y_n \equiv Lx_n \to y$, then, because $\exists x \in S \ni y = Lx$, by *definition* y is in this subspace. It should also be evident that if L has an inverse, then L has closed range.

At this point we give an example of an operator whose range is not closed. Let $x \in [-1, 1]$ and $f_n(x) \in \mathscr{L}_2[-1, 1]$, and define

$$g_n(x) \equiv \int_{-1}^{x} f_n(y)\, dy = \int_{-1}^{1} H(x-y) f_n(y)\, dy \equiv \int_{-1}^{1} k(x, y) f_n(y)\, dy \tag{2.181}$$

where the *Heaviside step function* is defined (almost everywhere) by

$$H(x) = \begin{cases} 1 & x > 0 \\ 0 & x < 0 \end{cases} \tag{2.182}$$

If we take a particular $f_n(x)$,

$$f_n(x) = \frac{n}{2} e^{-n|x|} \tag{2.183}$$

then $f_n(x) \in \mathscr{L}_2[-1, 1]$ since

$$\int_{-\infty}^{\infty} |f_n(x)|^2\, dx = 1$$

for all finite n. A simple calculation shows that

$$g_n(x) = \begin{cases} \frac{1}{2}(e^{-n|x|} - e^{-n}) & x \leq 0 \\ 1 - \frac{1}{2}(e^{-n|x|} + e^{-n}) & x \geq 0 \end{cases}$$

It is clear that

$$g(x) \equiv \lim_{n\to\infty} g_n(x) = H(x) \qquad x \neq 0 \tag{2.184}$$

so that

$$\int_{-1}^{1} |g(x)|^2\, dx = \int_0^1 dx = 1$$

Therefore $g(x) \in \mathscr{L}_2[-1, 1]$, although the limit of the sequence of functions in Eq. 2.183 does not exist. Nor does any $f(x) \in \mathscr{L}_2[-1, 1]$ exist such that $g(x) = \int_{-1}^{1} H(x-y) f(y)\, dy$. For reference later we note that

$$\int_{-1}^{1} dx \int_{-1}^{1} dy\, |k(x, y)|^2 = \int_{-1}^{1} dx \int_{-1}^{1} H(x-y)\, dy$$

$$= \int_{-1}^{1} dx \int_{-1}^{x} dy = \int_{-1}^{1} (x+1)\, dx = 2 \tag{2.185}$$

It is important to understand that the vector, x, referred to in Eq. 2.180 for an operator with closed range need not be the limit of the sequence $\{x_n\}$ in Eq. 2.179.

Example 2.8 Let

$$x_n = \begin{pmatrix} (-1)^n + \dfrac{1}{n} \\ (-1)^{n+1} + \dfrac{1}{n} \end{pmatrix} \qquad A = \begin{pmatrix} 1 & 1 \\ 1 & 1 \end{pmatrix}$$

so that

$$y_n \equiv A x_n = \begin{pmatrix} \dfrac{2}{n} \\ \dfrac{2}{n} \end{pmatrix} \xrightarrow[n\to\infty]{} 0 \equiv y$$

Although the sequence $\{x_n\}$ does not have a limit, there does exist a vector (i.e., $x = 0$) such that $y = Ax$.

We can easily see that all finite-dimensional square matrices have closed range. If $\det A \neq 0$, then A^{-1} exists and the statement is obvious. If $\det A = 0$, we are still given the condition that there are vectors x_m such that

$$y_m = A x_m \tag{2.186}$$

and when $y_m \to y$ we must prove the existence of an x such that $y = Ax$. However from Section 2.2 (cf. Eq. 2.27) we can write this as

$$\langle a_i, x_m \rangle = y_m^i \qquad i = 1, 2, \ldots, n$$

where *at most* $(n-1)$ of the $\{a_i\}$ can be linearly independent since det $A = 0$. This states that given y_m a set $\{\bar{x}_m\}$ is uniquely determined only in the subspace spanned by these $(n-1)$ linearly independent vectors and is completely arbitrary in the rest of the space. If the first $(n-1)$ of the $\{a_i\}$ are linearly independent, then we can use Cramer's rule to solve

$$\langle a_i, \bar{x}_m \rangle = y_m^i \qquad i = 1, 2, \ldots, (n-1)$$

where this will be true in a suitable basis. Since we are here *assuming* $\{x_m\}$ such that Eq. 2.186 is satisfied, the $\{\bar{x}_m\}$ so constructed will be consistent with

$$\langle a_n, \bar{x}_m \rangle = y_m^n$$

If we now simply take the nth component of $\bar{x}_m$ to be zero, then we have *an* $\bar{x}_m$ given in terms of the y_m and, since all sums involved are finite, this $\bar{x}_m$ will go to a limit as y_m does (i.e., $m \to \infty$).

The following theorem will now provide the necessary and sufficient condition for the existence of a solution to a singular system of inhomogeneous linear algebraic equations (as promised at the end of Section 2.2c). It also will have much wider general application.

THEOREM 2.14

Given an operator L with closed range, the necessary and sufficient condition that the equation

$$Lx = a$$

have a solution for a given a is that

$$\langle a, z \rangle = 0$$

for *every* solution to

$$L^\dagger z = 0$$

PROOF

a. It is trivial to prove that the condition stated is necessary since if we are given the existence of an x such that $Lx = a$ then

$$\langle z, Lx \rangle = \langle z, a \rangle = \langle L^\dagger z, x \rangle = 0$$

b. The proof of the sufficiency is a bit more subtle and can be accomplished by contradiction. We assume that $\langle z, a \rangle = 0$ for all z such that $L^\dagger z = 0$ but that a is not in the range of L (i.e., that there is no x in S such that $Lx = a$). Since the range of L is closed and, hence, a subspace, we know from Th. 1.1 that there must exist a nonzero vector $z \equiv a - a_r$, where a_r is the projection of a onto the range of L, such that z is orthogonal to the range of L; that is,

$$\langle z, Lx \rangle = 0 \qquad \forall x \in S$$

If we take $x=L^\dagger z$, then we have

$$\langle L^\dagger z, L^\dagger z\rangle \equiv \|L^\dagger z\|^2=0$$

so that

$$L^\dagger z=0$$

which states that $z=a-a_r$ is a solution to the homogeneous adjoint equation. However direct calculation shows that

$$\langle z, a\rangle=\langle z, (z+a_r)\rangle=\langle z, z\rangle+\langle z, a_r\rangle=\|z\|^2\neq 0$$

which yields a contradiction. Q.E.D.

Therefore we may summarize these very important results as follows:

i. $\exists x\neq 0 \ni Lx=0 \Leftrightarrow x\ni Lx=a$ cannot be unique.
ii. $\exists x\ni Lx=a \Leftrightarrow \langle z, a\rangle=0,\ \forall z\ni L^\dagger z=0$.

If we consider a self-adjoint operator with closed range,

$$L^\dagger=L$$

then uniqueness implies existence since

$$Lx_0\equiv L^\dagger x_0=0 \Rightarrow x_0=0 \qquad \text{(for uniqueness)}$$

so that

$$\langle a, x_0\rangle=0$$

or

$$\exists x\ni Lx=a$$

However the converse is not necessarily true (e.g., singular hermitian matrices). Such self-adjoint operators form the major class of operators met in physical applications.

2.10 COMPLETELY CONTINUOUS OPERATORS

The main purpose of this section is to reassure us that most of our labor in proving abstract theorems has not been in vain with regard to practical applications.

An operator L is said to be *completely continuous* if it can be uniformly approximated by separable operators. That is, if there exists a sequence of separable operators $\{L_n\}$ such that

$$\|Lx-L_nx\|<\varepsilon\,\|x\| \tag{2.187}$$

whenever $n>N$ for all vectors x and with ε independent of x, then L is a completely continuous operator.

THEOREM 2.15

If $\{e_j\}$ is an orthonormal basis for the space, and if L is of *trace class*,

$$\sum_{n=1}^{\infty} \|Le_n\|^2 < \infty \tag{2.188}$$

then L is completely continuous.

PROOF

We will also show that the trace class condition implies a uniform bound on L. Since the $\{e_j\}$ form an orthonormal basis, we can write

$$|x\rangle = \sum_j |e_j\rangle\langle e_j \mid x\rangle \tag{2.189}$$

for any vector in S. If we can prove the uniform convergence of

$$\sum_{j=1}^{\infty} L\, |e_j\rangle\langle e_j| \tag{2.190}$$

we will be finished since then

$$L = \sum_{j=1}^{\infty} L\, |e_j\rangle\langle e_j| \tag{2.191}$$

The norm of the remainder of (2.191) after n terms becomes, with repeated use of the Schwarz inequality,

$$\begin{aligned}
&\left\| L \sum_{j=1}^{n} |e_j\rangle\langle e_j \mid x\rangle - \sum_{j=1}^{\infty} L\, |e_j\rangle\langle e_j \mid x\rangle \right\|^2 \\
&\quad \equiv \left\| \sum_{j=n+1}^{\infty} L\, |e_j\rangle\langle e_j \mid x\rangle \right\|^2 = \left\langle \sum_{j=n+1}^{\infty} L\, |e_j\rangle\langle e_j \mid x\rangle,\ \sum_{l=n+1}^{\infty} L\, |e_l\rangle\langle e_l \mid x\rangle \right\rangle \\
&\quad = \sum_{j,l=n+1}^{\infty} \langle e_j \mid x\rangle^* \langle e_l \mid x\rangle\langle Le_j \mid Le_l\rangle \le \sum_{j,l=n+1}^{\infty} |\langle e_j \mid x\rangle|\, |\langle e_l \mid x\rangle\langle Le_j \mid Le_l\rangle| \\
&\quad \le \sum_{j=n+1}^{\infty} |\langle e_j \mid x\rangle| \left(\sum_{l=n+1}^{\infty} |\langle e_l \mid x\rangle|^2 \right)^{1/2} \left(\sum_{l=n+1}^{\infty} |\langle Le_j \mid Le_l\rangle|^2 \right)^{1/2} \\
&\quad \le \left(\sum_{j=n+1}^{\infty} |\langle e_j \mid x\rangle|^2 \right)^{1/2} \|x\| \left(\sum_{j=n+1}^{\infty} \sum_{l=n+1}^{\infty} |\langle Le_j \mid Le_l\rangle|^2 \right)^{1/2} \\
&\quad \le \|x\|^2 \left(\sum_{j,l=n+1}^{\infty} \|Le_j\|^2\, \|Le_l\|^2 \right)^{1/2}
\end{aligned} \tag{2.192}$$

so that

$$\begin{aligned}
\lim_{n\to\infty} \left\| L \sum_{j=1}^{n} |e_j\rangle\langle e_j \mid x\rangle - \sum_{j=1}^{\infty} L\, |e_j\rangle\langle e_j \mid x\rangle \right\| &\equiv \left\| Lx - \sum_{j=1}^{\infty} L\, |e_j\rangle\langle e_j \mid x\rangle \right\| \\
&\le \lim_{n\to\infty} \left(\sum_{j=n+1}^{\infty} \|Le_j\|^2 \right)^{1/2} \|x\| = 0 \quad \text{Q.E.D.}
\end{aligned}$$

Furthermore since

$$\|Lx\| = \left\|\sum_j L\,|e_j\rangle\langle e_j \mid x\rangle\right\| \le \sum_j \|Le_j\|\,|\langle e_j \mid x\rangle|$$

$$\le \left(\sum_j \|Le_j\|^2\right)^{1/2} \|x\| \tag{2.193}$$

we see that if an operator is of trace class, then it is bounded.

We also note for use later that for a bounded operator L the adjoint operator $L^\dagger$ has the same bound as L. From Th. 2.12 we know that $L^\dagger$ exists for a bounded L so that

$$|\langle y \mid Lx\rangle| = |\langle L^\dagger y \mid x\rangle| \le \|y\|\,\|Lx\| \le C\,\|y\|\,\|x\| \qquad \forall x \in S$$

If we let $x = L^\dagger y$ then

$$\|L^\dagger y\|^2 \le C\,\|L^\dagger y\|\,\|y\|$$

or

$$\|L^\dagger y\| \le C\,\|y\|$$

Example 2.9 As an instructive example, let us compute a bound on the norm of Lf for an integral operator. In the notation of Section 2.5 we write

$$\langle x \mid Lf\rangle \equiv Lf(x) = \int_a^b K(x, y) f(y)\, dy$$

Since

$$\|Lf\|^2 = \langle Lf \mid Lf\rangle = \langle f \mid L^\dagger Lf\rangle$$

$$= \int_a^b dx \int_a^b dy \int_a^b dz\, \langle f \mid x\rangle\langle x|\, L^\dagger\, |y\rangle\langle y|\, L\, |z\rangle\langle z \mid f\rangle$$

$$= \int_a^b dx \int_a^b dy \int_a^b dz\, f^*(x) K^*(y, x) K(y, z) f(z)$$

$$= \int_a^b dy \int_a^b K(y, z) f(z)\, dz \left(\int_a^b K(y, x) f(x)\, dx\right)^*$$

$$= \int_a^b dy \left|\int_a^b K(y, z) f(z)\, dz\right|^2 \le \int_a^b dy \int_a^b |K(y, z)|^2\, dz \int_a^b |f(x)|^2\, dx$$

$$= \left(\int_a^b dx \int_a^b dy\, |K(x, y)|^2\right) \|f\|^2$$

we conclude that

$$\|Lf\| \le \left(\int_a^b dx \int_a^b dy\, |K(x, y)|^2\right)^{1/2} \|f\| \tag{2.194}$$

Therefore, if $K(x, y) \in \mathscr{L}_2$, then L is a *bounded* operator.

Finally let us see what the trace class condition,

$$\sum_n \|Le_n\|^2 < \infty$$

becomes for an integral operator. We will see in Section 4.3a that

$$\varphi_n(x) \equiv \frac{1}{\sqrt{(b-a)}} \exp\left\{\frac{2\pi in[2x-(a+b)]}{(b-a)}\right\},$$

$n = 0, \pm 1, \pm 2, \ldots$, form a countable set of basis functions for $\mathscr{L}_2[a, b]$.

$$\langle x \mid \varphi_n \rangle = \varphi_n(x) \qquad \langle y \mid L\varphi_n \rangle = \int_a^b dx\, K(y, x)\varphi_n(x)$$

$$\langle \varphi_n \mid x \rangle = \varphi_n^*(x) \qquad \langle y | L | x \rangle = K(y, x)$$

$$\langle y | L^\dagger | x \rangle = K^*(x, y)$$

$$\begin{aligned}
\sum_n \langle L\varphi_n \mid L\varphi_n \rangle &= \sum_n \int_a^b dy\, |\langle y \mid L\varphi_n \rangle|^2 \\
&= \sum_n \int_a^b dy \left| \int_a^b dx\, K(y, x)\varphi_n(x) \right|^2 \\
&= \int_a^b dy \sum_{n=-\infty}^{\infty} \left| \int_a^b dx\, K(y, x)\varphi_n(x) \right|^2 \\
&= \int_a^b dy \int_a^b |K(y, x)|^2\, dx
\end{aligned} \tag{2.195}$$

Here the last step follows from the completeness of the $\{\varphi_n\}$. Therefore in this case L will be of trace class provided $K(x, y) \in \mathscr{L}_2[a, b]$; that is,

$$\int_a^b dx \int_a^b dy\, |K(x, y)|^2 < \infty \tag{2.196}$$

Of course, we would expect the trace class condition to imply boundedness for an integral operator since we have just proven such a result for a general linear operator.

In our study of integral equations in Chapter 5 we will be concerned with operators that will be of the form of the identity plus an operator of trace class. If we can convince ourselves that Th. 2.14 is applicable to such operators, then we shall be able to save ourselves a lot of algebra later. In Section A2.3 we establish the following theorem.

THEOREM 2.16

The operator $L = I + K_n$, where K_n is an n-term separable operator, is an operator with closed range.

We can now easily argue that a linear operator of the form $L = I + K$, where K

is a completely continuous operator, is an operator with closed range. Since K is completely continuous we may write

$$L=I+K_n+R_n=(I+R_n)+K_n \tag{2.197}$$

where K_n is an n-term separable operator and R_n is an operator with a bound as small as we please. We know that $(I+R_n)$ has an inverse (cf. Th. 2.11) so that it necessarily has closed range. Therefore if we define

$$Lx=y\equiv[(I+R_n)+K_n]x \tag{2.198}$$

then this implies

$$\begin{aligned}[I+(I+R_n)^{-1}K_n]x&=(I+R_n)^{-1}y\equiv a\\ &\equiv(I+K_n')x\end{aligned} \tag{2.199}$$

where $K_n'=(I+R_n)^{-1}K_n$ is again an n-term separable operator so that, by Th. 2.16, $(I+K_n')$ is an operator with closed range. This means that for any given y (or a) there exists an $x\in S$ that satisfies Eq. 2.199. But this x will also satisfy Eq. 2.198 so that $L=I+K$ has closed range. This establishes an important theorem.

THEOREM 2.17

The operator $L=I+K$, where K is a completely continuous linear operator, is an operator with closed range.

We will now apply the results of Theorems 2.13–2.17 to the integral equation

$$\phi(x)-\lambda\int_a^b K(x, y)\phi(y)\,dy=f(x)$$

where $\phi(x), f(x)\in\mathcal{L}_2[a, b]$.

THEOREM 2.18

If the kernel function $K(x, y)$ is of trace class,

$$\int_a^b dx\int_a^b dy\,|K(x, y)|^2<\infty \tag{2.200}$$

then the integral equation

$$\phi(x)-\lambda\int_a^b K(x, y)\phi(y)\,dy=f(x) \tag{2.201}$$

has a solution if and only if

$$\int_a^b f^*(x)\psi(x)\,dx=0 \tag{2.202}$$

for all solutions $\psi(x)$ of the homogeneous adjoint equation

$$\psi(x)-\lambda\int_a^b K^*(y, x)\psi(y)\, dy=0 \tag{2.203}$$

The necessary and sufficient condition that Eq. 2.201 have a unique solution is that there be no nontrivial solution to the homogeneous equation

$$\phi_0(x)-\lambda\int_a^b K(x, y)\phi_0(y)\, dy=0 \tag{2.204}$$

A2.1 A POLYNOMIAL EXPANSION FOR $\Delta(x)=\det\{a_{ij}+\delta_{ij}x\}$

We now derive an expansion for a determinant of the form

$$\Delta(x)\equiv\det\{a_{ij}+\delta_{ij}x\}=\begin{vmatrix} a_{11}+x & a_{12} & \cdots & a_{1n} \\ a_{21} & a_{22}+x & \cdots & a_{2n} \\ \cdot & \cdot & \cdot & \cdot \\ \cdot & \cdot & \cdot & \cdot \\ \cdot & \cdot & \cdot & \cdot \\ a_{n1} & a_{n2} & \cdots & a_{nn}+x \end{vmatrix} \tag{A2.1}$$

which is evidently a polynomial of degree n in x. If we consider the elements of each column to be the sum of two terms (i.e., $a_{ij}+\delta_{ij}x$), then we see from the basic definition, Eq. 2.6, that $\Delta(x)$ can be written as the sum of 2^n determinants, none of whose columns any longer contain sums. Since $\Delta(x)$ is an nth degree polynomial, it must have the form

$$\Delta(x)=x^n+S_1x^{n-1}+S_2x^{n-2}+\cdots+S_{n-1}x+S_n \tag{A2.2}$$

where $S_n=\Delta\equiv\det\{a_{ij}\}$.

We must now find explicit expressions for the coefficients, S_m. If we delete the first terms from $m\,(\leq n)$ columns of $\Delta(x)$ and keep only the first terms in the other $(n-m)$ columns, then the remaining terms in each of the m columns are all zero except for an x on the diagonal in each of these m columns. Each such determinant will reduce to x^m times a determinant of order $(n-m)$ obtained from Δ by striking these m columns and their corresponding rows and retaining the $(n-m)\times(n-m)$ determinant consisting only of the $\{a_{ij}\}$. These determinants with row-columns $i_1, i_2, \ldots, i_m$ struck we denote by $\Delta(i_1, i_2, \ldots, i_m)$. For example, if $n=4$ and $i_1=1$, $i_2=2$, we have

$$\begin{vmatrix} x & 0 & a_{13} & a_{14} \\ 0 & x & a_{23} & a_{24} \\ 0 & 0 & a_{33} & a_{34} \\ 0 & 0 & a_{43} & a_{44} \end{vmatrix}=x^2\begin{vmatrix} a_{33} & a_{34} \\ a_{43} & a_{44} \end{vmatrix}=x^2\,\Delta(1, 2)$$

Therefore in order to obtain S_{n-k} (i.e., the coefficient of x^k in Eq. A2.2) we must simply evaluate

$$S_{n-k} = \sum_{i_1<i_2<\cdots<i_k} \Delta(i_1, i_2, \ldots, i_k) \tag{A2.3}$$

Notice that since $\Delta(i_1, \ldots, i_k)$ is left invariant under the interchange of any two i_l's (by its very definition) and since none of the i_l's are the same, we can relax the inequalities on the sum in Eq. A2.3 and simply divide by $k!$ [and define $\Delta(i_1, \ldots, i_k)$ to be zero when any two of the i_l's are identical]. As a check that we do obtain all 2^n terms in this expansion of Eq. A2.1, we observe that there are

$$n(n-1)\cdots(n-k+1) = \frac{n!}{(n-k)!}$$

different ways of striking the k columns to obtain a $\Delta(i_1, \ldots, i_k)$ so that, by the binomial theorem,

$$\sum_{k=0}^{n} \frac{n!}{k!\,(n-k)!} = (1+1)^n = 2^n$$

We can put this into a better form for use later if we define a determinant of rank m as

$$A(j_1, j_2, \ldots, j_m) \equiv \begin{vmatrix} a_{j_1j_1} & a_{j_1j_2} & \cdots & a_{j_1j_m} \\ a_{j_2j_1} & a_{j_2j_2} & \cdots & a_{j_2j_m} \\ \cdot & \cdot & \cdot & \cdot \\ \cdot & \cdot & \cdot & \cdot \\ \cdot & \cdot & \cdot & \cdot \\ a_{j_mj_1} & a_{j_mj_2} & \cdots & a_{j_mj_m} \end{vmatrix} \tag{A2.4}$$

and write in place of Eq. A2.3

$$S_m = \sum_{j_1<j_2<\cdots<j_m} A(j_1, j_2, \ldots, j_m) \tag{A2.5}$$

Notice that $A(j_1, \ldots, j_m)$ vanishes if any two of the j_k's are identical and that the value of $A(j_1, \ldots, j_m)$ is left unchanged by the interchange of any pair of the j_k's (i.e., this is equivalent to *simultaneously* interchanging two rows *and* two columns). Since there are $m!$ different ways of ordering the $j_1, j_2, \ldots, j_m$, we may drop the inequalities in the summation indices in Eq. A2.5 and write simply

$$S_m = \frac{1}{m!} \sum_{j_1,\ldots,j_m=1}^{n} A(j_1, j_2, \ldots, j_m) \tag{A2.6}$$

where we define $S_0 \equiv 1$, to obtain

$$\Delta(x) = \sum_{m=0}^{n} S_m x^{n-m} \tag{A2.7}$$

A2.2 EXPONENTIATION OF THE TWO-DIMENSIONAL INFINITESIMAL SPATIAL ROTATIONS

Consider $R(d\theta)$ defined as

$$R(d\theta)=I+d\theta C \qquad C=\begin{pmatrix} 0 & 1 \\ -1 & 0 \end{pmatrix}$$

It is a simple matter to prove that

$$C^2=-I \qquad C^3=-C \qquad C^4=I, \ldots$$

so that

$$C^{2j}=(-1)^j I \qquad C^{2j+1}=(-1)^j C \qquad j=0, 1, 2, \ldots$$

This implies that

$$R^2(d\theta)\equiv R(d\theta)R(d\theta)=I+2d\theta C-I(d\theta)^2\simeq I+2d\theta C\simeq R(2d\theta) \tag{A2.8}$$

If we apply $R(d\theta)$ n times, then we obtain *exactly*

$$\begin{aligned} R^n(d\theta)&=(I+d\theta C)^n=n!\sum_{s=0}^{n}\frac{1}{(n-s)!\,s!}(d\theta C)^s \\ &=n!I\sum_{s=0}^{[n/2]}\frac{(-1)^s}{(n-2s)!\,(2s)!}(d\theta)^{2s}+n!\,C\sum_{s=0}^{[(n-1)/2]}\frac{(-1)^s}{(n-2s-1)!\,(2s+1)!}(d\theta)^{2s+1} \end{aligned} \tag{A2.9}$$

where $[N/2]$ means that the summation terminates at the *integer* less than or equal to $N/2$. But since

$$\frac{n!}{(n-2s)!}=\overbrace{n(n-1)(n-2)\cdots(n-2s+1)}^{2s\text{ terms}}=\begin{cases} n^{2s}\left[1\left(1-\frac{1}{n}\right)\left(1-\frac{2}{n}\right)\cdots\left(1-\frac{2s+1}{n}\right)\right] & s\geq 1 \\ 1 & s=0 \end{cases}$$

$$\frac{n!}{(n-2s-1)!}=\overbrace{n(n-1)(n-2)\cdots(n-2s)}^{2s+1\text{ terms}}=n^{2s+1}\left[1\left(1-\frac{1}{n}\right)\left(1-\frac{2}{n}\right)\cdots\left(1-\frac{2s}{n}\right)\right]$$

we have

$$\begin{aligned} R^n(d\theta)=I&\sum_{s=0}^{[n/2]}\frac{(-1)^s}{(2s)!}(n\,d\theta)^{2s}\left[1\left(1-\frac{1}{n}\right)\left(1-\frac{2}{n}\right)\cdots\left(1-\frac{2s-1}{n}\right)\right] \\ &+C\sum_{s=0}^{[(n-1)/2]}\frac{(-1)^s}{(2s+1)!}(n\,d\theta)^{2s+1}\left[1\left(1-\frac{1}{n}\right)\left(1-\frac{2}{n}\right)\cdots\left(1-\frac{2s}{n}\right)\right] \end{aligned} \tag{A2.10}$$

Now notice that

$$\begin{aligned} \lim_{n\to\infty}\sum_{s=1}^{[n/2]}\ln\left(1-\frac{2s-1}{n}\right)&=\lim_{n\to\infty}\left[\ln\left(1-\frac{1}{n}\right)+\ln\left(1-\frac{3}{n}\right)+\cdots+\ln\left(\frac{1}{n}\right)\right] \\ &=-\infty \end{aligned}$$

so that

$$\lim_{n\to\infty}\left\{\prod_{s=1}^{[n/2]} 1\left(1-\frac{1}{n}\right)\left(1-\frac{2}{n}\right)\cdots\left(1-\frac{2s-1}{n}\right)\right\}=0 \qquad \text{(A2.11)}$$

If we now let $d\theta \to 0$, $n \to \infty$ such that $n\,d\theta \to \theta$, a finite limit, then we finally arrive at

$$\lim_{n\to\infty} R^n(d\theta)=I\cos\theta+C\sin\theta=e^{C\theta}=R(\theta) \qquad \text{(A2.12)}$$

The methods of Section 3.7 allow a much simpler proof of this result.

A2.3 PROOF OF THEOREM 2.16

The proof that an operator of the form

$$L_N=I+\sum_{j=1}^{N}|a_j\rangle\langle b_j| \qquad \text{(A2.13)}$$

is an operator with closed range is straightforward but rather lengthy. In Eq. A2.13 let $N=1$,

$$L_1=I+|a_1\rangle\langle b_1|$$

so that

$$y_n=Lx_n=x_n+\langle b_1\mid x_n\rangle a_1$$

$$\langle b_1\mid y_n\rangle=\langle b_1\mid x_n\rangle[1+\langle b_1\mid a_1\rangle]$$

If $[1+\langle b_1\mid a_1\rangle\neq 0$, then

$$x_n=y_n-\frac{\langle b_1\mid y_n\rangle a_1}{1+\langle b_1\mid a_1\rangle}$$

so that

$$x\equiv\lim_{n\to\infty}x_n=y-\frac{\langle b_1\mid y\rangle a_1}{1+\langle b_1\mid a_1\rangle}$$

is such that

$$Lx\equiv y-\frac{\langle b_1\mid y\rangle a_1}{1+\langle b_1\mid a_1\rangle}+\langle b_1\mid y\rangle a_1-\frac{\langle b_1\mid y\rangle\langle b_1\mid a_1\rangle a_1}{1+\langle b_1\mid a_1\rangle}=y+\langle b_1\mid y\rangle a_1-\langle b_1\mid y\rangle a_1=y$$

If $1+\langle b_1\mid a_1\rangle=0$, then

$$\langle b_1\mid y_n\rangle=0 \qquad \forall n$$

which implies that $\langle b_1\mid y\rangle=0$. Let $x\equiv y\equiv\lim_{n\to\infty}y_n$ so that

$$Ly=y+\langle b_1\mid y\rangle a=y$$

Therefore L_1 is an operator with closed range (i.e., true for $N=1$). Also, since $L_1^\dagger=I+|b_1\rangle\langle a_1|$, both L_1 and $L_1^\dagger$ are necessarily operators with *closed range.*

Now assume that both L_N and $L_N^\dagger$ are operators with closed range and prove the same for L_N+1.

$$L_{N+1}=L_N+|a_{N+1}\rangle\langle b_{N+1}|$$
$$y_n=L_{N+1}x_n=L_Nx_n+\langle b_{N+1} \mid x_n\rangle a_{N+1}$$

α. $b_{N+1}\in\mathcal{R}(L_N^\dagger) \Rightarrow \exists b \ni b_{N+1}=L_N^\dagger b$

$$\therefore \quad y_n=L_Nx_n+\langle b \mid L_Nx_n\rangle a_{N+1}$$
$$\Rightarrow \langle b \mid y_n\rangle=\langle b \mid L_Nx_n\rangle[1+\langle b \mid a_{N+1}\rangle] \tag{A2.14}$$

i. *If* $1+\langle b \mid a_{N+1}\rangle\neq 0$, then

$$L_Nx_n=y_n-\frac{\langle b \mid y_n\rangle a_{N+1}}{1+\langle b \mid a_{N+1}\rangle}$$

Since L_Nx_n approaches a limit, then since L_N has closed range by assumption, there exists a vector x such that

$$L_Nx=y-\frac{\langle b \mid y\rangle a_{N+1}}{1+\langle b \mid a_{N+1}\rangle}$$

But this vector is also such that

$$L_{N+1}x\equiv L_Nx+\langle b_{N+1} \mid x\rangle a_{N+1}=L_Nx+\langle b \mid L_Nx\rangle a_{N+1}$$
$$=y-\frac{\langle b \mid y\rangle a_{N+1}}{1+\langle b \mid a_{N+1}\rangle}+\langle b \mid y\rangle a_{N+1}$$
$$-\frac{\langle b \mid y\rangle\langle b \mid a_{N+1}\rangle a_{N+1}}{1+\langle b \mid a_{N+1}\rangle}\equiv y$$

so that L_{N+1} has closed range in this case [i.e., (α.i)].

ii. If $1+\langle b \mid a_{N+1}\rangle=0$, then either $a_{N+1}\in\mathcal{R}(L_N)$ or $a_{N+1}\notin\mathcal{R}(L_N)$. If $a_{N+1}\in\mathcal{R}(L_N)$, there exists an $a \ni L_Na = a_{N+1}$. Since $\langle b \mid y\rangle=0$, Eq. A2.14 implies that

$$y_n=L_Nx_n+\langle b \mid L_Nx_n\rangle L_Na=L_N[x_n+\langle b \mid L_Nx_n\rangle a]$$

which states that $y\in\mathcal{R}(L_N)$. Since L_N is assumed to have closed range we know there exists an x such that

$$y=L_Nx$$

We conclude that

$$L_{N+1}x=L_Nx+\langle b_{N+1} \mid x\rangle a_{N+1}=L_Nx+\langle b \mid L_Nx\rangle a_{N+1}$$
$$=y+\langle b \mid y\rangle a_{N+1}=y$$

so that L_{N+1} has closed range in this case [i.e., (α.ii)].

iii. If $1+\langle b \mid a_{N+1}\rangle=0$ and $a_{N+1}\notin\mathcal{R}(L_N)$, then there is no solution to

$$a_{N+1}=L_Nw$$

so that (cf. Th. 2.14) there is a nontrivial solution to

$$L_N^\dagger z = 0$$

such that

$$\langle a_{N+1} \mid z \rangle \neq 0$$

From Eq. A2.14 we obtain

$$\begin{aligned}\langle (b+z) \mid y_n \rangle &= \langle (b+z) \mid L_N x_n \rangle + \langle (b+z) \mid L_N x_n \rangle \langle (b+z) \mid a_{N+1} \rangle \\ &= \langle (b+z) \mid L_N x_n \rangle [1 + \langle (b+z) \mid a_{N+1} \rangle] \\ &= \langle (b+z) \mid L_N x_n \rangle \langle z \mid \ a_{N+1} \rangle\end{aligned}$$

so that

$$\begin{aligned}y_n &= L_N x_n + \langle (b+z) \mid L_N x_n \rangle a_{N+1} \\ &= L_N x_n + \frac{\langle (b+z) \mid y_n \rangle}{\langle z \mid a_{N+1} \rangle} a_{N+1}\end{aligned}$$

Since L_N has closed range, there exists an x such that

$$L_N x = y - \frac{\langle (b+z) \mid y \rangle}{\langle z \mid a_{N+1} \rangle} a_{N+1}$$

This implies that

$$\begin{aligned}L_{N+1} x &= L_N x + \langle b_{N+1} \mid x \rangle a_{N+1} \\ &= y - \frac{\langle (b+z) \mid y \rangle a_{N+1}}{\langle z \mid a_{N+1} \rangle} + \langle (b+z) \mid L_N x \rangle a_{N+1} = y\end{aligned}$$

so that L_{N+1} has closed range in this case [i.e., (α.iii)].

β. $b_{N+1} \notin \mathcal{R}(L_N^\dagger)$

i. $a_{N+1} \in \mathcal{R}(L_N) \Rightarrow \exists a \ni$

$$\begin{aligned}a_{N+1} &= L_N a \\ y_n = L_{N+1} x_n &= L_N x_n + \langle b_{N+1} \mid x_n \rangle a_{N+1} \\ &= L_N [x_n + \langle b_{N+1} \mid x_n \rangle a]\end{aligned}$$

Since L_N has closed range, this requires the existence of a u such that

$$y = L_N u$$

Furthermore there is no solution to

$$b_{N+1} = L_N^\dagger w$$

so that a nontrivial z exists such that

$$L_N z = 0 \qquad \langle z \mid b_{N+1} \rangle \neq 0 \tag{A2.15}$$

Since

$$L_{N+1}u = L_N u + \langle b_{N+1} \mid u\rangle a_{N+1} = y + \langle b_{N+1} \mid u\rangle a_{N+1}$$
$$L_{N+1}z = L_N z + \langle b_{N+1} \mid z\rangle a_{N+1} = \langle b_{N+1} \mid z\rangle a_{N+1} \neq 0 \tag{A2.16}$$

let

$$x = u + \alpha z$$

so that

$$\begin{aligned} L_{N+1}x &= L_N u + \langle b_{N+1} \mid u\rangle a_{N+1} + \alpha\langle b_{N+1} \mid z\rangle a_{N+1} \\ &= L_N u + [\langle b_{N+1} \mid u\rangle + \alpha\langle b_{N+1} \mid z\rangle] a_{N+1} \equiv y \end{aligned}$$

or

$$\alpha = -\langle b_{N+1} \mid u\rangle / \langle b_{N+1} \mid z\rangle$$

Again, L_{N+1} has closed range in this case [i.e., (β.i)].

ii. Finally the only case left is $a_{N+1} \notin \mathcal{R}(L_N)$. Hence there is no solution to

$$a_{N+1} = L_N w$$

so that there exists a nontrivial v such that

$$L_N^\dagger v = 0 \qquad \langle a_{N+1} \mid v\rangle \neq 0$$

Since

$$y_n = L_N x_n + \langle b_{N+1} \mid x_n\rangle a_{N+1}$$

we obtain

$$\begin{aligned} \langle v \mid y_n\rangle &= \langle v \mid L_N x_n\rangle + \langle b_{N+1} \mid x_n\rangle\langle v \mid a_{N+1}\rangle \\ &= \langle b_{N+1} \mid x_n\rangle\langle v \mid a_{N+1}\rangle \end{aligned}$$

or

$$y_n = L_N x_n + \frac{\langle v \mid y_n\rangle}{\langle v \mid a_{N+1}\rangle} a_{N+1}$$

Since L_N has closed range, there exists a w such that

$$L_N w = y - \frac{\langle v \mid y\rangle}{\langle v \mid a_{N+1}\rangle} a_{N+1}$$

Let

$$x = w + \alpha z$$

Then from Eqs. A2.15 and A2.16 we have

$$L_{N+1}x = L_N w + \langle b_{N+1} \mid w \rangle a_{N+1} + \alpha L_N z + \alpha \langle b_{N+1} \mid z \rangle a_{N+1}$$

$$= y + \left[-\frac{\langle v \mid z \rangle}{\langle v \mid a_{N+1} \rangle} + \langle b_{N+1} \mid w \rangle + \alpha \langle b_{N+1} \mid z \rangle \right] a_{N+1}$$

so that

$$\alpha = \frac{1}{\langle b_{N+1} \mid z \rangle} \left[\frac{\langle v \mid y \rangle}{\langle v \mid a_{N+1} \rangle} - \langle b_{N+1} \mid w \rangle \right]$$

or

$$L_{N+1}x = y$$

Again L_{N+1} has closed range in this case [i.e., (β.ii)].

Hence in all possible cases if L_N and $L_N^\dagger$ have closed range for any N, then L_{N+1} also has closed range. However since (cf. Problem 2.16)

$$L_{N+1}^\dagger = L_N^\dagger + |b_{N+1}\rangle\langle a_{N+1}|$$

the proof just given demonstrates that $L_{N+1}^\dagger$ also has closed range. Since we have established this for $N=1$, we now know that

$$L_N \equiv I + \sum_{j=1}^{N} |a_j\rangle\ b_j|$$

has closed range for *all* N. Q.E.D.

A2.4 PROOF THAT $\mathcal{H} = \mathcal{R} \oplus \mathcal{N}$ FOR A BOUNDED, SELF-ADJOINT LINEAR OPERATOR DEFINED ON $\mathcal{H}$

Let L be a bounded, self-adjoint linear operator defined on a Hilbert space $\mathcal{H}$ and denote by $\mathcal{R}$ and $\mathcal{N}$ the range and null space of L, respectively. By choice we decompose $\mathcal{H}$ into the direct sum

$$\mathcal{H} = \mathcal{N} \oplus \mathcal{R}' \tag{A2.17}$$

where, by definition, $\mathcal{R}'$ is all of $\mathcal{H}$ not contained in $\mathcal{N}$. Since for all $g \in \mathcal{N}$

$$Lg = 0$$

then for all $f \in \mathcal{H}$, $\|f\| < \infty$, the boundedness of L requires that $\|Lf\| < \infty$ and we have

$$\langle f \mid Lg \rangle = 0 = \langle Lf \mid g \rangle \tag{A2.18}$$

which states that the range of a self adjoint operator is orthogonal to its null space. This allows us to decompose $\mathcal{R}'$ as

$$\mathcal{R}' = \mathcal{R} \oplus \mathcal{M}$$

so that

$$\mathscr{H}=\mathscr{N}\oplus\mathscr{R}\oplus\mathscr{M} \tag{A2.19}$$

or

$$f=g+h+w \tag{A2.20}$$

where $f \in \mathscr{H}$, $g \in \mathscr{N}$, $h \in \mathscr{R}$, $w \in \mathscr{M}$, and

$$\langle g \mid h\rangle=0=\langle g \mid w\rangle=\langle h \mid w\rangle \tag{A2.21}$$

We now need only establish that $\mathscr{M}$ contains no vectors not already in $\mathscr{R}$ or $\mathscr{N}$. Therefore we assume that

$$Lw\neq 0 \tag{A2.22}$$

By definition $Lw \in \mathscr{R}$ as is $L(Lw)$ so that Eqs. A2.19 and A2.20 imply

$$\langle w \mid L^2 w\rangle=0=\langle Lw \mid Lw\rangle=\|Lw\|^2$$

or

$$Lw=0 \tag{A2.23}$$

which contradicts Eq. A2.22. Thus $\mathscr{M}$ must be empty and

$$\mathscr{H}=\mathscr{R}\oplus\mathscr{N} \qquad \text{Q.E.D.} \tag{A2.24}$$

TABLE 2.2 Results on the Inversion of Separable Linear Operators

$L\,|x\rangle=|a\rangle$

1. $L=\sum_{j=1}^{n} |a_j\rangle\langle b_j|$
   a. *necessary conditions for a solution* $|x\rangle$ *to exist:*
      i. $|a\rangle=\sum_{j=1}^{n} \alpha_j\,|a_j\rangle$
      ii. $\exists\{\xi_j\} \ni \alpha_j=\sum_{k=1}^{n} \xi_k\langle b_j \mid b_k\rangle,\ j=1,2,\ldots,n$
      (This will be so if, for instance, the $\{b_j\}$ are linearly independent.)
   b. *the solution (generally nonunique):*
      $|x\rangle=\sum_{j=1}^{n} \xi_j\,|b_j\rangle+|c\rangle$
      where $|c\rangle$ is an arbitrary vector such that $\langle c \mid b_j\rangle=0,\ j=1,2,\ldots,n$

TABLE 2.2 (*continued*)

2. $L = I + \sum_{j=1}^{n} |a_j\rangle\langle b_j|$
   a. *necessary condition for a solution* $|x\rangle$ *to exist:*
   $\exists\{\langle b_j \mid x\rangle\} \ni \langle b_j \mid x\rangle + \sum_{k=1}^{n} \langle b_k \mid x\rangle\langle b_j \mid a_k\rangle = \langle b_j \mid a\rangle, j = 1, 2, \ldots, n$
   b. *the unique solution (if the* $\{\langle b_j \mid x\rangle\}$ *are uniquely determined):*
   $|x\rangle = |a\rangle - \sum_{j=1}^{n} \langle b_j \mid x\rangle |a_j\rangle$
   c. *if* $\langle b_j \mid a_k\rangle = \delta_{jk}\langle b_j \mid a_j\rangle$, *then:*
   i. if $1 + \langle b_j \mid a_j\rangle \neq 0$ for any j 1, 2, . . . , n, then the unique solution is
   $$|x\rangle = |a\rangle - \sum_{j=1}^{n} \frac{\langle b_j \mid a\rangle}{1 + \langle b_j \mid a_j\rangle} |a_j\rangle$$
   ii. if $1 + \langle b_j \mid a_j\rangle = 0$, $j = \{n_m\}$, there is no solution unless
   $\langle b_j \mid a\rangle = 0$, $j = \{n_m\}$
   in which case a nonunique solution is
   $$|x\rangle = |a\rangle - \sum_{\substack{j=1 \\ j \neq \{n_m\}}}^{n} \frac{\langle b_j \mid a\rangle}{1 + \langle b_j \mid a_j\rangle} |a_j\rangle + \sum_{j=\{n_m\}} \alpha_j |a_j\rangle$$
   where the $\{\alpha_j\}$ are arbitrary constants.

SUGGESTED REFERENCES

B. Friedman, *Principles and Techniques of Applied Mathematics.* Much of the material covered in the latter part of the present chapter is also treated in Chapter 1 of Friedman.

P. R. Halmos, *Finite-Dimensional Vector Spaces.*

For a treatment of linear algebra, matrices, and determinants, the reader may wish to consult the following textbooks.

G. Birkhoff and S. MacLane, *A Survey of Modern Algebra.*

F. W. Byron and R. W. Fuller, *Mathematics of Classical and Quantum Physics (Vol. 1).*

M. Marcus and H. Minc, *A Survey of Matrix Theory and Matrix Inequalities.*

D. C. Murdock, *Linear Algebra for Undergraduates.*

V. I. Smirnov, *Linear Algebra and Group Theory.* This is really an excellent text that treats many important topics from an elementary point of view. The reader will be well rewarded to acquaint himself with this work.

For a more general and abstract approach to operations in various spaces, the following are suggested.

R. Beals, *Topics in Operator Theory.*

F. Riesz and B. Sz-Nagy, *Functional Analysis.* This is a classic text that covers the theory of Lebesgue integration and then goes on to integral equations and linear transformations.

W. Schmeidler, *Linear Operators in Hilbert Space.* This little volume gives a very readable and compact summary of many topics in the theory of linear operators.

A. E. Taylor, *Introduction to Functional Analysis.*

PROBLEMS

2.1 Using the definitions of the indicated matrix operations, prove the following:

a. $AO = OA = O$

b. $IA = AI = A$

c. $A(B+C) = AB + AC$

d. $(A+B)C = AC + BC$

e. $A(BC) = (AB)C$

f. $A + B = B + A$

g. $(AB)^T = B^T A^T$

2.2 Add and multiply the following matrices and hence show that they do not commute with respect to multiplication but commute with respect to addition.

$$A = \begin{pmatrix} 1 & 2i \\ 3i & 2 \end{pmatrix} \qquad B = \begin{pmatrix} 0 & i \\ 2i & 1 \end{pmatrix}$$

2.3 a. Show that a diagonal matrix with identical elements will commute with any matrix.

b. Show that two matrices will commute if they are both diagonal.

2.4 If

$$A = \begin{pmatrix} 1 & -3 & 2 \\ 4 & 1 & -1 \\ -3 & 2 & 5 \end{pmatrix} \quad \text{and} \quad B = \begin{pmatrix} 2 & 1 & 5 \\ -1 & -2 & -2 \\ 3 & 1 & 2 \end{pmatrix}$$

find the matrices AB, BA, A^{-1}, B^{-1}, $(AB)^{-1}$. Check your results by verifying the relations

$$(AB)^{-1} = B^{-1}A^{-1}, \; AA^{-1} = I, \; B^{-1}B = I$$

2.5 Consider a collection of $n \times n$ square matrices $\{A, B, \ldots\}$. Suppose we define an operation,

$$(A, B) \equiv A^{\dagger}B$$

where ordinary matrix multiplication is indicated on the right, and "†" signifies the adjoint of a matrix. Would this operation be an acceptable inner product if we replace the requirement

$$\|x\| = 0 \Leftrightarrow x = 0$$

by

$$(A, A) = 0 \Leftrightarrow A = O$$

where O is the $n \times n$ zero matrix? If this is possible, are there any restrictions that must be imposed on A?

2.6 Let:

$$A = \begin{pmatrix} 1 & 0 & 0 \\ 0 & \cos\theta & -\sin\theta \\ 0 & \sin\theta & \cos\theta \end{pmatrix} \qquad S = \begin{pmatrix} \cos\phi & 0 & \sin\phi \\ 0 & 1 & 0 \\ -\sin\phi & 0 & \cos\phi \end{pmatrix}$$

a. Compute explicitly $SAS^{-1} = A_{(II)}$.

b. Compute $\mathbf{r}'_{(II)} = A_{(II)}\mathbf{r}_{(II)}$
and

$$\mathbf{r}_{(II)} = S_{(I)}\mathbf{r}_{(I)}$$

for a vector $\mathbf{r}_{(I)}$ initially contained in the y-z plane.

c. Show graphically the effect of A on $\mathbf{r}$ and of S on the coordinate axes.

2.7 Given the matrix

$$M = \begin{pmatrix} \alpha & \beta & \gamma & \delta \\ \beta & \alpha & \delta & \gamma \\ \gamma & \delta & \alpha & \beta \\ \delta & \gamma & \beta & \alpha \end{pmatrix}$$

where α, β, γ, and δ are complex quantities, prove that M can be decomposed into the sum of direct products

$$M = \alpha a \otimes a + \beta a \otimes b + \gamma b \otimes a + \delta b \otimes b$$

where

$$a = \begin{pmatrix} 1 & 0 \\ 0 & 1 \end{pmatrix} \qquad b = \begin{pmatrix} 0 & 1 \\ 1 & 0 \end{pmatrix}$$

Next show that

$$ab = ba$$

and find a 2×2 orthogonal matrix, u,

$$uu^{\mathrm{T}} = 1$$

such that

$$uau^{\mathrm{T}} = \begin{pmatrix} 1 & 0 \\ 0 & 1 \end{pmatrix} \equiv d_1 \qquad ubu^{\mathrm{T}} = \begin{pmatrix} 1 & 0 \\ 0 & -1 \end{pmatrix} \equiv d_2$$

where d_1 and d_2 are diagonal. Prove then that the 4×4 orthogonal matrix U,

$$U = u \otimes u, \; UU^{\mathrm{T}} = I$$

is such that

$$D \equiv UMU^{\mathrm{T}} = \alpha\, d_1 \otimes d_1 + \beta\, d_1 \otimes d_2 + \gamma\, d_2 \otimes d_1 + \delta\, d_2 \otimes d_2$$

where D is a diagonal matrix. Finally find the four roots $\lambda_1, \ldots, \lambda_4$, of the equation

$$\det|M - \lambda I| = 0$$

where I is the 4×4 identity matrix.

2.8 Let A be an $n \times n$ matrix. Suppose A has the form

$$A = \begin{pmatrix} R & O \\ O & S \end{pmatrix}$$

where R is $m \times m$ and S is $(n-m) \times (n-m)$. Prove that

$$\det A = \det R \det S$$

2.9 Either prove the following statement or find a counter example. If $C = A \otimes B$, then $\det C = \det A \det B$.

2.10 The final set of axes (x'', y'', z'') is obtained from the original set (x, y, z) by:

a. A counterclockwise rotation through an angle ϕ about the z axis to obtain the set (x', y', z');

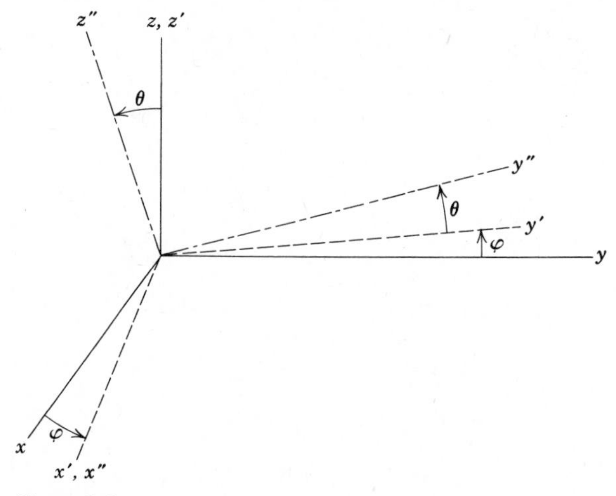

Figure 2.6

b. A counterclockwise rotation through an angle θ about the x' axis to obtain the set (x'', y'', z'').

Write down the rotation matrix $R(\phi, \theta)$ that will carry the set (x, y, z) directly into (x'', y'', z''), by writing $R(\phi)$ and $R(\theta)$ first. Consider $R(\phi, \theta)$ applied to an arbitrary vector $\mathbf{r}$ so that

$$\mathbf{r}' = R(\phi, \theta)\mathbf{r}$$

Show that the angle Φ between $\mathbf{r}'$ and $\mathbf{r}$ is given as

$$\cos\left(\frac{\Phi}{2}\right) = \cos\left(\frac{\phi}{2}\right)\cos\left(\frac{\theta}{2}\right)$$

Hint: First show that any real 3×3 orthogonal matrix has one real eigenvalue as well as a complex-conjugate pair of eigenvalues. Argue that two successive rotations must be equivalent to a single rotation of angle Φ about the vector left invariant by $R(\phi, \theta)$ so that there must exist a similarity transformation S such that

$$R(\Phi) = SR(\phi, \theta)S^{-1}$$

Now prove that the trace of a matrix A,

$$\text{Tr}\, A \equiv \sum_j a_{jj}$$

(i.e., the sum of the diagonal elements of A) is invariant under a similarity transformation.

2.11 Prove that if A and B are $n\times n$ hermitian matrices (i.e., $A^\dagger = A$, $B^\dagger = B$) then AB and BA are hermitian if and only if

$$[A, B] \equiv AB - BA = 0$$

That is, A and B must commute.

2.12 Given the matrix

$$A = \begin{pmatrix} \beta & 0 & 2 \\ 0 & 3 & \beta \\ 0 & \beta & 3 \end{pmatrix}$$

where β is a continuous parameter, find those values of β for which A^{-1} does not exist. Assuming that β is not one of these values, find A^{-1}.

2.13 Consider the equation

$$\mathbf{A}\mathbf{x}=0$$

where

$$A=\begin{pmatrix}1 & 0 & 1\\ 0 & 1 & 0\\ 1 & 1 & 1\end{pmatrix}$$

Prove that $\mathbf{x}$ is not unique by a geometrical argument that shows that $\mathbf{x}$ is determined only in a subspace. Display this subspace pictorially.

2.14 Consider the operator equation

$$\mathbf{A}\mathbf{x}=\mathbf{b}$$

where

$$A=\begin{pmatrix}\frac{1}{\sqrt{3}} & \frac{1}{\sqrt{2}} & \frac{1}{\sqrt{6}}\\ -\frac{1}{\sqrt{3}} & 0 & \frac{2}{\sqrt{6}}\\ \frac{1}{\sqrt{3}} & -\frac{1}{\sqrt{2}} & \frac{1}{\sqrt{6}}\end{pmatrix}\qquad \mathbf{x}=\begin{pmatrix}x_1\\ x_2\\ x_3\end{pmatrix}\qquad \mathbf{b}=\begin{pmatrix}1\\ 0\\ 1\end{pmatrix}$$

Find the vector $\mathbf{x}$. Is this solution unique?

2.15 Prove that, for a linear operator L, boundedness implies continuity and vice versa. Furthermore prove that if L is bounded, then

$$L\sum_{j=1}^{\infty} x_j=\sum_{j=1}^{\infty} Lx_j$$

2.16 Prove that if L is separable

$$L=\sum_{j=1}^{n} |a_j\rangle\langle b_j|$$

then its adjoint is

$$L^{\dagger}=\sum_{j=1}^{n} |b_j\rangle\, a_j|$$

2.17 You are given the integral equation

$$u(x)+\int_0^1 e^{-|x-y|}u(y)\,dy=f(x),\qquad x\in[0,1],\ f(x)\in\mathcal{L}_2[0,1]$$

Discuss as fully as you can the question of the existence and uniqueness of the solution to this problem.

Hint: For an integral operator recall that

$$\|Lu\|\leq\left(\int_0^1 dx\int_0^1 dy\,|K(x,y)|^2\right)^{1/2}\|u\|$$

2.18 Consider E_∞, the vector space with a countable infinity of basis vectors and defined over the field of all complex numbers. Take the basis vectors to be:

$$e_1 = (1, 0, 0, \ldots, 0, \ldots)$$
$$e_2 = (0, 1, 0, \ldots, 0, \ldots)$$
$$\vdots$$
$$e_n = (0, 0, 0, \ldots, 0, 1, 0, \ldots)$$
$$\longleftarrow \text{nth position} \longrightarrow$$
$$\vdots$$

{i.e., $e_n^{(j)} = \delta_{jn}$} so that any vector x may be represented as

$$x = (\xi_1, \xi_2, \ldots, \xi_n, \ldots) = \sum_{j=1}^{\infty} \xi_j e_j$$

We also require that $\|x\| < \infty$ for a vector to be in E_∞. Consider the linear operator, L, defined as

$$Lx = \left(\xi_1, \frac{1}{2}\xi_2, \frac{1}{3}\xi_3, \ldots, \frac{1}{n}\xi_n, \ldots\right)$$

Prove your answers to the following questions.

a. Is L bounded?
b. Is L of trace class?
c. Is L^{-1} a bounded linear operator?
d. Is $(I+L)^{-1}$ bounded? Need the Neumann series for it converge.
e. Does L have closed range? *Hint:* Consider the sequence of vectors

$$y_n = Lx_n \qquad x_n = \{\xi_k^{(n)}\} \qquad \xi_k^{(n)} = \left(1 + \frac{1}{n}\right)^{-k}$$

and investigate the limits of these sequences.

f. Does $(I+L)$ have closed range?

2.19 Consider the integral equation

$$u(x) + \lambda \int_{-1}^{1} e^{ikxy} u(y)\, dy = f(x)$$

Discuss the existence and uniqueness of $u(x)$. If a solution exists, find it.

2.20 Consider the integral equation

$$u(x) + \sqrt{\frac{3}{8}} \int_0^1 |x-y|^{-\alpha} u(y)\, dy = f(x) \qquad x \in [0, 1]$$

where $f(x) \in \mathscr{L}_2[0, 1]$.

a. For what range of values of α is the integral operator of trace class?
b. For what range of values of α will the Newmann series for $u(x)$ necessarily converge?

c. Assuming α is in the range found in (b), what can you say about the existence and uniqueness of $u(x)$? Why?

d. If $\alpha=0$, solve the integral equation explicitly. Is the solution unique?

e. Let $\alpha=\varepsilon>0$, where $\varepsilon \ll 1$. Write

$$u(x) \simeq u_0(x)+\varepsilon u_1(x)$$

where $u_0(x)$ is the solution found in (d). Take $f(x)$ to be a constant, for example, f_0. Expand the kernel of the integral equation about $\varepsilon=0$, insert $u(x)=u_0(x)+\varepsilon u_1(x)$, keep all terms of order ε, and compute $u_1(x)$.

3 | Spectral Analysis of Linear Operators

3.0 INTRODUCTION

We have seen previously (cf. Section 2.7a) that the diagonalization of an operator is tantamount to solving the linear equation

$$Lx = a \tag{3.1}$$

That is *if* we can find a basis $\{\hat{x}_j\}$ and its reciprocal $\{\hat{z}_j\}$ (which always exists once the $\{\hat{x}_j\}$ does) such that the operator L can be represented as

$$L = \sum_j \lambda_j \hat{x}_j\rangle\langle \hat{z}_j \tag{3.2}$$

then we can solve Eq. 3.1 as follows. Since $\langle \hat{x}_j, \hat{z}_k\rangle = \delta_{jk}$, we see that

$$L\hat{x}_j\rangle = \sum_k \lambda_k \langle \hat{z}_k \mid \hat{x}_j\rangle \, |\hat{x}_k\rangle = \lambda_j \hat{x}_j\rangle, \qquad j = 1, 2, \ldots \tag{3.3}$$

Here the $\{\lambda_j\}$ are the *eigenvalues* of L and the $\{\hat{x}_j\}$ are the *eigenvectors* of L corresponding to the $\{\lambda_j\}$. Since we have assumed the $\{\hat{x}_j\}$ form a *basis,* we may expand both a, which is known, and x, which is unknown, in terms of the $\{\hat{x}_j\}$ as

$$a = \sum_j \alpha_j \hat{x}_j \tag{3.4}$$

$$x = \sum_j \xi_j \hat{x}_j \tag{3.5}$$

so that Eq. 3.1 reduces to

$$Lx=\sum_j \xi_j L\hat{x}_j=\sum_j \xi_j\lambda_j\hat{x}_j=a=\sum_j \alpha_j\hat{x}_j$$

which yields

$$\lambda_j\xi_j=\alpha_j$$

If there are no zero eigenvalues (i.e., $\lambda_j\neq 0$ for *any* j), then we obtain the solution to our problem as

$$x=\sum_j \frac{\alpha_j}{\lambda_j}\hat{x}_j$$

Example 3.1 As a trivial illustration consider the following.

$$L=\begin{pmatrix}1 & 0\\ 0 & -1\end{pmatrix} \qquad \det L=-1\neq 0 \qquad e_1=\begin{pmatrix}1\\0\end{pmatrix} \qquad e_2=\begin{pmatrix}0\\1\end{pmatrix}$$

$$\langle e_j, e_k\rangle=\delta_{jk} \qquad j, k=1, 2$$

$$Le_1=e_1 \Rightarrow \lambda_1=+1$$

$$Le_2=-e_2 \Rightarrow \lambda_2=-1$$

$$x=\begin{pmatrix}\xi_1\\ \xi_2\end{pmatrix}=\xi_1 e_1+\xi_2 e_2 \qquad a=\begin{pmatrix}\alpha_1\\ \alpha_2\end{pmatrix}=\alpha_1 e_1+\alpha_2 e_2$$

$$x=\frac{\alpha_1}{1}+\frac{\alpha_2}{-1}=\begin{pmatrix}\alpha_1\\ -\alpha_2\end{pmatrix}$$

This, of course, checks with the solution obtained by constructing L^{-1} directly.

Most of this chapter will be concerned with finding conditions sufficient to insure the completeness of the eigenvectors of an operator in a finite dimensional space. In Chapters 5 and 8 we will return to this question for operators defined on an infinite dimensional separable Hilbert space.

3.1 INVARIANT MANIFOLDS

A manifold (or subspace) $\mathcal{M}$ is an *invariant manifold* (or subspace) of an operator L provided that whenever $x \in \mathcal{M}$ then $Lx \in \mathcal{M}$, $\forall x \in \mathcal{M}$. Some simple but important examples of invariant manifolds of an operator L defined on a space S are the entire space S, the space of the zero vector, the null space of L (i.e., all $y \ni Ly=0$), and the range of L. Furthermore the reader who has read Section A2.4 will realize that for any self-adjoint linear operator L in a finite-dimensional vector space the whole space can be decomposed into the direct sum of the null space of L plus the range of L. Also, if we are in a space having only vectors of finite norm, then any linear operator is bounded when restricted to its invariant manifolds so that, from Th. 2.8, L may be represented by a matrix on any of its finite-dimensional invariant subspaces.

THEOREM 3.1

If an n-dimensional space is the direct sum of two invariant subspaces $\mathcal{M}$ (of dimension l) and $\mathcal{N}$ (of dimension $n-l$) of a linear operator L, then L may be represented on this n-dimensional space by the block-diagonal matrix

$$L=\begin{pmatrix} A & 0 \\ 0 & B \end{pmatrix} \tag{3.6}$$

where A is an $l\times l$ matrix and B is an $(n-l)\times(n-l)$ matrix.

PROOF

Since the entire space is the direct sum of $\mathcal{M}$ and $\mathcal{N}$, we may choose $\{\hat{x}_j\}$, $j=1, 2, \ldots, l$, as a basis for $\mathcal{M}$ and $\{\hat{x}_j\}$, $j=l+1, \ldots, n$, as a basis for $\mathcal{N}$. Any x in this n-dimensional space can be represented as

$$x=\sum_{k=1}^{n} \xi_k \hat{x}_k$$

so that

$$y=Lx$$

or

$$\sum_{j=1}^{n} \eta_j \hat{x}_j = \sum_{k,j=1}^{n} \alpha_{jk} \xi_k \hat{x}_j$$

$$\eta_j = \sum_{k=1}^{n} \alpha_{jk} \xi_k \tag{3.7}$$

If we now choose an $x \in \mathcal{M}$, then $Lx \in \mathcal{M}$ also so that Eq. 3.7 becomes

$$0=\sum_{k=1}^{l} \alpha_{jk} \xi_k, \qquad j=l+1, \ldots, n$$

Since the $\{\xi_k\}$ are arbitrary we obtain

$$\alpha_{jk}=0, \qquad j=l+1, \ldots, n; \qquad k=1, 2, \ldots, l \tag{3.8}$$

Similarly, if we take $x \in \mathcal{N}$ so that $Lx \in \mathcal{N}$, then Eq. 3.7 becomes

$$0=\sum_{k=l+1}^{n} \alpha_{jk} \xi_k \qquad j=1, 2, \ldots, l$$

so that

$$\alpha_{jk}=0, \qquad j=1,2,\ldots,l \qquad k=l+1,\ldots,n \tag{3.9}$$

This is just Eq. 3.6 since $L=\{\alpha_{jk}\}$. Q.E.D.

If λ_j is an eigenvalue of L, then we say that the dimension of the null space of $(L-\lambda_j I)$ is the *multiplicity* of the eigenvalue λ_j.

It is evident that if $\mathcal{M}$ is a finite-dimensional subspace invariant under L, then there exists (at least) one nontrivial eigenvector of L in $\mathcal{M}$. Since L may be represented by a matrix $L=A=\{\alpha_{ij}\}$ in $\mathcal{M}$, we require a solution to

$$Lx \equiv Ax=\lambda x \qquad x \in \mathcal{M} \tag{3.10}$$

which reduces to

$$\sum_{k=1}^{l} \alpha_{jk}\xi_k=\lambda\xi_j \qquad j=1,2,\ldots,l \tag{3.11}$$

From Section 2.2b we know that Eq. (3.10) has a nontrivial solution if and only if

$$\det|A-\lambda I|=0 \tag{3.12}$$

This is simply an lth degree polynomial and therefore has exactly l (possibly redundant) roots $\{\lambda_j\}$, $j=1,2,\ldots,l$ (cf. Section 7.6). For any λ_j there will exist a nontrivial solution to Eq. 3.10.

We now give some examples of this last statement.

Example 3.2

$$L=\begin{pmatrix}5 & 1\\ 4 & 2\end{pmatrix} \qquad x=\begin{pmatrix}\xi_1\\ \xi_2\end{pmatrix} \qquad (L-\lambda I)\begin{pmatrix}\xi_1\\ \xi_2\end{pmatrix}=0$$

This yields the algebraic equations

$$\begin{aligned}(5-\lambda)\xi_1+\qquad \xi_2&=0\\ 4\xi_1+(2-\lambda)\xi_2&=0\end{aligned}$$

so that

$$\begin{vmatrix}5-\lambda & 1\\ 4 & 2-\lambda\end{vmatrix}=0$$

or $\lambda_1=1$, $\lambda_2=6$.

For $\lambda_1=1$ we obtain

$$\left.\begin{aligned}4\xi_1+\xi_2=0\\ 4\xi_1+\xi_2=0\end{aligned}\right\}$$

so that $\xi_2=-4\xi_1$, or

$$x_1=\alpha\begin{pmatrix}1\\ -4\end{pmatrix}$$

We cannot determine the scalar α (unless, for instance, we normalize x_1 to unity; i.e., $\langle x_1, x_1\rangle=1$).

For $\lambda_2=6$ we have

$$-\xi_1 + \xi_2 = 0$$
$$4\xi_1 - 4\xi_2 = 0$$

so that $\xi_2=\xi_1$, or

$$x_2 = \beta\begin{pmatrix}1\\1\end{pmatrix}$$

Here x_1 and x_2 are not orthogonal since $\langle x_1 | x_2\rangle = -3\alpha^*\beta \neq 0$, but they are linearly independent so that they span E_2.

Example 3.3 However there are operators whose eigenvectors do not span the space. Let

$$L=\begin{pmatrix}1 & 1 & 2\\0 & 1 & 3\\0 & 0 & 2\end{pmatrix}$$

so that

$$\det|L-\lambda I| = \begin{vmatrix}1-\lambda & 1 & 2\\0 & 1-\lambda & 3\\0 & 0 & 2-\lambda\end{vmatrix} = (2-\lambda)\begin{vmatrix}1-\lambda & 1\\0 & 1-\lambda\end{vmatrix} = (2-\lambda)(1-\lambda)^2 = 0$$

or

$$\lambda_1 = 2 \qquad \lambda_2=\lambda_3=1$$

$$\lambda_1=2 \qquad \left.\begin{aligned}-\xi_1+\xi_2+2\xi_3&=0\\0-\xi_2+3\xi_3&=0\\0+0+0&=0\end{aligned}\right\}\Rightarrow \left.\begin{aligned}\xi_2&=3\xi_3\\\xi_1&=5\xi_3\end{aligned}\right\}\Rightarrow x_1\propto\begin{pmatrix}5\\3\\1\end{pmatrix}$$

$$\lambda_2=\lambda_3=1 \qquad \left.\begin{aligned}\xi_2+2\xi_3&=0\\3\xi_3&=0\\\xi_3&=0\end{aligned}\right\}\Rightarrow \left.\begin{aligned}&\xi_2=\xi_3=0\\&\xi_1-\text{arbitrary}\end{aligned}\right\}\Rightarrow x_2 \quad \text{and} \quad x_3\propto\begin{pmatrix}1\\0\\0\end{pmatrix}$$

Although x_1 is linearly independent of x_2 or x_3, x_2 and x_3 are linearly dependent. We do not have three linearly independent eigenvectors so that these cannot be used as a basis for E_3.

We now have two different means of solving the equation

$$Lx = a$$

We may find the inverse of L (i.e., L^{-1}) or find the eigenvectors and eigenvalues of L and expand in terms of these (if they are complete). We will use both of these methods, as well as combinations of them, later.

However this method of using the eigenvectors of an operator to solve an inhomogeneous equation is applicable only if the eigenvectors span the space under consideration. Example 3.3 exhibits a matrix with repeated roots for which the three eigenvectors cannot be linearly independent so that they will not span E_3. Clearly any time none of the eigenvalues of a finite-dimensional matrix are repeated, there will always be enough eigenvectors to span the space.

However degeneracy of the eigenvalues need not necessarily imply lack of linear independence of the eigenvectors as shown by the following example.

Example 3.4

$$L = \begin{pmatrix} 1 & 0 & 0 \\ 0 & 2 & 0 \\ 0 & 0 & 2 \end{pmatrix} \qquad x = \begin{pmatrix} \alpha \\ \beta \\ \gamma \end{pmatrix}$$

$$Lx = \lambda x \Rightarrow \det|L - \lambda I| = 0 \Rightarrow \lambda = 1, 2 \text{ (repeated twice)}$$

$$\lambda = 1 \Rightarrow \left.\begin{matrix} \alpha = \alpha \\ 2\beta = \beta \\ 2\gamma = \gamma \end{matrix}\right\} \Rightarrow x_1 = \begin{pmatrix} 1 \\ 0 \\ 0 \end{pmatrix}$$

$$\lambda = 2 \Rightarrow \left.\begin{matrix} \alpha = 2\alpha \\ 2\beta = 2\beta \\ 2\gamma = 2\gamma \end{matrix}\right\} \Rightarrow \begin{matrix} \text{two possible} \\ \text{linearly independent} \\ \text{solutions are} \end{matrix} \; x_2 = \begin{pmatrix} 0 \\ 1 \\ 0 \end{pmatrix} \qquad x_3 = \begin{pmatrix} 0 \\ 0 \\ 1 \end{pmatrix}$$

The question of spanning the subspace of degenerate eigenvalues is extremely important in quantum mechanics, especially in perturbation theory. The problem of the linear independence of eigenvectors corresponding to degenerate eigenvalues turns on the possibility of being able to diagonalize L completely. In particular we will soon see that all self-adjoint matrices can be brought into diagonal form. For the most part hereafter we restrict our attention to such operators since nearly all operators of physical interest fall into this class.

3.2 CHARACTERISTIC EQUATION OF A MATRIX

We now consider actually finding the eigenvalues and eigenvectors of finite-dimensional matrices.

$$A = \{a_{ij}\} \qquad x = \begin{pmatrix} \xi_1 \\ \xi_2 \\ \cdot \\ \cdot \\ \cdot \\ \xi_n \end{pmatrix}$$

$$Ax = \lambda x \tag{3.13}$$

Since we are interested in nontrivial solutions to the system (3.13),

$$\sum_{j=1}^{n} (a_{ij}-\lambda\,\delta_{ij})\xi_j=0 \qquad i=1,2,\ldots,n \tag{3.14}$$

we know we must require that

$$c(\lambda)\equiv\det(A-\lambda I)=0 \tag{3.15}$$

as necessary and sufficient condition for nonzero ξ_j's to exist. Equation 3.15 is called the *characteristic equation* of the matrix, A. In other words, one first solves $c(\lambda)=0$ for the λ_k, $k=1,2,\ldots,n$ (i.e., an nth-degree polynomial equation in λ) and then for each λ_k one solves the set (3.14) for the $\{\xi_j^{(k)}\}$ to obtain the x_j such that

$$Lx_j=\lambda_j x_j \tag{3.16}$$

We will give an example of this procedure in Section 3.3.

It is easily seen that similar matrices possess the same characteristic equations since if

$$B=SAS^{-1} \tag{3.17}$$

then

$$\begin{aligned}\det|B-\lambda I| &= \det|SAS^{-1}-\lambda SIS^{-1}|=\det|S(A-\lambda I)S^{-1}|\\ &= \det S\det|A-\lambda I|\det S^{-1}=\det(SS^{-1})\det|A-\lambda I|\\ &= \det|A-\lambda I|\end{aligned} \tag{3.18}$$

The following theorem is true for any matrix A, but we will prove it only for those matrices that can be completely diagonalized by a similarity transformation.

THEOREM 3.2

Every matrix A that can be diagonalized by a similarity transformation satisfies its characteristic equation identically; that is

$$c(A)\equiv 0 \tag{3.19}$$

PROOF

We assume that A can be diagonalized by a similarity transformation as

$$B=SAS^{-1}$$

Since B is diagonal and since $c(\lambda)=0$ is invariant under a similarity transformation, it follows that the diagonal elements of B are just the eigenvalues $\{\lambda_j\}$ of A. Because $c(\lambda)$ is simply a finite-degree polynomial in λ, we could write out explicitly the matrix polynomial to show that

$$Sc(A)S^{-1}=c(B)$$

so that

$$c(A)=S^{-1}c(B)S \tag{3.20}$$

We now need only show that $c(B)\equiv 0$. The characteristic polynomial $c(\lambda)$ is just

$$c(\lambda)=\prod_{j=1}^{n}(\lambda-\lambda_j)$$

so that

$$c(B)=\prod_{j=1}^{n}(B-\lambda_j I)\equiv\prod_{j=1}^{n}C^{(j)} \tag{3.21}$$

where

$$C^{(j)}_{lm}=(\lambda_m-\lambda_j)\,\delta_{lm}$$

Let us now compute a matrix element of $c(B)$,

$$\begin{aligned}[c(B)]_{rs}&=\sum_{\substack{i_1,i_2,\ldots\\ \ldots,i_{n-1}=1}}^{n} C^{(1)}_{ri_1}C^{(2)}_{i_1i_2}C^{(3)}_{i_2i_3}\cdots C^{(n)}_{i_{n-1}s}\\
&=\sum_{i_1,\ldots,i_n=1}(\lambda_{i_1}-\lambda_1)\,\delta_{ri_1}(\lambda_{i_2}-\lambda_2)\,\delta_{i_1i_2}\cdots(\lambda_s-\lambda_n)\,\delta_{i_{n-1}s}\\
&=\delta_{rs}\prod_{j=1}^{n}(\lambda_s-\lambda_j)\equiv 0 \qquad \forall r,s\end{aligned} \tag{3.22}$$

since the product will contain at least one zero. This establishes that

$$c(B)\equiv 0$$

or that

$$c(A)=0$$

Q.E.D.

3.3 SELF-ADJOINT MATRICES AND COMPLETENESS

We begin with two simple but extremely important observations about self-adjoint (or hermitian) matrices. If A is hermitian (i.e., $A^{\dagger}=A$), then the eigenvalues of A are real. Let x_j be any eigenvector such that

$$Ax_j=\lambda_j x_j \tag{3.23}$$

Since

$$\begin{aligned}\langle x_j\mid Ax_j\rangle&=\lambda_j\langle x_j\mid x_j\rangle=\langle A^{\dagger}x_j\mid x_j\rangle=\langle Ax_j\mid x_j\rangle\\
&=\lambda_j^*\langle x_j\mid x_j\rangle\end{aligned} \tag{3.24}$$

we see that $\lambda_j^*=\lambda_j$. Furthermore, if A is real and hermitian, then

$$Ax_j=\lambda_j x_j$$
$$A^*x_j^*\equiv Ax_j^*=\lambda_j^*x_j^*=\lambda_j x_j^*$$

so that

$$A(x_j+x_j^*)=\lambda_j(x_j+x_j^*) \tag{3.25}$$

which states that in this case the eigenvectors may be chosen to be *real*.

Finally the eigenvectors corresponding to *distinct* eigenvalues are orthogonal for an hermitian matrix since

$$Ax_j=\lambda_j x_j \qquad Ax_k=\lambda_k x_k$$

so that

$$\langle x_j \mid Ax_k\rangle=\lambda_k\langle x_j \mid x_k\rangle=\langle Ax_j \mid x_k\rangle=\lambda_j\langle x_j \mid x_k\rangle$$

and

$$(\lambda_j-\lambda_k)\langle x_j \mid x_k\rangle=0 \tag{3.26}$$

Therefore if $\lambda_j\neq\lambda_k$, then $\langle x_j \mid x_k\rangle=0$

We will need the very important result that the eigenvectors of a self-adjoint matrix, $A=A^\dagger$, will always be sufficient to form an orthonormal basis for a finite-dimensional space. We will then be able to prove that any self-adjoint matrix can be brought into diagonal form.

First, if

$$Ax_j=\lambda_j x_j$$

and *all* n of the λ_j are distinct, then since

$$\langle x_j \mid x_k\rangle=\delta_{jk}\langle x_j \mid x_j\rangle$$

we may certainly use the x_j to set up an orthonormal basis for E_n.

Example 3.5

$$A=\begin{pmatrix}2&1&0\\1&3&1\\0&1&2\end{pmatrix} \qquad Ax=\lambda x \qquad x=\begin{pmatrix}\xi_1\\\xi_2\\\xi_3\end{pmatrix}$$

$$\det|A-\lambda I|=\begin{vmatrix}2-\lambda&1&0\\1&3-\lambda&1\\0&1&2-\lambda\end{vmatrix}\begin{array}{l}=(2-\lambda)[(3-\lambda)(2-\lambda)-1]-(2-\lambda)\\=(2-\lambda)[(3-\lambda)(2-\lambda)-2]\\=(2-\lambda)[6-5\lambda+\lambda^2-2]=(2-\lambda)(\lambda-1)(\lambda-4)=0\end{array}$$

$$\therefore\lambda_1=1 \qquad \lambda_2=2 \qquad \lambda_3=4$$

$$Ax_1=\lambda_1 x_1$$

$$\Rightarrow \left.\begin{array}{r}2\xi_1+\ \xi_2+\ 0=\xi_1\\ \xi_1+3\xi_2+\ \xi_3=\xi_2\\ 0+\ \xi_2+2\xi_3=\xi_3\end{array}\right\}\Rightarrow \left.\begin{array}{r}\xi_1+\ \xi_2\qquad=0\\ \xi_1+2\xi_2+\xi_3=0\\ \xi_2+\xi_3=0\end{array}\right\}\Rightarrow \xi_1=-\xi_2=\xi_3$$

$$\therefore x_1=\begin{pmatrix}1\\-1\\1\end{pmatrix}$$

$$Ax_2 = \lambda_2 x_2$$

$$\Rightarrow \left.\begin{array}{l} 2\xi_1 + \xi_2 + 0 = 2\xi_1 \\ \xi_1 + 3\xi_2 + \xi_3 = 2\xi_2 \\ 0 + \xi_2 + 2\xi_3 = 2\xi_3 \end{array}\right\} \Rightarrow \left.\begin{array}{l} \xi_2 \quad = 0 \\ \xi_1 + \xi_2 + \xi_3 = 0 \\ \xi_2 \quad = 0 \end{array}\right\} \Rightarrow \begin{array}{l} \xi_1 = -\xi_3 \\ \xi_2 = 0 \end{array} \qquad \therefore x_2 = \begin{pmatrix} 1 \\ 0 \\ -1 \end{pmatrix}$$

$$Ax_3 = \lambda_3 x_3$$

$$\Rightarrow \left.\begin{array}{l} 2\xi_1 + \xi_2 + 0 = 4\xi_1 \\ \xi_1 + 3\xi_2 + \xi_3 = 4\xi_2 \\ 0 + \xi_2 + 2\xi_3 = 4\xi_3 \end{array}\right\} \Rightarrow \left.\begin{array}{l} -2\xi_1 + \xi_2 \quad = 0 \\ \xi_1 - \xi_2 + \xi_3 = 0 \\ \xi_2 - 2\xi_3 = 0 \end{array}\right\} \Rightarrow \begin{array}{l} \xi_1 = \frac{1}{2}\xi_2 = \xi_3 \\ \xi_2 = 2\xi_3 \end{array} \qquad \therefore x_3 = \begin{pmatrix} 1 \\ 2 \\ 1 \end{pmatrix}$$

The vectors, x_1, x_2, x_3, are linearly independent since they are mutually orthogonal. Furthermore, the vectors

$$e_1 = \frac{1}{\sqrt{3}}\begin{pmatrix} 1 \\ -1 \\ 1 \end{pmatrix} \qquad e_2 = \frac{1}{\sqrt{2}}\begin{pmatrix} 1 \\ 0 \\ -1 \end{pmatrix} \qquad e_3 = \frac{1}{\sqrt{6}}\begin{pmatrix} 1 \\ 2 \\ 1 \end{pmatrix}$$

form an orthonormal basis for E_3.

However, suppose $\lambda = \lambda_0$ is a k-fold root to the characteristic equation, $c(\lambda) \equiv \det|A - \lambda I| = 0$. We must now show that there exist just k nontrivial linearly independent eigenvectors such that

$$Ax_j = \lambda_0 x_j \qquad j = 1, 2, \ldots, k$$

There is at least *one* x_j so that this is true. Assume now that there are m linearly independent $\{x_j\}$ such that

$$Ax_j = \lambda_0 x_j \qquad j = 1, 2, \ldots, m \qquad 1 \leq m \leq k$$

Define $\mathcal{M}$ to be that subspace that contains all these x_j; that is,

$$y \in \mathcal{M} \Rightarrow Ay \in \mathcal{M} \qquad \ni Ay = \lambda_0 y$$

We may always decompose S into a direct sum of $\mathcal{M}$ and the rest of S that is orthogonal to $\mathcal{M}$ such that,

$$S = \mathcal{M} \oplus \mathcal{N} \ni y \in \mathcal{M} \Rightarrow Ay = \lambda_0 y \qquad z \in \mathcal{N} \Rightarrow \langle y \mid z \rangle = 0$$

By construction $\mathcal{M}$ is an invariant subspace of A. We now show that $\mathcal{N}$ is also an invariant subspace of A (which is not true in general).

$$\left.\begin{array}{r} \langle y \mid z \rangle = 0 \\ Ay = \lambda_0 y \end{array}\right\} \Rightarrow \langle Ay \mid z \rangle = 0 = \langle y \mid Az \rangle$$

This implies that the vector Az is also orthogonal to $\mathcal{M}$ or in $\mathcal{N}$. Therefore, if $z \in \mathcal{N}$, then $Az \in \mathcal{N}$, so that $\mathcal{N}$ is also an invariant subspace and

$$S = \mathcal{M} \oplus \mathcal{N}$$

so that S can be written as the direct sum of two invariant subspaces. By Th. 3.1 we know that A can be written in block-diagonal form

$$A = \begin{pmatrix} B & 0 \\ 0 & C \end{pmatrix} \equiv \begin{pmatrix} B & 0 \\ 0 & I \end{pmatrix} \begin{pmatrix} I & 0 \\ 0 & C \end{pmatrix}$$

where B is an $m \times m$ matrix operating only on $\mathcal{M}$ and C is an $(n-m)\times(n-m)$ matrix operating only on $\mathcal{N}$.

The characteristic equation then becomes

$$c(\lambda) \equiv \det |A - \lambda I| = \det |B - \lambda I| \det |C - \lambda I|$$

Now by construction λ_0 is an m-fold root of the characteristic equation for B and not any root of that for C. Therefore from this expression for $c(\lambda)$ we have

$$c(\lambda) = (\lambda - \lambda_0)^m \mathcal{P}(\lambda)$$

where λ_0 is not a root of $\mathcal{P}(\lambda)$. However we were initially given that λ_0 was a k-fold root of $c(\lambda)$ so that

$$c(\lambda) = (\lambda - \lambda_0)^k \mathcal{R}(\lambda)$$

where λ_0 is not a root of $\mathcal{R}(\lambda)$. Since these two expressions for $c(\lambda)$ are identical, we have

$$(\lambda - \lambda_0)^m \mathcal{P}(\lambda) \equiv \mathcal{R}(\lambda)(\lambda - \lambda_0)^k$$

for *all* values of λ.

Therefore we can write

$$(\lambda - \lambda_0)^{k-m} \frac{\mathcal{R}(\lambda)}{\mathcal{P}(\lambda)} \equiv 1$$

for all values of λ; in particular, for $\lambda = \lambda_0$ so that $k = m$. We can, of course, make this argument for every repeated root of finite multiplicity. This gives us the very important result that for a self-adjoint matrix the number of linearly independent eigenvectors corresponding to a k-fold root λ_j of $c(\lambda) = 0$ is exactly k.

THEOREM 3.3

The eigenvectors of a self-adjoint operator can be chosen as an orthonormal basis for the space.

PROOF

If all the $\{\lambda_j\}$ are distinct, the corresponding $\{x_j\}$ are orthogonal and may therefore be normalized. If λ_j is repeated k times, then we know there are just

k $\{x_j\}$ belonging to λ_j, so that any linear combination,

$$z = \sum_{j=1}^{k} \alpha_j x_j$$

is such that $Lz = \lambda_j z$ and is also an eigenvector belonging to λ_j. By the Schmidt orthogonalization technique we may make these linear combinations orthonormal.

Q.E.D.

THEOREM 3.4

Any hermitian matrix may be diagonalized by a unitary similarity transformation; that is

$$D = U^{-1}AU = U^{\dagger}AU \tag{3.27}$$

PROOF

We will simply exhibit the required U. Consider the orthonormal eigenvectors of A, such that $Ax_j = \lambda_j x_j$. Construct U as that matrix with the x_j's as its columns:

$$U = \begin{pmatrix} x_1 & x_2 & \cdots & x_n \\ \Big| & \Big| & \cdots & \Big| \end{pmatrix} \tag{3.28}$$

That is, if we write the orthonormal eigenvectors of $\{x_j\}$ in terms of their components $\xi_l^{(j)}$ as $x_j = \{\xi_l^{(j)}\}$, then Eq. 3.28 is equivalent to $U = \{u_{ij}\}$ where

$$u_{ij} = \xi_i^{(j)} \tag{3.29}$$

The orthonormality condition on the $\{x_j\}$ is

$$\langle x_j \mid x_k \rangle \equiv \sum_{l=1}^{n} \xi_l^{(j)*} \xi_l^{(k)} = \delta_{jk} \tag{3.30}$$

It is evident that U is unitary since

$$(U^{\dagger}U)_{ij} = \sum_k u_{ki}^* u_{kj} = \sum_k \xi_k^{(i)*} \xi_k^{(j)} = \delta_{ij}$$

or

$$U^{\dagger}U = I \tag{3.31}$$

Now consider the matrix $U^{\dagger}AU$,

$$(U^{\dagger}AU)_{ij} = \sum_{k,l} u_{ki}^* a_{kl} u_{lj} = \sum_{k,l} \xi_k^{(i)*} a_{kl} \xi_l^{(j)}$$

$$= \sum_k \xi_k^{(i)*} \lambda_j \xi_k^{(j)} = \lambda_j \, \delta_{ij} \tag{3.32}$$

so that

$$U^\dagger AU = D = \{\lambda_j\, \delta_{ij}\} \qquad \text{Q.E.D.}$$

Obviously the elements of the diagonal matrix, D, are just the eigenvalues, $\{\lambda_j\}$.

3.4 QUADRATIC FORMS

If we have a hermitian matrix, $A = A^\dagger = \{a_{ij}\}$, and form the matrix element $\langle x|\, A\, |x\rangle$ of A with a vector x, we obtain a *hermitian quadratic form*

$$\langle x|\, A\, |x\rangle = \sum_{j,k=1}^{n} a_{jk} \xi_j^* \xi_k \tag{3.33}$$

If A and x are real, then we have *real quadratic form*. It is clear from Th. 3.4 that any hermitian quadratic form can be brought into diagonal form by a unitary similarity transformation U, where U is the matrix of the orthonormal eigenvectors of A appearing in Eq. 3.28. If we write

$$\begin{aligned} D &\equiv U^{-1}AU = U^\dagger AU = \{\delta_{ij}\lambda_j\} \\ y &\equiv U^\dagger x = \{\eta_i\} \end{aligned} \tag{3.34}$$

then we obtain

$$\begin{aligned} \langle x|\, A\, |x\rangle &= \langle x|\, UU^\dagger AUU^\dagger\, |x\rangle \\ &= \langle U^\dagger x|\, U^\dagger AU\, |U^\dagger x\rangle = \langle y|\, D\, |y\rangle = \sum_{j=1}^{n} \lambda_j\, |\eta_j|^2 \end{aligned} \tag{3.35}$$

We also note two useful invariants associated with these quadratic forms.

$$\text{i. } \langle x \mid x\rangle = \sum_{j=1}^{n} |\xi_j|^2 = \langle x \mid UU^\dagger x\rangle = \langle y \mid y\rangle = \sum_{j=1}^{n} |\eta_j|^2 \tag{3.36}$$

$$\text{ii. } \det A = \det\,(UDU^\dagger) = \det D = \prod_{j=1}^{n} \lambda_j \tag{3.37}$$

If we have a real quadratic form, then since the eigenvectors of A may be chosen to be real (cf. Eq. 3.25) and orthonormal, the diagonalizing transformation matrix may be chosen orthogonal and real (i.e., $U^* = U$, $U^T U = I$) rather than unitary. This matrix, U, will carry us from one orthonormal set of basis vectors to another.

As a simple example in real three-dimensional space, recall a theorem proved in elementary analytic geometry that states that an arbitrary second-degree surface

$$\sum_{j,k=1}^{3} \alpha_{jk} \xi_j \xi_k = c$$

expressed with reference to an arbitrary set of axes can always be expressed in canonical form,

$$\sum_{j=1}^{3} c_j \xi_j'^2 = c$$

in terms of a suitable set of axes. This transformation is sometimes referred to as the *pricipal-axis transformation.* We can now see just how trivial this once formidable-appearing statement is. In fact we recognize the $\{c_j\}$ as the $\{\lambda_j\}$ of $\det |A - \lambda I| = 0$ and the $x' \equiv \{\xi_j'\}$ as $x = Ux'$. Incidentally, if all three $\{\lambda_j\}$ are positive and $c > 0$, then the surface is necessarily a real ellipse whose canonical form is

$$\frac{x^2}{a^2} + \frac{y^2}{b^2} + \frac{z^2}{c^2} = 1$$

Similarly we can go on to classify all the second-degree surfaces. Again in classical mechanics, the principal-axis transformation is important with regard to the moment of inertia ellipsoid.

a. Minimax Principle

We will now discuss the minimax principle. In principal axes we have

$$\langle x| A |x\rangle = \sum_j \lambda_j |\xi_j|^2$$

where we use an orthonormal basis $\{e_j\}$

$$x = \sum_j \xi_j e_j$$

and since we want vectors of nonzero norm we choose

$$\langle x \mid x\rangle = \sum_j |\xi_j|^2 \equiv 1$$

Here $\{e_j\}$ is the complete set of orthonormal eigenfunctions of A,

$$Ae_j = \lambda_j e_j$$

and we order the eigenvalues $\{\lambda_j\}$ so that

$$\lambda_1 \leq \lambda_2 \leq \lambda_3 \leq \cdots \leq \lambda_n$$

Since

$$\langle x| A |x\rangle = \sum_j \lambda_j |\xi_j|^2 \geq \lambda_1 \sum_j |\xi_j|^2 = \lambda_1$$

and

$$\langle x| A |x\rangle \leq \lambda_n \sum_j |\xi_j|^2 = \lambda_n$$

we have our quadratic form bounded as

$$\lambda_1 \le \langle x|\, A\, |x\rangle \le \lambda_n \tag{3.38}$$

for all x such that $\|x\|=1$.

Suppose we now choose a vector x such that

$$\langle x \mid e_1\rangle = 0 \tag{3.39}$$

$$\|x\| = 1 \tag{3.40}$$

Then x has the expansion

$$x = \sum_{j=2}^{n} \xi_j e_j$$

$$\|x\|^2 = \sum_{j=2}^{n} |\xi_j|^2 = 1$$

We now obtain the bound

$$\langle x|\, A\, |x\rangle = \sum_{j=2}^{n} \lambda_j\, |\xi_j|^2 \ge \lambda_2 \sum_{j=2}^{n} |\xi_j|^2 = \lambda_2 \tag{3.41}$$

Notice that the result of Eq. 3.38 need not necessarily be true when A is not hermitian. A simple example will establish this. If we let

$$A = \begin{pmatrix} 1 & 0 \\ 2 & 1 \end{pmatrix} \tag{3.42}$$

then $\lambda_1 = \lambda_2 = 1$ and the eigenvector corresponding to

$$A\chi = \lambda\chi$$

is

$$\chi = \begin{pmatrix} 0 \\ 1 \end{pmatrix}$$

For the vector

$$x = \frac{1}{\sqrt{2}} \begin{pmatrix} 1 \\ 1 \end{pmatrix} \qquad \|x\| = 1 \tag{3.43}$$

we obtain

$$\langle x|\, A\, |x\rangle = 2 \tag{3.44}$$

which contradicts Eq. 3.38.

We now come to the *minimax principle* itself.

THEOREM 3.5 (Minimax principle)

Given the hermitian quadratic form $\langle x|\, A\, |x\rangle$ with $\|x\|=1$ and $\langle x \mid y\rangle = 0$ where y is any *fixed* vector and with the eigenvalues of A ordered as $\lambda_1 \le \lambda_2 \le \cdots \le \lambda_n$, vary x until $\langle x|\, A\, |x\rangle$ attains its *minimum* value. Now vary y until this minimum attains its maximum possible value. This maximum value of the minimum is just λ_2.

PROOF

We first claim that

$$\langle x|\, A\, |x\rangle_{\min} \leq \lambda_2 \tag{3.45}$$

This is proven by simply exhibiting an x that makes this true. Let

$$x = \xi_1 e_1 + \xi_2 e_2$$
$$y = \sum_{j=1}^{n} \eta_j e_j \tag{3.46}$$

$$\langle x \mid y\rangle = 0 = \sum_j \xi_j^* \eta_j = \xi_1^* \eta_1 + \xi_2^* \eta_2 = 0 \tag{3.47}$$

The important point is that for any *fixed* y (i.e., values of η_j, $j = 1, 2, \ldots, n$) it is *always* possible to choose an x of the form (3.46) subject to condition (3.47) and $\|x\| = 1$.

$$\begin{aligned}\langle x|\, A\, |x\rangle &= \lambda_1 |\xi_1|^2 + \lambda_2 |\xi_2|^2 = \lambda_2 - \lambda_2 |\xi_1|^2 + \lambda_1 |\xi_1|^2 \\ &= \lambda_2 - (\lambda_2 - \lambda_1)|\xi_1|^2 \leq \lambda_2\end{aligned}$$

This establishes (3.45). Notice that this is not necessarily *the* minimum value possible for $\langle x|\, A\, |x\rangle$. However when $\xi_1 = 0$ (i.e., $x = \xi_2 e_2$), which will be true for some y as we then vary y, $\langle x|\, A\, |x\rangle$ attains exactly the value λ_2, so that

$$\text{Max}_y \, \text{Min}_x \langle x|\, A\, |x\rangle = \lambda_2 \tag{3.48}$$

subject to the condition

$$\langle y \mid x\rangle = 0 \tag{3.49}$$ Q.E.D.

In Section A3.3 we verify this principle directly in a real three-dimensional space.

In general if we have a fixed set of vectors $\{y_j\}$, $j = 1, 2, \ldots, m$, then for $\|x\| = 1$ and $\langle x \mid y_j\rangle = 0$, $\forall j$,

$$\langle x\, |A|\, x\rangle \leq \lambda_{m+1}$$

Again this follows since we may take

$$x = \sum_{k=1}^{m+1} \xi_k e_k$$

and always satisfy

$$\langle x \mid y_j\rangle = 0 = \sum_{k=1}^{m+1} \xi_k^* \eta_k^{(j)} = 0 \qquad j = 1, 2, \ldots, m$$

This set of m equations in $(m+1)$ unknowns will always have at least one nontrivial solution. Therefore we obtain

$$\langle x|\, A\, |x\rangle = \sum_{k=1}^{m+1} \lambda_j |\xi_j|^2 = \lambda_{m+1} - \sum_{k=1}^{m} (\lambda_{m+1} - \lambda_k)|\xi_k|^2 \leq \lambda_{m+1} \tag{3.50}$$

as previously. However, if we now want to be guaranteed that the maximum value of (3.50), namely λ_{m+1}, is actually attained as we vary the $\{y_j\}$, then we must also require that the projections of the $\{y_j\}$ into the subspace spanned by the $(e_1, e_2, \ldots, e_{m+1})$ are all linearly independent. Under these conditions the minimax principle will hold for the $(m+1)$st eigenvalue.

b. Simultaneous Reduction of Two Quadratic Forms

A hermitian quadratic form is called positive definite if $\langle x| A |x\rangle > 0$ for all $x \neq 0$. An important example of a real positive-definite quadratic form is the kinetic energy of a mechanical system in some generalized coordinates $\{\zeta_j\}$

$$T = \frac{1}{2} \sum_{j,k=1}^{n} a_{jk} \dot{\zeta}_j \dot{\zeta}_k = \tfrac{1}{2} \langle \dot{z}| A |\dot{z}\rangle$$

Similarly an associated real (but not necessarily positive definite) quadratic form is the potential energy of a system of coupled harmonic oscillators in these generalized coordinates,

$$V = \frac{1}{2} \sum_{j,k=1}^{n} b_{jk} \zeta_j \zeta_k = \tfrac{1}{2} \langle z| B |z\rangle$$

We now show that these may be simultaneously diagonalized to yield

$$T = \frac{1}{2} \sum_{j=1}^{n} \dot{\eta}_j^2 = \tfrac{1}{2} \langle \dot{y} \mid \dot{y}\rangle$$

$$V = \frac{1}{2} \sum_{j=1}^{n} \lambda_j \eta_j^2 = \tfrac{1}{2} \langle y| D |y\rangle$$

These $\{\eta_j\}$ are the *normal coordinates* of the system. This makes the solution of the mechanical problem trivial since for a conservative system we have

$$T + V = E \text{ (a constant)}$$

so that $dE/dt = 0$ or

$$\ddot{\eta}_j(t) + \lambda_j \eta_j(t) = 0 \qquad j = 1, 2, \ldots, n$$

These equations are easy to integrate for $\eta_j = \eta_j(t)$.

THEOREM 3.6

If A and B are hermitian matrices and if A is positive definite, then there exists a nonsingular matrix R such that

$$R^\dagger A R = I, \qquad R^\dagger B R = D$$

where D is a diagonal matrix.

PROOF

Since A is hermitian and positive definite, we may solve

$$A\varphi_j = \alpha_j \varphi_j$$

for the complete orthonormal set of eigenvectors $\{\varphi_j\}$ and the real, *positive* eigenvalues $\{\alpha_j\}$. Just as in Th. 3.4, Eq. 3.28, we may construct a matrix S with the $\alpha_j^{-(1/2)}\varphi_j$ as its columns

$$S = \{\alpha_j^{-(1/2)}\varphi_j\}$$

so that

$$S^\dagger S = \{\alpha_j^{-1}\,\delta_{ij}\}$$

$$\det S^\dagger S \equiv |\det S|^2 = \prod_{j=1}^{n} \alpha_j^{-1} \neq 0$$

Therefore S is nonsingular so that S^{-1} exists. Furthermore direct calculation shows that

$$S^\dagger A S = I$$

Now consider the eigenvalue problem

$$(B - \lambda_j A)x_j = 0 \tag{3.51}$$

which is equivalent to

$$(S^\dagger B S - \lambda_j S^\dagger A S)S^{-1}x_j = 0$$

If we define

$$\begin{aligned} M &\equiv S^\dagger B S = M^\dagger \\ \psi_j &= S^{-1}x_j \end{aligned} \tag{3.52}$$

then (3.51) becomes

$$M\psi_j = \lambda_j \psi_j \tag{3.53}$$

That is, since M is hermitian we know there exists a unitary matrix U such that

$$U^\dagger M U = D_M$$

where D_M is a diagonal matrix. If we now construct an $R \equiv SU$, then

$$R^\dagger A R = U^\dagger S^\dagger A S U = UU^\dagger = I$$

$$R^\dagger B R = U^\dagger S^\dagger B S U = U^\dagger M U = D_M$$

which establishes the theorem. However we will also show explicitly how to construct the matrix R directly from A and B. Since M is hermitian, the $\{\psi_j\}$ form a complete orthonormal set and the $\{\lambda_j\}$ are all real. Because S is nonsingular, Eq. 3.52 implies that the $\{x_j\}$ of Eq. 3.51 are also linearly independent since if they were not there would exist a set of constants $\{c_j\}$ not all zero such that

$$\sum_{j=1}^{n} c_j x_j = 0$$

which would imply

$$\sum_{j=1}^{n} c_j \psi_j = 0$$

a contradiction. We see that the $\{x_j\}$ are orthonormal w.r.t. A since

$$\langle x_i | A | x_j \rangle = \langle S\psi_i | A | S\psi_j \rangle = \langle \psi_i | S^\dagger AS | \psi_j \rangle = \langle \psi_i | \psi_j \rangle = \delta_{ij}$$

Also we see that

$$\langle x_i | B | x_j \rangle = \lambda_j \langle x_i | A | x_j \rangle = \lambda_j \delta_{ij}$$

so that B is diagonal w.r.t. the $\{x_j\}$. Therefore in order to construct the R required we first solve for the $\{\lambda_j\}$ as the roots to

$$\det |B - \lambda A| = 0$$

We then solve the eigenvalue problem

$$(B - \lambda_j A) x_j = 0$$

and normalize these eigenvectors as

$$\langle x_i | A | x_j \rangle = \delta_{ij}$$

Matrix R, then, has these $\{x_j\}$ as its columns. That is, if $x_j = \{\xi_k^{(j)}\}$ and $R = \{r_{ij}\}$, then

$$r_{ij} = \xi_i^{(j)}$$

so that

$$(R^\dagger AR)_{ij} = \sum_{k,l} r_{ki}^* a_{kl} r_{lj} = \sum_{k,l} \xi_k^{(i)*} a_{kl} \xi_l^{(j)} = \delta_{ij}$$

$$(R^\dagger BR)_{ij} = \sum_{k,l} r_{ki}^* b_{kl} r_{lj} = \sum_{k,l} \xi_k^{(i)*} b_{kl} \xi_l^{(j)} = \lambda_j \delta_{ij} \qquad \text{Q.E.D.}$$

Notice that this proof is independent of the degeneracy of the eigenvalues $\{\lambda_j\}$ and that if A and B are both real self-adjoint matrices with A still positive definite, then R may be chosen real [since the eigenvectors for (3.51) may then be chosen real].

If we now define

$$z = Ry \tag{3.54}$$

then we obtain

$$\langle \dot{z} | A | \dot{z} \rangle = \langle R\dot{y} | A | R\dot{y} \rangle = \langle \dot{y} | R^\dagger AR | \dot{y} \rangle = \langle \dot{y} | \dot{y} \rangle$$

$$\langle z | B | z \rangle = \langle Ry | B | Ry \rangle = \langle y | R^\dagger BR | y \rangle = \langle y | D | y \rangle$$

That is, the transformation by R defined in (3.54) will carry us to normal coordinates as claimed.

3.5 SIMULTANEOUS DIAGONALIZATION OF COMMUTING HERMITIAN MATRICES

We will now prove a theorem that is of the utmost importance in quantum mechanics.

THEOREM 3.7

A necessary and sufficient condition that two hermitian matrices commute is that they possess a common complete set of orthonormal eigenvectors.

PROOF

i. *Sufficient.* We are given the existence of a complete set of orthonormal eigenvectors such that

$$Ax_j = \lambda_j x_j \qquad Bx_j = \tau_j x_j \qquad j = 1, 2, \ldots, n$$

Since this set is complete, we may expand any vector in our space as

$$x = \sum_{j=1}^{n} \xi_j x_j$$

Since

$$ABx = AB \sum_j \xi_j x_j = A \sum_j \xi_j \tau_j x_j = \sum_j \xi_j \lambda_j \tau_j x_j$$

$$= B \sum_j \xi_j \lambda_j x_j = BA \sum_j \xi_j x_j = BAx$$

we see that

$$[A, B]x = 0, \forall x$$

or that

$$[A, B] = 0$$

The expression $[A, B] = AB - BA$ is called the *commutator* of A and B. When $[A, B] = 0$, then the operators A and B commute. The last conclusion follows since if a matrix C is such that $Cx = 0$ for *all* x, then $C = 0$ since $Cx = 0$ is equivalent to

$$\sum_{j=1}^{n} c_{ij} \xi_j = 0 \qquad i = 1, 2, \ldots, n, \forall \xi_j$$

so that

$$c_{ij} = 0, \forall i, j$$

ii. *Necessary.* We are given that $[A, B] = 0$.

α. We first assume that all of the eigenvalues of at least one of the matrices (e.g., A) are distinct.

$$Ax_j = \lambda_j x_j .$$

Define

$$y_j \equiv Bx_j$$

so that

$$Ay_j = ABx_j = BAx_j = \lambda_j Bx_j = \lambda_j y_j$$

which states that y_j is also an eigenvector of A corresponding to eigenvalue λ_j since all the $\{\lambda_j\}$ are assumed distinct and the $\{x_j\}$ form an orthonormal basis. Therefore we have

$$y_j = \text{const. } x_j \equiv \tau_j x_j$$

so that

$$Bx_j \equiv y_j = \tau_j x_j$$

which states that the $\{x_j\}$ are a common complete set of orthonormal eigenvectors.

β. We now allow for degeneracy in the eigenvalues. Suppose that λ_1 is m-fold degenerate; that is

$$Ax_j = \lambda_1 x_j \qquad j = 1, 2, \ldots, m$$

Again define

$$y_k = Bx_k \qquad k = 1, 2, \ldots, m$$

Since

$$Ay_k = ABx_k = BAx_k = \lambda_1 Bx_k = \lambda_1 y_k$$

we now know that y_k is contained in the subspace $\mathcal{M}$ of λ_1 spanned by the $\{x_j\}$, $j = 1, 2, \ldots, m$ so that

$$y_k = \sum_{j=1}^{m} \alpha_j^{(k)} x_j = Bx_j$$

This states that the subspace $\mathcal{M}$ of λ_1 is an invariant subspace of B. In the subspace $\mathcal{M}_{\lambda_1}$ B can be represented by an $m \times m$ hermitian matrix. In this subspace we can solve the eigenvalue problem

$$Bz_l = \tau_l z_l \qquad l = 1, 2, \ldots, m$$

By Th. 3.3 we know that these $\{z_l\}$ can be chosen as an orthonormal basis for $\mathcal{M}_{\lambda_1}$. However, since the $\{x_j\}$, $j = 1, 2, \ldots, m$, also form a basis for $\mathcal{M}_{\lambda_1}$, we may expand the $\{z_l\}$ as

$$z_l = \sum_{j=1}^{m} \beta_j^{(l)} x_j$$

Since these are simply linear combinations of the $\{x_j\} \in \mathcal{M}_{\lambda_1}$, these $\{z_l\}$ are still eigenvectors of A with eigenvalue λ_1,

$$Az_l = \lambda_1 z_l \qquad l = 1, 2, \ldots, m \qquad \text{Q.E.D.}$$

Notice that this theorem only guarantees that common orthonormal sets of eigenvectors *can* be found. It does not state that an orthonormal set computed at random for one of two commuting hermitian operators will necessarily be a set of eigenvectors for the other operator.

THEOREM 3.8

Two commuting hermitian matrices may be simultaneously diagonalized by a single unitary transformation.

PROOF

This theorem is an immediate consequence of Th. 3.3 and Th. 3.7 if we simply take the unitary matrix to have the common orthonormal eigenvectors as its columns. Then if

$$Ax_j = \lambda_j x_j \qquad Bx_j = \tau_j x_j \qquad U = \{x_j\},$$

we conclude that

$$U^\dagger A U = D_A = \{\delta_{ij}\lambda_j\}$$

$$U^\dagger B U = D_B = \{\delta_{ij}\tau_j\} \qquad \text{Q.E.D.}$$

3.6 NORMAL MATRICES AND COMPLETENESS

We can, in fact, give a slightly more general condition under which a matrix can be diagonalized. A matrix M is called *normal* provided

$$[M, M^\dagger] = 0 \tag{3.55}$$

Define

$$A \equiv \tfrac{1}{2}(M + M^\dagger) = A^\dagger$$

$$B \equiv \frac{i}{2}(M - M^\dagger) = B^\dagger$$

Therefore A and B are hermitian and may be diagonalized. But by direct calculation we see that

$$[A, B] \equiv -\frac{i}{2}[M, M^\dagger] = 0$$

But by Th. 3.8 two commuting hermitian matrices may be diagonalized *simultaneously* by a common U as

$$D_A = U^\dagger A U$$

$$D_B = U^\dagger B U$$

so that

$$UMU^{\dagger} = D_A - iD_B \equiv D_M$$

It is also clear that any normal matrix possesses a complete set of orthonormal eigenvectors. This case includes the hermitian and unitary matrices.

3.7 FUNCTIONS OF AN OPERATOR†

We will restrict our few comments to linear operators whose eigenvectors form an orthonormal basis for our space S. Consider any holomorphic function that has a convergent Taylor series expansion

$$f(z) = \sum_{n=0}^{\infty} \alpha_n z^n$$

within some radius of convergence (cf. Section 7.4). If the operator L has a complete set of eigenvectors $\{x_j\}$,

$$Lx_j = \lambda_j x_j \tag{3.56}$$

then we may *define* $f(L)$ for any vector x

$$x = \sum_j \xi_j x_j$$

as

$$\begin{aligned} f(L)x &= \sum_n \alpha_n L^n \sum_j \xi_j x_j = \sum_{n,j} \alpha_n \lambda_j^n \xi_j x_j \\ &= \sum_j \xi_j \left(\sum_n \alpha_n \lambda_j^n \right) x_j = \sum_j \xi_j f(\lambda_j) x_j \end{aligned} \tag{3.57}$$

This technique is sometimes used to find the inverse of an operator as

$$[L-\lambda]^{-1} x = \sum_j (\lambda_j - \lambda)^{-1} \xi_j x_j \tag{3.58}$$

if $\lambda \neq \lambda_j$. In particular we can see how this is connected with the eigenvalue technique (cf. Section 3.0) of solving inhomogeneous problems,

$$Lx = a$$

If we take $\lambda = 0$ in Eq. 3.57, and if none of the eigenvalues of L given in Eq. 3.56 is zero, then we obtain

$$a = \sum_j \alpha_j x_j$$

$$x = L^{-1} a = \sum_j (\alpha_j / \lambda_j) x_j \equiv \sum_j \xi_j x_j$$

† This section is similar to the first part of the corresponding section in Chapter 3 of Friedman's *Principles and Techniques of Applied Mathematics.*

or

$$\xi_j = \alpha_j/\lambda_j$$

as previously.

What we have done so far simply tells us how to write down *formally* a holomorphic function of an operator. This is often not too useful in the actual evaluation of such a function of an operator.

We have seen in Th. 3.2 that a matrix (at least a self-adjoint one) satisfies its characteristic equation identically [i.e., $c(A)=0$]. Suppose that A is an $n \times n$ matrix and $f(z)$ is a holomorphic function of z whose Taylor series about $z=0$ has a radius of convergence R where $R>|\lambda_{\max}|$, $\lambda_{\max}$ being the eigenvalue of A with the largest absolute value. We now write

$$f(z)=c(z)g(z)+r(z) \tag{3.59}$$

where $r(z)$ is a polynomial of degree $(n-1)$. We first assume that all the roots $\{\lambda_j\}$ of

$$c(\lambda)\equiv \det|A-\lambda I|=0$$

are distinct and construct the polynomial $r(z)$ of degree $(n-1)$ by satisfying the conditions

$$f(\lambda_j)=r(\lambda_j) \qquad j=1, 2, \ldots, n \tag{3.60}$$

These n conditions uniquely determine the polynomial, $r(z)$. Since $f(z)$, $c(z)$, and $r(z)$ are given as being holomorphic and since $c(\lambda_j)=0$ and $f(\lambda_j)=r(\lambda_j)$, if follows that $g(z)$ is also a holomorphic function (at least for $|z|<R$). However we never need construct $g(z)$ explicitly. Therefore we obtain $f(A)$ as

$$f(A)=r(A) \tag{3.61}$$

This means that, at worst, we have to evaluate a matrix polynomial of degree $(n-1)$. The real advantage of this technique comes for evaluating operators such as e^A when A is a matrix of low-dimension.

Now suppose one of the eigenvalues, λ_j, for example, is m-fold degenerate. Then from Eq. 3.59 we obtain upon differentiation

$$f'(x)=c'(x)g(x)+g'(x)c(x)+r'(x) \tag{3.62}$$

so that

$$f'(\lambda_j)=r'(\lambda_j) \tag{3.63}$$

since $c(z)$ and its first $(m-1)$ derivatives will vanish at $x=\lambda_j$ because

$$c(z)=\prod_{j=1}^{n}(z-\lambda_j)$$

In fact we easily see that

$$f^{(k)}(\lambda_j)=r^{(k)}(\lambda_j) \qquad k=1, 2, 3, \ldots (m-1) \tag{3.64}$$

Therefore we still setermine $r(x)$ as a polynomial of degree $(n-1)$, but with λ_j as a root of degree $(m-1)$. The result (3.61) remains valid. Also for hermitian matrices since $U^\dagger AU = D_A$, we may evaluate a function of such matrix as

$$f(A) = U^\dagger f(D_A)U = U^\dagger\{f(\lambda_j)\,\delta_{ij}\}U \tag{3.65}$$

A3.1 LAGRANGE UNDETERMINED MULTIPLIERS

We wish to make a function $f(x, y)$ an extremum subject to the constraint

$$g(x, y) = \text{constant} \tag{A3.1}$$

We will first treat the problem geometrically to gain some insight. Let $z = f(x, y)$ be a two-dimensional surface in a real three-dimensional space and consider $g(x, y) = c$ as a cylinder of cross section $y = y(x)$ as depicted in Fig. 3.1. We want

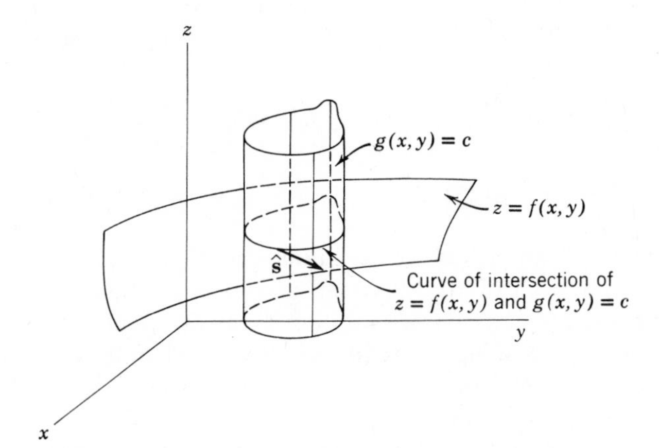

Figure 3.1

the extremal value of $f(x, y)$ along the curve of intersection. If we denote by $\hat{\mathbf{s}}$ the unit tangent vector along this curve of intersection, then the condition for an extremum will be the vanishing of the directional derivative

$$\frac{df}{ds} \equiv \nabla f \cdot \hat{\mathbf{s}} = 0 \tag{A3.2}$$

which implies that ∇f is orthogonal to $\hat{\mathbf{s}}$ at the extremum.

Since ∇g is a vector normal to the cylinder $g(x, y) = c$, it is also normal to $\hat{\mathbf{s}}$ so that ∇f and ∇g are contained in a plane normal to $\hat{\mathbf{s}}$ at the extremum. It is also clear geometrically that $\hat{\mathbf{s}}$ is parallel to the x-y plane at an extremum as ∇g *always* is. Therefore at the extremum ∇g and the component of ∇f parallel to the x-y plane are *collinear* so that there exists a scalar λ such that at (x_0, y_0)

$$\begin{aligned} \frac{\partial f}{\partial x}+\lambda\frac{\partial g}{\partial x}=0 \\ \frac{\partial f}{\partial y}+\lambda\frac{\partial g}{\partial y}=0 \end{aligned} \tag{A3.3}$$

Hence Eqs. A3.3 plus the constraint condition

$$g(x_0, y_0)=c \tag{A3.4}$$

give us three equations for the three unknowns x_0, y_0, and λ.

Therefore in practice we define a new function

$$h(x, y; \lambda)\equiv f(x, y)+\lambda g(x, y) \tag{A3.5}$$

and treat x and y as *independent* variables, set

$$\frac{\partial h}{\partial x}=0=\frac{\partial h}{\partial y} \tag{A3.6}$$

solve for $x_0=x_0(\lambda)$, $y_0=y_0(\lambda)$ and then put these back into the constraint condition

$$g[x_0(\lambda), y_0(\lambda)]=c \tag{A3.7}$$

to solve for λ, thus determining x_0 and y_0. Of course in our discussion we have assumed that the surface $z=f(x, y)$ and the cylinder $g(x, y)=c$ actually *do* intersect. If they in fact do not intersect, there is no solution to the problem.

Another way to treat this problem, which is really much shorter and more easily extended to any number of variables with several constraints, is the following. In principle we could use the constraint, Eq. A3.1, to solve for $y=y(x)$ and then write $f(x, y)=f[x, y(x)]$, which now becomes a function of the single independent variable, x, so that

$$\frac{df}{dx}=\frac{\partial f}{\partial x}+\frac{\partial f}{\partial y}\frac{dy}{dx}=\frac{\partial f}{\partial x}-\frac{\partial f}{\partial y}\left(\frac{\partial g/\partial x}{\partial g/\partial y}\right) \tag{A3.8}$$

If we now define $h(x, y; \lambda)$ as in Eq. A3.5 and set

$$\begin{aligned} \frac{\partial h}{\partial x}=0=\frac{\partial f}{\partial x}+\lambda\frac{\partial g}{\partial x} \\ \frac{\partial h}{\partial y}=0=\frac{\partial f}{\partial y}+\lambda\frac{\partial g}{\partial y} \end{aligned} \tag{A3.9}$$

and eliminate λ between these equations, we obtain the same result as gotten by setting $df/dx=0$ in Eq. A3.8.

A3.2 DERIVIATION OF EXTREMAL PROPERTIES OF HERMITIAN QUADRATIC FORMS USING LAGRANGE UNDETERMINED MULTIPLIERS

As an illustration of the method of Lagrange multipliers developed in Section A3.1, let us derive the bounds of Eqs. 3.38 and 3.41 for the hermitian quadratic form, $\langle x| A |x\rangle$, as a direct minimization problem subject to constraints. In the following we assume that all of the $\{\xi_j\}$ are real and that all of the $\{\lambda_j\}$ are distinct to simplify the algebra. These are not necessary restrictions.

$$A \equiv \langle x| A |x\rangle = \sum_j \lambda_j \xi_j^2 \qquad \sum_j \xi_j^2 = 1$$

For an extremum we require that

$$\delta A \equiv \sum_j \frac{\partial A}{\partial \xi_j} \delta\xi_j = 2 \sum_j \lambda_j \xi_j \, \delta\xi_j = 0$$

However not all the $\{\delta\xi_j\}$ are independent because of the constraint

$$2 \sum_j \xi_j \, \delta\xi_j = 0$$

If we introduce the undetermined multiplier μ as

$$\sum_j (\lambda_j + \mu)\xi_j \, \delta\xi_j = 0$$

we may treat all the $\{\delta\xi_j\}$ as independent since for some $j = k$ we may choose μ such that

$$(\lambda_k + \mu) = 0$$

or $\mu = -\lambda_k$. We then obtain

$$\sum_{j \neq k} (\lambda_j - \lambda_k)\xi_j^{(0)} \, \delta\xi_j = 0$$

where the $\{\xi_j^{(0)}\}$ are the values of $\{\xi_j\}$ we are expanding about.

Here the remaining $\{\delta\xi_j\}$ *are* independent so that their coefficients must vanish,

$$(\lambda_j - \lambda_k)\xi_j^{(0)} = 0 \qquad j \neq k$$

If all the eigenvalues are distinct, then

$$\xi_j^{(0)} = 0 \qquad j \neq k$$

$$\sum_j \xi_j^{(0)2} = 1 \Rightarrow \xi_k^{(0)} = 1$$

so that $\xi_j^{(0)} = \delta_{jk}$.

Therefore about $\xi_j = \xi_j^{(0)}$ and with $B \equiv A + \mu \sum_j \xi_j^2$ we have

$$\delta B \simeq \sum_{i,j} \frac{\partial^2 B}{\partial \xi_i \, \partial \xi_j} \delta\xi_i \, \delta\xi_j = \sum_j (\lambda_j - \lambda_k)(\delta\xi_j)^2 \tag{A3.10}$$

If we want a *minimum* then we require that as we move away from the $\{\xi_j^{(0)}\}$

$$\delta B > 0 \qquad \forall\ \delta\xi_j$$

which implies that

$$\lambda_j - \lambda_k \geq 0$$

or that $\lambda_k = \lambda_1$ so that $k=1$ and $\xi_j^{(0)} = \delta_{j1}$ which yields

$$\langle x|\ A\ |x\rangle_{\min} = \lambda_1 \tag{A3.11}$$

Similarly for a maximum we require that

$$\delta B < 0 \qquad \forall\ \delta\xi_j$$

so that the same type argument just given shows that $\lambda_k = \lambda_n$, or that $\xi_j^{(0)} = \delta_{jn}$. This implies that

$$\langle x|\ A\ |x\rangle_{\max} = \lambda_n \tag{A3.12}$$

Equations A3.11 and A3.12 together are the result stated in Eq. 3.38.
For the next problem we must minimize

$$A = \sum_j \lambda_j \xi_j^2$$

subject to the two constraints

$$\sum_j \xi_j^2 = 1 \qquad \sum_j \eta_j \xi_j = 0$$

where $\eta_j = \delta_{j1}$ so that

$$\xi_1 = 0 \tag{A3.13}$$

or

$$\delta\xi_1 = 0 \tag{A3.14}$$

Introducing two Lagrange multipliers we define

$$B \equiv A + \mu \sum_j \xi_j^2 + \nu \sum_j \eta_j \xi_j$$

so that $\delta B = 0$ requires

$$2(\lambda_j + \mu)\xi_j^{(0)} + \nu\eta_j = 0 \qquad \forall j \tag{A3.15}$$

For $j=1$, Eqs. A3.13 and A3.15 show that $\nu=0$ or that

$$(\lambda_j + \mu)\xi_j^{(0)} = 0 \qquad j = 2, 3, \ldots, n \tag{A3.16}$$

Since

$$\sum_{j=2}^{n} \xi_j^2 = 1$$

not all of the ξ_j vanish so that, by Eq. A3.16, $\mu = -\lambda_k$ for some *fixed* k, $2 \leq k \leq n$. We conclude that

$$\xi_j^{(0)} = \delta_{jk}$$

which allows us to write

$$\delta B \simeq \sum_{i,j} \frac{\partial^2 B}{\partial \xi_i\, \partial \xi_j}\, \delta\xi_i\, \delta\xi_j = \sum_{j=2}^{n} (\lambda_j - \lambda_k)(\delta\xi_j)^2 \tag{A3.17}$$

For a minimum we require that $\delta B > 0$ for all $\delta\xi_j$. Equation A3.17 implies that

$$\lambda_j - \lambda_k \geq 0$$

so that

$$\lambda_k = \lambda_2$$

or

$$\xi_j^{(0)} = \delta_{j2}$$

The final result becomes

$$\langle x|\, A\, |x\rangle_{\min} = \lambda_2 \tag{A3.18}$$

which is just Eq. 3.41.

A3.3 A DIRECT VERIFICATION OF THE MINIMAX PRINCIPLE IN A REAL THREE-DIMENSIONAL SPACE

In this section we establish the minimax principle, Th. 3.5, for a real three-dimensional space by first minimizing $\langle x|\, A\, |x\rangle$ subject to the constraint $\langle x \mid y\rangle = 0$ and then maximizing this result w.r.t. y.

Let y be any fixed vector in three space,

$$y = a(\sin\theta \cos\phi, \sin\theta \sin\phi, \cos\theta) \tag{A3.19}$$

We can obtain this y from a vector contained wholly along the z axis as

$$y = \begin{pmatrix} \cos\phi & -\sin\phi & 0 \\ \sin\phi & \cos\phi & 0 \\ 0 & 0 & 0 \end{pmatrix} \begin{pmatrix} \cos\theta & 0 & \sin\theta \\ 0 & 1 & 0 \\ -\sin\theta & 0 & \cos\theta \end{pmatrix} \begin{pmatrix} 0 \\ 0 \\ 1 \end{pmatrix}$$

$$= \begin{pmatrix} \cos\theta\cos\phi & -\sin\phi & \sin\theta\cos\phi \\ \cos\theta\sin\phi & \cos\phi & \sin\theta\sin\phi \\ -\sin\theta & 0 & \cos\theta \end{pmatrix} \begin{pmatrix} 0 \\ 0 \\ 1 \end{pmatrix} \equiv R(\theta, \phi) \begin{pmatrix} 0 \\ 0 \\ 1 \end{pmatrix} \tag{A3.20}$$

If we were to take $y' = a(0, 0, 1)$, then the most general vector of unit norm such that $\langle x' \mid y'\rangle = 0$ would be

$$x' = (\cos\phi', \sin\phi', 0) \tag{A3.21}$$

Therefore from Eqs. A3.20 and A3.21 we see that for y given by Eq. A3.19 the most general vector x such that

$$\|x\| = 1 \qquad \langle x, y\rangle = 0 \tag{A3.22}$$

is given by

$$x = \begin{pmatrix} \cos\theta\cos\phi\cos\phi' - \sin\phi\sin\phi' \\ \cos\theta\sin\phi\cos\phi' + \cos\phi\sin\phi' \\ -\sin\theta\cos\phi' \end{pmatrix} \tag{A3.23}$$

Now consider a 3×3 hermitian matrix A with eigenvalues λ_1, λ_2, λ_3 such that

$$\lambda_1 \leq \lambda_2 \leq \lambda_3$$

We take the directions of the orthogonal eigenvectors $\{x_j\}$,

$$Ax_j = \lambda_j x_j \qquad j = 1, 2, 3$$

to be along the x, y, and z axes, respectively, of our coordinate system. In principal axes we have

$$\begin{aligned} A(\theta, \phi; \phi') \equiv \langle x| A |x\rangle &= \lambda_1(\cos\theta \cos\phi \cos\phi' - \sin\phi \sin\phi')^2 \\ &\quad + \lambda_2(\cos\theta \sin\phi \cos\phi' + \cos\phi \sin\phi')^2 + \lambda_3 \sin^2\theta \cos^2\phi' \\ &\equiv \lambda_1\xi_1^2 + \lambda_2\xi_2^2 + \lambda_3\xi_3^2 \end{aligned}$$

$$\xi_1^2 + \xi_2^2 + \xi_3^2 = 1 \tag{A3.24}$$

We rewrite this as

$$\begin{aligned} A(\theta, \phi; \phi') &= \lambda_1\xi_1^2 + \lambda_2(1 - \xi_1^2 - \xi_3^2) + \lambda_3\xi_3^2 = \lambda_2 - (\lambda_2 - \lambda_1)\xi_1^2 + (\lambda_3 - \lambda_2)\xi_3^2 \\ &= \lambda_2 - (\lambda_2 - \lambda_1)(\cos\theta \cos\phi \cos\phi' - \sin\phi \sin\phi')^2 + (\lambda_3 - \lambda_2) \sin^2\theta \cos^2\phi' \\ &= \lambda_2 + [-(\lambda_2 - \lambda_1) \cos^2\theta \cos^2\phi + (\lambda_3 - \lambda_2) \sin^2\theta] \cos^2\phi' \\ &\quad + 2(\lambda_2 - \lambda_1) \cos\theta \cos\phi \sin\phi \cos\phi' \sin\phi' - (\lambda_2 - \lambda_1) \sin^2\phi \sin^2\phi' \\ &\equiv \lambda_2 + \alpha \cos^2\phi' + 2\beta \cos\phi' \sin\phi' - \gamma \sin^2\phi' \end{aligned} \tag{A3.25}$$

where

$$\alpha(\theta, \phi) \equiv -(\lambda_2 - \lambda_1) \cos^2\theta \cos^2\phi + (\lambda_3 - \lambda_2) \sin^2\theta$$

$$\beta(\theta, \phi) \equiv (\lambda_2 - \lambda_1) \cos\theta \cos\phi \sin\phi$$

$$\gamma(\theta, \phi) \equiv (\lambda_2 - \lambda_1) \sin^2\phi$$

Since we must first minimize this w.r.t. x, we set

$$\frac{\partial A}{\partial \phi'} = -2\alpha \cos\phi' \sin\phi' + 2\beta(\cos^2\phi' - \sin^2\phi') - 2\gamma \sin\phi' \cos{}' = 0$$

which implies that

$$(\alpha + \gamma) \sin\phi' \cos\phi' = \beta(\cos^2\phi' - \sin^2\phi')$$

or

$$\tfrac{1}{2}(\alpha + \gamma) \sin(2\phi') = \beta \cos(2\phi')$$

so that

$$\tan(2\phi') = \frac{2\beta}{(\alpha + \gamma)}$$

This vector is *not* necessarily contained in the x-y plane. The trigonometric identities

$$\sin^2(2\phi') = \frac{4\beta^2}{(\alpha + \gamma)^2}[1 - \sin^2(2\phi')]$$

$$\sin^2(2\phi')\left[\frac{(\alpha + \gamma)^2 + 4\beta^2}{(\alpha + \gamma)^2}\right] = \frac{4\beta^2}{(\alpha + \gamma)^2}$$

allow us to write

$$\sin(2\phi') = \frac{\pm 2\beta}{[(\alpha+\gamma)^2+4\beta^2]^{1/2}}$$
$$\cos(2\phi') = \frac{\pm(\alpha+\gamma)}{[(\alpha+\gamma)^2+4\beta^2]^{1/2}} \tag{A3.26}$$

Since we want a minimum (not a maximum), we compute the second derivative as

$$\begin{aligned}\frac{\partial^2 A}{\partial \phi'^2} &= -2\alpha(\cos^2\phi' - \sin^2\phi') - 8\beta(\sin\phi'\cos\phi') - 2\gamma(\cos^2\phi' - \sin^2\phi') \\ &= -2\alpha\cos(2\phi') - 4\beta\sin(2\phi') - 2\gamma\cos(2\phi') \\ &= -2[(\alpha+\gamma)\cos(2\phi') + 2\beta\sin(2\phi')] \\ &= \mp 2[(\alpha+\gamma)^2+4\beta^2]^{1/2}\end{aligned} \tag{A3.27}$$

and require this is to be positive so that we must choose the lower sign in Eq. A3.26.

$$\sin(2\phi') = \frac{-2\beta}{[(\alpha+\gamma)^2+4\beta^2]^{1/2}}$$
$$\cos(2\phi') = \frac{-(\alpha+\gamma)}{[(\alpha+\gamma)^2+4\beta^2]^{1/2}} \tag{A3.28}$$

Explicitly this minimum is given as

$$\begin{aligned}A(\theta, \phi; \phi')_{\min} &\equiv \bar{A}(\theta, \varphi) \\ &= \lambda_2 + \tfrac{1}{2}\{(\alpha-\gamma) - [(\alpha-\gamma)^2 + 4(\beta^2+\alpha\gamma)]^{1/2}\}\end{aligned}$$

Since

$$\beta^2 + \alpha\gamma = (\lambda_3-\lambda_2)(\lambda_2-\lambda_1)\sin^2\theta\sin^2\phi \geq 0 \tag{A3.29}$$

we see that

$$\{(\alpha-\gamma) - [(\alpha-\gamma)^2 + 4(\beta^2+\alpha\gamma)]^{1/2}\} \leq 0 \tag{A3.30}$$

This tells us that $\bar{A}(\theta, \phi)$ is bounded from above by λ_2 and that it will actually attain this maximum value provided the expression in Eq. A3.30 can vanish for some values of θ and ϕ. From Eqs. A3.29 and A3.30 we see that this will happen provided

$$(\lambda_3-\lambda_2)(\lambda_2-\lambda_1)\sin^2\theta\sin^2\phi = 0 \tag{A3.31}$$

If $\lambda_1 = \lambda_2$ or $\lambda_2 = \lambda_3$, then this condition can be satisfied for any y (i.e., θ and ϕ). If the eigenvalues are all distinct, then we would require that $\theta = 0$ and/or $\phi = 0, \pi$ (since $0 \leq \theta \leq \pi/2$). Since the values $\theta = 0$, $\phi = 0$ (or π) are possible values for which (cf. Eq. A3.27)

$$(\alpha+\gamma)^2 + 4\beta^2 = (\lambda_1-\lambda_2)^2 > 0$$

so that $\partial^2 A/\partial\phi'^2 < 0$, we finally obtain

$$\text{Max}_y[\text{Min}_x \langle x| A |x\rangle] = \lambda_2 \tag{A3.32}$$

The difficulty of this direct verification for even so simple a case should point up the beauty of the general proof given of Th. 3.5 since it avoids a direct calculation of $\text{Min}_x \langle x| A |x\rangle$.

SUGGESTED REFERENCES

R. Courant and D. Hilbert, *Methods of Mathematical Physics* (Vol. I). This classic text on mathematical physics gives a nice presentation of quadratic forms in Section 3 of Chapter I and a lucid discussion of the minimax principle in Section 4 of the same chapter. Many important topics are beautifully presented in this book.

B. Friedman, *Principles and Techniques of Applied Mathematics.* Chapter 2 of Friedman's book covers much of the same material as the present chapter. In addition pp. 67–91 contain a discussion of matrices whose eigenvectors do not form a complete set.

W. Kaplan, *Advanced Calculus.* Section 2.16 of Kaplan's book gives a very readable presentation of the method of Lagrange undetermined multipliers.

For some physical applications of techniques developed in this chapter, the reader might consult the following.

H. Goldstein, *Classical Mechanics.* Section 5.4 of Goldstein discusses the inertia tensor and the principal axis transformation, and Chapter 10 treats small oscillations and the transformation to normal modes.

F. Mandl, *Quantum Mechanics.* Section 19 gives an extremely clear discussion of the importance of the diagonalization of commuting hermitian operators and the simultaneous measurability of observables in quantum mechanics.

PROBLEMS

3.1 Let

$$A = \begin{pmatrix} 1 & 0 & 0 \\ 0 & 2 & i \\ 0 & 0 & 3 \end{pmatrix}$$

a. Does A^{-1} exist? If so, find it.
b. Find the eigenvalues and eigenvectors of A.
c. Do the eigenvectors span E_3? Can these eigenvectors be chosen orthonormal?
d. Solve the problem,

$$Ax = b \qquad b = \begin{pmatrix} 1 \\ 0 \\ 1 \end{pmatrix}$$

once by constructing A^{-1} (if it exists) and once by expanding both x and b in terms of the eigenvectors of A.

3.2 Consider the transformation

$$\mathbf{x}' = A\mathbf{x}$$

where

$$\mathbf{x}' = \begin{pmatrix} \xi_1' \\ \xi_2' \end{pmatrix} \qquad \mathbf{x} = \begin{pmatrix} \xi_1 \\ \xi_2 \end{pmatrix} \qquad A = \begin{pmatrix} 2 & 1 \\ 1 & 2 \end{pmatrix}$$

Take $\mathbf{x}'$ and $\mathbf{x}$ as two different vectors referred to one set of coordinate axes.

a. Show that the subspace defined by

$$\xi_1 + \xi_2 = 0$$

and the subspace defined by

$$\xi_1 - \xi_2 = 0$$

are each invariant subspaces of the operator, A.

b. Considering A as operating in a real two-dimensional space (i.e., E_2), show the geometrical significance of these two invariant subspaces.

c. Choose as a set of basis vectors two unit vectors, the first contained wholly in the first invariant subspace and the second contained wholly in the other invariant subspace. Now express A in this new basis. Could you have predicted the form of this result from general principles (i.e., theorems) we have studied? How?

3.3 Consider the operator defined by

$$\begin{aligned} x' &= 2x + y - 2z \\ y' &= -2x - y \\ z' &= \phantom{-2x - y - {}} 2z \end{aligned}$$

Show that the subspace defined by $x+y+z=0$ is invariant under this operator. Find a representation of this operator in this subspace and its eigenvalues. Find that transformation that will put two of the basis vectors in this invariant plane with the third basis vector orthogonal to the plane. Notice that $x'+y'+z'=0$ even when $x+y+z\neq 0$. What does this tell you geometrically about the operator?

3.4 Suppose that L can be written as

$$L = L_- L_+$$

where

$$[L_-, L_+] = I$$

Prove that if x is an eigenvector of L such that

$$Lx = \lambda x$$

then $y \equiv L_+ x$ is such that

$$Ly = (\lambda + 1)y$$

and $z \equiv L_- x$ is such that

$$Lz = (\lambda - 1)z$$

We will use such "raising" and "lowering" operators in Section 9.12a when we study angular momentum operators.

3.5 In the quantum-mechanical treatment of the harmonic oscillator the following eigenvalue problem arises,

$$L\varphi(x) = \lambda\varphi(x)$$

where

$$L = -\frac{d^2}{dx^2} + \frac{x^2}{4} + \frac{1}{2}$$

and $\varphi(x) \in \mathcal{L}_2(-\infty, \infty)$. Take

$$L_- = \frac{x}{2} + \frac{d}{dx}$$

$$L_+ = \frac{x}{2} - \frac{d}{dx}$$

and use the results of Problem 3.4 to find these eigenfunctions and eigenvalues. [*Hint:* Assume there does exist a *lowest* (finite) eigenvalue.]

3.6 Let

$$A = \begin{pmatrix} \frac{7}{4} & 0 & \frac{\sqrt{3}}{4} \\ 0 & 1 & 0 \\ \frac{\sqrt{3}}{4} & 0 & \frac{5}{4} \end{pmatrix} \qquad Ax_j = \lambda_j x_j$$

a. Find the eigenvalues, $\{\lambda_j\}$.
b. Find the eigenvectors, $\{x_j\}$.
c. Construct an orthonormal set of eigenvectors of A.
d. Find the orthogonal transformation, U, that diagonalizes A and write down the diagonal form of A.
e. Evaluate the matrix

$$B = A^3 - 4A^2 + 5A - 2I$$

f. Solve the equation

$$Ax = b = \begin{pmatrix} 1 \\ 0 \\ 1 \end{pmatrix}$$

by expanding the unknown vector x and the known vector b in terms of the orthonormal eigenvectors of A.

3.7 Prove that a matrix is hermitian if and only if its complete set of eigenvectors can all be chosen orthonormal and its eigenvalues are all real.

3.8 Prove that the necessary and sufficient condition that the inner product $\langle x, y\rangle$ be left invariant (for all x and y) under the transformation

$$x' = Ax \qquad y' = Ay$$

is that A be a unitary matrix. Assume that you are in a finite-dimensional vector space.

3.9 You are given that A, B, and C are $n \times n$ hermitian matrices, that A is positive definite, and that B and C commute. Prove that there exists a *nonsingular* matrix T such that

$$T^\dagger A T = I$$

$$T^\dagger B T = D_B$$

$$T^\dagger C T = D_C$$

where D_B and D_C are diagonal matrices. Use this result to solve the system of differential equations

$$A\frac{d^2z}{dt^2}+B\frac{dz}{dt}+Cz=0$$

where A, B, and C are hermitian matrices independent of the variable t and such that $\langle x|\, A\,|x\rangle>0$, $\|x\|\neq 0$, $[B, C]=0$, and z is a column vector $z=\{\zeta_j(t)\}$, $j=1, 2, \ldots, n$.

3.10 Obtain *explicitly* the general solution to the system of differential equations

$$\tfrac{3}{2}\ddot{x}_1(t)-\frac{i}{2}\ddot{x}_2(t)-i\dot{x}_2(t)-\tfrac{1}{2}x_1(t)+\tfrac{3}{2}ix_2(t)=0$$

$$\frac{i}{2}\ddot{x}_1(t)+\tfrac{3}{2}\ddot{x}_2(t)+i\dot{x}_1(t)-\tfrac{3}{2}ix_1(t)-\tfrac{1}{2}x_2(t)=0$$

3.11 If

$$A=\begin{pmatrix} 5 & 0 & \sqrt{3} \\ 0 & 3 & 0 \\ \sqrt{3} & 0 & 3 \end{pmatrix}$$

compute explicitly the matrix $\sin(\pi A)$.

3.12 Find the most general form for a two-dimensional orthogonal matrix. Compare it with the matrix for a rotation of the coordinate axes.

3.13 Prove that the determinant of an orthogonal transformation is ± 1. Show that the orthogonal transformation that diagonalizes a real self-adjoint matrix can be so chosen that its determinant is $+1$.

3.14 a. Consider the real, 3×3 orthogonal matrices,

$$AA^T=I$$

subject to the restriction

$$\det A=+1$$

How many real parameters are necessary to characterize such a matrix? Prove that these matrices form a group w.r.t. matrix multiplication. Do those matrices such that

$$\det A=-1$$

form a group?

b. Prove that a finite-dimensional $n\times n$ matrix S is unitary if and only if it can be written as

$$S=e^{iH} \qquad \text{where} \quad H^\dagger=H$$

That is, show that

$$SS^\dagger=I \Leftrightarrow H^\dagger=H \qquad S=e^{iH}$$

Also demonstrate that $\det S=\exp(i \operatorname{Tr} H)$.

(*Hint:* Think about diagonalizing S and H.)

3.15 Prove that the most general unitary symmetric (i.e., $S^\dagger S = I$, $S^T = S$) 2×2 matrix can be written as

$$S=\begin{pmatrix} \alpha e^{2i\beta} & i(1-\alpha^2)^{1/2}e^{i(\beta+\gamma)} \\ i(1-\alpha^2)^{1/2}e^{i(\beta+\gamma)} & \alpha e^{2i\gamma} \end{pmatrix}$$

where α, β, and γ are real parameters with $0\leq\alpha\leq 1$.

3.16 Given the matrices

$$A=\begin{pmatrix}1 & 0\\ 0 & 4\end{pmatrix} \qquad B=\begin{pmatrix}1 & 1\\ 1 & 1\end{pmatrix}$$

find a transformation that will bring both of them simultaneously into diagonal form. Then write down the diagonal forms of A and B.

3.17 Consider the two quadratic forms

$$(\dot{\xi}_1^2+\dot{\xi}_3^2)+2\dot{\xi}_2^2=\langle \dot{x}|\ A\ |\dot{x}\rangle$$

$$(\xi_1^2+2\xi_2^2+\xi_3^2-2\xi_1\xi_2-2\xi_2\xi_3)=\langle x|\ B\ |x\rangle$$

a. Show that A is positive definite.
b. Find the eigenvalues and eigenvectors of

$$(B-\lambda A)y=0$$

c. Find the transformation matrix which will simultaneously diagonalize A and B.
d. Write down the diagonal forms of A and B.

3.18

$$A=\begin{pmatrix}\frac{5}{4} & \frac{5}{4} & \frac{\sqrt{2}}{4}\\ \frac{5}{4} & \frac{5}{4} & \frac{\sqrt{2}}{4}\\ \frac{\sqrt{2}}{4} & \frac{\sqrt{2}}{4} & \frac{5}{2}\end{pmatrix} \qquad B=\begin{pmatrix}\frac{1}{2} & -\frac{1}{2} & -\frac{\sqrt{2}}{2}\\ -\frac{1}{2} & \frac{1}{2} & -\frac{\sqrt{2}}{2}\\ -\frac{\sqrt{2}}{2} & -\frac{\sqrt{2}}{2} & 0\end{pmatrix}$$

a. Show that $[A, B]=0$.
b. Find the eigenvalues of A and of B.
c. Find a common orthonormal set of eigenvectors for A and B.
d. Find the U that diagonalizes A and B simultaneously.
e. Diagonalize A and B.

3.19 You are given the following matrices

$$A=\begin{pmatrix}\frac{3}{2} & -\frac{i}{2}\\ \frac{i}{2} & \frac{3}{2}\end{pmatrix} \qquad B=\begin{pmatrix}1 & 0\\ 0 & 1\end{pmatrix}$$

$$C=\begin{pmatrix}0 & -i\\ i & 0\end{pmatrix} \qquad D=\begin{pmatrix}-\frac{1}{2} & \frac{3i}{2}\\ -\frac{3i}{2} & -\frac{1}{2}\end{pmatrix}$$

a. Prove that they all commute with each other.
b. Find the eigenvalues of each.
c. Construct a common orthonormal set of eigenvectors for A, B, C, and D.
d. Construct the unitary matrix, U, that will diagonalize A, B, C, and D simultaneously.
e. Write down the diagonal forms of A, B, C, and D.

3.20 Suppose that an $n^2 \times n^2$ matrix M can be decomposed into a sum of direct products as

$$M = \alpha a \otimes a + \beta a \otimes b + \gamma b \otimes a + \delta b \otimes b$$

where α, β, γ, and δ are real constants and a and b are each $n \times n$ matrices such that

$$a^{\dagger} = a \qquad b^{\dagger} = b \qquad [a, b] = 0$$

Denote the eigenvalues and common orthonormal set of eigenvectors of these matrices as

$$a\chi_j = \mu_j \chi_j, \; j = 1, 2, \ldots, n \qquad b\chi_k = \nu_k \chi_k, \; k = 1, 2, \ldots, n$$

Now prove that the eigenvectors and eigenvalues of M,

$$M\phi_i = \lambda_i \phi_i \qquad i = 1, 2, 3, \ldots, n^2$$

can be expressed as

$$\phi_i = \chi_j \otimes \chi_k \qquad j, k = 1, 2, 3, \ldots, n$$

$$\lambda_i = \alpha \mu_j \mu_k + \beta \mu_j \nu_k + \gamma \nu_j \mu_k + \delta \nu_j \nu_k, \qquad j, k = 1, 2, 3, \ldots, n$$

Furthermore prove that

$$\langle \phi_i \mid \phi_j \rangle = \delta_{ij} \qquad i, j = 1, 2, \ldots, n^2$$

3.21 Find the eigenvalues and an orthonormal set of eigenvectors for the following matrix.

$$A = \begin{pmatrix} 1 & 2 & 1 & -2 \\ 2 & 1 & -2 & 1 \\ 1 & -2 & 1 & 2 \\ -2 & 1 & 2 & 1 \end{pmatrix}$$

3.22 If

$$A \begin{pmatrix} \cos \theta & \sin \theta \\ -\sin \theta & \cos \theta \end{pmatrix}$$

find

a. A^n

b. $e^{\gamma A}$

3.23 Consider the "vector" matrix

$$\boldsymbol{\sigma} \equiv (\sigma_x, \sigma_y, \sigma_z)$$

where

$$\sigma_x = \begin{pmatrix} 0 & 1 \\ 1 & 0 \end{pmatrix} \qquad \sigma_y = \begin{pmatrix} 0 & -i \\ i & 0 \end{pmatrix} \qquad \sigma_z = \begin{pmatrix} 1 & 0 \\ 0 & -1 \end{pmatrix}$$

and the vector

$$\hat{\boldsymbol{n}} = (\sin \theta \cos \phi, \sin \theta \sin \phi, \cos \theta)$$

and define

$$\boldsymbol{\sigma} \cdot \hat{\boldsymbol{n}} \equiv n_x \sigma_x + n_y \sigma_y + n_z \sigma_z$$

a 2×2 matrix. Evaluate the matrix

$$\exp\left(\frac{i}{2}\psi\boldsymbol{\sigma}\cdot\hat{\boldsymbol{n}}\right)$$

where ψ is a real parameter. That is, write this matrix operator explicitly as a single 2×2 matrix.

3.24 If A and B are each $n\times n$ matrices, under what conditions is it true that

$$e^{(A+B)}=e^{A}e^{B}$$

4 Complete Sets of Functions

4.0 INTRODUCTION

Let us begin by emphasizing that the sense in which a set of functions $\{\varphi_j(x)\}$ is called *complete* is different from the use of that term when we discussed complete spaces (cf. Section 1.3d). Completeness for a set of functions has the same meaning as for a set of basis vectors $\{\hat{x}_j\}$ for a finite or infinite-dimensional space (cf. Section 3.3). That is, a set of basis vectors $\{\hat{x}_j\}$ for a space S is called *complete* provided that any vector $x \in S$ can be written as a linear combination of these $\{\hat{x}_j\}$ as

$$x = \sum_j \xi_j \hat{x}_j \tag{4.1}$$

Furthermore, if the $\{\hat{x}_j\}$ are chosen to be an orthonormal basis then the condition that x have finite norm reduces to the question of the convergence of the series

$$\|x\|^2 = \sum_j |\xi_j|^2 \tag{4.2}$$

In many applications we deal with functions in an infinite-dimensional separable Hilbert space and must therefore deal with an infinite set of vectors (i.e., functions) for a basis. That is, the analogues of Eqs. 4.1 and 4.2 now become the following. If we are given any $f(x) \in \mathscr{L}_2[a, b]$, we must decide whether or not there exists a countable infinity of basis functions $\{\varphi_j(x)\}$, $j = 1, 2, \ldots$, such that

$$f(x) \sim \sum_j c_j \varphi_j(x) \tag{4.3}$$

where, for the time being, "$\sim$" means "approximated by" in the sense that

$$\|f\|^2 \equiv \int_a^b |f(x)|^2 \, dx = \sum_j |c_j|^2 \tag{4.4}$$

In such a function space the question of the completeness of a set of functions now becomes a rather difficult one of the utmost importance.

We must begin by stating exactly what is meant by expanding a function $f(x)$ in terms of the set $\{\varphi_j(x)\}$.

4.1 CRITERION FOR COMPLETENESS

a. Bessel's Inequality

As a criterion for completeness we consider *Bessel's inequality* (cf. Problem 1.16). We assume the existence of an infinite, orthonormal set of functions $\{\varphi_j(x)\}$, $j=1, 2, \ldots$, such that

$$\langle \varphi_j \mid \varphi_k \rangle \equiv \int_a^b \varphi_j^*(x)\varphi_k(x) \, dx = \delta_{jk} \qquad j, k = 1, 2, \ldots \tag{4.5}$$

Note that we do not mean to imply that the set of functions $\{\varphi_j(x)\}$ is complete or spans the space since this is the point of the entire discussion.

THEOREM 4.1 (Bessel's inequality)

Given an infinite orthonormal set of functions $\{\varphi_j(x)\}$ and any $f(x) \in \mathscr{L}_2[a, b]$, then it follows that

$$\|f\|^2 \geq \sum_j |c_j|^2 \tag{4.6}$$

where the constants c_j are defined as

$$c_j \equiv \langle \varphi_j \mid f \rangle = \int_a^b \varphi_j^*(x) f(x) \, dx \tag{4.7}$$

PROOF

Even though we have seen one proof of this and of the next theorem for a general space with an inner product defined (cf. Problem 1.16), we will give here another proof for each that does not require differentiation. Notice that in the following we begin with just n terms in the sum.

$$0 \leq \left\| f - \sum_{j=1}^n c_j \varphi_j \right\|^2 \equiv \int_a^b \left| f(x) - \sum_{j=1}^n c_j \varphi_j(x) \right|^2 dx$$

$$= \|f\|^2 - 2\sum_j |c_j|^2 + \sum_j |c_j|^2 = \|f\|^2 - \sum_{j=1}^n |c_j|^2$$

We may now take the limit as $n \to \infty$ to obtain

$$\|f\|^2 \geq \sum_{j=1}^{\infty} |c_j|^2 \qquad \text{Q.E.D.}$$

b. Approximation in the Mean

Just as in fitting the "best" straight line to a set of points, we must now adopt some criterion by which to judge the "best" fit (e.g., mean square deviation). Suppose we approximate a square-integrable function $f(x)$ by a finite sum of n of our $\{\varphi_j(x)\}$ as

$$f(x) \sim \sum_{j=1}^{n} \alpha_j \varphi_j(x) \qquad x \in [a, b] \tag{4.8}$$

We take that fit (i.e., those values of α_j) as the best that makes the mean square error after n terms M_n,

$$\left\| f - \sum_{j=1}^{n} \alpha_j \varphi_j \right\|^2 \equiv M_n(b-a) \tag{4.9}$$

the least for any given value of n.

THEOREM 4.2

The mean square error, M_n,

$$M_n(b-a) \equiv \left\| f - \sum_{j=1}^{n} \alpha_j \varphi_j \right\|^2$$

is minimized when $\alpha_j = c_j \equiv \langle \varphi_j \mid f \rangle$.

PROOF

$$\begin{aligned}
0 \leq M_n(b-a) &\equiv \left\| f - \sum_{j=1}^{n} \alpha_j \varphi_j \right\|^2 = \left\| f - \sum_{j=1}^{n} (\alpha_j - c_j)\varphi_j - \sum_{j=1}^{n} c_j \varphi_j \right\|^2 \\
&= \|f\|^2 - \sum (\alpha_j - c_j) c_j^* - \sum |c_j|^2 - \sum (\alpha_j^* - c_j^*) c_j + \sum |\alpha_j - c_j|^2 \\
&\quad + \sum (\alpha_j^* - c_j^*) c_j - \sum |c_j|^2 + \sum c_j^* (\alpha_j - c_j) + \sum |c_j|^2 \\
&= \left[\|f\|^2 - \sum_{j=1}^{n} |c_j|^2 \right] + \sum_{j=1}^{n} |\alpha_j - c_j|^2
\end{aligned}$$

However from Th. 4.2 we know that the quantity in square brackets is nonnegative. Therefore M_n is minimized by the choice $\alpha_j = c_j$. Q.E.D.

We denote this minimum value of M_n by $\bar{M}_n$,

$$\bar{M}_n(b-a) \equiv \|f\|^2 - \sum_{j=1}^{n} |c_j|^2 \tag{4.10}$$

It is clear that $\bar{M}_n$ decreases (or possibly remains constant) as n increases. Therefore $\bar{M}_n$ attains its absolute minimum as $n \to \infty$,

$$\bar{M}_\infty(b-a) = \|f\|^2 - \sum_{j=1}^{\infty} |c_j|^2$$

If $\bar{M}_n \to 0$ as $n \to \infty$, then we have for all $f(x) \in \mathscr{L}_2[a, b]$

$$\|f\|^2 = \sum_{j=1}^{\infty} |c_j|^2 \tag{4.11}$$

and the set $\{\varphi_j(x)\}$ is called *complete*. Therefore our criterion for completeness, Eq. 4.11, is just approximation in the mean. If, *furthermore*, the series $\sum_{j=1}^{n} c_j \varphi_j(x)$ converges *uniformly* to a limit for all $x \in [a, b]$, then we may take the limit inside the integral (cf. Appendix I, Th. I.11) to obtain

$$\int_a^b \left| f(x) - \sum_{j=1}^{\infty} c_j \varphi_j(x) \right|^2 dx = 0 \tag{4.12}$$

From this we can then deduce that

$$f(x) = \sum_{j=1}^{\infty} c_j \varphi_j(x) \tag{4.13}$$

almost everywhere. Therefore in order to use an expansion like (4.13) we require that

i. $f(x) \in \mathscr{L}_2[a, b]$
ii. the completeness relation (4.11) is satisfied
iii. $\sum_{j=1}^{\infty} c_j \varphi_j(x)$ is uniformly convergent for $x \in [a, b]$

Notice that condition (iii) is something over and above (i) and (ii).

From a practical point of view, given a set $\{\varphi_j(x)\}$, one attempts to find the weakest conditions that must be imposed on $f(x)$ (e.g., piecewise continuity, differentiability, etc.) to insure the applicability of the expansion (4.13). For example in Section 8.3 we will prove the completeness of the solutions to the Sturm-Liouville differential equation.

4.2 WEIERSTRASS APPROXIMATION THEOREM

We will now prove a remarkable theorem due to Weierstrass that allows any continuous function on a finite interval to be uniformly approximated by a sequence of polynomials. Once we have demonstrated that such a function can be uniformly approximated by powers of x, we can construct linear combinations of such powers of x to form orthonormal polynomials (e.g., the Legendre polynomials, which are only orthogonal but not normalized to unity). We can also extend this to other sets of functions by relating these to powers of x.

THEOREM 4.3 (Weierstrass approximation theorem)

Any function continuous on the finite closed interval $x \in [a, b]$ can be uniformly approximated by polynomials in x on this interval; that is, for sufficiently large n and ε as small as we please, we have

$$|\mathscr{P}_n(x) - f(x)| < \varepsilon \qquad \forall x \in [a, b]$$

PROOF

We will take the closed interval on which $f(x)$ is continuous to be $[0, 1]$. We lose no generality in doing this since any finite interval $[a, b]$ can be mapped onto $[0, 1]$ by the linear transformation

$$y = \frac{x-a}{b-a}$$

In the present case we shall also take $f(0) = 0$ for convenience. If $f(0) \neq 0$, then our proof will hold for the function $[f(x) - f(0)]$, which is still a continuous function.

Since $f(x)$ is given as being continuous on the *closed* interval $[0, 1]$, it is uniformly continuous there (cf. Appendix I, Th. I.3), so that for *any* $x_1, x_2 \in [0, 1]$ we can choose a δ

$$|x_1 - x_2| < \delta \tag{4.14}$$

such that

$$|f(x_1) - f(x_2)| < \tfrac{1}{2}\varepsilon \tag{4.15}$$

for any ε we please. We now divide the interval $[0, 1]$ into m equal segments such that $m > 1/\delta$,

$$x_1 = 0 \qquad x_{m+1} = 1$$

and construct the function

$$g(f; x) = \sum_{k=1}^{m} g_k h(x_k; x) \tag{4.16}$$

where

$$g_k = \frac{f(x_{k+1}) - f(x_k)}{2(x_{k+1} - x_k)} \qquad k = 1, 2, \ldots, m \tag{4.17}$$

$$h(x_k; x) \equiv (x_{k+1} - x_k) + |x - x_k| - |x - x_{k+1}| = \begin{cases} 0, & x \leq x_k \\ 2(x - x_k), & x_k \leq x \leq x_{k+1} \\ 2(x_{k+1} - x_k), & x \geq x_{k+1} \end{cases} \tag{4.18}$$

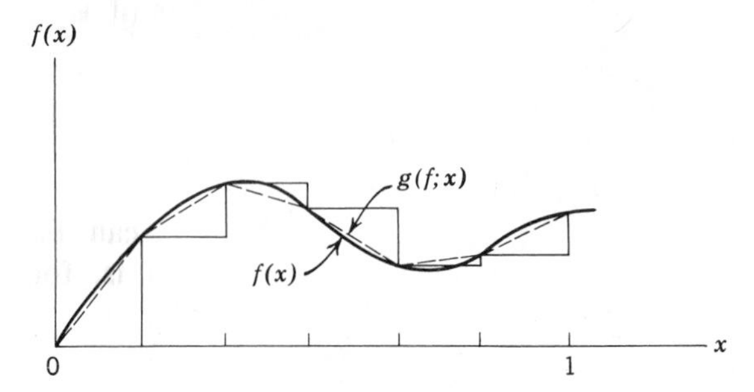

Figure 4.1

for $x \in [0, 1]$ as shown in Fig. 4.1. It is clear that $g(f; x_k) = f(x_k)$. From Eqs. 4.14–4.18 (cf. Fig. 4.1 also) we see that

$$|f(x) - g(f; x)| < \frac{1}{2}\varepsilon \qquad x \in [0, 1] \tag{4.19}$$

That is, let $x_k \le x \le x_{k+1}$ so that $|x - x_k| \le 1/m < \delta$ and

$$\begin{aligned}|f(x) - g(f; x)| &= |f(x) - f(x_k) - g_k h(x_k; x)| \\ &\le |f(x) - f(x_k)| + |g_k||h(x_k; x)| \\ &\le \varepsilon' + \frac{m}{2}\varepsilon' 2\delta \equiv \frac{\varepsilon}{2}\end{aligned}$$

Since $|x| \le 1$, we can write

$$|x| = \sqrt{1 - (1 - x^2)}$$

so that with $u \equiv (1 - x^2)$ we have from the binomial theorem

$$|x| = -\frac{\Gamma(\frac{3}{2})}{\pi} \sum_{s=0}^{\infty} \frac{\Gamma(s - \frac{1}{2})}{s!} u^s = \sum_{s=0}^{\infty} \frac{(2s)!\, u^s}{(s!)^2 (1 - 2s) 2^{2s}} \tag{4.20}$$

From the Weierstrass M test for the uniform convergence of a series of functions plus the integral test for the convergence of the series (cf. Appendix I, Th. I.9)

$$\sum_{s=0}^{\infty} \frac{\Gamma(s - \frac{1}{2})}{s!}$$

and the fact that (cf. Section 7.12b)

$$\frac{\Gamma(s - \frac{1}{2})}{\Gamma(s + 1)} \xrightarrow[s \to \infty]{} s^{-3/2}$$

we have

$$\int^{\infty} \frac{\Gamma(s - \frac{1}{2})}{\Gamma(s + 1)} ds < \infty$$

so that the series in Eq. 4.20 is *uniformly* convergent for $0 \leq u \leq 1$. This means that there exist polynomials, for example, of degrèe n, in $(1-x^2)$ such that $|x|$ can be uniformly approximated by them. Therefore the $g_k h(x_k; x)$ of Eq. 4.16 can be uniformly approximated arbitrarily closely by some polynomials $P_n^{(k)}(x)$ on $x \in [0, 1]$ as

$$|P_n^{(k)}(x) - g_k h(x_k; x)| < \frac{\varepsilon}{2m} \tag{4.21}$$

if we so choose. From Eqs. 4.16 and 4.21 we have

$$\begin{aligned}\left|g(f; x) - \sum_{k=1}^{m} P_n^{(k)}(x)\right| &= \left|\sum_{k=1}^{m} [g_k h(x_k; x) - P_n^{(k)}(x)]\right| \\ &\leq \sum_{k=1}^{m} |g_k h(x_k; x) - P_n^{(k)}(x)| < \sum_{k=1}^{m} \frac{\varepsilon}{2m} = \frac{\varepsilon}{2}\end{aligned} \tag{4.22}$$

Finally we obtain with the aid of Eqs. 4.19 and 4.22

$$\begin{aligned}\left|f(x) - \sum_{k=1}^{m} P_n^{(k)}(x)\right| &= \left|f(x) - g(f; x) + g(f; x) - \sum_{k=1}^{m} P_n^{(k)}(x)\right| \\ &\leq |f(x) - g(f; x)| + \left|g(f; x) - \sum_{k=1}^{m} P_n^{(k)}(x)\right| < \varepsilon\end{aligned}$$

This establishes the theorem since the finite sum

$$\sum_{k=1}^{m} P_n^{(k)}(x) \equiv \mathcal{P}_n(x)$$

is also a polynomial of degree n in x. Q.E.D.

Although this proof (due in essence to Lebesgue) is very simple and straightforward, it does not give any estimate of the rate at which $\mathcal{P}_n(x)$ approaches $f(x)$ as n increases. An alternative proof of this theorem is given in Section A4.1. The Weierstrass approximation theorem is easily extended to two (or more) variables for functions continuous in two (or more) variables in a closed finite region (cf. Section A4.2).

Now suppose we have an orthonormal set of polynomials in x, $\{\varphi_j(x)\}$ of order j, $j = 0, 1, 2, \ldots$, on the interval $[a, b]$,

$$\begin{gathered}\varphi_j(x) = \sum_{k=0}^{j} a_k^{(j)} x^k \qquad j = 1, 2, 3, \ldots \\ \langle \varphi_j \mid \varphi_k \rangle = \delta_{jk}\end{gathered} \tag{4.23}$$

Since powers of x are linearly independent as are the $\{\varphi_j(x)\}$, we could invert Eq. 4.23 and finally express the $\mathcal{P}_n(x)$ of Th. 4.3 as

$$\mathcal{P}_n(x) = \sum_{j=0}^{2n} \alpha_j^{(n)} \varphi_j(x)$$

We do *not* know that the $\{\alpha_j^{(n)}\}$ are the $\{c_j\}$. In fact generally they are not. However we know from the Weierstrass approximation theorem that for sufficiently large n and continuous $f(x)$, $\forall x \in [a, b]$,

$$\left|f(x)-\sum_{j=0}^{2n} \alpha_j^{(n)} \varphi_j(x)\right| < \varepsilon$$

where ε can be made to vanish as $n \to \infty$. Therefore we have

$$\left\|f-\sum_{j=1}^{2n} \alpha_j^{(n)} \varphi_j\right\|^2 < \int_a^b \varepsilon^2\, dx = \varepsilon^2(b-a) \tag{4.24}$$

so that the mean square error, Eq. 4.9, will vanish as $n \to \infty$. Since the sum in (4.24) is uniformly convergent, we may take the limit $n \to \infty$ inside the integral to deduce

$$\left\|f-\sum_{j=0}^{\infty} \alpha_j^{(\infty)} \varphi_j\right\|^2 = 0 \tag{4.25}$$

However we have seen from Th. 4.2 that Eq. 4.25 is fulfilled only when $\alpha_j^{(\infty)} \equiv c_j = \langle \varphi_j \mid f \rangle$.

The net result of all this is that any continuous $f(x)$, $x \in [a, b]$, can be approximated in the mean by the orthonormal polynomials $\{\varphi_j(x)\}$ as

$$\|f\|^2 = \sum_{j=0}^{\infty} |c_j|^2 \qquad c_j = \langle \varphi_j \mid f \rangle \tag{4.26}$$

This establishes the completeness of the $\{\varphi_j(x)\}$ for all continuous $f(x)$ (since these are obviously square integrable on any finite interval).

We can now show that if the set of polynomials $\{\varphi_j(x)\}$ is complete for all continuous $f(x)$, then it will also satisfy the completeness relation, Eq. 4.11, for any function with a finite number of finite discontinuities on the range $x \in [a, b]$. If $h(x)$ is discontinuous at $x = x_j$, $j = 1, 2, \ldots, m$, replace it by the straight-line segments between the points $[x_j - \delta, h(x_j - \delta)]$ and $[x_j + \delta, h(x_j + \delta)]$ in the neighborhoods of these discontinuities and call this new continuous function $f(x)$ (cf. Fig. 4.2). If M is the maximum absolute value of $h(x)$ on $[a, b]$, then its maximum discontinuity is $2M$ and

$$\|f-h\|^2 \equiv \int_a^b |f(x)-h(x)|^2\, dx = \sum_{j=1}^{m} \int_{x_j-\delta}^{x_j+\delta} |f(x)-h(x)|^2\, dx$$

$$\leq m \int_{-\delta}^{\delta} |2M|^2\, dx = 8M^2 m\delta$$

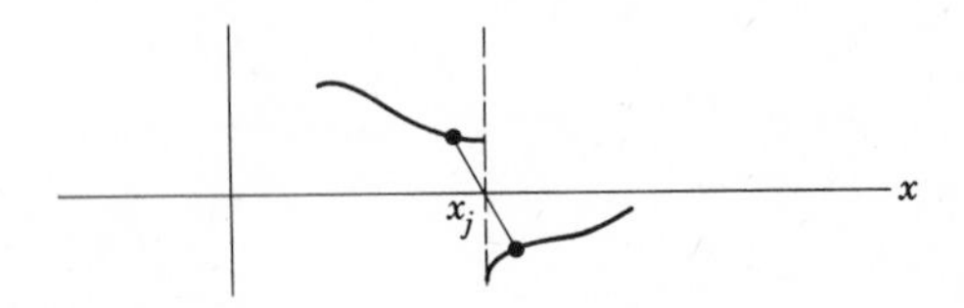

Figure 4.2

This may be made as small as we please by letting $\delta \to 0$. Since

$$\left\|h-\sum_{j=0}^{n} c_j\varphi_j\right\| = \left\|h-f+f-\sum_{j=0}^{n} c_j\varphi_j\right\| \le \|h-f\| + \left\|f-\sum_{j=0}^{n} c_j\varphi_j\right\| \tag{4.27}$$

and since each term on the right can be made arbitrarily small for sufficiently large n, we see that we may approximate any *piecewise* continuous function on $[a, b]$ in the mean by the orthonormal polynomials $\{\varphi_j(x)\}$, $j=0, 1, 2, \ldots, \infty$. However of course the series $\sum_j c_j\varphi_j(x)$ does not necessarily converge uniformly to $f(x)$ everywhere on $[a, b]$. In fact there is no simple criterion for $f(x)$ that will guarantee pointwise (or uniform) convergence for this series [cf. Hobson and Titchmarsh].

We have still not proven the completeness of these $\{\varphi_j(x)\}$ for *any* arbitrary $f(x) \in \mathscr{L}_2[a, b]$. If we were to make the plausible assumption that any $f(x) \in \mathscr{L}_2[a, b]$ can be approximated arbitrarily closely in the mean by a continuous function $g(x)$ so that

$$\|f-g\|^2 < \varepsilon$$

where ε can be made as small as we please, then we could use an argument like that given in Eq. 4.27 to prove completeness. However this is not such an easy matter and requires a knowledge of the theory of Lebesgue integration. The interested reader is referred to the book by Schwartz cited at the end of this chapter. (Also see Appendix II, Th. II.7.)

4.3 EXAMPLES OF COMPLETE SETS OF FUNCTIONS

a. Fourier Series

THEOREM 4.4

Any function $f(x)$ continuous on $x \in [-\pi, \pi]$ and such that $f(-\pi)=f(\pi)$ may be uniformly approximated by the series

$$\frac{\alpha_0}{2} + \sum_{n=1}^{N} (\alpha_n \cos nx + \beta_n \sin nx) \tag{4.28}$$

PROOF

Let us define the variables

$$r = \sigma \cos x \qquad s = \sigma \sin x$$

and the function

$$g(r, s) \equiv \sigma f(x)$$

As x ranges from $-\pi$ to $+\pi$ and σ from 0 to ∞, variables r and s take on all possible (real) values so that $g(r, s)$ is a continuous function of r and s in the entire r–s plane. On the unit circle in this plane, $r^2+s^2=1$, $g(r, s)$ coincides with $f(x)$. Therefore we could use a $\mathcal{P}_n(r, s)$ (cf. Section A4.2) for two variables to approximate $g(r, s)$ uniformly within a region containing the unit circle. On the unit circle the form of the uniformly approximating polynomial would be

$$\mathcal{P}_n(r, s) = \sum_{p,q=0}^{2n} \alpha_{pq} r^p s^q \tag{4.29}$$

Since

$$r^p = (\cos x)^p = \left(\frac{e^{ix}+e^{-ix}}{2}\right)^p$$

$$s^q = (\sin x)^q = \left(\frac{e^{ix}-e^{-ix}}{2i}\right)^q$$

we may use the binomial theorem to reduce (4.29) to the form of (4.28).

Q.E.D.

Since the mean square error will vanish as $n \to \infty$, we know that $\sin nx$ and $\cos nx$ form a complete set for the functions of this theorem. Hence the expansion coefficients must be the projections of $f(x)$ onto $\sin nx$ and $\cos nx$ (with proper normalization factors; cf. Problem 4.1). If the function $f(x)$ has a finite number of finite discontinuities on $x \in [-\pi, \pi]$, we can again use an argument similar to that in Section 4.2 to prove that the series (4.28) still converges in the mean to $f(x)$. Similarly if $f(-\pi) \neq f(\pi)$ we can *define* a new discontinuous function

$$g(x) \equiv f(x) \qquad -\pi \leq x < \pi$$

$$g(\pi) \equiv f(-\pi)$$

and still have convergence in the mean. It is also true that the series (4.28) converges uniformly to $f(x)$ on every closed interval on which $f(x)$ is continuous, as we will now prove in Th. 4.5. Therefore $\sin nx$ and $\cos nx$ are complete for every bounded, piecewise continuous function on $x \in [-\pi, \pi]$. We can state this as the fundamental theorem of Fourier series.

THEOREM 4.5

Every function $f(x)$ that is bounded and piecewise continuous (having at most a finite number of finite discontinuities) on $x \in [-\pi, \pi]$ may be approximated on this interval in the mean by the series

$$\frac{a_0}{2} + \sum_{n=1}^{\infty} (a_n \cos nx + b_n \sin nx) \tag{4.30}$$

where

$$a_n = \frac{1}{\pi}\int_{-\pi}^{\pi} f(x)\cos nx\,dx \qquad b_n = \frac{1}{\pi}\int_{-\pi}^{\pi} f(x)\sin nx\,dx \qquad \textbf{(4.31)}$$

Furthermore the convergence of the series (4.30) is uniform in every closed interval in which $f(x)$ is continuous and also at the end points if $f(\pi)=f(-\pi)$. At any discontinuity the series converges to the mean value as

$$\tfrac{1}{2}[f(x^+)+f(x^-)] = \frac{a_0}{2} + \sum_{n=1}^{\infty}(a_n\cos nx + b_n\sin nx)$$

PROOF

If we first assume that $f(x)$ is continuous, that $f(-\pi)=f(\pi)$, and that $f'(x)$ has at most a finite number of finite discontinuities so that $f'(x)$ is square integrable, then by Ths. 4.3 and 4.4 we may approximate $f'(x)$ *in the mean* as

$$\frac{1}{\pi}\|f'\|^2 = \sum_{n=0}^{\infty}(|\bar{a}_n|^2 + |\bar{b}_n|^2)$$

where

$$\bar{a}_n \equiv \frac{1}{\pi}\int_{-\pi}^{\pi} f'(x)\cos nx\,dx = \frac{1}{\pi}f(x)\cos nx\Big|_{-\pi}^{\pi} + \frac{n}{\pi}\int_{-\pi}^{\pi} f(x)\sin nx\,dx = nb_n$$

$$\bar{b}_n \equiv \frac{1}{\pi}\int_{-\pi}^{\pi} f'(x)\sin nx\,dx = \frac{1}{\pi}f(x)\sin nx\Big|_{-\pi}^{\pi} - \frac{n}{\pi}\int_{-\pi}^{\pi} f(x)\cos nx\,dx = -na_n$$

so that

$$\frac{\|f'\|^2}{\pi} = \sum_{n=0}^{\infty} n^2(|a_n|^2 + |b_n|^2) < \infty$$

Since for any complex scalars α and β

$$\begin{aligned}|\alpha+\beta|^2 &= |\alpha|^2 + |\beta|^2 + 2\,\mathrm{Re}\,(\alpha^*\beta) \le |\alpha|^2 + |\beta|^2 + 2\,|\alpha\beta| \\ &= |\alpha|^2 + |\beta|^2 + 2\,|\alpha||\beta| = (|\alpha| + |\beta|)^2 \le 4(|\alpha|^2 + |\beta|^2)\end{aligned}$$

we see that

$$\begin{aligned}\left|\sum_{n=1}^{\infty}(a_n\cos nx + b_n\sin nx)\right| &= \left|\sum_{n=1}^{\infty}\frac{1}{n}(na_n\cos nx + nb_n\sin nx)\right| \\ &\le \left(\sum_{n=1}^{\infty}\frac{1}{n^2}\right)^{1/2}\left(\sum_{n=1}^{\infty} n^2\,|a_n\cos nx + b_n\sin nx|^2\right)^{1/2} \\ &\le 2\left(\sum_{n=1}^{\infty}\frac{1}{n^2}\right)^{1/2}\left(\sum_{n=1}^{\infty} n^2[|a_n|^2\cos^2 nx + |b_n|^2\sin^2 nx]\right)^{1/2} \\ &\le 2\left(\sum_{n=1}^{\infty}\frac{1}{n^2}\right)^{1/2}\left(\sum_{n=1}^{\infty} n^2[|a_n|^2 + |b_n|^2]\right)^{1/2} = 2\left(\sum_{n=1}^{\infty}\frac{1}{n^2}\right)^{1/2}\frac{\|f'\|}{\sqrt{\pi}} < \infty\end{aligned}$$

This implies that $\sum_{n=1}^{\infty}(a_n \cos nx + b_n \sin nx)$ is uniformly convergent. Since a uniformly convergent sum of continuous functions is itself a continuous function (cf. Appendix I, Th. I.10, or Th. 7.5 of Section 7.4) and since the uniform convergence of the sum allows us to write the completeness relation as

$$\int_{-\pi}^{\pi}\left|f(x)-\frac{a_0}{2}-\sum_{n=1}^{\infty}(a_n \cos nx + b_n \sin nx)\right|^2 dx = 0$$

we conclude that

$$f(x) = \frac{a_0}{2} + \sum_{n=1}^{\infty}(a_n \cos nx + b_n \sin nx)$$

An alternate proof of this is outlined in Problem 4.2.

Furthermore, if $f(x)$ is discontinuous, we may replace it by a continuous $g(x)$ as in the discussion following Th. 4.3 (cf. Fig. 4.2) so that the series for $g(x)$ will converge uniformly to $f(x)$ on every closed subinterval of continuity. If we denote by a_n and a_n', respectively,

$$a_n = \frac{1}{\pi}\int_{-\pi}^{\pi} f(x) \cos nx\, dx$$

$$a_n' = \frac{1}{\pi}\int_{-\pi}^{\pi} g(x) \cos nx\, dx$$

then using the same argument and notation that led to Eq. 4.27 we have

$$|a_n - a_n'| = \left|\frac{1}{\pi}\sum_{j=1}^{m}\int_{x_j-\delta}^{x_j+\delta}[f(x)-g(x)] \cos nx\, dx\right|$$

$$\leq \frac{1}{\pi}\sum_{j=1}^{m}\int_{x_j-\delta}^{x_j+\delta}|f(x)-g(x)|\, dx \leq \frac{8}{\pi}M^2 m\,\delta$$

so that $a_n' \to a_n$ as $\delta \to 0$. A similar argument holds for the $\{b_n\}$.

Suppose now that $f(x)$ has just one discontinuity Δ at $x = x_0$. (The extension to m discontinuities is immediate and is left to the reader). The function defined as [Δ is positive if $f(x)$ decreases at x_0, negative otherwise]

$$F(x) \equiv \begin{cases} f(x) - \Delta & x \leq x_0 \\ f(x) & x > x_0 \end{cases}$$

is continuous so that from the previous part of this proof we have

$$F(x) = \frac{\alpha_0}{2} + \sum_{n=1}^{\infty}(\alpha_n \cos nx + \beta_n \sin nx)$$

where

$$\alpha_n = \frac{1}{\pi}\int_{-\pi}^{\pi} F(x)\cos nx\,dx = \frac{1}{\pi}\int_{-\pi}^{\pi} f(x)\cos nx\,dx - \frac{\Delta}{\pi}\int_{-\pi}^{x_0}\cos nx\,dx \qquad n\geq 1$$

$$\beta_n = \frac{1}{\pi}\int_{-\pi}^{\pi} F(x)\sin nx\,dx = \frac{1}{\pi}\int_{-\pi}^{\pi} f(x)\sin nx\,dx - \frac{\Delta}{\pi}\int_{-\pi}^{x_0}\sin nx\,dx \qquad n\geq 1$$

or

$$\alpha_n = a_n - \frac{\Delta}{\pi n}\sin(nx_0),\ n\geq 1 \qquad \alpha_0 = \frac{1}{\pi}\int_{-\pi}^{\pi} f(x)\,dx - \frac{\Delta}{\pi}(x_0+\pi)$$

$$\beta_n = b_n + \frac{\Delta}{\pi n}[\cos(nx_0) - (-1)^n]$$

If we take the limits

$$\lim_{x\to x_0^-} F(x) \equiv F(x_0) = f(x_0^-) - \Delta = \frac{\alpha_0}{2} + \sum_{n=1}^{\infty}(\alpha_n\cos nx_0 + \beta_n\sin nx_0)$$

$$\lim_{x\to x_0^+} F(x) = F(x_0) = f(x_0^+) = \frac{\alpha_0}{2} + \sum_{n=1}^{\infty}(\alpha_n\cos nx_0 + \beta_n\sin nx_0)$$

and add, we obtain

$$\begin{aligned}
\tfrac{1}{2}[f(x_0^+)+f(x_0^-)] &= \frac{\Delta}{2} + \frac{\alpha_0}{2} + \sum_{n=1}^{\infty}(\alpha_n\cos nx_0 + \beta_n\sin nx_0)\\
&= \frac{a_0}{2} + \sum_{n=1}^{\infty}(a_n\cos nx_0 + b_n\sin nx_0) - \frac{\Delta}{2\pi}(x_0+\pi) + \frac{\Delta}{2}\\
&\quad + \sum_{n=1}^{\infty}\left\{-\frac{\Delta}{\pi n}\sin(nx_0)\cos(nx_0) + \frac{\Delta}{\pi n}[\cos(nx_0)-(-1)^n]\sin(nx_0)\right\}\\
&= \frac{a_0}{2} + \sum_{n=1}^{\infty}(a_n\cos nx_0 + b_n\sin nx_0) - \frac{\Delta}{2\pi}x_0 - \sum_{n=1}^{\infty}\frac{\Delta}{\pi n}(-1)^n\sin(nx_0)
\end{aligned}$$

However from the standard expansion

$$\sum_{n=1}^{\infty}\frac{(-1)^n x^n}{n} = -\ln(1+x) \qquad |x|\leq 1 \qquad x\neq -1$$

we deduce that

$$\begin{aligned}
\sum_{n=1}^{\infty}\frac{(-1)^n}{n}\sin(nx_0) &= \frac{1}{2i}\sum_{n=1}^{\infty}\frac{(-1)^n}{n}[e^{inx_0} - e^{-inx_0}]\\
&= -\frac{1}{2i}[\ln(1+e^{ix_0}) - \ln(1+e^{-ix_0})] = -\frac{1}{2i}\ln\left(\frac{1+e^{ix_0}}{1+e^{-ix_0}}\right)\\
&\equiv -\frac{1}{2i}\ln[\rho e^{i\phi}] = -\frac{1}{2i}\{\ln 1 + i\tan^{-1}[\tan(x_0)]\} = -\frac{x_0}{2}
\end{aligned}$$

so that

$$\tfrac{1}{2}[f(x_0^+)+f(x_0^-)]=\frac{a_0}{2}+\sum_{n=1}^{\infty}(a_n \cos nx+b_n \sin nx)$$

as claimed in the theorem. Q.E.D.

Notice that if a function is not periodic with period 2π, then the Fourier series (4.30) will give a faithful representation of it only in the region $x\in[-\pi, \pi]$. If we use the series outside this basic interval, the series simply reproduces a periodic extension of $f(x)$ since $\sin nx$ and $\cos nx$ have period 2π (cf. Fig. 4.3).

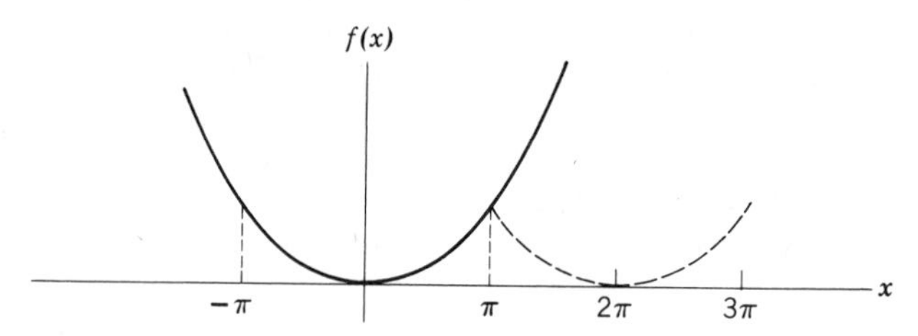

Figure 4.3

Note that if $f(x)$ is odd on $x\in[-\pi, \pi]$, then the series reduces to one involving only sines, and if even on $x\in[-\pi, \pi]$, then to one involving only cosines.

Example 4.1

$$f(x)=\begin{cases} x & 0\leq x\leq \pi \\ -x & -\pi\leq x\leq 0 \end{cases}$$

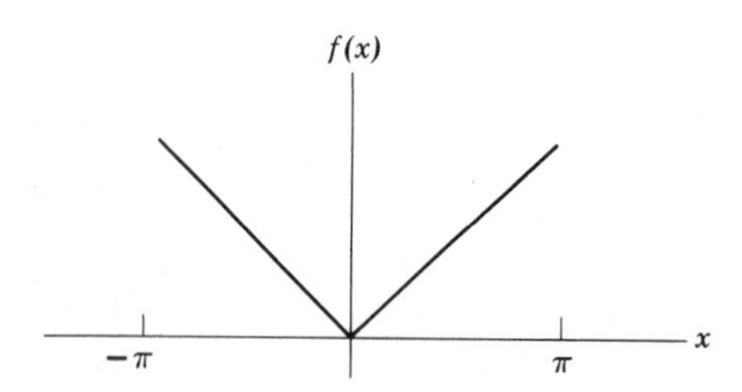

Figure 4.4

Since $f(x)$ is even, we have

$$a_0=\frac{1}{\pi}\int_{-\pi}^{\pi} f(x)\,dx=\frac{2}{\pi}\int_0^{\pi} f(x)\,dx=\frac{2}{\pi}\int_0^{\pi} x\,dx=\frac{x^2}{\pi}\bigg|_0^{\pi}=\pi$$

$$a_n=\frac{1}{\pi}\int_{-\pi}^{\pi} f(x)\cos nx\,dx=\frac{2}{\pi}\int_0^{\pi} x\cos nx\,dx=\frac{2}{\pi}\left[\frac{1}{n^2}\cos nx+\frac{1}{n}x\sin nx\right]\bigg|_0^{\pi}$$

$$=\frac{2[(-1)^n-1]}{\pi n^2}=\begin{cases} 0 & n \text{ even} \\ -\dfrac{4}{\pi n^2} & n \text{ odd}\end{cases}$$

$$b_n=\frac{1}{\pi}\int_{-\pi}^{\pi} f(x)\sin nx\,dx=0$$

so that

$$f(x)=\frac{\pi}{2}-\frac{4}{\pi}\sum_{n=0}^{\infty}\frac{\cos[(2n+1)x]}{(2n+1)^2}$$

This series is *uniformly* convergent since

$$\left|\sum_{n=0}^{\infty}\frac{\cos[(2n+1)x]}{(2n+1)^2}\right|\le\sum_{n=0}^{\infty}\frac{1}{(2n+1)^2}<\infty$$

Example 4.2

$$f(x)=\begin{cases}-1 & -\pi<x<0 \\ +1 & 0<x<\pi\end{cases}$$

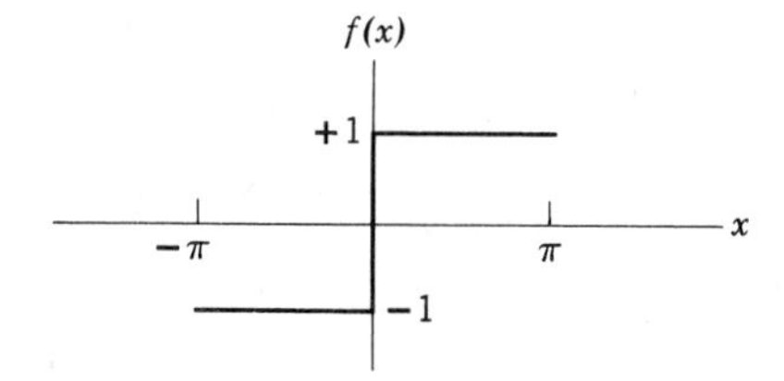

Figure 4.5

Since this is an odd function we have

$$a_0=0=a_n$$

$$b_n=\frac{1}{\pi}\int_{-\pi}^{\pi} f(x)\sin nx\,dx=\frac{2}{\pi}\int_0^{\pi}\sin nx\,dx=-\frac{2}{\pi}\frac{\cos nx}{n}\bigg|_0^{\pi}=\frac{-2}{\pi n}[(-1)^n-1]$$

so that

$$f(x)=\frac{4}{\pi}\sum_{n=0}^{\infty}\frac{\sin[(2n+1)x]}{(2n+1)}$$

This is a nonuniformly convergent series.

---

b. Legendre Polynomials

From the Weierstrass approximation theorem we know that an orthonormal set of polynomials on the interval $[-1,+1]$ forms a complete set. That is, using 1, x, $x^2,\ldots,x^n,\ldots$, we may construct an orthonormal set of functions $\{\varphi_n(x)\}$ on $-1,+1$. Some of these were given in Problem 1.14. The *Legendre polynomials*, which we will define shortly, are simply proportional to these orthonormal $\{\varphi_n(x)\}$. The Legendre polynomials are extremely important in physical applications, and we will treat them extensively later in complex variable theory (cf. Sections 7.13b and A7.4). Here we simply exhibit some of their more elementary properties for real values of $x\in[-1,1]$.

There are several different ways of defining the Legendre polynomials. We choose the following:

$$P_0(x)=1 \qquad P_n(x)=\frac{1}{2^n n!}\frac{d^n(x^2-1)^n}{dx^n} \qquad n=1,2,3,\ldots \tag{4.32}$$

We must prove that these are polynomials of degree n and that they are orthogonal on $[-1,+1]$. From the binomial expansion, we have

$$(-1+x^2)^n=\sum_{r=0}^{n}\frac{n!\,(-1)^{n-r}}{(n-r)!\,r!}x^{2r}$$

so that

$$P_n(x)\equiv\frac{1}{2^n n!}\frac{d^n(-1+x^2)^n}{dx^n}$$

$$=\frac{1}{2^n}\sum_{r=0}^{n}\frac{(-1)^{n-r}2r(2r-1)(2n-2)\cdots(2r-n+1)x^{2r-n}}{(n-r)!\,r!}$$

$$\times\frac{(2r-n)(2r-n-1)\cdots3\cdot2\cdot1}{(2r-n)(2r-n-1)\cdots3\cdot2\cdot1}$$

$$=\frac{1}{2^n}\sum_{r=0}^{n}\frac{(-1)^{n-r}(2r)!\,x^{2r-n}}{(n-r)!\,r!\,(2r-n)!}=\sum_{r=0}^{n}\frac{(-1)^{n-r}1\cdot3\cdot5\cdots(2r-1)x^{2r-n}}{(n-r)!\,(2r-n)!\,2^{n-r}} \tag{4.33}$$

Since $(-m)!\equiv\infty$, we have for the first term,

$$\frac{(-1)^{(n-1)/2}1\cdot3\cdot5\cdots n}{2\cdot4\cdot6\cdots(n-1)}x$$

if n is odd, since the leading term arises from $2r=n+1$, and

$$\frac{(-1)^{n/2}1\cdot 3\cdot 5\cdots(n-1)}{2\cdot 4\cdot 6\cdots n}$$

if n is even, since the leading term arises from $2r=n$. It also follows directly from Eq. 4.32 by writing out the leading term that

$$P_n(1)=1 \qquad \forall n\geq 0 \tag{4.34}$$

Obviously these are polynomials of degree n and, in fact, are even or odd depending on whether n is even or odd so that

$$P_n(x)=(-1)^n P_n(-x) \tag{4.35}$$

The first few Legendre polynomials are

$$P_0(x)=1 \qquad P_1(x)=x \qquad P_2(x)=\tfrac{3}{2}x^2-\tfrac{1}{2}$$
$$P_3(x)=\tfrac{5}{2}x^3-\tfrac{3}{2}x \qquad P_4(x)=\tfrac{35}{8}x^4-\tfrac{15}{4}x^2+\tfrac{3}{8}$$

We must now show that the $\{P_n(x)\}$ are orthogonal. If $m<n$ then integration by parts yields

$$\int_{-1}^{1} P_n(x)x^m\,dx\equiv\frac{1}{2^n n!}\int_{-1}^{1} dx\,\frac{d^n[(-1+x^2)^n]}{dx^n}\,x^m=0 \qquad m<n$$

which implies

$$\int_{-1}^{1} P_m(x)P_n(x)\;dx=0 \qquad m\neq n$$

Finally it will be shown in Section A4.4 that

$$\int_{-1}^{1} P_m(x)P_n(x)\,dx=\frac{2}{2n+1}\,\delta_{mn} \tag{4.36}$$

so that the required orthogonal set is given as

$$\varphi_n(x)=\sqrt{\frac{2n+1}{2}}\,P_n(x) \qquad \langle\varphi_n\,|\,\varphi_m\rangle=\delta_{mn}$$

A very important application of these functions is in the so-called *Coulomb expansion* (cf. Section A4.3),

$$\frac{1}{\sqrt{1-2\mu x+\mu^2}}=\sum_{n=0}^{\infty}\mu^n P_n(x) \qquad |\mu|<1 \qquad -1\leq x\leq +1 \tag{4.37}$$

Geometrically this allows us to find the distance between a point at r_0 on the polar axis and any other point (r,θ) not on the sphere of radius r_0.

i. $r<r_o$ $\quad \dfrac{1}{R}\equiv\dfrac{1}{r_0\sqrt{1+(r/r_0)^2-2(r/r_0)\cos\theta}}=\dfrac{1}{r_0}\displaystyle\sum_{n=0}^{\infty}\left(\frac{r}{r_0}\right)^n P_n(\cos\theta)$

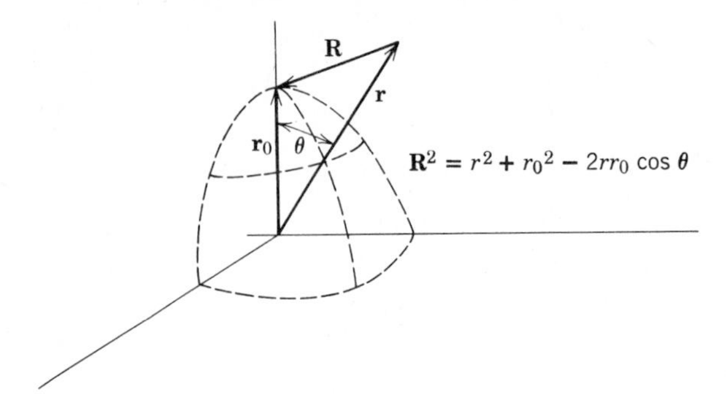

Figure 4.6

ii. $r > r_0$ $$\frac{1}{R} \equiv \frac{1}{r\sqrt{1+(r_0/r)^2 - 2(r_0/r)\cos\theta}} = \frac{1}{r}\sum_{n=0}^{\infty}\left(\frac{r_0}{r}\right)^n P_n(\cos\theta)$$

These are *not* valid expansions for $r = r_0$.

Finally we list two simple properties of these functions (cf. Section A4.4).

i. Recursion formula:

$$(n+1)P_{n+1}(x) - (2n+1)xP_n(x) + nP_{n-1}(x) = 0 \qquad n \geq 1 \tag{4.38}$$

ii. Differential equation:

$$(x^2-1)\frac{d^2}{dx^2}P_n(x) + 2x\frac{d}{dx}P_n(x) - n(n+1)P_n(x) = 0 \tag{4.39}$$

In Section A4.5 it is shown that the *associated Legendre polynomials* defined as

$$P_n^m(x) \equiv (1-x^2)^{m/2}\frac{d^m P_n(x)}{dx^m} \qquad 0 \leq m \leq n$$
$$P_n^0(x) \equiv P_n(x) \tag{4.40}$$

form an infinite orthogonal set of functions over $[-1, 1]$; that is,

$$\int_{-1}^{1} P_n^m(x)P_{n'}^m(x)\,dx = \frac{2}{(2n+1)}\frac{(n+m)!}{(n-m)!}\delta_{nn'} \tag{4.41}$$

Clearly if $m > n$ then $P_n^m(x) \equiv 0$ and, because of $(n-m)!$ in the denominator, Eq. 4.41 remains valid. We now show that for any fixed m these $\{P_n^m(x)\}$ also form a complete set of functions, at least for bounded piecewise continuous functions on $[-1, 1]$.

THEOREM 4.6

Any given orthonormal set $\{\varphi_n(x)\}$ is complete if and only if $\langle \varphi_n \mid f \rangle = 0$ for all n implies that $\|f\| = 0$.

PROOF

i. *Sufficient.* If the set is complete, then

$$\|f\|^2 = \sum_{n=0}^{\infty} |c_n|^2$$

so that if $c_n \equiv \langle \varphi_n \mid f \rangle = 0$, $\forall n$, then $\|f\| = 0$.

ii. *Necessary.* Suppose that $c_n \equiv \langle \varphi_n \mid f \rangle = 0$ $\forall n$ implies that $\|f\| = 0$, but that the set is *not* complete. Since the set is assumed not complete, then Bessel's inequality states that

$$\|f\|^2 > \sum_{n=0}^{\infty} |c_n|^2 = 0$$

so that $\|f\| > 0$, which is a contradiction. Q.E.D.

Therefore, if we can show that

$$\int_{-1}^{1} P_n^m(x) f(x)\, dx = 0 \qquad \forall n \geq 0 \tag{4.42}$$

implies that $f(x) = 0$, almost everywhere, then we will have proven the completeness of these $\{P_n^m(x)\}$ for all positive integer m, $0 \leq m \leq n$. Also in Section A4.5 the recursion relation

$$x(2n+1)P_n^m(x) = (n-m+1)P_{n+1}^m(x) + (n+m)P_{n-1}^m(x) \qquad n \geq 1 \tag{4.43}$$

is derived. This relation simply becomes an empty identity (i.e., $0=0$) when $m > n$. If $n = 0$, we replace (4.43) by $xP_0^0(x) = P_1(x)$. If we multiply (4.43) by $f(x)$, integrate, and use assumption (4.42), we obtain

$$\int_{-1}^{1} P_n^m(x) x f(x)\, dx = 0 \qquad \forall n \geq 0 \tag{4.44}$$

If we repeat this process k times we deduce that

$$\int_{-1}^{1} P_n^m(x) x^k f(x)\, dx = 0 \qquad \forall n,\ k = 0, 1, 2, \ldots \tag{4.45}$$

Now we can multiply (4.45) for $k = 0, 1, 2, \ldots, l$ by the appropriate coefficients given in Eq. 4.33 and add to obtain

$$\int_{-1}^{1} P_n^m(x) P_l(x) f(x)\, dx = 0 \qquad \forall n,\ l = 0, 1, \ldots \tag{4.46}$$

Since the $\{P_l(x)\}$ form a complete set, we may use Th. 4.6 to conclude that

$$\|P_n^m f\| = 0$$

or that

$$P_n^m(x) f(x) = 0$$

almost everywhere. However we see from the definition of the $P_n^m(x)$, Eq. 4.40, that if $P_n^m(x)$ were to vanish identically on some continuum then

$$\frac{d^m}{dx^m} P_n(x)=0$$

would also hold there. Since $P_n(x)$ is a polynomial of degree n in x, this would be a polynomial of degree $(n-m)\geq 0$ in x. If this polynomial of degree $(n-m)$ were to vanish identically on some continuum, then all of its coefficients would have to vanish, which would imply that $P_n(x)$ were a polynomial of degree less than n, a contradiction. We also see from this argument that $P_n^m(x)$ can vanish only on a *finite* set of discrete points. Therefore if $P_n^m(x)f(x)=0$ almost everywhere then $f(x)=0$ almost everywhere so that the $\{P_n^m(x)\}$ form a complete set on $[-1, 1]$.

If we combine this result with Th. 4.5, we have the important result that the *spherical harmonics* defined as

$$Y_l^m(\theta, \varphi)\equiv(-1)^{(1/2)(m+|m|)}\sqrt{\frac{(2l+1)}{4\pi}\frac{(l-|m|)!}{(l+|m|)!}}\, e^{im\varphi}P_l^{|m|}(\cos\theta) \tag{4.47}$$

$$m=-l, -l+1, \ldots, l-1, l \qquad l\geq 0$$

are a complete orthonormal set over the unit sphere, $0\leq\varphi\leq 2\pi$, $0\leq\theta\leq\pi$

$$\langle Y_{l'}^{m'} \mid Y_l^m\rangle\equiv\int_0^{2\pi} d\varphi\int_{-1}^{1} d(\cos\theta)\, Y_{l'}^{m'*}(\theta, \varphi)\, Y_l^m(\theta, \varphi)=\delta_{mm'}\,\delta_{ll'} \tag{4.48}$$

4.4 RIEMANN-LEBESGUE LEMMA

The following two theorems are useful in the theory of Fourier integrals as well as in the formal theory of quantum-mechanical scattering.

THEOREM 4.7 (Riemann-Lebesgue lemma)

If $f(x)$ is absolutely integrable on $(-\infty, \infty)$, that is if

$$\int_{-\infty}^{\infty} |f(x)|\, dx<\infty \tag{4.49}$$

then

$$\lim_{\lambda\to\infty}\int_{-\infty}^{\infty} f(x)e^{i\lambda x}\, dx=0 \tag{4.50}$$

where λ is real.

PROOF

We can choose an R large enough so that

$$\int_R^{\infty} |f(x)|\, dx<\varepsilon \qquad \int_{-\infty}^{-R} |f(x)|\, dx<\varepsilon \tag{4.51}$$

We will restrict ourselves in this proof to functions with a finite number of integrable discontinuities on $[-R, R]$, although the theorem is true as long as $f(x) \in \mathscr{L}(-\infty, \infty)$. First let us assume that $f(x)$ is bounded on $[-R, R]$. Then we can define a continuous once-differentiable function $g(x)$ such that

$$\int_{-R}^{R} |f(x) - g(x)|\, dx < \varepsilon$$

which implies

$$\left| \int_{-R}^{R} [f(x) - g(x)] e^{i\lambda x}\, dx \right| < \varepsilon \tag{4.52}$$

where ε can be made arbitrarily small. Then it follows that

$$\int_{-R}^{R} g(x) e^{i\lambda x}\, dx = \frac{g(R) e^{i\lambda R}}{i\lambda} - \frac{g(-R) e^{-i\lambda R}}{i\lambda} - \frac{1}{i\lambda} \int_{-R}^{R} g'(x) e^{i\lambda x}\, dx$$

so that for λ large enough

$$\left| \int_{-R}^{R} g(x) e^{i\lambda x}\, dx \right| < \varepsilon \tag{4.53}$$

From Eqs. 4.51–4.53 we have

$$\left| \int_{-\infty}^{\infty} f(x) e^{i\lambda x}\, dx \right| < 4\varepsilon$$

where ε may be made as small as we please for λ sufficiently large.

Now suppose that $f(x)$ becomes unbounded at a finite number of points $\{x_j\}$, $j = 1, 2, \ldots, m$, on $[-R, R]$. We may exclude these points by small intervals γ_j, $j = 1, 2, \ldots, m$, and apply the argument just given to the remaining portions of $[-R, R]$. By definition of an improper integral that is absolutely convergent we have

$$\lim_{\delta \to 0} \left\{ \int_{x_j - \gamma}^{x_j - \delta} |f(x)|\, dx + \int_{x_j + \delta}^{x_j + \gamma} |f(x)|\, dx \right\} < \infty \qquad 0 < \delta < \gamma$$

when

$$|f(x_j)| \to \infty$$

That is, if $\lim_{x \to x_j} |f(x)|$ becomes undefined, the improper integral, when it exists in the Riemann sense, is defined to be

$$\int_{x_j}^{R} |f(x)|\, dx = \lim_{\delta \to 0} \int_{x_j + \delta}^{R} |f(x)|\, dx$$

so that, for such an improper integral,

$$\lim_{\delta \to 0} \int_{x_j}^{x_j + \delta} |f(x)|\, dx = 0$$

As a simple example of this consider

$$\lim_{\delta \to 0} \int_{-\delta}^{\delta} \frac{dx}{|x|^{\alpha}} = \lim_{\delta \to 0} \frac{2\delta^{1-\alpha}}{(1-\alpha)} = 0 \qquad 0<\alpha<1$$

Therefore we may take

$$\sum_{j=1}^{m} \int_{\gamma_j} |f(x)|\, dx < \varepsilon'$$

where ε' can again be made as small as we please. Q.E.D.

If the integration range is finite and, if we restrict $f(x)$ somewhat, we can obtain a sharper limit.

THEOREM 4.8

If $f(x)$ has limited total fluctuation on $[a, b]$, then

$$\lim_{\lambda \to \infty} \int_a^b f(x) e^{i\lambda x}\, dx = O\left(\frac{1}{\lambda}\right) \tag{4.54}$$

Here $O(1/\lambda)$ means that for large λ the integral vanishes *at least* as rapidly as $1/\lambda$.

PROOF

We begin by defining the *fluctuation* of a function $f(x)$ on a range $[a, b]$. If we subdivide the interval $[a, b]$ with a set of points $\{x_j\}$, $j=0, 1, \ldots, n+1$, such that $x_0=a$, $x_{n+1}=b$, then we define the fluctuation of $f(x)$ for this particular subdivision as

$$|f(a)-f(x_1)|+|f(x_1)-f(x_2)|+\cdots+|f(x_n)-f(b)| = \sum_{j=1}^{n+1} |f(x_{j-1})-f(x_j)| \equiv \sum_{j=1}^{n+1} |\Delta f(x_j)|$$

If for all possible subdivisions of $[a, b]$ there exists *some* upper bound $F(a, b)$ on the fluctuation, then $f(x)$ is said to be of *limited total fluctuation.* (Many modern texts use the term *bounded variation* rather than *limited total fluctuation.*) For example, if $f(x)$ is monotonic on $[a, x]$, then we have very simply

$$F(a, x) = |f(x) - f(a)|$$

An example of a bounded function not having limited total fluctuation would be $f(x) = \sin(1/x)$, $x \in (0, 1)$. We see that $F(a, x)$ is a nonnegative, nondecreasing function of $x \in [a, b]$. We may always choose $F(a, x)$ to be positive and may adjust its minimum (positive) value arbitrarily. We assume that $f(x)$ is not identically zero on $[a, b]$ since then the theorem is already trivially true. Since $f(x)$ must be bounded for $F(a, x)$ to exist, this allows us to define two positive nondecreasing bounded functions as

$$g(x)=\tfrac{1}{2}[F(a, x)+f(x)]$$
$$h(x)=\tfrac{1}{2}[F(a, x)-f(x)]$$

such that

$$f(x)=g(x)-h(x)$$

By the mean value theorem for integrals (cf. Appendix I, Th. 1.5b), if $k(x)$ is an integrable positive nondecreasing function on $[a, b]$ and $f(x)$ is integrable, then

$$\int_a^b f(x)k(x)\,dx=k(b)\int_\xi^b f(x)\,dx \qquad a\leq\xi\leq b$$

Therefore we have

$$\left|\int_a^b g(x)e^{i\lambda x}\,dx\right|=\left|g(b)\int_\xi^b e^{i\lambda x}\,dx\right|\leq\frac{2g(b)}{\lambda}$$
$$\left|\int_a^b h(x)e^{i\lambda x}\,dx\right|\leq\frac{2h(b)}{\lambda}$$

so that finally

$$\left|\int_a^b f(x)e^{i\lambda x}\,dx\right|\leq\left|\int_a^b g(x)e^{i\lambda x}\,dx\right|+\left|\int_a^b h(x)e^{i\lambda x}\,dx\right|$$
$$\leq\frac{2[g(b)+h(b)]}{\lambda}$$

Q.E.D.

4.5 FOURIER INTEGRALS

a. A Heuristic Approach

We begin with a heuristic argument that will allow us to extend the Fourier series, Eq. 4.30, to an integral when $x\in(-\infty, \infty)$. It is a simple matter to see from (4.30) that we can expand a function on $x\in[-l, l]$ as

$$f(x)=\frac{a_0}{2}+\sum_{n=1}^{\infty}\left[a_n\cos\left(\frac{n\pi x}{l}\right)+b_n\sin\left(\frac{n\pi x}{l}\right)\right] \tag{4.55}$$

where

$$a_n=\frac{1}{l}\int_{-l}^{l} f(t)\cos\left(\frac{n\pi t}{l}\right)dt \qquad b_n=\frac{1}{l}\int_{-l}^{l} f(t)\sin\left(\frac{n\pi t}{l}\right)dt \tag{4.56}$$

If we make use of the identities

$$\cos x=\frac{e^{ix}+e^{-ix}}{2} \qquad \sin x=\frac{e^{ix}-e^{-ix}}{2i}$$

then we can rewrite Eqs. 4.55 and 4.56 as

$$f(x) = \sum_{n=-\infty}^{\infty} \alpha_n e^{in\pi x/l} \tag{4.57}$$

$$\alpha_n = \frac{1}{2l} \int_{-l}^{l} f(t) e^{-in\pi t/l}\, dt \tag{4.58}$$

If we now define

$$k_n \equiv \frac{n\pi}{l} \qquad \Delta k \equiv \frac{\pi}{l} = k_{n+1} - k_n$$

then Eq. 4.57 becomes

$$f(x) = \frac{1}{2\pi} \sum_{n=-\infty}^{\infty} \Delta k \int_{-l}^{l} f(t) e^{ik_n(x-t)}\, dt \tag{4.59}$$

If we pass to the limit and if all the integrals converge (a question we will study in the next section), we obtain

$$f(x) = \frac{1}{2\pi} \int_{-\infty}^{\infty} dk \int_{-\infty}^{\infty} dt\, f(t) e^{ik(x-t)} \tag{4.60}$$

If the *Fourier transform* of $f(x)$ is defined as

$$f(k) = \frac{1}{\sqrt{2\pi}} \int_{-\infty}^{\infty} e^{ikx} f(x)\, dx \tag{4.61}$$

then the reciprocal relation is

$$f(x) = \frac{1}{\sqrt{2\pi}} \int_{-\infty}^{\infty} e^{-ikx} f(k)\, dk \tag{4.62}$$

Of course the actual functional forms of $f(x)$ and $f(k)$ are in general quite different and we should use different symbols for each [e.g., $f(x)$ and $F(k)$]. However the notation of Eqs. 4.61 and 4.62 is quite common and we will continue to use it because it lends itself well to the interpretation of $f(x)$ and $f(k)$ being two different representations of the same abstract function (vector) f (cf. Section 2.5). No confusion need arise.

Such Fourier transforms, when they exist, are appropriate for nonperiodic functions defined on $(-\infty, \infty)$.

b. Fourier Integral Theorem

We now give a rigorous proof of the *Fourier integral theorem*, Eq. 4.60, under the assumption that $f(x)$ is of limited total fluctuation and that $\int_{-\infty}^{\infty} |f(x)|\, dx$ exists.

THEOREM 4.9 (Fourier integral theorem)

If $f(x)$ is absolutely integrable on $(-\infty, \infty)$

$$\int_{-\infty}^{\infty} |f(x)|\, dx < \infty$$

and of limited total fluctuation, then

$$\tfrac{1}{2}[f(x+0)+f(x-0)] = \frac{1}{2\pi}\int_{-\infty}^{\infty} dk \int_{-\infty}^{\infty} dt\, f(t)e^{ik(x-t)} \tag{4.63}$$

PROOF

We begin by observing that since

$$\int_{-\infty}^{\infty} |f(t)|\, dt < \infty$$

then the integral

$$\int_{-\infty}^{\infty} f(t)e^{ik(x-t)}\, dt \tag{4.64}$$

is uniformly convergent with respect to k over any finite interval in k. That is,

$$\left|\int_{R}^{\infty} f(t)e^{ik(x-t)}\, dt\right|$$

can be made as small as we please for R sufficiently large *independent* of the value of k. This follows from the absolute integrability of $f(x)$ since this requires that

$$\int_{R}^{\infty} |f(t)|\, dt < \varepsilon$$

where ε can be made arbitrarily small for sufficiently large R and

$$\left|\int_{R}^{\infty} f(t)e^{ik(x-t)}\, dt\right| \leq \left|\int_{R}^{\infty} f(t)\, dt\right| \leq \int_{R}^{\infty} |f(t)|\, dt < \varepsilon$$

We need this result shortly.

Now consider the integral

$$I(\lambda) \equiv \int_{-\lambda}^{\lambda} dk \int_{-\infty}^{\infty} f(t)e^{ik(x-t)}\, dt \qquad x \in [-R, R] \tag{4.65}$$

We want to interchange the orders of the integrations in Eq. 4.65. Let us write

$$I(\lambda) = \int_{-\lambda}^{\lambda} dk h(k)$$

where

$$h(k) = \int_{-\infty}^{\infty} f(t) e^{ik(x-t)} \, dt$$

$$|h(k)| \leq \int_{-\infty}^{\infty} |f(t)| \, dt$$

We can see that $h(k)$ is continuous as follows.

$$\begin{aligned}
|h(k) - h(k')| &= \left| \int_{-\infty}^{\infty} f(t)[e^{ik(x-t)} - e^{ik'(x-t)}] \, dt \right| \\
&\leq \int_{-\infty}^{-R} |f(t)||e^{ik(x-t)} - e^{ik'(x-t)}| \, dt + \int_{R}^{\infty} |f(t)||e^{ik(x-t)} - e^{ik'(x-t)}| \, dt \\
&\quad + \int_{-R}^{R} |f(t)||e^{ik(x-t)} - e^{ik'(x-t)}| \, dt \\
&\leq 2\left[\int_{-\infty}^{-R} |f(t)| \, dt + \int_{R}^{\infty} |f(t)| \, dt \right] + \int_{-R}^{R} |f(t)||x-t||k-k'| \, dt
\end{aligned}$$

We have used the mean value theorem as

$$e^{ik(x-t)} - e^{ik'(x-t)} = i(k-k')(x-t)e^{i\bar{k}(x-t)} \qquad k \leq \bar{k} \leq k'$$

to obtain the last integral above. Now for any ε we choose, no matter how small, we can take R large enough so that

$$2\int_{-\infty}^{-R} |f(t)| \, dt < \frac{\varepsilon}{3}$$

$$2\int_{R}^{\infty} |f(t)| \, dt < \frac{\varepsilon}{3}$$

Furthermore since

$$\int_{-\infty}^{\infty} |f(x)| \, dx < \infty$$

and since $|x-t| < 2R$, once we have chosen R, we can choose a δ such that

$$|k-k'| < \delta < \frac{\varepsilon}{3} \Big/ \left(2R \int_{-\infty}^{\infty} |f(x)| \, dx \right)$$

which implies that

$$|k-k'| \int_{-R}^{R} |x-t||f(t)| \, dt < \delta 2R \int_{-\infty}^{\infty} |f(t)| \, dt < \frac{\varepsilon}{3}$$

so that finally

$$|h(k) - h(k')| < \varepsilon$$

whenever $|k-k'| < \delta$. Since this states that $h(k)$ is continuous, then $I(\lambda)$ is a continuous, differentiable function of λ having a continuous first derivative such that

$$\frac{d}{d\lambda} I(\lambda) = h(\lambda) - h(-\lambda) = \int_{-\infty}^{\infty} f(t)[e^{i\lambda(x-t)} - e^{-i\lambda(x-t)}]\, dt$$

$$= 2i \int_{-\infty}^{\infty} f(t) \sin[\lambda(x-t)]\, dt$$

If we now consider the integral

$$\bar{I}(\lambda) = \int_{-\infty}^{\infty} dt f(t) \int_{-\lambda}^{\lambda} dk e^{ik(x-t)}$$

then we may differentiate it as (cf. Appendix I, Th. I.15)

$$\frac{d}{d\lambda} \bar{I}(\lambda) = \int_{-\infty}^{\infty} dt\, f(t)[e^{i\lambda(x-t)} - e^{-i\lambda(x-t)}]$$

$$= 2i \int_{-\infty}^{\infty} dt\, f(t) \sin[\lambda(x-t)] = \frac{d}{d\lambda} I(\lambda)$$

since the resultant integral is uniformly convergent w.r.t. λ as follows from the uniform convergence of (4.64). Finally since

$$\bar{I}(\lambda = 0) = I(\lambda = 0) = 0$$

we see that

$$\bar{I}(\lambda) \equiv I(\lambda) \qquad \forall\lambda$$

so that we may write Eq. 4.65 as

$$I(\lambda) = \int_{-\infty}^{\infty} f(t)\, dt \int_{-\lambda}^{\lambda} e^{ik(x-t)}\, dk = 2 \int_{-\infty}^{\infty} f(t) \frac{\sin[\lambda(x-t)]}{(x-t)}\, dt \tag{4.66}$$

However by Th. 4.7 we have

$$\lim_{\lambda\to\infty} I(\lambda) = 2 \lim_{\lambda\to\infty} \int_{x-\delta}^{x+\delta} f(t) \frac{\sin[\lambda(x-t)]}{(x-t)}\, dt$$

$$= 2 \lim_{\lambda\to\infty} \int_{0}^{\delta} [f(x+y) + f(x-y)] \frac{\sin \lambda y}{y}\, dy \tag{4.67}$$

where δ is some fixed, small, positive quantity. Also by use of the standard definite integral

$$\int_0^{\infty} \frac{\sin x}{x}\, dx = \frac{\pi}{2}$$

we have

$$\lim_{\lambda\to\infty} \int_0^{\delta} \frac{\sin \lambda y}{y}\, dy = \lim_{\lambda\to\infty} \int_0^{\lambda\delta} \frac{\sin x}{x}\, dx = \frac{\pi}{2} \tag{4.68}$$

If we now combine Eqs. 4.65, 4.67, and 4.68, we obtain

$$\lim_{\lambda\to\infty}\left\{\frac{1}{2\pi}\int_{-\lambda}^{\lambda} dk\int_{-\infty}^{\infty} f(t)e^{ik(x-t)}\,dt-\tfrac{1}{2}[f(x+0)+f(x-0)]\right.$$

$$=\lim_{\lambda\to\infty}\frac{1}{\pi}\int_0^{\delta}[f(x+y)+f(x-y)-f(x+0)-f(x-0)]\frac{\sin\lambda y}{y}\,dy \quad \textbf{(4.69)}$$

However the function

$$g(y)\equiv f(x+y)+f(x-y)-f(x+0)-f(x-0)$$

is of limited total fluctuation so that, as we have shown in the proof of Th. 4.8, it can be written as the difference of two positive nondecreasing bounded functions $h_1(y)$ and $h_2(y)$

$$g(y)\equiv h_1(y)-h_2(y)$$

In this case we may choose $h_1(0)=\varepsilon'=h_2(0)$, where ε' is an arbitrarily small positive quantity, so that for sufficiently small y each can be made as small as we please. That is, we can find a δ' such that

$$\varepsilon'<|h_j(y)|<\varepsilon \qquad |y|\le\delta'<\delta \qquad j=1,2 \quad \textbf{(4.70)}$$

so that for $j=1, 2$

$$\left|\int_0^{\delta} h_j(y)\frac{\sin\lambda y}{y}\,dy\right|\le\left|\int_0^{\delta'} h_j(y)\frac{\sin\lambda y}{y}\,dy\right|+\left|\int_{\delta'}^{\delta} h_j(y)\frac{\sin\lambda y}{y}\,dy\right|$$

$$=\left|h_j(\delta')\int_{\xi}^{\delta'}\frac{\sin\lambda y}{y}\,dy\right|+\left|\int_{\delta'}^{\delta} h_j(y)\frac{\sin\lambda y}{y}\,dy\right| \qquad (0\le\xi\le\delta')$$

$$=\left|h_j(\delta')\int_{\lambda\xi}^{\lambda\delta'}\frac{\sin x}{x}\,dx\right|+\left|\int_{\delta'}^{\delta} h_j\frac{\sin\lambda y}{y}\,dy\right| \quad \textbf{(4.71)}$$

If $\xi>0$, we may apply Th. 4.7 directly to both integrals in Eq. 4.71. If $\xi=0$, then the first integral in Eq. 4.71 tends to $\pi/2$ as $\lambda\to\infty$ so that the first term is bounded by $(\pi/2)\varepsilon$. Here we have again used the mean value theorem for integrals and Eqs. 4.68 and 4.70. Therefore in any case we can make (4.69) as small as we please. Q.E.D.

It is clear that, when $f(x)$ is also continuous, Eq. 4.63 reduces to either of Eqs. 4.60 or 4.62. Furthermore, in this case, $f(k)$ as defined in (4.61) is not only bounded and uniformly convergent but also continuous. By the same type argument as given for the interchange of the order of integration in Eq. 4.65 we can justify the following interchange of integrations.

$$\int_{-\infty}^{\infty} f(k)g^*(k)\,dk=\int_{-\infty}^{\infty} f(k)\left(\frac{1}{\sqrt{2\pi}}\int_{-\infty}^{\infty} g^*(x)e^{-ikx}\,dx\right)dk$$

$$=\int_{-\infty}^{\infty} g^*(x)\left(\frac{1}{\sqrt{2\pi}}\int_{-\infty}^{\infty} f(k)e^{-ikx}\,dk\right)dx=\int_{-\infty}^{\infty} f(x)g^*(x)\,dx \quad \textbf{(4.72)}$$

This relation is known as *Parseval's formula.* In particular if $g(x)=f(x)$ we obtain the completeness relation for Fourier integrals

$$\int_{-\infty}^{\infty} |f(k)|^2\, dk = \int_{-\infty}^{\infty} |f(x)|^2\, dx \tag{4.73}$$

The Fourier integral theorem can be proven under much weaker assumptions than we have made [cf. Titchmarsh where the only restriction is that $f(x) \in \mathscr{L}_2(-\infty, \infty)$]. In fact if we had been willing to assume Fubini's theorem (cf. Appendix II, Th. II.2), on the interchange of the order of integration, then we could have arrived at Eqs. 4.72 and 4.73 immediately.

A4.1 AN ALTERNATIVE PROOF OF THE WEIERSTRASS APPROXIMATION THEOREM IN ONE VARIABLE

We will show directly that the polynomial of degree $2n$ defined as

$$\mathscr{P}_n(x) \equiv \frac{\int_\alpha^\beta f(s)[1-(s-x)^2]^n\, ds}{\int_{-1}^{1} (1-s^2)^n\, ds} \qquad n=0, 1, 2, \ldots \tag{A4.1}$$

$$0<\alpha<a\le x\le b<\beta<1$$

will serve to approximate uniformly any function continuous on the *closed* interval $[a, b]$, where for convenience and without loss of generality we take $0<a<b<1$ for this proof. Since $f(x)$ is given as being continuous on the closed interval $[a, b]$, we can make a continuous extension of it to $x\in[\alpha, \beta]$.

If we let $s=x+y$, then we have

$$\mathscr{P}_n(x) = \frac{\int_{\alpha-x}^{\beta-x} f(x+y)(1-y^2)^n\, dy}{2\int_0^1 (1-y^2)^n\, dy} \tag{A4.2}$$

so that with $0<\delta<1$

$$\begin{aligned}
|\mathscr{P}_n(x)-f(x)| &\\
= \frac{1}{2\int_0^1 (1-y^2)^n\, dy} &\left\{\left| \int_{\alpha-x}^{-\delta} f(x+y)(1-y^2)^n\, dy + \int_{-\delta}^{\delta} f(x+y)(1-y^2)^n\, dy \right.\right.\\
&\left.\left. + \int_\delta^{\beta-x} f(x+y)(1-y^2)^n\, dy - f(x)\int_{-1}^{1}(1-y^2)^n\, dy \right|\right\}\\
= \frac{1}{2\int_0^1 (1-y^2)^n\, dy} &\left\{\left| \int_{\alpha-x}^{\delta} f(x+y)(1-y^2)^n\, dy + \int_\delta^{\beta-x} f(x+y)(1-y^2)^n\, dy \right.\right.\\
&\left.\left. -2f(x)\int_\delta^1 (1-y^2)^n\, dy + \int_{-\delta}^{\delta} [f(x+y)-f(x)](1-y^2)^n\, dy \right|\right\}
\end{aligned} \tag{A4.3}$$

If we now define the quantities

$$A_n \equiv \int_0^1 (1-y^2)^n \, dy \tag{A4.4}$$

$$B_n \equiv \int_\delta^1 (1-y^2)^n \, dy \tag{A4.5}$$

from which it follows that

$$A_n > \int_0^1 (1-y)^n \, dy = \frac{1}{(n+1)}$$

$$B_n < (1-\delta^2)^n (1-\delta) < (1-\delta^2)^n$$

then we have

$$\frac{B_n}{A_n} < (1-\delta^2)^n (n+1) \xrightarrow[n\to\infty]{} 0 \tag{A4.6}$$

Let M be the maximum value of $|f(x)|$ and use the (uniform) continuity of $f(x)$ as

$$|f(x+y)-f(x)| < \varepsilon \qquad |y| \leq \delta$$

in Eq. A4.3 to obtain

$$|\mathscr{P}_n(x)-f(x)| \leq \frac{2M\int_\delta^1 (1-y^2)^n \, dy}{\int_0^1 (1-y^2)^n \, dy} + \frac{\varepsilon\int_0^\delta (1-y^2)^n \, dy}{\int_0^1 (1-y^2)^n \, dy} \leq 2M\frac{B_n}{A_n} + \varepsilon \tag{A4.7}$$

This bound in Eq. A4.7 is a uniform one since it is independent of x. But from Eq. A4.6 we see that $\lim_{n\to\infty} (B_n/A_n) = 0$ so that for *any* $\varepsilon > 0$ we choose there exists a sufficiently large value of N such that

$$|\mathscr{P}_n(x)-f(x)| < \varepsilon \qquad n > N \tag{A4.8}$$

which states that $\mathscr{P}_n(x)$ uniformly approximates $f(x)$ arbitrarily well.

A4.2 TWO PROOFS OF THE WEIERSTRASS APPROXIMATION THEOREM IN TWO VARIABLES

We will first give a proof of the Weierstrass approximation theorem in two variables following the method used in establishing Th. 4.3 of Section 4.2. We assume that $f(x, y)$ is continuous in x and y in the region $x \in [0, 1]$, $y \in [0, 1]$. We cover this region with a square grid, each grid being of area $1/m \times 1/m$. Begin by holding y = constant and expand $\bar{f}(x) \equiv f(x, y = \text{const.})$ as in the proof of Th. 4.3, by defining

$$\bar{g}(f; x, y) \equiv \sum_{k=1}^{m} g_k(y)h(x_k; x) \tag{A4.9}$$

where $h(x_k, x)$ has been defined in Eq. 4.18, $g_k(y)$ is a continuous function of y given as

$$g_k(y) = \frac{f(x_{k+1}, y) - f(x_k, y)}{2(x_{k+1} - x_k)} \tag{A4.10}$$

Now define

$$g[g_k; y] \equiv \sum_{l=1}^{m} g_{kl}h(y_l; y) \tag{A4.11}$$

with

$$g_{kl} = \frac{g_k(y_{l+1}) - g_k(y_l)}{2(y_{l+1} - y_l)} \tag{A4.12}$$

Using the same type argument by which we arrived at Eq. 4.19, we see that we may take

$$|g_k(y) - g[g_k; y]| < \frac{1}{3}\frac{\varepsilon}{m} \tag{A4.13}$$

where we have used the continuity of $f(x, y)$ as

$$|f(x, y) - f(x_0, y_0)| < \varepsilon'$$

whenever

$$\sqrt{(x - x_0)^2 + (y - y_0)^2} < \delta$$

with $\delta < \sqrt{2}/m$. If we define an approximating function of two variables as

$$\begin{aligned} g(f; x, y) &\equiv \sum_{k,l=1}^{m} g_{kl}h(y_l; y)h(x_k; x) \\ &= \sum_{k=1}^{m} g[g_k; y]h(x_k; x) \end{aligned} \tag{A4.14}$$

then we have

$$\begin{aligned} |g(f; x, y) - \bar{g}(f; x, y)| &\leq \sum_{k=1}^{m} |g_k(y) - g[g_k; y]||h(x_k; x)| \\ &< \sum_{k=1}^{m} |g_k(y) - g[g_k; y]| < \frac{\varepsilon}{3} \end{aligned} \tag{A4.15}$$

For any fixed $y \in [0, 1]$, it follows from Th. 4.3 that

$$|f(x, y) - \bar{g}(f; x, y)| \equiv \left| f(x, y) - \sum_{k=1}^{m} g_k(y)h(x_k; x) \right| < \tfrac{1}{3}\varepsilon \tag{A4.16}$$

since we may use the largest of the $\varepsilon(y)$ for ε in Eq. A4.16. As in the proof of Th. 4.3 we may expand $|x|=\sqrt{1-(1-x^2)}$ and $|y|=\sqrt{1-(1-y^2)}$ each as uniformly convergent polynomials, say of degree n, in $(1-x^2)$ and in $(1-y^2)$, respectively, and write

$$|P_n^{(k,l)}(x, y)-g_{kl}h(x_k; x)h(y_l; y)|<\frac{\varepsilon}{3m^2} \tag{A4.17}$$

The inequalities

$$\left|g(f; x, y)-\sum_{k,l=1}^{n} P_n^{(k,l)}(x, y)\right|$$

$$=\left|\sum_{k,l}[g_{kl}h(x_l; x)h(y_k, y)-P_n^{(k,l)}(x, y)\right|$$

$$\leq\sum_{k,l}|g_{kl}h(x_l; x)h(y_k; y)-P_n^{(k,l)}(x, y)|<\sum_{k,l=1}^{m}\frac{\varepsilon}{3m^2}=\frac{\varepsilon}{3} \tag{A4.18}$$

and

$$\left|f(x, y)-\sum_{k,l=1}^{m} P_n^{(k,l)}(x, y)\right|$$

$$=\left|f(x, y)-\bar{g}(f; x, y)-g(f; x, y)+\bar{g}(f; x, y)+g(f; x, y)-\sum_{k,l} P_n^{(k,l)}(x, y)\right|<\varepsilon \tag{A4.19}$$

are now easily established. If we define

$$\mathcal{P}_n(x, y)\equiv\sum_{k,l=1}^{m} P_n^{(k,l)}(x, y) \tag{A4.20}$$

then we obtain

$$|f(x, y)-\mathcal{P}_n(x, y)|<\varepsilon \tag{A4.21}$$

which states that we may uniformly approximate $f(x, y)$ by a polynomial of degree n in x and y.

We can also establish this result by the same type argument as used in Section A4.1. Define a polynomial as

$$\mathcal{P}_n(x, y)\equiv\frac{\int_\alpha^\beta du\int_\alpha^\beta dv f(u, v)[1-(u-x)^2]^n[1-(v-y)^2]^n}{\left[\int_{-1}^{1}(1-u^2)^n\,du\right]^2} \qquad n=0, 1, 2, \ldots$$

$$0<\alpha<a\leq x\leq b<\beta<1 \qquad 0<\alpha<a\leq y\leq b<\beta<1 \tag{A4.22}$$

Suppose we again let $u=s+x$ and write the s integral as

$$\int_{\alpha-x}^{-\delta}+\int_{-\delta}^{\delta}+\int_{\delta}^{\beta-x}$$

Just as for one variable the first and last integrals will contribute to a factor bounded by a constant times B_n. That is, we write

$$\begin{aligned}
\mathcal{P}_n(x, y)(2A_n)^2 &= \int_\alpha^\beta dv\left\{\int_{\alpha-x}^{\beta-x} f(s+x, v)(1-s^2)^n\, ds\right\}[1-(v-y)^2]^n \\
&= \int_{\alpha-y}^{\beta-y} dt \int_{\alpha-x}^{\beta-x} ds f(s+x, t+y)(1-s^2)^n(1-t^2)^n \\
&= \int_{\alpha-y}^{\beta-y} (1-t^2)^n\, dt\left\{\int_{\alpha-x}^{-\delta} + \int_{-\delta}^{\delta} + \int_{\delta}^{\beta-x}\right\}
\end{aligned}$$

so that

$$\begin{aligned}
|\mathcal{P}_n(x, y)-f(x, y)| &\le \frac{1}{(2A_n)^2}\left|\int_{\alpha-y}^{\beta-y}(1-t^2)^n\, dt\left[\int_{\alpha-x}^{-\delta} + \int_{\delta}^{\beta-x} + \int_{-\delta}^{\delta}\right] - (2A_n)^2 f(x, y)\right| \\
&< \frac{1}{(2A_n)^2}\left\{2\int_{\alpha-y}^{\beta-y}(1-t^2)^n\, dt MB_n\right. \\
&\quad + \left|\int_{\alpha-y}^{\beta-y}(1-t^2)^n\, dt\int_{-\delta}^{\delta} f(s-x, t+y)(1-s^2)^n\, ds - (2A_n)^2 f(x, y)\right| \\
&< \frac{1}{(2A_n)^2}\left\{2MB_n[2B_n+2(A_n-B_n)]+2MB_n 2(A_n-B_n)\right. \\
&\quad \left. + \left|\int_{-\delta}^{\delta}(1-t^2)^n\, dt\int_{-\delta}^{\delta}(1-s^2)^n\, ds f(s+x, t+y)-(2A_n)^2 f(x, y)\right|\right\} \\
&< \frac{1}{(2A_n)^2}\left\{4MB_n(2A_n-B_n)\right. \\
&\qquad + \left|\int_{-\delta}^{\delta}(1-t^2)^n\, dt\int_{-\delta}^{\delta}(1-s^2)^n\, ds[f(s+x, t+y)-f(x, y)]\right. \\
&\quad \left.\left. + f(x, y)\int_{-\delta}^{\delta}(1-t^2)^n\, dt\int_{-\delta}^{\delta}(1-s^2)^n\, ds - (2A_n)^2 f(x, y)\right|\right\} \\
&< \frac{1}{(2A_n)^2}\left\{4MB_n(2A_n-B_n)+|4(A_n-B_n)^2-4A_n^2||f(x, y)|\right. \\
&\quad \left. + \varepsilon\int_{-1}^{1}(1-t^2)^n\, dt\int_{-1}^{1}(1-s^2)^n\, ds\right\} \\
&\le \frac{1}{(2A_n)^2}\{4MB_n(2A_n-B_n)+M(8A_nB_n+4B_n^2)+\varepsilon(2A_n)^2\}
\end{aligned}$$

This implies that

$$|\mathcal{P}_n(x, y)-f(x, y)| < 4M\frac{B_n}{A_n}+\varepsilon \tag{A4.23}$$

and the desired result follows from Eq. A4.6. In fact for m variables we obtain

$$|\mathcal{P}_n(x_1, x_2, \ldots, x_m)-f(x_1, x_2, \ldots, x_m)| < 2^m M\frac{B_n}{A_n}+\varepsilon$$

A4.3 A PROOF OF THE COULOMB EXPANSION

The following is a direct but somewhat lengthy verification of the Coulomb expansion stated in Eq. 4.37. From the identities

$$(1-2\mu x+\mu^2)(1-2\nu x+\nu^2)\equiv\alpha+\beta x+\gamma x^2$$
$$=(1+\nu^2+\mu^2+\mu^2\nu^2)-2(\nu+\mu+\mu\nu^2+\nu\mu^2)x+4\mu\nu x^2$$
$$\alpha\pm\beta+\gamma=(1+\mu\nu\mp\mu\mp\nu)^2$$
$$\pm\gamma+\tfrac{1}{2}\beta=\pm4\mu\nu-(\nu+\mu+\mu\nu^2+\nu\mu^2)$$

we can evaluate the integral

$$I(\mu,\nu)\equiv\int_{-1}^{1}\frac{dx}{\sqrt{1-2\mu x+\mu^2}\sqrt{1-2\nu x+\nu^2}}$$
$$=\frac{1}{2\sqrt{\mu\nu}}\ln\left(\frac{\gamma+\frac{1}{2}\beta+2\sqrt{\mu\nu}\sqrt{\alpha+\beta+\gamma}}{-\gamma+\frac{1}{2}\beta+2\sqrt{\mu\nu}\sqrt{\alpha-\beta+\gamma}}\right)=\frac{1}{\sqrt{\mu\nu}}\ln\left(\frac{1+\sqrt{\mu\nu}}{1-\sqrt{\mu\nu}}\right)$$
$$=2\sum_{n=0}^{\infty}\frac{(\mu\nu)^n}{(2n+1)} \tag{A4.24}$$

If we consider a Taylor expansion of $(1-2\mu x+\mu^2)^{-1/2}$ about $\mu=0$, we obtain

$$\frac{1}{\sqrt{1-2\mu x+\mu^2}}=\sum_{n=0}^{\infty}\mu^n Q_n(x) \tag{A4.25}$$

where $Q_n(x)$ is some polynomial of degree n since

$$Q_n(x)=\frac{1}{n!}\frac{d^n}{d\mu^n}[(1-2\mu x+\mu^2)^{-1/2}]\Big|_{\mu=0}$$

Eqs. A4.24 and A4.25 together imply that

$$\int_{-1}^{1}Q_n(x)Q_m(x)\,dx=\frac{2}{(2n+1)}\delta_{nm} \tag{A4.26}$$

We may expand the Legendre polynomial $P_n(x)$ in terms of the $\{Q_m(x)\}$ as

$$P_n(x)=\sum_{m=0}^{n}\alpha_{nm}Q_m(x) \tag{A4.27}$$

so that

$$\langle P_n\,|\,Q_r\rangle=\frac{2}{(2r+1)}\alpha_{nr}\qquad r=0,1,2,\ldots,n \tag{A4.28}$$

Since $P_n(x)$ is orthogonal to all x^r for $r<n$ (cf. Section 4.3b), this implies that

$$\alpha_{nr}=0\qquad\forall r<n \tag{A4.29}$$

or that

$$\alpha_{nn}Q_n(x)=P_n(x) \tag{A4.30}$$

If we consider Eq. A4.25 for $x=1$ we obtain

$$\frac{1}{\sqrt{1-2\mu+\mu^2}}=\frac{1}{1-\mu}=\sum_{n=0}^{\infty}\mu^n=\sum_{n=0}^{\infty}\mu^n Q_n(1) \qquad \textbf{(A4.31)}$$

Since $Q_n(1)=1$, Eq. 4.34 requires that $\alpha_{nn}=1$ or that $Q_n(x)=P_n(x)$. Equation A4.25 is, then, just the Coulomb expansion.

A4.4 DERIVATION OF SOME IMPORTANT RELATIONS SATISFIED BY THE LEGENDRE POLYNOMIALS, $P_n(x)$

In this section we will derive some of the more important relations satisfied by the Legendre polynomials including the results stated in Eqs. 4.36, 4.38, and 4.39.

If we differentiate the Coulomb expansion, Eq. 4.37, w.r.t. μ, we obtain

$$\frac{(x-\mu)}{(1-2\mu x+\mu^2)^{3/2}}=\sum_{n=0}^{\infty}n\mu^{n-1}P_n(x) \qquad \textbf{(A4.32)}$$

so that use of Eq. 4.37 again yields

$$\sum_{n=0}^{\infty}(x-\mu)\mu^n P_n(x)=\sum_{n=0}^{\infty}(1-2\mu x+\mu^2)n\mu^{n-1}P_n(x) \qquad \textbf{(A4.33)}$$

If we now equate the coefficients of μ^n on both sides of this equation, we obtain the useful recursion relation

$$(n+1)P_{n+1}(x)-(2n+1)xP_n(x)+nP_{n-1}(x)=0 \qquad \textbf{(A4.34)}$$

We can use the power series expression for $P_n(x)$ given in Eq. 4.33 to see that

$$\begin{aligned}
&(x^2-1)\frac{d^2}{dx^2}P_n(x)+2x\frac{d}{dx}P_n(x)-n(n+1)P_n(x)\\
&\quad\equiv\sum_{r=0}^{n}\frac{(-1)^{n-r}1\cdot3\cdot5\cdots(2r-1)}{(n-r)!\,(2r-n)!\,2^{n-1}}\\
&\qquad\times[-n^2-n+4r-2n+4r^2-2rn-2r-2nr+n^2+n]x^{2r-n}\\
&\qquad+\sum_{r=0}^{n-1}\frac{2(-1)^{n-r}1\cdot3\cdot5\cdots(2r-1)(2r+1)(2r-n+2)(2r-n+1)}{(n-r-1)!\,(2r+2-n)!\,2^{n-r}}x^{2r-n}\\
&\quad=\sum_{r=0}^{n-1}\frac{(-1)^{n-r}1\cdot3\cdot5\cdots(2r-1)}{(n-r)!\,(2r-n)!\,2^{n-r}}x^{2r-n}\\
&\qquad\times\left[2(1+2r)(r-n)+\frac{2(2r+1)(2r-n+2)(2r-n+1)(n-r)}{(2r+2-n)(2r+1-n)}\right]\\
&\quad\equiv 0
\end{aligned} \qquad \textbf{(A4.35)}$$

If we use Eq. A4.34 with n replaced by $(n-1)$ and the orthogonality of the

$\{P_n(x)\}$ established in Section 4.3b, we can show that

$$\int_{-1}^{1} P_n(x)P_n(x)\,dx = \frac{(2n-1)}{n}\int_{-1}^{1} xP_n(x)P_{n-1}(x)\,dx + \frac{(n-1)}{n}\int_{-1}^{1} P_n(x)P_{n-2}(x)\,dx$$

$$= \frac{(2n-1)}{n}\int_{-1}^{1}\left[\left(\frac{n+1}{2n+1}\right)P_{n+1}(x) + \frac{n}{(2n+1)}P_{n-1}(x)\right]P_{n-1}(x)\,dx$$

$$= \left(\frac{2n-1}{2n+1}\right)\int_{-1}^{1} P_{n-1}(x)P_{n-1}(x)\,dx \tag{A4.36}$$

We easily verify directly that

$$\int_{-1}^{1} [P_0(x)]^2\,dx = \int_{-1}^{1} dx = 2 \tag{A4.37}$$

$$\int_{-1}^{1} [P_1(x)]^2\,dx = \int_{-1}^{1} x^2\,dx = \tfrac{2}{3} \tag{A4.38}$$

Eqs. A4.36–A4.38 plus mathematical induction show that

$$\int_{-1}^{1} [P_n(x)]^2\,dx = \frac{2}{(2n+1)} \qquad n = 0, 1, 2, \ldots \tag{A4.39}$$

Use of the Coulomb expansion, Eq. 4.37, plus the bionomial theorem allow us to write

$$\frac{1}{1\pm\mu} = \sum_{n=0}^{\infty} (\mp 1)^n \mu^n = \sum_{n=0}^{\infty} \mu^n P_n(\mp 1)$$

so that

$$P_n(\pm 1) = (\pm 1)^n \tag{A4.40}$$

and

$$\frac{1}{\sqrt{1+\mu^2}} = \sum_{n=0}^{\infty} (-1)^n \frac{1\cdot 3\cdot 5\cdots(2n-1)}{2\cdot 4\cdot 6\cdots(2n)}\mu^{2n} = \sum_{n=0}^{\infty} \mu^n P_n(0)$$

so that

$$P_{2n+1}(0) = 0 \tag{A4.41}$$

$$P_{2n}(0) = \frac{(-1)^n 1\cdot 3\cdot 5\cdots(2n-1)}{2\cdot 4\cdot 6\cdots(2n)} = \frac{(-1)^n(2n)!}{2^{2n}(n!)^2} \tag{A4.42}$$

In Section A7.3 we will see that the properties of $P_n(x)$ derived here are simply special cases of those satisfied by the Jacobi polynomials.

A4.5 DERIVATION OF SOME IMPORTANT RELATIONS SATISFIED BY THE ASSOCIATED LEGENDRE POLYNOMIALS, $P_n^m(x)$

The associated Legendre polynomials are defined as

$$P_n^m(x) = (1-x^2)^{m/2}\frac{d^m}{dx^m}P_n(x) \qquad m = 0, 1, 2, \ldots, n;\ n \geq 0 \tag{A4.43}$$

$$P_n^0(x) = P_n(x) \tag{A4.44}$$

If we use Leibnitz' rule for differentiation,

$$\frac{d^n}{dx^n}[f(x)g(x)] = \sum_{s=0}^{n} \frac{n!}{s!(n-s)!} \frac{d^{n-s}}{dx^{n-s}} f(x) \frac{d^s}{dx^s} g(x) \tag{A4.45}$$

plus the definition of the $P_n(x)$ given in Eq. 4.32, we easily establish the identities

$$\frac{d^m}{dx^m}[(x^2-1)P_n''(x)] = m!\sum_{s=0}^{m} \frac{1}{s!(m-s)!} \frac{d^{m-s}}{dx^{m-s}}(x^2-1)\frac{d^s}{dx^s}P_n''(x)$$

$$= m!\left\{\frac{1}{m!}(x^2-1)\frac{d^m}{dx^m}P_n''(x) + \frac{1}{(m-1)!}2x\frac{d^{m-1}}{dx^{m-1}}P_n''(x) + \frac{1\cdot 2}{(m-2)!\,2!}\frac{d^{m-2}}{dx^{m-2}}P_n''(x)\right\}$$

$$\frac{d^m}{dx^m}[2xP_n'(x)] = m!\left\{\frac{1}{m!}2x\frac{d^m}{dx^m}P_n'(x) + \frac{1}{(m-1)!}2\frac{d^{m-1}}{dx^{m-1}}P_n'(x)\right\}$$

so that from Eq. A4.35

$$\frac{d^m}{dx^m}[(x^2-1)P_n''(x) + 2xP_n'(x) - n(n+1)P_n(x)] \equiv 0$$

$$= (x^2-1)\frac{d^2}{dx^2}\left(\frac{d^m}{dx^m}P_n(x)\right) + 2x(m+1)\frac{d}{dx}\left(\frac{d^m}{dx^m}P_n(x)\right) - (n-m)(n+m+1)\left(\frac{d^m}{dx^m}P_n(x)\right) \tag{A4.46}$$

If we now define

$$f(x) \equiv (1-x^2)^{m/2}\frac{d^m}{dx^m}P_n(x) \equiv (1-x^2)^{m/2}g(x)$$

then the expressions

$$g'(x) = \left[f'(x) + \frac{mxf(x)}{1-x^2}\right](1-x^2)^{-m/2}$$

$$g''(x) = \left[f''(x) + \frac{2mxf'(x)}{1-x^2} + \frac{mf(x)}{1-x^2} + \frac{m(m+2)x^2f(x)}{(1-x^2)^2}\right](1-x^2)^{-m/2}$$

when substituted into Eq. A4.46 yield

$$(1-x^2)\frac{d^2}{dx^2}P_n^m(x) - 2x\frac{d}{dx}P_n^m(x) + \left[n(n+1) - \frac{m^2}{1-x^2}\right]P_n^m(x) = 0 \tag{A4.47}$$

The property

$$P_n^m(\pm 1) = 0 \qquad m > 0 \tag{A4.48}$$

follows directly from the definition of $P_n^m(x)$ in Eq. A4.43.

We will now use mathematical induction to establish the orthogonality condition given in Eq. 4.41. It is true for $r=0$ from Eq. 4.36. From Eq. A4.43 we have

$$P_n^{m+1}(x) \equiv (1-x^2)^{(m+1)/2} \frac{d}{dx}\left[\frac{d^m}{dx^m} P_n(x)\right]$$
$$= (1-x^2)^{1/2} \frac{d}{dx} P_n^m(x) + \frac{mx}{(1-x^2)^{1/2}} P_n^m(x)$$

so that integration by parts and use of the differential equation (A4.47) imply that

$$\int_{-1}^{1} P_n^{m+1}(x) P_{n'}^{m+1}(x)\, dx = \int_{-1}^{1} (1-x^2) \frac{d}{dx} P_n^m(x) \frac{d}{dx} P_{n'}^m(x)\, dx$$
$$+ m \int_{-1}^{1} x\left[P_n^m(x) \frac{d}{dx} P_{n'}^m(x) + P_{n'}^m(x) \frac{d}{dx} P_n^m(x)\right] dx$$
$$+ m^2 \int_{-1}^{1} \left(\frac{1}{1-x^2} - 1\right) P_n^m(x) P_{n'}^m(x)\, dx$$
$$= \left[\frac{n(n+1)}{2} + \frac{n'(n'+1)}{2} - m^2\right] \int_{-1}^{1} P_n^m(x) P_{n'}^m(x)\, dx$$
$$+ m \int_{-1}^{1} x\left[P_n^m(x) \frac{d}{dx} P_{n'}^m(x) + P_{n'}^m(x) \frac{d}{dx} P_n^m(x)\right] dx$$
$$= \left[\frac{n(n+1)}{2} + \frac{n'(n'+1)}{2} - m^2\right] \int_{-1}^{1} P_n^m(x) P_{n'}^m(x)\, dx$$
$$- m\left\{\int_{-1}^{1} P_{n'}^m(x)\left[P_n^m(x) + x \frac{d}{dx} P_n^m(x)\right] dx\right.$$
$$\left. + \int_{-1}^{1} P_n^m(x)\left[P_{n'}^m(x) + x \frac{d}{dx} P_{n'}^m(x)\right] dx\right\}$$
$$= \left[\frac{n(n+1)}{2} + \frac{n'(n'+1)}{2} - m^2 - m\right] \int_{-1}^{1} P_n^m(x) P_{n'}^m(x)\, dx$$
$$= (n+m+1)(n-m) \frac{2}{(2n+1)} \frac{(n+m)!}{(n-m)!} \delta_{nn'}$$
$$= \frac{2}{(2n+1)} \frac{(n+m+1)!}{(n-m-1)!} \delta_{nn'}$$

Therefore if Eq. 4.41 is true for m it is true for $(m+1)$, which completes the proof.

The recursion relation given in Eq. 4.43 is established by differentiating the Coulomb expansion, Eq. 4.37, m times w.r.t. x to obtain

$$(1-x^2)^{m/2} \frac{d^m}{dx^m} (1-2\mu x + \mu^2)^{-1/2}$$
$$= 1 \cdot 3 \cdot 5 \cdots (2m-1) \mu^m (1-2\mu x + \mu^2)^{-(2m+1)/2} (1-x^2)^{m/2}$$
$$= \sum_{n=0}^{\infty} \mu^n P_n^m(x) \qquad \textbf{(A4.49)}$$

If we now differentiate both sides of Eq. A4.49 w.r.t. μ we find

$$1\cdot 3\cdot 5\cdots(2-m-1)(1-x^2)^{m/2}\{m\mu^{m-1}(1-2\mu x+\mu^2)^{-(2m+1)/2} +(2m+1)(1-2\mu x+\mu^2)^{-(2m+3)/2}(x-\mu)\}=\sum_{n=0}^{\infty} n\mu^{n-1}P_n^m(x)$$

so that

$$\frac{m}{\mu}\sum_{n=0}^{\infty}\mu^n P_n^m(x)+\frac{1}{(1-2\mu x+\mu^2)}(2m+1)(x-\mu)\sum_{n=0}^{\infty}\mu^n P_n^m(x)=\sum_{n=0}^{\infty} n\mu^{n-1}P_n^m(x)$$

TABLE 4.1 Some Relations Satisfied by the Legendre Polynomials

$$P_n(x)=\frac{1}{2^n n!}\frac{d^n}{dx^n}[(x^2-1)^n]=\sum_{r=0}^{n}\frac{(-1)^{n-r}1\cdot 3\cdot 5\cdots(2r-1)x^{2r-n}}{(n-r)!(2r-n)!2^{n-r}} \qquad n\geq 0$$

$$P_n(1)=1$$

$$P_n(-x)=(-1)^n P_n(x)$$

$$P_0(x)=1 \qquad P_1(x)=x \qquad P_2(x)=\tfrac{3}{2}x^2-\tfrac{1}{2}$$

$$P_3(x)=\tfrac{5}{2}x^3-\tfrac{3}{2}x \qquad P_4(x)=\tfrac{35}{8}x^4-\tfrac{15}{4}x^2+\tfrac{3}{8}$$

$$P_n^m(x)=(1-x^2)^{m/2}\frac{d^m}{dx^m}P_n(x) \qquad 0\leq m\leq n,\ n\geq 0$$

$$P_n^0(x)=P_n(x)$$

$$\int_{-1}^{1}P_n^m(x)P_{n'}^m(x)\,dx=\frac{2}{(2n+1)}\frac{(n+m)!}{(n-m)!}\delta_{nn'}$$

$$(1-x^2)\frac{d^2}{dx^2}P_n^m(x)-2x\frac{d}{dx}P_n^m(x)+\left[n(n+1)-\frac{m^2}{1-x^2}\right]P_n^m(x)=0$$

$$(n-m+1)P_{n+1}^m(x)=(2n+1)xP_n^m(x)-(n+m)P_{n-1}^m(x) \qquad n\geq 1$$

$$P_{2n+m+1}^m(0)=0 \qquad P_{2n+m}^m(0)=\frac{(-1)^n(2m+2n)!}{2^{m+2n}n!(m+n)!}$$

$$P_n^m(\pm 1)=0 \qquad m>0$$

$$\frac{2^m\Gamma(m+\frac{1}{2})(1-x^2)^{m/2}}{\sqrt{\pi}[1-2\mu x+\mu^2]^{m+\frac{1}{2}}}=\sum_{n=0}^{\infty}\mu^n P_{n+m}^m(x)$$

$$(1-x^2)\frac{d}{dx}P_n^m(x)=(n+1)xP_n^m(x)-(n-m+1)P_{n+1}^m(x)$$

$$=-nxP_n^m(x)+(n+m)P_{n-1}^m(x)$$

$$\int_0^1 x^m P_n(x)\,dx=\frac{\sqrt{\pi}\,m!}{2^{m+1}\Gamma(1+\frac{1}{2}m-\frac{1}{2}n)\Gamma(\frac{3}{2}+\frac{1}{2}m+\frac{1}{2}n)} \qquad m\geq 0$$

$$\int_{-1}^{1}\frac{1}{\sqrt{1-x}}P_n(x)\,dx=\frac{2^{3/2}}{(2n+1)}$$

$$\int_{-1}^{1}(1+x)^{m+n}P_m(x)P_n(x)\,dx=\frac{2^{m+n+1}[(m+n)!]^4}{(m!n!)^2(2m+2n+1)!}$$

$$\int_{-1}^{1}\frac{1}{(1-x^2)}[P_n^m(x)]^2\,dx=\frac{(n+m)!}{m(n-m)!}$$

or

$$m(1-2\mu x+\mu^2)\sum_{n=0}^{\infty}\mu^n P_n^m(x)+(2m+1)(x-\mu)\sum_{n=0}^{\infty}\mu^{n+1}P_n^m(x)$$
$$=(1-2\mu x+\mu^2)\sum_{n=0}^{\infty}n\mu^n P_n^m(x)$$

If we rewrite this as

$$\sum_{n=0}^{\infty}\mu^n[mP_n^m(x)-2mxP_{n-1}^m(x)+mP_{n-2}^m(x)$$
$$+(2m+1)xP_{n-1}^m(x)-(2m+1)P_{n-2}^m(x)-nP_n^m(x)+2x(n-1)P_{n-1}^m(x)$$
$$-(n-2)P_{n-2}^m(x)]\equiv 0$$

TABLE 4.2 A Short Table of Fourier Transforms

$f(x)$		$f(k)=\frac{1}{\sqrt{2\pi}}\int_{-\infty}^{\infty} e^{ikx}f(x)\,dx$	
$f(ax+b)$		$\frac{1}{a}e^{-i(b/a)}f(k/a)$	
$f(ax)e^{ibx}$		$\frac{1}{a}f\left(\frac{k+b}{a}\right)$	
$i^n x^n f(x)$		$\frac{d^n}{dk^n}f(k)$	
$\frac{d^n}{dx^n}f(x)$		$(-i)^n k^n f(k)$	
$(1+x^2)^{-1}$		$\sqrt{\frac{\pi}{2}}e^{-\lvert k\rvert}$	
$e^{-(1/2)x^2}$		$e^{-(1/2)k^2}$	
$(a+ix)^{-\nu}$	$\nu>0$	$\sqrt{2\pi}\,k^{\nu-1}e^{-ak}/\Gamma(\nu)$	$k>0$
	$a>0$	0	$k<0$
$(a-ix)^{-\nu}$	$\nu>0$	0	$k>0$
	$a>0$	$-\sqrt{2\pi}\,(-k)^{\nu-1}e^{-ak}/\Gamma(\nu)$	$k<0$
$\frac{e^{-\lambda x}}{(a+e^{-x})}$	$a>0$	$\sqrt{\frac{\pi}{2}}a^{\lambda-1-ik}\csc(\pi\lambda-i\pi k)$	
	$0<\lambda<1$		

and equate to zero the coefficient of μ^n, we see that

$$(m-n)P_n^m(x)+(2n-1)xP_{n-1}^m(x)+(-m-n+1)P_{n-2}^m(x)=0$$

or with $n \rightarrow n+1$,

$$x(2n+1)P_n^m(x)=(n-m+1)P_{n+1}^m(x)+(n+m)P_{n-1}^m(x) \qquad \textbf{(A4.50)}$$

SUGGESTED REFERENCES

F. W. Byron and R. W. Fuller, *Mathematics of Classical and Quantum Physics* (Vols. I and II). Chapter 5 gives a very nice presentation of complete sets of functions. In particular Section 5.8 discusses the completeness of the spherical harmonics directly from the Weierstrass approximation theorem for three variables. Section 9.1 proves the completeness of the eigenvectors of completely continuous operators in Hilbert space.

R. V. Churchill, *Fourier Series and Boundary Value Problems.* This gives an elementary introduction to Fourier series with many detailed examples and applications.

R. Courant and D. Hilbert, *Methods of Mathematical Physics.* Chapter II gives a classic presentation of expansions of functions in terms of different complete sets. Section 9.6 of Chapter II gives a proof of the completeness of the Laguerre and Hermite polynomials on the infinite intervals $[0, \infty)$ and $(-\infty, \infty)$, respectively.

H. Lass, *Elements of Pure and Applied Mathematics.* Chapter 6 presents an elementary and rigorous treatment of Fourier series and integrals, including the expansion theorems.

L. Schwartz, *Mathematics for the Physical Sciences.* This is a beautiful work which discusses many of the techniques of applied mathematics in great generality and at a high level of mathematical rigor. Chapters 4 and 5 cover Fourier series and integrals for functions belonging to $\mathscr{L}_2$.

I. N. Sneddon, *Fourier Transforms.* This is an exhaustive treatment on the applications of Fourier transforms to problems in the physical sciences.

E. C. Titchmarsh, *Introduction to the Theory of Fourier Integrals.* This is one of the standard references on the theory of Fourier integrals and gives proofs of all the important theorems in great detail with all of the δ's and ε's displayed.

For the difficult question of the uniform convergence of a series $\sum_j \langle \phi_j \mid f \rangle \phi_j(x)$ see either of the following.

E. W. Hobson, *Theory of Spherical and Ellipsoidal Harmonics.* Section 205 establishes uniform convergence for the Legendre polynomials provided $f(x)$ is continuous, of bounded variation, and $f(x)/(1-x^2)^{1/4} \in \mathscr{L}[-1, 1]$.

E. C. Titchmarsh, *Eigenfunction Expansions* (Part I). Chapters II and IX discuss the convergence of expansions in terms of the eigenfunctions of the Sturm-Liouville equation.

PROBLEMS

4.1 Prove that the set of functions

$$\frac{1}{\sqrt{2\pi}} \qquad \frac{\cos nx}{\sqrt{\pi}} \qquad \frac{\sin nx}{\sqrt{\pi}}$$

is orthonormal on the range $[-\pi, \pi]$.

4.2 Assume that you are given a continuous bounded function $f(x)$, $x \in [-\pi, \pi]$, such that $f(-\pi) = f(\pi)$ and such that df/dx is bounded and of limited total fluctuation on $x \in [-\pi, \pi]$. Use one of the Riemann-Lebesgue lemmas to prove that the Fourier series for $f(x)$,

$$f(x) = \frac{a_0}{2} + \sum_{n=1} (a_n \cos nx + b_n \sin nx)$$

$$a_n = \frac{1}{\pi} \int_{-\pi}^{\pi} f(x) \cos nx \, dx \qquad b_n = \frac{1}{\pi} \int_{-\pi}^{\pi} f(x) \sin nx \, dx$$

is *uniformly* convergent on $x \in [-\pi, \pi]$.

For the next five problems sketch graphs of each of the functions given and expand them in Fourier series.

4.3

$$f(x) = \begin{cases} -\pi - x & -\pi \le x \le -\pi/2 \\ x & -\pi/2 \le x \le \pi/2 \\ \pi - x & \pi/2 \le x \le \pi \end{cases}$$

Verify by the Weierstrass comparison test that the above Fourier series converges uniformly on $[-\pi, \pi]$.

4.4

$$f(x) = x^2 \qquad -\pi \le x \le \pi$$

4.5

$$f(x) = e^x \qquad -\pi \le x \le \pi$$

4.6

$$f(x) = \begin{cases} \dfrac{\pi}{2} - x & -\pi < x < -\dfrac{\pi}{2} \\ x & -\dfrac{\pi}{2} < x < \dfrac{\pi}{2} \\ \dfrac{\pi}{2} - x & \dfrac{\pi}{2} < x < \pi \end{cases}$$

4.7

$$f(x) = \begin{cases} 0 & -\pi \le x \le 0 \\ x & 0 \le x \le \pi \end{cases}$$

4.8 You are given only the following information about a function

$$f(x) = x \qquad 0 \le x \le \pi$$

a. Construct a Fourier cosine series on $-\pi < x < \pi$ that will converge to $f(x)$ for $0 < x < \pi$.

b. Construct a Fourier sine series on $-\pi<x<\pi$ that will converge to $f(x)$ for $0<x<\pi$.

c. Which of these series would you expect to converge uniformly at $x=\pi$?

4.9 A famous and useful series in mathematics is

$$\sum_{n=1}^{\infty}\frac{1}{n^2}$$

which is known to converge. Evaluate this series by expanding the function $f(x)=x^2$ on the interval $-\pi\leq x\leq\pi$ in a Fourier series. Now evaluate

$$\sum_{n=1}^{\infty}\frac{1}{n^4} \quad \text{and} \quad \sum_{n=1}^{\infty}\frac{(-1)^n}{n^4}$$

4.10 Prove that

$$\int_{-1}^{1} x^n P_n(x)\,dx=\frac{2^{n+1}(n!)^2}{(2n+1)!}$$

4.11 Expand x^n as a series of Legendre polynomials, $x\in[-1,1]$.

4.12 Expand the function

$$f(x)=\begin{cases}-1 & -1\leq x<0\\ +1 & 0<x\leq 1\end{cases}$$

in a series of Legendre polynomials $P_n(x)$. That is, evaluate *explicitly* the expansion coefficients.

4.13 Prove that

$$1-2\mu\cos\theta+\mu^2=(1-\mu e^{i\theta})(1-\mu e^{-i\theta})$$

and then expand $(1-2\mu\cos\theta+\mu^2)^{-1/2}$ in terms of a product of two series to show that

$$P_n(\cos\theta)=\frac{1\cdot 3\cdot 5\cdots(2n-1)}{2\cdot 4\cdot 6\cdots(2n)}\left[2\cos(n\theta)+\frac{1\cdot(2n)}{2\cdot(2n-1)}2\cos(n-2)\theta\right.$$
$$\left.+\frac{1\cdot 3}{2\cdot 4}\frac{(2n-2)(2n)}{(2n-3)(2n-1)}2\cos(n-4)\theta+\cdots\right]$$

where for n even, the constant term (i.e., the last one for n even) is

$$\left[\frac{1\cdot 3\cdot 5\cdots(n-1)}{2\cdot 4\cdot 6\cdots n}\right]^2$$

rather than that given in this series. Then deduce that $|P_n(\cos\theta)|\leq 1$ for all (real) values of θ.

4.14 Prove the following properties of the Fourier transform

$$f(k)=\frac{1}{\sqrt{2\pi}}\int_{-\infty}^{\infty} f(x)e^{ikx}\,dx$$

a. $$\frac{d}{dk}f(k)=\frac{1}{\sqrt{2\pi}}\int_{-\infty}^{\infty}[ixf(x)]e^{ikx}\,dx$$

b. $\frac{1}{\sqrt{2\pi}}\int_{-\infty}^{\infty}\frac{d}{dx}f(x)e^{ikx}\,dx=-ikf(k)$

c. $\frac{1}{\sqrt{2\pi}}\int_{-\infty}^{\infty}[f(x)e^{ik_0x}]e^{ikx}\,dx=f(k+k_0)$

d. $\frac{1}{\sqrt{2\pi}}\int_{-\infty}^{\infty}f(x+x_0)e^{ikx}\,dx=e^{-ikx_0}f(k)$

4.15 The *convolution* of two functions, $f(x)$ and $g(x)$, is defined as

$$\frac{1}{\sqrt{2\pi}}\int_{-\infty}^{\infty}f(y)g(x-y)\,dy$$

Prove that the Fourier transform of this convolution is just $f(k)g(k)$. Use the result to find a solution $f(x)$ to the integral equation

$$h(x)=\int_{-\infty}^{\infty}f(y)g(x-y)\,dy$$

where $h(x)$ and $g(y)$ are given functions whose Fourier transforms are known. Is the solution thus obtained always valid?

4.16 Compute the Fourier transforms of the following functions.

a. $f(x)=\begin{cases}1 & |x|<a\\ 0 & |x|>a\end{cases}$

b. $f(x)=e^{-x^2/2}$

4.17 Consider the differential equation

$$\frac{d^2}{dx^2}\psi(x)-V(x)\psi(x)+k^2\psi(x)=0$$

for the function $\psi(x)$ where k^2 is a parameter. Define the Fourier transform of $\psi(x)$ as $\tilde{\psi}(p)$,

$$\tilde{\psi}(p)=\frac{1}{\sqrt{2\pi}}\int_{-\infty}^{\infty}\psi(x)e^{ipx}\,dx$$

Show that the integral equation satisfied by $\tilde{\psi}(p)$ is

$$p^2\tilde{\psi}(p)+\frac{1}{\sqrt{2\pi}}\int_{-\infty}^{\infty}dp'\,\tilde{V}(p-p')\tilde{\psi}(p')=k^2\tilde{\psi}(p)$$

5 Integral Equations

5.0 INTRODUCTION

We have already proven some very important theorems about integral equations in Section 2.10. These covered the existence and uniqueness of the solutions to those integral equations whose kernels were of trace class. In this chapter we extend this theory and develop techniques for actually finding these solutions. Finally we return to the question of the completeness of the eigenfunctions of these integral equations.

We now point out an important distinction in the general assumptions which will be made in proving many of the theorems in this chapter. Some of the arguments will require only that the norm, $\|K\|$, of the linear operator be finite while others will require that a particular representation of K, for example, $K(x, y)$, be continuous or bounded. The latter representation-dependent restrictions are clearly much more stringent then the former, representation-independent norm restrictions. Since statements about norms of operators or vectors do not depend on any particular representation, the results derived under these assumptions alone will be very general in their applicability and will be valid for any operator in a Hilbert space satisfying them. Naturally, if we are willing to make stronger assumptions about K in some particular representation, then we should obtain a sharper result. Generally speaking the sharper result we obtain is uniform convergence of certain series rather than simple convergence and pointwise continuity of solutions rather than just existence. Of course representation-dependent results true in one representation need not be true in another representation. For example, the discontinuous function

$$f(x)=\begin{cases} 0 & |x|>1 \\ \dfrac{1}{\sqrt{2}} & |x|<1 \end{cases}$$

has a continuous Fourier transformation

$$f(k)=\frac{1}{\sqrt{\pi}}\frac{\sin k}{k}$$

while both have the finite norm $\|f\|=1$. This abstract representation-independent treatment is developed further in Sections A5.4 and A5.5 at the end of the chapter.

5.1 VOLTERRA EQUATIONS

a. Equations of the First and Second Kind

The *Volterra integral equations* of the *first* and *second* kind are defined, respectively, as

$$\int_a^x K(x, y)\phi(y)\, dy=f(x) \qquad x\in[a, b] \tag{5.1}$$

and

$$\phi(x)-\lambda\int_a^x K(x, y)\phi(y)\, dy=f(x) \qquad x\in[a, b] \tag{5.2}$$

We begin by studying Eq. (5.2). We now prove that the Neumann series (cf. Section 2.7b) for this equation will converge to a solution provided $K(x, y)$ satisfies certain conditions.

THEOREM 5.1

If $K(x, y)$ is a continuous function of both x and y on $[a, b]$ and if $f(x)$ is also continuous here, then the Volterra integral equation of the second kind

$$\phi(x)-\lambda\int_a^x K(x, y)\phi(y)\, dy=f(x) \qquad x\in[a, b]$$

has one and only one solution that is given by the uniformly convergent series

$$\phi(x)=f(x)+\sum_{n=1}^{\infty} \lambda^n\int_a^x K_n(x, y)f(y)\, dy \tag{5.3}$$

where

$$K_1(x, y) \equiv K(x, y)$$

$$K_{n+1}(x, y) \equiv \int_y^x K(x, s) K_n(s, y)\, ds \qquad n \geq 1 \tag{5.4}$$

PROOF

If we rewrite Eq. 5.2 as

$$(I - \lambda K)\phi(x) = f(x)$$

then *formally* the Neumann series would be

$$\phi(x) = f(x) + \sum_{n=1}^{\infty} \lambda^n K^n f(x) \tag{5.5}$$

We must first prove the convergence of the series in Eq. 5.5. We define $\kappa_m^2(x)$ as

$$\begin{aligned}
\kappa_m^2(x) &\equiv |K^m f(x)|^2 \equiv |KK^{m-1} f(x)|^2 = \left| \int_a^x K(x, y)[K^{m-1} f(y)]\, dy \right|^2 \\
&\leq \left[\int_a^x dy |K(x, y)|^2 \right] \left[\int_a^x dy |K^{m-1} f(y)|^2 \right] \\
&= k^2(x) \left[\int_a^x dy\, \kappa_{m-1}^2(y) \right]
\end{aligned} \tag{5.6}$$

where by definition

$$k^2(x) = \int_a^x |K(x, y)|^2\, dy \leq \int_a^b |K(x, y)|^2\, dy$$

We next establish a bound on $\kappa_m^2(x)$ for $m = 1$.

$$\begin{aligned}
\kappa_1^2(x) &\equiv |Kf(x)|^2 = \left| \int_a^x dy\, K(x, y) f(y) \right|^2 \\
&\leq \int_a^x dy |K(x, y)|^2 \int_a^x dy |f(y)|^2 \leq k^2(x) \int_a^b dy |f(y)|^2
\end{aligned}$$

To obtain the last inequality we have used the fact that $\int_a^x dy|f(y)|^2$ is a nondecreasing function of x. Now let $m = 2$.

$$\begin{aligned}
\kappa_2^2(x) &= \left| \int_a^x dy\, K(x, y) \int_a^y ds\, K(y, s) f(s) \right|^2 \leq \int_a^x dy\, |k(x, y)|^2 \int_a^x dy \left| \int_a^y ds\, K(y, s) f(s) \right|^2 \\
&\leq k^2(x) \int_a^x dy \left[\int_a^y ds\, |K(y, s)|^2 \right] \left[\int_a^y |f(s)|^2\, ds \right] \\
&\leq k^2(x) \int_a^x dy\, k^2(y) \int_a^y |f(s)|^2\, ds \leq k^2(x) \int_a^x dy\, k^2(y) \int_a^b |f(y)|^2\, dy
\end{aligned} \tag{5.8}$$

We can now use induction to prove that

$$\kappa_m^2(x) \leq \frac{k^2(x)}{(m-1)!} \left[\int_a^x dy\, k^2(y) \right]^{m-1} \int_a^b |f(y)|^2\, dy \tag{5.9}$$

Since we have just shown this to be true for $m=1$ and 2, we assume it is true for $m-1$ and then prove it for m. We begin with Eq. (5.6).

$$\kappa_m^2(x) \leq k^2(x) \int_a^x dy\, \kappa_{m-1}^2(y) \leq k^2(x) \int_a^x dy \frac{k^2(y)}{(m-2)!} \left[\int_a^y ds\, k^2(s) \right]^{m-2} \int_a^b |f(s)|^2\, ds$$

$$\equiv \frac{k^2(x)}{(m-2)!} \int_a^x \left[\int_a^y ds\, k^2(s) \right]^{m-2} d\left[\int_a^y ds\, k^2(s) \right] \int_a^b |f(s)|^2\, ds$$

$$\equiv \frac{k^2(x)}{(m-1)!} \left[\int_a^x k^2(s)\, ds \right]^{m-1} \int_a^b |f(y)|^2\, dy \tag{5.10}$$

We now establish a bound for the general term $\lambda^n K^n f(x)$ in Eq. (5.5).

$$|\lambda^n K^n f(x)| = |\lambda|^n\, |K^n f(x)| \leq |\lambda|^n \left\{ \frac{k^2(x)}{(n-1)!} \left[\int_a^x k^2(y)\, dy \right]^{n-1} \right\}^{1/2} \|f\|$$

$$\leq |\lambda|^n k(x) \left[\frac{\mathbf{|K|}^{n-1}}{\sqrt{(n-1)!}} \right] \|f\|$$

where $\mathbf{|K|}^2 \equiv \int_a^b dx \int_a^b dy\, |K(x, y)|^2$. Since

$$\frac{|\lambda|^{n+1}\, \mathbf{|K|}^n}{\sqrt{n!}} \bigg/ \frac{|\lambda|^n\, \mathbf{|K|}^{n-1}}{\sqrt{(n-1)!}} = \frac{|\lambda|\, \mathbf{|K|}}{\sqrt{n}} \xrightarrow[n\to\infty]{} 0$$

we see that the series

$$\sum_{n=1}^{\infty} \frac{|\lambda|^n\, \mathbf{|K|}^{n-1}}{\sqrt{(n-1)!}}$$

is absolutely convergent, which implies by the Weierstrass M test that the series

$$\sum_{n=1}^{\infty} \lambda^n K^n f(x)$$

is uniformly convergent. Since every term in the series (5.5) is a continuous function of x, so is the function defined by this sum (cf. Appendix I, Th. I.10).

As a technical point we must now show that Eq. 5.5 actually *is* a solution to Eq. 5.2. This is rather simple since we can integrate a uniformly convergent series term by term (cf. Appendix I, Th. I.11).

$$f(x) + \sum_{n=1}^{\infty} \lambda^n K^n f(x) - \lambda K f(x) - \sum_{n=1}^{\infty} \lambda^{n+1} K^{n+1} f(x)$$

$$= f(x) + \sum_{n=2}^{\infty} \lambda^n K^n f(x) - \sum_{n=1}^{\infty} \lambda^{n+1} K^{n+1} f(x) = f(x) \tag{5.11}$$

Since this is valid for *any* $f(x)$, we see that $\left[I + \sum_{n=1}^{\infty} \lambda^n K^n \right]$ is the inverse of $(I - \lambda K)$ and *always* exists. That is, the operator $(I - \lambda K)$ defined here is a nonsingular operator. Therefore we see at once that there can be no other solution to Eq. 5.2 since if there were the difference of the two solutions would satisfy Eq. 5.2 with

$f(x)=0$. Then Eq. 5.5 would state that the difference of these two solutions would be zero.

We can also establish directly the uniqueness of the solution to Eq. 5.2 as follows. Consider the homogeneous version of Eq. 5.2 [i.e., set $f(x)=0$]. It then follows that

$$\begin{aligned}|\phi(x)|^2 &= |\lambda|^2\left|\int_a^x K(x,y)\phi(y)\,dy\right|^2 \\ &\le |\lambda|^2\int_a^x |K(x,y)|^2\,dx\int_a^x |\phi(y)|^2\,dy \\ &= |\lambda|^2 k^2(x)\int_a^x |\phi(y)|^2\,dy \\ &\le |\lambda|^2 k^2(x)\|\phi\|^2\end{aligned}$$

so that

$$\begin{aligned}|\phi(x)|^2 &\le |\lambda|^2 k^2(x)\int_a^x [|\lambda|^2 k^2(y)\|\phi\|^2]\,dy = |\lambda|^4 k^2(x)\int_a^x k^2(y)\,dy\,\|\phi\|^2 \\ &\le |\lambda|^4 k^2(x)\mathbf{|K|}^2\|\phi\|^2\end{aligned}$$

and

$$\begin{aligned}|\phi(x)|^2 &\le |\lambda|^2 k^2(x)\int_a^x |\lambda|^4\,dy\left[k^2(y)\int_a^y k^2(z)\,dz\right]\|\phi\|^2 \\ &= |\lambda|^6 k^2(x)\int_a^x d\left[\int_a^y k^2(z)\,dz\right]\int_a^y k^2(z)\,dz\|\phi\|^2 \\ &= |\lambda|^6 k^2(x)\left[\frac{1}{2}\int_a^y k^2(z)\,dz\right]\Bigg|_a^x\|\phi\|^2 = \frac{|\lambda|^6 k^2(x)}{2!}\left[\int_a^x k^2(z)\,dz\right]^2\|\phi\|^2 \\ &\le \frac{|\lambda|^6 k^2(x)}{2!}[\mathbf{|K|}^2]\|\phi\|\end{aligned}$$

Continuing successively we may use induction, just as we did for Eq. (5.9), to establish that

$$|\phi(x)|^2 \le |\lambda|^2 k^2(x)\frac{[|\lambda|^2\mathbf{|K|}^2]^n}{n!} \qquad \forall n=0,1,2,\ldots$$

If we now take the limit as $n\to\infty$, we see that

$$|\phi(x)|=0$$

since $k^2(x)$ is everywhere finite by assumption. This establishes the uniqueness of the solution to the Volterra integral equation of the second kind since it states that the *only* solution to the homogeneous equation is the trivial one.

Finally it remains to show that Eq. 5.5 is identical to Eqs. 5.3–5.4. This is most easily done if we introduce the Heaviside step function (cf. Eq. 2.182) into the multiple integrals. For instance for $n=2$ we have

$$K^2f(x)=\int_a^x K(x, y)\, dy \int_a^y K(y, s)f(s)\, ds$$
$$=\int_a^b \int_a^b dy\, ds\, H(x-y)H(y-s)K(x, y)K(y, s)f(s)$$
$$=\int_a^b \int_a^b dy\, ds\, H(x-s)H(s-y)K(x, s)K(s, y)f(y)$$

so that the integrals are zero unless $x>s>y$. Therefore as claimed for $n=2$ we obtain

$$K^2f(x)=\int_a^x dy\left[\int_y^x dsK(x, s)K(s, y)\right]f(y)$$
$$\equiv\int_a^x K_2(x, y)f(y)\, dy$$

For arbitrary $n\geq 1$ we have

$$K^nf(x)\equiv\int_a^x dy_1K(x, y_1)\int_a^{y_1} dy_2K(y_1, y_2)\cdots\int_a^{y_{n-1}} dy_nK(y_{n-1}, y_n)f(y_n)$$
$$=\int_a^b dy_1\cdots\int_a^b dy_nH(x-y_1)$$
$$\times H(y_1-y_2)\cdots H(y_{n-1}, y_n)K(x, y_1)\cdots K(y_{n-1}, y_n)f(y_n)$$
$$=\int_a^b dy_1\cdots\int_a^b dy_nH(x-y_2)$$
$$\times H(y_2-y_3)\cdots H(y_n-y_1)K(x, y_2)K(y_2, y_3)K(y_3, y_4)\cdots K(y_n, y_1)f(y_1)$$

where we have made the change of variables $y_n\to y_1\to y_2\to y_3\to\cdots\to y_{n-1}\to y_n$ in the last step. Since the product of step functions requires that $x>y_2>y_3>\cdots>y_{n-1}>y_n>y_1$, we can rewrite this as

$$K^nf(x)\equiv\int_a^x dy_1f(y_1)K_n(x, y_1)$$
$$=\int_a^x f(y_1)\, dy_1\int_{y_1}^x dy_2\, K(x, y_2)\cdots\int_{y_1}^{y_{n-2}} dy_{n-1}\, K(y_{n-2}, y_{n-1})$$
$$\times\int_{y_1}^{y_{n-1}} dy_n\, K(y_{n-1}, y_n)K(y_n, y_1)$$
$$=\int_a^x dy_1\, f(y_1)\int_{y_1}^x dy_2\, K(x, y_2)\left[\int_{y_1}^{y_2} dy_3\, K(y_2, y_3)\cdots\right.$$
$$\left.\times\int_{y_1}^{y_{n-2}} dy_{n-1}\, K(y_{n-2}, y_{n-1})\int_{y_1}^{y_{n-1}} dy_n\, K(y_{n-1}, y_n)K(y_n, y_1)\right]$$
$$=\int_a^x dy_1\, f(y_1)\left[\int_{y_1}^x dy_2\, K(x, y_2)K_{n-1}(y_2, y_1)\right]$$

Q.E.D.

We have proven this important theorem under the assumption that $K(x, y)$ and $f(x)$ are *continuous* on the closed interval $[a, b]$. In fact a weaker form of the theorem still holds provided only that $K(x, y)$ and $f(x)$ are $\mathscr{L}_2[a, b]$; that is, if

$$\int_a^b dx \int_a^b dy\, |K(x, y)|^2 < \infty$$

$$\int_a^b |f(x)|^2\, dx < \infty$$

We will simply state the necessary results from the theory of Lebesgue integration cited in Appendix II. *Fubini's theorem* (Th. II.2) states that if $K(x, y)$ is $\mathscr{L}_2[a, b]$ then the function

$$k^2(x) \equiv \int_a^x |K(x, y)|^2\, dy$$

exists almost everywhere and is also $\mathscr{L}_2[a, b]$. Next since $k^2(x) \in \mathscr{L}_2[a, b]$ we may write

$$\frac{d}{dy} \int_a^y ds\, k^2(s) = k^2(y)$$

almost everywhere and then perform the crucial integration by parts in Eq. 5.10 (cf. Th. II.5). This allows us to demonstrate the absolute convergence of the series in Eq. 5.5 almost everywhere. Since the majorant of the terms of the series in Eq. 5.5, that is, $\kappa_m^2(x)$ of Eq. 5.10, is $\mathscr{L}_2[a, b]$, the series may be integrated term by term (cf. Th. II.3) so that the argument given in Eq. 5.11 is valid.

Notice also that the proof we gave of Th. 5.1 did not depend on the range $[a, b]$ being of finite extent. That is, the proof given will remain valid even for the infinite range provided all integrals converge (i.e., $\|K\| < \infty$, $\|f\| < \infty$).

In later work, especially for Fredholm equations, we will use the *resolvent* defined as

$$\mathscr{R}(x, y; \lambda) \equiv \sum_{n=1}^{\infty} \lambda^{n-1} K_n(x, y) \quad \textbf{(5.12)}$$

Since Eq. 5.12 is a uniformly convergent series of continuous functions, $\mathscr{R}(x, y; \lambda)$ is a continuous function of x and y. We can then express the solution to Eq. 5.2 as

$$\phi(x) = f(x) + \lambda \int_a^x \mathscr{R}(x, y; \lambda) f(y)\, dy \quad \textbf{(5.13)}$$

It is a simple matter to find the integral equation satisfied by this resolvent. If we substitute Eq. 5.13 into Eq. 5.2 we obtain

$$\int_a^x [\mathcal{R}(x, y; \lambda) - K(x, y)]f(y)\, dy$$

$$= \lambda \int_a^x dy\, K(x, y) \int_a^y \mathcal{R}(y, s; \lambda) f(s)\, ds$$

$$= \lambda \int_a^b dy \int_a^b ds\, H(x-y)H(y-s)K(x, y)\mathcal{R}(y, s; \lambda)f(s)\, ds$$

$$= \lambda \int_a^b dy \int_a^b ds\, H(x-s)H(s-y)K(x, s)\mathcal{R}(s, y; \lambda)f(y)\, dy$$

$$= \lambda \int_a^x dy \left[\int_y^x ds\, K(x, s)\mathcal{R}(s, y; \lambda) \right] f(y)\, dy \qquad \textbf{(5.14)}$$

Since $K(x, y)$ and $\mathcal{R}(x, y; \lambda)$ do not depend on $f(x)$ and since Eq. 5.2 has solutions for all $f(x) \in \mathcal{L}_2[a, b]$, we may choose $f(x)$ to be, successively, all the functions of a complete set and use Th. 4.6 to conclude that

$$\mathcal{R}(x, y; \lambda) - \lambda \int_y^x K(x, s)\mathcal{R}(s, y; \lambda)\, ds = K(x, y) \qquad \textbf{(5.15)}$$

Later we will need another version of Eq. 5.15. To this end we prove that

$$\int_y^x K_n(x, s)K(s, y)\, ds = K_{n+1}(x, y) \qquad \textbf{(5.16)}$$

For $n = 1$ this is trivially true by Eq. 5.4. Now assume this is true for $n-1$. Then the proof by induction is simple.

$$\int_y^x K_n(x, s)K(s, y)\, ds$$

$$= \int_y^x ds \int_s^x ds'\, K(x, s')K_{n-1}(s', s)K(s, y)$$

$$= \int_a^b \int_a^b ds\, ds'\, H(x-s)H(s-y)H(x-s')H(s'-s)K(x, s')K_{n-1}(s', s)K(s, y)$$

$$= \int_y^x ds'\, K(x, s') \int_y^{s'} K_{n-1}(s', s)K(s, y)\, ds = \int_y^x ds'\, K(x, s')K_n(s', y)$$

$$\equiv K_{n+1}(x, y)$$

From Eqs. 5.12 and 5.16 we deduce at once that

$$\mathcal{R}(x, y; \lambda) - \lambda \int_y^x \mathcal{R}(x, s; \lambda)K(s, y)\, ds = K(x, y) \qquad \textbf{(5.17)}$$

Although we will not deal with Volterra equations of the first kind, Eq. 5.1, we can easily show that these often can be converted into Volterra equations of the second kind by differentiation to yield

$$\phi(x)+\int_a^x \frac{1}{K(x, x)}\frac{\partial}{\partial x} K(x, y)\phi(y)\, dy=\frac{f'(x)}{K(x, x)} \tag{5.18}$$

Although this requires that the derivatives indicated exist (at least almost everywhere) and be $\mathscr{L}_2[a, b]$, the really crucial assumption is that $K(x, x)\neq 0$ anywhere on $x\in[a, b]$. We will not now treat the special case when $K(x, x)=0$ at some discrete set of points on $[a, b]$ since we shall see this can often be treated as a Fredholm equation with a weak singularity (cf. Section 5.7) or as a singular integral equation (cf. Section 7.10).

b. Connection with Ordinary Differential Equations

We will now show that solving the nonsingular second-order linear inhomogeneous ordinary differential equation

$$a(x)\frac{d^2u(x)}{dx^2}+b(x)\frac{du(x)}{dx}+c(x)u(x)=f(x) \tag{5.19}$$

subject to the initial conditions $u(x_0)=c_1$, $u'(x_0)=c_2$, $x_0\in[a, b]$, is equivalent to solving a Volterra integral equation of the second kind. If we assume that $a(x)$ does not vanish anywhere on $[a, b]$, then the integral equation

$$\phi(x)+\int_{x_0}^{x} K(x, y)\phi(y)\, dy=F(x) \tag{5.20}$$

with

$$\phi(x)\equiv\frac{d^2u(x)}{dx^2}$$

$$K(x, y)\equiv\frac{b(x)}{a(x)}+\frac{c(x)}{a(x)}(x-y)$$

$$F(x)\equiv\frac{1}{a(x)}\{f(x)-c_2b(x)-c(x)[c_1+c_2(x-x_0)]\}$$

$$u(x_0)\equiv c_1$$

$$u'(x_0)\equiv c_2$$

is satisfied by virtue of Eq. 5.19. Conversely, if we construct from $\phi(x)$, the solution to Eq. 5.20, a function

$$u(x)\equiv c_1+c_2(x-x_0)+\int_{x_0}^{x}(x-y)\phi(y)\, dy \tag{5.21}$$

then this $u(x)$ satisfies the differential equation (5.19) and the initial conditions,

$$u(x_0)=c_1 \qquad u'(x_0)=c_2 \tag{5.22}$$

However from Th. 5.1 we know that if $a(x)$, $b(x)$, $c(x)$, and $f(x)$ are continuous and $a(x)\neq 0$ for any $x\in[a, b]$, then the solution of Eq. 5.20 exists and is unique. Furthermore if $a(x)$ is of one sign (e.g., positive) everywhere on $[a, b]$ while $b(x)$, $c(x)$, and $f(x)$ are piecewise continuous and bounded, then $\phi(x)$ will exist and, in fact, be continuous except possibly at the points of discontinuity of $b(x)$, $c(x)$, and $f(x)$ [where $\phi(x)$ may have finite discontinuities]. Even in this case however, Eq. 5.21 and

$$\frac{du(x)}{dx}\equiv\int_{x_0}^{x}\phi(y)\,dy+c_2$$

imply that both $u(x)$ and $u'(x)$ are continuous for all $x\in[a, b]$. We therefore have the following important existence and uniqueness theorem for differential equations that we will return to in Section 8.2.

THEOREM 5.2

If for $x\in[a, b]$ the functions $a(x)$, $b(x)$, $c(x)$, and $f(x)$ are piecewise continuous and bounded while $a(x)$ is of one sign there, then for the differential equation

$$a(x)\frac{d^2u(x)}{dx^2}+b(x)\frac{du(x)}{dx}+c(x)u(x)=f(x)$$

subject to the initial conditions

$$u(x_0)=c_1$$
$$u'(x_0)=c_2 \qquad x_0\in[a, b]$$

there exists a unique solution such that $u(x)$ and $du(x)/dx$ are continuous for all $x\in[a, b]$.

We now discuss briefly a beautiful technique, in essence due originally to Fubini, of treating the general differential equation

$$a(x)u''(x)+b(x)u'(x)+c(x)u(x)=f(x)$$
$$u(x_0)=c_1 \qquad u'(x_0)=c_2 \tag{5.23}$$

when we know the solutions to the equation

$$A(x)v''(x)+B(x)v'(x)+C(x)v(x)=0 \tag{5.24}$$

We assume for the following that neither $a(x)$ nor $A(x)$ vanish anywhere on $x\in[a, b]$. Actually what is really sufficient is that none of the ratios $b(x)/a(x)$, $c(x)/a(x)$, $f(x)/a(x)$, $B(x)/A(x)$, $C(x)/A(x)$ is singular for any $x\in[a, b]$. We will denote the two linearly independent solutions to the homogeneous Eq. 5.24 by

$v(x)$ and $w(x)$ (cf. Section 8.2) whose *Wronskian* $W[v(x), w(x)]$ is defined as

$$W[v, w] \equiv \begin{vmatrix} v(x) & w(x) \\ v'(x) & w'(x) \end{vmatrix} = v(x)w'(x) - v'(x)w(x) \tag{5.25}$$

This cannot vanish on $[a, b]$ since

$$W[v(x), w(x)] = W[v(x_0), w(x_0)] \exp\left[-\int_{x_0}^{x} \frac{B(x')}{A(x')} dx'\right]$$

as will be shown in Section 8.2a.

In Section A5.1 we show that the solution to Eq. 5.23 is given as

$$\begin{aligned} u(x) = {} & \gamma_1\left[v(x) + \int_{x_0}^{x} H(x, y)g_1(y)\, dy\right] + \gamma_2\left[w(x) + \int_{x_0}^{x} H(x, y)g_2(y)\, dy\right] \\ & + \int_{x_0}^{x} H(x, y)g_3(y)\, dy - \int_{x_0}^{x} \frac{H(x, y)f(y)\, dy}{a(y)W[v, w]} \end{aligned} \tag{5.26}$$

where

$$H(x, y) \equiv v(x)w(y) - w(x)v(y) \tag{5.27}$$

with γ_1 and γ_2 determined as the solutions to

$$\begin{aligned} \gamma_1 v(x_0) + \gamma_2 w(x_0) &= c_1 \\ \gamma_1 v'(x_0) + \gamma_2 w'(x_0) &= c_2 \end{aligned} \tag{5.28}$$

In Eq. 5.26 the $g_j(x)$, $j = 1, 2, 3$ are the solutions to the Volterra integral equations

$$g_j(x) - \int_{x_0}^{x} K(x, y)g_j(y)\, dy = F_j(x) \qquad j = 1, 2, 3 \tag{5.29}$$

where

$$F_1(x) = \frac{1}{W[v, w]}\left\{\left[\frac{b(x)}{a(x)} - \frac{B(x)}{A(x)}\right]v'(x) + \left[\frac{c(x)}{a(x)} - \frac{C(x)}{A(x)}\right]v(x)\right\} \tag{5.30}$$

$$F_2(x) = \frac{1}{W[v, w]}\left\{\left[\frac{b(x)}{a(x)} - \frac{B(x)}{A(x)}\right]w'(x) + \left[\frac{c(x)}{a(x)} - \frac{C(x)}{A(x)}\right]w(x)\right\} \tag{5.31}$$

$$F_3(x) = F_1(x)G_1(x) + F_2(x)G_2(x) \tag{5.32}$$

$$G_1(x) = -\int_{x_0}^{x} \frac{f(y)w(y)\, dy}{a(y)W[v, w]} \tag{5.33}$$

$$G_2(x) = \int_{x_0}^{x} \frac{f(y)v(y)\, dy}{a(y)W[v, w]} \tag{5.34}$$

$$K(x, y) = \begin{vmatrix} F_1(x) & F_2(x) \\ v(y) & w(y) \end{vmatrix} \tag{5.35}$$

There are many important special cases of Eq. 5.26, some of which we will study later. For example, when we have the solutions to the homogeneous version of Eq. 5.23 [i.e., $f(x)=0$], and we want the solution to the inhomogeneous problem, then since $a(x)=A(x)$, $b(x)=B(x)$, $c(x)=C(x)$ in this case, $g_j(x)=0$, $j=1, 2, 3$, and $H(x, y)$ in Eq. 5.27 becomes the *Green's function* for the problem (cf. Section 8.6a). When $f(x)=0$ in Eq. 5.23 and, usually, $a(x)=A(x)$, $b(x)=B(x)$, but $c(x)\neq C(x)$, then we have a *perturbation expansion* in terms of the function $[C(x)-c(x)]$.

It should be clear that Eqs. 5.26 and 5.29 together always give a convergent expansion by Th. 5.1. However suppose that $F_1(x)$ and $F_2(x)$ both depend on a parameter λ and all vanish uniformly as $O(\lambda^\alpha)$, $\alpha>0$, for all $x\in[a, b]$ as $\lambda\to 0$. Then from Eqs. 5.26 and 5.29 we see that

$$u(x;\lambda)\xrightarrow[\lambda\to 0]{}\gamma_1\left[v(x)+\int_{x_0}^{x}H(x, y)F_1(y)\,dy\right]+\gamma_2\left[w(x)+\int_{x_0}^{x}H(x, y)F_2(y)\,dy\right]$$
$$+\int_{x_0}^{x}H(x, y)F_3(y)\,dy-\int_{x_0}^{x}\frac{H(x, y)f(y)\,dy}{a(y)W[v, w]}+O(\lambda^{2\alpha})$$

Example 5.1 Perhaps an example using these results will help clarify the general discussion just presented. This example is related to Problem 5.1. If we consider

$$\left[\frac{d^2}{dx^2}+k^2+\lambda V(x)\right]u(x)=0 \qquad x\in[0, \infty)$$

and

$$\left[\frac{d^2}{dx^2}+k^2\right]v(x)=0 \qquad x\in[0, \infty)$$

with k^2 real and positive at first and where $\lambda V(x)$ is assumed real and bounded such that

$$|V(x)|<\frac{\text{const.}}{x^{1+\varepsilon}} \qquad \varepsilon>0 \qquad \text{as } x\to\infty$$

then we have in terms of our notation above

$$a(x)=A(x)=1 \qquad b(x)=B(x)=0 \qquad f(x)=0$$
$$C(x)=k^2 \qquad c(x)=k^2+\lambda V(x)$$
$$v(x)=e^{ikx} \qquad v(x)=e^{-ikx}$$

Simple calculations show that

$$F_1(x)=\frac{i\lambda}{2k}V(x)v(x)$$

$$F_2(x)=\frac{i\lambda}{2k}V(x)w(x)$$

$$K(x, y)=-\frac{\lambda V(x)}{k}\sin[k(x-y)]$$

$$G_1(x)=G_2(x)=F_3(x)=g_3(x)=0$$

$$G(x)=\frac{i\lambda}{2k}V(x)[\gamma_1 v(x)+\gamma_2 w(x)]$$

so that, with $x_0=0$,

$$g_1(x)+\frac{\lambda V(x)}{k}\int_0^x \sin[k(x-y)]g_j(y)\,dy=F_j(x) \qquad j=1,2,3$$

If we define as usual

$$k^2(x)\equiv\int_0^x |K(x,y)|^2\,dy$$

then we can establish

$$\left|\int_0^x k^2(x)\,dx\right|\leq\frac{\lambda^2}{k^2}\left[\frac{1}{2}\int_0^\infty yV^2(y)\,dy+\frac{1}{4k}\int_0^\infty V^2(y)\,dy\right]$$

so that the Neumann series for $g_j(x)$, $j=1, 2$, will converge. (The reader should convince himself that this is true.)

Therefore we obtain

$$H(x,y)=2i\sin[k(x-y)]$$

so that

$$u(x)=\gamma_1\left[v(x)+\int_0^x H(x,y)g_1(y)\,dy\right]+\gamma_2\left[w(x)+\int_0^x H(x,y)g_2(y)\,dy\right]$$

which is exact. If we now assume that λ is very small, then to first order in λ we see that

$$u(x)\simeq\left[\gamma_1+\frac{i\gamma_1\lambda}{2k}\int_0^x V(y)\,dy+\frac{i\gamma_2\lambda}{2k}\int_0^x e^{-2iky}V(y)\,dy\right]e^{ikx}$$
$$+\left[\gamma_2-\frac{i\gamma_1\lambda}{2k}\int_0^x e^{2iky}V(y)\,dy-\frac{i\gamma_2\lambda}{2k}\int_0^x V(y)\,dy\right]e^{-ikx}\xrightarrow[x\to 0]{}(\gamma_1+\gamma_2)$$

If we require, for example, that $u(x=0)=0$, then $\gamma_2=-\gamma_1$. We leave the completion of the solution for $k^2>0$ for Problem 5.1 and now set $k^2=-\kappa^2<0$ in which case $k=i\kappa$, $\kappa>0$. Since $\gamma_2=-\gamma_1$ the reader should have no trouble in showing that we can obtain a solution $u(x)$ that vanishes as $x\to\infty$ if and only if

$$-1-\frac{\lambda}{2\kappa}\int_0^\infty e^{-2\kappa y}V(y)\,dy+\frac{\lambda}{2\kappa}\int_0^\infty V(y)\,dy=0$$

which is an eigenvalue equation for the allowed (discrete) values of κ. This shows explicitly how the boundary conditions determine the allowed eigenvalues.

If we consider the particularly simple case

$$V(x)=\begin{cases}V_0, & 0\leq x<a \\ 0, & x>a\end{cases} \qquad V_0>0$$

then the eigenvalue equation becomes (for $\kappa\neq 0$)

$$\frac{2\kappa}{\lambda V_0}=a-\frac{1}{2\kappa}(1-e^{-2\kappa a})$$

Let the reader show that $\kappa=0$ is a root if and only if $a^2\lambda V_0=2$. If we define the dimensionless parameters

$$\varepsilon=2\kappa a \qquad \alpha=\lambda a^2 V_0$$

then the eigenvalue equation can be written as

$$\varepsilon^2-\alpha\varepsilon+\alpha=\alpha e^{-\varepsilon}$$

There are no solutions to this when $\lambda<0$. For $\lambda>0$ there is one root $\varepsilon>0$ if and only if $\lambda a^2 V_0>2$.

5.2 CLASSIFICATION OF FREDHOLM EQUATIONS

As for Volterra integral equations we first classify *Fredholm* integral equations as either the *first kind*,

$$\int_a^b K(x, y)\phi(y)\,dy=f(x)$$

or the *second kind*,

$$\phi(x)-\lambda\int_a^b K(x, y)\,\phi(y)\,dy=f(x)$$

We will be concerned with equations of the second kind.

We further classify this equation according to its kernel.

i. If $K(x, y)$ is continuous for $a\leq x\leq b$, $a\leq y\leq b$ or if the discontinuities are such that

$$\int_a^b dx\int_a^b dy|K(x, y)|^2<\infty$$

then the equation is one of the *Fredholm type*.

ii. If

$$k(x, y)=\frac{B(x, y)}{|x-y|^\alpha}$$

where $B(x, y)$ is bounded and

$$0<\alpha<1$$

then the equation has a *weak singularity*.

iii. If

$$K(x, y)=\frac{C(x, y)}{(x-y)}$$

where $C(x, y)$ satisfies a *Hölder condition* in x and y, then the equation has a *Cauchy singularity*. A function $f(x)$ is said to satisfy a Hölder condition on $x\in[a, b]$ provided that for all x_1, $x_2\in[a, b]$ there exists a constant A and an index

μ, $0<\mu\leq 1$, such that

$$|f(x_1)-f(x_2)|\leq A\,|x_1-x_2|^{\mu}$$

We will study this type of singular integral equation in Section 7.10.

Although this classification is not complete or exhaustive in that there are Fredholm equations that do not come under any of the categories listed above, we will consider only these cases since they are the most important ones in practice.

It is clear that the Volterra integral equations of Section 5.1 can be considered as special cases of Fredholm integral equations with kernel $H(x-y)K(x, y)$ where $H(x-y)$ is the Heaviside step function [cf. Eq. (2.182)].

5.3 SUCCESSIVE APPROXIMATIONS

Given the Fredholm integral equation of the second kind,

$$\phi(x)-\lambda\int_a^b K(x, y)\phi(y)\,dy=f(x) \tag{5.36}$$

the most obvious thing we might try for a solution is a scheme of successive approximations as in Th. 5.1 for the corresponding Volterra equation. Let:

$$k^2(x)\equiv\int_a^b |K(x, y)|^2\,dy \tag{5.37}$$

$$\|K\|^2\equiv\int_a^b dx\int_a^b dy\,|K(x, y)|^2<\infty \tag{5.38}$$

$$K_{n+1}(x, y)\equiv\int_a^b K(x, s)K_n(s, y)\,ds \tag{5.39}$$

$$K_1(x, y)\equiv K(x, y)$$

$$\kappa_m^2(x)\equiv|K^n f(x)|^2 \tag{5.40}$$

$$\int_a^b |f(x)|^2\,dx<\infty \tag{5.41}$$

It then follows immediately that

$$\begin{aligned} K^n f(x) &\equiv \int_a^b dy_1\int_a^b dy_2\cdots\int_a^b dy_n K(x, y_1)K(y_1, y_2)\cdots K(y_{n-1}, y_n)f(y_n) \\ &= \int_a^b K_n(x, y)f(y)\,dy \end{aligned} \tag{5.42}$$

We now ask when the infinite series

$$\begin{aligned} \phi(x) &\equiv f(x)+\sum_{n=1}^{\infty}\lambda^n K^n f(x) \\ &= f(x)+\sum_{n=1}^{\infty}\lambda^n\int_a^b K_n(x, y)f(y)\,dy \end{aligned} \tag{5.43}$$

converges to a solution of Eq. 5.36. If we look at the proof for the convergence of the corresponding series in Th. 5.1, we see that the crucial step is the bound on $\kappa_n^2(x)$ given in Eq. 5.9. In fact we were able to prove this inequality in Eq. 5.10 only because we could write

$$k^2(x)=\frac{d}{dx}\int_a^x k^2(s)\,ds$$

and then integrate. Since we now have fixed limits of integration, we obtain the result (cf. Problem 5.3)

$$\kappa_n^2(x)\leq k^2(x)\left[\int_x^b k^2(y)\,dy\right]^{n-1}\|f\|^2\equiv k^2(x)|\mathbf{K}|^{n-1}\|f\|^2 \tag{5.44}$$

Therefore when we majorize the series in Eq. 5.43 we now obtain the bound

$$k(x)\sum_{n=1}^{\infty}|\lambda|^n\left[\int_a^b k^2(y)\,dy\right]^{(n-1)/2}\|f\|\equiv k(x)\left(\sum_{n=1}^{\infty}|\lambda|^n|\mathbf{K}|^{n-1}\right)\|f\| \tag{5.45}$$

This series will be uniformly convergent for all values of λ such that

$$|\lambda|<\frac{1}{|\mathbf{K}|} \tag{5.46}$$

provided $k(x)$ is bounded for all $x\in[a, b]$. The rest of the proof that Eq. 5.43 actually is the only solution to Eq. 5.36 is the same as the corresponding proof in Th. 5.1 and is left to the reader. Therefore, we have the following theorem.

THEOREM 5.3

If

$$|\mathbf{K}|^2\equiv\int_a^b dx\int_a^b dy|K(x, y)|^2<\infty$$

$$\|f\|^2\equiv\int_a^b dx|f(x)|^2<\infty$$

and if

$$k^2(x)\equiv\int_a^b|K(x, y)|^2\,dy$$

is bounded for all $x\in[a, b]$, then for those values of λ such that

$$|\lambda|<\frac{1}{|\mathbf{K}|}$$

the Fredholm integral equation

$$\phi(x)-\lambda\int_a^b K(x, y)\phi(y)\,dy=f(x)$$

has one and only one solution which is given by the uniformly convergent series

$$\phi(x)=f(x)+\sum_{n=1}^{\infty}\lambda^n\int_a^b K_n(x, y)f(y)\,dy$$

Just as for the corresponding Volterra equation, as discussed after the proof of Th. 5.1, conditions (5.38) and (5.41) alone [still with (5.46), of course] are sufficient to guarantee the uniqueness and convergence of (5.43) almost everywhere.

Finally let us observe that condition (5.46) for the convergence of the Neumann series in (5.43) is exactly what we would have expected from Th. 2.11 since

$$\begin{aligned}\|\lambda Kf\|^2&\equiv|\lambda|^2\int_a^b dx\int_a^b ds'\,K^*(x, s')f^*(s')\int_a^b ds\,K(x, s)f(s)\\&\leq|\lambda|^2\int_a^b dx\left(\int_a^b ds'\,|K(x, s')|^2\right)^{1/2}\|f\|\left(\int_a^b ds\,|K(x, s)|^2\right)^{1/2}\|f\|\\&=|\lambda|^2\int_a^b dx\int_a^b ds\,|K(x, s)|^2\|f\|^2\equiv|\lambda|^2\mathbf{I}K\mathbf{I}^2\|f\|^2\end{aligned}$$

For values of λ satisfying Eq. 5.46 we define the *resolvent* as

$$\mathscr{R}(x, y; \lambda)\equiv\sum_{n=1}^{\infty}\lambda^{n-1}K_n(x, y) \qquad \textbf{(5.47)}$$

Therefore at least for values of λ such that $|\lambda|<\mathbf{I}K\mathbf{I}^{-1}$, once we have constructed the resolvent for a given kernel $K(x, y)$, we obtain the solution to Eq. 5.36 for any $f(x)$ as

$$\phi(x)=f(x)+\lambda\int_a^b \mathscr{R}(x, y; \lambda)f(y)\,dy \qquad \textbf{(5.48)}$$

We can also obtain an integral equation satisfied by the resolvent, at least for $|\lambda|<\mathbf{I}K\mathbf{I}^{-1}$. From Eqs. 5.39 and 5.47 we easily see that

$$\begin{aligned}&\mathscr{R}(x, y; \lambda)-\lambda\int_a^b K(x, s)\mathscr{R}(s, y; \lambda)\,ds\\&\qquad=\sum_{n=1}^{\infty}\lambda^{n-1}K_n(x, y)-\lambda\int_a^b\sum_{n=1}^{\infty}K(x, s)\lambda^{n-1}K_n(s, y)\,ds\\&\qquad=K(x, y)+\sum_{n=2}^{\infty}\lambda^{n-1}K_n(x, y)-\sum_{n=1}^{\infty}\lambda^nK_{n+1}(x, y)\end{aligned}$$

so that

$$\mathscr{R}(x, y; \lambda)-K(x, y)=\lambda\int_a^b K(x, s)\mathscr{R}(s, y; \lambda)\,ds=\lambda\int_a^b \mathscr{R}(x, s; \lambda)K(s, y)\,ds \qquad \textbf{(5.49)}$$

In Sections A5.2 and 5.6 we will show that this equation is valid for *all* finite values of λ, except when λ is an eigenvalue.

5.4 DEGENERATE AND COMPLETELY CONTINUOUS KERNELS

We must now extend our results to the case in which $|\lambda| \geq \|K\|$. We do this by considering *degenerate* or separable (cf. Section 2.7a) kernels of the form

$$K(x, y) = \sum_{j=1}^{n} a_j(x) b_j^*(y) \tag{5.50}$$

where, without loss of generality, we can always assume the $\{a_j(x)\}$ [or the $\{b_j(x)\}$] to be linearly independent among themselves if we wish. The Fredholm equation, Eq. 5.36, then becomes

$$\phi(x) - \lambda \sum_{j=1}^{n} \langle b_j \mid \phi \rangle a_j(x) = f(x) \tag{5.51}$$

If we now project this onto the $\{b_j(x)\}$, we obtain

$$\langle b_j \mid \phi \rangle - \lambda \sum_{k=1}^{n} \langle b_k \mid \phi \rangle \langle b_j \mid a_k \rangle = \langle b_j \mid f \rangle \qquad j = 1, 2, \ldots, n \tag{5.52}$$

This is simply a set of inhomogeneous linear algebraic equations for the unknowns $\langle b_j \mid \phi \rangle$ of the type studied in Section 2.2. We must examine the $n \times n$ determinant

$$D(\lambda) \equiv \det \{\delta_{ij} - \lambda \langle b_i \mid a_k \rangle|$$

$$= \begin{vmatrix} 1-\lambda\langle b_1 \mid a_1\rangle & -\lambda\langle b_1 \mid a_2\rangle & \cdots & -\lambda\langle b_1 \mid a_n\rangle \\ -\lambda\langle b_2 \mid a_1\rangle & 1-\lambda\langle b_2 \mid a_2\rangle & \cdots & -\lambda\langle b_2 \mid a_n\rangle \\ \vdots & \vdots & \ddots & \vdots \\ -\lambda\langle b_n \mid a_1\rangle & -\lambda\langle b_n \mid a_2\rangle & \cdots & 1-\lambda\langle b_n \mid a_n\rangle \end{vmatrix} \tag{5.53}$$

This determinant is a polynomial of degree n in λ and not identically zero since $D(0) = 1$. The equation

$$D(\lambda) = 0$$

has n roots, $\lambda = \lambda_j$, $j = 1, 2, \ldots, n$. If the λ in Eq. 5.52 does not coincide with one of these $\{\lambda_j\}$, then there exists a unique solution for the $\langle b_j \mid \phi \rangle$ (cf. Th. 2.4) and the unique solution to Eq. 5.51 is given as

$$\phi(x) = f(x) + \lambda \sum_{j=1}^{n} \langle b_j \mid \phi \rangle a_j(x)$$

If $\lambda = \lambda_j$, an eigenvalue, the system in Eq. 5.52 is either inconsistent or else has an infinite number of solutions (cf. Section 2.2c and Th. 2.14). We will return to the case when λ is an eigenvalue.

We can now use these results to study the general case of completely continuous kernels that are the cornerstone of Fredholm theory. We have proven in Section 2.10 that condition (5.38), that is,

$$\|K\|^2 \equiv \int_a^b dx \int_a^b dy\, |K(x, y)|^2 < \infty$$

guarantees that $K(x, y)$ is completely continuous so that it can be approximated arbitrarily closely in the mean by a degenerate kernel (cf. Eq. 2.187). In fact, if we take $[a, b]$ to be a finite range, then we can expand (at *least* in the mean) $K(x, y)$ in a double series of orthonormal functions $\{\varphi_j(x)\}$ (e.g., a Fourier series, Section 4.3a) as

$$K(x, y) = \sum_{j,k=0}^{\infty} \alpha_{jk}\varphi_j(x)\varphi_k^*(y) \tag{5.54}$$

such that

$$\|K\|^2 = \sum_{j,k=0}^{\infty} |\alpha_{jk}|^2 \tag{5.55}$$

The basic idea now is to break off a separable piece of (5.54), which we can handle essentially as we did (5.50), and have a "small" remainder to which we can apply Th. 5.3. To this end we let

$$K_n(x, y) \equiv \sum_{j,k=0}^{n} \alpha_{jk}\varphi_j(x)\varphi_k^*(y) \tag{5.56}$$

$$\begin{aligned} R_n(x, y) &\equiv K(x, y) - K_n(x, y) \\ &= \sum_{j=0}^{n}\sum_{k=n+1}^{\infty} \alpha_{jk}\varphi_j(x)\varphi_k^*(y) + \sum_{j=n+1}^{\infty}\sum_{k=0}^{\infty} \alpha_{jk}\varphi_j(x)\varphi_k^*(y) \end{aligned} \tag{5.57}$$

such that

$$\|R_n\|^2 \equiv \sum_{j=0}^{n}\sum_{k=n+1}^{\infty} |\alpha_{jk}|^2 + \sum_{j=n+1}^{\infty}\sum_{k=0}^{\infty} |\alpha_{jk}|^2 < \frac{1}{|\lambda|^2} \tag{5.58}$$

We can now make the necessary arguments very compactly as follows if we write the original Fredholm equation, Eq. 5.36, in the abstract form

$$(I - \lambda K)\phi = f$$

$$K = K_n + R_n$$

$$K_n = \sum_{j=1}^{n} |a_j\rangle\langle b_j|$$

where, for any given value of λ, we choose n large enough so that

$$|\lambda|\, \|R_n\| < 1$$

which implies that the resolvent operator for $\mathcal{R}_n(\lambda)$ is given by the uniformly convergent series

$$\mathcal{R}_n(\lambda) = \sum_{m=1}^{\infty} \lambda^{m-1} R_n^m$$

We can then rewrite the original Fredholm equation as

$$(I-\lambda R_n)\phi=f+\lambda K_n\phi \tag{5.59}$$

so that Eq. 5.59 implies

$$\phi=\lambda\mathcal{R}_n(\lambda)f+\lambda^2\mathcal{R}_n(\lambda)K_n\phi+f+\lambda K_n\phi$$

or

$$[I-\lambda K_n-\lambda^2\mathcal{R}_n(\lambda)K_n]\phi=[f+\lambda\mathcal{R}_n(\lambda)f] \tag{5.60}$$

If we define the new separable kernel, K'_n, as

$$K'_n\equiv K_n+\lambda\mathcal{R}_n(\lambda)K_n=\sum_{j=1}^{n}[|a_j\rangle+\lambda\mathcal{R}_n(\lambda)\,|a_j\rangle]\langle b_j|\equiv\sum_{j=1}^{n}|\bar{a}_j(\lambda)\rangle\langle b_j|$$

and a known quantity $g(\lambda)$ as

$$g(\lambda)\equiv f+\lambda\mathcal{R}_n(\lambda)f$$

then Eq. 5.60 can be written as

$$[I-\lambda K'_n(\lambda)]\phi=g(\lambda) \tag{5.61}$$

We see that since Eq. 5.61 has a degenerate kernel (i.e., of the form of Eq. 5.50), all of the results true for Fredholm equations with degenerate kernels hold for the general Fredholm equation subject only to Eqs. 5.38 and 5.41. In fact the linear algebraic equations obtained from Eq. 5.61 are just

$$\langle b_k\,|g\rangle=\sum_{j=1}^{n}[\delta_{kj}-\lambda\langle b_k\mid\bar{a}_j(\lambda)\rangle]\langle b_j\mid\phi\rangle\equiv\sum_{j=1}^{n}\alpha_{kj}(\lambda)\langle b_j\mid\phi\rangle \tag{5.62}$$

so that the $n\times n$ determinant whose roots determine the eigenvalues is

$$D(\lambda)=\det\{\delta_{kj}-\lambda\langle b_k\mid\bar{a}_j(\lambda)\rangle\}\equiv\prod_j(\lambda-\lambda_j) \tag{5.63}$$

Since $D(0)=1$, we see that $D(\lambda)$ is a not identically zero holomorphic function (cf. Section 7.1) of λ (i.e., an $n\times n$ determinant each of whose elements is a holomorphic function of λ). Each value of λ such that $D(\lambda_j)=0$ is an eigenvalue. Since a holomorphic function of λ can have, at most, a *finite* number of zeros (each of finite multiplicity) in a finite region of the λ plane (cf. Th. 7.6), there can be but a finite number of eigenvalues λ_j (although of course there may be none).

Notice finally if we write Eq. 5.61 for the homogeneous case (i.e., $f=0$) *as*

$$A(\lambda_j)\phi_j=0$$

so that the eigenvalues are given as

$$\det A(\lambda_j)=0$$

then if λ_j is an eigenvalue of

$$(I-\lambda_jK)\phi_j=0$$

it follows that λ_j^* will be an eigenvalue of the adjoint equation

$$(I-\lambda_j^*K^\dagger)\psi_j=0$$

Of course the ϕ_j and ψ_j will in general be different vectors.

These results, plus Th. 2.18, lead to the theorems of the next section.

5.5 FREDHOLM'S THEOREMS

We term *eigenfunctions* nontrivial solutions $\phi_j(x)$ to the homogeneous Fredholm equation

$$\phi_j(x)-\lambda_j\int_a^b K(x, y)\phi_j(y)\, dy=0 \tag{5.64}$$

It is clear from our discussion in Section 5.4 that there exists a nontrivial solution to Eq. 5.64 if and only if λ_j is such that $D(\lambda_j)=0$. The following theorems either require no further proof or only a few comments.

THEOREM 5.4

In any finite portion of the λ plane the Fredholm equation

$$\phi(x)-\lambda\int_a^b K(x, y)\phi(y)\, dy=f(x)$$

has at most a finite number of eigenvalues so that the total number of eigenvalues is either finite or denumerably infinite, in which latter case the magnitudes of the eigenvalues increase without limit.

THEOREM 5.5

If $\lambda=\lambda_j$ is an eigenvalue, then there exist m nontrivial linearly independent eigenfunctions $\{\phi_j^{(s)}(x)\}$, $s=1, 2, \ldots, m$ (where m is a finite integer greater than or equal to unity) such that

$$\phi_j^{(s)}(x)-\lambda_j\int_a^b K(x, y)\phi_j^{(s)}(y)\, dy$$

PROOF

This follows from considering the set of n homogeneous linear algebraic equations (i.e., Eq. 5.62 with $f=0$) for the coefficients $\langle b_k \mid \phi_j^{(s)}\rangle$. From Section 2.2b we know that there is at least one (and not more than m) nontrivial (linearly independent) solution.

Q.E.D.

THEOREM 5.6: (Fredholm alternative theorem)

If, for a completely continuous kernel, the homogeneous Fredholm equation

$$\phi(x)-\lambda\int_a^b K(x, y)\phi(y)\, dy=0 \tag{5.65}$$

has only the trivial solution, then the corresponding inhomogeneous equation

$$\phi(x)-\lambda\int_a^b K(x, y)\phi(y)\, dy=f(x) \tag{5.66}$$

always has exactly one solution. If the homogeneous equation has some nontrivial solutions, then the inhomogeneous equation has no solution unless

$$\int_a^b f^*(x)\psi(x)\, dx=0 \tag{5.67}$$

for all solutions to the homogeneous adjoint equation

$$\psi(x)-\lambda^*\int_a^b K^*(y, x)\psi(y)\, dy=0 \tag{5.68}$$

in which case the inhomogeneous equation has infinitely many solutions.

PROOF

This follows from and is essentially a restatement of Th. 2.18. If and only if Eq. 5.68 has a nontrivial solution, then so does Eq. 5.65, and conversely. Therefore, if Eq. 5.65 has no nontrivial solutions, then neither does Eq. 5.68 and Eq. 5.67 is identically fulfilled so that Eq. 5.66 has a solution by Th. 2.18. Furthermore the existence of only the trivial solution to Eq. 5.65 guarantees the uniqueness of this solution to Eq. 5.66. The second part of this theorem is just a restatement of Th. 2.18. Q.E.D.

We can now define the resolvent for Eq. 5.36 for any value of λ for which there is a unique solution as

$$\phi(x)=f(x)+\lambda\int_a^b \mathcal{R}(x, y; \lambda)f(y)\, dy$$

The resolvent is given explicitly by the Neumann series, Eq. 5.47, for those values of λ such that $|\lambda|<1/\|K\|$. In the next section we will develop an explicit expression for $\mathcal{R}(x, y; \lambda)$ valid for all regular values of λ (i.e., $\lambda\neq\lambda_j$, an eigenvalue).

5.6 FREDHOLM'S RESOLVENT

In this section we will discuss an elegent expression derived by Fredholm for the resolvent operator $\mathcal{R}(x, y; \lambda)$. These results are of great importance in general discussions although they are of little use in constructing closed form solutions (when this is possible) or in obtaining solutions numerically. Since the results contained in this section do not play an essential role in what follows in the text, some readers may wish to omit it.

In Section A5.2 it is proven that if the kernel $K(x, y)$ of the Fredholm equation (5.36) satisfies the bound

$$|K(x, y)| \leq N < \infty$$

for all x and y on the finite range $[a, b]$, then the resolvent operator can be written as

$$\mathcal{R}(x, y; \lambda) = \frac{N(x, y; \lambda)}{D(\lambda)} \tag{5.69}$$

for all values of λ for which $D(\lambda) \neq 0$. Furthermore it is shown that neither $N(x, y; \lambda)$ nor $D(\lambda)$ has any singularities for any finite value of λ. In fact one might expect from the limiting argument that leads to Eq. A5.21 that $D(\lambda_j) = 0$ if and only if λ_j is an eigenvalue of the Fredholm equation. Let us assume that $\mathcal{R}(x, y; \lambda)$ were holomorphic in the neighborhood of $\lambda = \lambda_j$ an eigenvalue; that is when

$$\phi_j(x) - \lambda_j \int_a^b K(x, y)\phi_j(y)\, dy = 0$$

From Th. 5.6 we know that the adjoint equation

$$\psi_j(x) - \lambda_j^* \int_a^b K^*(y, x)\psi_j(y)\, dy = 0$$

has a nontrivial solution $\psi_j(x)$. But then also by Th. 5.6 the equation

$$\phi(x) - \lambda_j \int_a^b K(x, y)\phi(y)\, dy = \psi_j(x) \tag{5.70}$$

can have no solution. Let λ be a regular point (i.e., not an eigenvalue) close to λ_j so that the equation

$$\bar{\phi}(x) - \lambda \int_a^b K(x, y)\bar{\phi}(y)\, dy = \psi_j(x)$$

has a solution given by

$$\bar{\phi}(x) = \psi_j(x) + \lambda \int_a^b \mathcal{R}(x, y; \lambda)\psi_j(y)\, dy \tag{5.71}$$

Now if $\mathcal{R}(x, y; \lambda)$ were not singular at $\lambda = \lambda_j$, we could take the limit $\lambda \to \lambda_j$ in Eq. 5.71 to obtain a solution to Eq. 5.70, which would be a contradiction. Therefore $\mathcal{R}(x, y; \lambda)$ must have a pole at $\lambda = \lambda_j$ [i.e., $D(\lambda_j) = 0$].

Now suppose $D(\lambda_j) = 0$. From Eq. A5.23 it follows that

$$N(x, y; \lambda_j) - \lambda_j \int_a^b K(x, s)N(s, y; \lambda_j)\, ds = 0 \tag{5.72}$$

so that $N(x, y; \lambda_j)$ is a nontrivial solution to the eigenvalue problem, Eq. 5.64, provided it does not vanish. From Eqs. A5.21 and A5.24 we see that

$$\frac{d}{d\lambda} D(\lambda) = -\sum_{m=1}^{\infty} \frac{(-\lambda)^{m-1}}{(m-1)!} \int_a^b dx_1 \cdots dx_m K\begin{pmatrix} x_1, x_2, \ldots, x_m \\ x_1, x_2, \ldots, x_m \end{pmatrix}$$

$$= -\sum_{m=0}^{\infty} \frac{(-\lambda)^m}{m!} \int_a^b dx dx_1 \cdots dx_m K\begin{pmatrix} x, x_1, x_2, \ldots, x_m \\ x, x_1, x_2, \ldots, x_m \end{pmatrix}$$

$$= -\int_a^b N(x, x; \lambda)\, dx \tag{5.73}$$

If $D(\lambda)$ has a *simple* (or first order) *zero* (cf. Section 7.5) at $\lambda = \lambda_j$ so that

$$\frac{d}{d\lambda} D(\lambda)|_{\lambda_j} \neq 0$$

then $N(x, y; \lambda_j)$ cannot vanish identically. Therefore $N(x, y; \lambda)$ is a nontrivial eigenfunction. When $D(\lambda)$ has a zero of order $m > 1$ [i.e., $D(\lambda) \xrightarrow[\lambda \to \lambda_j]{} \text{const.}\, (\lambda - \lambda_j)^m$], then a nontrivial eigenfunction is given by $d^{m-1}N(x, y; \lambda)/d\lambda^{m-1}$, although the proof is quite involved. Since a complete discussion of this point would carry us rather far afield and cover several pages, we refer the interested reader to Lovitt's book (Chapter 3, Sec. 21) and state the following theorem, nearly all of which we have established.

THEOREM 5.7

If the kernel satisfies the bound $|K(x, y)| < N$ on the finite range $[a, b]$, then the Fredholm resolvent can be written as the ratio of two functions holomorphic in the entire λ plane,

$$\mathcal{R}(x, y; \lambda) = \frac{N(x, y; \lambda)}{D(\lambda)} \tag{5.74}$$

where $N(x, y; \lambda)$ is defined in Eq. A5.24 and $D(\lambda)$ in Eq. A5.21. Furthermore $D(\lambda_j) = 0$ if and only if λ_j is an eigenvalue of the Fredholm equation (Eq. 5.64).

5.7 WEAK SINGULARITIES

In Section 5.2 we defined a kernel with a *weak singularity* as one such that

$$K(x, y) = \frac{B(x, y)}{|x-y|^{\alpha}} \qquad 0 < \alpha < 1 \tag{5.75}$$

where $B(x, y)$ is bounded on $[a, b]$. We will first show that if we iterate the kernel of Eq. 5.75 a sufficient number of times according to formula (5.39), then

$K_n(x, y)$ will finally become $\mathscr{L}_2[a, b]$. Assume $x<y$ and consider the integral

$$I(x, y) \equiv \int_a^b \frac{B(x, s)B'(s, y)\,ds}{|x-s|^\alpha\,|s-y|^{\alpha'}}$$

$$= \int_a^x \frac{B(x, s)B'(s, y)\,ds}{(x-s)^\alpha(y-s)^{\alpha'}} + \int_x^y \frac{B(x, s)B'(s, y)\,ds}{(s-x)^\alpha(y-s)^{\alpha'}} + \int_y^b \frac{B(x, s)B'(s, y)\,ds}{(s-x)^\alpha(s-y)^{\alpha'}}$$

Now for each of these integrals in succession let

$$s = x-(y-x)t$$

$$s = x+(y-x)t$$

$$s = y+(y-x)t$$

so that

$$I(x, y) = (y-x)^{1-\alpha-\alpha'}\left\{\int_0^{(x-a)/(y-x)} \frac{B[x, x-(y-x)t]B'[x-(y-x)t, y]\,dt}{t^\alpha(1+t)^{\alpha'}}\right.$$

$$+\int_0^1 \frac{B[x, x+(y-x)t]B'[x+(y-x)t, y]\,dt}{t^\alpha(1-t)^{\alpha'}}$$

$$\left.+\int_0^{(b-y)/(y-x)} \frac{B[x, y+(y-x)t]B'[y+(y-x)t, y]\,dt}{(1+t)^\alpha t^{\alpha'}}\right\}$$

$$\equiv (y-x)^{1-\alpha-\alpha'}\bar{B}(x, y) \tag{5.76}$$

where $\bar{B}(x, y)$ is necessarily a bounded function since $B(x, y)$ and $B'(x, y)$ are assumed bounded and since

$$\int_0^1 \frac{dt}{t^\alpha(1-t)^{\alpha'}}$$

$$\int_0^R \frac{dt}{t^\alpha(1+t)^{\alpha'}}$$

both exist because $0<\alpha<1$, $0<\alpha'<1$. Specifically, if we compute $K_2(x, y)$, we find

$$K_2(x, y) \equiv \int_a^b \frac{B(x, s)B(s, y)\,ds}{|x-s|^\alpha\,|s-y|^\alpha} \equiv |x-y|^{1-2\alpha}B_2(x, y) \tag{5.77}$$

If $0<\alpha\leq\frac{1}{2}$, then $K_2(x, y)$ is no longer singular on $[a, b]$. Furthermore we easily see by repeated application of Eq. 5.76 that

$$K_n(x, y) = |x-y|^{n(1-\alpha)-1}B_n(x, y) \tag{5.78}$$

where $B_n(x, y)$ is bounded, so that even with $\frac{1}{2}<\alpha<1$ we can always iterate enough times until $n(1-\alpha)-1>0$, in which case $K_n(x, y)$ will be bounded on $[a, b]$ as, of course, will be all the further iterated kernels. Since $K_n(x, y)$ eventually becomes *bounded* on $[a, b]$, the Fredholm resolvent of Section 5.6 is applicable.

What remains to be done is to examine the relation between the original Fredholm equation

$$\phi(x)-\lambda\int_a^b K(x, y)\phi(y)\,dy=f(x) \tag{5.79}$$

and an equation with an iterated kernel $K_n(x, y)$. If we operate on Eq. 5.79 with $(I+\lambda K)$ we find that

$$\phi(x)-\lambda^2\int_a^b K_2(x, y)\phi(y)\,dy=f_2(x) \tag{5.80}$$

where

$$f_2(x)=f(x)+\lambda\int_a^b K(x, y)f(y)\,dy$$

In fact one easily shows that

$$\phi(x)-\lambda^n\int_a^b K_n(x, y)\phi(y)\,dy=f_n(x) \qquad n=1, 2, \ldots \tag{5.81}$$

where

$$f_1(x)\equiv f(x)$$

$$f_{n+1}(x)=f(x)+\lambda\int_a^b K(x, y)f_n(y)\,dy \qquad n\geq 1 \tag{5.82}$$

It is clear that all the solutions of (5.79) are solutions of Eq. 5.81, but the converse is not necessarily true. Since all the solutions of (5.79) are contained in those of (5.81), we may find all the solutions of (5.81) for some large enough n [i.e., $n>1/(1-\alpha)$] by the methods of the previous sections of this chapter and then substitute each of these back into Eq. 5.79 to see which ones actually are the solutions. Notice that the integral in Eq. 5.79 will always exist in this process since $\phi(y)$ will be bounded and the singularity of $K(x, y)$ is integrable because $0<\alpha<1$. The theorems of Section 5.5 (i.e., Th. 5.4–5.6, inclusive) are easily extended to the class of kernels of Eq. 5.75. The interested reader can show this for himself.

We will now show that all of the eigenvalues $\{\mu_k\}$ of

$$\phi_k(x)-\mu_k\int_a^b K_2(x, y)\phi_k(y)\,dy=0 \tag{5.83}$$

are contained in the set $\{\lambda_j^2\}$ of the eigenvalues of

$$\phi_j(x)-\lambda_j\int_a^b K(x, y)\phi(y)\,dy=0 \tag{5.84}$$

It is clear from what we have already said above that $\mu_k=\lambda_j^2$ is necessarily an eigenvalue of Eq. 5.83. We demonstrate the converse by writing Eq. 5.83 as

$$(I-\sqrt{\mu_k}K)(I+\sqrt{\mu_k}K)\phi_k=0 \tag{5.85}$$

If $\sqrt{\mu_j}=\lambda_j$, the statement is true. If $\sqrt{\mu_j}\neq\lambda_j$, then

$$(I-\sqrt{\mu_b}K)\psi=0$$

has only the trivial solution $\psi=0$ so that Eq. 5.85 implies that

$$(I+\sqrt{\mu_k}K)\phi_k=0$$

Since ϕ_k was assumed nontrivial, $-\sqrt{\mu_k}$ must be an eigenvalue of K (i.e., $-\sqrt{\mu_k}=\lambda_j$) and the claim is established. Therefore, as long as λ_2 is not an eigenvalue of (5.83), then both (5.80) and (5.79) have unique solutions and they must be identical. However if $\lambda^2=\mu_k$, an eigenvalue of (5.83), then (5.80) may have no solutions or infinitely many. Of course if (5.80) has no solution, then neither does (5.79). If (5.80) has infinitely many solutions, we do not know which, if any, of these are solutions to (5.79).

If we now observe that any polynomial of degree n can always be written as a product of factors of its roots x_j as

$$\sum_{s=0}^{n} a_s x^s=\prod_{j=1}^{n}(x-x_j), \qquad a_n=1$$

then

$$\begin{aligned}1-a^n &= -\prod_{s=1}^{n}(a-\eta_n^{-s})\\ &= -\prod_{s=1}^{n}\eta_n^{-s}(\eta_n^{s}a-1)=\left(\prod_{s=1}^{n}\eta_n^{-s}\right)(-1)^{n-1}\prod_{s=1}^{n}(1-\eta_n^{s}a)=\prod_{s=1}^{n}(1-\eta_n^{s}a)\end{aligned}$$

where $\eta_n=e^{2\pi i/n}$, the n roots of unity, and

$$\prod_{s=1}^{n}\eta_n^{-s}=e^{-\frac{2\pi i}{n}\sum_{s=1}^{n}s}=e^{-\frac{2\pi i}{n}\frac{n(n+1)}{2}}=(-1)^{n+1}=(-1)^{n-1}$$

Therefore we can rewrite Eq. 5.81 in abstract form as

$$(I-\lambda^nK^n)\phi=\prod_{s=1}^{n}(I-\eta_n^{s}\lambda K)\phi=f_n \qquad n=1,2,\ldots \tag{5.86}$$

We can now repeat the argument given for Eq. 5.83 to deduce that *all* of the eigenvalues of

$$\phi_k-\mu_k\int_a^b K_n(x,y)\phi_k(y)\,dy=0$$

are contained in the $\{\lambda_j^n\}$. That is, finally for some $s=1,2,\ldots,n$,

$$\eta_n^{s}\mu_k^{1/n}=\lambda_j$$

where λ_j is an eigenvalue of Eq. 5.79. Therefore we have the following theorem.

THEOREM 5.8

The set of eigenvalues of the nth iterated kernel, $K_n(x, y)$, coincides with the set of nth powers, $\{\lambda_j^n\}$, of the eigenvalues $\{\lambda_j\}$ of the kernel $K(x, y)$.

Finally, once we have a nontrivial solution of

$$(I-\lambda_j^n K^n)\psi_j=0 \tag{5.87}$$

we can always write a solution of

$$(I-\lambda_j K)\phi_j=0 \tag{5.88}$$

as

$$\phi_j=\prod_{s=1}^{n-1}(I-\eta_n^s\lambda_j K)\psi_j \tag{5.89}$$

This ϕ_j will necessarily be nontrivial unless at least one of the $\eta_n^s\lambda_j$, $s=1, 2, \ldots, n-1$, is in the eigenvalue spectrum of Eq. 5.88. Let $s=\{n_i\}$, $i=1, 2, \ldots, m$, label the $\eta_n^s\lambda_j$ in the eigenvalue spectrum of Eq. 5.88. Since all of the solutions of Eq. 5.88 are solutions of Eq. 5.87, it follows that there exist nontrivial linearly independent eigenfunctions $\psi_j^{(i)}$, 1, 2, ..., m, of Eq. 5.87 such that the functions

$$\phi_j^{(i)}=\prod_{\substack{s=1\\ s\neq n_i}}^{n-1}(I-\eta_n^s\lambda_j K)\psi_j^{(i)} \qquad i=1, 2, \ldots, m \tag{5.90}$$

are nontrivial eigenfunctions of

$$(I-\eta_n^{n_i}\lambda_j K)\phi_j^{(i)}=0 \qquad i=1, 2, \ldots, m$$

We need only begin with a nontrivial solution ψ_j of Eq. 5.87 and construct in succession

$$(I-\eta_n\lambda_j K)\psi_j,\ (I-\eta_n\lambda_j K)(I-\eta_n^2\lambda_j K)\psi_j, \ldots,\ (I-\eta_n\lambda_j K)(I-\eta_n^2\lambda_j K)\cdots(I-\eta_n^{n-1}\lambda_j K)\psi_j$$

If one of these should vanish, then we have found another eigenvalue of Eq. 5.88.

Therefore once we have the solutions to the iterated Fredholm equations, Eqs. 5.81 or 5.87, we can obtain the solutions to the original equations with weak singularities, Eqs. 5.79 or 5.88.

5.8 HILBERT-SCHMIDT THEORY

This section is concerned with *symmetric* or self-adjoint kernels,

$$K(x, y)=K^*(y, x) \tag{5.91}$$

As we might expect from our study of self-adjoint matrices, we should be able to say considerably more about Fredholm equations with symmetric kernels than we have been able to about ones with an arbitrary kernel. As usual we shall assume that $K(x, y)$ is $\mathscr{L}_2[a, b]$. In particular we will discuss the conditions under which the eigenfunctions $\{\phi_j(x)\}$ of

$$\phi_j(x)-\lambda_j\int_a^b K(x, y)\phi_j(y)\,dy=0 \tag{5.92}$$

form a complete set.

There are a few important facts about the eigenvalues and eigenfunctions of symmetric kernels that we simply list here. If truth of these statements is not obvious to the reader, he can supply the proofs easily enough.

i. The eigenvalues of a symmetric kernel are real.
ii. The eigenfunctions of a symmetric kernel corresponding to different eigenvalues are orthogonal.
iii. The sequence of eigenfunctions $\{\phi_j(x)\}$ of a symmetric kernel can be made orthonormal on the range $[a, b]$.

Since Th. 5.4 guarantees that the eigenvalue spectrum of Eq. 5.92 is discrete, we agree to label our eigenvalues $\{\lambda_j\}$ such that

$$|\lambda_1|\leq|\lambda_2|\leq|\lambda_3|\leq\cdots\leq|\lambda_n|\leq\cdots \tag{5.93}$$

THEOREM 5.9

For a symmetric kernel, the magnitude of the smallest eigenvalue is given as

$$\frac{1}{|\lambda_1|}=\text{Max}\,|\langle\varphi \mid K\varphi\rangle|\equiv\text{Max}\left|\int_a^b dx\int_a^b dy\,K(x, y)\varphi^*(x)\varphi(y)\right| \tag{5.94}$$

where the minimum is taken with respect to all $\varphi(x)$ such that $\|\varphi\|=1$.

PROOF

From Eq. 2.187 for a completely continuous kernel,

$$\lim_{n\to\infty}\|(K-K_n)\varphi\|=0 \tag{5.95}$$

and Eqs. 5.54 and 5.56, where $\alpha_{jk}=\alpha_{kj}^*$ now, we see from the Schwarz inequality that

$$|\langle\varphi|\,(K-K_n)\,|\varphi\rangle|\leq\|\varphi\|\;\|(K-K_n)\varphi\|$$

so that Eq. 5.95 implies

$$\lim_{n\to\infty}|\langle\varphi|\,K_n\,|\varphi\rangle|=|\langle\varphi|\,K\,|\varphi\rangle| \tag{5.96}$$

We can then use the bounds on $\langle x|\,A\,|x\rangle$ established for self-adjoint matrices in Eq. 3.38 to show that the modulus of the minimum eigenvalue of K_n is given as

$$|\lambda_1^{(n)}|^{-1} = \text{Max}\, |\langle\varphi|\, K_n\, |\varphi\rangle| \qquad \|\varphi\| = 1$$

where K_n is the approximate kernel of Eq. 5.56 and $\lambda_1^{(n)}$ is the root of smallest norm of

$$D_n(\lambda) \equiv \det\{\delta_{kl} - \lambda\alpha_{kl}\} = 0 \tag{5.97}$$

From Sections A5.2 and 5.6 (cf. especially Eqs. A5.16 and A5.21) we see that the roots of Eq. 5.97 become the exact eigenvalues of Eq. 5.92 in the limit $n \to \infty$ under the assumption that $K(x, y)$ is bounded {which is a stronger assumption than $K(x, y) \in \mathscr{L}_2[a, b]$}. Under this assumption on $K(x, y)$ we can then use Eq. 5.96 to write

$$|\lambda_1|^{-1} = \lim_{n\to\infty} |\lambda_1^{(n)}|^{-1} = \lim_{n\to\infty} \underset{\|\varphi\|=1}{\text{Max}}\, |\langle\varphi|\, K_n\, |\varphi\rangle| = \underset{\|\varphi\|=1}{\text{Max}}\, |\langle\varphi|\, K\, |\varphi\rangle| \tag{5.98}$$

which would establish the desired result. More generally, if we compare Eq. 5.97 with Eq. 5.63, which gives the *exact* eigenvalues as long as $\mathbf{|}K\mathbf{|} < \infty$, we see that we must estimate the magnitude of the term

$$\sum_{j=1}^{n} \alpha_{jk} \langle\varphi_j|\, \mathscr{R}_n(\lambda)\, |\varphi_j\rangle \tag{5.99}$$

since

$$D(\lambda) = \det\left\{\delta_{kl} - \lambda\alpha_{kl} - \lambda \sum_{j=0}^{n} \alpha_{jk} \langle\varphi_l|\, \mathscr{R}_n(\lambda)\, |\varphi_j\rangle\right\} \tag{5.100}$$

Here we are computing both determinants in the orthonormal basis $\{\varphi_j(x)\}$ of Eq. 5.54. In Section A5.3 it is shown that the expression in Eq. 5.99 vanishes in the limit $n \to \infty$. This implies that the roots of Eq. 5.97 approach the roots of Eq. 5.100 as $n \to \infty$ so that we may again establish Eq. 5.98.

We can also give a more elegant proof of this if we use the results of Sections A5.4 and A5.5. These two sections taken together show that we may equivalently define a completely continuous operator K as one that takes every bounded sequence $\{\chi_n\}$ in the Hilbert space into a sequence $\{K\chi_n\}$ that contains a strongly convergent subsequence. Let us begin by assuming that K is a symmetric (i.e., self-adjoint) operator on our Hilbert space. Its norm $\|K\|$ is defined as the smallest number that provides an upper bound (i.e., a least upper bound, written l.u.b.) on $\|K\chi\|$, $\forall \chi \ni \|\chi\| = 1$,

$$\text{l.u.b.}\, \|K\chi\| \equiv \|K\| \tag{5.101}$$

Notice that the norm defined here, $\|K\|$, and that of Eq. 5.38, that is, $\mathbf{|}K\mathbf{|}$, are *not* the same. In fact we have

$$\|K\varphi\|^2 \equiv \int_a^b \left|\int_a^b dy\, K(x, y)\varphi(y)\right|^2 dx \leq \int_a^b dx \int_a^b dy\, |K(x, y)|^2\, \|\varphi\|^2 \equiv \mathbf{|}K\mathbf{|}^2\, \|\varphi\|^2$$

so that

$$\|K\| \leq |K| \tag{5.102}$$

If we again restrict ourselves to those χ such that $\|\chi\|=1$ we have

$$|\langle\chi| K |\chi\rangle| \leq \|K\chi\| \leq \|K\| \tag{5.103}$$

We define M as the least upper bound of $|\langle\chi| K |\chi\rangle|$,

$$\underset{\|\chi\|=1}{\text{l.u.b.}} |\langle\chi| K |\chi\rangle| \equiv M$$

We now establish that $M=\|K\|$. We already see that $M \leq \|K\|$. For any real value of α we have the following identity.

$$\|K\chi\|^2 \equiv \frac{1}{4}\left[\left\langle K\left(\alpha\chi+\frac{1}{\alpha} K\chi\right), \alpha\chi+\frac{1}{\alpha} K\chi\right\rangle - \left\langle K\left(\alpha\chi-\frac{1}{\alpha} \chi\right), \alpha\chi-\frac{1}{\alpha} K\chi\right\rangle\right]$$

$$\leq \frac{M}{4}\left(\left\|\alpha\chi+\frac{1}{\alpha} K\chi\right\|^2 + \left\|\alpha\chi-\frac{1}{\alpha} K\chi\right\|^2\right)$$

$$= \frac{M}{2}\left(\alpha^2+\frac{1}{\alpha^2}\|K\chi\|^2\right)$$

If $\|K\chi\| \neq 0$ (and we assume this since $K=0$ is an uninteresting case), then the final expression above attains its minimum value when

$$\alpha^2 = \|K\chi\|$$

so that

$$\|K\chi\|^2 \leq M\|K\chi\|$$

or

$$\|K\chi\| \leq M$$

which implies that

$$\|K\| \leq M$$

Therefore the two inequalities established between M and $\|K\|$ show that

$$\underset{\|\chi\|=1}{\text{l.u.b.}} |\langle\chi| K |\chi\rangle| = \underset{\|\chi\|=1}{\text{l.u.b.}} \|K\chi\| = \|K\| \tag{5.104}$$

Now define a bounded sequence of vectors $\{\chi_n\}$, $\|\chi_n\|=1$, such that

$$\lim_{n\to\infty} |\langle\chi_n| K |\chi_n\rangle| = \|K\| \tag{5.105}$$

Since

$$|\langle\chi_n| K |\chi_n\rangle| \leq \|K\|$$

we know (cf. Section A5.5 or the Bolzano-Weierstrass theorem) that there exists a convergent subsequence of the $\{\langle\chi_n \mid K\chi_n\rangle\}$ that we now consider such that

$$\lim_{n\to\infty} \langle\chi_n \mid K\chi_n\rangle \equiv \mu$$

so that $\mu = \pm\|K\|$. We then have

$$0 \leq \|K\chi_n - \mu\chi_n\|^2 = \|K\chi_n\|^2 - 2\mu\langle K\chi_n, \chi_n\rangle + \mu^2$$
$$\leq 2\mu(\mu - \langle K\chi_n, \chi_n\rangle)$$

If we take the limit as $n \to \infty$ we obtain

$$\lim_{n\to\infty} \|K\chi_n - \mu\chi_n\|^2 = 0$$

or

$$\lim_{n\to\infty} (K\chi_n - \mu\chi_n) = 0 \tag{5.106}$$

We have not yet used the fact that K is completely continuous. If we do, then we know there exists a strongly convergent subsequence of $\{K\chi_n\}$, for example, $\{K\bar{\chi}_n\}$, such that

$$\lim_{n\to\infty} K\bar{\chi}_n = K\bar{\chi} \tag{5.107}$$

which implies that

$$\lim_{n\to\infty} (\mu\bar{\chi}_n) = K\bar{\chi} = \mu\bar{\chi} \qquad \|\bar{\chi}\| = 1 \tag{5.108}$$

Therefore Eq. 5.108 states that every completely continuous operator K (not identically zero) has at least one nontrivial solution for

$$\phi_1 - \lambda_1 K\phi_1 = 0$$

corresponding to $\lambda_1 = \pm\|K\|^{-1}$. Furthermore there can be no other eigenvalue, for example, $\bar{\lambda}$, such that $|\bar{\lambda}| < |\lambda_1|$ since if there were we would have

$$\bar{\phi} - \bar{\lambda}K\bar{\phi} = 0 \qquad \|\bar{\phi}\| = 1$$

which would imply

$$|\langle\bar{\phi}|\, K \,|\bar{\phi}\rangle| = \frac{1}{|\bar{\lambda}|} > \frac{1}{|\lambda_1|} \equiv \|K\|$$

which would contradict the fact that $\|K\|$ is the *least* upper bound of $|\langle\phi|\, K \,|\phi\rangle|$ for all ϕ, $\|\phi\| = 1$, in the Hilbert space.

We can now use Bessel's inequality (cf. Section 4.1a) to deduce from Eq. 5.92 that

$$k^2(x) = \int_a^b |K(x, y)|^2\, dy \geq \sum_{j=1}^{\infty} \left|\int_a^b K(x, y)\phi_j^*(y)\, dy\right|^2 = \sum_{j=1}^{\infty} \frac{|\phi_j(x)|^2}{\lambda_j^2} \tag{5.109}$$

so that this infinite sum converges and is bounded for all values of x provided $k^2(x) < \infty$. [In any event, since $\mathbf{|K|} < \infty$ this series will converge almost everywhere (cf. Appendix II, Th. II.3).] If we normalize our eigenfunctions to unity, then Eq. 5.109 states that

$$\sum_{j=1}^{\infty} \frac{1}{\lambda_j^2} \leq \mathbf{|K|}^2 < \infty \tag{5.110}$$

so that there can be only *finite* degeneracy of the $\{\phi_i\}$ associated with any (finite) λ_j.

Also, since $\langle\varphi|\,K\,|\varphi\rangle\equiv 0$, $\forall\varphi\ni\|\varphi\|=1$, would require $\mathbf{|K|}=0$ or that $K(x, y)=0$ almost everywhere (which we assume is not the case) and since

$$\left|\int_a^b dx\int_a^b dy\, K(x, y)\varphi^*(x)\varphi(y)\right|\le\left(\int_a^b dx\left|\int_a^b dy\, K(x, y)\varphi(y)\right|^2\right)^{1/2}\|\varphi\|$$

$$\le\left(\int_a^b dx\int_a^b dy|K(x, y)|^2\right)^{1/2}=\mathbf{|K|}$$

we see from Eq. 5.94 that a nonzero symmetric kernel always has a finite, nonzero lowest eigenvalue. We can now use the bound of Eq. 3.41 and the proof of Th. 5.9 to see that

$$|\lambda_2|^{-1}=\operatorname{Max}|\langle\varphi|\,K\,|\varphi\rangle| \qquad \|\varphi\|=1 \qquad \langle\varphi\mid\phi_1\rangle=0 \tag{5.111}$$

where ϕ_1 is the normalized eigenfunction of Eq. 5.92 corresponding to eigenvalue λ_1. Furthermore we can effectively remove eigenvalue λ_1 from the spectrum of $K(x, y)$ if we define another symmetric kernel $K^{(2)}(x, y)$ as

$$K(x, y)\equiv K^{(2)}(x, y)+\frac{\phi_1(x)\phi_1^*(y)}{\lambda_1} \tag{5.112}$$

so that

$$K^{(2)}\phi_1=K\phi_1-\langle\phi_1\mid\phi_1\rangle\frac{\phi_1}{\lambda_1}=0 \tag{5.113}$$

Since $K^{(2)}(x, y)$ is a symmetric kernel, then either it is identically zero (almost everywhere) or the reciprocal of its lowest modulus, nonzero eigenvalue is given by

$$\frac{1}{|\lambda_2|}=\operatorname{Max}|\langle\varphi|\,K^{(2)}\,|\varphi\rangle| \qquad \ni\|\varphi\|=1 \tag{5.114}$$

However the value obtained from Eq. 5.114 is the same as that gotten from Eq. 5.111 since we may write any φ as

$$\varphi=\alpha\phi_1+\chi \qquad \ni\langle\phi_1\mid\chi\rangle=0$$

so that Eq. 5.111 becomes

$$\langle\varphi|\,K\,|\varphi\rangle=\langle\varphi\mid K^{(2)}\,|\varphi\rangle+\frac{|\langle\phi_1\mid\varphi\rangle|^2}{\lambda_1}=\langle\chi|\,K^{(2)}\,|\chi\rangle\equiv\langle\varphi|\,K^{(2)}\,|\varphi\rangle$$

This tells us that for a symmetric kernel either there exists a denumerable infinity of eigenvalues (and, of course, *at least* one) or the kernel

$$K^{(n)}(x, y)\equiv K(x, y)-\sum_{j=1}^{n-1}\frac{\phi_j(x)\phi_j^*(y)}{\lambda_j} \qquad n\ge 2 \tag{5.115}$$

vanishes identically once n is large enough so that $K(x, y)$ is, in fact, degenerate. We also easily see that $K^{(n)}(x, y)$ has *only* the eigenvalues $\lambda_n, \lambda_{n+1}, \ldots$, and eigenfunctions $\phi_n, \phi_{n+1}, \ldots$, and *no* others. Clearly if

$$\phi_j - \lambda_j K\phi_j = 0 \qquad j = n, n+1, \ldots$$

then

$$\phi_j - \lambda_j K^{(n)}\phi_j = \lambda_j \sum_{l=1}^{n-1} \frac{\langle \phi_l \mid \phi_j \rangle}{\lambda_l} \phi_l = 0$$

We need not consider ϕ_j for $j < n$, since $K^{(n)}\phi_j \equiv 0$, $j < n$. If

$$\psi - \mu K^{(n)}\psi = 0$$

then

$$\psi - \mu K\psi = -\mu \sum_{l=1}^{n-1} \frac{\langle \phi_l \mid \psi \rangle}{\lambda_l} \phi_l \tag{5.116}$$

or

$$\langle \phi_k \mid \psi \rangle = 0 \qquad k < n \tag{5.117}$$

$$\langle \phi_k \mid \psi \rangle \left(1 - \frac{\mu}{\lambda_k}\right) = 0 \qquad k \geq n$$

If we substitute Eq. 5.117 back into Eq. 5.116, we find that

$$\psi - \mu K\psi = 0$$

But by Th. 5.6 we see that this requires that the ψ be an eigenfunction and μ an eigenvalue.

We can now establish the main theorem of this section.

THEOREM 5.10 (Hilbert-Schmidt theorem)

If $\{\lambda_j\}$ are the eigenvalues and $\{\phi_j(x)\}$ the orthonormal eigenfunctions of a symmetric kernel $K(x, y)$ satisfying

$$k^2(x) \equiv \int_a^b |K(x, y)|^2 \, dy < C \tag{5.118}$$

(i.e., a uniform bound independent of x) and $h(x) \in \mathscr{L}_2[a, b]$, then the function

$$f(x) \equiv Kh(x) = \int_a^b K(x, y)h(y) \, dy \tag{5.119}$$

may be expanded in the absolutely and uniformly convergent series

$$f(x) = \sum_{j=1}^{\infty} \frac{\langle h \mid \phi_j \rangle^*}{\lambda_j} \phi_j(x) \tag{5.120}$$

PROOF

It is a simple matter to establish the uniform and absolute convergence of the series in Eq. 5.120 by use of the Schwarz and Bessel inequalities as (cf. Eq. 5.109)

$$\left|\sum_{j=N}^{\infty} \langle h \mid \phi_j\rangle^* \frac{\phi_j(x)}{\lambda_j}\right| \le \left(\sum_{j=N}^{\infty} |\langle h \mid \phi_j\rangle|^2\right)^{1/2} \left(\sum_{j=N}^{\infty} \frac{|\phi_j(x)|^2}{\lambda_j^2}\right)^{1/2}$$
$$\le \left(\sum_{j=N}^{\infty} |\langle h \mid \phi_j\rangle|^2\right)^{1/2} k(x)$$
$$\le \left(\sum_{j=N}^{\infty} |\langle h \mid \phi_j\rangle|^2\right)^{1/2} \sqrt{C} \le \|h\|\sqrt{C}$$

We also know that if $K(x, y)$ is *continuous* then so are the $\{\phi_j(x)\}$ and $f(x) = Kh(x)$. Finally we must show that the sequence of partial sums

$$S_n(x) = \sum_{j=1}^{n} \frac{\langle h \mid \phi_j\rangle^*}{\lambda_j} \phi_j(x)$$

actually does converge to the $f(x)$ of Eq. 5.119.

$$\|f - S_n\|^2 = \|K^{(n+1)}h\|^2 = |\langle h \mid K^{(n+1)^2}h\rangle|$$
$$= |\langle h \mid K_2^{(n+1)} \, |h\rangle| \le \frac{\|h\|^2}{\lambda_{n+1}^2}$$

Here we have used Eq. 5.114, generalized to arbitrary n, and Ths. 5.8 and 5.9. Therefore we obtain

$$\lim_{n\to\infty} \|f - S_n\| = 0$$

which is not yet a statement of the uniform (or pointwise) convergence of the series in Eq. 5.120 to $f(x)$. However we can obtain pointwise convergence to $f(x)$ as follows. Since K is a completely continuous self-adjoint operator we know from Section A2.4 that our Hilbert space $\mathcal{H}$ can be decomposed into a direct sum of the range $\mathcal{R}$ of K and the null space $\mathcal{N}$ of K,

$$\mathcal{H} = \mathcal{R} \oplus \mathcal{N} \tag{5.121}$$

That is, we can take any $h(x) \in \mathcal{L}_2$ and express it as

$$h(x) = w(x) + g(x) \tag{5.122}$$

such that

$$Kw(x) = 0$$
$$\langle w \mid g\rangle = 0$$

Since we have just shown that any f that can be written as

$$f = Kh$$

is such that

$$\|f\|^2 = \|Kh\|^2 = \sum_{j=1}^{\infty} \frac{|\langle h \mid \phi_j \rangle|^2}{\lambda_j^2} = \sum_{j=1}^{\infty} |\langle f | \phi_j \rangle|^2$$

we see that the eigenfunctions $\{\phi_j\}$ form a complete set for the range $\mathcal{R}$ of K. Therefore for any $h(x)$ we can use Eq. 5.122 where $g(x)$ will be in the range $\mathcal{R}$ so that

$$\|h\|^2 = \|w\|^2 + \|g\|^2 = \|w\|^2 + \sum_{j=1}^{\infty} |\langle \phi_j \mid g \rangle|^2$$

$$= \|w\|^2 + \sum_{j=1}^{\infty} |\langle \phi_j \mid h \rangle|^2$$

We may define the sequence of functions

$$g_n(x) = \sum_{j=1}^{n} \langle h \mid \phi_j \rangle \phi_j(x)$$

such that $\|g_n\| \leq \|h\| < \infty$ for all n and such that $\|g_n\| \xrightarrow[n\to\infty]{} \|g\|$. We now have, for any $f(x) = Kh(x)$,

$$\begin{aligned}
|f(x) - S_n(x)|^2 &= \left| Kh(x) - \sum_{j=1}^{n} \langle h \mid \phi_j \rangle K\phi_j(x) \right|^2 \\
&= \left| K\left[h(x) - \sum_{j=1}^{n} \langle h \mid \phi_j \rangle \phi_j(x) \right] \right|^2 = |K[h(x) - g_n(x)]|^2 \\
&= |K[w(x) + g(x) - g_n(x)]|^2 = |K[g(x) - g_n(x)]|^2 \\
&= \left| \int_a^b K(x, y)[g(y) - g_n(y)]\, dy \right|^2 \\
&\leq \int_a^b |K(x, y)|^2\, dy \|g - g_n\|^2 \leq C\|g - g_n\|^2 \xrightarrow[n\to\infty]{} 0
\end{aligned}$$

which completes the proof of this theorem. Q.E.D.

It should be clear that if we have only $\|K\| < \infty$, rather than Eq. 5.118, we still have convergence of the series (5.120) in the mean (cf. discussion immediately following the proof of Th. 5.1).

It is important to note that Th. 5.10 does *not* yet guarantee that the eigenfunctions $\{\phi_j(x)\}$ of Eq. (5.92) form a complete set for the entire Hilbert space $\mathcal{L}_2[a, b]$. However these $\{\phi_j\}$ will form a complete set provided there exist only trivial solutions to

$$Kw = 0 \tag{5.123}$$

that is, when the null space of K contains only the zero vector. If the null space of K contains nonzero vectors, then we can always add a denumerable orthonormal set, say $\{\chi_i\}$, such that

$$K\chi_i = 0 \tag{5.124}$$

since any incomplete orthonormal set in a separable Hilbert space can be made complete by the addition of a denumerable orthonormal set. That is, if $\mathcal{H}$ is separable [as $\mathcal{L}_2(-\infty, \infty)$ is (cf. Schmeidler, pp. 26–29) or Section A8.3], then since $\mathcal{R}$ has just been proven separable, so must $\mathcal{N}$ be separable [cf. Eq. (5.121)]. We can simply call these basis vectors $\{\chi_j\}$. In the present case these additional vectors must be in the null space of K since $\mathcal{H} = \mathcal{R} \oplus \mathcal{N}$ and the $\{\phi_j\}$ already form a complete set for $\mathcal{R}$.

As a specific example of the application of Th. 5.10, consider the integral equation (with symmetric, completely continuous kernel)

$$\phi_n(x) - \frac{\lambda_n}{2\pi}\left[\int_{-\pi}^{x} (\pi - x)(\pi + y)\phi_n(y)\, dy + \int_{x}^{\pi} (\pi + x)(\pi - y)\phi_n(y)\, dy\right]$$
$$\equiv \phi_n(x) - \lambda_n \int_{-\pi}^{\pi} K(x, y)\phi_n(y)\, dy = 0 \quad \textbf{(5.125)}$$

Let us first demonstrate that the null space of K contains only the zero vector. Define

$$I(x) \equiv Kf(x) = \frac{1}{2\pi}\left[\int_{-\pi}^{x} (\pi - x)(\pi + y)f(y)\, dy + \int_{x}^{\pi} (\pi + x)(\pi - y)f(y)\, dy\right]$$

so that

$$\frac{dI(x)}{dx} = \frac{1}{2\pi}\left[-\int_{\pi}^{x} (\pi + y)f(y)\, dy + \int_{x}^{\pi} (\pi - y)f(y)\, dy\right]$$

$$\frac{d^2I(x)}{dx^2} = -f(x)$$

Therefore $Kf(x) \equiv 0$ requires that $f(x) \equiv 0$ so that the null space of K contains only the zero vector. We conclude from Th. 5.10 that these $\{\phi_n(x)\}$ form a complete set of functions. (We will discuss this further in a moment.)

If we differentiate Eq. 5.125 directly we easily verify that

$$\frac{d^2\phi_n(x)}{dx^2} + \lambda_n\phi_n(x) = 0 \quad \textbf{(5.126)}$$

while it follows directly from Eq. 5.125 that

$$\phi_n(-\pi) = 0 = \phi_n(\pi) \quad \textbf{(5.127)}$$

Conversely, all $\{\phi_n(x)\}$ satisfying Eqs. 5.126 and 5.127 are solutions to Eq. 5.125. It is a simple matter to verify that the eigenvalues and orthonormal eigenfunctions to Eqs. 5.126 and 5.127 are, respectively,

$$\sqrt{\lambda_n} = \frac{n}{2} \qquad n = 0, 1, 2, \ldots \quad \textbf{(5.128)}$$

$$\phi_n(x) = \begin{cases} \dfrac{1}{\sqrt{\pi}} \sin\left(\dfrac{n}{2}x\right) & n = 0, 2, 4, \ldots \\[2ex] \dfrac{1}{\sqrt{\pi}} \cos\left(\dfrac{n}{2}x\right) & n = 1, 3, 5, \ldots \end{cases} \quad \textbf{(5.129)}$$

The kernel $K(x, y)$ defined in Eq. 5.125 brings out an interesting point. We know that this kernel has the property that $\mathcal{H}$ {i.e., $\mathcal{L}_2[-\pi, \pi]$ here} can be decomposed as a direct sum of the range $\mathcal{R}$ and the null space $\mathcal{N}$ of K. However it is obvious from the definition of K in Eq. 5.125 that for *any* $h(x)$ and $f(x)$ defined as

$$f(x) \equiv Kh(x)$$

is such that

$$f(-\pi) = 0 = f(\pi)$$

That is, all functions in the range of K have this property. On the other hand, a function $h(x) \equiv 1$, $x \in [-\pi, \pi]$ is not in this range, nor is it in the null space of K since

$$Kh(x) = \pi(\pi^2 - x^2) \neq 0$$

What has happened to the general property that $\mathcal{H} = \mathcal{R} \oplus \mathcal{N}$? We can see the source of the paradox, if we consider a function $\bar{h}(x)$,

$$\bar{h}(x) = \begin{cases} 0 & x = \pm\pi \\ 1 & -\pi < x < \pi \end{cases}$$

Since $h(x)$ and $\bar{h}(x)$ differ only on a set of measure zero (i.e., at $+\pi$ and $-\pi$) then they are essentially the same function since for any complete set $\{\chi_j\}$

$$\langle h \mid \chi_j \rangle = \langle \bar{h} \mid \chi_j \rangle \qquad \forall j$$

Therefore, of course, the set $\{\phi_n(x)\}$ of Eq. 5.129 forms a complete set as claimed in Th. 5.10, but the convergence of the expansion cannot be uniform for functions that do not satisfy $f(-\pi) = 0 = f(\pi)$.

Let us end this chapter with a few important observations. We have proven most of our theorems under the condition that the integral kernel, $K(x, y)$, satisfies

$$\|K\|^2 \equiv \int_a^b dx \int_a^b dy |K(x, y)|^2 < \infty$$

However it should be clear that the really crucial assumption is that the operators be completely continuous (i.e., able to be uniformly approximated by separable operators). In fact under the assumption that

$$\|(K - K_n)f\| \xrightarrow[n \to \infty]{} 0$$

we can prove Th. 5.3–5.6 and 5.8–5.10 for

$$\phi - \lambda K \phi = f$$

or for

$$\phi_j - \lambda_j K \phi_j = 0$$

at least in the mean (cf. Problems 5.8, 5.9, and 5.11). However we do lose some important results if we use only this abstract approach. For instance the induction proof of Th. 5.1 (and subsequently Th. 5.2) depended explicitly on the form $\int_a^x K(x, y)f(y)\,dy$ for the Volterra kernel. Also another representation-dependent assumption

$$k^2(x) \equiv \int_a^b |K(x, y)|^2\,dy < C$$

allowed us to prove *uniform* convergence of certain series expansions, which is more than convergence in the mean.

Finally, we will establish an important property of the resolvent operator for a symmetric kernel. Let us rewrite the inhomogeneous Fredholm equation as

$$\phi(x) - f(x) = \lambda \int_a^b K(x, y)\phi(y)\,dy \qquad \lambda \neq \lambda_i \neq 0 \tag{5.130}$$

It is clear that the function $[\phi(x) - f(x)]$ satisfies the conditions of Th. 5.10 so that it may be expressed as

$$\phi(x) - f(x) = \sum_{j=1}^{\infty} c_j \phi_j(x) \tag{5.131}$$

where

$$c_j = \langle \phi_j \mid (\phi - f) \rangle = \langle \phi_j \mid \phi \rangle - \langle \phi_j \mid f \rangle = \lambda \langle \phi_j \mid K\phi \rangle = \frac{\lambda}{\lambda_j} \langle \phi_j \mid \phi \rangle$$

We then have

$$(\lambda_j - \lambda)\langle \phi_j \mid \phi \rangle = \lambda_j \langle \phi_j \mid f \rangle$$

$$\langle \phi_j \mid \phi \rangle = \frac{\lambda_j \langle \phi_j \mid f \rangle}{(\lambda_j - \lambda)}$$

or

$$c_j = \frac{\lambda \langle \phi_j \mid f \rangle}{(\lambda_j - \lambda)} \tag{5.132}$$

If we substitute Eq. 5.132 into Eq. 5.131 and use the definition of the resolvent, $\mathcal{R}(x, y; \lambda)$, for Eq. 5.130, we obtain

$$\begin{aligned} \phi(x) &= f(x) + \lambda \sum_{j=1}^{\infty} \frac{\langle \phi_j \mid f \rangle}{(\lambda_j - \lambda)} \phi_j(x) \\ &\equiv f(x) + \lambda \int_a^b \mathcal{R}(x, y; \lambda) f(y)\,dy \end{aligned} \tag{5.133}$$

If the series in Eq. 5.133 converges almost everywhere, then we can conclude that

$$\mathcal{R}(x, y; \lambda) = \sum_{j=1}^{\infty} \frac{\phi_j^*(y)\phi_j(x)}{(\lambda_j - \lambda)}$$

In any case, however, Eq. 5.49 for the resolvent, plus Th. 5.10 applied again, yield

$$\mathcal{R}(x, y; \lambda) = K(x, y) + \lambda \sum_{k=1}^{\infty} \int_a^b \frac{\mathcal{R}(s, y; \lambda)\phi_k^*(s)\, ds}{\lambda_k} \phi_k(x)$$

$$= K(x, y) + \lambda \sum_{j=1}^{\infty} \frac{\phi_j^*(y)\phi_j(x)}{\lambda_j(\lambda_j - \lambda)} \tag{5.134}$$

Equation 5.134 shows that for completely continuous symmetric kernels the resolvent has only *simple* poles.

A5.1 A DERIVATION OF FUBINI'S METHOD GIVEN IN EQS. 5.26 AND 5.29

We may write the solution to Eq. 5.23 as

$$u(x) = f_1(x)v(x) + f_2(x)w(x) \tag{A5.1}$$

where $f_1(x)$ and $f_2(x)$ are subject to the condition

$$f_1'(x)v(x) + f_2'(x)w(x) = 0 \tag{A5.2}$$

That is, it is *always* possible to represent a given differentiable function $u(x)$ (which we here know to be uniquely determined by Eq. 5.23 according to Th. 5.2) in terms of two linearly independent functions $v(x)$ and $w(x)$ as in Eq. A5.1 where $f_1(x)$ and $f_2(x)$ are constrained to satisfy Eq. A5.2 (cf. Problem 5.2). It is clear that Eqs. 5.23, 5.24, A5.1, and A5.2 together determine a coupled set of first order linear differential equations for $f_1(x)$ and $f_2(x)$. If we define the known functions

$$F_1(x) \equiv \frac{1}{W[v, w]} \left\{ \left[\frac{b(x)}{a(x)} - \frac{B(x)}{A(x)} \right] v'(x) + \left[\frac{c(x)}{a(x)} - \frac{C(x)}{A(x)} \right] v(x) \right\}$$

$$F_2(x) \equiv \frac{1}{W[v, w]} \left\{ \left[\frac{b(x)}{a(x)} - \frac{B(x)}{A(x)} \right] w'(x) + \left[\frac{c(x)}{a(x)} - \frac{C(x)}{A(x)} \right] w(x) \right\}$$

then we obtain for $f_1(x)$ and $f_2(x)$

$$\begin{aligned} \frac{df_1(x)}{dx} &= [f_1(x)F_1(x) + f_2(x)F_2(x)]w(x) - \frac{f(x)w(x)}{a(x)W[v, w]} \\ \frac{df_2(x)}{dx} &= -[f_1(x)F_1(x) + f_2(x)F_2(x)]v(x) + \frac{f(x)v(x)}{a(x)W[v, w]} \end{aligned} \tag{A5.3}$$

From Eqs. 5.23, A5.1, and A5.2 we have

$$u(x_0) \equiv c_1 \equiv \gamma_1 v(x_0) + \gamma_2 w(x_0)$$

$$u'(x_0) \equiv c_2 \equiv \gamma_1 v'(x_0) + \gamma_2 w'(x_0)$$

which allows us to solve for γ_1 and γ_2 in terms of c_1 and c_2. If we define two more known functions as

$$G_1(x) \equiv -\int_{x_0}^{x} \frac{f(y)w(y)\,dy}{a(y)W[v, w]}$$

$$G_2(x) \equiv \int_{x_0}^{x} \frac{f(y)v(y)\,dy}{a(y)W[v, w]}$$

then Eqs. A5.3 with the initial conditions $f_j(x_0) \equiv \gamma_j$, $j=1, 2$, are equivalent to

$$\begin{aligned} f_1(x) &= \gamma_1 + \int_{x_0}^{x} [f_1(y)F_1(y) + f_2(y)F_2(y)]w(y)\,dy + G_1(x) \\ f_2(x) &= \gamma_2 - \int_{x_0}^{x} [f_1(y)F_1(y) + f_2(y)F_2(y)]v(y)\,dy + G_2(x) \end{aligned} \tag{A5.4}$$

If we define

$$g(x) \equiv f_1(x)F_1(x) + f_2(x)F_2(x)$$

then

$$\begin{aligned} g(x) = &\int_{x_0}^{x} g(y)[F_1(x)w(y) - F_2(x)v(y)]\,dy \\ &+ \gamma_1 F_1(x) + \gamma_2 F_2(x) + F_1(x)G_1(x) + F_2(x)G_2(x) \end{aligned}$$

or

$$g(x) - \int_{x_0}^{x} K(x, y)g(y)\,dy = G(x) \tag{A5.5}$$

where

$$K(x, y) \equiv \begin{vmatrix} F_1(x) & F_2(x) \\ v(y) & w(y) \end{vmatrix}$$

$$G(x) \equiv \gamma_1 F_1(x) + \gamma_2 F_2(x) + F_1(x)G_1(x) + F_2(x)G_2(x)$$

We can display explicitly the dependence of the solutions on the γ_j as follows. Because of the structure of the necessarily convergent iterative solution of Eq. A5.5 given by Eq. 5.3, we see that $g(x)$ must be of the form

$$g(x) = \gamma_1 g_1(x) + \gamma_2 g_2(x) + g_3(x) \tag{A5.6}$$

where the $g_j(x)$, $j=1, 2, 3$, are *independent of* γ_1 and γ_2. Since we may vary the initial conditions at will, we can solve the following three integral equations in place of Eq. A5.5.

$$g_j(x) - \int_{x_0}^{x} K(x, y)g_j(y)\,dy = F_j(x) \qquad j=1, 2, 3 \tag{A5.7}$$

where

$$F_3(x) \equiv F_1(x)G_1(x) + F_2(x)G_2(x)$$

Once we have solved Eq. A5.7 for the $g_j(x)$, we obtain $u(x)$ from Eqs. A5.1,

A5.4, and A5.6 as

$$u(x) \equiv f_1(x)v(x) + f_2(x)w(x)$$
$$= \gamma_1 v(x) + \gamma_2 w(x) + \int_{x_0}^{x} H(x, y)g(y)\, dy - \int_{x_0}^{x} \frac{H(x, y)f(y)\, dy}{a(y)W[v, w]}$$
$$= \gamma_1\left[v(x) + \int_{x_0}^{x} H(x, y)g_1(y)\, dy\right] + \gamma_2\left[w(x) + \int_{x_0}^{x} H(x, y)g_2(y)\, dy\right]$$
$$+ \int_{x_0}^{x} H(x, y)g_3(g)\, dy - \int_{x_0}^{x} \frac{H(x, y)f(y)\, dy}{a(y)W[v, w]} \tag{A5.8}$$

where

$$H(x, y) \equiv v(x)w(y) - w(x)v(y) \tag{A5.9}$$

A5.2 A DERIVATION OF FREDHOLM'S EXPRESSION FOR THE RESOLVENT, $\mathcal{R}(x, y; \lambda)$

The basic motivation behind this discussion is that the resolvent, $\mathcal{R}(x, y; \lambda)$, for the Fredholm equation,

$$\phi(x) - \lambda \int_a^b K(x, y)\phi(y)\, dy = f(x) \tag{A5.10}$$

should be obtainable as

$$\mathcal{R}(x, y; \lambda) = \lim_{n\to\infty} \mathcal{R}_n(x, y; \lambda) \tag{A5.11}$$

where $\mathcal{R}_n(x, y; \lambda)$ is the resolvent for

$$\phi^{(n)}(x) - \lambda^{(n)} \int_a^b K_n(x, y)\phi^{(n)}(y)\, dy = f(x) \tag{A5.12}$$

Here $K_n(x, y)$ is an n-term separable operator. That Eq. A5.11 is a reasonable expectation follows since, in some sense, $K(x, y) = \lim_{n\to\infty} K_n(x, y)$ for $\|K\| < \infty$. (For a precise statement see Eq. 2.187.) We now outline the proof. One constructs $D_n(\lambda)$ (cf. Eq. 5.53) and defines

$$D(\lambda) = \lim_{n\to\infty} D_n(\lambda) \tag{A5.13}$$

to obtain $D(\lambda)$ in terms of $K(x, y)$ only. The fact that $D(\lambda)$ actually exists and is a holomorphic function of λ for all finite values of λ is established next. A quantity $N(x, y; \lambda)$ is defined as

$$N(x, y; \lambda) \equiv D(\lambda)\mathcal{R}(x, y; \lambda) \tag{A5.14}$$

and then demonstrated to exist and be holomorphic for all finite λ. We now present the details of this proof.

We begin our argument by approximating the kernel, $K(x, y)$, by a separable kernel $K_n(x, y)$,

$$K_n(x, y) = \sum_{j=1}^{n} a_j(x) b_j^*(y) \tag{A5.15}$$

According to Eq. 5.53 we must examine the determinant

$$D_n(\lambda) \equiv \det\{D_{ij}^{(n)}(\lambda)\} = \det\{\delta_{ij} - \lambda\langle b_i \mid a_j\rangle\} \tag{A5.16}$$

However, according to Eqs. A2.1 and A2.7 of Section A2.1, we may expand (A5.16) as

$$D_n(\lambda) = 1 + \sum_{m=1}^{n} (-1)^m \lambda^m S_m^{(n)} \tag{A5.17}$$

with

$$S_m^{(n)} = \frac{1}{m!} \sum_{j_1=1}^{n} \sum_{j_2=1}^{n} \cdots \sum_{j_m=1}^{n} \begin{vmatrix} \langle b_{j_1} \mid a_{j_1}\rangle & \langle b_{j_1} \mid a_{j_2}\rangle & \cdots & \langle b_{j_1} \mid a_{j_m}\rangle \\ \langle b_{j_2} \mid a_{j_1}\rangle & \langle b_{j_2} \mid a_{j_2}\rangle & \cdots & \langle b_{j_2} \mid a_{j_m}\rangle \\ \cdot & \cdot & \cdot & \cdot \\ \cdot & \cdot & \cdot & \cdot \\ \cdot & \cdot & \cdot & \cdot \\ \langle b_{j_m} \mid a_{j_1}\rangle & \langle b_{j_m} \mid a_{j_2}\rangle & \cdots & \langle b_{j_m} \mid a_{j_m}\rangle \end{vmatrix} \tag{A5.18}$$

$$\equiv \frac{1}{m!} \sum_{j_1=1}^{n} \cdots \sum_{j_m=1}^{n} \Delta^{(n)}(j_1, j_2, \ldots, j_m)$$

where the last equation *defines* $\Delta^{(n)}(j_1, j_2, \ldots, j_m)$. We define

$$K_n\begin{pmatrix} x_1, x_2, \ldots, x_m \\ y_1, y_2, \ldots, y_m \end{pmatrix} \equiv \begin{vmatrix} K_n(x_1, y_1) & K_n(x_1, y_2) & \cdots & K_n(x_1, y_m) \\ K_n(x_2, y_1) & K_n(x_2, y_2) & \cdots & K_n(x_2, y_m) \\ \cdot & \cdot & & \cdot \\ \cdot & \cdot & \cdot & \cdot \\ \cdot & \cdot & \cdot & \cdot \\ K_n(x_m, y_1) & K_n(x_m, y_2) & \cdots & K_m(x_m, y_m) \end{vmatrix} \tag{A5.19}$$

If we now use the basic definition and properties of determinants in Section 2.1b, we can prove the following identity.

$$\begin{aligned} &\int_a^b dx_1 \int_a^b dx_2 \cdots \int_a^b dx_m K_n\begin{pmatrix} x_1, x_2, \ldots, x_m \\ x_1, x_2, \ldots, x_m \end{pmatrix} \\ &= \int_a^b dx_1 \cdots dx_m \varepsilon_{i_1 i_2 \ldots i_m} K_n(x_1, x_{i_1}) K_n(x_2, x_{i_2}) \cdots K_n(x_m, x_{i_m}) \\ &= \sum_{j_1=1}^{n} \sum_{j_2=1}^{n} \cdots \sum_{j_m=1}^{n} \int_a^b dx_1 \cdots dx_m \varepsilon_{i_1 i_2 \cdots i_m} a_{j_1}(x_1) b_{j_1}^*(x_{i_1}) a_{j_2}(x_2) b_{j_2}^*(x_{i_2}) \cdots a_{j_m}(x_m) b_{j_m}^*(x_{i_m}) \\ &= \sum_{j_1=1}^{n} \cdots \sum_{j_m=1}^{n} \varepsilon_{i_1 i_2 \cdots i_m} \langle b_{j_{i_1}} \mid a_{j_1}\rangle \langle b_{j_{i_2}} \mid a_{j_2}\rangle \cdots \langle b_{j_{i_m}} \mid a_{j_m}\rangle \\ &\equiv m!\, S_m^{(n)} \end{aligned} \tag{A5.20}$$

It may help the reader if we carry out explicitly all of the operations leading to Eq. A5.20 for $n=2$.

$$\int_a^b dx_1 \int_a^b dx_2 K_2\begin{pmatrix} x_1, x_2 \\ x_1, x_2 \end{pmatrix}$$

$$= \int_a^b dx_1 \int_a^b dx_2 \varepsilon_{i_1 i_2} K_2(x_1, x_{i_1}) K_2(x_2, x_{i_2})$$

$$= \sum_{j=1}^{2} \sum_{k=1}^{2} \int_a^b dx_1 \int_a^b dx_2 \varepsilon_{i_1 i_2} a_j(x_1) b_j^*(x_{i_1}) a_k(x_2) b_k^*(x_{i_2})$$

$$= \sum_{j=1}^{2} \sum_{k=1}^{2} \int_a^b dx_1 \int_a^b dx_2 [a_j(x_1) b_j^*(x_1) a_k(x_2) b_k^*(x_2) - a_j(x_1) b_k^*(x_1) a_k(x_2) b_j^*(x_2)]$$

$$= \sum_{j=1}^{2} \sum_{k=1}^{2} [\langle a_j \mid k_j\rangle^* \langle a_k \mid b_k\rangle^* - \langle a_j \mid b_k\rangle^* \langle a_k \mid b_j\rangle^*]$$

$$= \sum_{\substack{j_1=1 \\ j_2=1}}^{2} [\langle b_{j_1} \mid a_{j_1}\rangle \langle b_{j_2} \mid a_{j_2}\rangle - \langle b_{j_2} \mid a_{j_1}\rangle \langle b_{j_1} \mid a_{j_2}\rangle]$$

$$= \sum_{j_1, j_2=1}^{2} \varepsilon_{i_1 i_2} \langle b_{j_{i_1}} \mid a_{j_1}\rangle \langle b_{j_{i_2}} \mid a_{j_2}\rangle$$

Suppose we now *formally* let $n \to \infty$ and simply replace $K_n(x, y)$ by $K(x, y)$. We then obtain

$$D(\lambda) \equiv 1 + \sum_{m=1}^{\infty} \frac{(-\lambda)^m}{m!} \int_a^b dx_1 \cdots \int_a^b dx_m K\begin{pmatrix} x_1, x_2, \ldots, x_m \\ x_1, x_2, \ldots, x_m \end{pmatrix}$$

$$\equiv 1 + \sum_{m=1}^{\infty} \frac{(-\lambda)^m}{m!} D^{(m)} \quad \textbf{(A5.21)}$$

If we make the assumption that $K(x, y)$ is *bounded* on $[a, b]$ (here taken to be a finite range) as

$$|K(x, y)| \leq N < \infty \quad \textbf{(A5.22)}$$

then we can prove that the series (A5.21) defining $D(\lambda)$ is convergent for *all* finite values of λ. Hadamard's inequality (cf. Section 2.3b) implies that

$$\left| K\begin{pmatrix} x_1, x_2, \ldots, x_m \\ x_1, x_2, \ldots, x_m \end{pmatrix} \right| \leq m^{m/2} N^m$$

so that

$$\left| \int_a^b dx_1 \cdots dx_m K\begin{pmatrix} x_1, x_2, \ldots, x_m \\ x_1, x_2, \ldots, x_m \end{pmatrix} \right| \leq \left(\prod_{j=1}^{m} \int_a^b dx_j \right) m^{m/2} N^m$$

$$= (b-a)^m m^{m/2} N^m$$

or

$$|D(\lambda)| \leq 1 + \sum_{m=1}^{\infty} \frac{[(b-a)|\lambda| N]^m m^{1/2m}}{m!}$$

Since we have (cf. Sterling's approximation in Section 7.12b)

$$\lim_{m\to\infty}\frac{[(b-a)|\lambda|N]^m m^{m/2}}{m!}=\lim_{m\to\infty}\frac{[(b-a)|\lambda|N]^m m^{m/2}}{\sqrt{2\pi}m^{m+1/2}e^{-m}}$$

$$=\lim_{m\to\infty}\frac{1}{\sqrt{2\pi m}}\exp\{m[\ln[(b-a)|\lambda|N]+1-\tfrac{1}{2}\ln m]\}=0$$

and

$$\lim_{m\to\infty}\frac{(b-a)|\lambda|N(m+1)^{(m+1)/2}}{(m+1)m^{m/2}}=\lim_{m\to\infty}\frac{(b-a)|\lambda|N}{(m+1)^{1/2}}\left(1+\frac{1}{m}\right)^{m/2}$$

$$=0$$

where by definition

$$\lim_{n\to\infty}\left(1+\frac{1}{n}\right)^n\equiv e$$

we see that the series (A5.21) defining $D(\lambda)$ converges uniformly for all finite values of λ.

We now wish to define a quantity $N(x, y; \lambda)$ by Eq. A5.14 that must satisfy the following integral equation by Eq. 5.49,

$$N(x, y; \lambda)=D(\lambda)K(x, y)+\lambda\int_a^b K(x, s)N(s, y; \lambda)\,ds \qquad \textbf{(A5.23)}$$

at least for $|\lambda|<\|K\|^{-1}$. From Eqs. A5.19 and A5.21 it is not difficult to see that $N(x, y; \lambda)$ is given as

$$N(x, y; \lambda)=\sum_{m=0}^{\infty}\frac{(-\lambda)^m}{m!}\int_a^b dx_1\cdots\int_a^b dx_m K\begin{pmatrix}x, x_1, x_2, \ldots, x_m\\ y, x_1, x_2, \ldots, x_m\end{pmatrix}$$

$$\equiv\sum_{m=0}^{\infty}\frac{(-\lambda)^m}{m!}N^{(m)}(x, y) \qquad \textbf{(A5.24)}$$

since if we expand the integral in Eq. A5.24 by cofactors across the first row we obtain

$$N^{(m)}(x, y)=\int_a^b dx_1\cdots\int_a^b dx_m\left[K(x, y)K\begin{pmatrix}x_1, x_2, \ldots, x_m\\ x_1, x_2, \ldots, x_m\end{pmatrix}\right.$$

$$\left.+\sum_{j=1}^{m}(-1)^j K(x, x_j)K\begin{pmatrix}x_1, x_2, \ldots, & \ldots, x_m\\ y, x_1, \ldots, x_{j-1}, x_{j+1}, & \ldots, x_m\end{pmatrix}\right]$$

$$=K(x, y)D^{(m)}-\sum_{j=1}^{m}\int_a^b dx_j K(x, x_j)\int_a^b dx_1\cdots\int_a^b dx_{j-1}\int_a^b dx_{j+1}\cdots\int_a^b dx_m$$

$$\times K\begin{pmatrix}x_j, x_1, \ldots, x_{j-1}, x_{j+1}, \ldots, x_n\\ y, x_1, \ldots, x_{j-1}, x_{j+1}, \ldots, x_n\end{pmatrix}$$

$$=K(x, y)D^{(m)}-\sum_{j=1}^{m}\int_a^b dx_j K(x, x_j)N^{(m-1)}(x_j, y)$$

$$=K(x, y)D^{(m)}-m\int_a^b K(x, s)N^{(m-1)}(s, y)\,ds \qquad \textbf{(A5.25)}$$

If we now multiply each term in Eq. A5.25 by $(-\lambda)^m/m!$ and sum on m from zero to infinity, we obtain Eq. A5.23. Finally we can prove the convergence of the series in Eq. A5.24 for all finite λ by majorizing each term in the series as

$$\left|\int_a^b dx_1 \cdots dx_m K\begin{pmatrix} x, x_1, \ldots x_m \\ y, x_1, \ldots, x_m \end{pmatrix}\right| \leq (b-a)^m N^{m+1}(m+1)^{(m+1)/2}$$

and then proceeding as we did for $D(\lambda)$.

Since $D(\lambda)$ is holomorphic for all finite values of λ it can have at most a finite number of zeros in the finite part of the λ-plane (cf. Th. 7.8). Since $N(x, y; \lambda)$ is holomorphic for all finite λ, it can have no singularities in the finite portion of the λ-plane. Therefore the resolvent, which can be written as

$$\mathcal{R}(x, y; \lambda) = \frac{N(x, y; \lambda)}{D(\lambda)} \qquad \textbf{(A5.26)}$$

can have only poles (cf. Section 7.5) at a finite, discrete set of points in the λ-plane. By the principle of analytic continuation (cf. Section 7.11) it follows that the integral equation of Eq. 5.49, that is,

$$\mathcal{R}(x, y; \lambda) - \lambda \int_a^b K(x, s)\mathcal{R}(s, y; \lambda)\, ds = K(x, y)$$

is satisfied for all finite values of λ for which $\mathcal{R}(x, y; \lambda)$ is not singular. It is also evident that $N(x, y; \lambda)$ satisfies Eq. A5.23 for *all* finite values of λ.

A5.3 A DIRECT PROOF THAT $\lim_{n\to\infty} \sum_{j=0}^{n} \alpha_{jk}\langle\varphi_l \mid \mathcal{R}_n(\lambda)|\varphi_j\rangle = 0$

If we use Eq. 5.57 and the power series expansion for $\mathcal{R}_n(\lambda)$, we see that

$$\langle\varphi_l \mid \mathcal{R}_n(\lambda) \mid \varphi_j\rangle = \sum_{m=1}^{\infty} \lambda^{m-1}\langle\varphi_l \mid R_n^m \mid \varphi_j\rangle$$

where

$$\langle\varphi_r | R | \varphi_s\rangle = \alpha_{rs} \qquad \begin{cases} r \leq n & s \geq n+1 \\ & \text{or} \\ r \geq n+1 & s \geq 0 \end{cases}$$

so that $\langle\varphi_l| R |\varphi_j\rangle = 0$ since l, $j \leq n$. Now for $m \geq 2$ we may use the Schwarz inequality and Eq. 5.58 to obtain (e.g., by induction) the following inequalities.

$$|\langle\varphi_r| R_n^2 |\varphi_s\rangle| = \left|\sum_{l=0}^{\infty} \langle\varphi_r| R |\varphi_l\rangle\langle\varphi_l| R |\varphi_s\rangle\right| = \left|\sum_{l=n+1}^{\infty} \alpha_{rl}\alpha_{ls}\right|$$

$$\leq \left(\sum_{l=n+1}^{\infty} |\alpha_{rl}|^2\right)^{1/2} \left(\sum_{l=n+1}^{\infty} |\alpha_{sl}|^2\right)^{1/2}$$

$$|\langle\varphi_r| R_n^m |\varphi_s\rangle| \leq \left(\sum_{l=n+1}^{\infty} |\alpha_{rl}|^2\right)^{1/2} \left(\sum_{l=n+1}^{\infty} |\alpha_{sl}|^2\right)^{1/2} \left(\sum_{p,q=n+1}^{\infty} |\alpha_{pq}|^2\right)^{(m-2)/2}$$

$$\leq \left(\sum_{l=n+1}^{\infty} |\alpha_{rl}|^2\right)^{1/2} \left(\sum_{l=n+1}^{\infty} |\alpha_{sl}|^2\right)^{1/2} \| R_n \|^{m-2}$$

Therefore we see that

$$|\langle\varphi_r|\,\mathcal{R}_n(\lambda)\,|\varphi_s\rangle| \le \sum_{m=2}^{\infty} |\lambda|^{m-1}\|R_n\|^{m-2}\left(\sum_{l=n+1}^{\infty} |\alpha_{rl}|^2\right)^{1/2}\left(\sum_{l=n+1}^{\infty} |\alpha_{sl}|^2\right)^{1/2}$$

$$= \frac{|\lambda|}{1-|\lambda|\,\|R_n\|}\left(\sum_{l=n+1}^{\infty} |\alpha_{rl}|^2\right)^{1/2}\left(\sum_{l=n+1}^{\infty} |\alpha_{sl}|^2\right)^{1/2}$$

or that

$$\left|\sum_{j=0}^{n} \alpha_{jk}\langle\varphi_l|\,\mathcal{R}_n(\lambda)\,|\varphi_j\rangle\right| \le \left[\frac{|\lambda|\,\|R_n\|}{1-|\lambda|\,\|R_n\|}\right]\left(\sum_{r=0}^{\infty} |\alpha_{rk}|^2\right)^{1/2}\left(\sum_{p=n+1}^{\infty} |\alpha_{lp}|^2\right)^{1/2}$$

which implies the desired result.

A5.4 AN EQUIVALENT DEFINITION OF COMPLETELY CONTINUOUS OPERATORS IN TERMS OF STRONG CONVERGENCE

A sequence of vectors $\{\chi_j\}$ is said to converge *weakly* to zero provided

$$\langle h \mid \chi_j\rangle \xrightarrow[j\to\infty]{} 0 \qquad \forall h \ni \|h\| < \infty \tag{A5.27}$$

and to converge *strongly* (or in the norm) to zero provided

$$\|\chi_j\| \xrightarrow[j\to\infty]{} 0$$

One can define a completely continuous operator K as one that takes every weakly convergent sequence into a strongly convergent sequence. We can show that this definition of complete continuity for an operator is equivalent to our definition in terms of the limit of a sequence of separable operators (cf. Eq. 2.187) as follows.

We begin by assuming that K is completely continuous as defined in Eq. 2.187; that is,

$$\lim_{n\to\infty} \|(K-K_n)f\| = 0 \qquad \forall f \ni \|f\| < \infty \tag{A5.28}$$

where K_n is a separable operator,

$$K_n = \sum_{k=1}^{n} |a_k\rangle\langle b_k| \tag{A5.28}$$

Let $\{\chi_j\}$, $\|\chi_j\| < \infty$, be a weakly convergent sequence of vectors and consider the sequence $\{K_n\chi_j\}$. We easily see from Eq. A5.27 that

$$\|K_n\chi_j\| = \left\|\sum_{k=1}^{n} |a_b\rangle\langle b_k \mid \chi_j\rangle\right\| \le \sum_{j=1}^{n} |\langle b_k \mid \chi_j\rangle|\,\|a_k\| \xrightarrow[j\to\infty]{} 0 \tag{A5.29}$$

since this sum is finite. Hence K_n takes every weakly convergent sequence of vectors into a strongly convergent one. If we use Eq. A5.28 (or Eq. 2.187) and Eq. A5.29, we have for any weakly convergent sequence $\{\chi_j\}$

$$\|K\chi_j\| \le \|(K-K_n)\chi_j\| + \|K_n\chi_j\| \le \varepsilon_n \|\chi_j\| + \|K_n\chi_j\| < \varepsilon \qquad \text{as } j\to\infty \tag{A5.30}$$

where ε may be made as small as we please for n sufficiently large. This establishes the fact that any operator K satisfying Eq. A5.28 will take every weakly convergent sequence of vectors into a strongly convergent sequence of vectors.

Now let us assume that K is such that

$$\|K \mid \chi_n\rangle\| \xrightarrow[n\to\infty]{} 0$$

whenever the $\{\chi_n\}$ are weakly convergent. If we choose a complete orthonormal set of basis vectors $\{e_l\}$, then for

$$|\chi_n\rangle \equiv |f\rangle - \sum_{l=0}^{n} \langle e_l \mid f\rangle \mid e_l\rangle$$

we have for any $h \ni \|h\| < \infty$

$$\langle h \mid \chi_n\rangle \equiv \langle h \mid f\rangle - \sum_{l=0}^{n} \langle e_l \mid f\rangle\langle h \mid e_l\rangle \xrightarrow[n\to\infty]{} 0 \tag{A5.31}$$

which is simply a statement of the completeness of the $\{e_l\}$. Therefore these $\{\chi_i\}$ form a weakly convergent sequence of vectors so that, by assumption, the sequence $\{K \mid \chi_i\rangle\}$ must be strongly convergent. If we now define the separable operator

$$K_n \equiv \sum_{l=0}^{n} K\,|e_l\rangle\langle e_l|$$

then it follows from Eq. A5.30 and the fact that $\{K \mid \chi_n\rangle\}$ is strongly convergent that for any f, $\|f\| < \infty$,

$$\|(K-K_n)f\| = \left\|\left(K - \sum_{l=0}^{n} K\,|e_l\rangle\langle e_l|\right)f\right\| = \left\|K \mid f\rangle - \sum_{l=0}^{n} \langle e_l \mid f\rangle K \mid e_l\rangle\right\|$$

$$= \|K\,|\chi_n\rangle\| \xrightarrow[n\to\infty]{} 0 \tag{A5.32}$$

which is just Eq. A5.28. Therefore we have proven the equivalence of these two definitions of completely continuous operators.

A5.5 PROOF THAT EVERY BOUNDED SEQUENCE OF VECTORS IN A HILBERT SPACE HAS A WEAKLY CONVERGENT SUBSEQUENCE

We consider a bounded sequence of vectors, each vector in the sequence having finite norm. Since we may always normalize a vector of finite norm to one of unit norm, we choose our sequence $\{x_n\}$ to be such that $\|x_n\|=1$ for all n. The basic thrust of the following proof is to construct nested subsequences of vectors and finally obtain a weakly convergent subsequence.

For any x_n and x_j from the original sequence we have

$$|\langle x_j \mid x_n \rangle| \leq 1$$

so that $\{\langle x_j \mid x_n \rangle\}$, $n=1, 2, \ldots$, for fixed j, is a bounded sequence. By the Bolzano-Weierstrass theorem (cf. Th. I.1) we know that there exists at least one limit point of this sequence. Notice that if the sequence $\{x_n\}$ contains only a finite number N of distinct vectors, then we may take (for example) x_N to be the limit of the sequence and we are finished with our proof. Therefore the only case requiring further discussion is that in which there are infinitely many *distinct* vectors in the sequence, as we now assume.

Hence, if the subsequence $\{\langle x_j \mid x_k \rangle\}$ (fixed j), $k \in \{n_j\}$, is convergent to one of these limit points, we consider the subsequence of vectors $\{x_k\}$, $k \in \{n_j\}$, which we hereafter denote by $\{y_n^{(j)}\}$, $n=1, 2, \ldots$. By construction the sequence $\{\langle x_1 \mid y_n^{(1)} \rangle\}$ forms a convergent subsequence to one of these limit points as $n \to \infty$, where $\{y_n^{(1)}\} \subset \{x_n\}$. But since

$$|\langle x_2 \mid y_n^{(1)} \rangle| \leq 1$$

the sequence $\{\langle x_2 \mid y_n^{(1)} \rangle\}$ is bounded and by the argument just given we may choose a subsequence $\{y_n^{(2)}\} \subset \{y_n^{(1)}\} \subset \{x_n\}$ such that $\{\langle x_2 \mid y_n^{(2)} \rangle\}$ converges to a limit as $n \to \infty$. We may continue this process indefinitely and choose nested subsequences

$$\{y_n^{(j)}\} \subset \{y_n^{(j-1)}\} \subset \cdots \subset \{y_n^{(2)}\} \subset \{y_n^{(1)}\} \subset \{x_n\}$$

such that the sequence $\{\langle x_j \mid y_n^{(j)} \rangle\}$ approaches a limit as $n \to \infty$. If we now take a sequence of vectors $\{z_j\} \subset \{x_n\}$, where $z_j \equiv y_j^{(j)}$, then the sequence $\{\langle x_n \mid z_j \rangle\}$ converges as $j \to \infty$ so that $\lim_{j \to \infty} \langle x_n \mid z_j \rangle$ exists for all n.

The entire Hilbert space $\mathcal{H}$ can be written as the direct sum of those vectors expressible as linear combinations of the $\{x_n\}$ or as strong limits (cf. Section A5.4 for the definition of a strong limit or of strong convergence) of Cauchy sequences of the $\{x_n\}$, which subspace we denote by $\mathcal{S}$, and of the rest of the space, say $\mathcal{W}$, as

$$\mathcal{H} = \mathcal{S} \oplus \mathcal{W}$$

or

$$h = s + w \qquad \langle s \mid w \rangle = 0 \tag{A5.33}$$

for all $h \in \mathcal{H}$. By construction $\{z_j\} \in \mathcal{S}$ so that the limits

$$\lim_{j \to \infty} \langle s_n \mid z_j \rangle \equiv \lambda_n \qquad \forall n \tag{A5.34}$$

exist for any Cauchy sequence $\{s_n\}$ of linear combinations of the $\{x_n\}$. That these vectors $\{s_n\}$ form a Cauchy sequence means that (cf. Eq. 1.16)

$$\|s_n - s_m\| < \varepsilon \qquad \forall n, m > N \tag{A5.35}$$

From Eqs. A5.34 and A5.35 we see that the $\{\lambda_n\}$ form a Cauchy sequence of scalars since

$$|\lambda_n - \lambda_m| = \lim_{j\to\infty} |\langle (s_n - s_m) | z_j \rangle| \le \|s_n - s_m\| < \varepsilon,\ n, m > N \tag{A5.36}$$

Therefore we have the existence of the limit

$$\lambda \equiv \lim_{n\to\infty} \lambda_n = \lim_{j\to\infty} \langle s \mid z_j \rangle = \lim_{j\to\infty} \langle h \mid z_j \rangle,\ \forall h \in \mathcal{H} \tag{A5.37}$$

where we have used Eq. A5.33 to obtain the last equality.

If we now define the linear functional $f(h)$ as

$$f(h) \equiv \lim_{j\to\infty} \langle z_j \mid h \rangle \tag{A5.38}$$

then $f(h)$ is bounded since $|\langle z_j \mid h \rangle| \le \|h\|$. From Th. 1.2 we are guaranteed the existence of a unique vector z such that

$$f(h) = \langle z \mid h \rangle \tag{A5.39}$$

Equations A5.38 and A5.39 together imply that

$$\lim_{j\to\infty} \langle z_j \mid h \rangle = \langle z \mid h \rangle \qquad \forall h \in \mathcal{H} \tag{A5.40}$$

so that the sequence $\{z_j\} \subset \{x_n\}$ has z as its weak limit.

SUGGESTED REFERENCES

F. W. Byron and R. W. Fuller, *Mathematics of Classical and Quantum Physics* (Vol. II). Chapters 8 and 9 treat the standard topics in linear integral equations. In particular Sections 8.7 and 9.1 are good examples of an abstract treatment of the eigenvalue problem for a completely continuous operator in a Hilbert space.

R. Courant and D. Hilbert, *Methods of Mathematical Physics* (Vol. I). Chapter III contains an elementary and very readable exposition on linear integral equations.

J. W. Dettman, *Mathematical Methods in Physics and Engineering*. This fine advanced undergraduate text contains a lucid and concise discussion of completely continuous operators in Section 2.8.

W. V. Lovitt, *Linear Integral Equations*. This book gives an elementary but thorough treatment of the Fredholm theory of integral equations. In particular Section 21 is one of the few places that the case of repeated zeros of $D(\lambda)$ is covered completely.

S. G. Mikhlin, *Integral Equations.* This text gives an exhauistive discussion of linear integral equations relying mainly on techniques of linear algebra. As a result the discussions are usually elementary but often extremely tedious. Many engineering applications are covered in great detail.

F. Riesz and B. Sz. Nagy, *Functional Analysis.* The second part of this book gives a rigorous and sophisticated treatment of integral equations based on the Lebesgue theory of integration. Chapter IV covers much of the material of the present chapter. Also Section 113 of Chapter VII gives the proof of a beautiful theorem that contains the Fourier integral theorem for $f \in \mathcal{L}_2(-\infty, \infty)$ as a special case.

F. G. Tricomi, *Integral Equations.* This is a rigorous, readable, and masterful presentation of many topics in integral equations including singular and some nonlinear equations as examples.

PROBLEMS

5.1 You are given the differential equation

$$\left[\frac{d^2}{dx^2}+k^2+\lambda V(x)\right]u(x)=0 \qquad x\in[0,\infty)$$

where $V(x)$ is bounded real function such that

$$|V(x)|<\frac{\text{const.}}{x^{1+\varepsilon}} \qquad \varepsilon>0$$

for large x, and k is a real quantity. Using the two linearly independent solutions to

$$\left[\frac{d^2}{dx^2}+k^2\right]v(x)=0$$

that is,

$$v_1(x)=e^{ikx} \qquad v_2(x)=e^{-ikx}$$

apply the results of Section 5.1b being careful to examine the question of convergence on the infinite domain to show that

$$u(x)\xrightarrow[x\to\infty]{}\alpha v_1(x)+\beta v_2(x)$$

where α and β are constants. Next require that $u(x=0)=0$ and assume that λ is small and compute the ration (β/α). Show that if $\beta/\alpha\equiv -e^{-2i\delta}$, then to first order in λ

$$\delta=\frac{\lambda}{k}\int_0^\infty V(y)\sin^2(ky)\,dy$$

Also complete the discussion of the case $k^2<0$ given at the end of Section 5.1b.

5.2 Prove that if you are given a known differentiable function $u(x)$ and any two linearly independent functions $v(x)$ and $w(x)$ of Eq. 5.24, then it is always possible to find an $f_1(x)$ and an $f_2(x)$ such that

$$u(x)=f_1(x)v(x)+f_2(x)w(x)$$

while

$$f_1'(x)v(x)+f_2'(x)w(x)=0$$

[*Hint:* Simply construct explicity in terms of $u(x)$, $v(x)$, and $w(x)$ the pair of functions $f_1(x)$ and $f_2(x)$ that satisfy both of these conditions.]

5.3 Prove (e.g., by induction) the inequality stated in Eq. 5.44.

5.4 Given the integral equation

$$\phi(x)+\lambda x^2\int_{-\infty}^{\infty}\frac{\phi(y)\,dy}{(1+y^2)^2}=\frac{\cos x}{x^2+4} \qquad x\in(-\infty,\infty)$$

discuss the existence and uniqueness of a solution. Find a solution when it exists.

5.5 Given the kernel

$$K(x,y)=x+y$$

on the range [0, 1], find the resolvent. Now solve

$$\phi(x)-\lambda\int_0^1(x+y)\phi(y)\,dy=f(x)$$

when λ is not an eigenvalue. Also find the eigenvalues and eigenfunctions.

5.6 Solve the *Abel equation*

$$\int_0^x\frac{\phi(y)\,dy}{\sqrt{(x-y)}}=f(x) \qquad x\in[0,1]$$

[*Hint:* Multiply by $(z-x)^{-1/2}$ and integrate w.r.t. x.] Then use the result (cf. Section 7.12a)

$$\int_0^1\frac{dx}{(1-x)^{\alpha}x^{1-\alpha}}=\frac{\pi}{\sin(\pi\alpha)} \qquad 0<\alpha<1$$

to solve the more general equation

$$\int_0^x\frac{\phi(y)\,dy}{(x-y)^{\alpha}}=f(x) \qquad 0<\alpha<1$$

5.7 Consider the Fredholm equation of the *first kind*

$$\int_a^b K(x,y)\phi(y)\,dy=f(x)$$

for the symmetric (Fredholm) kernel $K(x, y)$. Assume that the null space of K contains only the zero vector. Solve this equation. Is this solution unique? What happens when the null space of K contains nonzero vectors?

5.8 Assuming only that K is completely continuous and that $|\lambda|<\|K\|^{-1}$ where the *norm* of K is defined as the smallest M such that

$$\|Kf\|\leq M\,\|f\|$$

(cf. Eq. 5.101) for all $f\ni\|f\|<\infty$, prove the convergence of the Neumann series

$$\sum_{n=1}^{\infty}\lambda^nK^nf$$

Then prove that in the mean

$$\phi\equiv f+\sum_{n=1}^{\infty}\lambda^nK^nf$$

is the solution to

$$\phi - \lambda K\phi = f$$

(See proof for convergence of the Neumann series in Section 2.7b.)

5.9 Prove Th. 5.4 assuming only complete continuity for K.

5.10 Construct the resolvent operator $\mathcal{R}(x, y; \lambda)$ for the Fredholm equation

$$\phi - \lambda \int_0^1 K(x, y)\phi(y)\, dy = f(x) \qquad x \in [0, 1]$$

where $K(x, y) = 1 + xy$. Show explicitly for this example that

$$\mathcal{R}(x, y; \lambda) = \frac{N(x, y; \lambda)}{D(\lambda)}$$

and that *all* of the eigenvalues are given as the roots to $D(\lambda) = 0$, while $N(x, y; \lambda)$ has *no* singularities as a function of λ. Explicitly compute $\mathbf{|K|}$ and show that $|\lambda_j|^{-1} \leq \mathbf{|K|}$, $j = 1, 2$.

5.11 Prove that if K is completely continuous, $K^\dagger = K$, and if

$$\phi_j - \lambda_j K\phi_j = 0 \qquad j = 1, 2, \ldots$$

$$\langle \phi_j \mid \phi_k \rangle = \delta_{jk}$$

$$f \equiv Kh \qquad \|h\| < \infty$$

then

$$\|f\|^2 = \sum_{j=1}^{\infty} \frac{|\langle h \mid \phi_j \rangle|^2}{\lambda_j^2}$$

where the series is to be proved convergent. That is, prove the Hilbert-Schmidt theorem in the mean using the norm condition $\|K\| < \infty$. Can you still prove that any finite eigenvalue λ_j can have only finite degeneracy?

5.12 Use the result of Th. 5.9 to prove directly that the largest eigenvalue of $\det |A - \lambda I| = 0$ is $|\lambda_{\max}| = n$ when

$$A = \{\alpha_{ij}\} \qquad \alpha_{ij} \equiv 1 \qquad \forall i, j = 1, 2, \ldots \quad n$$

(*Hint:* Use a Lagrange multiplier and explicitly maximize the quadratic form $\langle x | A | x \rangle$ subject to the constraint $\|x\| = 1$). In fact, show that $\lambda = 0$ is an $(n-1)$-fold root and that $\lambda = n$ is the other root.

5.13 Prove directly that the eigenfunctions given in Eq. 5.129 form a complete orthogonal set for $f \in \mathcal{L}_2[-\pi, \pi]$. [*Hint:* Use the completeness of the Fourier expansion functions, $\sin(nx)$ and $\cos(nx)$.]

6 | Calculus of Variations

6.0 INTRODUCTION

In ordinary differential calculus we are used to finding extrema of functions $f(x, y)$. We now consider the problem of finding a *function* which will make an *integral* an extremum. Let

$$I=\int_a^b f\left[x, y, \frac{dy}{dx}\right] dx$$

where $y=y(x)$. As we vary the choice (i.e., functional form) of $y(x)$, we vary the value of I. We want that particular $y=y(x)$ that will make I a minimum or a maximum. Some of the classic problems in the calculus of variations were the shortest path between two points on a given surface, the path of quickest descent between two fixed points in a gravitational field, and the surface of minimum area enclosing a given volume.

Just as in the differential calculus the vanishing of the first derivative is a *necessary*, but not sufficient (e.g., saddle point), condition for a maximum or a minimum, so in the calculus of variations we speak of the first and second variations. In our brief treatment we will work only with the first variation and often rely on geometrical or physical reasoning as to maximum or minimum for the extremum found.

The material contained in the last two Sections of this Chapter (Secs. 6.4 and 6.5) is of a somewhat specialized nature, having application mainly in the Lagrangian formulation of mechanics and of field theories. This material is not required later in the text and may be omitted without interrupting the continuity of topics presented.

6.1 EXTREMUM OF AN INTEGRAL WITH FIXED END POINTS

a. Euler-Lagrange Conditions

We will often need the results of the following theorem.

THEOREM 6.1

If $F(x)$ is a continuous real function for $x \in [a, b]$ and if

$$\int_a^b \eta(x)F(x)\,dx = 0$$

for all real, continuous, once-differentiable functions $\eta(x)$ such that

$$\eta(a) = 0 = \eta(b)$$

then $F(x)$ vanishes identically for $x \in [a, b]$.

PROOF

Notice that we require $\eta(x)$ to be continuous *and* differentiable, while $F(x)$ need only be continuous, not necessarily differentiable. In fact if $\eta(x)$ were only required to be continuous (and not vanish at a and b), then the theorem would be trivial since we could take $\eta(x) = F(x)$ so that

$$\int_a^b |F(x)|^2\,dx = 0$$

or

$$F(x) = 0$$

If $F(x)$ should be complex, then we may make the following argument for the real and imaginary parts of the function separately. However to establish the theorem as stated, we need only consider the following specific case of $\eta(x)$.

$$\eta(x) = \begin{cases} 0 & a \le x \le x_1 \\ (x - x_1)^2(x - x_2)^2 & x_1 \le x \le x_2 \\ 0 & x_2 \le x \le b \end{cases} \tag{6.1}$$

Then we have

$$\int_a^b \eta(x)F(x)\,dx = \int_{x_1}^{x_2} (x - x_1)^2(x - x_2)^2 F(x)\,dx \tag{6.2}$$

where we have chosen x_1 and x_2 such that $F(x) \ge 0$, $x_1 \le x \le x_2$, which can always be done since $F(x)$ is continuous. That is, we can divide the entire range $a \le x \le b$ into segments such that $F(x) \ge 0$ or $F(x) \le 0$. This is easily seen as follows. Assume that $F(x_0) > 0$, $x_0 \in [a, b]$. By the definition of continuity,

$$|F(x_0 + \delta) - F(x_0)| < \varepsilon$$

where ε can be made *arbitrarily* small for δ sufficiently small. Therefore we must have (cf. Problem 1.7)

$$|F(x_0+\delta)-F(x_0)|<\tfrac{1}{2}F(x_0)$$

or

$$\tfrac{1}{2}F(x_0)<F(x_0+\delta)<\tfrac{3}{2}F(x_0+\delta)$$

Since this must be true for all x_0 and δ, we can indeed so divide $a\leq x\leq b$ into regions where $F(x)$ is either positive or negative, as claimed. Now assume that the integrand in Eq. 6.2 is positive everywhere on the range of integration. However the integral is to be *zero*. Hence the contradiction implies that $F(x)\equiv 0$, $x_1\leq x\leq x_2$. Similarly this argument can be extended to the entire range $a\leq x\leq b$.

Q.E.D.

Several extensions are rather easy. In the proof just given $F(x)$ can have a finite number of finite discontinuities on the range of definition. Then $F(x)=0$ *almost everywhere*. Also an $\eta(x)$ can be constructed that is n-fold differentiable. Finally the proof for several independent variables is straightforward.

We now consider

$$I=\int_a^b f[x, y(x), y'(x)]\,dx \tag{6.3}$$

such that the numbers $y(a)=y_1$, $y(b)=y_2$, a, and b are given. We want the differential equation satisfied by that $y=y(x)$ that renders I an extremum. Let $y(x)$ be that function that extremizes I and define a set of comparison functions

$$Y(x;\,\varepsilon)\equiv y(x)+\varepsilon\eta(x)$$

where $\eta(x)$ is an arbitrary differentiable function for which

$$\eta(a)=0=\eta(b)$$

and ε is an arbitrary parameter. This family of curves will represent any continuous $Y(x)$ having the required end-point values y_1 and y_2 and such that $Y(x)=y(x)$ for $\varepsilon=0$. We can speak of the neighborhood of the minimizing curve $y=y(x)$ by choosing ε sufficiently small so that $|\varepsilon\eta(x)|$ is arbitrarily small for $a\leq x\leq b$ as shown graphically in Fig. 6.1.

From the comparison curves, $Y(x)$, we can form a comparison integral

$$I(\varepsilon)=\int_a^b f[x, Y(x), Y'(x)]\,dx$$

such that $I(0)$ is the extremum sought.

$$Y'(x)=y'(x)+\varepsilon\eta'(x)$$

It is important to note that $I(\varepsilon=0)$ is the extremum for *any* choice of $\eta(x)$. We

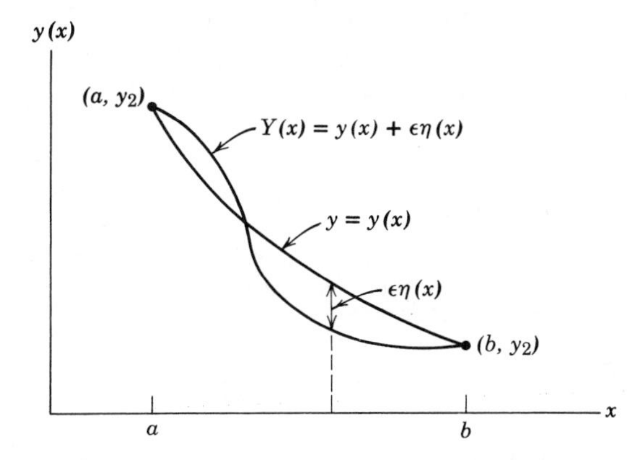

Figure 6.1

now have a function $I(\varepsilon)$ that we wish to make an extremum. The condition is just

$$\frac{d}{d\varepsilon} I(\varepsilon)=0$$

where we already know that the extremum will occur for $\varepsilon=0$. Therefore we can write

$$\frac{dI}{d\varepsilon}=\int_a^b \left\{\frac{\partial f}{\partial Y}\frac{\partial Y}{\partial \varepsilon}+\frac{\partial f}{\partial Y'}\frac{\partial Y'}{\partial \varepsilon}\right\} dx=\int_a^b \left\{\frac{\partial f}{\partial Y}\eta(x)+\frac{\partial f}{\partial Y'}\eta'(x)\right\} dx$$

so that

$$I'(0)=\int_a^b \left\{\frac{\partial f}{\partial y}\eta(x)+\frac{\partial f}{\partial y'}\eta'(x)\right\} dx=0$$

Since $\eta'(x)$ has been assumed differentiable we have

$$\int_a^b \frac{\partial f}{\partial y'}\eta'(x)\, dx=\frac{\partial f}{\partial y'}\eta(x)\bigg|_a^b-\int_a^b \frac{d}{dx}\left(\frac{\partial f}{\partial y}\right)\eta(x)\, dx$$

which implies

$$\int_a^b \left\{\frac{\partial f}{\partial y}-\frac{d}{dx}\left(\frac{\partial f}{\partial y'}\right)\right\}\eta(x)\, dx=0$$

From Th. 6.1 we conclude that

$$\frac{\partial f}{\partial y}-\frac{d}{dx}\left(\frac{\partial f}{\partial y'}\right)=0 \tag{6.4}$$

This is the *Euler-Langrange* differential equation. We now have proven the following theorem.

THEOREM 6.2

If $f(x, y, y')$ is a continuous twice-differentiable function of its arguments, then a necessary condition that the integral

$$I \equiv \int_a^b f(x, y, y')\, dx$$

be an extremum for a continuous twice-differentiable curve such that $y(a) = y_1$, $y(b) = y_2$ is that the function $y(x)$ satisfy the differential equation

$$\frac{\partial f}{\partial y} - \frac{d}{dx}\left(\frac{\partial f}{\partial y'}\right) = 0$$

This differential equation is one for $y = y(x)$, *not* for $f = f(x, y, y')$, whose functional form is given in the problem. Furthermore the $y(x)$ that this equation yields will be *twice* differentiable in x. Before extending the formalism to more general situations, we shall formulate and discuss a few simple problems.

Example 6.1 Suppose we consider two points in the plane and ask for the curve connecting these two points such that the distance along this curve is the shortest possible. An element of arc in the plane is given as

$$ds^2 = dx^2 + dy^2 = \left[1 + \left(\frac{dy}{dx}\right)^2\right] dx^2$$

The length of this arc is just

$$I = \int_{x_1}^{x_2} ds = \int_{x_1}^{x_2} \sqrt{1 + y'^2}\, dx$$

We wish to minimize this. For this example we have

$$f(x, y, y') = \sqrt{1 + y'^2}$$

$$\frac{\partial f}{\partial y} = 0$$

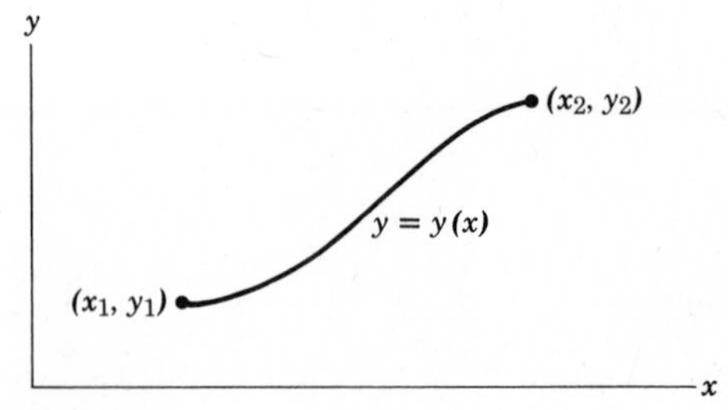

Figure 6.2

so that

$$\frac{d}{dx}\left(\frac{\partial f}{\partial y'}\right)=0$$

or

$$\frac{\partial f}{\partial y'}=c_1=\frac{y'}{\sqrt{1+y'^2}}$$

The Euler-Lagrange equation then reduces to

$$(1-c_1^2)y'^2=c_1^2$$

or to

$$\frac{dy}{dx}=\frac{c_1}{\sqrt{1-c_1^2}}\equiv a$$

This is just the equation of a straight line, of course.

An obvious extension of this problem is the question of the shortest distance between two points on a general two-dimensional surface (e.g., a sphere). Such curves are known as *geodesics*.

Example 6.2 A problem that originally gave impetus to the calculus of variations was that of the *brachistochrone*. We wish to find that path (i.e., the shape of a smooth wire) of quickest descent between two points, assuming that the point mass m begins at rest and that the only external force acting is a uniform gravitational field. The time taken to travel an infinitesimal distance ds is simply ds/v where v is the instantaneous speed of the particle. Therefore the total time for descent is

$$I=\int_{x_1}^{x_2}\frac{ds}{v}=\int_{x_1}^{x_2}\frac{\sqrt{1+y'^2}\,dx}{v}$$

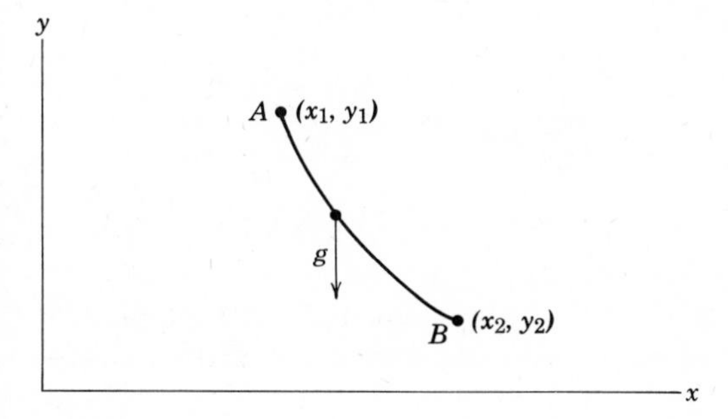

Figure 6.3

We must now be able to express v in terms of x. This can be done by conserving energy as

$$\tfrac{1}{2}mv^2+mgy=mgy_1$$

so that

$$v=\sqrt{2g(y_1-y)}$$

or

$$I=\int_{x_1}^{x_2}\sqrt{\frac{1+y'^2}{2g(y_1-y)}}\,dx$$

and

$$\frac{\partial f}{\partial y}-\frac{d}{dx}\left(\frac{\partial f}{\partial y'}\right)=0$$

Since x does not appear explicitly in f, we have $\partial f/\partial x=0$ and as an identity

$$\begin{aligned}\frac{d}{dx}\left(y'\frac{\partial f}{\partial y'}-f\right)&=y'\frac{d}{dx}\left(\frac{\partial f}{\partial y'}\right)+y''\frac{\partial f}{\partial y'}-\frac{\partial f}{\partial x}-y'\frac{\partial f}{\partial y}-y''\frac{\partial f}{\partial y'}\\&=y'\left[\frac{d}{dx}\left(\frac{\partial f}{\partial y'}\right)-\frac{\partial f}{\partial y}\right]-\frac{\partial f}{\partial x}=-\frac{\partial f}{\partial x}\end{aligned}$$

In the present case

$$f(x,y,y')=\sqrt{\frac{1+y'^2}{y_1-y}}$$

so that $\partial f/\partial x=0$ and therefore

$$y'\frac{\partial f}{\partial y'}-f=\text{const.}$$

$$\frac{\partial f}{\partial y'}=\frac{y'}{\sqrt{(1+y'^2)(y_1-y)}}$$

This yields

$$\frac{-1}{\sqrt{(y_1-y)(1+y'^2)}}=\frac{1}{\sqrt{c_1}}\qquad\text{or}\qquad 1+y'^2=\frac{c_1}{y_1-y}$$

which implies that

$$y'\equiv\frac{dy}{dx}=-\sqrt{\frac{c_1-y_1+y}{y_1-y}}$$

Here we have taken account of the fact that $dy/dx<0$ as we have depicted the geometry in Fig. 6.3. If this is integrated one obtains an arc of a cycloid. It is obvious that the extremum thus found is not a maximum since the time of descent could be made arbitrarily long.

Example 6.3 A more interesting example is furnished by the surface of revolution of minimum area passing through two points in space. In the previous two examples it was evident from geometrical or physical reasoning that the extremum curves obtained actually corresponded to minima. However in the present case we shall see that things need not be so simple.

An element of area for a surface of revolution is just

$$dA = 2\pi y\, ds$$

as shown in Fig. 6.4.

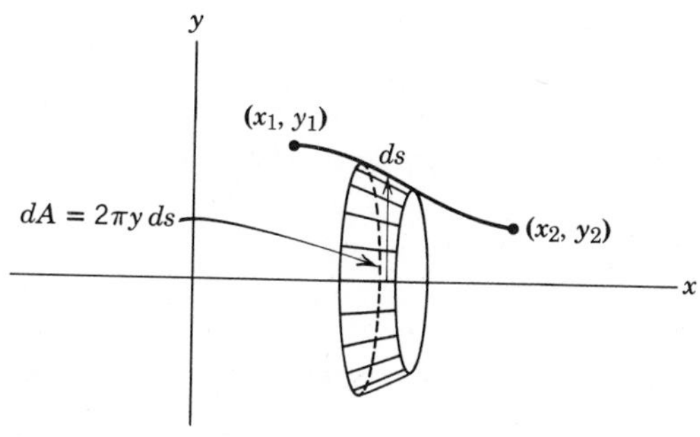

Figure 6.4

$$I = 2\pi \int_{x_1}^{x_2} y\sqrt{1+y'^2}\, dx$$

Again $\partial f/\partial x = 0$ so that

$$y'\frac{\partial f}{\partial y'} - f = \text{const.}$$

We then obtain

$$\frac{yy'^2}{\sqrt{1+y'^2}} - \frac{y(1+y'^2)}{\sqrt{1+y'^2}} = a = \frac{-y}{\sqrt{1+y'^2}}$$

$$a^2 = \frac{y^2}{1+y'^2}$$

or

$$y' \equiv \frac{dy}{dx} = \frac{\sqrt{y^2-a^2}}{a}$$

so that

$$x = a\cosh^{-1}(y/a) + b$$

and

$$y = a\cosh\left(\frac{x-b}{a}\right)$$

the equation of a *catenary*.

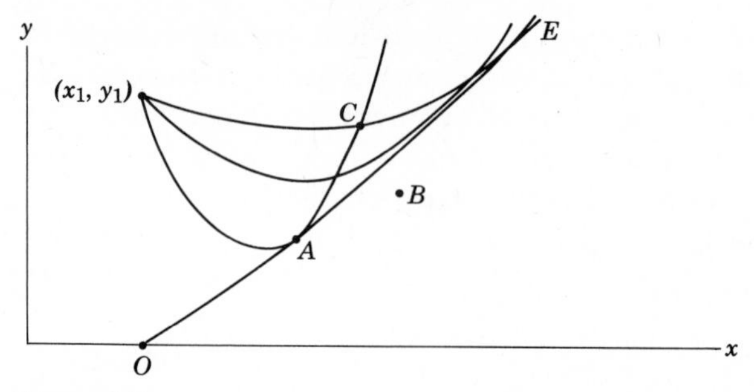

Figure 6.5

The constants, a and b, are adjusted so that the curve passes through points (x_1, y_1) and (x_2, y_2), if possible. Constants a and b are connected by

$$y_1 = a \cos h\left(\frac{x_1 - b}{a}\right)$$

for a catenary through (x_1, y_1). This graph represents some members of the one-parameter family of curves. Every member of this family is tangent to the envelope OE and cannot cross it. Hence point B cannot have any members pass through it. Only one member can pass through any point A on the envelope. Through any point C pass exactly two members of the family. Hence, if point (x_2, y_2) is B, no member $y(x)$ of this twice-differentiable class can pass through it and this $y(x)$ cannot even yield an extremum. In fact the minimum area is generated by the *discontinuous solution,*

$$x = x_1 (0 \leq y \leq y_1) \qquad y = 0 (x_1 \leq x \leq x_2) \qquad x = x_2 (0 \leq y \leq y_2) \tag{6.5}$$

Again even if (x_2, y_2) is on the envelope, for example, point A, the catenary does not yield a minimum or even a relative minimum. It is the discontinuous solution, Eq. 6.5, which gives the minimum. If two catenaries pass through (x_2, y_2), say point C, the upper one provides a *relative* minimum, unless (x_2, y_2) lies sufficiently far above the envelope. Then this upper one will provide an absolute minimum rather than the discontinuous solution. In every case the discontinuous solution, Eq. 6.5, whose area is $\pi(y_1^2 + y_2^2)$, affords a relative minimum, at least. This example is studied further in Problem 6.5.

This discussion points up the fact that the Euler-Lagrange equation is merely a *necessary*, and not a sufficient, condition for an extremum and then only for a *twice differentiable* function $y = y(x)$.

A very nice experimental study of the minimum surface area problem for a fixed boundary curve is provided by forming thin soap films on a wire frame. Under the action of surface tension the film will form the minimum surface area on the frame. As an example, if two circular wires are held close together, and a soap film produced between them, the surface is a catenary of revolution. As the rings are separated the film finally ruptures and reforms as two disjoint discs, one on each ring (cf. Problem 6.5b).

There is a beautiful paper by Courant and Roberts on this subject.

b. Several Dependent Variables

We now consider the case in which there are several dependent variables $\{y_j(t)\}$ that are functions of *one* independent variable t. We then have

$$I=\int_{t_1}^{t_2} f(y_1, y_2, \ldots, y_n; \dot{y}_1, \dot{y}_2, \ldots, \dot{y}_n; t)\, dt \qquad \dot{y}_j \equiv \frac{dy_j(t)}{dt} \tag{6.6}$$

We again construct a set of comparison functions

$$Y_j(t) \equiv y_j(t) + \varepsilon\eta_j(t)$$

$$\eta_j(t_1) = 0 = \eta_j(t_2) \qquad j=1, 2, \ldots, n$$

Proceeding as before we have

$$I(\varepsilon)=\int_{t_1}^{t_2} f(Y_1, Y_2, \ldots, Y_n; Y_1', Y_2', \ldots, Y_n'; t)\, dt$$

so that

$$I'(0)=\int_{t_1}^{t_2} \sum_{j=1}^{n} \left\{\frac{\partial f}{\partial y_j}\eta_j + \frac{\partial f}{\partial \dot{y}_j}\dot{\eta}_j\right\} dt = \int_{t_1}^{t_2} \sum_{j=1}^{n} \left\{\frac{\partial f}{\partial y_j} - \frac{d}{dt}\left(\frac{\partial f}{\partial \dot{y}_j}\right)\right\}\eta_j(t)\, dt = 0$$

as previously, for all $\eta_j(t)$. If we take, in turn, all these to be zero except one, we obtain

$$\frac{\partial f}{\partial y_j} - \frac{d}{dt}\left(\frac{\partial f}{\partial \dot{y}_j}\right) = 0 \qquad j=1, 2, \ldots, n \tag{6.7}$$

These are again the Euler-Lagrange equations.

6.2 VARIABLE END POINTS

We now consider the case in which no value is prescribed at one of the end points, for example, (b, y_2). Again we have

$$I=\int_a^b f(x, y, y')\, dx$$

and

$$I'(0)=\int_a^b \left\{\frac{\partial f}{\partial y}\eta + \frac{\partial f}{\partial y'}\eta'\right\} dx = 0$$

$$= -\frac{\partial f}{\partial y'}\eta(x)\Big|_b + \int_a^b \left\{\frac{\partial f}{\partial y} - \frac{d}{dx}\left(\frac{\partial f}{\partial y'}\right)\right\}\eta(x)\, dx = 0$$

Since this must be true for *all* $\eta(x)$, we may choose those $\eta(b)$ such that $\eta(b)=0$ and recover the Euler-Lagrange equation as before. Then choose $\eta(b)=$ const. $\neq 0$. Finally we have

$$\frac{\partial f}{\partial y}-\frac{d}{dx}\left(\frac{\partial f}{\partial y'}\right)=0$$

and

$$\left.\frac{\partial f}{\partial y'}\right|_b=0$$

as an additional condition to be fulfilled. Of course this is readily extended to several dependent variables $y_j=y_j(t)$.

An example of such a problem is the curve of quickest descent from a fixed point to a given vertical straight line. Again the curve is an arc of a cycloid but this time that one with a horizontal tangent at $x=b$.

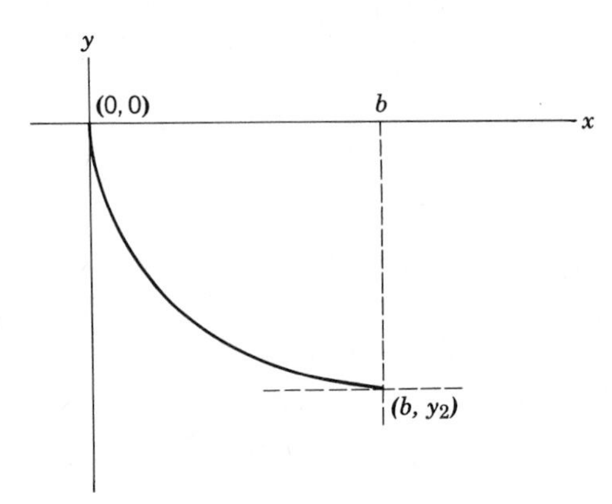

Figure 6.6

6.3 ISOPERIMETRIC PROBLEMS

Here we consider minimizing a given integral

$$I=\int_a^b f(x, y, y')\,dx \tag{6.8}$$

when a second integral

$$J=\int_a^b g(x, y, y')\,dx \tag{6.9}$$

has a prescribed value as a side condition. We now require a *two-parameter* family of comparison functions,

$$Y(x)=y(x)+\varepsilon_1\eta_1(x)+\varepsilon_2\eta_2(x)$$
$$\eta_1(a)=\eta_1(b)=0=\eta_2(a)=\eta_2(b)$$

The comparison problem then becomes

$$I(\varepsilon_1, \varepsilon_2)=\int_a^b f(x, Y, Y')\,dx$$

$$J(\varepsilon_1, \varepsilon_2)=\int_a^b g(x, Y, Y')\,dx$$

Of course ε_1 and ε_2 are not independent since they are constrained by the relation

$$J(\varepsilon_1, \varepsilon_2)=\text{const.}$$

Since $y(x)$ is assumed to be the actual extremizing function, the extremum is given by $\varepsilon_1=0=\varepsilon_2$. By the method of *Lagrange undetermined multipliers* we may treat ε_1 and ε_2 as independent parameters (cf. Section A3.1). Let

$$H(\varepsilon_1, \varepsilon_2; \lambda)\equiv I(\varepsilon_1, \varepsilon_2)+\lambda J(\varepsilon_1, \varepsilon_2)=\int_a^b h(x, Y, Y')\,dx$$

where

$$h=f+\lambda g$$

where λ is the Lagrange multiplier to be determined later. The equations

$$\frac{\partial H}{\partial \varepsilon_1}=0=\frac{\partial H}{\partial \varepsilon_2} \qquad J(\varepsilon_1, \varepsilon_2)=\text{const.}$$

provide three conditions for the three unknowns, ε_1, ε_2, and λ. (Of course we already know that $\varepsilon_1=0=\varepsilon_2$ here.) As before we have

$$\left.\frac{\partial H}{\partial \varepsilon_j}\right|_{\varepsilon_j=0}=\int_a^b \left\{\frac{\partial h}{\partial y}\eta_j+\frac{\partial h}{\partial y'}\eta_j'\right\}dx=\int_a^b \left\{\frac{\partial h}{\partial y}-\frac{d}{dx}\left(\frac{\partial h}{\partial y'}\right)\right\}\eta_j\,dx \qquad j=1, 2$$

Since the η_j are arbitrary, we have

$$\frac{\partial h}{\partial y}-\frac{d}{dx}\left(\frac{\partial h}{\partial y'}\right)=0 \tag{6.10}$$

The solution of this second-order equation for $y=y(x)$ contains, generally, two integration constants as well as the parameter λ. The λ is determined by the condition J=const. and the integration constants by $y(a)=y_1$, $y(b)=y_2$.

We may again readily extend this to the case of several dependent variables subject to several side conditions.

$$I=\int_{t_1}^{t_2} f(y_1, \ldots, y_n; \dot{y}_1, \ldots, \dot{y}_n; t)\, dt \qquad j=1, 2, \ldots, n \tag{6.11}$$

$$J_k=\int_{t_1}^{t_2} g_k(y_1, \ldots, y_n; \dot{y}_1, \ldots, \dot{y}_n; t)\, dt \qquad k=1, 2, \ldots, m \tag{6.12}$$

$$h=\sum_{k=1}^{m} \lambda_k g_k + f \tag{6.13}$$

$$\frac{\partial h}{\partial y_j}-\frac{d}{dt}\left(\frac{\partial h}{\partial \dot{y}_j}\right)=0 \tag{6.14}$$

Example 6.4 As a simple application consider the closed plane curve that encloses the maximum area for a given perimeter. Let

$$x=x(t) \qquad y=y(t)$$

$$I=\frac{1}{2}\int_{t_1}^{t_2} (x\dot{y}-y\dot{x})\, dt$$

which is easily seen from Green's theorem or from the identity

$$\oint \mathbf{A}\cdot d\mathbf{l}=\int_{\Sigma} (\nabla\times\mathbf{A})\cdot d\mathbf{S}$$

Therefore the problem becomes the following

$$J=\int_{t_1}^{t_2} \sqrt{\dot{x}^2+\dot{y}^2}\, dt=L \qquad \text{a constant}$$

$$h=\tfrac{1}{2}(x\dot{y}-y\dot{x})+\lambda\sqrt{\dot{x}^2+\dot{y}^2}$$

$$-\tfrac{1}{2}\dot{x}-\frac{d}{dt}\left(\tfrac{1}{2}x+\frac{\lambda\dot{y}}{\sqrt{\dot{x}^2+\dot{y}^2}}\right)=0 \tag{6.15}$$

$$\tfrac{1}{2}\dot{y}-\frac{d}{dt}\left(-\tfrac{1}{2}y+\frac{\lambda\dot{x}}{\sqrt{\dot{x}^2+\dot{y}^2}}\right)=0 \tag{6.16}$$

If we integrate Eqs. 6.15 and 6.16 w.r.t. t we obtain

$$x+\frac{\lambda\dot{y}}{\sqrt{\dot{x}^2+\dot{y}^2}}=c_1 \qquad y-\frac{\lambda\dot{x}}{\sqrt{\dot{x}^2+\dot{y}^2}}=c_2 \tag{6.17}$$

Equations 6.17 can be written as

$$(y-c_2)^2=\frac{\lambda^2\dot{x}^2}{\dot{x}^2+\dot{y}^2} \qquad (x-c_1)^2=\frac{\lambda^2\dot{y}^2}{\dot{x}^2+\dot{y}^2}$$

so that the solution becomes

$$(x-c_1)^2+(y-c_2)^2=\lambda^2=\left(\frac{L}{2\pi}\right)^2$$

This is simply a circle of circumference L.

6.4 LAGRANGIAN FIELD THEORIES

We consider a set of fields, $\{\phi^\alpha(x_\mu)\}\equiv\{\phi^\alpha(x)\}$, $\alpha=1, 2, \ldots, N$, with the independent variables x_μ, $\mu=1, 2, 3, 4$. We could of course take any number of independent variables. We have a problem of several dependent and several independent variables. We will use the compact notation

$$\phi^\alpha{}_{,\nu}\equiv\frac{\partial\phi^\alpha}{\partial x_\nu}$$

for the gradient of the fields. Here for simplicity $\mathscr{L}$, the *Lagrangian density*,

$$\mathscr{L}=\mathscr{L}(\phi^\alpha, \phi^\alpha{}_{,\nu})$$

is assumed to depend upon the fields, $\{\phi^\alpha\}$, and their *first* derivatives $\{\phi^\alpha{}_{,\nu}\}$ only and not on x_μ explicitly. Let

$$I\equiv\int_V \mathscr{L}(\phi^\alpha, \phi^\alpha{}_{,\nu})\, d^4x \tag{6.18}$$

where $d^4x=dx_1\, dx_2\, dx_3\, dx_4$. We are now integrating over a four-dimensional volume enclosed by a three-dimensional surface Σ. We prescribe the values of the fields ϕ^α on Σ as

$$\delta\phi^\alpha|_\Sigma=0 \tag{6.19}$$

If we set the first variation of I equal to zero, we obtain

$$\begin{aligned}\delta I=0&=\int_V \left\{\frac{\partial\mathscr{L}}{\partial\phi^\alpha}\,\delta\phi^\alpha+\frac{\partial\mathscr{L}}{\partial\phi^\alpha{}_{,\nu}}\,\delta\phi^\alpha{}_{,\nu}\right\} d^4x\\ &\equiv\int_V \left\{\frac{\partial\mathscr{L}}{\partial\phi^\alpha}-\frac{\partial}{\partial x_\nu}\left(\frac{\partial\mathscr{L}}{\partial\phi^\alpha{}_{,\nu}}\right)\right\}\delta\phi^\alpha\, d^4x+\int_V \frac{\partial}{\partial x_\nu}\left(\frac{\partial\mathscr{L}}{\partial\phi^\alpha{}_{,\nu}}\,\delta\phi^\alpha\right) d^4x\end{aligned} \tag{6.20}$$

Notice here that

$$\delta\phi^\alpha{}_{,\nu}\equiv\frac{\partial}{\partial x_\nu}(\delta\phi^\alpha) \tag{6.21}$$

since we are using $\delta\phi^\alpha$ as a shorthand notation for

$$\delta\phi^\alpha(x)\equiv\Phi^\alpha(x, \varepsilon)-\phi^\alpha(x) \tag{6.22}$$

where $\{\Phi^\alpha(x, \varepsilon)\}$ is the set of comparison fields

$$\Phi^\alpha(x, \varepsilon) \equiv \phi^\alpha(x) + \varepsilon\eta^\alpha(x) \tag{6.23}$$

such that

$$\eta^\alpha(x)|_\Sigma \equiv 0 \tag{6.24}$$

We will use the Einstein summation convention according to which repeated indices are summed over. We may use Gauss' theorem for the last integral in Eq. 6.20 to obtain

$$\int_V \frac{\partial}{\partial x_\nu}\left(\frac{\partial \mathcal{L}}{\partial \phi^\alpha{}_{,\nu}}\,\delta\phi^\alpha\right) d^4x = \int_\Sigma \frac{\partial \mathcal{L}}{\partial \phi^\alpha{}_{,\nu}}\,\delta\phi^\alpha\, d\sigma_\nu = 0$$

where the last step follows from Eq. 6.19. Therefore the Euler-Lagrange equations become a set of partial differential equations for the fields $\{\phi^\alpha(x)\}$

$$\frac{\partial \mathcal{L}}{\partial \phi^\alpha} - \frac{\partial}{\partial x_\nu}\left(\frac{\partial \mathcal{L}}{\partial \phi^\alpha{}_{,\nu}}\right) = 0 \qquad \alpha = 1, 2, \ldots, N \tag{6.25}$$

This Lagrangian density is not unique since if we add to it the divergence of a four-vector depending only on the fields $\phi^\alpha(x)$ the field equations are left unchanged.

$$\mathcal{L} \to \mathcal{L}' = \mathcal{L} + \frac{\partial \Lambda_\nu}{\partial x_\nu} \qquad \Lambda_\nu = \Lambda_\nu(\phi^\alpha)$$

This follows since

$$\delta \int_V \frac{\partial \Lambda_\nu}{\partial x_\nu}\, d^4x = \delta \int_\Sigma \Lambda_\nu\, d\sigma_\nu = \int_\Sigma \frac{\partial \Lambda_\nu}{\partial \phi^\alpha}\,\delta\phi^\alpha\, d\sigma_\nu = 0$$

again by Eq. 6.19.

Example 6.5 Let

$$\mathcal{L} = \tfrac{1}{2}(-\phi_{,\mu}\phi_{,\mu} + m^2\phi^2)$$

so that

$$\frac{\partial \mathcal{L}}{\partial \phi_{,\mu}} = -\phi_{,\mu}$$

with the Euler-Lagrange equation

$$\phi_{,\mu\mu} + m^2\phi = 0$$

This is the *Klein–Gordon* equation whose solution is

$$\phi(x) = Ae^{ik_\mu x_\mu}$$

subject to the condition

$$k_\mu k_\mu = m^2$$

6.5 NOETHER'S THEOREM

We will see that to every invariance group (cf. Chapter 9) of the system there corresponds a conserved quantity, and we will see how to construct this quantity from the Lagrangian density. This is similar to the situation in the Lagrangian formulation of classical mechanics (cf. Courant and Hilbert or Goldstein). Consider a system of fields $\{\phi^\alpha(x)\}$ characterized by the integral

$$I=\int_V \mathscr{L}(\phi^\alpha, \phi^\alpha{}_{,\nu})\, d^4x \tag{6.26}$$

If the coordinates are subjected to a certain infinitesimal transformation

$$x_\mu \longrightarrow x'_\mu \equiv x_\mu + \delta x_\mu \tag{6.27}$$

then the fields $\{\phi^\alpha(x)\}$ representing this same group will undergo the induced transformation

$$\phi^\alpha(x) \rightarrow \phi'^\alpha(x') \equiv \phi^\alpha(x) + \delta\phi^\alpha(x) \tag{6.28}$$

Notice that this $\delta\phi^\alpha$ is very *different* from the $\delta\phi^\alpha$ of Eq. 6.22.

Example 6.6 Consider the spatial rotations in two dimensions and the basis functions $e^{\pm i\varphi}$. It is convenient to diagonalize $R(\theta)$ by means of a unitary transformation to obtain the similar matrix

$$\mathscr{R}(\theta)=\begin{pmatrix} e^{i\theta} & 0 \\ 0 & e^{-i\theta} \end{pmatrix}$$

Let

$$f^{(1)}(\varphi)=e^{-i\varphi} \qquad\qquad f^{(2)}(\varphi)=e^{i\varphi}$$

so that

$$f'^{(\alpha)}(\varphi') \equiv f^{(\alpha)}(\varphi') = \mathscr{R}^{\alpha\beta}(\theta) f^{(\beta)}(\varphi)$$

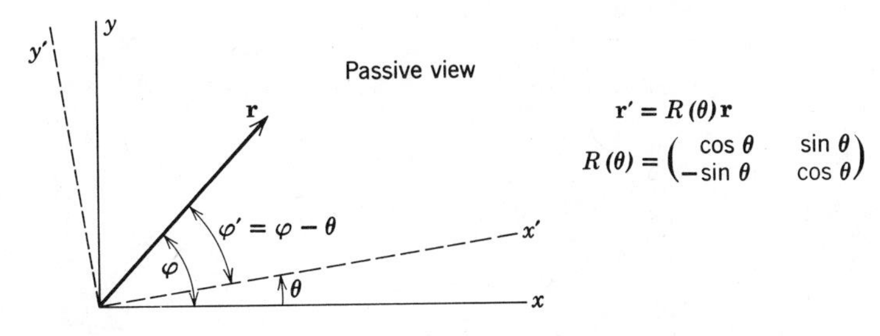

Figure 6.7

or

$$f'^{(1)}(\varphi')=e^{-i(\varphi-\theta)}=e^{i\theta}f^{(1)}(\varphi)$$
$$f'^{(2)}(\varphi')=e^{i(\varphi-\theta)}=e^{-i\theta}f^{(2)}(\varphi)$$

To first order we would have

$$\delta f^{(1)}(\varphi)\equiv f'^{(1)}(\varphi')-f^{(1)}(\varphi)=e^{i\theta}f^{(1)}(\varphi)-f^{(1)}(\varphi)\simeq i\theta f^{(1)}(\varphi)$$

In general the $\phi'^{\alpha}(x')$ will be expressible as some linear combination of the $\phi^{\alpha}(x)$ as

$$\phi'^{\alpha}(x')=\mathscr{D}^{\alpha\beta}\phi^{\beta}(x)$$

We will discuss this more in Section 9.2.

We now assume that the integral of Eq. 6.26 is left *form invariant* under the set of transformations Eq. 6.27 and 6.28. That is, for all V we have the relation

$$\int_{V'} \mathscr{L}[\phi'^{\alpha}(x'), \phi'^{\alpha}{}_{,\nu}(x')]\, d^4x' = \int_V \mathscr{L}[\phi^{\alpha}(x), \phi^{\alpha}{}_{,\nu}(x)]\, d^4x \tag{6.29}$$

Of course we already realize that in general there could be an additional term

$$\int_V \frac{\partial \Lambda_{\mu}(\phi^{\alpha})}{\partial x_{\mu}}\, d^4x$$

on the right. Therefore from Eq. 6.29 we obtain

$$\int_{V'} \mathscr{L}[\phi^{\alpha}+\delta\phi^{\alpha}, \phi^{\alpha}{}_{,\nu}+\delta\phi^{\alpha}{}_{,\nu}]\, d^4x' - \int_V \mathscr{L}[\phi^{\alpha}, \phi^{\alpha}{}_{,\nu}]\, d^4x \equiv 0 \tag{6.30}$$

We must first transform from V' to V as

$$d^4x' = \left|\frac{\partial(x')}{\partial(x)}\right| d^4x \tag{6.31}$$

where the Jacobian of transformation is

$$\left|\frac{\partial(x')}{\partial(x)}\right| \equiv \begin{vmatrix} \frac{\partial x_1'}{\partial x_1} & \frac{\partial x_1'}{\partial x_2} & \frac{\partial x_1'}{\partial x_3} & \frac{\partial x_1'}{\partial x_4} \\ \frac{\partial x_2'}{\partial x_1} & \cdots & \cdot & \cdot \\ \cdot & \cdot & \cdot & \cdot \\ \cdot & \cdot & \cdot & \cdot \\ \frac{\partial x_4'}{\partial x_1} & \cdots & \cdot & \frac{\partial x_4'}{\partial x_4} \end{vmatrix} = \begin{vmatrix} 1+\frac{\partial \delta x_1}{\partial x_1} & \frac{\partial \delta x_1}{\partial x_2} & \frac{\partial \delta x_1}{\partial x_3} & \frac{\partial \delta x_1}{\partial x_4} \\ \frac{\partial \delta x_2}{\partial x_1} & 1+\frac{\partial \delta x_2}{\partial x_2} & \cdots & \cdot \\ \cdot & \cdot & \cdot & \cdot \\ \cdot & \cdot & \cdot & \cdot \\ \frac{\partial \delta x_4}{\partial x_1} & \cdots & \cdot & 1+\frac{\partial \delta x_4}{\partial x_4} \end{vmatrix}$$

since

$$\frac{\partial x'_\mu}{\partial x_\nu} \simeq \delta_{\mu\nu} + \frac{\partial \delta x_\mu}{\partial x_\nu}$$

to first order in δx_μ. Again to first order in δx_μ we have

$$\left|\frac{\partial(x')}{\partial(x)}\right| = 1 + \frac{\partial \delta x_\mu}{\partial x_\mu} \tag{6.32}$$

Using a Taylor series to first order in $\delta\phi^\alpha$ we see that

$$\mathcal{L}(\phi^\alpha + \delta\phi^\alpha, \phi^\alpha{}_{,\nu} + \delta\phi^\alpha{}_{,\nu}) \simeq \mathcal{L}(\phi^\alpha, \phi^\alpha{}_{,\nu}) + \frac{\partial \mathcal{L}}{\partial \phi^\alpha}\, \delta\phi^\alpha + \frac{\partial \mathcal{L}}{\partial \phi^\alpha{}_{,\nu}}\, \delta\phi^\alpha{}_{,\nu} \tag{6.33}$$

Hence to first order we have from Eqs. 6.30–6.33

$$\int_V \left\{ \frac{\partial \mathcal{L}}{\partial \phi^\alpha}\, \delta\phi^\alpha + \frac{\partial \mathcal{L}}{\partial \phi^\alpha{}_{,\nu}}\, \delta\phi^\alpha{}_{,\nu} + \mathcal{L}\frac{\partial \delta x_\nu}{\partial x_\nu} \right\} d^4x = 0 \tag{6.34}$$

We will now transform the integrand into a four divergence. We have the following expressions at our disposal:

i. $\dfrac{\partial \mathcal{L}}{\partial \phi^\alpha} - \dfrac{\partial}{\partial x_\nu}\left(\dfrac{\partial \mathcal{L}}{\partial \phi^\alpha{}_{,\nu}}\right) = 0$

ii. $\dfrac{\partial \mathcal{L}}{\partial x_\mu} \equiv \dfrac{\partial \mathcal{L}}{\partial \phi^\alpha}\, \phi^\alpha{}_{,\mu} + \dfrac{\partial \mathcal{L}}{\partial \phi^\alpha{}_{,\sigma}}\, \dfrac{\partial \phi^\alpha{}_{,\sigma}}{\partial x_\mu}$ —an identity

iii.
$$\delta\phi^\alpha{}_{,\mu} \equiv \delta\left(\frac{\partial \phi^\alpha}{\partial x_\mu}\right) \equiv \frac{\partial \phi'^\alpha(x')}{\partial x'_\mu} - \frac{\partial \phi^\alpha(x)}{\partial x_\mu}$$
$$= \frac{\partial[\phi^\alpha(x) + \delta\phi^\alpha(x)]}{\partial x_\sigma}\, \frac{\partial x_\sigma}{\partial x'_\mu} - \phi^\alpha{}_{,\mu} = \left[\phi^\alpha{}_{,\sigma}(x) + \frac{\partial \delta\phi^\alpha}{\partial x_\sigma}\right]\left[\delta_{\sigma\mu} - \frac{\partial \delta x_\sigma}{\partial x_\mu}\right] - \phi^\alpha{}_{,\mu}$$
$$\simeq \phi^\alpha{}_{,\mu} - \phi^\alpha{}_{,\sigma}\, \frac{\partial \delta x_\sigma}{\partial x_\mu} + \frac{\partial \delta\phi^\alpha}{\partial x_\mu} - \phi^\alpha{}_{,\mu}$$

We can successively transform each of the three terms in the integrand as follows:

iv.
$$\frac{\partial \mathcal{L}}{\partial \phi^\alpha}\, \delta\phi^\alpha = \frac{\partial}{\partial x_\mu}\left(\frac{\partial \mathcal{L}}{\partial \phi^\alpha{}_{,\mu}}\right) \delta\phi^\alpha \qquad \text{(by Eq. 6.25)}$$
$$= \frac{\partial}{\partial x_\mu}\left(\frac{\partial \mathcal{L}}{\partial \phi^\alpha{}_{,\mu}}\, \delta\phi^\alpha\right) - \frac{\partial \mathcal{L}}{\partial \phi^\alpha{}_{,\mu}}\, \frac{\partial \delta\phi^\alpha}{\partial x_\mu}$$

v.
$$\frac{\partial \mathcal{L}}{\partial \phi^\alpha{}_{,\mu}}\, \delta(\phi^\alpha{}_{,\mu}) = \frac{\partial \mathcal{L}}{\partial \phi^\alpha{}_{,\mu}}\left[-\phi^\alpha{}_{,\sigma}\, \frac{\partial(\delta x_\sigma)}{\partial x_\mu} + \frac{\partial \delta\phi^\alpha}{\partial x_\mu}\right]$$
$$= -\frac{\partial}{\partial x_\mu}\left(\frac{\partial \mathcal{L}}{\partial \phi^\alpha{}_{,\mu}}\, \phi^\alpha{}_{,\sigma}\, \delta x_\sigma\right) + \frac{\partial}{\partial x_\mu}\left(\frac{\partial \mathcal{L}}{\partial \phi^\alpha{}_{,\mu}}\right)\phi^\alpha{}_{,\sigma}\, \delta x_\sigma$$
$$+ \frac{\partial \mathcal{L}}{\partial \phi^\alpha{}_{,\mu}}\, \frac{\partial}{\partial x_\mu}(\phi^\alpha{}_{,\sigma})\, \delta x_\sigma + \frac{\partial \mathcal{L}}{\partial \phi^\alpha{}_{,\mu}}\, \frac{\partial \delta\phi^\alpha}{\partial x_\mu}$$

$$= -\frac{\partial}{\partial x_\mu}\left(\frac{\partial \mathcal{L}}{\partial \phi^\alpha_{,\mu}} \phi^\alpha_{,\sigma}\, \delta x_\sigma\right) + \frac{\partial \mathcal{L}}{\partial x_\nu}\, \delta x_\nu$$

$$+ \frac{\partial \mathcal{L}}{\partial \phi^\alpha_{,\mu}} \frac{\partial \delta\phi^\alpha}{\partial x_\mu}; \text{ [by Eqs. 6.25 and (ii)]}$$

vi. $\mathcal{L}\dfrac{\partial \delta x_\mu}{\partial x_\mu} = \dfrac{\partial}{\partial x_\mu}(\mathcal{L}\delta_{\mu\nu}\, \delta x_\nu) - \dfrac{\partial \mathcal{L}}{\partial x_\nu}\, \delta x_\nu$

Therefore the integral Eq. 6.34 becomes

$$\int_V \left\{\frac{\partial}{\partial x_\mu}\left[\left(\mathcal{L}\delta_{\mu\nu} - \frac{\partial \mathcal{L}}{\partial \phi^\alpha_{,\mu}} \phi^\alpha_{,\nu}\right) \delta x_\nu + \frac{\partial \mathcal{L}}{\partial \phi^\alpha_{,\mu}}\, \delta\phi^\alpha\right]\right\} d^4x = 0$$

Since V is arbitrary we have (just as in the proof of Th. 6.1)

$$\frac{\partial f_\mu}{\partial x_\mu} = 0 \tag{6.35}$$

where

$$f_\mu \equiv \left(\mathcal{L}\delta_{\mu\nu} - \frac{\partial \mathcal{L}}{\partial \phi^\alpha_{,\mu}} \phi^\alpha_{,\nu}\right) \delta x_\nu + \frac{\partial \mathcal{L}}{\partial \phi^\alpha_{,\mu}}\, \delta\phi^\alpha \tag{6.36}$$

THEOREM 6.3 (Noether's theorem)

If the group of transformations

$$x_\mu \to x'_\mu \equiv x_\mu + \delta x_\mu$$
$$\phi^\alpha(x) \to \phi'^\alpha(x') \equiv \phi^\alpha(x) + \delta\phi^\alpha(x)$$

is such that

$$\int_{V'} \mathcal{L}[\phi'^\alpha(x'), \phi'^\alpha_{,\nu'}(x')]\, d^4x' = \int_V \mathcal{L}[\phi^\alpha(x), \phi^\alpha_{,\nu}(x)]\, d^4x$$

for all volumes V, then

$$\frac{\partial f_\mu}{\partial x_\mu} \equiv 0$$

where

$$f_\mu = \left(\mathcal{L}\delta_{\mu\nu} - \frac{\partial \mathcal{L}}{\partial \phi^\alpha_{,\mu}} \phi^\alpha_{,\nu}\right) \delta x_\nu + \frac{\partial \mathcal{L}}{\partial \phi^\alpha_{,\mu}}\, \delta\phi^\alpha$$

We mentioned at the beginning of this section that the invariance of Eq. 6.26 (i.e., Eq. 6.29) under the group of transformations, Eqs. 6.27–6.28, would imply the conservation of some quantity. This will follow from Eq. 6.35 under the following conditions. We have considered four independent variables. (Of course as far as the mathematics is concerned, we could have taken any number.) If we let x_1, x_2, x_3 be spatial coordinates and x_4 be a time coordinate (actually

$x_4 = it$), we can write Eq. 6.35 as

$$\frac{\partial f_4}{\partial x_4}+\sum_{j=1}^{3}\frac{\partial f_j}{\partial x_j}\equiv\frac{\partial f_4}{\partial x_4}+\nabla\cdot\mathbf{f}=0 \tag{6.37}$$

If we now integrate Eq. 6.37 over a three-dimensional spatial volume V_3 (bounded by Σ_2), we obtain

$$\int_{V_3}\frac{\partial f_4}{\partial x_4}\,d^3v\equiv\frac{d}{dx_4}\int_{V_3}f_4\,d^3v=-\int_{V_3}\nabla\cdot\mathbf{f}\,d^3v=-\int_{\Sigma_2}\mathbf{f}\cdot d\mathbf{S}$$

If we assume that the fields $\phi^\alpha(x)$ and their derivatives (and hence f_μ) vanish at spatial infinity (i.e., when x_1, x_2, or $x_3 \rightarrow \pm\infty$, but not necessarily as $x_4 \rightarrow \pm\infty$), then for an infinite spatial region V_3 we obtain

$$\frac{d}{dx_4}\int_{\substack{\text{all}\\\text{space}}}f_4\,d^3v=0$$

This states that the quantity

$$F(t)\equiv\int_{\substack{\text{all}\\\text{space}}}f_4(\mathbf{r},t)\,d^3v \tag{6.38}$$

is independent of time,

$$\frac{dF}{dt}\equiv 0$$

or a constant (i.e., a *conserved quantity*).

Suppose now that the Lagrangian density depends *explicitly* on the coordinates x_μ but that the integral

$$I\equiv\int_V\mathcal{L}(\phi^\alpha,\phi^\alpha_{,\mu};x_\mu)\,d^4x \tag{6.39}$$

is still such that it is left invariant (i.e., $\delta I=0$) under the transformations of Eqs. 6.27 and 6.28. Then the result of Th. 6.3 remains valid (cf. Problem 6.9). If, however, $\mathcal{L}$ can be decomposed as

$$\mathcal{L}=\mathcal{L}_0+\mathcal{L}_I \tag{6.40}$$

such that, under the transformations of Eqs. 6.27 and 6.28,

$$\delta\int_V\mathcal{L}_0\,d^4x=0$$

$$\delta\int_V\mathcal{L}_I\,d^4x\neq 0$$

then

$$\frac{\partial f_\mu}{\partial x_\mu}=\delta\mathcal{L}_I \tag{6.41}$$

where f_μ is still given by Eq. 6.36 and where we have assumed $|\partial(x')/\partial(x)|=1$ for simplicity.

Example 6.7 Consider the following Lagrangian density

$$\mathcal{L}=-\tfrac{1}{2}\phi_{,\sigma}\phi_{,\sigma}+\tfrac{1}{2}m^2\phi^2+\lambda A_\sigma\phi_{,\sigma}$$

where $A_\sigma(x)$ is a *given* external field. Take the transformations to be

$$\phi'(x')=\phi(x)$$

$$x'_\mu=x_\mu+\varepsilon_\mu$$

where the $\{\varepsilon_\mu\}$, $\mu=1, 2, 3, 4$, are infinitesimal constants.

$$A_\sigma(x)\rightarrow A'_\sigma(x')\equiv A_\sigma(x')=A_\sigma(x_\mu+\delta x_\mu)$$

$$\simeq A_\sigma(x)+\frac{\partial A_\sigma}{\partial x_\mu}\,\delta x_\mu$$

According to Eq. 6.41 we should have

$$\frac{\partial f_\mu}{\partial x_\mu}=\delta\mathcal{L}_I=\left.\frac{\partial\mathcal{L}}{\partial x_\mu}\right|_{\text{explicit}}\delta x_\mu=\lambda\frac{\partial A_\sigma}{\partial x_\mu}\,\phi_{,\sigma}\,\delta x_\mu$$

where $\mathcal{L}_I=\lambda A_\sigma\phi_{,\sigma}$. Let us now compute every term explicitly.

$$\frac{\partial\mathcal{L}}{\partial\phi}=m^2\phi \qquad \frac{\partial\mathcal{L}}{\partial\phi_{,\nu}}=-\phi_{,\nu}+\lambda A_\nu$$

The field equation then becomes

$$\phi_{,\mu\mu}+m^2\phi=\lambda\frac{\partial A_\mu}{\partial x_\mu}$$

so that

$$f_\mu=\{[-\tfrac{1}{2}\phi_{,\sigma}\phi_{,\sigma}+\tfrac{1}{2}m^2\phi^2+\lambda A_\sigma\phi_{,\sigma}]\,\delta_{\mu\nu}+\phi_{,\mu}\phi_{,\nu}-\lambda A_\mu\phi_{,\nu}\}\varepsilon_\nu$$

and

$$\frac{\partial f_\mu}{\partial x_\mu}\equiv\{-\phi_{,\sigma}\phi_{,\sigma\nu}+m^2\phi\phi_{,\nu}+\lambda A_\sigma\phi_{,\sigma\nu}+\lambda A_{\sigma,\nu}\phi_{,\sigma}$$

$$+\phi_{,\mu\mu}\phi_{,\nu}+\phi_{,\mu}\phi_{,\mu\nu}-\lambda A_{\mu,\mu}\phi_{,\nu}-\lambda A_\mu\phi_{,\nu\mu}\}\varepsilon_\nu$$

$$=\left\{\phi_{,\nu}\lambda\frac{\partial A_\sigma}{\partial x_\sigma}-\lambda A_{\mu,\mu}\phi_\nu+\lambda\frac{\partial A_\sigma}{\partial x_\nu}\,\phi_{,\sigma}\right\}\varepsilon_\nu$$

$$=\lambda\frac{\partial A_\sigma}{\partial x_\nu}\,\phi_{,\sigma}\,\delta x_\nu\equiv\left.\frac{\partial\mathcal{L}}{\partial x_\mu}\right|_{\text{explicit}}\delta x_\mu=\delta\mathcal{L}_I$$

as expected.

SUGGESTED REFERENCES

H. Sagan, *Introduction to the Calculus of Variations.* This gives a quite advanced and thorough treatment of the problem of extremizing an integral.

R. Weinstock, *Calculus of Variations.* This is an elementary and very readable treatment of variational problems with many applications to physics and engineering.

The following discuss physical applications of variational techniques.

R. Courant and D. Hilbert, *Methods of Mathematical Physics.* Section 12.8 of Chapter IV discusses Noether's theorem for classical particle mechanics.

R. Courant and H. Robbins, "Plateau's Problem," in *The World of Mathematics* (ed. J. R. Newman) (pp. 901–909). This beautiful and delightful article discusses how thin soap films on wire boundaries assume the shapes of minimum surface areas under the influence of surface tension.

W. Yourgrau and S. Mandelstam, *Variational Principles in Dynamics and Quantum Theory.*

P. Roman, *Theory of Elementary Particles.* Section 1 of Chapter IV applies Noether's theorem to physical systems to demonstrate the intimate connection between geometrical invariances of a system and conserved quantities.

PROBLEMS

6.1 Suppose you are given a plane sheet of metal whose surface density is $\rho(x, y)$. You are now asked to find that closed curve of given length such that the mass,

$$m = \iint\limits_R \rho(x, y)\, dx\, dy$$

assumes the largest possible value, where R is the interior of the curve of given length. Find the differential equation satisfied by this extremal curve.

6.2 Show that the geodesics of a spherical surface are great circles (i.e., circles on the sphere whose centers lie at the center of the sphere).

6.3 Complete the solution of the brachistochrone problem of Example 6.2 and show that the required curve is a cycloid. Show that this time of descent is less than that for the path $(0, 0) \to (0, y_2) \to (b, y_2)$ along straight-line segments (cf. Fig. 6.6). Then, rather than finding the curve of fastest descent between two points in a vertical plane, find the corresponding curve between a fixed point and a given vertical line.

6.4 Show that if a particular function $y = y(x)$ renders I an extremum subject to the side condition $J = \text{const.}$, then the same function renders J an extremum subject to $I = \text{const.}$, provided the Lagrange multiplier, λ, is not zero for the first problem. What would be the significance of $\lambda = 0$? Using this result show that, of all the simple closed curves enclosing a given area, the least perimeter is possessed by a circle.

6.5 a. Find the parametric equation for the envelope of the family of catenaries shown in Fig. 6.5. Prove that if point C of Fig. 6.5 lies sufficiently far above this envelope, then a catenary passing through this point will provide the absolute minimum of surface area rather than the discontinuous solution discussed in the text. [*Hint:* prove that if a one-parameter family of curves defined by

$$g(x, y; \alpha) = 0$$

has an envelope, then this curve $y = y(x)$ is the solution to

$$g(x, y; \alpha) = 0$$

$$\frac{\partial}{\partial \alpha} g(x, y; \alpha) = 0$$

once the parameter, α, has been eliminated from these two equations.]

b. Now consider the special case in which $y_1 = y_2$ (i.e., $x = 0 \Rightarrow y = y_1$, $x = x_2 \Rightarrow y = y_2$). Show that if $(x_2/y_1) < 1.32$, two catenaries pass through the specified points, if $(x_2/y_1) = 1.32$, just one catenary, and if $(x_2/y_1) > 1.32$, no catenaries. Show that the upper catenary provides the absolute minimum area of revolution until $(x_2/y_1) = 1.02$, after which the discontinuous solution provides the absolute minimum.

6.6 Assume you are given a Lagrangian density $\mathscr{L}' = \mathscr{L}'(\phi, \phi_{,\nu}, \phi_{,\mu\nu})$ that depends on the field $\phi(x_\mu)$ and its first and second derivatives. Show that the condition

$$\delta \int_V \mathscr{L}' \, d^4x = 0$$

with the boundary conditions

$$\delta\phi|_\Sigma = 0 \qquad \delta\phi_{,\mu}|_\Sigma = 0$$

where Σ is the surface enclosing the integration volume, V, implies the field equation

$$\frac{\partial \mathscr{L}'}{\partial \phi} - \frac{\partial}{\partial x_\nu}\left(\frac{\partial \mathscr{L}'}{\partial \phi_{,\nu}}\right) + \frac{\partial^2}{\partial x_\mu \, \partial x_\nu}\left(\frac{\partial \mathscr{L}'}{\partial \phi_{,\mu\nu}}\right) = 0$$

Furthermore prove that, if

$$\mathscr{L}'(\phi, \phi_{,\nu}, \phi_{,\mu\nu}) = \mathscr{L}(\phi, \phi_{,\nu}) + \frac{\partial}{\partial x_\sigma} \Lambda_\sigma(\phi, \phi_{,\nu})$$

then these field equations reduce to the simpler form

$$\frac{\partial \mathscr{L}}{\partial \phi} - \frac{\partial}{\partial x_\nu}\left(\frac{\partial \mathscr{L}}{\partial \phi_{,\nu}}\right) = 0$$

6.7 Assume that the Lagrangian density, $\mathscr{L} = \mathscr{L}(\phi^\alpha, \phi^\alpha_{,\nu}, \phi^\alpha_{,\mu\nu})$, has its action

$$I = \int_V \mathscr{L}(\phi^\alpha, \phi^\alpha_{,\nu}, \phi^\alpha_{,\mu\nu}) \, d^4x$$

left invariant under the continuous infinitesimal transformations,

$$x_\mu \to x'_\mu \equiv x_\mu + \delta x_\mu$$

$$\phi^\alpha(x) \to \phi'^\alpha(x') \equiv \phi^\alpha(x) + \delta\phi^\alpha(x)$$

Generalize Noether's theorem to this case and show that the f_μ such that

$$\frac{\partial f_\mu}{\partial x_\mu}=0$$

by virtue of the field equations for the $\phi^\alpha(x)$ is given by

$$f_\mu=\mathscr{L}\,\delta x_\mu+\frac{2\mathscr{L}}{\partial\phi^\alpha{}_{,\mu}}(\delta\phi^\alpha-\phi^\alpha{}_{,\sigma}\,\delta x_\sigma)+\frac{\partial\mathscr{L}}{\partial\phi^\alpha{}_{,\mu\nu}}\frac{\partial}{\partial x_\nu}(\delta\phi^\alpha-\phi^\alpha{}_{,\sigma}\,\delta x_\sigma)$$
$$-\frac{\partial}{\partial x_\nu}\left(\frac{\partial\mathscr{L}}{\partial\phi^\alpha{}_{,\mu\nu}}\right)(\delta\phi^\alpha-\phi^\alpha{}_{,\sigma}\,\delta x_\sigma)$$

6.8 Given the Lagrangian density

$$\mathscr{L}=-\frac{1}{2}\frac{\partial A_\mu}{\partial x_\nu}\frac{\partial A_\mu}{\partial x_\nu}-(\phi_{,\mu}\phi^*{}_{,\mu}-m^2\phi\phi^*)$$
$$+i\lambda A_\mu(\phi^*{}_{,\mu}\phi-\phi_{,\mu}\phi^*)-\lambda^2A_\mu A_\mu\phi\phi^*$$

where λ is a real constant and A_μ a real field, use the Euler-Lagrange equations to find the field equations satisfied by the fields $\phi(x)$, $\phi^*(x)$, and $A_\mu(x)$. Here, as usual, x stands for x_μ, $\mu=1, 2, 3, 4$. Prove that the integral I of Eq. (6.26) is left invariant under the transformation

$$x_\mu\to x'_\mu=x_\mu$$
$$\phi\to\phi'=\phi+i\lambda\phi\;\delta\alpha$$
$$\phi^*\to\phi^{*\prime}=\phi^*-i\lambda\phi^*\;\delta\alpha$$
$$A_\mu\to A'_\mu=A_\mu$$

where $\delta\alpha$ is a real infinitesimal parameter, and use Noether's theorem to construct the conserved f_μ. Then show explicitly that $\partial f_\mu/\partial x_\mu\equiv 0$ by virtue of the equations of motion. Repeat this for the infinitesimal transformation,

$$x_\mu\to x'_\mu=x_\mu+\varepsilon_\mu$$
$$\phi\to\phi'=\phi$$
$$\phi^*\to\phi^{*\prime}=\phi^*$$
$$A_\mu\to A'_\mu=A_\mu$$

where the $\{\varepsilon_\mu\}$ are real infinitesimal parameters.

6.9 Prove that if $\mathscr{L}=\mathscr{L}(\phi^\alpha,\phi^\alpha{}_{,\mu};x_\mu)$ but $\delta I=0$, then Th. 6.3 remains valid. Realize that if $\mathscr{L}$ depends upon x_μ *explicitly*, then

$$\frac{\partial\mathscr{L}}{\partial x_\mu}\equiv\frac{\partial\mathscr{L}}{\partial\phi^\alpha}\phi^\alpha{}_{,\mu}+\frac{\partial\mathscr{L}}{\partial\phi^\alpha{}_{,\sigma}}\frac{\partial\phi^\alpha{}_{,\sigma}}{\partial x_\mu}+\left.\frac{\partial\mathscr{L}}{\partial x_\mu}\right|_{\text{explicit}}$$

Next prove the result stated in Eq. 6.41.

6.10 You are given the Lagrangian density

$$\mathscr{L}=\frac{i}{2}\psi^*\dot{\psi}-\frac{i}{2}\dot{\psi}^*\psi-\frac{1}{2m}\nabla\psi^*\cdot\nabla\psi-V(\mathbf{r},t)\psi^*\psi$$

where $\psi(\mathbf{r}, t)$ and $\psi^*(\mathbf{r}, t)$ are independent fields, $V(\mathbf{r}, t)$ is a given real function and $\dot{\psi} \equiv \partial\psi/\partial t$.

a. Use the Euler-Lagrange equations to obtain the partial differential equations satisfied by ψ and ψ^*.

b. If the Lagrangian is given as being left invariant under the infinitesimal transformation

$$\psi(x) \to \psi'(x') = \psi(x)$$
$$\psi^*(x) \to \psi^{*\prime}(x') = \psi^*(x)$$
$$x_j \to x_j' = x_j, \; j = 1, 2, 3$$
$$t \to t' = t + \varepsilon \qquad (\varepsilon\text{—an infinitesimal parameter})$$

find what quantity is a constant of the motion from Noether's theorem. What assumptions must you make about $V(\mathbf{r}, t)$? What conditions must be imposed on the field $\psi(\mathbf{r}, t)$ as $|\mathbf{r}| \to \infty$?

c. Repeat (b) for the transformations

$$\psi(x) \to \psi'(x') = \psi(x)$$
$$x_j \to x_j' = x_j + \varepsilon_j$$
$$t \to t' = t$$

What restrictions are necessary on $V(\mathbf{r}, t)$?

d. Prove that $\mathcal{L}$ is left invariant under the phase transformation

$$x_\mu \to x_\mu' = x_\mu \qquad \mu = 1, 2, 3, 4$$
$$\psi(x) \to \psi'(x') = e^{i\alpha}\psi(x)$$
$$\psi^*(x) \to \psi^{*\prime}(x') = e^{-i\alpha}\psi^*(x)$$

where α is a real infinitesimal parameter. Construct the conserved quantity.

7 | Complex Variables

7.0 INTRODUCTION

We will assume some basic knowledge of what a complex number is. In applied mathematics we often find consideration of functions defined on the real axis alone (i.e., $-\infty<x<+\infty$) not sufficient. We must consider functions of the complex variable, z, defined as

$$z \equiv x+iy \equiv \text{Re } z+i \text{ Im } z \equiv \rho e^{i\varphi} \tag{7.1}$$

$$\rho=|z|=\sqrt{x^2+y^2} \geq 0 \qquad \varphi=\tan^{-1}(y/x) \qquad i^2=-1$$

The complex conjugate is defined as

$$z^* \equiv x-iy \tag{7.2}$$

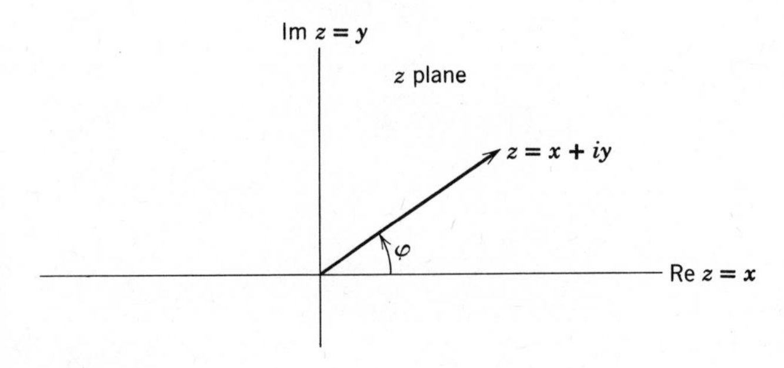

Figure 7.1

That is, we have let $i \to -i$ or $\varphi \to -\varphi$. For use with multiple-valued functions later it is better to state the operation of taking the complex conjugate in a modified form such as

$$\varphi \to 2\pi - \varphi \tag{7.3}$$

We will discuss this further in Section 7.9.

7.1 DEFINITION OF A HOLOMORPHIC FUNCTION

Since any complex quantity may be decomposed into its real and imaginary parts, we may always write

$$f(x, y) = u(x, y) + iv(x, y) \tag{7.4}$$

where $u(x, y)$ and $v(x, z)$ are real functions of the real variables, x and y. Initially our study of complex variables will be restricted to those single-valued functions that depend on z *only*, and not on both z and z^*. In a short while we will enlarge our class to include multiple-valued functions of z as well, but the study of functions of two or more complex variables is extremely difficult, and we will have nothing further to say about it. The f written above is, without further restrictions, simply a complex function of the two independent variables, x and y. Such functions are much more general than those depending upon z alone, since a knowledge of z is not equivalent to that of x and y separately. Given the directed line segment, z (i.e., a vector), in the plane, we do not know the relative direction of the x axis. However given z and z^*, the x axis bisects the angle between z and z^* (cf. Fig. 7.2). This is similar to a change of variables

$$s = x + y \qquad t = x - y$$

Therefore the complex function

$$u(x, y) + iv(x, y) \equiv f(z, z^*) \tag{7.5}$$

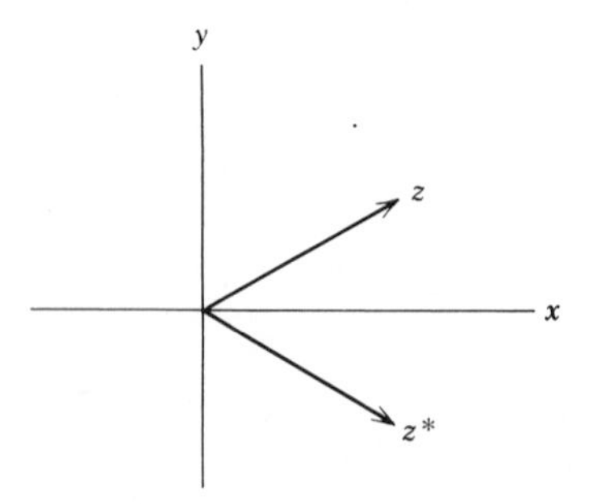

Figure 7.2

will generally depend on both z and z^*, and this turns out to be too broad a class of functions. We must impose restrictions on f.

We say that a function $f(z)$, single-valued in a region, is *holomorphic* if it possesses a *unique* derivative at every point in the region. That is, given an $f(z)$, the ratio

$$\frac{f(z)-f(z_0)}{z-z_0} \tag{7.6}$$

approaches a (unique) limit as $z \to z_0$ along *any* path in the z plane. (Notice that if a function is not single-valued in R it cannot have a unique derivative everywhere there.) Since we have a function defined in a plane, we must approach a given point (e.g., z_0) along a path, which is not now restricted to the real axis as in the theory of functions of one real variable. Holomorphy is an extremely restrictive requirement to make and, in fact, will not usually be satisfied by an arbitrary complex function of x and y. For example let

$$f(x, y)=x+2iy=\tfrac{3}{2}z-\tfrac{1}{2}z^*$$

If we compute the ratio of Eq. 7.6 for an approach along the y axis (i.e., $x=0$) for $z_0=0$, then we obtain

$$\frac{f(y)-f(y_0)}{iy-iy_0}=\frac{2iy}{iy}\xrightarrow[y\to y_0=0]{} 2$$

If we approach the origin along the real axis (i.e., $y=0$), then we have

$$\frac{f(x)-f(x_0)}{x-x_0}=\frac{x}{x}\xrightarrow[x\to x_0=0]{} 1$$

Therefore, the derivative, defined as

$$\lim_{\Delta z\to 0}\frac{f(z+\Delta z)-f(z)}{\Delta z}$$

is not unique for this example.

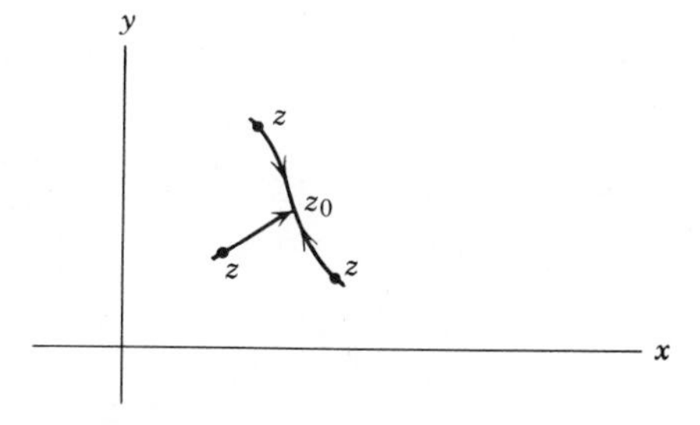

Figure 7.3

7.2 CAUCHY-RIEMANN CONDITIONS

As the discussion above shows holomorphic functions of z do not encompass all complex functions of two real variables (x, y), but only a very special class of them. Now since any complex function of two real variables can be written as

$$f(x, y) = u(x, y) + iv(x, y)$$

we expect that the condition of holomorphy on $f(z)$ will impose some relations between u and v.

THEOREM 7.1

If in a given region R the two real, single-valued functions $u(x, y)$ and $v(x, y)$ are defined, then the necessary and sufficient condition that the complex function

$$f = u(x, y) + iv(x, y)$$

be holomorphic in R is that here all first partial derivatives of u and v exist, be continuous, and satisfy the *Cauchy-Riemann conditions*

$$\frac{\partial u}{\partial x} = \frac{\partial v}{\partial y} \qquad \frac{\partial u}{\partial y} = -\frac{\partial v}{\partial x} \tag{7.7}$$

PROOF

a. Necessary

Assume that there exists a *unique* $f'(z)$, independent of the direction of approach of Δz to z, defined as

$$f'(z) = \frac{df(z)}{dz} \equiv \lim_{\Delta z \to 0} \frac{f(z+\Delta z) - f(z)}{\Delta z}$$

Since

$$\frac{f(z+\Delta z) - f(z)}{\Delta z} \equiv \frac{[u(x+\Delta x, y+\Delta y) - u(x, y)] + i[v(x+\Delta x, y+\Delta y) - v(x, y)]}{\Delta x + i\,\Delta y} \tag{7.8}$$

and since the limit $\Delta z \to 0$ is given to exist for *all* Δz, we may first take $\Delta y = 0$ to conclude that

$$\lim_{\Delta x \to 0} \left[\frac{u(x+\Delta x, y) - u(x, y)}{\Delta x} + i\,\frac{v(x+\Delta x, y) - v(x, y)}{\Delta x} \right] \equiv \frac{\partial u}{\partial x} + i\,\frac{\partial v}{\partial x}$$

exists so that $\partial u/\partial x$ and $\partial v/\partial x$ exist everywhere in R (since the real and imaginary parts of the limit of a complex quantity must exist separately). Similarly, if we take $\Delta x=0$, we deduce that $\partial u/\partial y$ and $\partial v/\partial y$ exist everywhere. The existence of these unique limits establishes the existence of all of the first partial derivatives as claimed. We shall see in Section 7.3b (cf. Eq. 7.16) that the derivative of any holomorphic function is itself holomorphic so that df/dz must be continuous. (Of course, this result will be established independently of the present theorem.) From this it follows that all of the first partial derivatives of $u(x, y)$ and $v(x, y)$ are also continuous.

We can now return to Eq. 7.8 and compute $f'(z)$, once for $\Delta y=0$ and once for $\Delta x=0$, to obtain

$$\frac{df}{dz}\equiv \lim_{\Delta z\to 0}\frac{\Delta f}{\Delta z}=\frac{\partial u}{\partial x}+i\frac{\partial v}{\partial x}=\frac{1}{i}\left(\frac{\partial u}{\partial y}+i\frac{\partial v}{\partial y}\right) \tag{7.9}$$

from which follows Eq. 7.7.

b. Sufficient

If we realize that $dy/dx=\lim_{\Delta x\to 0}\Delta y/\Delta x$ is the slope of the path of approach to the point (x, y) in the z plane, we may use the mean value theorem for two variables (cf. Appendix I, Th. I.6) and Eq. 7.8 to write

$$\frac{df}{dz}\equiv \lim_{\Delta z\to 0}\frac{\Delta f}{\Delta z}=\frac{\dfrac{\partial u}{\partial x}+i\dfrac{\partial v}{\partial x}+\left(\dfrac{\partial u}{\partial y}+i\dfrac{\partial v}{\partial y}\right)\dfrac{dy}{dx}}{1+i\,dy/dx}$$

$$=\frac{\dfrac{\partial u}{\partial x}+i\dfrac{\partial v}{\partial x}+\left(-\dfrac{\partial v}{\partial x}+i\dfrac{\partial u}{\partial x}\right)\dfrac{dy}{dx}}{1+i\,dy/dx}=\frac{\partial u}{\partial x}+i\frac{\partial v}{\partial x} \tag{7.10}$$

where we have used Eq. 7.7. The final expression in Eq. 7.10 is independent of dy/dx so that df/dz is unique (i.e., independent of the direction of approach to z). (Clearly df/dz as given by Eq. 7.10 exists and is continuous since $\partial u/\partial x$ and $\partial v/\partial x$ are given as having these properties.) Q.E.D.

We have defined a general complex function of two variables

$$f(z, z^*)\equiv u(x, y)+iv(x, y)$$

to be a *holomorphic function* of $z=x+iy$ provided u and v satisfy the Cauchy-Riemann equations

$$\frac{\partial u}{\partial x}=\frac{\partial v}{\partial y} \qquad \frac{\partial u}{\partial y}=-\frac{\partial v}{\partial x}$$

The equivalence of the requirement that there exist a unique

$$\lim_{\Delta z\to 0}\frac{\Delta f}{\Delta z}$$

and that a continuous, differentiable function f depend only on z can be made explicit by computing $\partial f/\partial z^*$ and setting it equal to zero. Since

$$z = x + iy$$
$$z^* = x - iy$$

or

$$x = \tfrac{1}{2}(z + z^*)$$
$$y = -\frac{i}{2}(z - z^*)$$

we have

$$\frac{\partial f}{\partial z^*} = \frac{\partial f}{\partial x}\frac{\partial x}{\partial z^*} + \frac{\partial f}{\partial y}\frac{\partial y}{\partial z^*} = \frac{1}{2}\left\{\frac{\partial u}{\partial x} + i\frac{\partial v}{\partial x} + i\frac{\partial u}{\partial y} - \frac{\partial v}{\partial y}\right\} \equiv 0$$

which requires

$$\frac{\partial u}{\partial x} = \frac{\partial v}{\partial y} \qquad \frac{\partial u}{\partial y} = -\frac{\partial v}{\partial x}$$

Again this is both necessary and sufficient.

Notice that a complex quantity being equal to zero implies that the real and imaginary parts are each separately zero since with

$$f = u + iv = \sigma e^{i\varphi} \qquad \sigma = \sqrt{u^2 + v^2} \qquad \varphi = \tan^{-1}(v/u)$$

the condition

$$f = 0$$

implies

$$ff^* = 0 = \sigma^2$$

from which we conclude that

$$u = 0 = v$$

Therefore we may demand that $f'(z)$ be unique or that $\partial f/\partial z^* = 0$. If $\lim_{\Delta z \to 0} (\Delta f/\Delta z)$ is not independent of the direction of Δz, then it is merely $\partial f/\partial z$ and should not be termed *the* derivative.

7.3 CAUCHY'S THEOREMS

We can now show that our definition of a holomorphic function can be used to deduce two representations of such a function. The second one we will consider is analogous to one in real variable theory in which a function is called holomorphic provided it can be expanded in a Taylor series in a region. However we will first see that a holomorphic function in a region may be expressed in terms of its values on a contour enclosing the region.

We define contour integrals in the complex plane in terms of ordinary contour integrals familiar from real variable theory

$$\oint f(z)\,dz \equiv \oint [u(x, y) + iv(x, y)][dx + i\,dy]$$

$$= \oint \{[u(x, y)\,dx - v(x, y)\,dy] + i[v(x, y)\,dx + u(x, y)\,dy]\} \quad \textbf{(7.11)}$$

where $y = y(x)$ is a closed contour C in the x-y plane (which we will often refer to as the z plane). We mention here a technical point about the type curves we will allow in our theorems. We assume that the curve C is made up of a finite number of arcs such that each arc is defined by parametric equations of the form $x = \phi(t)$, $y = \psi(t)$ where the derivatives $\phi'(t)$ and $\psi'(t)$ are continuous over each closed arc. We call such curves *ordinary curves*. The reason for this restriction is that it simplifies arguments about the existence of contour integrals since then

$$\oint f(z)\,dz = \int_{t_1}^{t_2} \{u[\phi(t), \psi(t)] + iv[\phi(t), \psi(t)]\}\{\phi'(t) + i\psi'(t)\}\,dt$$

This class of curves is sufficiently broad for any application we shall study.

a. Cauchy-Goursat Theorem

THEOREM 7.2 (Cauchy-Goursat theorem)

If $f(z)$ is holomorphic in a given finite region R enclosed by a curve C lying wholly within R and enclosing only points of R, then

$$\int_C f(z)\,dz = 0 \quad \textbf{(7.12)}$$

PROOF

Since $f(z)$ is holomorphic in R, we know that it has a uniquely defined (i.e., single valued, finite) first derivative at every point within R. Notice that we have said that C can enclose only points of R. We have in mind a *simply-connected* region. That is, we could shrink C continuously to a point without leaving region R. By convention the positive direction around the contour is always *counterclockwise* (i.e., the region of integration is on the left as one traverses C).

We divide the area within C into N squares and partial squares. Therefore we can write

$$\int_C f(z)\,dz = \sum_j \int_{\gamma_j} f(z)\,dz$$

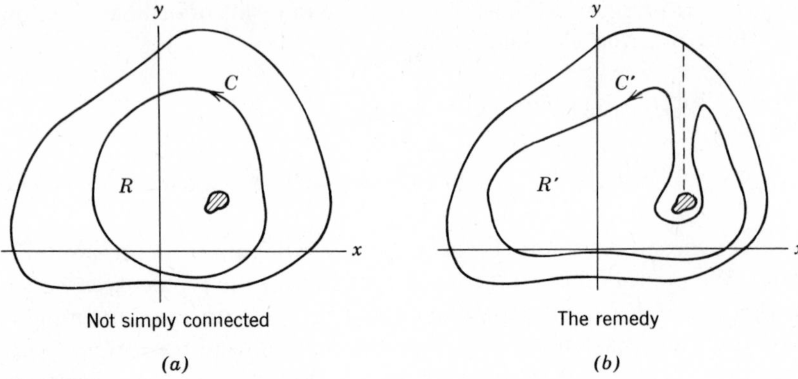

Figure 7.4

where the γ_j's are these new small contours. Notice that the contributions from internal squares cancel since (cf. Fig. 7.6)

$$\int_A^B f(z)\,dz = -\int_B^A f(z)\,dz$$

Consider a typical integral around one of the γ_j as shown in Fig. 7.7. Here z is on γ_j and the area, A_j, enclosed by γ_j, is $A_j = a^2$. Since the $f'(z)$ is given to exist everywhere in R, for sufficiently small squares we can always write exactly

$$f(z) = f(z_j) + f'(z_j)(z - z_j) + \eta(z)(z - z_j) \tag{7.13}$$

where $|\eta(z)| < \varepsilon$ and ε can be made arbitrarily small. Furthermore we easily see explicitly that

$$\int_{\gamma_j} dz = 0 \qquad \int_{\gamma_j} z\,dz = 0$$

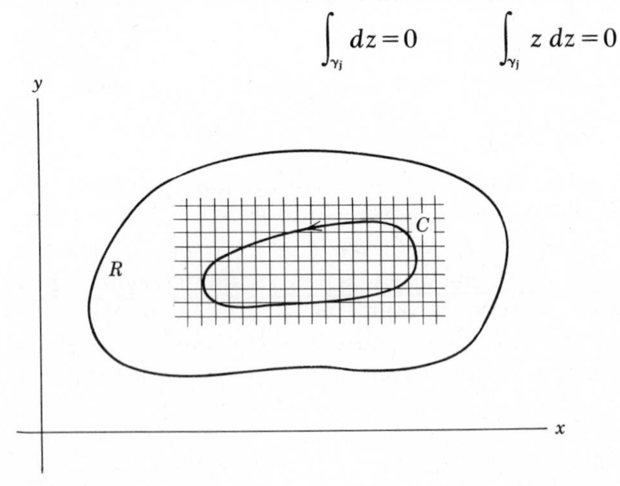

Figure 7.5

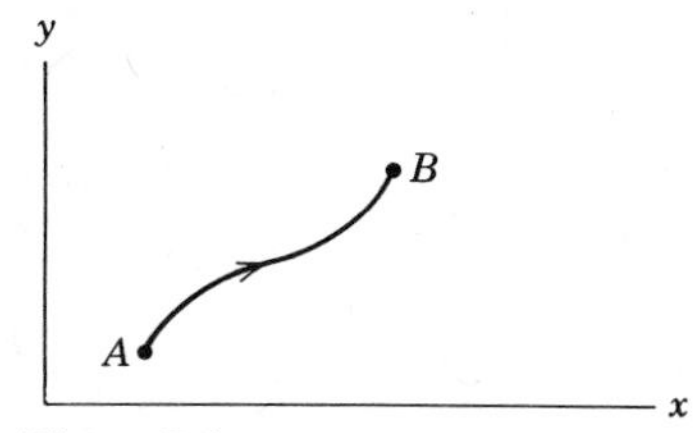

Figure 7.6

since

$$\int_{\gamma_j} dz = \int_{\gamma_j} (dx + i\,dy) = x|_1^1 + iy|_1^1 = 0$$

(cf. Fig. 7.8) and

$$\int_{\gamma_j} z\,dz = \int_{\gamma_j} (x+iy)(dx+i\,dy) = \int_{\gamma_j} [(x\,dx - y\,dy) + i(y\,dx + x\,dy)] = 0$$

In fact these would vanish for *any* closed contour in the z plane [i.e., simply apply Green's integral theorem (cf. Section 7.8)]. Therefore we have

$$\left|\int_{\gamma_j} f(z)\,dz\right| = \left|\int_{\gamma_j} \eta(z)(z-z_j)\,dz\right| < \varepsilon \int_{\gamma_j} |z-z_j||dz|$$

$$< \varepsilon\,|z-z_j|_{\max} \int_{\gamma_j} |dz| = \frac{\varepsilon a}{\sqrt{2}} 4a = 2\sqrt{2}\,a^2 = \varepsilon 2\sqrt{2}\,A_j$$

or

$$\left|\int_C f(z)\,dz - \sum_{\substack{\text{squares}\\ \text{only}}} \int_{\gamma_j} f(z)\,dz\right| = \left|\sum_{\substack{\text{partial}\\ \text{squares}}} \int_{\gamma_j} f(z)\,dz\right| \xrightarrow[N\to\infty]{} 0$$

so that

$$\left|\int_C f(z)\,dz\right| < 2\sqrt{2}\,\varepsilon A_C \equiv \varepsilon'$$

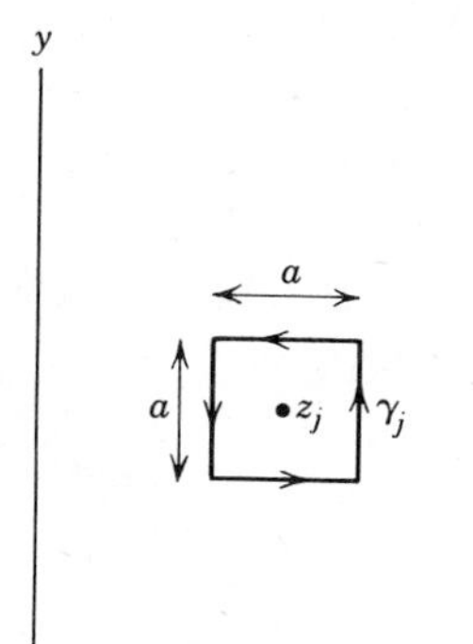

Figure 7.7

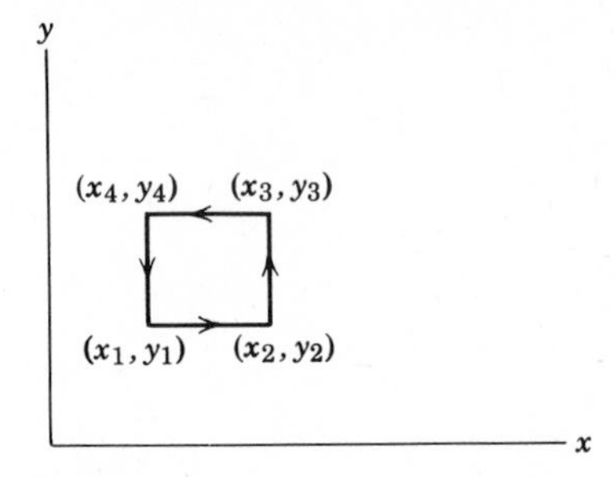

Figure 7.8

where ε' can be made arbitrarily small. Since we may choose ε arbitrarily small by taking a finer and finer grid, we finally arrive at

$$\oint f(z)\, dz = 0$$

Q.E.D.

In Section 7.8 we will give a proof of this theorem that is valid even when contour C extends to infinity. If region R is multiply connected, we must connect the various parts of the contour by cross cuts, the contributions along which cancel. That is, we construct a new region that is simply connected.

We can now extend this result so that C may actually be the boundary of R, provided $f(z)$ converges uniformly to its value on C (when approached from the interior). We may already take any contour arbitrarily close to the boundary. Since $f(z)$ converges *uniformly* on C, its value at a point displaced by δ within C will differ arbitrarily little from its value at z. Therefore the difference between the integral along C and along C' can be made as small as we please.

It should be obvious from this Cauchy-Goursat theorem and a knowledge of line integrals that an integral between any two points in such a region R has a value that is the same for all paths connecting these two points, provided only that this curve lies wholly within R and encloses only points of R. An extremely

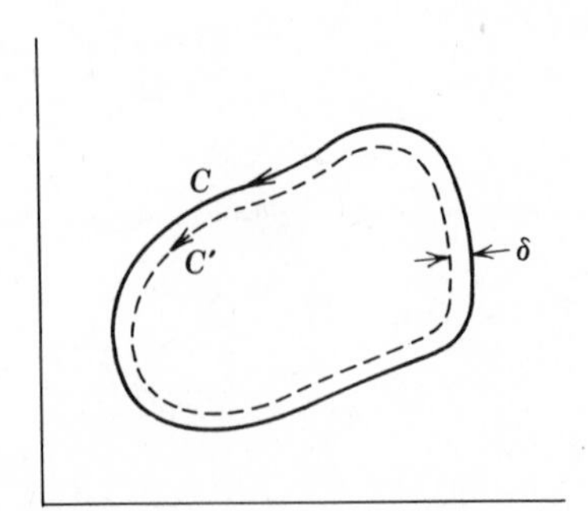

Figure 7.9

basic and important concept to grasp in complex variable theory is that the contour, C, may be distorted *arbitrarily within* R without changing the value of this integral (i.e., it remains equal to zero).

The following result is essentially the converse of the Cauchy-Goursat theorem.

THEOREM 7.3 (Morera's theorem)

If $f(z)$ is a continuous function in a region R and if

$$\oint f(z)\, dz = 0$$

when taken around *all* closed contours C lying wholly within R, then $f(z)$ is holomorphic everywhere in R.

PROOF

We may assume R to be simply connected without loss of generality since, if it were not, we could introduce a system of cross cuts to make it so. Since

$$\oint f(z)\, dz = 0$$

the value of

$$\int_{z_0}^{z_0+\Delta z} f(z)\, dz$$

is independent of the path chosen between z_0 and $(z_0+\Delta z)$. Therefore we may choose a straight line between these two points (cf. Fig. 7.10). We define

$$F(z_0) \equiv \int_{\alpha}^{z_0} f(z)\, dz$$

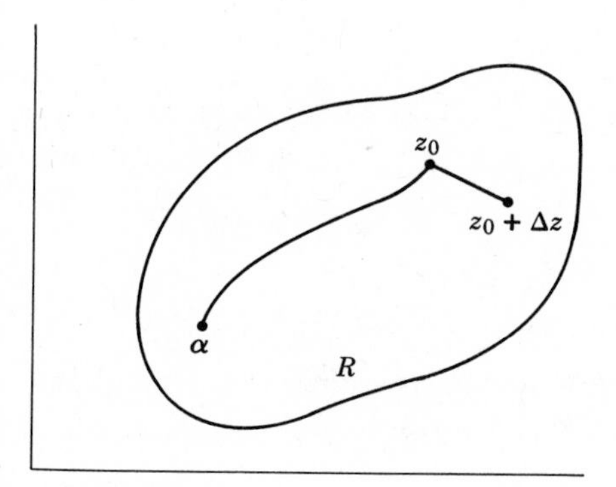

Figure 7.10

so that

$$F(z_0+\Delta z)=\int_\alpha^{z_0+\Delta z} f(z)\,dz$$

or

$$\Delta F(z_0)\equiv F(z_0+\Delta z)-F(z_0)=\int_{z_0}^{z_0+\Delta z} f(z)\,dz$$

Since $f(z)$ is continuous in R we have, as in Eq. 7.13,

$$f(z)=f(z_0)+\eta(z)$$

where $|\eta(z)|<\varepsilon$, an arbitrarily small quantity, whenever δ is small enough with $|\Delta z|<\delta$, $z\equiv z_0+\Delta z$, so that

$$\frac{\Delta F}{\Delta z}=f(z_0)+\frac{1}{\Delta z}\int_{z_0}^{z_0+\Delta z}\eta(z)\,dz$$

However we easily see that

$$\left|\frac{1}{\Delta z}\int_{z_0}^{z_0+\Delta z}\eta(z)\,dz\right|\leq\frac{1}{|\Delta z|}\int_{z_0}^{z_0+\Delta z}|\eta(z)||dz|<\frac{\varepsilon}{|\Delta z|}\int_{z_0}^{z_0+\Delta z}|dz|=\varepsilon$$

where the last step follows because we have been able to choose a straight-line path from z_0 to $(z_0+\Delta z)$. Since

$$\left|\frac{\Delta F}{\Delta z}-f(z_0)\right|<\varepsilon \qquad |\Delta z|<\delta$$

we have the existence of the following limit

$$\lim_{\Delta z\to 0}\frac{\Delta F}{\Delta z}\equiv F'(z_0)=f(z_0)$$

which limit is independent of the direction of approach of Δz to z_0 for all z_0 in R. This establishes that $F(z)$ is holomorphic in R (cf. Sec. 7.1), from which it follows that $F'(z)$ is also holomorphic here (cf. Eq. 7.16). Q.E.D.

b. Cauchy's Integral Formula

This formula that we now derive is the most important and often-used tool in the entire theory of complex variables.

THEOREM 7.4 (Cauchy integral theorem)

If R is a finite region bounded by a closed curve C consisting of a finite number of ordinary curves, and if $f(z)$ is holomorphic within R and converges uniformly to a value along C, then for any point α interior to R we can write

$$f(\alpha)=\frac{1}{2\pi i}\int_C\frac{f(z)\,dz}{(z-\alpha)} \qquad \textbf{(7.14)}$$

PROOF

Figure 7.11 may aid in following the proof. By means of a cross cut we join a small circle γ of radius ρ about point α to main contour C. The contribution along the cross cut cancels. Within the new region (i.e., excluding a neighborhood about $z=\alpha$) the function $f(z)/(z-\alpha)$ is holomorphic so that we may apply Th. 7.2 to obtain

$$0=\int_C \frac{f(z)\,dz}{(z-\alpha)}-\int_\gamma \frac{f(z)\,dz}{(z-\alpha)} \tag{7.15}$$

The change of sign on the second integral comes about as a result of the direction of the integration on γ originally indicated in Fig. 7.11. In Eq. 7.15 we are now taking both integrals counterclockwise around C and γ. To evaluate the second term on the right-hand side of Eq. 7.15 we set $z-\alpha=\rho e^{i\varphi}$, which puts z on circle of radius ρ about point α (ρ=constant). Since

$$dz=i\rho e^{i\varphi}\,d\varphi=i(z-\alpha)\,d\varphi$$

we can write

$$\int_\gamma \frac{f(z)\,dz}{z-\alpha}=i\int_0^{2\pi} f(\alpha+\rho e^{i\varphi})\,d\varphi$$

Since $f(z)$ is holomorphic about $z=\alpha$, we have (cf. Eq. 7.13)

$$f(\alpha+\rho e^{i\varphi})=f(\alpha)+[f'(\alpha)+\eta(\alpha+\rho e^{i\varphi})]\rho e^{i\varphi}$$

so that as $\rho \to 0$,

$$\int_\gamma \frac{f(z)\,dz}{(z-\alpha)}=2\pi i f(\alpha)$$

or

$$f(\alpha)=\frac{1}{2\pi i}\int_C \frac{f(z)\,dz}{(z-\alpha)} \qquad \text{Q.E.D.}$$

Notice that again we may arbitrarily deform the contour, C, within the region of holomorphy of $f(z)$. This amazing theorem tells us that if we know the values of a holomorphic function at every point on the boundary of its domain of holomorphy, then we know its value at all points interior to C.

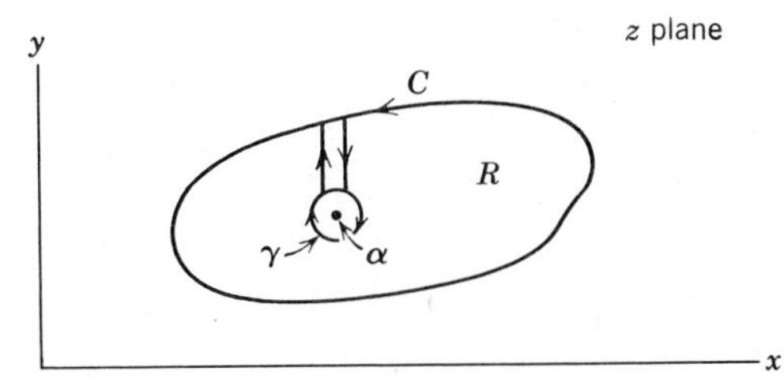

Figure 7.11

However we may easily prove another remarkable property of holomorphic functions. Since the point, α, is restricted to lie within but *not on* the contour, C, we may take a limit inside the integral sign. Therefore it follows that

$$f'(z) \equiv \lim_{\Delta z \to 0} \frac{f(z+\Delta z)-f(z)}{\Delta z} = \lim_{\Delta z \to 0} \frac{1}{2\pi i} \int_C \frac{f(t)\,dt}{\Delta z} \left[\frac{1}{t-z-\Delta z} - \frac{1}{t-z} \right]$$

$$= \frac{1}{2\pi i} \int_C \frac{dt f(t)}{(t-z)} \frac{d}{dw} \left[\frac{1}{1-w/(t-z)} \right] \Bigg|_{w=0} = \frac{1}{2\pi i} \int_C \frac{f(t)\,dt}{(t-z)^2}$$

This function is also holomorphic within R. In fact it is easily proven in this fashion that

$$f^{(n)}(z) = \frac{n!}{2\pi i} \int_C \frac{f(t)\,dt}{(t-z)^{n+1}} \tag{7.16}$$

Therefore we know that if $f(z)$ is holomorphic in R, then *all* its derivatives exist and are holomorphic in R.

7.4 TAYLOR SERIES

Consider a function $f(z)$ holomorphic in a region R. We may write

$$\frac{1}{w-z} = \frac{1}{(w-\alpha)-(z-\alpha)} = \frac{1}{(w-\alpha)\left[1-\left(\frac{z-\alpha}{w-\alpha}\right)\right]}$$

$$= \sum_{n=0}^{\infty} \frac{(z-\alpha)^n}{(w-\alpha)^{n+1}} \tag{7.17}$$

provided only that $|z-\alpha|<|w-\alpha|$ from the binomial theorem (cf. geometrical progression). The point is that by induction we can prove

$$S_n \equiv \sum_{s=0}^{n} r^s = \frac{1-r^{n+1}}{1-r}$$

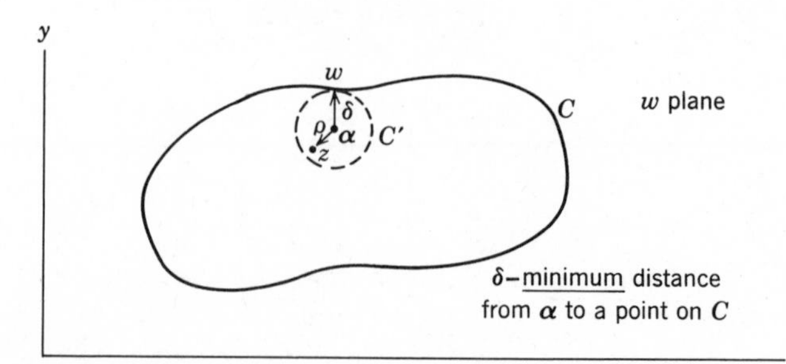

Figure 7.12

for any r, real or complex. If furthermore we want the limit

$$S_\infty \equiv \lim_{n\to\infty} S_n$$

to exist, we must require that $|r|<1$ so that $|r|^{n+1} \xrightarrow[n\to\infty]{} 0$. Therefore from Eqs. 7.14, 7.16, and 7.17 we obtain

$$\frac{1}{2\pi i}\int_C \frac{f(w)\,dw}{w-z} \equiv f(z) = \frac{1}{2\pi i}\sum_{n=0}^{\infty}(z-\alpha)^n \int_C \frac{f(w)\,dw}{(w-\alpha)^{n+1}}$$

$$= \sum_{n=0}^{\infty} \frac{f^{(n)}(\alpha)}{n!}(z-\alpha)^n$$

The interchange of summation and integration is justified since a uniformly convergent series of continuous functions may be integrated term by term (cf. Appendix I, Th. I.11). The radius of convergence of the series is any circle about the point α such that $|z-\alpha| \leq \rho < \delta$. Since we can deform the contour, C, outward until we come to a point at which $f(z)$ is no longer holomorphic, we see that the radius of convergence of $f(z)$ about the point, $z=\alpha$, is the smallest distance from α to a point at which $f(z)$ is not holomorphic. We state this result as the following theorem.

THEOREM 7.5 (Taylor series)

If z_0 is a point interior to the domain of holomorphy of $f(z)$, then $f(z)$ may be represented by the *Taylor series*

$$f(z) = \sum_{n=0}^{\infty} \alpha_n (z-z_0)^n \tag{7.18}$$

where

$$\alpha_n = \frac{f^{(n)}(z_0)}{n!} = \frac{1}{2\pi i}\int_{C'} \frac{f(z)\,dz}{(z-z_0)^{n+1}} \tag{7.19}$$

with C' any closed curve about the point, z_0, enclosing only points of holomorphy of $f(z)$. Furthermore this series will converge for all points inside a circle centered about z_0 whose radius is the distance from z_0 to the closest point at which $f(z)$ ceases to be holomorphic.

The simplest and most often used test for the convergence of a series is the ratio test. That is, the series of nonvanishing $\{\beta_n\}$

$$\sum_{n=0}^{\infty} \beta_n$$

will converge provided

$$l \equiv \lim_{n\to\infty} \frac{\beta_{n+1}}{\beta_n} < 1$$

If we let $r=l+(1-l)/2$ and $\gamma_n \equiv \beta_{n+1}/\beta_n$, then since $\lim_{n\to\infty} \gamma_n = l$ and $l<r<1$, we see from the definition of the convergence of a sequence that $0<\gamma_n<r$ for all $n>N$, once N is sufficiently large. This implies that

$$\left|\sum_{n=N+1}^{\infty} \beta_n\right| \leq \sum_{n=N+1}^{\infty} |\beta_n| = |\beta_{N+1}| \sum_{n=N+1}^{\infty} \left|\frac{\beta_n}{\beta_{N+1}}\right|$$
$$= |\beta_{N+1}| \left\{1 + \left|\frac{\beta_{N+2}}{\beta_{N+1}}\right| + \left|\frac{\beta_{N+2}}{\beta_{N+1}}\right| \left|\frac{\beta_{N+3}}{\beta_{N+2}}\right| + \cdots\right\}$$
$$< |\beta_{N+1}| \sum_{n=0}^{\infty} r^n = \frac{|\beta_{N+1}|}{1-r}$$

which states that $\sum_{n=0}^{\infty} \beta_n$ must converge. If we apply this to a Taylor series we see that for a given value of z_0 the radius of convergence will be the radius of that circle about z_0 containing z for which

$$\lim_{n\to\infty} \left|\frac{\alpha_{n+1}(z-z_0)^{n+1}}{\alpha_n (z-z_0)^n}\right| = |z-z_0| \lim_{n\to\infty} \left|\frac{\alpha_{n+1}}{\alpha_n}\right| < 1$$

which may be restated as

$$|z-z_0| < \lim_{n\to\infty} \left|\frac{\alpha_{n+1}}{\alpha_n}\right|^{-1} \tag{7.20}$$

We now obtain the *Cauchy inequality* for the Taylor coefficients. If $f(z)$ is holomorphic in a region then, since $f'(z)$ exists everywhere there, $f(z)$ must be continuous (cf. Eq. 7.13). Any function continuous in a closed region (e.g., on C and in its interior) is *uniformly* continuous there and hence bounded (cf. Appendix I, Th. I.3). Let $|f(z)|$ be bounded by M interior to and on a circle of radius r' about $z=z_0$. We will set $z'-z_0=r'e^{i\varphi}$ with $r<r'<\delta$.

$$|\alpha_n| = \left|\frac{1}{2\pi i} \int_C \frac{f(z')\,dz'}{(z'-z_0)^{n+1}}\right| \leq \frac{M(r')}{2\pi} \oint \left|\frac{dz'}{(z'-z_0)^{n+1}}\right|$$
$$= \frac{M(r')}{2\pi} \int_0^{2\pi} \left|\frac{r'e^{i\varphi}\,d\varphi}{r'^{n+1}e^{i(n+1)\varphi}}\right| = \frac{M(r')}{2\pi r'^n} \int_0^{2\pi} |d\varphi| = \frac{M(r')}{r'^n} \tag{7.21}$$

Since for $z-z_0=re^{i\varphi}$,

$$\left|\sum_{n=0}^{\infty} \alpha_n (z-z_0)^n\right| \leq \sum_{n=0}^{\infty} |\alpha_n||z-z_0|^n \leq \sum_{n=0}^{\infty} \frac{M(r')}{r'^n} |z-z_0|^n = M(r') \sum_{n=0}^{\infty} \left(\frac{r}{r'}\right)^n = \frac{M(r')}{1-(r/r')}$$

we see that, by the Weierstrass M test, the series will necessarily converge provided that

$$\frac{|z-z_0|}{r'} < 1$$

or that

$$|z-z_0| < \delta$$

This, in fact, establishes that the Taylor series, Eq. 7.18, is uniformly convergent within the domain of holomorphy of $f(z)$.

We may equivalently define a holomorphic function as one that can be expanded in a uniformly convergent power series in $(z-z_0)$. This will be a special case of the following theorem.

THEOREM 7.6

If the functions $\{u_n(z)\}$ are all holomorphic within a region R and if the series

$$\sum_{n=0}^{\infty} u_n(z)$$

is uniformly convergent everywhere interior to R, then the function

$$f(z) \equiv \sum_{n=0}^{\infty} u_n(z) \tag{7.22}$$

is holomorphic within R.

PROOF

From Th. 7.4 we have

$$u_n(z) = \frac{1}{2\pi i} \int_C \frac{u_n(z')\, dz'}{(z'-z)}$$

for any curve C interior to R with z interior to C. Since $1/(z'-z)$ converges uniformly to a value everywhere on C, we may multiply the uniformly convergent series $\sum u_n(z')$ by $(z'-z)^{-1}$ and integrate, term by term, to obtain

$$\int_C \frac{f(z')\, dz'}{(z'-z)} = \int_C \frac{dz'}{(z'-z)} \sum_{n=0}^{\infty} u_n(z') = \sum_{n=0}^{\infty} \int_C \frac{u_n(z')\, dz'}{(z'-z)}$$

$$= 2\pi i \sum_{n=0}^{\infty} u_n(z) = 2\pi i f(z)$$

Now, just as in Section 7.3 (cf. Eq. 7.16), we can deduce that $f'(z)$ exists so that $f(z)$ does indeed define a holomorphic function of z. Q.E.D.

Therefore, if we are given a uniformly convergent power series

$$\sum_{n=0}^{\infty} \alpha_n (z-z_0)^n$$

we may take $u_n(z) = \alpha_n (z-z_0)^n$ in Th. 7.6 to see that this power series defines a holomorphic function within its circle of convergence.

We now prove another remarkable theorem for holomorphic functions. This result will be needed in Section 7.12b on asymptotic expansions.

THEOREM 7.7 (Maximum modulus theorem)

For any holomorphic function $f(z)$ the modulus, $|f(z)|$, attains its maximum value somewhere on the boundary of the domain of holomorphy but can never attain a maximum at an interior point [unless $f(z)$ is a constant].

PROOF

Let R be the region of holomorphy bounded by a closed curve C. Assume that $|f(z)|$ attained a maximum value at some point z_0 interior to C. There would then exist a circular region of radius ρ centered about z_0 and lying wholly within R such that the Taylor series

$$f(z)=\sum_{n=0}^{\infty}\alpha_n(z-z_0)^n$$

would be uniformly convergent. If we let $z-z_0=\rho e^{i\varphi}$, then under these assumptions we would have

$$|\alpha_0|^2\equiv\frac{1}{2\pi}\int_0^{2\pi}|\alpha_0|^2\,d\varphi=\frac{1}{2\pi}\int_0^{2\pi}|f(z_0)|^2\,d\varphi\geq\frac{1}{2\pi}\int_0^{2\pi}|f(z_0+\rho e^{i\varphi})|^2\,d\varphi$$
$$=\frac{1}{2\pi}\sum_{n,m=0}^{\infty}\alpha_n\alpha_m^*\rho^{n+m}\int_0^{2\pi}e^{i(n-m)\varphi}\,d\varphi=\sum_{n=0}^{\infty}|\alpha_n|^2\rho^{2n}$$

This implies that $\alpha_n=0$ for *all* $n\geq1$ so that there is a contradiction unless $f(z)$ is a constant within this circle. We may now repeat this argument at any interior point, or use the Cauchy-Riemann conditions of Eq. 7.7, to conclude that $f(z)$ is a constant everywhere within R. Therefore either $f(z)$ is a constant within and on C or $|f(z)|$ has no maximum within C. Another way to state this is that $|f(z_0)|$ for any interior point z_0 cannot be at least equal to the value $|f(z)|$ takes on everywhere else in the region. Q.E.D.

It follows immediately that neither $\operatorname{Re} f(z)\equiv u(x, y)$ nor $\operatorname{Im} f(z)\equiv v(x, y)$ can have either maxima or minima interior to a region of holomorphy of $f(z)$. This is simply seen by contradiction. If $u(x, y)$ attained a maximum at $z_0=(x_0, y_0)$, then so would $e^{u(x,y)}\equiv|e^{f(z)}|$ since $e^{f(z)}$ is a holomorphic function of z. Unless $u(x, y)=$ constant, Th. 7.7 will not allow this. Similar arguments for $e^{-u(x,z)}=|e^{-f(z)}|$, $e^{-iv(x,z)}=|e^{if(z)}|$, and $e^{iv(x,y)}=|e^{-if(z)}|$ will complete the proof of the stated result.

7.5 ZEROS AND SINGULARITIES

So far we have considered only functions of z that are holomorphic in a given region. We must now treat the various classes of singularities that $f(z)$ may have.

However, first we will discuss the zeros of $f(z)$. Suppose that $f(z_0)=0$. We say that $f(z)$ has a *zero of order* k at $z=z_0$ if there exists a positive integer k such that the function

$$f(z)/(z-z_0)^k$$

is holomorphic in the neighborhood of z_0 and different from zero there. Notice that any function that is holomorphic in the neighborhood of z_0 and that also vanishes at z_0 (not being identically zero everywhere) will have this behavior since it is expressible as a Taylor series about this point. If $k=1$ we say $f(z)$ has a *simple zero* at z_0. We now show that in any finite region of the z plane a holomorphic function can have at most a finite number of zeros (unless the function is identically zero).

THEOREM 7.8

If $f(z)$ is holomorphic in a region R and if

$$f(z_j)=0 \qquad j=1,2,\ldots$$

at a set of points $\{z_j\}$ having an accumulation point z_∞ within R, then $f(z)\equiv 0$ everywhere within R.

PROOF

For convenience we take $z_\infty=0$ (i.e., the origin). If $f(z)$ does not vanish identically, then there is some first nonvanishing coefficient, say α_m, in the Taylor series, Eq. 7.18, so that

$$f(z)=z^m\sum_{n=0}^{\infty}\alpha_{m+n}z^n \tag{7.23}$$

If we take a circle of radius ρ within R centered about $z=0$, we may use the Cauchy inequality, Eq. 7.21,

$$|\alpha_n|\,\rho^n\leq M \tag{7.24}$$

where M is some constant. The following inequality holds for any complex scalars α and β,

$$|\alpha|+|\beta|\geq|\alpha+\beta|\geq|\alpha|-|\beta| \tag{7.25}$$

since

$$\begin{aligned}|\alpha+\beta|^2&=|\alpha|^2+|\beta|^2+2\,\text{Re}\,\alpha^*\beta\leq|\alpha|^2+|\beta|^2+2\,|\alpha^*\beta|\\&=|\alpha|^2+|\beta|^2+2\,|\alpha||\beta|=[|\alpha|+|\beta|]^2\\|\alpha+\beta|^2&\geq|\alpha|^2+|\beta|^2-2\,\text{Re}\,|\alpha^*\beta|\geq|\alpha|^2+|\beta|^2-2\,|\alpha^*\beta|\\&=|\alpha|^2+|\beta|^2-2\,|\alpha||\beta|=[|\alpha|-|\beta|]^2\end{aligned}$$

Equations 7.23–7.25 imply that

$$
\begin{aligned}
|f(z)| &= |z|^m \left| \alpha_m + \sum_{n=1}^{\infty} \alpha_{m+n} z^n \right| \\
&\geq |z|^m \left[|\alpha_m| - \sum_{n=1}^{\infty} |\alpha_{m+n}| |z|^n \right] \\
&\geq |z|^m \left[|\alpha_m| - \frac{M}{\rho^m} \sum_{n=1}^{\infty} \left| \frac{z}{\rho} \right|^n \right] \\
&= |z|^m \left[|\alpha_m| - \frac{M|z|}{\rho^m(\rho - |z|)} \right]
\end{aligned}
\tag{7.26}
$$

For z sufficiently small (but not zero) we see that $|f(z)|>0$, which contradicts the assumption that $z=0$ is an accumulation point of zeros of $f(z)$, since if $z=0$ were such an accumulation point there would have to be zeros *arbitrarily close* to $z=0$.

Q.E.D.

Since the Bolzano-Weierstrass theorem (cf. Appendix I, Th. I.1) states that every bounded infinite set has at least one accumulation point, we see that any holomorphic function having infinitely many zeros in a finite region would necessarily have an accumulation point of zeros somewhere in that region. But by the theorem we have just established this function would have to vanish identically here.

We say that a function $f(z)$ has a *singular point* at $z=z_0$ if it ceases to be holomorphic at that point [i.e., there does not exist a (unique) derivative of $f(z)$ there]. Point $z=z_0$ is a *pole* or *nonessential singularity* of function $f(z)$ if there exists a positive integral value of k such that the product

$$(z-z_0)^k f(z) \tag{7.27}$$

is holomorphic in the neighborhood of z_0 and different from zero for $z=z_0$. The integer, k, is called the *order* of the pole. If $k=1$, we have a *simple pole.* If we cannot so remove the singularity of $f(z)$, then z_0 is said to be an *essential singular point.* (Remember, thus far we are considering only those functions that are single valued at every point where they are defined.)

We can easily see that the poles of $f(z)$ cannot be dense on any line since the functions $(z-z_0)^k f(z)$ must be holomorphic in the neighborhood of z_0, which implies continuity there. Also, by Th. 7.8 if $f(z)$ is not identically zero, it cannot vanish in the immediate neighborhood of a zero. [Notice furthermore that, if a holomorphic function $f(z)$ is identically zero along any arbitrarily small arc, we can compute its derivatives along this arc and conclude that $f^{(n)}(z)\equiv 0$, $\forall n$, here. Then, using a Taylor series and the Cauchy-Riemann equations, we would see that $f(z)\equiv 0$ within its domain of holomorphy.]

For the sake of clarity, and for future reference, we give here some definitions often used in complex variable theory. A function $f(z)$ is said to be:

i. *Holomorphic* in a region if it is single valued and has a uniquely determined derivative in every neighborhood of R.
ii. *Analytic* in a region if it is holomorphic there with the possible exception of some points that do not interrupt the continuity of R (in the literature today the terms *holomorphic* and *analytic* are often used interchangeably).
iii. *Meromorphic* in a region if it is single valued in R and has (at most) a finite number of poles but no essential singular points.

Finally, in order to consider the nature of a function in the neighborhood of point $z=\infty$, we make the substitution $\zeta = 1/z$ and study $\tilde{f}(\zeta) \equiv f(1/\zeta)$ as $\zeta \to 0$.

Example 7.1

$$f(z) = \frac{z^2+1}{(z-2)^2} = \frac{(z+i)(z-i)}{(z-2)^2}$$

i. Pole of order 2 at $z=2$.
ii. Zeros of order 1 at $z=\pm i$.
iii. $\zeta = 1/z \Rightarrow f(1/\zeta) = \dfrac{1+\zeta^2}{(1-2\zeta)^2} \xrightarrow[\zeta\to 0]{} 1$

$$\Rightarrow f(z) \xrightarrow[z\to\infty]{} 1 \Rightarrow \text{no singularity at } z=\infty$$

iv. *Meromorphic* in entire z plane, including $z=\infty$.
v. *Holomorphic* beyond circle $|z|=2$.
vi. *Analytic* in the entire z plane.

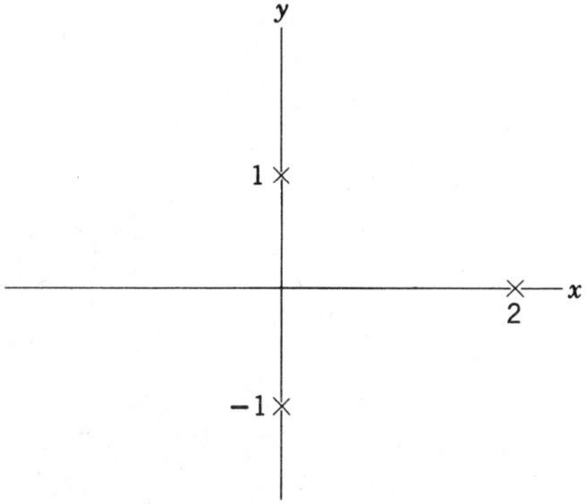

Figure 7.13

Example 7.2 An essential singularity may be isolated or not:

a. $e^{1/z}$, $\sin(1/z)$, $\cos(1/z)$ all have an *isolated* essential singular point at $z=0$.
b. $\csc(1/z)$ has a nonisolated essential singular point at $z=0$ since $z=0$ is a limit point of the poles at $z=1/(n\pi)$.

7.6 LIOUVILLE'S THEOREM

We will now prove a theorem that is used to demonstrate the fundamental theorem of algebra that every finite-degree polynomial in x has at least one root that makes it vanish (cf. Problem 7.12).

THEOREM 7.9 (Liouville's theorem)

Any holomorphic function that has no singularity either in the finite portion of the plane or at infinity is a constant.

PROOF

Since $f(z)$ is assumed to have no singularity anywhere, it must be bounded by some positive constant, for example, M, everywhere. Otherwise there would exist a point z_0 at which $f(z)$ would exceed all finite bounds. From our assumption we know that $f(z)$ allows a Taylor expansion about the origin,

$$f(z)=\sum_{n=0}^{\infty}\alpha_n z^n \qquad \alpha_n=\frac{f^{(n)}(0)}{n!}=\frac{1}{2\pi i}\int_C\frac{f(z)\,dz}{z^{n+1}}$$

where we may take C to be a circle of radius ρ about the origin as center, where ρ may be arbitrarily large. Therefore, from Cauchy's inequality, Eq. 7.21, we have

$$|\alpha_n|\le\frac{1}{2\pi}\int_C\frac{M\,|dz|}{\rho^{n+1}}=\frac{M}{\rho^n}<\varepsilon \qquad n>0$$

Since ρ can be made as large as we please, we see that

$$\alpha_n=0 \qquad n=1,2,3,\ldots$$

which implies

$$f(z)=\alpha_0$$

a constant. Q.E.D.

This theorem implies that every single-valued analytic function that is not a constant must have at least one singular point either in the finite portion of the plane or at infinity.

7.7 LAURENT SERIES

We have seen that an analytic function $f(z)$ can be expanded in a Taylor series about a regular point (i.e., one at which a unique derivative exists). We will now consider the series expansion of a function about a singular point. We will see that this involves negative powers of $(z-\alpha)$ and is called a *Laurent series.*

THEOREM 7.10

If $f(z)$ is holomorphic in a region R bounded by two circles concentric about the point, α, then within this region $f(z)$ can be represented by the series

$$f(z)=\sum_{n=-\infty}^{\infty}\alpha_n(z-\alpha)^n \tag{7.28}$$

where

$$\alpha_n=\frac{1}{2\pi i}\int_C (t-\alpha)^{-n-1}f(t)\,dt \tag{7.29}$$

and C is any ordinary curve interior to R and enclosing the inner circle.

PROOF

Suppose we take a region R that contains a singular point α of $f(z)$. We may obtain a region R' of holomorphy by deforming the original contour C' into C_1 and C_2 as shown in Fig. 7.14. Within the annular region, R', between C_1 and C_2 $f(z)$ is holomorphic. Notice that the contours are traversed in *opposite* directions. By Cauchy's integral formula we have

$$f(z)=\frac{1}{2\pi i}\int_{C_1}\frac{f(t)\,dt}{t-z}-\frac{1}{2\pi i}\int_{C_2}\frac{f(t)\,dt}{t-z} \tag{7.30}$$

where both contours are now taken counterclockwise. For values of t on the

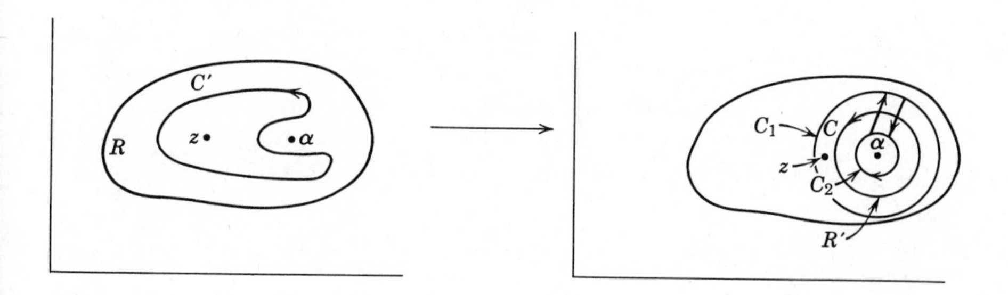

Figure 7.14

circles, C_1 and C_2, respectively, we have the following inequalities satisfied:

$$t\in C_1: |t-\alpha|>|z-\alpha|$$

$$t\in C_2: |t-\alpha|<|z-\alpha|$$

where z is confined to the annular region R'. Therefore, just as with the Taylor series of Section 7.4, we may write:

$$t\in C_1: \frac{1}{t-z}=\frac{1}{(t-\alpha)-(z-\alpha)}=\frac{1}{(t-\alpha)[1-(z-\alpha)/(t-\alpha)]}=\sum_{n=0}^{\infty}\frac{(z-\alpha)^n}{(t-\alpha)^{n+1}} \tag{7.31}$$

$$t\in C_2: \frac{1}{t-z}=\frac{-1}{(z-\alpha)-(t-\alpha)}=\frac{-1}{(z-\alpha)[1-(t-\alpha)/(z-\alpha)]}=\frac{-1}{(z-\alpha)}\sum_{n=0}^{\infty}\frac{(t-\alpha)^n}{(z-\alpha)^n}$$

$$=\frac{-1}{(z-\alpha)}\sum_{n=1}^{\infty}\frac{(t-\alpha)^{n-1}}{(z-\alpha)^{n-1}}=-\sum_{n=1}^{\infty}\frac{(t-\alpha)^{n-1}}{(z-\alpha)^n} \tag{7.32}$$

We may again integrate term by term and obtain

$$f(z)=\sum_{n=0}^{\infty}\frac{(z-\alpha)^n}{2\pi i}\int_{C_1}\frac{f(t)\,dt}{(t-\alpha)^{n+1}}+\sum_{n=1}^{\infty}\frac{(z-\alpha)^{-n}}{2\pi i}\int_{C_2}f(t)\,dt(t-\alpha)^{n-1}$$

$$=\sum_{n=-\infty}^{\infty}\alpha_n(z-\alpha)^n$$

$$\alpha_n\equiv\frac{1}{2\pi i}\int_C(t-\alpha)^{-n-1}f(t)\,dt$$

where C is any curve contained between C_1 and C_2, since the *integrands* are holomorphic everywhere here. We recognize the terms $0\le n<\infty$ as the usual Taylor series for α a nonsingular point, while the terms of negative powers of $(z-\alpha)$ are called the *principal part* of the Laurent expansion. Q.E.D.

The size of the annular region is easily determined. C_2 must exclude the singularity at $z=\alpha$ (which is not necessarily isolated) and the radius of C_1 will be the distance from α to the nearest other singularity of $f(z)$ (since beyond this circle we may no longer apply Cauchy's integral formula along C_1). If the principal part of the series terminates after m terms, then $f(z)$ has a pole of order m at $z=\alpha$. If this part of the series does not terminate, then $f(z)$ has an essential singularity at $z=\alpha$. We see that a Laurent series converges *uniformly* within this annular region by essentially the same argument as given for the uniform convergence of the Taylor series in Section 7.4.

We will give only a few specific examples of Laurent series since their main interest for us will be an abstract one in the application to residues that we will see soon.

Example 7.3 If $f(z)$ has a pole of order k at $z=z_0$, then we may write $f(z)=(z-z_0)^{-k}\phi(z)$, where $\phi(z)$ is holomorphic about z_0 and in a circle up to the next singularity of $f(z)$. We may expand $\phi(z)$ in a Taylor series as

$$\phi(z)=\sum_{n=0}^{\infty} \alpha_n(z-z_0)^n$$

and then simply divide through by $(z-z_0)^k$ to obtain the Laurent series for $f(z)$. Specifically consider

$$f(z)=\frac{1}{z^2(1-z)}$$

The Laurent expansion about $z=0$ and within the circle of radius 1 about the origin is

$$f(z)=\frac{1}{z^2}\sum_{n=0}^{\infty} z^n=\sum_{n=-2}^{\infty} z^n=\frac{1}{z^2}+\frac{1}{z}+1+z+z^2+\cdots$$

where the principal part is

$$\left(\frac{1}{z^2}+\frac{1}{z}\right)$$

Example 7.4 Let us apply directly the formulas for the Laurent expansion to the function above.

$$f(z)=\sum_{n=-\infty}^{\infty} \alpha_n(z-z_0)^n \qquad \alpha_n=\frac{1}{2\pi i}\int_C (t-z_1)^{-n-1}f(t)\,dt$$

Here we have

$$f(z)=\frac{1}{z^2(1-z)} \qquad z_0=0$$

Let us take C as a circle of radius ρ, $0<\rho<1$, and set

$$t=\rho e^{i\varphi} \qquad dt=i\rho e^{i\varphi}\,d\varphi=it\,d\varphi$$

We see directly that

$$\begin{aligned}\alpha_n&=\frac{1}{2\pi i}\int_C t^{-n-1}f(t)\,dt=\frac{1}{2\pi i}\int_C \frac{t^{-n-3}it\,d\varphi}{(1-\rho e^{i\varphi})}\\&=\frac{1}{2\pi}\int_0^{2\pi}\frac{\rho^{-n-2}e^{-(n+2)i\varphi}\,d\varphi}{(1-\rho e^{i\varphi})}=\frac{\rho^{-n-2}}{2\pi}\sum_{m=0}^{\infty}\int_0^{2\pi} e^{-(n+2)i\varphi}\rho^m e^{im\varphi}\,d\varphi\\&=\frac{\rho^{-n-2}}{2\pi}\sum_{m=0}^{\infty}\rho^m\int_0^{2\pi} e^{-(n-m+2)i\varphi}\,d\varphi\end{aligned}$$

But since

$$\int_0^{2\pi} e^{-(n-m+2)i\varphi}\,d\varphi=\begin{cases}0 & m\neq n+2\\ 2\pi & m=n+2\end{cases}$$

we obtain

$$\alpha_n = \begin{cases} \dfrac{1}{2\pi} 2\pi = 1 & n = -2, -1, 0, 1, 2, 3, \ldots, \infty \\ 0 & n < -2 \end{cases}$$

or

$$f(z) = \sum_{n=-2}^{\infty} z^n$$

This result is the same as that of Example 7.3, of course.

Use of the Laurent series allows a very simple and elegant proof of the completeness of the functions $e^{in\varphi}$, $n = 0, \pm 1, 2, \ldots$, for all functions that are holomorphic in an annular region that includes the unit circle (cf. Problem 7.46).

7.8 THEORY OF RESIDUES

We have seen from the Cauchy-Goursat theorem (Th. 7.2) that the contour integral of a function around a closed path encircling only a domain of holomorphy of $f(z)$ vanishes;

$$\oint f(z)\, dz = 0$$

We now consider the value of such an integral when the contour encloses a *finite* number of isolated singularities. If the original contour encloses several isolated singularities, then we may replace it by a set of equivalent ones such that each encircles but one singularity (i.e., use cross cuts), as indicated in Fig. 7.15. Therefore we need consider only the case in which there is just *one* singular point within the contour. In such a region (i.e., about an isolated singular or essential singular point) $f(z)$ has a Laurent expansion

$$f(z) = \sum_{n=-\infty}^{\infty} \alpha_n (z - z_0)^n$$

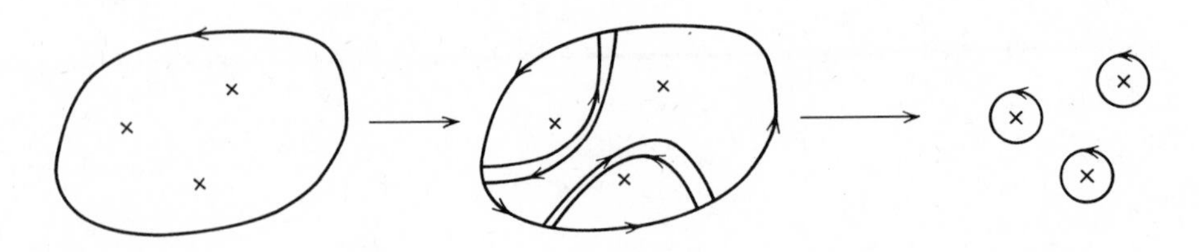

Figure 7.15

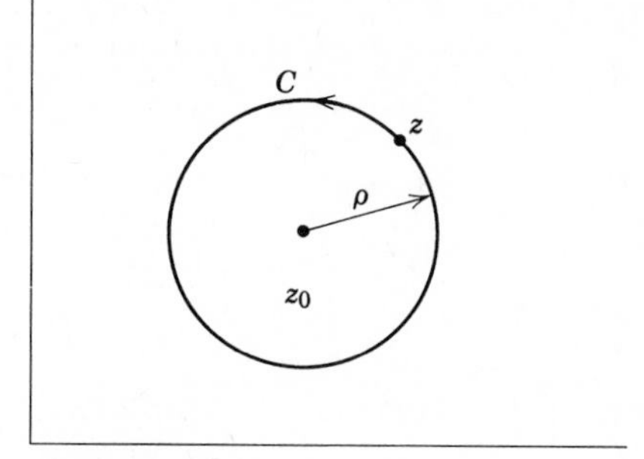

Figure 7.16

Consider now

$$\oint f(z)\,dz = \sum_{n=-\infty}^{\infty} \alpha_n \oint (z-z_0)^n\,dz \tag{7.33}$$

and let

$$(z-z_0) = \rho e^{i\theta} \qquad dz = i\rho e^{i\theta}\,d\theta = i(z-z_0)\,d\theta$$

Again the interchange of summation and integration is allowed since the series is uniformly convergent, as can be seen by computing bounds on the $|\alpha_n|$ as done in Eq. 7.21 for the Taylor series, once on the circle C_1 of Fig. 7.14 for $n \geq 0$ and once on C_2 of this same figure for $n<0$.

If we now take C to be a circle of radius ρ about z_0, then Eq. 7.33 becomes

$$\oint f(z)\,dz = \sum_{n=-\infty}^{\infty} i\alpha_n \rho^{n+1} \int_0^{2\pi} e^{i(n+1)\theta}\,d\theta$$

But, as we have seen before,

$$\int_0^{2\pi} e^{i(n+1)\theta}\,d\theta = \begin{cases} 0 & n \neq -1 \\ 2\pi & n = -1 \end{cases} \tag{7.34}$$

so that

$$\oint f(z)\,dz = 2\pi i \alpha_{-1} \tag{7.35}$$

This also follows immediately from Eq. 7.29. That is, the contour integral is equal to $2\pi i$ times the coefficient of $(z-z_0)^{-1}$ in the Laurent expansion of $f(z)$. We call this coefficient the *residue* of $f(z)$.

$$\mathscr{R} \equiv \operatorname{Res}\,[f(z), z=z_0] = \alpha_{-1}$$

Obviously, if there had been m isolated singularities enclosed by C, then we would have

$$\oint f(z)\,dz = 2\pi i \sum_{j=1}^{m} \mathscr{R}_j \tag{7.36}$$

The important and surprising point is that the value of the contour integral depends on the value of α_{-1} in the Laurent expansion, but not upon any other of the α_n's.

Example 7.5 Consider the function

$$f(z) = \frac{1}{z^2(1-z)}$$

and the integral $\oint f(z)\,dz$ taken around the path of radius $|z| = \frac{1}{2}$ (for example) about $z=0$. We have already seen that the Laurent expansion about $z=0$ is

$$f(z) = \frac{1}{z^2} + \frac{1}{z} + 1 + z + z^2 + \cdots$$

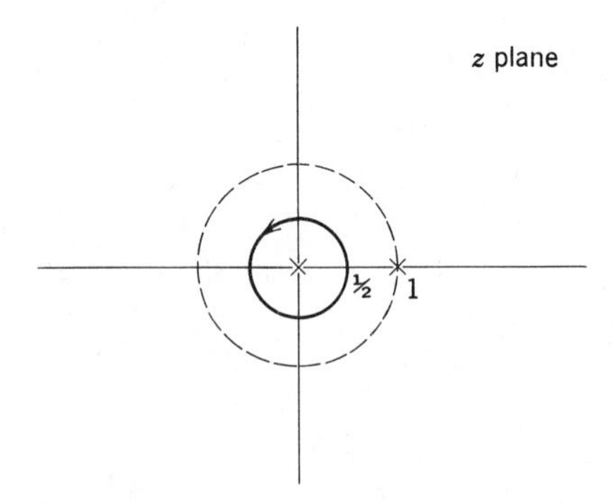

Figure 7.17

so that the residue is

$$\mathcal{R} \equiv \alpha_{-1} = 1$$

This implies that

$$\oint_{\rho=1/2} f(z)\,dz = 2\pi i$$

In general, if $f(z)$ has a pole of finite order k at $z = z_0$, then the residue is equal to (cf. Eq. 7.16)

$$\alpha_{-1} = \frac{1}{2\pi i} \int_C \frac{\phi(t)\,dt}{(t-z_0)^k} \equiv \frac{\phi^{(k-1)}(z_0)}{(k-1)!} \tag{7.37}$$

where $f(z) = \phi(z)/(z-z_0)^k$. We can also see this in another way since

$$f(z) = \frac{1}{(z-z_0)^k}\,\phi(z) = \frac{1}{(z-z_0)^k} \sum_{n=0}^{\infty} \frac{\phi^{(n)}(z_0)}{n!}(z-z_0)^n = \sum_{n=0}^{\infty} \frac{\phi^{(n)}(z_0)}{n!}(z-z_0)^{n-k}$$

We want the coefficient for $n-k=-1$, or $n=k-1$, so that

$$\mathcal{R}=\frac{\phi^{(k-1)}(z_0)}{(k-1)!}$$

Example 7.6 Consider again

$$f(z)=\frac{1}{z^2(1-z)}$$

only this time let $\rho=2$. We have already seen that $R_1=1$ at $z=0$. About $z_0=1$, $\phi(z)=-1/z^2$, so that, for this simple pole, $R_2=-1$, and

$$\oint_{\rho=2} f(z)\,dz=2\pi i-2\pi i=0$$

One of the most important applications of the calculus of residues is the evaluation of integrals along the real axis and of trigonometric integrals. First, however, we must extend the proof of Cauchy's two integral theorems so that they hold for an infinite domain. We are interested in the contour integral

$$\oint f(z)\,dz$$

where

$$f(z)=u(x, y)+iv(x, y) \qquad dz=dx+i\,dy$$

so that

$$\oint f(z)\,dz=\oint[u(x, y)\,dx-v(x, y)\,dy]+i\oint[v(x, y)\,dx+u(x, y)\,dy]$$

where each contour integral in the x-y plane is now real. We must remember that x and y are constrained to lie on contour C.

We can make use of the following theorem from real variable theory. In a simply connected region in which $P(x, y)$, $Q(x, y)$, and their first partial derivatives are continuous, the necessary and sufficient condition that the

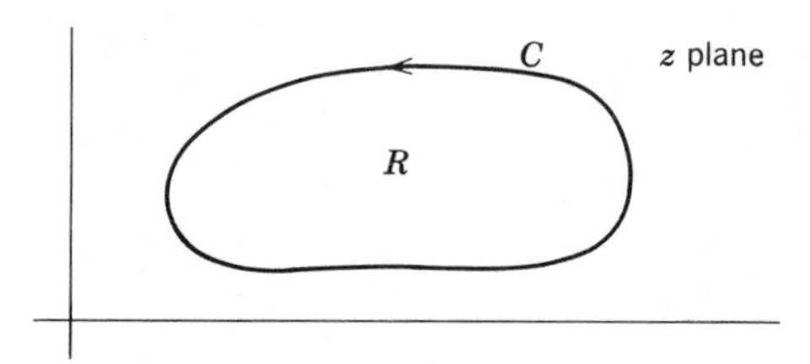

Figure 7.18

integral (assumed to exist)

$$\oint [P(x, y)\,dx + Q(x, y)\,dy]$$

around a closed path should be zero and that, therefore, the integral along a path connecting two points should be independent of the path is

$$\frac{\partial}{\partial y} P(x, y) \equiv \frac{\partial}{\partial x} Q(x, y)$$

everywhere within and on this contour. This follows from an application of Green's theorem in the plane,

$$\iint_R \left(\frac{\partial P}{\partial y} - \frac{\partial Q}{\partial x}\right) dx\,dy = -\int_C (P\,dx + Q\,dy)$$

or equivalently from Stokes' theorem,

$$\int_\Sigma (\nabla \times \mathbf{A}) \cdot d\mathbf{S} = \int_C \mathbf{A} \cdot d\mathbf{l}$$

This is easily proved as follows for a two-dimensional plane region. Consider the integral

$$\iint_R \frac{\partial P}{\partial y}\,dx\,dy = \int_a^b dx \int_{y_1(x)}^{y_2(x)} \frac{\partial P}{\partial y}\,dy$$

$$= \int_a^b dx[P(x, y_2) - P(x, y_1)] = -\int_a^b P(x, y_1)\,dx - \int_b^a P(x, y_2)\,dx$$

$$\equiv -\int_C P[x, y(x)]\,dx$$

where we must remember that dx is constrained to move on the contour, C. Similarly we can show that

$$\iint_R \frac{\partial Q}{\partial x}\,dx\,dy = \int_C Q\,dy$$

from which Green's theorem follows.

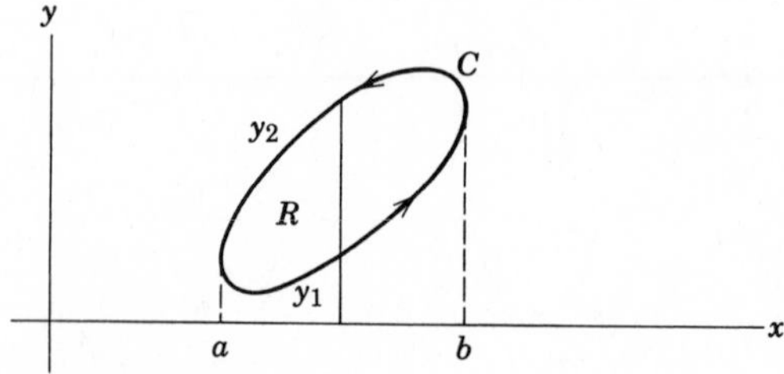

Figure 7.19

However for our $f(z)$ we have the Cauchy-Riemann conditions

$$\frac{\partial u}{\partial x}=\frac{\partial v}{\partial y} \qquad \frac{\partial u}{\partial y}=-\frac{\partial v}{\partial x}$$

so that

$$\oint f(z)\,dz=0$$

independent of the size of the region, as long as $f(z)$ is holomorphic within and on C and provided all integrals exist. Whenever we let part of C approach infinity, we must always demonstrate the existence of the integral along this portion of C. Finally we can also prove

$$\frac{1}{2\pi i}\oint \frac{f(t)\,dt}{t-z}=f(z)$$

as before from $\oint f(z)\,dz=0$ within the corresponding cross-cut region. We then prove, again independent of the size of the region, that

$$\oint f(z)\,dz=2\pi i \sum_{j=1}^{N} \text{Res}\,[f(z), z_j]$$

We now consider some applications of the calculus of residues.

a. Rational Algebraic Integrands

Consider

$$\int_{-\infty}^{\infty} dx\,\frac{P(x)}{Q(x)}$$

where $P(x)$ and $Q(x)$ are polynomials in x and such that $Q(x)$ is at least *two* degrees higher than $P(x)$. Let

$$f(z)\equiv\frac{P(z)}{Q(z)}$$

On the circle of radius R of Fig. 7.20 set

$$z=\text{Re}^{i\theta} \qquad dz=iz\,d\theta$$

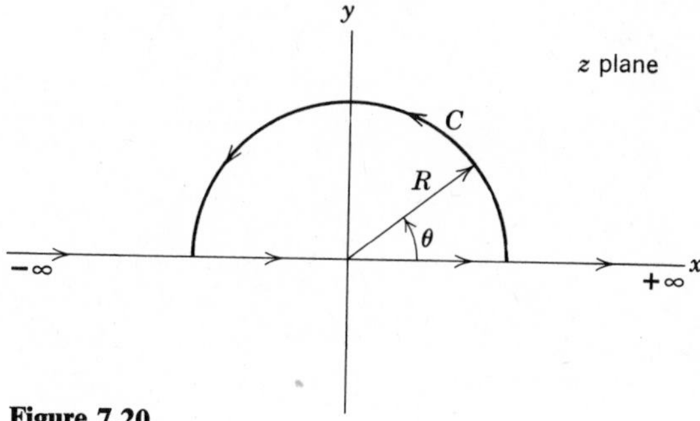

Figure 7.20

Therefore on the perimeter of the semicircle

$$\left| dz \frac{P(z)}{Q(z)} \right| = \left| iz \frac{P(z)}{Q(z)} \, d\theta \right| \xrightarrow[R\to\infty]{} \left| iz \frac{z^m}{z^n} \, d\theta \right| = \left| \frac{R^{m+1}}{R^n} \right| |d\theta| \xrightarrow[R\to\infty]{} 0$$

if $n>m+1$. Hence the contribution to

$$\oint f(z) \, dz$$

from the large semicircle vanishes and we have

$$\int_{-\infty}^{\infty} \frac{P(x)}{Q(x)} \, dx = \oint f(z) \, dz = 2\pi i \sum_{j=1}^{N} \text{Res}\,[f(z), z_j] \tag{7.38}$$

where the $\{z_j\}$ are just the zeros of

$$Q(z)=0$$

lying *within* the contour, C. We have assumed, of course, that $f(x)$ has no poles on the real axis since, if it had, then Cauchy's theorem would not apply. We will consider the necessary modifications for this case later. In general, one can evaluate integrals of functions on the range $-\infty<x<+\infty$ provided only that $|zf(z)| \xrightarrow[|z|\to\infty]{} 0$ and that $f(z)$ is single-valued for all directions of z.

Example 7.7 We take first a case that can easily be done by standard means as well.

$$\int_0^{\infty} \frac{dx}{1+x^2} = \tan^{-1}(x) \Big|_0^{\infty} = \frac{\pi}{2}$$

$$\int_0^{\infty} \frac{dx}{1+x^2} = \frac{1}{2} \int_{-\infty}^{\infty} \frac{dx}{1+x^2} = \frac{1}{2} \int_C \frac{dz}{(z+i)(z-i)} = \frac{2\pi i}{2} \left(\frac{1}{2i} \right) = \frac{\pi}{2}$$

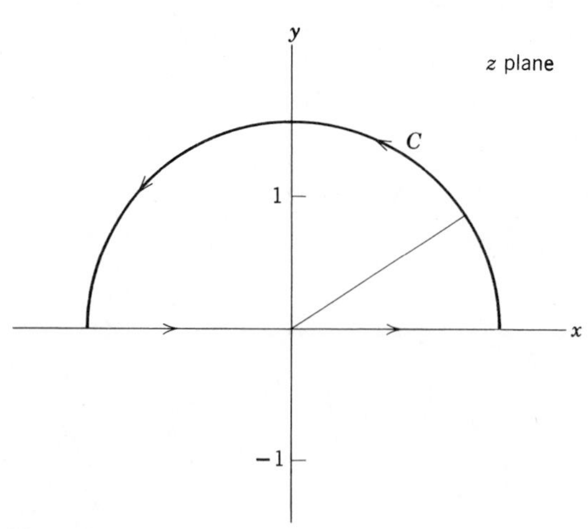

Figure 7.21

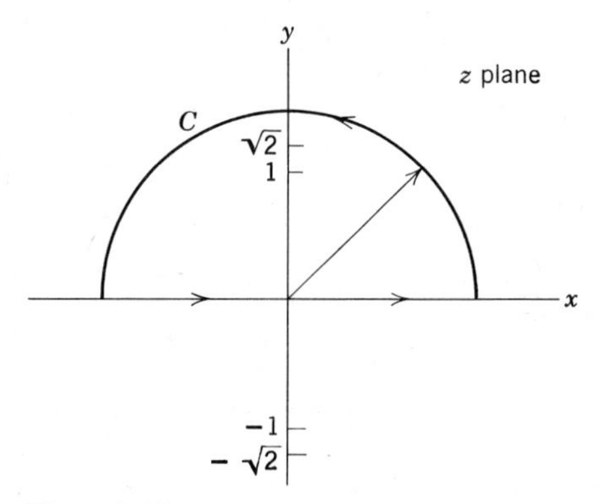

Figure 7.22

Example 7.8

$$\int_0^\infty \frac{dx}{(x^2+1)(x^2+2)} = \frac{1}{2}\int_{-\infty}^{\infty} \frac{dx}{(x^2+1)(x^2+2)}$$

$$= \frac{1}{2}\int_C \frac{dz}{(z+i)(z-i)(z+i\sqrt{2})(z-i\sqrt{2})}$$

$$= \frac{2\pi i}{2}\left\{\frac{1}{(i+i)(i+i\sqrt{2})(i-i\sqrt{2})} + \frac{1}{(i\sqrt{2}+i)(i\sqrt{2}-i)(i\sqrt{2}+i\sqrt{2})}\right\}$$

$$= \pi\left(\frac{1}{2} - \frac{1}{2\sqrt{2}}\right) = \frac{\pi}{4}(2-\sqrt{2})$$

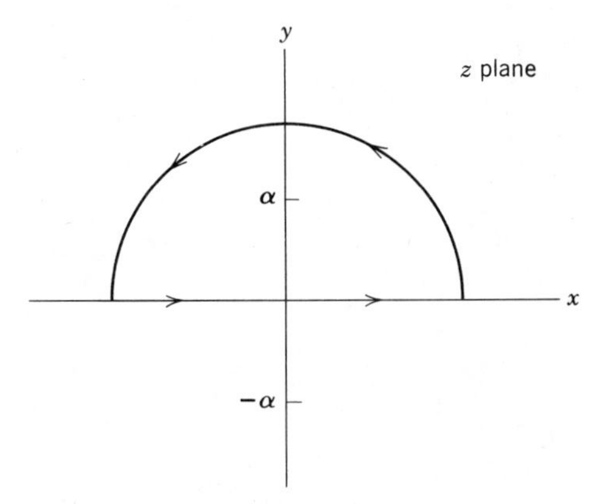

Figure 7.23

Example 7.9

$$I=\int_{-\infty}^{\infty}\frac{x^2\,dx}{(x^2+\alpha^2)^3}$$

This has poles of order *three* at $z=\pm i\alpha$.

$$z^2\equiv(z-i\alpha)^2+2i\alpha(z-i\alpha)-\alpha^2$$

$$\frac{1}{(z+i\alpha)^3}=\sum_{n=0}^{\infty}\frac{(-1)^n(n+3)!}{n!\,3!\,(2i\alpha)^{n+3}}(z-i\alpha)^n$$

This last expression is the Taylor series expansion of $(z+i\alpha)^{-3}$ about $z=i\alpha$. Therefore the Laurent series for $f(z)$ about $z=i\alpha$ is

$$f(z)\equiv\frac{z^2}{(z-i\alpha)^3(z+i\alpha)^3}\equiv\frac{\phi(z)}{(z-i\alpha)^3}$$

$$=[(z-i\alpha)^2+2i\alpha(z-i\alpha)-\alpha^2]\sum_{n=0}^{\infty}\frac{(-1)^n(n+1)(n+2)(n+3)}{3!\,(2i\alpha)^{n+3}}(z-i\alpha)^{n-3}$$

from which it follows that

$$\text{Res}\,[f(z),\,z=i\alpha]=\frac{1}{(2i\alpha)^3}-\frac{2i\alpha(4)}{(2i\alpha)^4}-\frac{\alpha^2(10)}{(2i\alpha)^5}=\frac{1}{16i\alpha^3}$$

and

$$\int_{-\infty}^{\infty}\frac{x^2\,dx}{(x^2+\alpha^2)^3}=2\pi i\left(\frac{1}{16i\alpha^3}\right)=\frac{\pi}{8\alpha^3}$$

We could also have calculated this residue from the expression given by Eq. 7.37,

$$\text{Res}\,[f(z),\,z_0]=\frac{\phi''(z_0)}{2!}$$

We find that

$$\phi'(z)=\frac{z(2i\alpha-z)}{(z+i\alpha)^4}$$

$$\phi''(z)=\frac{2(z^2-4i\alpha z-\alpha^2)}{(z+i\alpha)^5}$$

Therefore we obtain

$$\frac{\phi''(z=i\alpha)}{2}=\frac{1}{16i\alpha^3}$$

which yields the same result as gotten previously.

---

b. Trigonometric Integrands

We now apply residue calculus to integrals with trigonometric integrands. One substitution that allows us to do these integrals is

$$e^{iz}=\cos z+i\sin z \qquad \textbf{(7.39)}$$

The technique is best illustrated with some examples.

Example 7.10

$$I=\int_0^{\infty}\frac{\sin x}{x}\,dx=\frac{1}{2}\int_{-\infty}^{\infty}\frac{\sin x}{x}\,dx$$

We must be careful with this case since $\sin z$ is no longer bounded by unity in absolute value for complex z (cf. Problem 7.7). However we make use of the identity

$$\sin z=\frac{e^{iz}-e^{-iz}}{2i}$$

and begin by considering the integral

$$I'\equiv\frac{1}{2i}\int_C\frac{e^{iz}\,dz}{z}=0$$

where contour C is shown in Fig. 7.24. Since

$$|e^{iz}|=|e^{iR\cos\varphi}e^{-R\sin\varphi}|=e^{-R\sin\varphi}$$

we see that for $0<\varphi<\pi$ the contribution from the large semicircle will vanish as $R\to\infty$. Furthermore we can explicitly evaluate the integral around the small semicircle in the limit $\rho\to 0$ as

$$\int_{C_1}\frac{e^{iz}\,dz}{z}=\int_{\pi}^{0}e^{i[\rho e^{i\varphi}]}i\,d\varphi=i\int_{\pi}^{0}d\varphi[1+i\rho e^{i\varphi}+\cdots]$$

$$=-\pi i+O(\rho)\xrightarrow[\rho\to 0]{}-\pi i$$

We then obtain

$$\int_{-R}^{-\rho}\frac{e^{ix}}{x}\,dx+\int_{C_1}\frac{e^{iz}\,dz}{z}+\int_{\rho}^{R}\frac{e^{ix}}{x}\,dx+\int_{R}\frac{e^{iz}\,dz}{z}=0$$

so that

$$\int_{\rho}^{R}\frac{(e^{ix}-e^{-ix})}{x}\,dx=-\int_{R}\frac{e^{iz}\,dz}{z}-\int_{C_1}\frac{e^{iz}\,dz}{z}$$

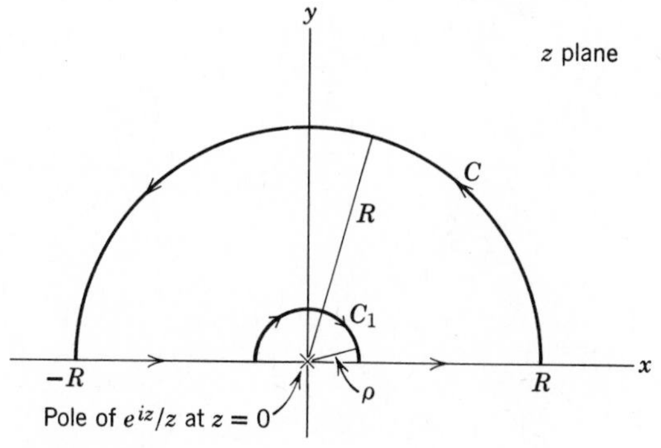

Figure 7.24

and

$$\lim_{\substack{R\to\infty\\ \rho\to 0}} \int_{\rho}^{R} \frac{(e^{ix}-e^{-ix})}{x}\,dx \equiv 2i\int_{0}^{\infty} \frac{\sin x}{x}\,dx = \pi i$$

which finally yields

$$\int_0^\infty \frac{\sin x}{x}\,dx = \frac{\pi}{2}$$

This example brings out a very important point. Whenever the singularities occur *on* the path of integration care must be taken. One cannot simply evaluate the antiderivative at the end points but must go through a suitable limiting process in the neighborhood of the singularity. Integrals such as those through $z=0$ in this example are known as *principal-value integrals*, and we will see more of these shortly.

Example 7.11

$$I = \int_0^{2\pi} \frac{\sin^2 \varphi \, d\varphi}{(1+\varepsilon \cos \varphi)^2} \qquad 0<|\varepsilon|<1$$

If we use the substitution

$$z = e^{i\varphi} = \cos \varphi + i \sin \varphi$$

$$dz = iz\, d\varphi \Rightarrow d\varphi = -i\frac{dz}{z}$$

then we can write

$$\sin \varphi = \frac{z-(1/z)}{2i} \qquad \cos \varphi = \frac{z+(1/z)}{2}$$

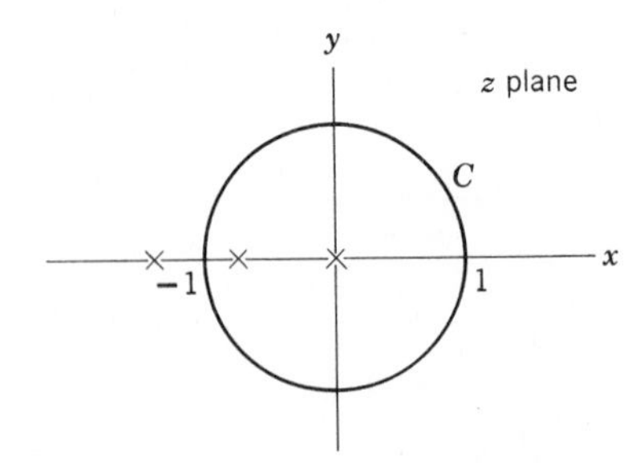

Figure 7.25

so that

$$I = -\frac{1}{4}\int_C \frac{[z-(1/z)]^2[-(i\,dz/z)]}{[1+[\varepsilon z+(\varepsilon/z)]/2]^2} = i\int_C \frac{dz}{z}\frac{(z^2-1)^2}{(\varepsilon z^2+2z+\varepsilon)^2}$$

$$= \varepsilon^2 i \int_C \frac{dz}{z}\frac{(z^2-1)^2}{\{[\varepsilon z+1-\sqrt{1-\varepsilon^2}][\varepsilon z+1+\sqrt{1-\varepsilon^2}]\}^2}$$

Let

$$f(z) = \frac{\phi(z)}{\varepsilon^2[z+(1-\sqrt{1-\varepsilon^2})/\varepsilon]^2}$$

$$\phi(z) = \frac{(z^2-1)^2}{z[\varepsilon z+1+\sqrt{1-\varepsilon^2}]^2}$$

so that

$$\phi'(z) = \frac{4(z^2-1)}{[\varepsilon z+1+\sqrt{1-\varepsilon^2}]^2} - \frac{(z^2-1)^2}{z^2[\varepsilon z+1+\sqrt{1-\varepsilon^2}]^2} - \frac{2(z^2-1)^2\varepsilon}{z[\varepsilon z+1+\sqrt{1-\varepsilon^2}]^3}$$

and

$$z_0 = \frac{-1+\sqrt{1-\varepsilon^2}}{\varepsilon}$$

$$\phi'(z_0) = \frac{\sqrt{1-\varepsilon^2}-1}{\varepsilon^2\sqrt{1-\varepsilon^2}} - \frac{1}{\varepsilon^2}$$

$$\text{Res}\,[f(z),\, z=z_0] = \phi'(z_0)/\varepsilon^2$$

$$\text{Res}\,[f(z),\, z=0] = 1/\varepsilon^4$$

The final result is (since only *two* poles lie inside the unit circle)

$$I = \varepsilon^2 i(2\pi i)\left\{\frac{\sqrt{1-\varepsilon^2}-1}{\varepsilon^4\sqrt{1-\varepsilon^2}} - \frac{1}{\varepsilon^4} + \frac{1}{\varepsilon^4}\right\} = \frac{2\pi}{\varepsilon^2}\left(\frac{1}{\sqrt{1-\varepsilon^2}} - 1\right)$$

This integral, which can also be done with integration by parts, occurs in Sommerfeld's semiclassical relativistic theory of the fine structure of the hydrogen atom. In fact this integral can be performed as

$$\int_0^{2\pi} \frac{\sin^2\varphi\, d\varphi}{(1+\varepsilon\cos\varphi)^2} = -\frac{1}{\varepsilon}\int_0^{2\pi}\frac{\cos\varphi\, d\varphi}{(1+\varepsilon\cos\varphi)} = -\frac{2}{3}\int_{-1}^{1}\frac{y\,dy}{(1-y^2)^{1/2}(1+\varepsilon y)}$$

$$= \frac{2\pi}{\varepsilon^2}\left(\frac{1}{\sqrt{1-\varepsilon^2}} - 1\right)$$

7.9 MULTIPLE-VALUED FUNCTIONS

So far we have considered *single-valued* functions only. This means that $f(z)$ has a unique value at every point z. As a working definition or test for single-valuedness in a region we could say that if we begin at a point z_0, where $f(z)$ has a value $f(z_0)$, and travel along *any* closed curve in the region back finally to z_0, then $f(z)$ must *always* return to its original value $f(z_0)$. The following discussion will be somewhat qualitative and informal but sufficient for our purposes.

a. Branch Points and Branch Cuts

Consider a simple and often-quoted example of a multiple-valued function

$$f(z)=\sqrt{z}$$

First take $z_0=0$ and set $z=\rho e^{i\varphi}$ so that

$$f(z)=\sqrt{z}=\sqrt{\rho}\, e^{i\varphi/2}$$

Begin at $z=\rho$ (i.e., $\varphi=0$) and proceed around a circle of radius ρ counterclockwise back to $z=\rho$ (i.e., $\varphi=2\pi$). Then z has returned to its original value, but $f(z)$ has gone from

$$f(z=\rho,\, \varphi=0)=+\sqrt{\rho}$$

to

$$f(z=\rho,\, \varphi=2\pi)=-\sqrt{\rho}$$

Therefore $f(z)$ does not return to its original value as z executes a closed circuit in the z plane enclosing $z=0$. However, if we now traverse the path again so that $2\pi \rightarrow \varphi \rightarrow 4\pi$, then $f(z)$ returns to its original value. Hence $f(z)$ is double valued and has two *branches*. In order to study the other singular point, we set $\zeta=1/z$ so that

$$\tilde{f}(\zeta)\equiv f(1/\zeta)=1/\sqrt{\zeta}=\rho^{-1/2}e^{-i\varphi/2}$$

Again we see that the behavior of $\tilde{f}(\zeta)$ is the same as for $f(z)$ above about $z=0$. The points $z=0$ and $z=\infty$ are *branch points* of $f(z)$ where two (or more) branches of the function become equal [i.e., $f(z)=\pm\sqrt{z}$].

If $(k+1)$ branches of a multiple-valued function become equal at a point, this point is a branch point of *order* k. In our example $z=0$ and $z=\infty$ are branch points of order one (or *simple* branch points). An example of a function with m branches is

$$f(z)=\sqrt[m]{z}=z^{1/m}$$

Here points $z=0$ and $z=\infty$ are branch points of order $(m-1)$. The branch points of a function are fixed singularities just as are poles.

As a slightly different example consider

$$f(z)=\sqrt{z^2-1}=\sqrt{(z+1)(z-1)}$$

As we might guess the points $z=\pm 1$ are branch points, since if

$$z-1=\rho e^{i\varphi} \qquad \rho \ll 1$$

then

$$f(z)=\sqrt{\rho}\,e^{i\varphi/2}\sqrt{2+\rho e^{i\varphi}} \rightarrow \sqrt{2\rho}\,e^{i\varphi/2}$$

As before, when $0 \rightarrow \varphi \rightarrow 2\pi$, $f(z)$ does not return to its initial value. The same is true for $z \simeq -1$. It takes two circuits around just *one* of the branch points to regain the original value of $f(z)$. Let us now consider point $z=\infty$. Here

$$\tilde{f}(\zeta)=\sqrt{\frac{1}{\zeta^2}-1}=\frac{\sqrt{1-\zeta^2}}{\zeta} \xrightarrow[\zeta\to 0]{} \frac{1}{\zeta}$$

so that $\tilde{f}(\zeta)$ is single-valued in the neighborhood of $\zeta=0$. Therefore point $z=\infty$ is *not* a branch point but only a simple pole.

In discussing multiple-valued functions we have denoted the various single-valued, continuous "pieces" of such a function

$$w=f(z)$$

as the *branches* of the function. This concept may become clearer if we plot two separate planes: one for z and one for w.

Example 7.12 Consider

$$w=f(z)=z^{1/3}=\rho^{1/3}e^{i\varphi/3}\equiv\sigma e^{i\theta}$$

so that

$$\sigma=\rho^{1/3} \qquad \theta=\varphi/3$$

As φ goes from 0 to 6π, we obtain, in succession, the three branches:

i. $\varphi=0$ $\qquad w=\rho^{1/3}$

ii. $\varphi=2\pi$ $\qquad w=\frac{1}{2}(-1+\sqrt{3})\rho^{1/3}$

iii. $\varphi=4\pi$ $\qquad w=-\frac{1}{2}(1+i\sqrt{3})\rho^{1/3}$

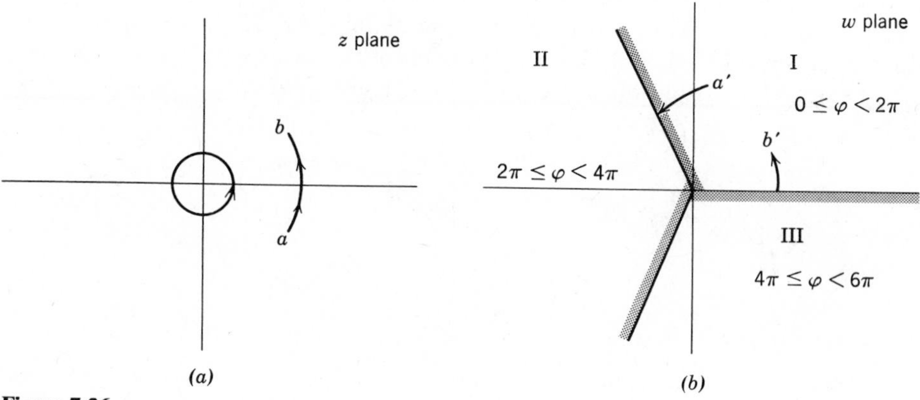

Figure 7.26

In other words for each point in the z plane there correspond three points in the w plane. This function has three branches (i.e., each is a "piece" of the function).

We can also look upon the function

$$w = z^{1/3}$$

as a mapping of the points of the z plane onto the w plane. It is as though the z plane were cut along the positive x axis, for example, and contracted upon itself into one of the *fundamental regions* in the w plane (e.g., region I). Hence a continuous line in the z plane, $\widehat{ab}$ for example, can become discontinuous in the w plane. We denote the various regions in the w plane as $w_1(z)$, $w_2(z)$, and $w_3(z)$ and refer to them as the various *branches* of $w = f(z) = z^{1/3}$. Each of these is a single-valued analytic function of z and taken together they constitute the entire function.

In general we can give the following formal definition of the *branch* of a multiple-valued function, $w = f(z)$. A collection of values (w, z) is said to define a branch w_k of a function if:

i. The set of all z points of the region of existence fills a region R_k just once.
ii. Only one value of w corresponds to each point of R_k.
iii. The set of w points corresponding to points in R_k represents a continuous function of z.

A point is called a branch point of the given function if some of the branches interchange as the independent variable describes a closed path about it. At branch points (some) branches of the function become equal. We can now see that if we begin at some point in the z plane, for example, z_0 (which is assumed not to be a singular point of the function), and travel on a closed contour encircling *all* the branch points of $w = f(z)$, *provided* there is only a finite number of branch points and that each is of finite order, then we end on the same branch on which we began. This is evident from the definition of the branch points. If a curve encloses *all* of them, then in the region of the plane exterior to this curve the function is by definition single-valued, so that if we traverse the curve in the

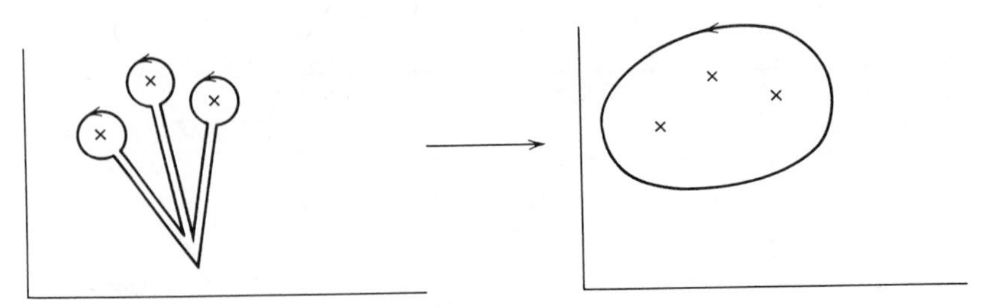

Figure 7.27

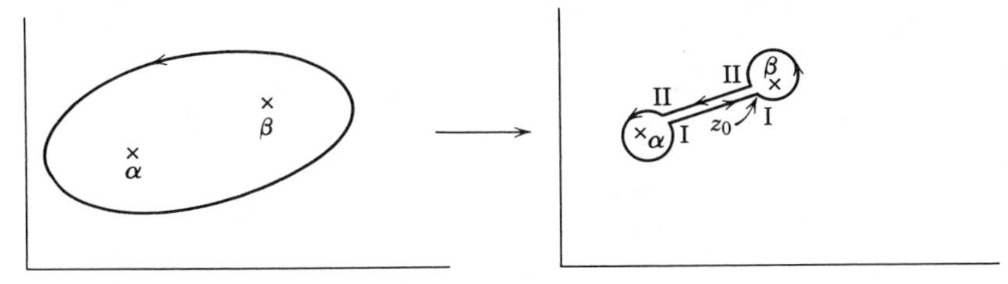

Figure 7.28

opposite direction the function is single-valued within C. Furthermore the effect of encircling a finite number of branch points with one curve is equivalent to circling each one separately. The equivalence of these two ways of evaluating the change of a function around a closed contour is shown in Fig. 7.27.

To be specific consider a $w=f(z)$, which has just two simple branch points. By the definition of a branch point the branches of a function must interchange as we go around it once. If we start at z_0 and go around β back to z_0, we go from branch I to branch II. In going along the cross cut we encircle no branch points and, hence, remain on the same branch (i.e., II). In going around α we go from II to I and, therefore, arrive back at z_0 on branch I where we began.

Example 7.13 An example of a function that has two simple branch points and one second-order branch point is

$$w^3-3w+4z-2=0$$

This means that there are three roots, w_1, w_2, and w_3, for each value of z. We could use techniques from the theory of multiple-valued functions or simply Cardan's method of resolution of a cubic equation to find these values of z for which two or more roots become equal. However, since we are interested in this only as an example, let us simply investigate the nature of the points, $z=0, 1, \infty$.

For $z=0$ we have

$$0=w^3-3w-2=(w+1)(w^2-w-2)=(w+1)^2(w-2)$$

so that two branches become equal here, which implies that $z=0$ is a simple branch point. For $z=1$ we have

$$0=w^3-3w+2=(w-1)(w^2+w-2)=(w-1)^2(w+2)$$

so that two branches become equal here, which makes $z=1$ a simple branch point also. If we let $z=\infty$ (actually, $z=1/\zeta$, $w=1/\omega$, $\zeta \to 0$), we see that $w=\infty$ is a triple root so that three branches become equal here. Therefore $z=\infty$ is a branch point of order *two*.

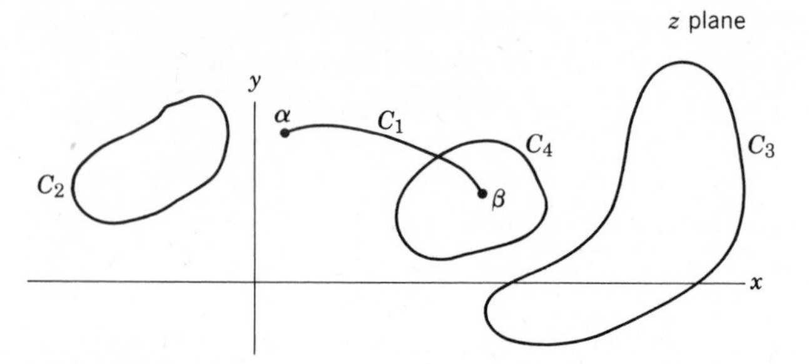

Figure 7.29

We now introduce the extremely useful concept of a *branch cut.* A branch cut is simply a (non self intersecting) line connecting two branch points (*distinct* ones of course). The lines are arbitrary in shape and need not be straight lines (although they are often taken as such for convenience). The only thing definite about them is that they begin and end on distinct branch points. Their usefulness consists of the following. Consider a function $f(z)$ that has just two branch points, α and β, and connect these by branch cut C_1. Then $f(z)$ will be single valued in the region excluding the branch cut, C_1. That is, $f(z)$ will return to its original value after one circuit around C_2 or C_3 but not around C_4. Thus we could apply Cauchy's theorem along any curve C not crossing or encircling branch cut C_1.

In fact, the line along which the discontinuity occurs is intimately connected with the choice of phase made for the function. If, as in Fig. 7.1, we agree to measure the phase φ of $z=\rho e^{i\varphi}$ up from the positive x axis, increasing counterclockwise, then the discontinuity of $f(z)=\sqrt{z}$ appears along the positive x axis. That is, if we compare the values of $f(\rho, \varphi)$ with those of $f(\rho, 2\pi-\varphi)$ as $\varphi \to 0$, these values do *not* approach each other but suffer a discontinuity as shown in Fig. 7.30. On the other hand if we had agreed to let the phase φ run

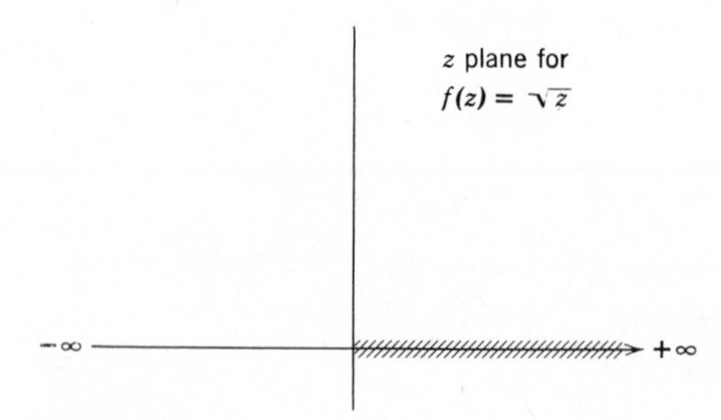

Figure 7.30

from $-\pi \to \varphi \to +\pi$, then the discontinuity would have appeared across the negative x axis. In our applications we will choose our branch cuts to run along the line of discontinuity of the function. We will also measure the phase of z as indicated in Fig. 7.1.

It will prove useful for reference later [cf. Section 7.13b and the Legendre function of the second kind, $Q_\alpha(z)$] to study in some detail the cut structure and phase of the following function.

Example 7.14 Let

$$f(z)=(z^2-1)^\alpha=(z-1)^\alpha(z+1)^\alpha$$

where α is any (real) noninteger number. In general the points, $z=\pm 1, \infty$ will be branch points. (Notice that in the special case of $\alpha=\frac{1}{2}$ the point at infinity is *not* a branch point.) With the phase convention discussed above, the cut z plane is shown in Fig. 7.31. The phases indicated in this figure correspond to the quantities

$$z=\rho e^{i\varphi}$$

$$w_1=\sigma_1 e^{i\theta_1}=(z+1)^\alpha$$

$$w_2=\sigma_2 e^{i\theta_2}=(z-1)^\alpha$$

so that with

$$f(z)=w=\sigma e^{i\theta}=(z^2-1)^\alpha$$

we have

$$\theta=\theta_1+\theta_2$$

It is clear from the geometry of this figure that, if we begin on the positive x axis with $x<1$ and complete one full circuit about the origin (e.g., on a circle), then the phase θ begins at $\theta=\alpha\pi$ and returns to this same value. However, if we begin

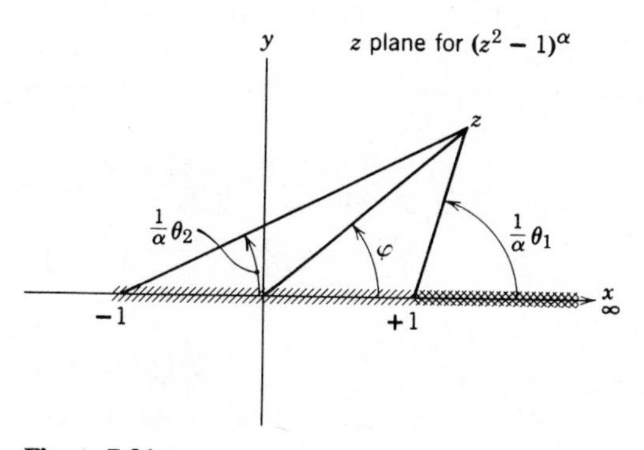

Figure 7.31

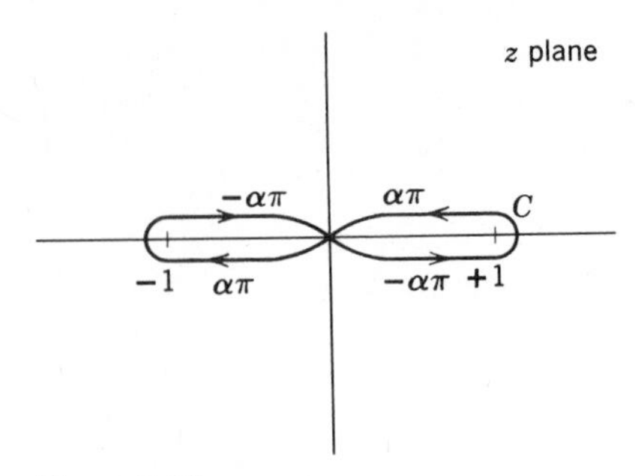

Figure 7.32

on the positive real axis with $x>1$, where $\theta=0$, and complete a similar circuit, then the final value of θ is $\theta=4\alpha\pi$ so that, unless $\alpha=\frac{1}{2}$, $f(z)$ does not return to its original value. Finally for this example let us consider the circuit shown in Fig. 7.32. From the geometry of Fig. 7.31 we see that θ returns to its initial value as we transverse C once in the direction shown in Fig. 7.32. The values of the phase θ are indicated on the various horizontal segments of C in the figure.

From this discussion we can now see that branch cuts, which simply join branch points, are a means of effectively isolating a given branch of a function of $w=f(z)$. If we never cross a branch cut, then we can never encircle a single branch point in order to change branches. We can only encircle all of them, if there is a finite number of them, and, as we have seen above, we end on the same branch we began on.

As a final example of a multiple-valued function, we consider

$$f(z)=\log z=\log \rho+i\varphi \qquad z=\rho e^{i\varphi}$$

We see that $z=0$ is a branch point of this function and that each circuit around $z=0$ increases $f(z)$ by $2\pi i$ so that the function will never return to its initial value. For $z=\infty$ we have

$$\tilde{f}(\zeta)=\log(1/\zeta)=-\log|\zeta|-i\theta$$

so that this is also a branch point. This function has *infinitely* many branches so that the points, $z=0$ and $z=\infty$, are branch points of infinitely high order.

b. Riemann Sheets

We have seen that branch points are convenient artifices by which we exclude troublesome points in the z plane. Here we consider an associated artifice known as *Riemann surfaces.* To begin let us take again the simple example of the

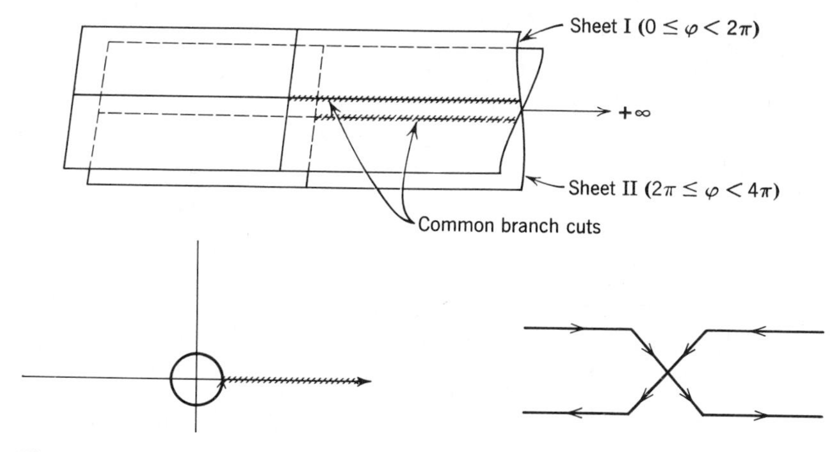

Figure 7.33

double-valued function

$$f(z)=\sqrt{z}$$

We saw previously that, as z ran through all of its values, $z=\rho e^{i\varphi}$, $0 \to \varphi \to 2\pi$, $f(z)$ did not (necessarily) return to its original value so that for (almost) every point in the z plane we could assign two values to $f(z)$. One obvious way to overcome this ambiguity would be to employ *two* z planes, one associated with values $0 \le \varphi < 2\pi$ and the other with values $2\pi \le \varphi < 4\pi$ and each plane having a branch cut (e.g., from $0 \to +\infty$ along the positive real axis). Then $f(z)$ would be *single valued* on each of these sheets and we could apply all the techniques developed for single-valued functions (e.g., Cauchy's theorem, etc.) provided, of course, we avoid the branch cut. Now a Riemann surface is simply a convenient means of treating these two separate surfaces as one, in a sense. Instead of picturing them as two disjoint regions, we take them as stacked one atop the other with their branch cuts coinciding. That is, the two sheets are literally "cut" along their positive real axes and joined together as shown in Fig. 7.33. The artifice of employing a Riemann surface has the effect of making the z plane simply connected for a multiple-valued function $w=f(z)$ by enlarging the z plane to be the entire Riemann surface. We see that there are as many Riemann sheets as there are branches of the function and that the union of all these sheets is the Riemann surface for $f(z)$. In a simply connected region any piecewise smooth nonintersecting (closed) contour can always be continuously contracted to a point without leaving the region. If we consider a singly cut z plane, it is obviously not simply connected (cf. Fig. 7.34).

Cauchy's theorems cannot be applied to such regions since their proofs require that $f(z)$ be holomorphic within C and continuous on the smooth contour

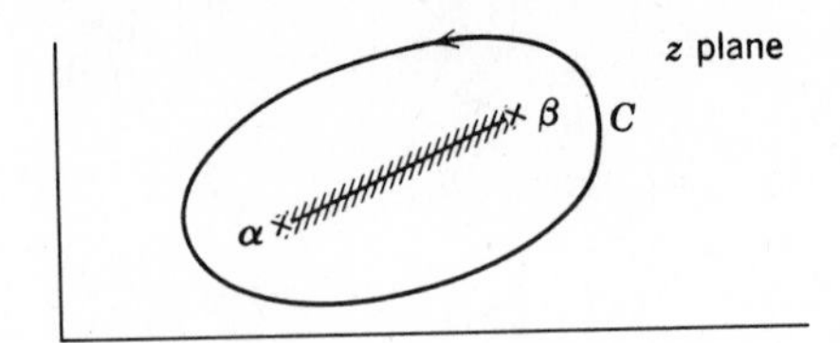

Figure 7.34

C. However, if we consider the entire Riemann surface, then the function is single valued by construction on this surface. Furthermore, if there is only a finite number of sheets, then the surface is simply connected. We may apply Cauchy's integral formula on such a surface with very little modification.

Example 7.15 Let

$$f(z)=\sqrt{z^2+1}=\sqrt{(z+i)(z-i)}$$

and consider the integral

$$I \equiv \int_C f(z)\, dz$$

Since $f(z)$ is single valued and continuous within and on C of Fig. 7.35 (in fact holomorphic with the exception of $z=i$ where the derivative is not defined), we *might expect* that

$$\int_C f(z)\, dz=0$$

To prove this conjecture we may deform this contour so that both circles are of

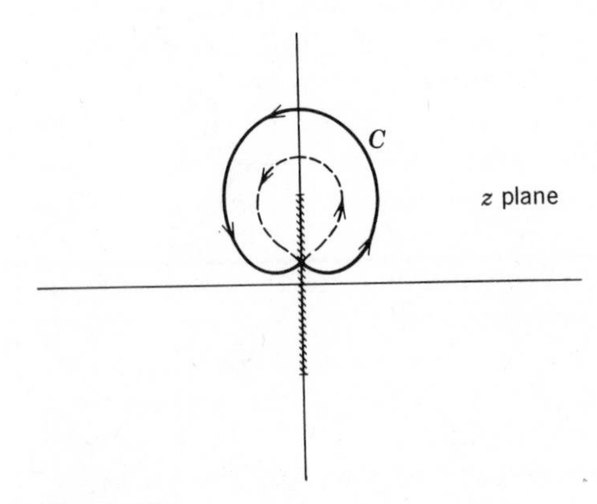

Figure 7.35

radius ρ about $z=i$ where ρ may be as small as we please. With the usual substitutions

$$-i+z=\rho e^{i\varphi} \qquad dz=i\rho e^{i\varphi}\,d\varphi$$

and the fact that

$$f(z)\xrightarrow[\rho\to 0]{}\sqrt{2i}\sqrt{\rho}\,e^{i\varphi/2}$$

we see that

$$I\xrightarrow[\rho\to 0]{} i\sqrt{2i}\,\rho^{3/2}\int_0^{4\pi} e^{i(3/2)\varphi}\,d\varphi=\frac{2}{3}\sqrt{2i}\,\rho^{3/2}e^{i(3/2)\varphi}\Big|_0^{4\pi}=0$$

In general, if we wanted to extend such an integral theorem to a Riemann surface for a contour enclosing a branch point, we would simply exclude the branch point with a cross cut and a small circle about the branch point, apply Cauchy's theorem to the remaining annular region, and then evaluate the integral around the branch point explicitly in the limit $\rho\to 0$. Since this is a branch point, all the branches have the same value as $\rho\to 0$. If this common value is finite, we obtain

$$\int_C f(z)\,dz=0$$

as before. If the value of the function becomes unbounded as $\rho\to 0$, we obtain a modified form of Cauchy's residue theorem. The interested reader can develop this for himself. Since we do not need the result, we will pursue it no further here.

c. Integral Along a Branch Cut

In physical applications we often require the evaluation of an integral *along* a branch cut (i.e., all on *one* Riemann sheet). We will begin by considering an example,

$$I=\int_0^{\infty} x^{a-1}f(x)\,dx \qquad \textbf{(7.40)}$$

where:

i. $f(x)$ is any *rational* function of x with *no poles* on the *positive* real axis.
ii. a—*any* noninteger number.
iii. $\lim_{|z|\to\infty} f(z)z^a=0$.
iv. $\lim_{|z|\to 0} f(z)z^a=0$.

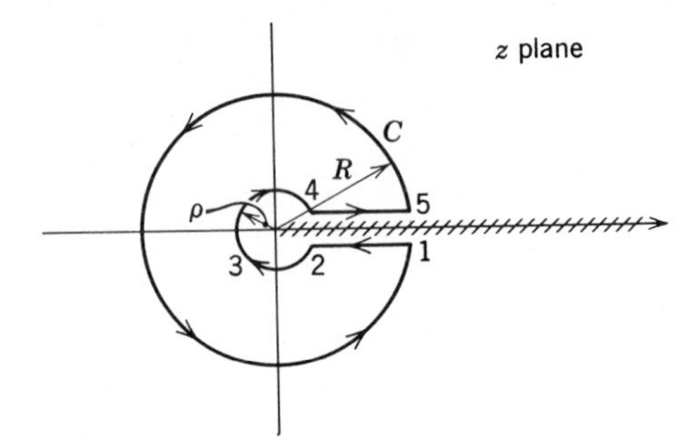

Figure 7.36

We begin by considering the contour integral

$$\int_C z^{a-1} f(z)\, dz \tag{7.41}$$

where C is shown in Fig. 7.36. First let us establish that $[z^{a-1}f(z)]$ has branch points at $z=0$ and at $z=\infty$. Since $f(z)$ is rational (i.e., a ratio of two polynomials in z), it is single valued and, therefore, returns to its initial value as $0 \to \varphi \to 2\pi$. Hence we need consider only z^{a-1}. With

$$z = \rho e^{i\varphi}$$

$$z^{a-1} = \rho^{(a-1)} e^{i(a-1)\varphi}$$

we see that for $\phi = 0$

$$z^{a-1} = \rho^{(a-1)}$$

while for $\phi = 2\pi$

$$z^{a-1} = \rho^{(a-1)} e^{i2\pi(a-1)}$$

Only if a is some *integer* will z^{a-1} be single valued (i.e., have the same *phase* at $\varphi = 0$ and at $\varphi = 2\pi$). Since by assumption a is not integer, z^{a-1} is multiple valued. By making the substitution $z \to 1/\zeta$, we see that $z = \infty$ is also a branch point. In fact if a is not real and rational, then there are infinitely many sheets for $[z^{a-1}f(z)]$.

We can choose the branch cut to run along the positive x axis as indicated in Fig. 7.36. Since our contour C avoids the branch cut and branch points, we may apply Cauchy's integral theorem as

$$\int_C z^{a-1} f(z)\, dz = 2\pi i \sum_j R_j \tag{7.42}$$

We may write this as

$$\int_C z^{a-1}f(z)\,dz = \int_R z^{a-1}f(z)\,dz + \int_{1\infty}^{2\rho} z^{a-1}f(z)\,dz + \int_{\rho}^{234} z^{a-1}f(z)\,dz$$
$$+ \int_{4\rho}^{5\infty} z^{a-1}f(z)\,dz \tag{7.43}$$

If we set

$$z = Re^{i\varphi}$$

on the large circle of radius R and

$$z = \sigma e^{i\varphi}$$

on the small circle of radius σ, then the first and third integrals vanish since, by assumption,

$$\lim_{|z|\to \substack{0\\ \infty}} |z^a f(z)| = 0$$

Along the path $1 \to 2$, $\varphi = 2\pi$ and the phase of z is

$$z = \rho e^{2\pi i} = xe^{2\pi i}$$

and along $4 \to 5$, $\varphi = 0$, so that

$$z = \rho = x$$

Therefore Eq. 7.43 reduces to

$$\int_C z^{a-1}f(z)\,dz = \int_{\infty}^{0} e^{2\pi i(a-1)}x^{a-1}f(x)\,dx e^{2\pi i} + \int_0^{\infty} x^{a-1}f(x)\,dx$$
$$= \int_0^{\infty} x^{a-1}f(x)\,dx[-e^{2\pi a i}+1]$$
$$= e^{\pi(a-1)i}2i\sin \pi a \int_0^{\infty} x^{a-1}f(x)\,dx$$

from which we obtain

$$\int_0^{\infty} x^{a-1}f(x)\,dx = \pi e^{-\pi(a-1)i}\csc(\pi a)\sum_j \text{Res}\,[z^{a-1}f(z)]$$
$$= \pi\csc(\pi a)\sum_j \text{Res}\,[(e^{-\pi i}z)^{a-1}f(z)] \tag{7.44}$$

Finally suppose $f(z)$ has a finite number of *simple* poles on the positive real axis. Then we should modify the contour as shown in Fig. 7.37. If these poles on the real axis are located at $z = a$, b, c, and so forth, consider the integral about one of these located at $z = b$ as shown in Fig. 7.37.

$$I_b \equiv \int_{b_1}^{b_2} z^{a-1}f(z)\,dz + \int_{b_3}^{b_4} z^{a-1}f(z)\,dz$$

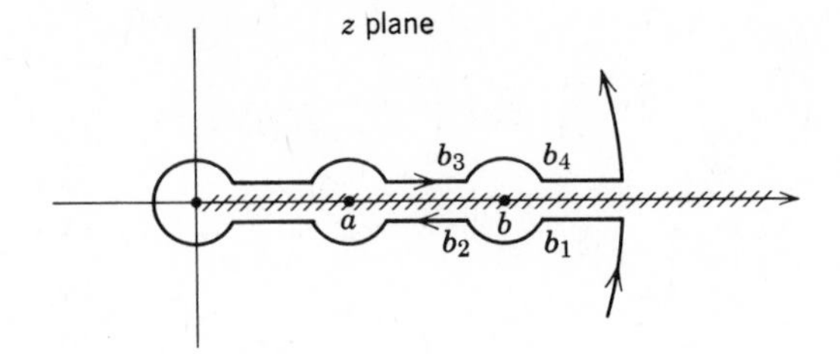

Figure 7.37

On the semicircular arc, $\widehat{b_1 b_2}$, of Fig. 7.37 we set

$$z - be^{2\pi i} = \sigma e^{i\theta}$$

while on the arc, $\widehat{b_3 b_4}$, we set

$$z - b = \sigma e^{i\theta}$$

We then obtain

$$\begin{aligned} I_b &= \int_{2\pi}^{\pi} (e^{2\pi i}b + \sigma e^{i\theta})^{a-1} f(e^{2\pi i b} + \sigma e^{i\theta})\sigma e^{i\theta} i \, d\theta \\ &\quad + \int_{\pi}^{0} (b + \sigma e^{i\theta})^{a-1} f(b + \sigma e^{i\theta})\sigma e^{i\theta} i \, d\theta \\ &\xrightarrow[\sigma \to 0]{} e^{2\pi i(a-1)} b^{a-1} R_b' \int_{2\pi}^{\pi} i \, d\theta + b^{a-1} R_b' \int_{\pi}^{0} i \, d\theta \\ &= -\pi i b^{a-1} R_b' [e^{2\pi i(a-1)} + 1] = \pi b^{a-1} R_b' e^{\pi i(a-1)} 2i \left(\frac{e^{\pi i a} + e^{-\pi i a}}{2} \right) \\ &= \pi b^{a-1} R_b' e^{\pi i(a-1)} 2i \cos(\pi a) \end{aligned}$$

where R_b' is the residue of $f(z)$ at $z = b$. This term will occur on the left [i.e., as part of the contour integral $\int_C z^{a-1} f(z) \, dz$] of the expression for the first case treated above (i.e., the case in which there were no poles on the positive real axis). Therefore the final expression becomes

$$\begin{aligned} \int_0^\infty x^{a-1} f(x) \, dx = \pi \Big\{ &\csc(\pi a) \sum_j \operatorname{Res}[(e^{-\pi i} z_j)^{a-1} f(z_j)] \\ &- \cot(\pi a) \sum_j \operatorname{Res}[x_j^{a-1} f(x_j)] \Big\} \end{aligned} \tag{7.45}$$

It is important to realize that the first sum in Eq. 7.45 is over all the residues of the poles of $f(z)$ not on the positive real axis. If there are poles of higher order than the first on the positive real axis, then the integral along this axis is not defined. Notice that these formulas fail when a is an integer since then there is no discontinuity across the positive x axis. In such a case if $x^{a-1} f(x)$ were an even function of x, we might choose a contour from $-\infty \to x \to +\infty$ and a large semicircle as previously.

d. Dispersion Representation

We now consider a general representation of functions having poles (assumed simple for convenience) and a branch cut. This treatment can easily be extended to more than one branch cut. Take a function $f(z)$ that has the following singularities and asymptotic properties:

i. A branch point at $z=\alpha$ (α – *real* and nonnegative).
ii. A branch point at $z=\infty$.
iii. Simple poles at $z=z_n$, $n=1, 2, \ldots, N$; (none of the z_n lie on the branch cut $\alpha \leq x < +\infty$).
iv. A branch cut chosen along the positive real axis from α to ∞.
v. $f(z)$ a *real* function [i.e., $f^*(z^*)=f(z)$—*definition* of a real function of z] on this Riemann sheet.
vi. $|z|\, f(z) \xrightarrow[|z|\to\infty]{} 0$.
vii. $f(\alpha)$ finite.

We take a contour C that excludes all the singularities of $f(z)$ so that we may apply Cauchy's integral theorem. The contour can then be deformed as shown in Fig. 7.38. Since $|z|\, f(z) \xrightarrow[|z|\to\infty]{} 0$, the integral around the large outer contour

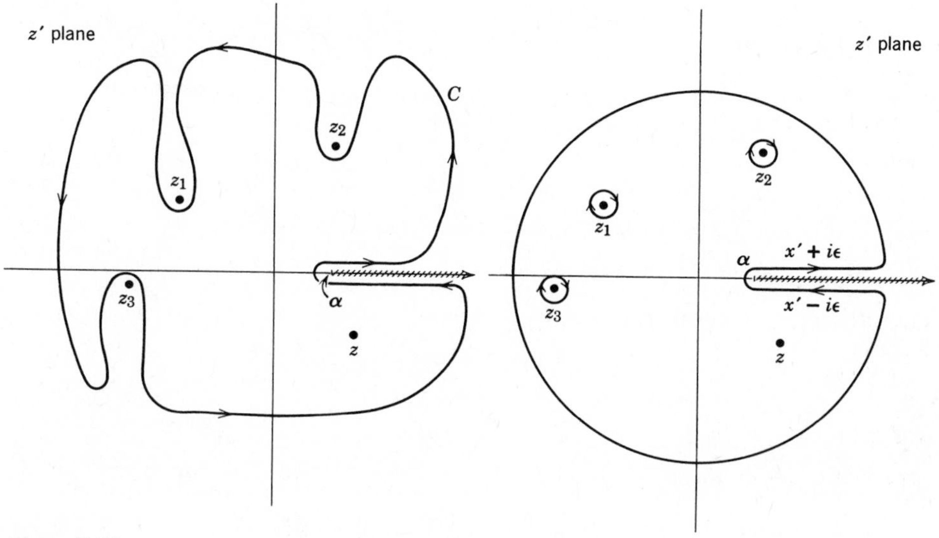

Figure 7.38

vanishes. Therefore we have

$$f(z)=\frac{1}{2\pi i}\int_C \frac{f(z')\,dz'}{(z'-z)}=-\frac{1}{2\pi i}\sum\oint_{\text{poles}}+\frac{1}{2\pi i}\left\{\int_{\infty-i\varepsilon}^{\alpha}+\int_{\alpha}^{\infty+i\varepsilon}\right\} \tag{7.46}$$

The integral around the small circle about $z=\alpha$ vanishes in the limit that the radius of the circle vanishes since $f(z=\alpha)$ is assumed finite. Consider next an integral about a small circle enclosing a pole.

$$\oint \frac{f(z')\,dz'}{(z'-z)} \xrightarrow[\rho\to 0]{} \int_0^{2\pi}\frac{R_n\,dz'}{(z'-z_n)(z'-z)} \to i\int_0^{2\pi}\frac{R_n\,d\varphi}{(z_n-z)}=\frac{2\pi i R_n}{(z_n-z)}=-\frac{2\pi i R_n}{(z-z_n)}$$

Here as usual we have set

$$(z'-z_n)=\rho e^{i\varphi}$$

to do the integral. Finally for the integral along the cut (i.e., just above and below it) we set

$$z'=x'\pm i\varepsilon$$

Then we obtain

$$\int_{\infty}^{\alpha}\frac{f(x'-i\varepsilon)\,dx'}{(x'-i\varepsilon-z)}+\int_{\alpha}^{\infty}\frac{f(x'+i\varepsilon)\,dx'}{(x'+i\varepsilon-z)}=\int_{\alpha}^{\infty}dx'\left\{-\frac{f(x'-i\varepsilon)}{x'-i\varepsilon-z}+\frac{f(x'+i\varepsilon)}{x'+i\varepsilon-z}\right\} \tag{7.47}$$

As long as z is kept off the branch cut on the portion of the positive real axis between α and ∞, the denominators $(x'-z\pm i\varepsilon)$ approach $(x'-z)$ uniformly as $\varepsilon\to 0$. Of course the numerators do *not* approach each other as $\varepsilon\to 0$ since there is a discontinuity of $f(z)$ along the cut. Therefore, remembering that z cannot be on the branch cut, we may write the r.h.s. of Eq. 7.47 as

$$\int_{\alpha}^{\infty}\frac{dx'}{(x'-z)}[f(x'+i\varepsilon)-f(x'-i\varepsilon)] \tag{7.48}$$

However we have assumed that $f(z)$ is a *real* analytic function so that

$$\begin{aligned}
f(x'+i\varepsilon)-f(x'-i\varepsilon)&\equiv f(x'+i\varepsilon)-f[(x'+i\varepsilon)^*]\\
&=f(x'+i\varepsilon)-f^*(x'+i\varepsilon)=\text{Re}\,f(x'+i\varepsilon)\\
&\quad+i\,\text{Im}\,f(x'+i\varepsilon)-\text{Re}\,f(x'+i\varepsilon)+i\,\text{Im}\,f(x'+i\varepsilon)\\
&=2i\,\text{Im}\,f(x'+i\varepsilon)
\end{aligned}$$

Therefore the integral of (7.48) becomes

$$2i\int_{\alpha}^{\infty}\frac{dx'\,\text{Im}\,f(x'+i\varepsilon)}{(x'-z)}$$

We can now take the limit as $\varepsilon\to 0$ and obtain the value of $\text{Im}\,f(x')$ as the cut is approached from *above*. We finally obtain from Eqs. 7.46–7.48 the expression

$$f(z)=\sum_{n=1}^{N}\frac{R_n}{(z-z_n)}+\frac{1}{\pi}\int_{\alpha}^{\infty}\frac{\text{Im}\,f(x')\,dx'}{(x'-z)} \tag{7.49}$$

Notice that for all values of z off the cut the denominator $(x'-z)$ is never zero and therefore the integral is well defined. We will treat the limiting case as z approaches the cut when we discuss principal value integrals.

The important fact about this relation is that a knowledge of the residues of the poles (if any at all) and of the imaginary part of the function along the cut *alone*, as well as of the asymptotic behavior, determine the entire function everywhere. It almost appears as though one is getting something for nothing. This should again make clear the stringency of the requirement that a function be analytic. In physical applications expressions of this kind are known as *dispersion relations* and are widely used in theoretical physics today, especially for functions of two (or more) complex variables. (See Section A7.1 for an example of a dispersion-theory problem.) As an interesting point, suppose we have an analytic function that has no poles or cuts and that vanishes at infinity. Then we are told that

$$f(z) \equiv 0$$

Of course this also follows directly from Liouville's theorem (Th. 7.9).

e. Principal-Value Integrals

We have seen before, for example in Section 7.9c, integrals in which a singularity occurs for the integrand *on* the path of integration. We know from real variable theory that if we are considering

$$\int_a^b q(x)\,dx$$

and $q(x)$ "blows up" at, for example, the lower limit, then we define an *improper integral* as

$$\lim_{\delta \to 0} \int_{a+\delta}^b q(x)\,dx$$

if the limit exists. If such a singularity occurs for some $c \ni a < c < b$, then we define the improper integral as

$$\left\{ \lim_{\delta \to 0} \int_a^{c-\delta} q(x)\,dx + \lim_{\delta' \to 0} \int_{c+\delta'}^b q(x)\,dx \right\}$$

again, if this limit exists. Note that δ and δ' are *independent*. It may happen however that neither of these limits exists when $\delta, \delta' \to 0$ independently but that the limit

$$\lim_{\delta \to 0} \left\{ \int_a^{c-\delta} q(x)\,dx + \int_{c+\delta}^b q(x)\,dx \right\} \equiv P \int_a^b q(x)\,dx \tag{7.50}$$

does exist. This is called the Cauchy *principal-value* integral.

Before we proceed to consider these in detail, let us show the difference between a principal-value integral and an ordinary one. If we take

$$\int_{-1}^{1} \frac{dx}{x}$$

we see that it is not well defined. However, if we proceed formally, we obtain

$$\int_{-1}^{1} \frac{dx}{x} = \ln x \Big|_{-1}^{1} = \ln(1) - \ln(-1) = \pi i$$

This result doesn't make much sense since this integral, if it exists, must be real. If we now attempt to define an improper Riemann integral as above, we have

$$\begin{aligned}\int_{-1}^{1} \frac{dx}{x} &= \lim_{\delta\to 0} \ln x \Big|_{-1}^{-\delta} + \lim_{\delta'\to 0} \ln x \Big|_{\delta'}^{+1} \\ &= \lim_{\delta\to 0} \ln(-\delta) - \ln(-1) + \ln(1) - \lim_{\delta'\to 0} \ln(\delta') \\ &= -\pi i + \lim_{\delta\to 0} \ln(-\delta) - \lim_{\delta'\to 0} \ln(\delta')\end{aligned}$$

which does not exist. Therefore the improper integral in the usual Riemann sense does not exist. However if we use Eq. 7.50 we find that

$$P\int_{-1}^{1} \frac{dx}{x} \equiv \lim_{\delta\to 0} \left\{ \int_{-1}^{-\delta} \frac{dx}{x} + \int_{\delta}^{1} \frac{dx}{x} \right\} = \lim_{\delta\to 0} \left\{ -\int_{\delta}^{1} \frac{dx}{x} + \int_{\delta}^{1} \frac{dx}{x} \right\} = 0$$

as one might guess naïvely from the oddness of $1/x$. Of course this oddness argument is *not* a proper proof here.

A very important case for us is a Cauchy integral with z on the contour

$$P\oint \frac{f(z')\,dz'}{(z'-z)}$$

where z is some point *on* the contour of integration. We consider only the case in which $f(z)$ is nonsingular on the contour. From the definition of a principal-value integral we are considering the contour shown in Fig. 7.39. We can add and subtract the well-defined line integral along γ to obtain

$$P\oint \frac{f(z')\,dz'}{(z'-z)} = \int_{C+\gamma} \frac{f(z')\,dz'}{(z'-z)} - \int_{\gamma} \frac{f(z')\,dz'}{(z'-z)}$$

We may now apply Cauchy's integral theorem to the first integral since the limiting process, $\rho \to 0$, is defined as taken *after* the integral has been evaluated. This integral yields simply

$$\int_{C+\gamma} \frac{f(z')\,dz'}{(z'-z)} = 2\pi i f(z)$$

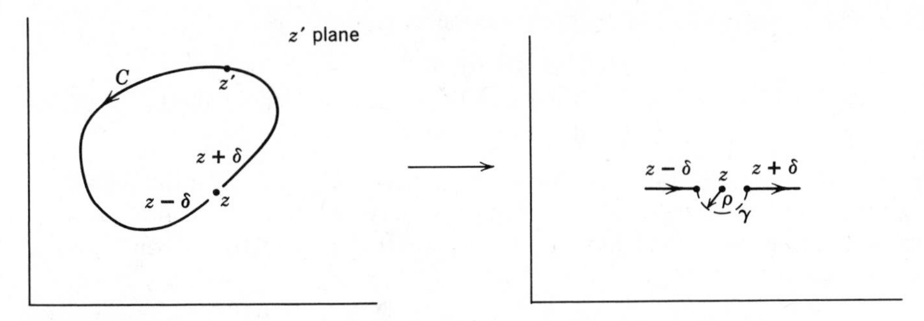

Figure 7.39

The integral along γ can be evaluated as usual by setting

$$(z'-z)=\rho e^{i\varphi} \qquad \frac{dz'}{(z'-z)}=i\,d\varphi$$

and by making a Taylor series expansion of $f(z')$ about $z'=z$ as

$$\lim_{\rho\to 0}\int_{\gamma}\frac{f(z')\,dz'}{z'-z}=\lim_{\rho\to 0} f(z)\int_{\pi}^{2\pi} i\,d\varphi=\pi i f(z)$$

Therefore we obtain

$$P\oint\frac{f(z')\,dz'}{(z'-z)}=2\pi i f(z)-\pi i f(z)=\pi i f(z) \tag{7.51}$$

Notice that the artifice of bulging the contour out below the singularity is equivalent to raising the singularity above the contour and then taking the limit (cf. Fig. 7.40). However *neither* of these is equivalent to a principal-value integral. We get the same result whether we displace the contour over the singularity or displace the singularity below the contour. We may summarize this symbolically as

$$P\int_C\frac{f(z)\,dz}{(z-\alpha)}=\int_C\frac{f(z)\,dz}{(z-\alpha\mp i\varepsilon)}\mp\pi i f(\alpha) \qquad \varepsilon\to 0^+ \tag{7.52}$$

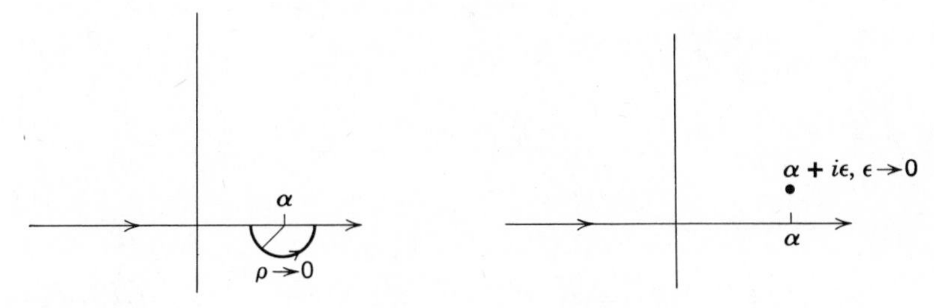

Figure 7.40

This is equivalent to stating Cauchy's integral formula as

$$\oint \frac{f(z)\,dz}{z-\alpha} = 2\pi i f(\alpha) \begin{cases} 1, & \text{if } \alpha \text{ is within } C \\ \frac{1}{2}, & \text{if } \alpha \text{ is on } C \text{ (i.e., the principal value)} \\ 0, & \text{if } \alpha \text{ is outside } C \end{cases} \tag{7.53}$$

We shall now consider the question of the *existence* of principal-value integrals. Let us first recall the definition of a Hölder (or H-) *condition* for a function (cf. Section 5.2). A function $f(z)$ is said to satisfy a Hölder condition for $z \in C$ provided

$$|f(z_2) - f(z_1)| \le A\,|z_2 - z_1|^{\mu} \qquad 0 < \mu \le 1 \tag{7.54}$$

for all $(z_1, z_2) \in C$ where A is the *Hölder constant* and μ the *Hölder index*. The case with $\mu > 1$ is of no interest since in that case

$$\left|\frac{df}{dz}\right| = \lim_{\Delta z \to 0} \left|\frac{f(z+\Delta z) - f(z)}{\Delta z}\right| \le A \lim_{\Delta z \to 0} |\Delta z|^{\mu-1} = 0$$

which states that $f(z)$ is merely a constant. Often different segments of C have different A's and μ's. Furthermore it is easy to show that a *sufficient* condition that $f(z)$ satisfy a Hölder condition on the real axis is that $f'(x)$ exist on the real axis. To see this we use the mean value theorem that states that

$$|f(x_2) - f(x_1)| = |f'(\bar{x})||x_2 - x_1| \le A\,|x_2 - x_1|$$

where we take $A = \max|f'(x)|$, $x_1 \le x \le x_2$, and $\mu = 1$. In fact this is true for all values of z lying on an ordinary curve C since on such a curve, where $x = x(t)$, $y = y(t)$ (cf. Appendix I, Th. I.6),

$$\begin{aligned} |f(z+\Delta z) - f(z)| &= |f[x + \dot{x}\,\Delta t + i(y + \dot{y}\,\Delta t)] - f(x+iy)| \\ &\equiv |f(t+\Delta t) - f(t)| = |f'(\bar{t})\,\Delta t| = |f'(\bar{z})\dot{z}\,\Delta t| \\ &= |f'(\bar{z})||\Delta x + i\,\Delta y| = |f'(\bar{z})||\Delta z| \end{aligned}$$

We also see that a function that is Hölder continuous must also be continuous in the usual sense (cf. the definition of continuity).

We now prove a theorem that is useful in the application of dispersion relations and of principal value integrals.

THEOREM 7.11

If we define a function $F(z)$ as

$$F(z) = P\int_0^{\infty} \frac{g(t)\,dt}{(t-z)} \tag{7.55}$$

then $F(z)$ will exist for all values of z and vanish as $z \to \infty$ provided:

i. $g(t)$ satisfies a Hölder condition on $0 \le t < \infty$.
ii. $g(t)$ is bounded by Ct^{ε}, $\varepsilon > 0$, as $t \to 0$.
iii. $g(t)$ is bounded by C'/t as $t \to \infty$ and $[tg(t)]$ is H-continuous for sufficiently large t.

PROOF

We begin by considering a principal-value integral with finite limits $[0, b]$ and with $0 < z < b$ but z not an end point. Let us write

$$P\int_0^b \frac{g(t)\,dt}{t-z} = P\int_0^b \frac{[g(t)-g(z)]\,dt}{t-z} + g(z)P\int_0^b \frac{dt}{t-z} \tag{7.56}$$

and evaluate the last integral explicitly

$$P\int_0^b \frac{dt}{t-z} \equiv \lim_{\delta\to 0}\left\{-\int_0^{z-\delta}\frac{dt}{(z-t)} + \int_{z+\delta}^{b}\frac{dt}{(t-z)}\right\}$$

$$= \lim_{\delta\to 0}\{\ln(\delta) - \ln(z) + \ln(b-z) - \ln(\delta)\}$$

$$= \ln\left(\frac{b-z}{z}\right)$$

Equation 7.56 then becomes

$$P\int_0^b \frac{g(t)\,dt}{t-z} = \int_0^b \frac{[g(t)-g(z)]\,dt}{t-z} + g(z)\ln\left(\frac{b-z}{z}\right)$$

There is no need for a P-symbol for the integral on the r.h.s. since it is well defined in the ordinary Riemann sense by virtue of the H-condition, (i), on $g(t)$. We can take $0 \le z < b$ on the real axis since, by (ii), as $z \to 0^+$

$$\lim_{z\to 0^+}\left|g(z)\ln\left(\frac{b-z}{z}\right)\right| = \lim_{z\to 0^+}|g(z)\ln z| \le |C| \lim_{z\to 0^+}|z^{\varepsilon}\ln z|$$

$$= C\lim_{z\to 0^+}\frac{|z^{\varepsilon}|}{\varepsilon} = 0$$

The only point remaining is the limit as $z \to \infty$. We handle this by taking $|z| > R$ where R is large enough that property (iii) can be used for $g(t)$. If we write

$$P\int_0^\infty \frac{g(t)\,dt}{t-z} = \int_0^R \frac{g(t)\,dt}{t-z} + P\int_R^\infty \frac{g(t)\,dt}{t-z}$$

then the first integral vanishes as $z \to \infty$ since

$$\left|\int_0^R \frac{g(t)\,dt}{t-z}\right| \le \int_0^R \frac{|g(t)|\,dt}{|t-z|} < \frac{1}{|R-z|}\int_0^R |g(t)|\,dt \xrightarrow[z\to\infty]{} 0$$

If we now change variables as $\zeta = 1/z$, $y = 1/t$ and define $h(y) \equiv (1/y)g(1/y)$, we obtain

$$P\int_R^\infty \frac{g(t)\,dt}{(t-z)} = -\zeta P\int_0^{1/R} \frac{h(y)\,dy}{(y-\zeta)}$$
$$= -\zeta\left\{\int_0^{1/R} \frac{[h(y)-h(\zeta)]}{(y-\zeta)}\,dy + h(\zeta)P\int_0^{1/R} \frac{dy}{(y-\zeta)}\right\}$$
$$= -\zeta\left\{\int_0^{1/R} \frac{[h(y)-h(\zeta)]}{(y-\zeta)}\,dy + h(\zeta)\ln\left(\frac{1/R-\zeta}{\zeta}\right)\right\}$$

where as previously the first integral exists since $h(y)$ satisfies an H-condition by (iii). Also by (iii) we have

$$\lim_{\zeta\to 0}\left|\zeta h(\zeta)\ln\left[\frac{1/R-\zeta}{\zeta}\right]\right| = \lim_{\zeta\to 0}|\zeta||h(\zeta)||\ln \zeta|$$
$$\leq C' \lim_{\zeta\to 0}|\zeta||\ln \zeta| = 0$$

Of course if $z \to \infty$ along any direction other than the positive real axis, then $F(z)$ vanishes even more rapidly. Q.E.D.

It is evident that the real axis $[0, \infty)$ is a branch cut for $F(z)$, Eq. 7.55, since we may use Eq. 7.52 to compute the discontinuity of $F(z)$ as

$$\lim_{\varepsilon\to 0}[F(x+i\varepsilon)-F(x-i\varepsilon)] = 2\pi i g(x) \tag{7.57}$$

In general the end points of a principal value integral are branch points of a function $F(z)$ defined as

$$F(z) = \int_a^b \frac{g(t)\,dt}{(t-z)} \tag{7.58}$$

7.10 SINGULAR INTEGRAL EQUATIONS

In this section we study a class of singular integral equations with Cauchy kernels (cf. Section 5.2) that are important in many applications. The particular form we will study is often referred to as the *Omnès equation* after the author who applied it to a famous problem in theoretical physics (cf. reference to Omnès' paper at end of chapter). This presentation closely parallels Omnès' original work. In contrast to much of the discussion in Chapter 5 on integral equations, the following is, in essence, representation dependent. Consider

$$\phi(x) = f(x) + \frac{1}{\pi} P\int_1^\infty \frac{\sigma(x')\phi(x')\,dx'}{(x'-x)} \qquad x \in [1, \infty) \tag{7.59}$$

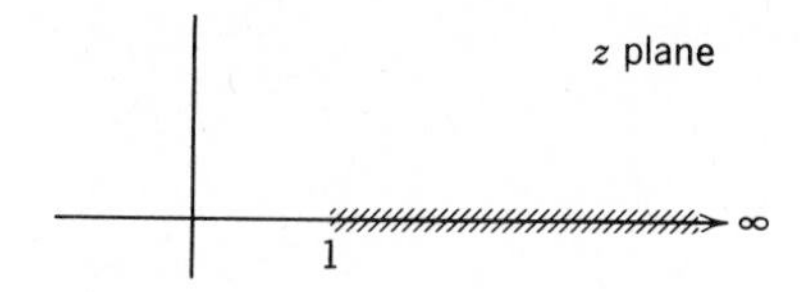

Figure 7.41

where $\sigma(x)$ and $f(x)$ satisfy Hölder conditions and a $\phi(x)$ is sought which is also H-continuous. For simplicity we assume that $\sigma(x) \neq 0$, $x \in [1, \infty)$. That is, $\phi(x)$ is defined on the cut $1 \leq x < +\infty$ in the z plane by this integral equation. We invert this equation as follows. Here $\sigma(x)$ and $f(x)$ are given *real* functions. We define an $F(z)$ holomorphic in the cut z plane as

$$F(z) \equiv \frac{1}{2\pi i} \int_1^\infty \frac{\sigma(x')\phi(x')\,dx'}{(x'-z)} \tag{7.60}$$

From Eq. 7.52 we have

$$F(z^+) - F(z^-) = \sigma(x)\phi(x) \qquad x > +1 \tag{7.61}$$

and

$$F(z^+) + F(z^-) = \frac{1}{\pi i} P \int_1^\infty \frac{\sigma(x')\phi(x')\,dx'}{(x'-x)} \tag{7.62}$$

where

$$z^\pm \equiv \lim_{\varepsilon \to 0^+} (x \pm i\varepsilon)$$

If we substitute these into the original equation, (7.59), we obtain

$$\frac{1}{\sigma(x)}[F(z^+) - F(z^-)] = i[F(z^+) + F(z^-)] + f(x) \tag{7.63}$$

Therefore we can write

$$[1 - i\sigma(x)]F(z^+) - [1 + i\sigma(x)]F(z^-) = f(x)\sigma(x) \tag{7.64}$$

Let us decompose $F(z)$ into a product of two other functions holomorphic in the cut z plane as

$$F(z) \equiv \Phi(z)\Omega(z) \tag{7.65}$$

where $\Omega(z)$ satisfies the homogeneous version of the *Hilbert problem* above, Eq. 7.64; that is,

$$[1 - i\sigma(x)]\Omega(z^+) - [1 + i\sigma(x)]\Omega(z^-) = 0 \tag{7.66}$$

We can easily solve this by taking the logarithm to obtain

$$\ln \Omega(z^+) - \ln \Omega(z^-) = \ln \left[\frac{1 + i\sigma(x)}{1 - i\sigma(x)}\right] \tag{7.67}$$

It is convenient to define

$$\sigma(x) \equiv \tan \delta(x)$$

since then

$$\ln\left[\frac{1+i\sigma(x)}{1-i\sigma(x)}\right] = \ln\left[\frac{1-\sigma^2(x)+2i\sigma(x)}{1+\sigma^2(x)}\right]$$

$$= \ln\left\{\frac{[1-\sigma^2(x)]^2+4\sigma^2(x)}{[1+\sigma^2(x)]^2}\right\} + i\tan^{-1}\left[\frac{2\sigma(x)}{1-\sigma^2(x)}\right]$$

$$= \ln 1 + i\tan^{-1}\left[\frac{2\tan\delta(x)}{1-\tan^2\delta(x)}\right] = i\tan^{-1}[\tan 2\delta(x)] = 2i\,\delta(x)$$

Equation 7.67 becomes

$$\ln \Omega(z^+) - \ln \Omega(z^-) = 2i\,\delta(x)$$

This defines a function $\ln \Omega(z)$ that is holomorphic in the z plane with cut $1 \le x < +\infty$ and with discontinuity $2i\,\delta(x)$ on this cut. It is simple to see that all of these properties are possessed by the function

$$\ln \Omega(z) = \frac{1}{\pi}\int_1^\infty \frac{\delta(x')\,dx'}{(x'-z)}$$

or

$$\Omega(z) = \exp\left\{\frac{1}{\pi}\int_1^\infty \frac{\delta(x')\,dx'}{(x'-z)}\right\} \tag{7.68}$$

where, of course, we have assumed $\delta(x)$ sufficiently well behaved at infinity so that the integral exists (cf. Th. 7.11). If we define

$$\rho(x) \equiv \frac{1}{\pi} P\int_1^\infty \frac{\delta(x')\,dx'}{(x'-x)}$$

then we have

$$\Omega(z^\pm) = \exp[\rho(x) \pm i\,\delta(x)]$$

Now Eq. 7.64 becomes

$$[1-i\sigma(x)]\Phi(z^+)\Omega(z^+) - [1+i\sigma(x)]\Phi(z^-)\Omega(z^-) = f(x)\sigma(x)$$

which implies that (cf. Eq. 7.66)

$$[1+i\sigma(x)]\Omega(z^-)[\Phi(z^+) - \Phi(z^-)] = f(x)\sigma(x)$$

But

$$1+i\sigma(x) = 1+i\tan\delta(x) = \frac{\cos\delta(x)+i\sin\delta(x)}{\cos\delta(x)} = \frac{e^{i\,\delta(x)}}{\cos\delta(x)}$$

so that

$$\Phi(z^+) - \Phi(z^-) = \frac{f(x)\sigma(x)}{[1+i\sigma(x)]\Omega(z^-)} = f(x)e^{-\rho(x)}\sin\delta(x)$$

The same reasoning that led to Eq. 7.68 implies that

$$\Phi(z)=\frac{1}{2\pi i}\int_1^\infty \frac{dx' f(x')e^{-\rho(x')}\sin\delta(x')}{(x'-z)} \tag{7.69}$$

Therefore we obtain

$$F(z)\equiv\Phi(z)\Omega(z)=\frac{1}{2\pi i}\exp\left[\frac{1}{\pi}\int_1^\infty \frac{\delta(x')\,dx'}{(x'-z)}\right]\int_1^\infty \frac{dx'\,f(x')e^{-\rho(x')}\sin\delta(x')}{(x'-z)} \tag{7.70}$$

Finally a solution to Eq. 7.62 is given as

$$\begin{aligned}
\phi(x)&=\frac{1}{\sigma(x)}[F(z^+)-F(z^-)]=\frac{1}{\sigma(x)}[\Omega(z^+)\Phi(z^+)-\Omega(z^-)\Phi(z^-)]\\
&=\frac{\cos\delta(x)}{\sin\delta(x)}\frac{1}{2\pi i}\left\{e^{\rho(x)}e^{i\delta(x)}\left[P\int_1^\infty \frac{dx' f(x')e^{-\rho(x')}\sin\delta(x')}{(x'-x)}\right.\right.\\
&\qquad\left.+\pi i f(x)e^{-\rho(x)}\sin\delta(x)\right]\\
&\qquad\left.-e^{\rho(x)}e^{-i\delta(x)}\left[P\int_1^\infty \frac{dx' f(x')e^{-\rho(x')}\sin\delta(x')}{(x'-x)}-\pi i f(x)e^{-\rho(x)}\sin\delta(x)\right]\right\}\\
&=\frac{\cos\delta(x)}{\sin\delta(x)}\left\{\frac{[e^{i\delta(x)}+e^{-i\delta(x)}]}{2}f(x)\sin\delta(x)\right.\\
&\qquad\left.+\frac{e^{\rho(x)}[e^{i\delta(x)}-e^{-i\delta(x)}]}{2\pi i}P\int_1^\infty \frac{dx' f(x')e^{-\rho(x')}\sin\delta(x')}{(x'-x)}\right\}
\end{aligned}$$

so that

$$\phi(x)=\cos\delta(x)\left\{f(x)\cos\delta(x)+e^{\rho(x)}\frac{1}{\pi}P\int_1^\infty \frac{dx' f(x')e^{-\rho(x')}\sin\delta(x')}{(x'-x)}\right\} \tag{7.71}$$

Of course, as usual, we can add to this particular solution any solution to the homogeneous equation

$$\phi_0(x)=\frac{1}{\pi}P\int_1^\infty \frac{\sigma(x')\phi_0(x')\,dx'}{(x'-x)} \tag{7.72}$$

As in Eq. 7.65 we set

$$F_0(z)\equiv\Phi_0(z)\Omega(z)=\frac{1}{2\pi i}\int_1^\infty \frac{\sigma(x')\phi_0(x')\,dx'}{(x'-z)}$$

where $\Omega(z)$ is still given by Eq. 7.68 so that $\Phi_0(z)$ satisfies

$$\Phi_0(z^+)-\Phi_0(z^-)=0$$

which states that $\Phi_0(z)$ is holomorphic in the entire complex z plane (except possibly at $x=1$ or at $x=\infty$). That is, $\Phi_0(z)$ cannot have poles (or essential singularities) on the open interval $1<x<\infty$ since the integral equation for $\phi_0(x)$

must be well defined. If we exclude essential singularities, then the only solution for $\Phi_0(z)$ is

$$\Phi_0(z)=\frac{P_m(z)}{(z-1)^n}$$

where $P_m(z)$ is a polynomial of degree m in z and n is an integer. The solutions to the homogeneous equation are of the form

$$\phi_0(x)=e^{\rho(x)}e^{i\delta(x)}\frac{P_m(x)}{(x-1)^n} \tag{7.73}$$

Therefore the original mathematical problem as posed in Eq. 7.59 has solutions. but there is no unique solution since the polynomial $P_m(x)$ and the integer n in Eq. 7.73 are arbitrary. If there are subsidiary conditions which $\phi(x)$ must satisfy as $x \to 1^+$ and as $x \to +\infty$, these can be used to remove this arbitrariness.

Finally this treatment may be extended to singular integral equations of the form

$$\phi(x)=f(x)+\frac{1}{\pi}P\int_1^\infty \left[\frac{\sigma(x')}{(x'-x)}+K(x',x)\right]\phi(x')\,dx \tag{7.74}$$

where $K(x, x')$ is a Fredholm kernel. One simply applies the methods of this section to the singular part and obtains a Fredholm equation that can be solved by the techniques of Chapter 5. This will yield a linear Fredholm equation since Eq. 7.71 is linear in $f(x)$. It should be evident that the cut need not be $1 \le x < \infty$ but that any cut (or set of cuts) would do equally well.

7.11 ANALYTIC CONTINUATION

We now come to one of the most important topics in complex variable theory as used today in many physical applications. When we are given an explicit algebraic expression for a function,

$$\sum_{j=0}^{n} g_j(z)[f(z)]^{n-j}=0$$

where the $g_j(z)$ are rational integral functions (i.e., polynomials) of z, so that (by definition) $f(z)$ is an algebraic function of z, then it is a relatively simple matter to find the singularities, asymptotic behavior, and so on, of the function. However we most often deal with functions that are initially represented by a power series or by an integral (as we will see shortly). We have found that a power series fails to converge beyond a certain radius of convergence (which is usually finite). The fact that a power series fails to converge beyond a certain radius leaves completely open the question of what the behavior of the function itself is like beyond this radius.

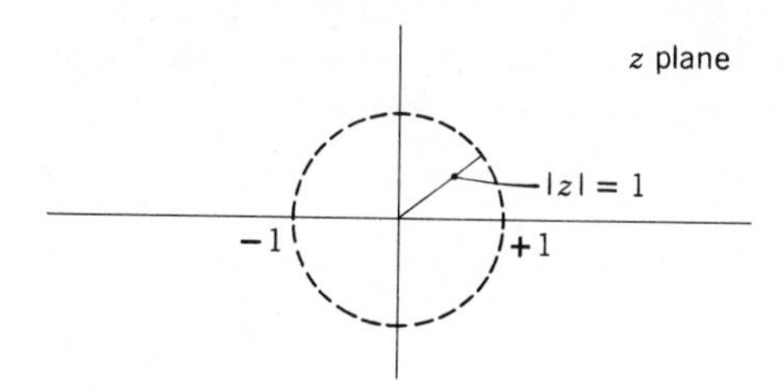

Figure 7.42

Example 7.16 Consider the function

$$f(z) \equiv \frac{1}{1-z} = \sum_{n=0}^{\infty} z^n = -\int_0^{\infty} e^{t(1-z)}\, dt$$

This Taylor series converges only within the circle of radius unity about the origin. On the other hand the integral representation on the right is defined (i.e., converges) only for Re $z > 1$. Each of the representations has a different region of convergence and, in this case, they do *not* overlap. However the actual function itself, that is,

$$\frac{1}{1-z}$$

is defined everywhere throughout the z plane with the exception of the point, $z = +1$, where there is a simple pole. This behavior of an integral representation converging in an infinite region or one side of a line is typical.

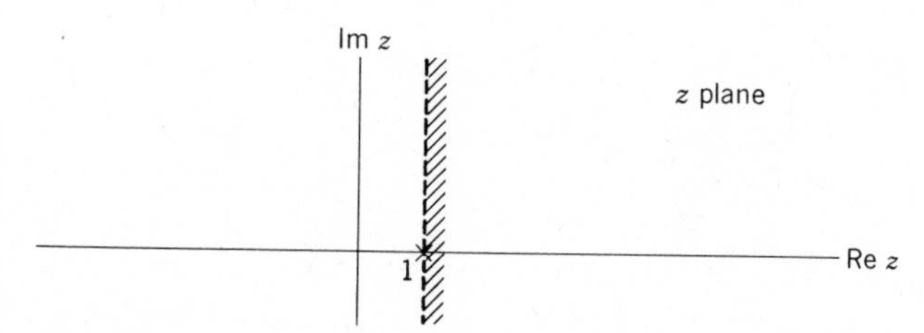

Figure 7.43

a. Power Series

We will find in general that the power series representation of a function is most useful for abstract discussions and for proving theorems but that the integral ones are best suited to applications. We first prove some general theorems about analytic continuation.

THEOREM 7.12

A holomorphic function, $f(z)$, defined in a connected region R whose boundary contains no multiple points, is uniquely determined for all values of z in R by its values along any arbitrarily small arc of an ordinary curve in R.

PROOF

Let α be an end point of an arc in R and β be any other point of R. Let $\mathscr{L}$ be a simple arc lying entirely in R and connecting α and β. We must prove that if we are given $f(z)$ along γ, then we know its value at β. Since $f(z)$ is holomorphic in the neighborhood of $z=\alpha$, so are all of its derivatives (cf., Section 7.3 and Eq. 7.16). Since the values of these derivatives must be independent of the direction of approach of Δz, we may evaluate the $f^{(m)}(\alpha)$ by approaching $z=\alpha$ along γ. Therefore we have the Taylor series for $f(z)$ about $z=\alpha$

$$f(z)\equiv\phi_0(z)=\sum_{n=0}^{\infty}\frac{f^{(n)}(\alpha)}{n!}(z-\alpha)^n$$

where we have used $\phi_0(z)$ to denote that *element* of $f(z)$ represented by the series within its radius of convergence. Since $f(z)$ is holomorphic in R, $\phi_0(z)$ will represent $f(z)$ within any circle C_0 lying wholly within R and centered at $z=\alpha$. Let this largest possible circle be C_0 as in Fig. 7.44. If β lies within C_0 then we are

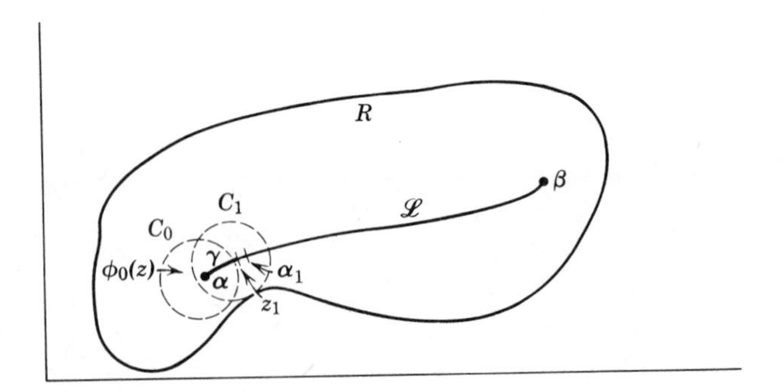

Figure 7.44

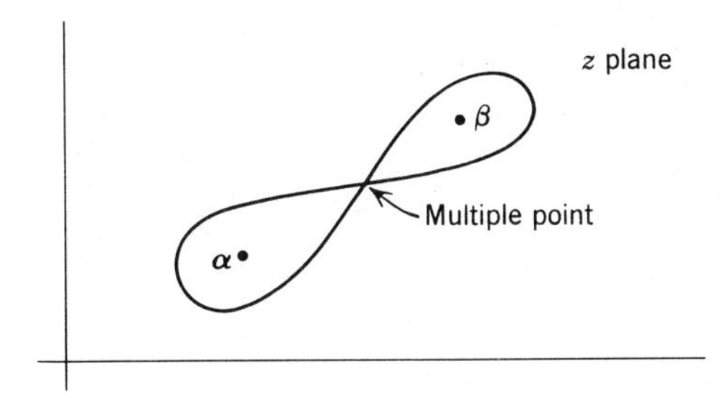

Figure 7.45

finished since, in such a case, $f(\beta)\equiv\phi_0(\beta)$. If β does not lie within C_0, then let α_1 denote the point of intersection of C_0 and $\mathscr{L}$. Choose z_1 as an interior point of C_0 but close to α_1. By means of the element $\phi_0(z)$ we may evaluate all the $f^{(n)}(z_1)$ and so obtain a Taylor series about $z=z_1$

$$\phi_1(z)=\sum_{n=0}^{\infty}\frac{f^{(n)}(z_1)}{n!}(z-z_1)^n\equiv\sum_{n=0}^{\infty}\frac{\phi_0^{(n)}(z_1)}{n!}(z-z_1)^n$$

We may continue this process n times until $z=\beta$ finally falls within some circle C_n. We will then have obtained the value of $f(z)$ at $z=\beta$ as

$$f(\beta)=\phi_n(\beta)$$

from a knowledge of $f(z)$ along the arc, γ, *alone.* It is important that the boundary curve of R contain no multiple points since, if it did, then we might not be able to get from α to β and still remain wholly within R. Consider, for example, Fig. 7.45. This is the reason we have inserted the terms *connected region* and *no multiple points* (i.e., a simple curve) in the theorem. Q.E.D.

The following theorem is a direct consequence of this theorem.

THEOREM 7.13

If two functions are holomorphic in a region R and are equal for all values of z in the neighborhood of a point $z=\alpha$ of R, or for all values of z along an arbitrarily small arc through α, then the two functions are equal for all values of z in R.

PROOF

This follows simply by defining a $g(z)$ such that

$$g(z)\equiv f_1(z)-f_2(z)=0$$

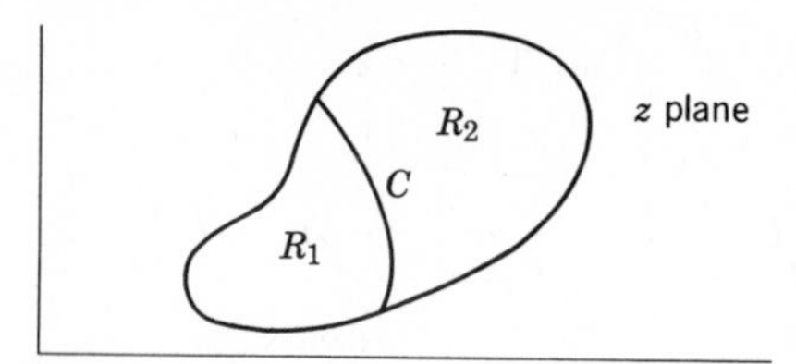

Figure 7.46

for z in the neighborhood of α, from which we see that all the coefficients of the Taylor expansion of $g(z)$ about $z=\alpha$, that is, $g^{(n)}(\alpha)/n!$, are identically zero. Therefore all the subsequent expansions about $z_1, z_2, \ldots, z_n$ will also be identically zero so that $g(z)\equiv 0$ *throughout* R. Q.E.D.

In the above theorems we have simply assumed that a function is holomorphic throughout a region. We have not said anything about the maximum possible size of this region. In practice we are interested in taking a region and extending this domain of holomorphy. We will see that an analytic function is completely and uniquely determined once it is known for any region, however small.

Before we actually prove one of the central theorems on analytic continuation, let us consider the general nature of what is to be done. Suppose we have two neighboring regions, R_1 and R_2, with a common boundary, C, as shown in Fig. 7.46. Let $\phi_1(z)$ be defined in R_1 and be holomorphic there. Suppose $\phi_2(z)$ is holomorphic in R_2 and that $\phi_1(z)$ and $\phi_2(z)$ are continuous along C and equal to each other here. Then $\phi_1(z)$ and $\phi_2(z)$, taken together, define a function $f(z)$ in the region $R=R_1+R_2+C$, and this function is holomorphic in R (cf. Th. 7.14). The function $\phi_2(z)$ is called the *analytic continuation* of $\phi_1(z)$ and the process of obtaining $\phi_2(z)$ from $\phi_1(z)$ is called analytic continuation. It follows from Th. 7.13 that the $f(z)$ so determined is *unique*. Suppose that a $\Phi_1(z)$ could also be found such that it had a value other than $\phi_2(z)$ in R_2. Then we should have another holomorphic function, $F(z)$, defined in R_1+R_2+C, as well as the holomorphic function, $f(z)$. However both $f(z)$ and $F(z)$ have identical values in R_1 so that they are identical throughout R_1+R_2+C, by Th. 7.13. We call functions $\phi_1(z)$ and $\phi_2(z)$ *elements* of the function $f(z)$.

The following theorem gives the necessary and sufficient condition for one function to be the analytic continuation of the other.

THEOREM 7.14

If $\phi_1(z)$ is holomorphic in a region R_1 and $\phi_2(z)$ is holomorphic in R_2 and if R_1 and R_2 have an ordinary curve C as part of their common boundary, then $\phi_1(z)$ and $\phi_2(z)$ are analytic continuations of each other if and only if they converge uniformly to common values along C.

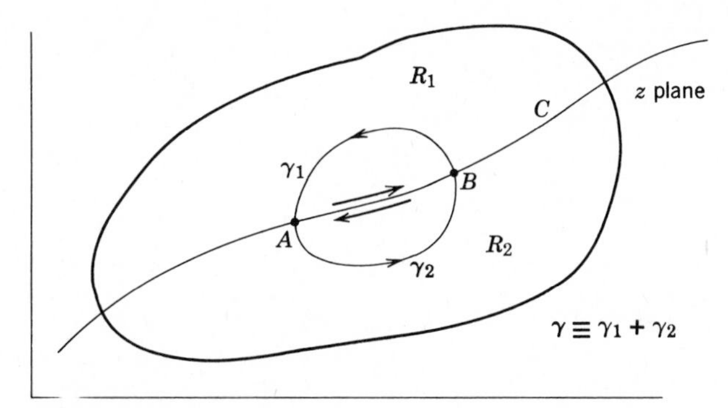

Figure 7.47

PROOF

That the condition is necessary follows from the definition of analytic continuation. We need only show it to be sufficient. Define $f(z)$ as:

$$f(z) = \begin{cases} \phi_1(z) \text{ in } R_1 \\ \phi_2(z) \text{ in } R_2 \\ \phi_1(z) = \phi_2(z) \text{ along } C \end{cases}$$

We must show that $f(z)$ is holomorphic in region $R = R_1 + R_2 + C$. Let γ_1 and γ_2 be any two ordinary curves joining two points A and B of C and lying wholly within R_1 and R_2, respectively (cf. Fig. 7.47). By the Cauchy-Goursat theorem (Th. 7.2) we have

$$\int_{\gamma_1} \phi_1(z)\,dz + \int_{AB} \phi_1(z)\,dz = 0 \qquad \int_{\gamma_2} \phi_2(z)\,dz + \int_{BA} \phi_2(z)\,dz = 0$$

Since $\phi_1(z) = \phi_2(z)$ along C we have

$$\int_{\gamma} f(z)\,dz \equiv \int_{\gamma_1} \phi_1(z)\,dz + \int_{\gamma_2} \phi_2(z)\,dz = 0$$

for any γ lying wholly within $R_1 + R_2$. As we have seen in Morera's theorem (Th. 7.3), this implies that $f(z)$ is holomorphic everywhere in $R = R_1 + R_2 + C$. From the proof of Th. 7.12 we see that $\phi_1(z)$ and $\phi_2(z)$ are uniquely determined from one another. Q.E.D.

We have already seen in our proof of Th. 7.12 the general argument by which we can take an element, for example, $\phi_0(z)$, and extend it so as to construct the entire holomorphic $f(z)$ (cf. Fig. 7.48). We can cover the entire region of holomorphy of $f(z)$ if we are given an element $\phi_0(z)$ in C_0

$$\phi_0(z) = \sum_{n=0}^{\infty} \alpha_n (z - z_0)^n$$

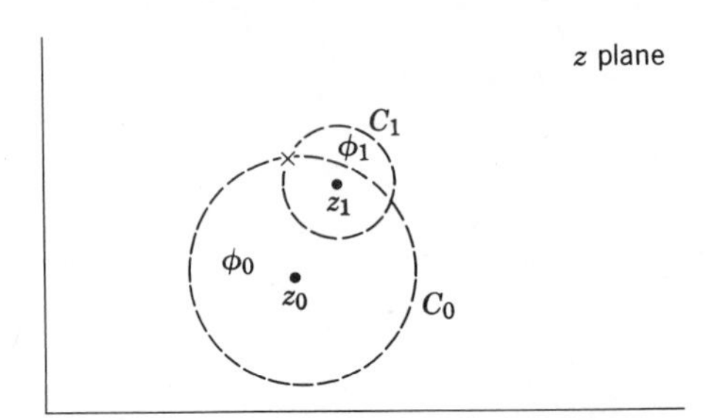

Figure 7.48

simply by constructing elements $\{\phi_j(z)\}$ and extending these circular domains $\{C_j\}$.

One should not conclude, however, that we can always extend the domain of holomorphy to the entire z plane. The "function" has certain intrinsic analytic properties that cannot be changed. First, if we attempt to extend the domain beyond a pole singularity, then we obtain the same value of the function independent of the path taken around the pole (cf. Fig. 7.49). This is simply because the function is holomorphic in the union of all these regions if the neighborhood of the pole is deleted. This uniqueness of the analytic continuation then follows from the theorems above.

Now if a function is continued along two different routes from z_0 to z_1 and two different values are obtained for the function at z_1, then the function must have a singularity (of *some* type) between the two routes. If there were no such singularity between the two routes, then we could certainly fill the region between them with convergent power series expansions and obtain enough overlap to apply the uniqueness theorem (Th. 7.12).

We have *not* stated that, if there is a singularity between these paths, then the two continuations *cannot* have the same value. The particular type of singularity

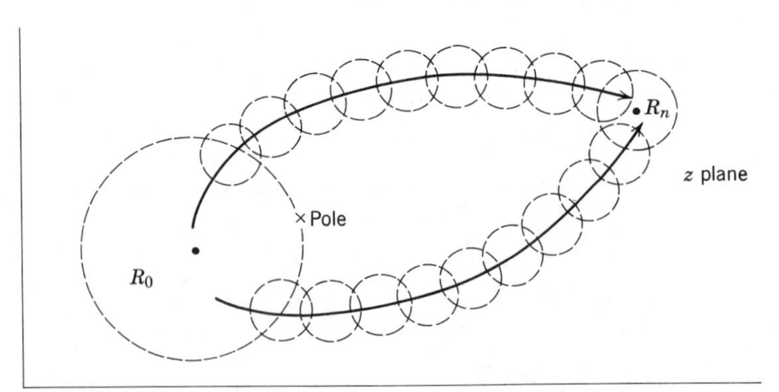

Figure 7.49

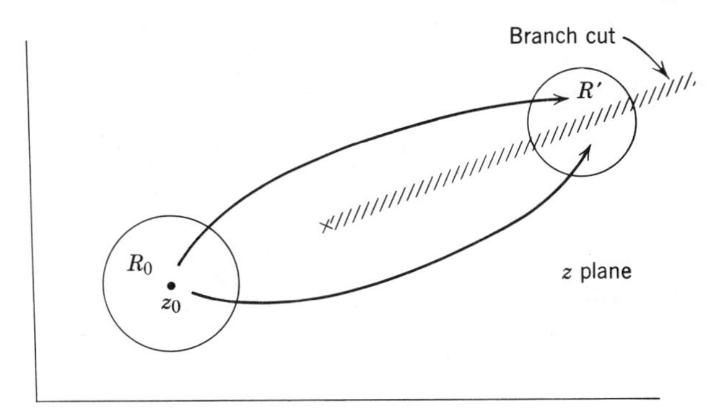

Figure 7.50

that can produce such nonunique continuations is a *branch point* that we have met before. The values of the function obtained for the two parts of R' (cf. Fig. 7.50) will differ since they are on opposite sides of the branch cut. However we still term this an *analytic* function since the values on the various branches can be obtained from a knowledge of a single element, $\phi_0(z)$, on one branch. This is a multiple-valued analytic function.

It can also happen that we encounter a closed curve of singularities beyond which the analytic continuation cannot be carried. Such a curve of singularities is known as a *natural boundary* of the domain of analyticity of the function.

Example 7.17 The function

$$f(z)=\sum_{n=0}^{\infty} z^{n!}$$

has a natural boundary. The series clearly converges for $|z|<1$ and is, therefore, a representation of an analytic [in fact, holomorphic (cf. Th. 7.6)] function there. Let

$$z=re^{2\pi ip/q} \qquad \phi=2\pi p/q$$

where p and q are integers and consider the behavior of $f(z)$ as $r\rightarrow 1$ through real values.

$$f(z)=\sum_{n=0}^{q-1} z^{n!}+\sum_{n=q}^{\infty} z^{n!}\equiv f_1(z)+f_2(z)$$

Since $f_1(z)$ is simply a polynomial in z, it tends to a finite limit as $r\rightarrow 1$. When $n\geq q$, q is a divisor of $n!$ so that

$$z^{n!}\equiv r^{n!} \qquad n\geq q$$

and

$$f_2(z)=\sum_{n=q}^{\infty} r^{n!}$$

This tends to infinity as $r \to 1$ and $f(z) \to \infty$ so that $z=e^{2\pi ip/q}$ is a singularity of $f(z)$. But points of this kind are dense everywhere around the unit circle so that there is no arc, however small, on which $f(z)$ is regular on the unit circle. Therefore it is impossible to continue $f(z)$ across the unit circle which constitutes a natural boundary for $f(z)$.

---

b. Poisson Integral Formula

We now obtain an integral representation that will allow us to continue a function analytically.

THEOREM 7.15 (Poisson integral formula)

If $f(z)=u(x, y)+iv(x, y)$ is holomorphic within a simply connected region that contains a circle of radius R in its interior, and if $u(x, y)$ is given everywhere on this circle, then $f(z)$ is given inside the circle as

$$f(z_0)=\frac{1}{2\pi}\int_0^{2\pi}\left(\frac{z+z_0}{z-z_0}\right)u(z)\,d\varphi+iv(0) \tag{7.75}$$

where

$$z_0=r_0e^{i\varphi_0} \qquad z=Re^{i\varphi}$$

PROOF

We may apply Cauchy's integral formula as (cf. Fig. 7.51)

$$f(z_0)=\frac{1}{2\pi i}\int_{C_R}\frac{f(z)\,dz}{(z-z_0)} \tag{7.76}$$

If we take a point z_1 exterior to C_R defined by

$$z_1z_0^*=R^2 \qquad |z_1|=\left(\frac{R}{r_0}\right)R>R$$

then the integrand is everywhere holomorphic within C_R and

$$\frac{1}{2\pi i}\int_{C_R}\frac{f(z)\,dz}{(z-z_1)}=0 \tag{7.77}$$

Let $z=Re^{i\varphi}$ so that Eq. 7.76 becomes

$$f(z_0)=\frac{1}{2\pi}\int_0^{2\pi}\frac{z(u+iv)\,d\varphi}{(z-z_0)} \tag{7.78}$$

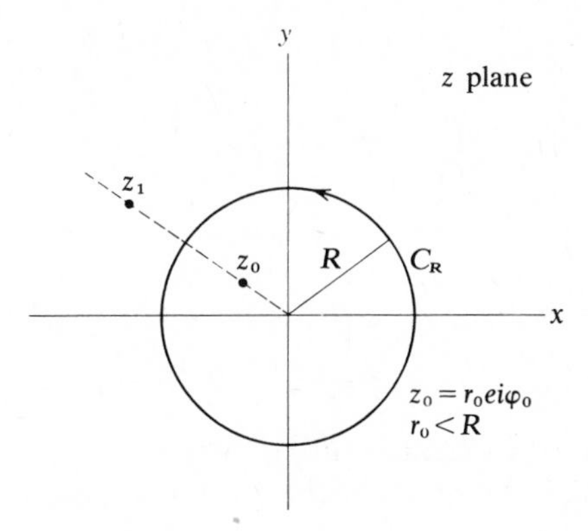

Figure 7.51

Also Eq. 7.77 can be written as

$$0=\frac{1}{2\pi}\int_0^{2\pi}\frac{z^*(u-iv)\,d\varphi}{(z^*-z_1^*)}=\frac{1}{2\pi}\int_0^{2\pi}\frac{R^2(u-iv)\,d\varphi}{z\left(\frac{R^2}{z}-\frac{R^2}{z_0}\right)} \tag{7.79}$$

$$=\frac{z_0}{2\pi}\int_0^{2\pi}\frac{(u-iv)\,d\varphi}{(z-z_0)}$$

If we add Eqs. 7.78 and 7.79, we have

$$f(z_0)=\frac{1}{2\pi}\int_0^{2\pi}\left(\frac{z+z_0}{z-z_0}\right)u\,d\varphi+\frac{i}{2\pi}\int_0^{2\pi}v\,d\varphi\equiv u+iv \tag{7.80}$$

or

$$u(z_0)\equiv u(x_0, z_0)=\text{Re } f(z_0)=\frac{1}{2\pi}\int_0^{2\pi}\text{Re}\left(\frac{z+z_0}{z-z_0}\right)u\,d\varphi \tag{7.81}$$

Since

$$\frac{z+z_0}{z-z_0}=\frac{(z+z_0)(z^*-z_0^*)}{|z-z_0|^2}=\frac{R^2-r_0^2+2ir_0R\sin(\varphi_0-\varphi)}{R^2+r_0^2-2Rr_0\cos(\varphi_0-\varphi)}$$

then Eq. 7.81 becomes

$$u(r_0e^{i\varphi_0})=\frac{1}{2\pi}\int_0^{2\pi}\frac{(R^2-r_0^2)v(Re^{i\varphi})\,d\varphi}{R^2+r_0^2-2Rr_0\cos(\varphi_0-\varphi)} \tag{7.82}$$

However $-if(z)\equiv-i(u+iv)=v-iu$ is also a holomorphic function of z with real part $v(x, y)$. Therefore immediately from Eq. 7.82 it follows that

$$v(z_0)=\frac{1}{2\pi}\int_0^{2\pi}\frac{(R^2-r_0^2)}{R^2+r_0^2-2Rr_0\cos(\varphi_0-\varphi)}v(Re^{i\varphi})\,d\varphi \tag{7.83}$$

For $z_0=0$ Eq. 7.83 reduces to

$$v(0)=\frac{1}{2\pi}\int_0^{2\pi} v(Re^{i\varphi})\, d\varphi$$

so that Eq. 7.80 yields the desired result

$$f(z_0)=\frac{1}{2\pi}\int_0^{2\pi}\left(\frac{z+z_0}{z-z_0}\right)u\, d\varphi+iv(0) \qquad \text{Q.E.D.}$$

Therefore, if we are given a real function, for example, $h(\varphi)$, defined on the circle, $|z|=R$, then we can construct a function $f(z_0)$ holomorphic in the interior of C_R as

$$f(z_0)=\frac{1}{2\pi}\int_0^{2\pi}\left(\frac{Re^{i\varphi}+r_0e^{i\varphi_0}}{Re^{i\varphi}-r_0e^{i\varphi_0}}\right)h(\varphi)\, d\varphi \qquad \textbf{(7.84)}$$

Furthermore, up to an additive purely imaginary constant, this is the unique holomorphic extension of $h(\varphi)$ to the interior of C_R.

In general, given the Cauchy-Riemann conditions, Eq. 7.7, and $u(x, y)$, we can construct the $v(x, y)$ as

$$v(x, y)=\int_{z_0}^{z}\left[-\frac{\partial u(x', y')}{\partial y'}\, dx'+\frac{\partial u(x', y')}{\partial x'}\, dy'\right] \qquad \textbf{(7.85)}$$

since

$$dv\equiv\frac{\partial v}{\partial x}\, dx+\frac{\partial v}{\partial y}\, dy=-\frac{\partial u}{\partial y}\, dx+\frac{\partial u}{\partial x}\, dy$$

The arbitrary additive constant is accounted for in Eq. 7.85 by the lower limit, z_0, in the integral. If the domain is *simply connected*, then the Cauchy-Riemann conditions imply that this integral is independent of path. In this case the $f(z)$ constructed as $f=u+iv$ from $u(x, y)$ and the $v(x, y)$ of Eq. 7.85 will be single valued.

c. Dirichlet Problem and Conformal Mapping

The Poisson integral formula furnishes the solution to a particular case of the *Dirichlet problem*. Before we state the general form of this famous problem, we define a *harmonic function*. A function $u(x, y)$ is said to be harmonic in a region R provided the partial derivatives

$$\frac{\partial u}{\partial x} \quad \frac{\partial u}{\partial y} \quad \frac{\partial^2 u}{\partial x^2} \quad \frac{\partial^2 u}{\partial y^2}$$

exist and are continuous and satisfy

$$\nabla^2 u \equiv \frac{\partial^2 u}{\partial x^2} + \frac{\partial^2 u}{\partial y^2} = 0 \tag{7.86}$$

everywhere in R. Equation 7.86 is *Laplace's equation* in two dimensions. From Th. 7.1 it is clear that both the real and imaginary parts of a holomorphic function satisfy Eq. 7.86. That is, if we apply Th. 7.1 to $f(z)$ and to $f'(z)$, then we know that $\partial^2 u/\partial x\,\partial y$ and $\partial^2 v/\partial x\,\partial y$ both exist so that by use of the Cauchy-Riemann conditions we see that

$$\frac{\partial^2 v}{\partial x\,\partial y} = \frac{\partial^2 v}{\partial y\,\partial x}$$

or that

$$\frac{\partial^2 u}{\partial x^2} = -\frac{\partial^2 u}{\partial y^2}$$

and similarly that

$$\frac{\partial^2 u}{\partial x\,\partial y} = \frac{\partial^2 u}{\partial y\,\partial x} \tag{7.87}$$

or

$$-\frac{\partial^2 v}{\partial x^2} = \frac{\partial^2 v}{\partial y^2}$$

In its simplest form the Dirichlet problem for a plane two-dimensional region R is the following. We are given that a real function $u(x, y)$ is harmonic in a simply-connected region R whose boundary is an ordinary closed curve C. We are also given the values $u_C(x, y)$ on this boundary curve C. The problem is then from $u_C(x, y)$ alone to determine $u(x, y)$ everywhere in R.

It is a simple matter to see that *if* a solution to this problem exists, then it is unique. If we take *Stokes' theorem* from elementary vector analysis

$$\oint \mathbf{A}\cdot d\mathbf{l} = \int_\Sigma (\nabla\times\mathbf{A})\cdot d\mathbf{S}$$

and let $\mathbf{A}=\mathbf{F}\times\mathbf{C}$, where $\mathbf{C}$ is some arbitrary constant vector, then we obtain with the aid of some vector identities

$$\oint \mathbf{C}\cdot d\mathbf{l}\times\mathbf{F} = \int_\Sigma \mathbf{C}\cdot(d\mathbf{S}\times\nabla)\times\mathbf{F}$$

Since $\mathbf{C}$ is arbitrary we deduce that

$$\oint d\mathbf{l}\times\mathbf{F} = \int_\Sigma (d\mathbf{S}\times\nabla)\times\mathbf{F}$$

If we now choose $\mathbf{F}=\frac{1}{2}\nabla(u^2)$ where $u(x, y)$ is harmonic and restrict ourselves to

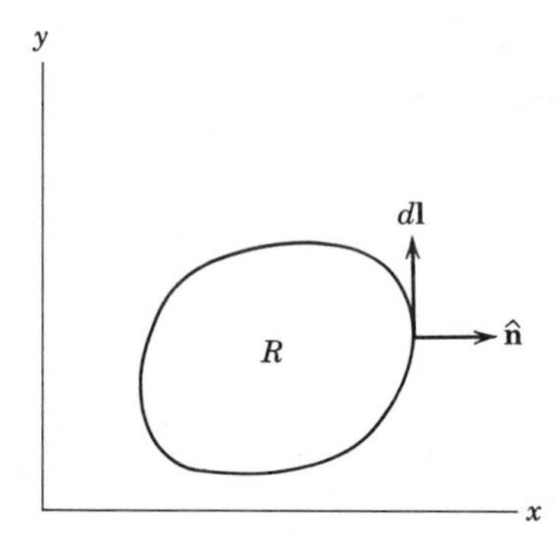

Figure 7.52

the x-y plane, we obtain (cf. Fig. 7.52)

$$\begin{aligned}\frac{1}{2}\oint d\mathbf{l}\times\boldsymbol{\nabla}(u^2) &= \oint d\mathbf{l}\times(u\,\boldsymbol{\nabla}u) = -\hat{\mathbf{k}}\oint u_C\frac{du}{dn}\,dl \\ &= \int_\Sigma (\hat{\mathbf{k}}\times\boldsymbol{\nabla})\times[\tfrac{1}{2}\boldsymbol{\nabla}(u^2)]\,dx\,dy \\ &= \int_\Sigma [(\hat{\mathbf{k}}\cdot\boldsymbol{\nabla})\boldsymbol{\nabla}-(\boldsymbol{\nabla}\cdot\boldsymbol{\nabla})\hat{\mathbf{k}}](\tfrac{1}{2}u^2)\,dx\,dy \\ &= -\hat{\mathbf{k}}\int_\Sigma \boldsymbol{\nabla}\cdot(u\,\boldsymbol{\nabla}u)\,dx\,dy \\ &= -\hat{\mathbf{k}}\int_\Sigma [(\boldsymbol{\nabla}u\cdot\boldsymbol{\nabla}u)+u\boldsymbol{\nabla}^2u]\,dx\,dy\end{aligned}$$

so that

$$\int_\Sigma \left[\left(\frac{\partial u}{\partial x}\right)^2+\left(\frac{\partial u}{\partial y}\right)^2\right]dx\,dy=\oint u_C\frac{du}{dn}\,dl \tag{7.88}$$

Since Laplace's equation, (7.86), is linear, if there were two different solutions to the Dirichlet problem having the same value on C, then their difference would be a solution having $u_C=0$. From Eq. 7.88 we could then conclude that $u(x, y)=\text{const.}=0$ everywhere in R and on C. This same conclusion can also be gotten a bit more elegantly from the fact that a nonconstant harmonic function cannot have a maximum or a minimum at a point interior to its region of definition (cf. Th. 7.7 and the discussion following it).

We now outline a beautiful solution to the Dirichlet problem in two dimensions by means of *conformal mapping*. If we take a function $w=f(z)$, holomorphic in some region, we may consider this as a conformal mapping of the z plane onto the w plane

$$w\equiv u(x, y)+iv(x, y) \tag{7.89}$$

Furthermore, we know from advanced calculus (cf. Th. I.17) that this transformation will be invertible, that is, we can solve for x and y as

$$x = X(u, v)$$
$$y = Y(u, v)$$

provided the *Jacobian of transformation*,

$$\frac{\partial(u, v)}{\partial(x, y)} \equiv \begin{vmatrix} \dfrac{\partial u}{\partial x} & \dfrac{\partial u}{\partial y} \\ \dfrac{\partial v}{\partial x} & \dfrac{\partial v}{\partial y} \end{vmatrix} = \left(\frac{\partial u}{\partial x}\right)^2 + \left(\frac{\partial v}{\partial x}\right)^2 = |f'(z)|^2 \tag{7.90}$$

does not vanish. However Eq. 7.90 shows that $\partial(u, v)/\partial(x, y)$ can vanish if and only if $f'(z) = 0$. Therefore, as long as $f'(z)$ does not vanish, we can map reversibly from the z plane to the w plane. It is also straightforward to show that a holomorphic function of a holomorphic function is itself holomorphic. If $g(w)$ is a holomorphic function of w and if $w = f(z)$ is a holomorphic function of z, then

$$\frac{d}{dz} g[w(z)] = \frac{dg}{dw}\frac{dw}{dz}$$

so that dg/dz exists and is single valued.

One approach to the solution of the Dirichlet problem is now fairly evident. If we are given a region R bounded by a curve C in the z plane, and if we can find a conformal mapping $w = f(z)$ that will take C into a circle and R into its interior in the w plane, then we can use the Poisson integral formula, Eq. 7.84, to solve the problem in the w plane. We then need only invert the transformation from the w to the z plane to obtain the solution to the original problem. In fact the *Riemann mapping theorem*, which we will not attempt to prove, guarantees that a simply-connected domain can be mapped conformally onto the interior of the unit circle (cf. Nehari). We will not say much more about the extensive and beautiful subject of conformal mapping although we will treat one case in some detail when we discuss the convergence of Tchebichef polynomial expansions in the complex plane in Section A7.5.

We now give a more complete treatment of the Dirichlet problem in two dimensions using the techniques that we developed for Fredholm integral equations in Chapter 5. As we have seen from the previous discussion of this section, solving the Dirichlet problem is equivalent to finding a holomorphic function $f(z) = u + iv$ whose real part reduces to the prescribed values, $u_C(x, y)$, on the boundary curve C. Let us now demonstrate that the required $f(z)$ can be written as

$$f(z) = \frac{1}{2\pi i} \int_C \frac{\mu(z')\, dz'}{z' - z} \tag{7.91}$$

where we are taking C to bound a finite region R and $\mu(z)$ is assumed real. Let $z \to z_0$, a point on C, and use Eq. 7.52 to obtain

$$f(z_0) = \tfrac{1}{2}\mu(z_0) + \frac{1}{2\pi i} P \int_C \frac{\mu(z')\,dz'}{(z'-z_0)} \tag{7.92}$$

Since, by definition, $\text{Re}\,[f(z_0)] = u(z_0)$, we obtain from Eq. 7.92

$$\mu(z_0) + \frac{1}{\pi} P \int_C \mu(z')\,\text{Im}\left(\frac{dz'}{z'-z_0}\right) = 2u(z_0) \tag{7.93}$$

If, as usual, we set $(z'-z_0) = re^{i\theta}$, then we have

$$\text{Im}\left(\frac{dz'}{z'-z_0}\right) = \text{Im}\,[d \ln (z'-z_0)] = d\theta$$

Suppose we now begin at some arbitrary origin on C and measure the arc length, s, along C from this origin so that as s goes from 0 to l we traverse C once. That is, we are going to assume that on C we can write $x = x(s)$, $y = y(s)$ and that C has a tangent everywhere (i.e., dx/ds and dy/ds exist) and *finite curvature* κ

$$\kappa(s) \equiv \frac{d^2y/dx^2}{[1+(dy/dx)^2]^{3/2}} = \frac{\dot{x}\ddot{y} - \dot{y}\ddot{x}}{(\dot{x}^2 + \dot{y}^2)^{3/2}} \tag{7.94}$$

As indicated in Fig. 7.53 we set

$$z' - z_0 = [x(s) - x(s_0)] + i[y(s) - y(s_0)] = r(s, s_0)e^{i\theta(s, s_0)}$$

so that

$$\theta = \tan^{-1}\left[\frac{y(s) - y(s_0)}{x(s) - x(s_0)}\right]$$

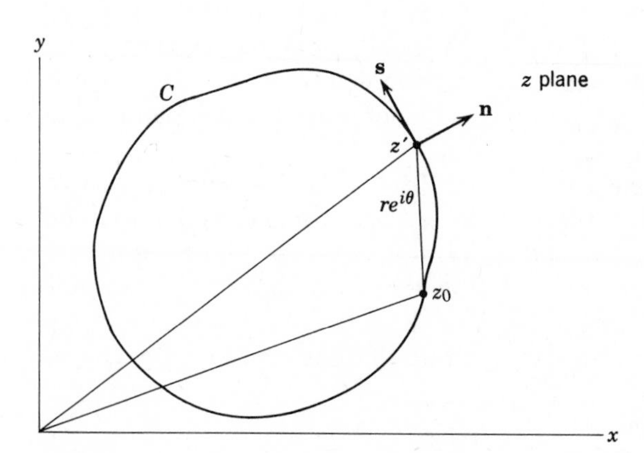

Figure 7.53

and

$$d\theta = \frac{\{[x(s)-x(s_0)]\dot{y}(s)-[y(s)-y(s_0)]\dot{x}(s)\}}{[x(s)-x(s_0)]^2+[y(s)-y(s_0)]^2}\,ds$$
$$\equiv K(s_0, s)\,ds \tag{7.95}$$

From Eqs. 7.94 and 7.95 we easily see that

$$\lim_{s\to s_0} K(s_0, s) = \lim_{s\to s_0}\left\{\frac{\frac{1}{2}(s-s_0)^2[\ddot{x}(s_0)\dot{y}(s_0)-\ddot{y}(s_0)\dot{x}(s_0)]}{(\dot{x}^2+\dot{y}^2)(s-s_0)^2}\right\}$$
$$= -\tfrac{1}{2}\kappa(s_0)[\dot{x}^2(s_0)+\dot{y}^2(s_0)]^{1/2} \tag{7.96}$$

This establishes that $K(s, s_0)$ is everywhere finite (under our assumptions on the boundary curve C) so that we do not need the P symbol in Eq. 7.93, which now becomes the Fredholm equation

$$\mu(s_0)+\frac{1}{\pi}\int_0^l \mu(s)K(s_0, s)\,ds = 2u(s_0) \tag{7.97}$$

If we can establish the existence of a solution to Eq. 7.97, we will have demonstrated that the Dirichlet problem has a solution under the assumptions we have made about C. The existence of such a solution will follow from the Fredholm alternative (Th. 5.6) if we can show that the homogeneous equation

$$\mu_0(s_0)+\frac{1}{\pi}\int_0^l \mu_0(s)K(s_0, s)\,ds = 0 \tag{7.98}$$

has only the trivial solution. If, as in Eq. 7.91, we define

$$f_0(z) = \frac{1}{2\pi i}\int_C \frac{\mu_0(z')\,dz'}{(z'-z)} \tag{7.99}$$

then comparison with Eq. 7.97 shows that Re $f_0(z_0)\equiv 0$ everywhere on C. But the uniqueness argument we gave following Eq. 7.88 implies that Re $f_0(z)\equiv 0$ everywhere within C also so that, by the Cauchy-Riemann equations, $f_0(z)$ can be at most an imaginary constant $f_0(z)=i\alpha$, which means that $f_0(z)$ has an analytic extension beyond C. It is also clear that the function, $g_0(z)$, defined for z *outside* C

$$g_0(z) \equiv \frac{1}{2\pi i}\int_C \frac{\mu_0(z')\,dz'}{(z'-z)} \tag{7.100}$$

is a function holomorphic exterior to C and vanishing at infinity. Notice that, in general, $f_0(z)\not\equiv g_0(z)$ since their regions of definition are separated by a closed curve C, which may be a branch line. Again, using Eq. 7.52 and letting $z \to z_0 \in C$ from the *outside*, we obtain

$$g_0(z_0) = \frac{1}{2\pi i}P\int_C \frac{\mu_0(z')\,dz'}{z'-z_0} - \frac{1}{2}\mu_0(z_0) \tag{7.101}$$

while from Eq. 7.99 and the fact that $f(z)=i\alpha$ within and on C

$$i\alpha=\frac{1}{2\pi i}P\int_C\frac{\mu_0(z')\,dz'}{z'-z_0}+\frac{1}{2}\mu_0(z_0) \qquad \textbf{(7.102)}$$

Equations 7.101 and 7.102 together imply that

$$g_0(z_0)-i\alpha=-\mu_0(z_0)$$
$$\text{Im } g_0(z_0)=\alpha$$
$$\text{Re } g_0(z_0)=-\mu_0(z_0)$$

If we now take z *exterior* to C, then from Cauchy's theorem we have

$$\frac{1}{2\pi i}\int_C\frac{\alpha\,dz'}{(z'-z)}=0$$

so that

$$g_0(z)=\frac{1}{2\pi i}\int_C\frac{[\mu_0(z')-\alpha]}{(z'-z)}\,dz'$$

If we let $z\to z_0$ from the exterior of C, we find that

$$g_0(z_0)=\frac{1}{2\pi i}P\int_C\frac{[\mu_0(z')-\alpha]\,dz'}{(z'-z_0)}-\tfrac{1}{2}[\mu_0(z_0)-\alpha]$$

or

$$\text{Re } g_0(z_0)=\frac{1}{2\pi}P\int_C[\mu_0(z')-\alpha]\,\text{Im}\left(\frac{dz'}{z'-z_0}\right)-\tfrac{1}{2}[\mu_0(z_0)-\alpha]$$
$$=-\mu_0(z_0)$$

Since this implies that

$$\mu_0(z_0)=-\alpha-\frac{1}{\pi}\int_C[\mu_0(z')-\alpha]\,d\theta$$

we can use Eq. 7.98 to deduce that

$$0=-\alpha+\frac{\alpha}{\pi}\int_0^{2\pi}d\theta=\alpha$$

This requires that $\text{Im } g_0(z_0)=0\ \forall z_0\in C$ and we already know that $\text{Im } g_0(\infty)=0$. However, as we have shown in Th. 7.7, neither the real nor imaginary parts of a holomorphic function can have a maximum or a minimum within the domain of holomorphy of a function unless the function is a constant. Therefore $\text{Im } g(z)\equiv 0$ for z outside C so that we may use the Cauchy-Riemann equations here to conclude that $g_0(z)=\text{constant}$. Since $g_0(\infty)=0$ we see that $g_0(z)\equiv 0$, z exterior to and on C so that $\mu_0\equiv-\text{Re } g_0(z_0)=0$ as claimed, which establishes the existence of a (unique) solution to Eq. 7.97.

d. Schwarz Principle of Reflection

We now state another useful method of analytic continuation before considering specific examples.

THEOREM 7.16 (Schwarz principle of reflection)

If $f(z)$ is holomorphic within a region R containing a portion of the real axis and is real on the real axis (in this region), then one obtains for conjugate values of z conjugate values of f; that is,

$$f(z^*) = f^*(z) \tag{7.103}$$

PROOF

Since $f(z)$ is holomorphic in R, it may be expanded in a Taylor series as

$$f(z) = \sum_{n=0}^{\infty} \alpha_n (z - x_0)^n$$

where x_0 is taken to be some point on the real axis. Since $f(x)$ is real here, then all the $\{\alpha_n\}$ are real. That is, we have

$$f^*(x) = f(x)$$

or

$$\sum_{n=0}^{\infty} \alpha_n^* (x - x_0)^n = \sum_{n=0}^{\infty} \alpha_n (x - x_0)^n$$

which implies

$$\alpha_n^* = \alpha_n \qquad \forall n$$

Also we can write

$$f(z^*) = \sum_{n=0}^{\infty} \alpha_n (z^* - x_0)^n$$

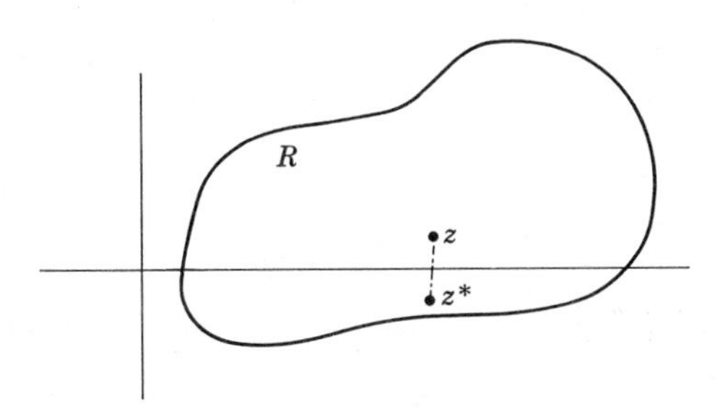

Figure 7.54

while

$$f^*(z)=\left\{\sum_{n=0}^{\infty}\alpha_n(z-x_0)^n\right\}^*=\sum_{n=0}^{\infty}\alpha_n(z^*-x_0)^n$$

so that

$$f(z^*)=f^*(z) \qquad \forall z\in R$$

Another way to state this is

$$f(z^*)=u(x,-y)+iv(x,-y)$$
$$f^*(z)=u(x,y)-iv(x,y)$$

so that

$$u(x,y)=u(x,-y) \qquad v(x,y)=-v(x,-y) \qquad \text{Q.E.D.}$$

Notice also that if a cut extends along the real axis but leaves a finite part of the real axis without a branch cut, then the theorem can still be applied to points above and below the cut provided only that $f(z)$ is real on the real axis where there is no cut. The basic argument is indicated in Fig. 7.55. We are given that

$$\phi_0^*(z)=\phi_0(z^*)$$

We can now extend this to circular regions above and below the cut, each of which partially overlaps C_0, and obtain

$$f(z)=\sum_{n=0}^{\infty}\frac{\phi_0^{(n)}(z_1)}{n!}(z-z_1)^n$$

$$f(z^*)=\sum_{n=0}^{\infty}\frac{\phi_0^{(n)}(z_1^*)}{n!}(z^*-z_1^*)^n=\left\{\sum_{n=0}^{\infty}\frac{\phi_0^{(n)}(z_1)}{n!}(z-z_1)^n\right\}^*=f^*(z)$$

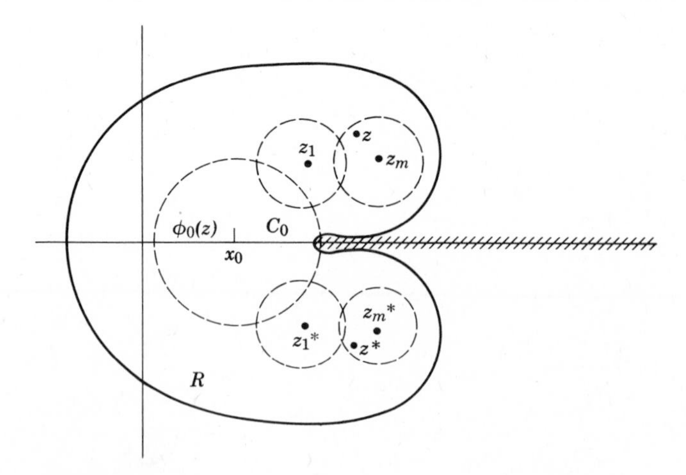

Figure 7.55

We may now repeat this argument to reach any points (say z_m and z_m^*) above and below the cut.

Example 7.18 First consider the function

$$f(z)=\sqrt{z}=\rho^{1/2}e^{i\varphi/2}$$

with the branch cut along the positive real axis. This function is *not* real on the negative real axis below the cut so that we would not expect Th. 7.16 to apply. In fact since the operation of complex conjugation here amounts to $\varphi \to 2\pi-\varphi$, we see directly that

$$f(z)=\rho^{1/2}[\cos(\varphi/2)+i\sin(\varphi/2)]$$
$$f(z^*)=\rho^{1/2}e^{-i\varphi/2}e^{i\pi}=\rho^{1/2}[-\cos(\varphi/2)+i\sin(\varphi/2)]$$

so that

$$f^*(z)=-f(z^*)\neq f(z^*)$$

However the function

$$\begin{aligned} g(z)&=e^{-i\pi/2}\sqrt{z}=e^{-i\pi/2}\rho^{1/2}e^{i\varphi/2}\\ &=\rho^{1/2}[-i\cos(\varphi/2)+\sin(\varphi/2)]\end{aligned}$$

is real for real negative z (i.e., $\varphi=\pi$) and has a cut along the positive real axis. Furthermore we have

$$\begin{aligned} g(z^*)&=e^{-i\pi/2}\rho^{1/2}e^{-i\varphi/2}e^{i\pi}\\ &=\rho^{1/2}[i\cos(\varphi/2)+\sin(\varphi/2)]\\ &\equiv[g(z)]^*\end{aligned}$$

as we expect from Th. 7.16.

7.12 INTEGRAL REPRESENTATIONS

In this and the following section we consider a few important examples of analytic continuation for the famous functions of classical mathematics.

a. Γ(z)—The Gamma Function

The gamma function is often defined as

$$\Gamma(z)=\int_0^\infty e^{-t}t^{z-1}\,dt \tag{7.104}$$

and is simply a generalization of $n!$ for noninteger positive n. In fact one readily proves by induction and integration by parts that

$$\Gamma(n+1)=n! \tag{7.105}$$

for n a positive integer on zero. We can see that $\Gamma(z)$ as defined by Eq. 7.104 is a holomorphic function of z for all finite z such that Re $z>0$ as follows from examining the integrand in the neighborhood of $t=0$ and at $t=\infty$ where the damped exponential insures convergence for all finite z. If we choose any point z_0, Re $z_0>0$, we could let $z=z_0+\rho e^{i\varphi}$ and execute a closed circuit about this point as $0 \to \varphi \to 2\pi$. Since the integral would converge for all points z on the circle (as long as ρ is not too large), we see that $\Gamma(z)$ is defined and single-valued for all finite z, Re $z>0$. Next, since we may differentiate this function under the integral sign (by the uniform convergence of the integral; cf. Appendix I, Th. I.15), we see that $\partial\Gamma/\partial x$ and $\partial\Gamma/\partial y$ exist and are continuous for Re $z>0$. Finally since $\partial\Gamma/\partial z^*\equiv 0$, it follows from Section 7.2 that the Cauchy-Riemann conditions are satisfied. This establishes the holomorphy of $\Gamma(z)$ for Re $z>0$ (cf. Fig. 7.56).

If we now simply integrate $\Gamma(z)$ by parts for Re $z>0$, we find

$$\Gamma(z)=\frac{1}{z}\,e^{-t}t^{z}\Big|_0^\infty+\frac{1}{z}\int_0^\infty e^{-t}t^{z}\,dt=\frac{1}{z}\int_0^\infty e^{-t}t^{(z+1)-1}=\frac{1}{z}\Gamma(z+1)$$

so that

$$\Gamma(z+1)=z\Gamma(z) \qquad \text{Re } z>0 \tag{7.106}$$

If we define

$$f(z)\equiv\Gamma(z+1)-z\Gamma(z) \tag{7.107}$$

then since $f(z)\equiv 0$, Re $z>0$, we see from Th. 7.13 that $f(z)\equiv 0$ everywhere in its domain of holomorphy. Furthermore, since the integral representation for $\Gamma(z+1)$, Eq. 7.104, converges for Re $z>-1$, we may take the limit of Eq. 7.106 as $z \to 0$ to obtain

$$\Gamma(z)\xrightarrow[z\to 0]{}\frac{1}{z} \tag{7.108}$$

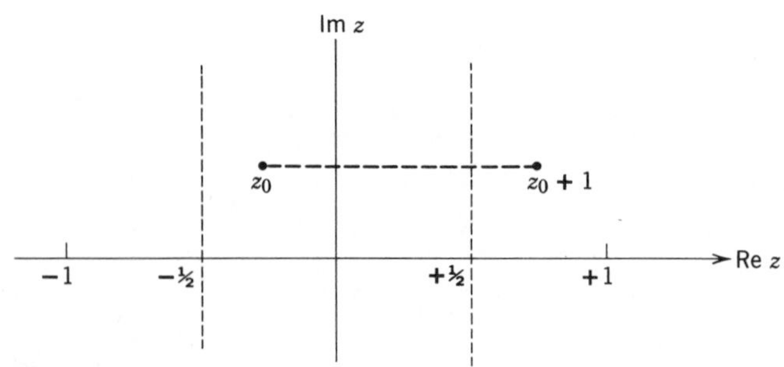

Figure 7.56

In fact as $\text{Re}\, z \to 0$, $\text{Im}\, z \neq 0$, we see from Eqs. 7.104 and 7.106 that $\Gamma(z)$ remains well defined. This implies that $\Gamma(z)$ has only a simple pole at $z=0$ so that the line $\text{Re}\, z=0$ is not a natural boundary of the $f(z)$ defined by Eq. 7.107. Therefore, by means of the functional relation (7.106), we can extend the definition of $\Gamma(z)$ to values of z such that $\text{Re}\, z<0$, but excluding $\text{Re}\, z=$negative integer or zero. From the values of $z \ni \text{Re}\, z>0$ we obtain those for $\text{Re}\, z>-1$ with a common region of definition for $\text{Re}\, z>0$, so that from Th. 7.13 these two elements are analytic continuations of each other.

We may apply our recursion relation, Eq. 7.106, n times to obtain

$$\Gamma(z)=\frac{\Gamma(z+n+1)}{(z+n)(z+n-1)(z+n-2)\cdots(z)} \tag{7.109}$$

Since $\text{Re}\,(z+n+1)>0$, $\Gamma(z+n+1)$ is finite and the denominator is nonzero *except* when $z=0$ or a negative integer. We can continue $\Gamma(z)$ to the negative half z plane with the exception of the neighborhoods of the negative integers and zero. Since $\Gamma(1)=1$ we see that $\Gamma(z)$ has simple poles at zero and the negative integers. Therefore $\Gamma(z)$ is an analytic (in fact, *meromorphic*) function in the entire finite z plane and the residue of $\Gamma(z)$ at $z=-n$ is just $(-1)^n/n!$ from Eq. 7.109.

Now the function $\Gamma(1-z)$ has simple poles at $z=+n$, $n=1, 2, 3, \ldots$, the positive integers. Therefore the function $\Gamma(z)\Gamma(1-z)$ has simple poles at all the integers. Also $\sin \pi z$ has simple zeros at $z=n$ since

$$\sin \pi z=\sin[\pi(z-n)+n\pi]=(-1)^n \sin[\pi(z-n)] \xrightarrow[z\to n]{} (-1)^n\pi(z-n)$$

It follows that the function $\sin \pi z\Gamma(z)\Gamma(1-z)$ is holomorphic everywhere in the finite z plane. We will now prove that this is just a constant. (Notice that we cannot use Liouville's Theorem, Th. 7.9, here since we do not know the behavior as $z \to \infty$.) First let $0<\text{Re}\, z<1$. Then we can write

$$\begin{aligned}
\Gamma(z)\Gamma(1-z)&=\int_0^\infty e^{-x^2}x^{2(z-1)}2x\,dx\int_0^\infty e^{-y^2}y^{-2z}2y\,dy\\
&=4\int_0^\infty dx\int_0^\infty dy e^{-(x^2+y^2)}x^{(2z-1)}y^{-(2z-1)}\\
&=4\int_0^\infty e^{-r^2}r\,dr\int_0^{\pi/2} d\theta(\cot\theta)^{(2z-1)}\\
&=2\int_0^{\pi/2} d\theta(\cot\theta)^{(2z-1)}=2\int_0^\infty \frac{s^{(2z-1)}\,ds}{(1+s^2)}=2\int_0^\infty \frac{s^{(2z-1)}\,ds}{(s+i)(s-i)}\\
&=2\pi\csc(2\pi z)\left[\frac{(-i)^{(2z-1)}}{2i}-\frac{(i)^{(2z-1)}}{2i}\right]=2\pi\csc(2\pi z)\left[\frac{e^{-\pi i z}+e^{\pi i z}}{2}\right]\\
&=\pi\csc(\pi z)
\end{aligned}$$

by the result given in Eq. 7.44. Therefore we obtain

$$\Gamma(z)\Gamma(1-z)=\pi \csc(\pi z) \qquad 0<\text{Re } z<1 \tag{7.110}$$

However all the functions involved are analytic ones in the entire finite z plane so that this relationship must hold for *all* finite z if it holds for those such that $0<\text{Re } z<1$. Also this implies that

$$\frac{1}{\Gamma(z)}=\frac{\sin(\pi z)\Gamma(1-z)}{\pi}$$

where the left-hand side can have a singularity only if $\Gamma(z)=0$ for some finite z. Since $\sin(\pi z)$ is finite for all finite z, $\Gamma(1-z)$ would have to be infinite at these values of z. However $\Gamma(1-z)$ has simple poles only at z=positive integer and here $\Gamma(n)=(n-1)!$, which is not zero. Therefore $[\Gamma(z)]^{-1}$ has no singularities anywhere in the finite z plane. A function that is holomorphic in every finite region of the z plane is called an *integral* or *entire* function of z.

Above, we have taken the original integral representation for $\Gamma(z)$, which was valid only for $\text{Re } z>0$, and managed to learn that $\Gamma(z)$ is an analytic function in the entire z plane. However we do not yet have a representation valid for all values of z but only for those such that $\text{Re } z>0$. We already know where $\Gamma(z)$ is defined. Our only problem is to find a defining representation for all finite z. It must reduce to Eq. 7.104

$$\int_0^{\infty} e^{-t}t^{(z-1)}\, dt$$

for $\text{Re } z>0$. Therefore we attempt

$$\Gamma(z)=A(z)\int_C e^{-t}(e^{-\pi i}t)^{(z-1)}\, dt \tag{7.111}$$

where both $A(z)$ and contour C are as yet unspecified. We try such a representation in analogy to the results derived in Section 7.9c. We notice that for arbitrary (i.e., noninteger) values of z the integrand has a branch point at $t=0$ (and also at $t=\infty$). We may take the branch cut along $0\leq t<+\infty$. The

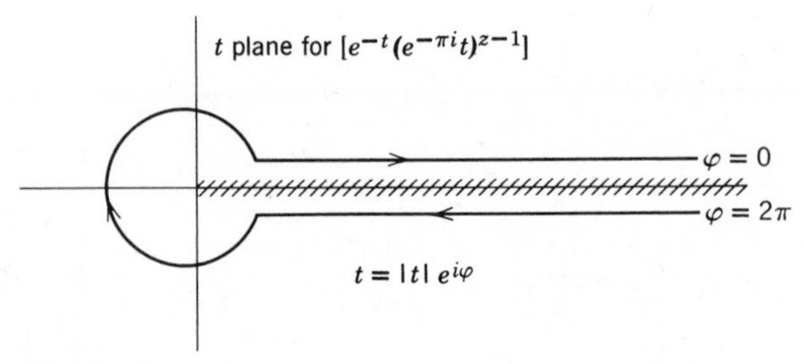

Figure 7.57

reason for our choice of phase $(e^{-\pi i}t)^{(z-1)}$ rather than $t^{(z-1)}$ is to make $(e^{-\pi i}t)^{(z-1)}$ real for real z when $t<0$. A contour C, which will make the integral well defined for all finite z, is shown in Fig. 7.57. On this contour $e^{-t}(e^{-\pi i}t)^{(z-1)}$ vanishes at the end points (i.e., $+\infty \pm i\varepsilon$) for *any* finite value of z. Furthermore the troublesome point, $t=0$, for Re $z<0$ is not on C so that this integral representation is defined for all values of z. We need only evaluate the factor $A(z)$ multiplying the integral. We do this for Re $z>0$ as follows.

$$\Gamma(z)=A(z)\int_C e^{-t}(e^{-\pi i}t)^{(z-1)}\,dt=A(z)\left\{\int_\infty^0 e^{-t}(e^{-\pi i}e^{2\pi i}t)^{(z-1)}\,dte^{2\pi i}\right.$$
$$\left.+\int_0^\infty e^{-t}(e^{-\pi i}t)^{(z-1)}\,dt\right\}=A(z)\left\{-\int_0^\infty e^{-t}t^{(z-1)}e^{\pi i(z-1)}\,dt\right.$$
$$\left.+\int_0^\infty e^{-t}t^{(z-1)}e^{-\pi i(z-1)}\,dt\right\}$$
$$=A(z)(e^{\pi iz}-e^{-\pi iz})\int_0^\infty e^{-t}t^{(z-1)}\,dt=2iA(z)\sin(\pi z)\int_0^\infty e^{-t}t^{(z-1)}\,dt \quad \textbf{(7.112)}$$

Comparing this with Eq. 7.104 we see that

$$A(z)=\frac{1}{2i\sin(\pi z)}$$

Since this $A(z)$ is an analytic function and is equal to $[2i\sin\pi z]^{-1}$ for Re $z>0$, it is equal to this for all finite z. Therefore we obtain

$$\Gamma(z)=\frac{1}{2i\sin(\pi z)}\int_C e^{-t}(e^{-\pi i}t)^{(z-1)}\,dt \quad \textbf{(7.113)}$$

If we recall Eq. 7.110, then we also have

$$\frac{1}{\Gamma(z)}\equiv\frac{\sin(\pi z)\Gamma(1-z)}{\pi}=\frac{1}{2\pi i}\int_C e^{-t}(e^{-\pi i}t)^{-z}\,dt \quad \textbf{(7.114)}$$

Again we should stress that these last integral representations are valid for *all* finite value of z for the contour shown above. Such integral representations are useful because they remain valid as the contour is deformed as long as C does not pass through any singular points of the integrand in the t plane. Just to be certain this point about deforming C is clear, realize that the contour C in this case is *not* a closed contour. We are arguing that the value of the *line integral* from $+\infty-i\varepsilon$ to $+\infty+i\varepsilon$ along C is independent of path (as long as we do not cross the branch cut). This follows since we *could* close this contour with another C' as shown in Fig. 7.58. Within the closed contour, $C+C'$, the function is holomorphic so that

$$\int_{C+C'} e^{-t}(e^{-\pi i}t)^{(z-1)}\,dt=0$$

from which path independence follows.

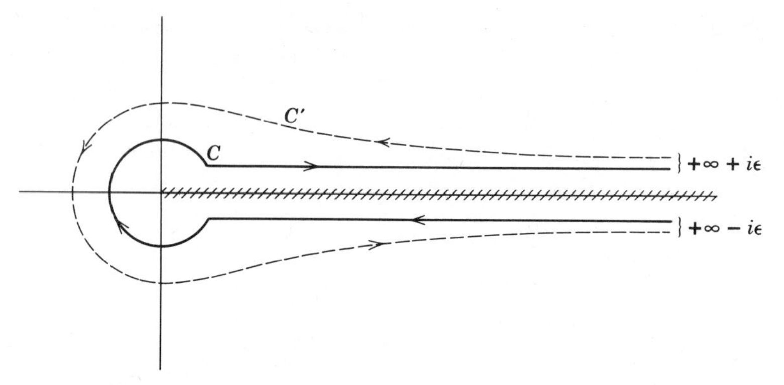

Figure 7.58

b. Method of Steepest Descent

We are often interested in the value of functions given by integral representations for large values of z. This is the *asymptotic* form of the function. The method of steepest descent, or saddle point method, is applicable to representations of the form

$$F(z) = \int_C \exp[zf(t)]w(t)\, dt \tag{7.115}$$

We can bring the representation for $\Gamma(z+1)$, Eq. 7.104, for Re $z > -1$ into this form by the substitution $\tau = tz$, where τ now denotes the real integration variable of Eq. 7.104. That is, with

$$t = |t|\, e^{i\theta} \qquad z = |z|\, e^{i\phi} \qquad \tau = |\tau|\, e^{i\alpha} = |t||z|\, e^{i(\theta+\phi)}$$

we choose $\alpha = 0$ so that $\theta = -\phi$. We must be careful to transform the contour when we change variables.

$$\Gamma(z+1) \equiv \int_0^\infty e^{-\tau}\tau^z\, d\tau = \int_0^\infty e^{-tz}(tz)^z z\, dt$$

$$= z^{z+1} \int_0^\infty e^{z(\ln t - t)}\, dt \tag{7.116}$$

The corresponding contours are shown in Fig. 7.59.

Before we can develop the method of steepest descent we must exhibit some important geometrical properties of the functions $u(x, y)$ and $v(x, y)$ of $f(z) = u(x, y) + iv(x, y)$. We have already seen in Sections 7.4 (cf. especially Th. 7.7) and 7.11c that functions $u(x, y)$ and $v(x, y)$ are both harmonic and that neither can have a maximum or a minimum within the domain of holomorphy of $f(z)$. However a saddle point (or minimax) is possible for $u(x, y)$ or for $v(x, y)$

provided that at (x_0, y_0) the conditions

$$\left.\frac{\partial u}{\partial x}\right|_{x_0,y_0}=0=\left.\frac{\partial u}{\partial y}\right|_{x_0,y_0}$$

$$\left.\left[\left(\frac{\partial^2 u}{\partial x\,\partial y}\right)^2-\left(\frac{\partial^2 u}{\partial x^2}\right)\left(\frac{\partial^2 u}{\partial y^2}\right)\right]\right|_{x_0,y_0}>0$$

are satisfied, as shown in Appendix I, Th. I.4. From the Cauchy-Riemann conditions of Eq. 7.7 we see that

$$\frac{\partial u}{\partial x}=0=\frac{\partial u}{\partial y}\Leftrightarrow\frac{\partial v}{\partial x}=0=\frac{\partial v}{\partial y}$$

Furthermore as an identity we have

$$\frac{df}{dz}=\frac{\partial u}{\partial x}+\frac{\partial v}{\partial y}+i\left(\frac{\partial v}{\partial x}-\frac{\partial u}{\partial y}\right)$$

If we again use the fact that u and v are harmonic and satisfy the Cauchy-Riemann conditions, we can easily verify the relation

$$\begin{aligned}\frac{d^2f}{dz^2}&=\frac{\partial^2 u}{\partial x^2}-\frac{\partial^2 u}{\partial y^2}+2\frac{\partial^2 v}{\partial x\,\partial y}+i\left(\frac{\partial^2 v}{\partial x^2}-\frac{\partial^2 v}{\partial y^2}-2\frac{\partial^2 u}{\partial x\,\partial y}\right)\\&=4\frac{\partial^2 u}{\partial x^2}+4i\frac{\partial^2 v}{\partial x^2}\end{aligned}$$

If either $u(x, y)$ or $v(x, y)$ has a saddle point at $z_0=(x_0, y_0)$, then they both must simultaneously and the condition for this is

$$\left.\left[\left(\frac{\partial^2 u}{\partial x\,\partial y}\right)^2-\left(\frac{\partial^2 u}{\partial x^2}\right)\left(\frac{\partial^2 u}{\partial y^2}\right)\right]\right|_{x_0,y_0}\equiv\left.\left[\left(\frac{\partial^2 v}{\partial x^2}\right)^2+\left(\frac{\partial^2 u}{\partial x^2}\right)^2\right]\right|_{x_0,y_0}>0$$

Therefore unless $f''(z_0)=0$ then z_0 will be a saddle point of both $u(x, y)$ and $v(x, y)$ whenever $f'(z_0)=0$.

Furthermore, if we have an analytic function of z, then the curves $u(x, y)=$ constant and $v(x, y)=$constant are everywhere orthogonal since the condition for orthogonality is that the tangents to the curves in question be the negative

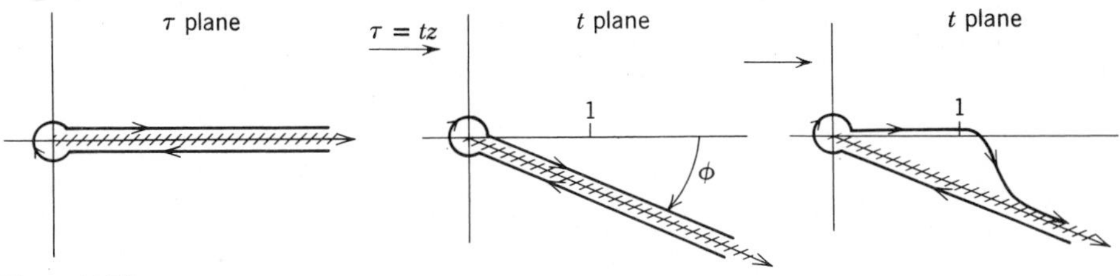

Figure 7.59

reciprocals of each other. If we are on a curve defined by $u(x, y) = \text{const.}$, then along this curve

$$du \equiv \frac{\partial u}{\partial x}\, dx + \frac{\partial u}{\partial y}\, dy = 0$$

so that

$$\left(\frac{dy}{dx}\right)_1 = -\frac{\partial u/\partial x}{\partial u/\partial y}$$

and

$$\left(\frac{dy}{dx}\right)_2 = -\frac{\partial v/\partial x}{\partial v/\partial y}$$

The condition for orthogonality is just

$$\left(\frac{dy}{dx}\right)_1 = -\left[\left(\frac{dy}{dx}\right)_2\right]^{-1}$$

or

$$\frac{\partial u}{\partial x}\frac{\partial v}{\partial x} + \frac{\partial u}{\partial y}\frac{\partial v}{\partial y} = 0$$

as indicated in Fig. 7.60. However this condition is identically satisfied by virtue of the Cauchy-Riemann equations. Now if we proceed along the curve

$$v(x, y) = \text{const.}$$

then on this curve $u(x, y)$ changes at its maximum rate.

If we consider the function $zf(t)$ for fixed z as a function of the complex variable $t = \xi + i\eta$, then a point of t_0 such that $f'(t_0) = 0$ will be a minimax or saddle point for $\mathrm{Re}\,[zf(t)]$. This is shown in Fig. 7.62. The condition $\mathrm{Im}\,[zf(t)] = \text{const.}$ then corresponds to the curves (or paths) AB and CD in this figure since $\mathrm{Re}\,[zf(t)]$ changes at a maximum rate along these. Since we want the integral

$$F(z) = \int_C e^{zf(t)} w(t)\, dt$$

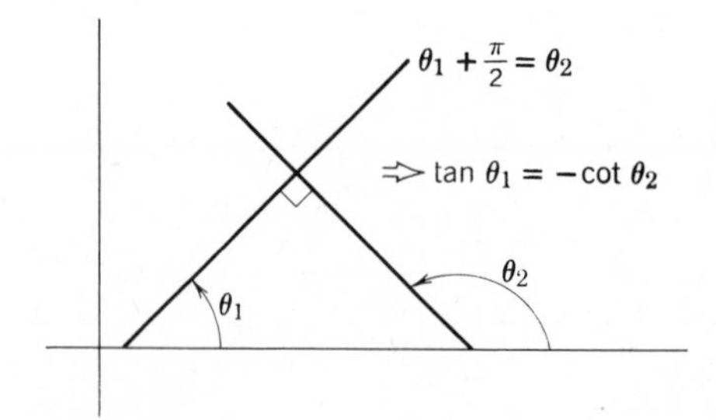

Figure 7.60

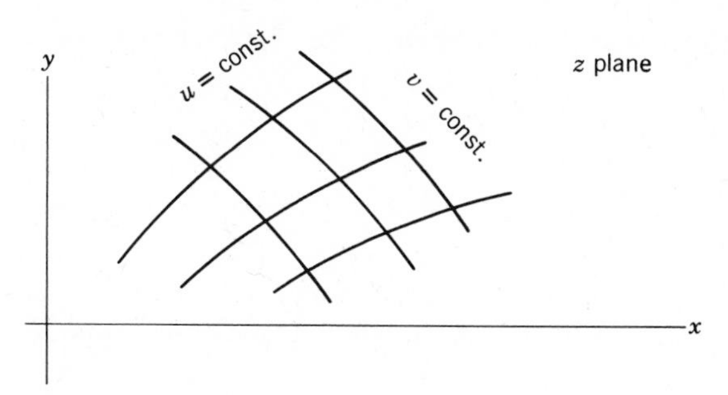

Figure 7.61

to converge as C goes to its end points even as $|z| \to \infty$, we must choose the path AB along which $\text{Re}\,[zf(t)]$ *decreases* at a maximum rate from its largest value at t_0. Therefore along AB $e^{zf(t)}$ goes to its end-point values along the path of steepest descent and passes through the minimax point, t_0.

In the rest of the discussion we will assume that the curve of steepest descent of $\text{Re}\,[zf(t)]$ corresponding to

$$\begin{aligned} \text{Im}\,[zf(t)] &= \text{Im}\,[zf(t_0)] \\ f'(t_0) &= 0 \end{aligned} \qquad \textbf{(7.117)}$$

has been found and that the contour C of the integral representation of Eq. 7.115 can be continuously deformed into this path of steepest descent without passing through any of the singularities of $f(t)$ or of $w(t)$. These conditions must always be checked in any given application of this method. Because of the

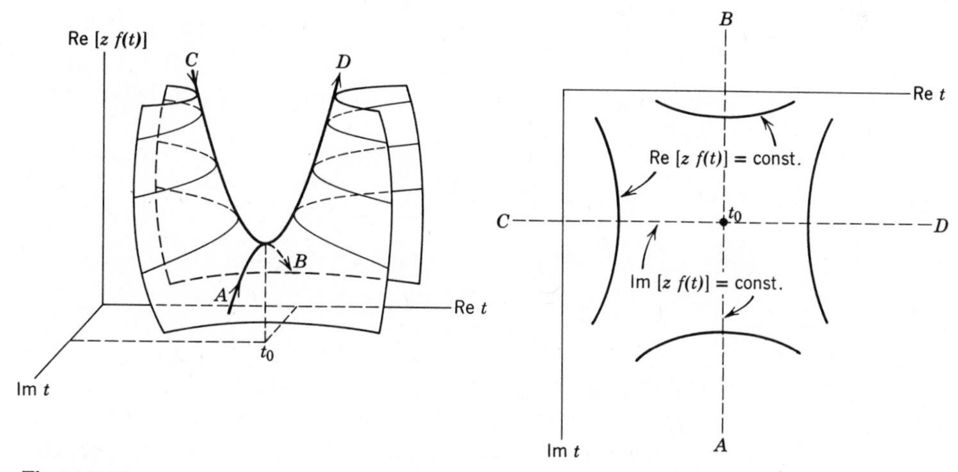

Figure 7.62

condition (7.117) satisfied everywhere on C we can rewrite Eq. 7.115 exactly as

$$F(z)=e^{i\,\mathrm{Im}[zf(t_0)]}\int_C e^{\mathrm{Re}[zf(t)]}w(t)\,dt$$

$$=e^{zf(t_0)}\int_C e^{\mathrm{Re}\{z[f(t)-f(t_0)]\}}w(t)\,dt \tag{7.118}$$

It is important to realize that the function $z[f(t)-f(t_0)]$ is *real* everywhere on C (cf. Eq. 7.117) and *negative* except at $t=t_0$ where it vanishes. If we now define a real function $v(t)$ as

$$e^{i\phi}[f(t)-f(t_0)]\equiv -v^2(t) \tag{7.119}$$

where we write $z=|z|\,e^{i\phi}$, then we obtain

$$F(z)=e^{zf(t_0)}\int_C e^{-|z|v^2(t)}w(t)\,dt$$

$$=e^{zf(t_0)}\int_L e^{-|z|v^2}w(t)\left(\frac{dt}{dv}\right)dv \tag{7.120}$$

where L is the image of C in the v plane. It runs along the *real* axis in the v plane (often from $-\infty$ to $+\infty$) and passes through the origin (i.e., $v=0$). Equation 7.120 is exact since no approximations have been made. As $|z|\to\infty$ the exponential in Eq. 7.120 becomes very steeply damped and, since we are assuming that $w(t)$ does not destroy the convergence properties of the integral, we may make the following approximations that become better and better as $|z|$ becomes larger.

$$F(z)=e^{zf(t_0)}\int_L e^{-|z|v^2}w(t)\left(\frac{dt}{dv}\right)dv$$

$$\xrightarrow[|z|\to\infty]{} e^{zf(t_0)}\int_{-\infty}^{\infty} e^{-|z|v^2}w[t(v)]\left(\frac{dt}{dv}\right)dv \tag{7.121}$$

In order to carry out the final integral in Eq. 7.121 we must solve for (dt/dv) and w as functions of v. Let

$$g(t)\equiv e^{-i\phi/2}v(t)=\sqrt{e^{-\pi i}[f(t)-f(t_0)]}=\sum_{n=1}^{\infty}\alpha_n(t-t_0)^n \tag{7.122}$$

Notice that $g(t)$ is indeed holomorphic in the neighborhood of t_0 in spite of the apparent square-root singularity in its definition. This follows since $f'(t_0)=0$ so that

$$f(t)-f(t_0)\simeq -f''(t_0)\frac{(t-t_0)^2}{2!}+\cdots$$

Since we know the function, $f(t)$, we can compute its Taylor series, extract the square root, and obtain the expansion coefficients, $\{\alpha_n\}$, which we hereafter consider known. We can next invert this expansion as

$$(t-t_0)=\sum_{n=0}^{\infty}\frac{a_n}{(n+1)}[g(t)]^{n+1}=\sum_{n=0}^{\infty}\frac{e^{-[i(n+1)/2]\phi}a_n}{(n+1)}v^{n+1} \tag{7.123}$$

and solve for the $\{a_n\}$ in terms of the $\{\alpha_n\}$ (cf. Problem 7.41). Similarly we write

$$w(t)=\sum_{m=0}^{\infty}\beta_m(t-t_0)^m=\sum_{m=0}^{\infty}\gamma_m v^m \tag{7.124}$$

If we then use Eqs. 7.123 and 7.124 in Eq. 7.121 we obtain

$$F(z)\xrightarrow[|z|\to\infty]{} e^{zf(t_0)}\sum_{m,n=0}^{\infty}\gamma_m a_n e^{-i(n+1)\phi/2}\int_{-\infty}^{\infty}e^{-|z|v^2}v^{n+m}\,dv \tag{7.125}$$

Since the interchange of these summations and the integration is not always justifiable, we do not necessarily produce a convergent series but often only an *asymptotic series* as we discuss below. If we observe that the symmetric integral in Eq. 7.125 vanishes whenever $n+m$ is an odd integer and make use of the standard integral

$$\int_{-\infty}^{\infty}e^{-\alpha x^2}x^{2n}\,dx=\frac{\Gamma(2n+1)}{\Gamma(n+1)2^{2n}\alpha^n}\sqrt{\frac{\pi}{\alpha}} \tag{7.126}$$

we arrive at

$$F(z)\xrightarrow[|z|\to\infty]{} e^{zf(t_0)}\sqrt{\frac{\pi}{z}}\left\{\gamma_0 a_0+\frac{1}{2|z|^2}[\gamma_1 a_1 e^{-i\phi/2}+\gamma_0 a_2 e^{-i\phi}+\gamma_2 a_0]+\cdots\right\} \tag{7.127}$$

where we have used the fact that $z=|z|\,e^{i\phi}$. We can write explicitly the leading term in the expansion if we realize that $\gamma_0=w(t_0)$ (cf. Eq. 7.124) and $a_0=1/\alpha_1=[e^{i\pi}f''(t_0)/2]^{-1/2}$ [cf. Eqs. 7.122 and 7.123] so that

$$F(z)\xrightarrow[|z|\to\infty]{} e^{zf(t_0)}w(t_0)\sqrt{\frac{2\pi}{e^{i\pi}f''(t_0)z}} \tag{7.128}$$

Another important case is that for which $w(t)=1$. If we use the relation

$$\Gamma(2n+1)=\frac{2^{2n}\Gamma(n+\frac{1}{2})\Gamma(n+1)}{\Gamma(\frac{1}{2})} \tag{7.129}$$

$$\Gamma(\tfrac{1}{2})=\sqrt{\pi}$$

which is easily proved by induction, we can write

$$F(z)\xrightarrow[|z|\to\infty]{} e^{zf(t_0)}\sqrt{\frac{\pi}{z}}\,a_0\sum_{n=0}^{\infty}\left(\frac{a_{2n}}{a_0}\right)\frac{\Gamma(n+\frac{1}{2})}{\Gamma(\frac{1}{2})}\left(\frac{1}{z}\right)^n \tag{7.130}$$

Notice that we have recovered the phase, $e^{i\phi}$, of z in every term of Eq. 7.130 since only the terms with a_{2n} of Eq. 7.123 survive in this case.

We now apply these results in some detail to the gamma function defined in Eq. 7.116

$$\Gamma(z+1) = z^{z+1} \int_0^\infty e^{z(\ln t - t)}\, dt$$

where the integration path is a straight line from 0 to ∞ of slope $-\phi$ as shown in Fig. 7.59. Since $f(t)$ is given as

$$f(t) = \ln t - t$$

we see that

$$f'(t) = \frac{1}{t} - 1$$

$$f''(t) = -\frac{1}{t^2}$$

so that $t_0 = 1$ and

$$f(t_0) = -1$$

$$f''(t_0) = -1 \tag{7.131}$$

We must next find the path such that on it

$$\operatorname{Im}[zf(t)] = \operatorname{Im}[zf(t_0)] = -\operatorname{Im} z = -y = -|z| \sin\phi$$

If we set $z = |z|\, e^{i\phi}$ and $t - 1 = \rho e^{i\theta}$, then we have

$$zf(t) = |z| \left[\ln\sqrt{1+\rho^2+2\rho\cos\theta}\, e^{i\phi} + ie^{i\phi} \tan^{-1}\left(\frac{\rho\sin\theta}{1+\rho\cos\theta}\right) - e^{i\phi} - \rho e^{i(\theta+\phi)} \right]$$

This implies that

$$\operatorname{Re}[zf(t)] = |z| \left[\ln\sqrt{1+\rho^2+2\rho\cos\theta}\cos\phi - \tan^{-1}\left(\frac{\rho\sin\theta}{1+\rho\cos\theta}\right)\sin\phi - \cos\phi - \rho\cos(\theta+\phi) \right]$$

and that the path of constant $\operatorname{Im}[zf(t)]$ is given by

$$\ln\sqrt{1+\rho^2+2\rho\cos\theta}\sin\phi + \tan^{-1}\left(\frac{\rho\sin\theta}{1+\rho\cos\theta}\right)\cos\phi - \rho\sin(\theta+\phi) = 0 \tag{7.132}$$

We can readily check that at the end points of the original contour of Fig. 7.59 (i.e., at $\rho = 1$, $\theta = \pi$ and at $\rho = \infty$, $\theta = -\phi$) we have

$$\operatorname{Re}[zf(t)] \xrightarrow[\theta=\pi]{\rho\to 1} |z| \ln(1-\rho)\cos\phi \to -\infty \qquad -\frac{\pi}{2} < \phi < \frac{\pi}{2}$$

$$\operatorname{Re}[zf(t)] \xrightarrow[\theta=-\phi]{\rho\to\infty} |z|[\ln\rho - \rho] \to -\infty$$

In fact $\text{Re}\,[zf(t)] \to -\infty$ as $\rho \to \infty$ as long as $\cos(\theta+\phi) > 0$ which requires that

$$-\frac{\pi}{2}-\phi<\theta<\frac{\pi}{2}-\phi$$

so that the integral would always converge for $\text{Re}\,z>0$ if the positive real axis were taken as the integration path, although this would *not* coincide with the path of steepest descent unless $\phi=0$ (i.e., z purely real). Since the saddle point $t_0=1$ (i.e., $\rho=0$) lies on the curve of Eq. 7.132, we can deform the integration path of Fig. 7.59 to lie on this path of steepest descent. In particular in the neighborhood of t_0 Eq. 7.132 reduces to

$$\rho^2 \sin(2\theta+\phi)=0$$

and

$$\text{Re}\,[zf(t)] \simeq -|z|\,[\cos\phi + \tfrac{1}{2}\rho^2 \cos(2\theta+\phi)]$$

Since we require that $\text{Re}\,[zf(t)] \xrightarrow[|z|\to\infty]{} -\infty$, these two conditions imply that $\theta=-\phi/2$ and $\theta=-\phi/2+\pi$, as shown in the qualitative sketch of the path of steepest descent in Fig. 7.63.

It is now a simple matter to use Eqs. 7.131 and 7.128 to obtain the leading term in the asymptotic expansion for $\Gamma(z+1)$. If the first few terms of Eq. 7.130 are computed, one finds

$$\Gamma(z) \xrightarrow[\substack{z\to\infty \\ \text{Re}\,z>0}]{} \sqrt{2\pi}\, z^{z-(1/2)} e^{-z}\left[1+\frac{1}{12z}+\frac{1}{288z^2}-\frac{139}{51840z^3}+O(z^{-4})\right] \quad \textbf{(7.133)}$$

which is known as *Stirling's approximation* for $\Gamma(z)$. Although it is by no means

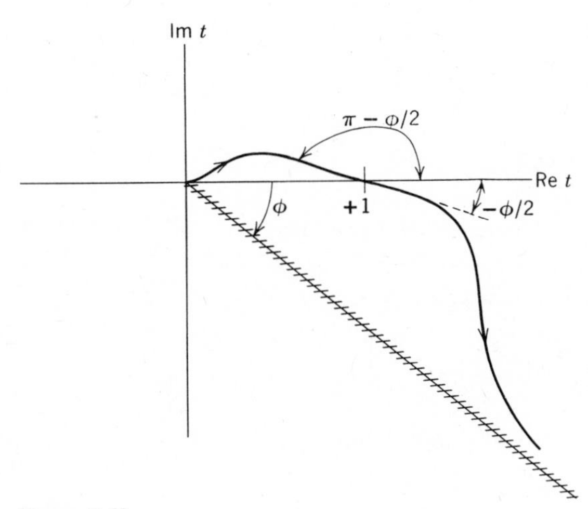

Figure 7.63

evident from an examination of the first few terms in Eq. 7.133, this series diverges as the number of terms is increased without limit for fixed z (cf. Whittaker and Watson, p. 252). It is an example of an *asymptotic series* about which we shall now say a few words in general. Suppose that as $|z| \to \infty$ we have

$$f(z) = \varphi(z)\left[A_0 + \frac{A_1}{z} + \frac{A_2}{z^2} + \cdots\right] \tag{7.134}$$

where $\varphi(z)$ is known. If $[f(z)/\varphi(z)]$ has an essential singularity as $z \to \infty$, then the series $\sum A_n z^{-n}$ will diverge. This follows simply from the definition of an essential singularity. Nevertheless this series is useful for calculating $f(z)$ for large values of z provided the difference between $f(z)/\varphi(z)$ and the first $(n+1)$ terms of the series can be made of order $1/z^{n+1}$ so that for large enough z this difference can be made quite small.

More precisely, we say that the first $(n+1)$ terms $\sum_{p=0}^{n} A_p z^{-p}$ form an *asymptotic series* for $[f(z)/\varphi(z)]$ if

$$\lim_{|z|\to\infty}\left\{z^n\left[\frac{f(z)}{\varphi(z)} - \sum_{p=0}^{n}\frac{A_p}{z^p}\right]\right\} = 0 \qquad n \text{ fixed} \tag{7.135}$$

even though

$$\lim_{n\to\infty}\left\{z^n\left[\frac{f(z)}{\varphi(z)} - \sum_{p=0}^{n}\frac{A_p}{z^p}\right]\right\} = \infty \qquad z \text{ fixed} \tag{7.136}$$

Obviously, if the second statement did not hold, then $f(z)/\varphi(z)$ would *not* have an essential singularity at $z = \infty$.

Let us consider a simple example

$$f(x) \equiv \int_x^\infty \frac{1}{t} e^{x-t}\, dt \qquad x > 0$$

If we let $y = t - x$, then we obtain

$$f(x) = \int_0^\infty \frac{e^{-y}\, dy}{(y+x)} < \frac{1}{x}\int_0^\infty e^{-y}\, dy = \frac{1}{x} \qquad x > 0$$

Repeated integration by parts shows that

$$f(x) = \frac{1}{x}\left[1 - \frac{1}{x} + \frac{2!}{x^2} + \cdots + \frac{(-1)^{n-1}(n-1)!}{x^{n-1}}\right] + (-1)^n n! \int_x^\infty \frac{e^{x-t}\, dt}{t^{n+1}}$$

Since the ratio of the $(n+1)$st to the nth term is

$$\left|\frac{u_{n+1}(x)}{u_n(x)}\right| = \left|\frac{n!}{x^{n+1}}\frac{x^n}{(n-1)!}\right| = \left|\frac{n}{x}\right| \xrightarrow[x \text{ fixed}]{n\to\infty} \infty$$

we see that this series is divergent for all finite values of x. If we let,

$$S_n(x) \equiv \sum_{m=1}^{n} u_m(x) \qquad u_m(x) = \frac{(-1)^m m!}{x^m}$$

then with $\varphi(x)=1/x$ we have

$$\left|\frac{f(x)}{\varphi(x)}-S_n(x)\right|\equiv|R_n(x)|=(n+1)!\int_x^\infty \frac{e^{x-t}\,dt}{t^{n+2}}<(n+1)!\int_x^\infty \frac{dt}{t^{n+2}}=\frac{n!}{x^{n+1}}$$

Therefore for fixed n we can find an x sufficiently large so that $|[f(x)/\varphi(x)]-S_n(x)|<\varepsilon$. Suppose that we hold x fixed (large) and ask for that n that will minimize the upper bound on the error (i.e., where to terminate the series). If we define

$$\frac{\Gamma(n+1)}{x^{n+1}}\equiv g(n;x) \qquad \ni|R_n(x)|<g(n;x) \qquad x \text{ fixed}$$

then we have

$$\frac{dg}{dn}=\frac{x^{n+1}\psi(n+1)\Gamma(n+1)-\Gamma(n+1)(n+1)x^n\ln x}{x^{2n+2}}$$

$$=\frac{x^n\Gamma(n+1)[x\psi(n+1)-(n+1)\ln x]}{x^{2n+2}}$$

where

$$\psi(z)\equiv\frac{d}{dz}\ln\Gamma(z)=\frac{1}{\Gamma(z)}\frac{d\Gamma(z)}{dz}\xrightarrow[z\to\infty]{}\ln z-\frac{1}{2z}-\frac{1}{12z^2}+\frac{1}{120z^4}+\cdots$$

We see that, for any fixed x, $dg(n;x)/dn=0$ implies that

$$\psi(n+1)=\frac{(n+1)}{x}\ln x$$

or that $x\simeq n+1$. Clearly, for fixed x, this is the optimal value for n. Also, however, as $x\to\infty$ for *fixed* n the error vanishes. Therefore we arrive at our definition of an asymptotic series given above in Eqs. 7.135 and 7.136. It is the condition stated in Eq. 7.135 that makes asymptotic series useful for calculations.

Finally for the Γ-function we have

$$\frac{\Gamma(z+1)}{z^{z+1}}=\int_0^\infty e^{z(\ln t-t)}\,dt$$

Clearly for $z=\rho e^{i\phi}$, $-(\pi/2)<\phi<(\pi/2)$, the integral on the right vanishes as $|z|\to\infty$. The Stirling approximation is an asymptotic expansion for this integral.

Notice that the region in the z plane (i.e., range of ϕ) for which the asymptotic expansion holds is crucial. That is, Stirling's formula need not be valid outside this range. (In fact it obviously cannot be valid along the negative real z axis!). This is an example of *Stokes' phenomenon.* Such nonuniqueness of a limiting value for an $f(z)$ is characteristic of an essential singularity (e.g., e^{-z} as $z\to\infty$).

c. Analytic Properties of Fourier Integrals

In Section 4.5 we considered Fourier transforms for *real* values of x and k only.

$$f(k)=\frac{1}{\sqrt{2\pi}}\int_{-\infty}^{\infty} e^{ikx}f(x)\,dx \tag{7.137}$$

$$f(x)=\frac{1}{\sqrt{2\pi}}\int_{-\infty}^{\infty} e^{-ikx}f(k)\,dk \tag{7.138}$$

It is often useful or necessary in applications to study the extension of these transforms to the complex k plane.

Suppose that $f(z)$ is holomorphic in a strip in the z plane defined by (cf. Fig. 7.64),

$$-\alpha<y<\beta \qquad \alpha>0,\ \beta>0$$

$$-\infty<x<\infty$$

If $|f(z)|$ is also bounded uniformly by some $M(x)$ as $x\to\pm\infty$ within this strip and $M(x)\to 0$ as $x\to\pm\infty$, then we can shift the path of integration of Eq. 7.137 to any line parallel to the x axis within this strip. For a rectangular path C as shown in the figure Cauchy's integral theorem states that

$$\frac{1}{\sqrt{2\pi}}\int_C e^{ikz}f(z)\,dz=0 \tag{7.139}$$

If we now let $X\to\infty$ then the integrals along the vertical strips vanish since, for instance,

$$\lim_{X\to\infty}\left|\int_{X-i\gamma_1}^{X+i\gamma_2} e^{ikz}f(z)\,dz\right|\le \lim_{X\to\infty}\int_{\gamma_1}^{\gamma_2} e^{-ky}\,|f(X+iy)|\,dy$$

$$\le \lim_{X\to\infty} M(X)\int_{-\gamma_1}^{\gamma_2} e^{-ky}\,dy=0$$

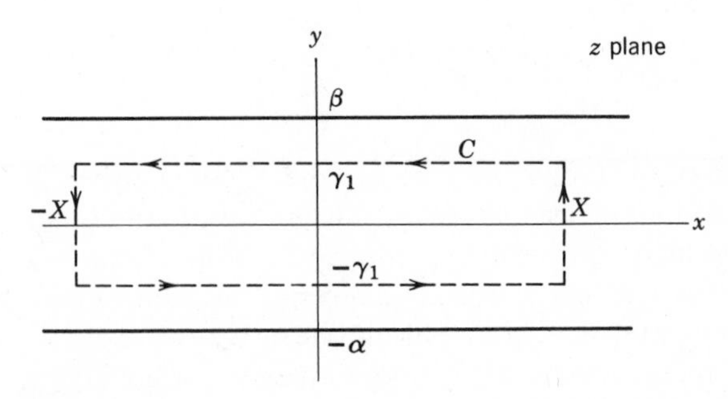

Figure 7.64

Therefore, if we take both paths of integration to run from left to right in Fig. 7.64, Eq. 7.139 implies that

$$f(k) \equiv \frac{1}{\sqrt{2\pi}} \int_{-\infty}^{\infty} e^{ikx} f(x)\, dx = \frac{1}{\sqrt{2\pi}} \int_{-\infty+i\gamma}^{+\infty+i\gamma} e^{ikz} f(z)\, dz \qquad -\alpha<\gamma<\beta \quad \textbf{(7.140)}$$

Also if $f(x)=O(e^{\mu x})$, $\mu>0$ as $x \to +\infty$, then $f(x)$ will not have a Fourier transform in the usual sense since it cannot be absolutely integrable on $(-\infty, \infty)$ (cf. Th. 4.9). However we can define a new function

$$g(x) \equiv e^{-\sigma x} f(x) \qquad \sigma>\mu \quad \textbf{(7.141)}$$

which will vanish as $x \to +\infty$. If $g(x)$ has proper behavior as $x \to -\infty$, then it will have a Fourier transform given as

$$\begin{aligned} g(k) &= \frac{1}{\sqrt{2\pi}} \int_{-\infty}^{\infty} e^{-\sigma x} f(x) e^{ikx}\, dx = \frac{1}{\sqrt{2\pi}} \int_{-\infty}^{\infty} e^{i(k+i\sigma)x} f(x)\, dx \\ &= f(k+i\sigma) \qquad \sigma>\mu \end{aligned} \quad \textbf{(7.142)}$$

This Fourier transform of $f(x)$ is defined by the integral in Eq. 7.142 only for those complex k such that $\operatorname{Im} k>\mu$. From Eqs. 7.138, 7. 141, and 7.142 we obtain

$$\begin{aligned} f(x) &= \frac{1}{\sqrt{2\pi}} \int_{-\infty}^{\infty} e^{-i(k+i\sigma)x} f(k+i\sigma)\, dk \\ &= \frac{1}{\sqrt{2\pi}} \int_{-\infty+i\sigma}^{\infty+i\sigma} e^{-ikx} f(k)\, dk \qquad \sigma>\mu \end{aligned} \quad \textbf{(7.143)}$$

where it must be realized that k is a *complex* integration variable in the last integral.

In fact it is easy to see that if, within the strip indicated in Fig. 7.64, $f(z)$ satisfies

$$|f(z)| \to \begin{cases} C_1 e^{-\alpha_1 x} & \text{as} \quad x \to +\infty \quad \alpha_1>0 \\ C_2 e^{\alpha_2 x} & \text{as} \quad x \to -\infty \quad \alpha_2>0 \end{cases} \quad \textbf{(7.144)}$$
$$-\alpha<y<\beta$$

then the $f(k)$ defined by Eq. 7.137 is bounded as $(k=r+is)$

$$|f(k)| \to \begin{cases} C_3 e^{-\beta r} & \text{as} \quad r \to +\infty \\ C_4 e^{\alpha r} & \text{as} \quad r \to -\infty \end{cases} \quad \textbf{(7.145)}$$
$$-\alpha_1<s<\alpha_2$$

where the α and β are the boundaries of the strip of holomorphy of $f(z)$ indicated in the figure. We prove this by using Eq. 7.140. First let $0 \leq \gamma<\beta$ so that

$$f(k) \equiv f(r+is) = \frac{1}{\sqrt{2\pi}} \int_{-\infty+i\gamma}^{\infty+i\gamma} e^{ikz} f(z)\, dz = \frac{1}{\sqrt{2\pi}} \int_{-\infty}^{\infty} e^{i(r+is)(x+i\gamma)} f(x+i\gamma)\, dx$$

$$= \frac{e^{-r\gamma} e^{-is\gamma}}{\sqrt{2\pi}} \int_{-\infty}^{\infty} e^{-sx} e^{irx} f(x+i\gamma)\, dx$$

from which we obtain

$$|f(k)| \xrightarrow[r\to\infty]{} e^{-r\gamma} \left| \frac{1}{\sqrt{2\pi}} \int_{-\infty}^{\infty} e^{-sx} e^{irx} f(x+i\gamma)\, dx \right|$$

$$\leq e^{-r\gamma} \frac{1}{\sqrt{2\pi}} \int_{-\infty}^{\infty} e^{-sx}\, |f(x+i\gamma)|\, dx \qquad \textbf{(7.146)}$$

If we examine the integrand as $x \to \pm\infty$, we find

$$e^{-sx}\, |f(x+i\gamma)| \xrightarrow[x\to+\infty]{} e^{-x(s+\alpha_1)}$$

so that

$$s > -\alpha_1$$

and

$$e^{-sx}\, |f(x+i\gamma)| \xrightarrow[x\to-\infty]{} e^{|x|(s-\alpha_2)}$$

so that

$$s < \alpha_2$$

However in Eq. 7.146 we can make γ as close to β as we please. A similar argument holds for $-\alpha < \gamma \leq 0$, which establishes Eq. 7.145.

If the function $f(x)$ has bad asymptotic properties as both $x \to +\infty$ and $x \to -\infty$, then the artifice of Eq. 7.141 cannot be used to define a Fourier transform. However suppose that we can find two different convergence parameters $\sigma_1 > 0$ and $\sigma_2 > 0$ such that both

$$\int_0^{\infty} |f(x)|\, e^{-\sigma_1 x}\, dx < \infty$$

and

$$\int_{-\infty}^{0} |f(x)|\, e^{\sigma_2 x}\, dx < \infty$$

If we now define

$$f_+(x) = \begin{cases} f(x) & x \geq 0 \\ 0 & x < 0 \end{cases}$$

$$f_-(x) = \begin{cases} 0 & x > 0 \\ f(x) & x \leq 0 \end{cases}$$

so that

$$f(x) \equiv f_+(x) + f_-(x)$$

we can apply Eq. 7.142 to obtain

$$f_+(k)=\frac{1}{\sqrt{2\pi}}\int_0^{\infty} f(x)e^{ikx}\,dx \qquad \text{Im } k>\sigma_1 \tag{7.147}$$

$$f_-(k)=\frac{1}{\sqrt{2\pi}}\int_{-\infty}^{0} f(x)e^{ikx}\,dx \qquad \text{Im } k<-\sigma_2 \tag{7.148}$$

Clearly $f_+(k)$ is analytic in the upper half k plane, Im $k>\sigma_1$, while $f_-(k)$ is analytic in the lower half k plane, Im $k<-\sigma_2$. We can apply the Fourier inversion theorem (cf. Eq. 7.143) to obtain

$$f(x)=\frac{1}{\sqrt{2\pi}}\int_{-\infty+i\sigma_1}^{\infty+i\sigma_1} f_+(k)e^{-ikx}\,dk+\frac{1}{\sqrt{2\pi}}\int_{-\infty-i\sigma_2}^{\infty-i\sigma_2} f_-(k)e^{-ikx}\,dk \tag{7.149}$$

It is important to realize that in general $f_+(k)$ and $f_-(k)$ need not be related to any common function in the k plane.

Example 7.19 Let $f(x)=e^{|x|}$ so that

$$f_+(k)=\frac{1}{\sqrt{2\pi}}\int_0^{\infty} e^{x}e^{ikx}\,dx=-\frac{1}{\sqrt{2\pi}}\left(\frac{1}{1+ik}\right)$$

and

$$f_-(k)=\frac{1}{\sqrt{2\pi}}\int_{-\infty}^{0} e^{-x}e^{ikx}\,dx=\frac{1}{\sqrt{2\pi}}\left(\frac{1}{-1+ik}\right)$$

If the conditions of Eq. 7.144 apply to $f(x)$, then we may decompose $f(k)$ as

$$f(k)=f_+(k)+f_-(k) \tag{7.150}$$

where $f_+(k)$ and $f_-(k)$ are still given by Eqs. 7.147 and 7.148. In this case $f_+(k)$ is holomorphic for Im $k>-\alpha_1$ while $f_-(k)$ is holomorphic for Im $k<\alpha_2$, which allows us to write Eq. 7.150 (cf. Eq. 7.149 with $\sigma_1=\sigma_2=0$).

We now present a rather loose and incomplete discussion of the asymptotic behavior of the Fourier transform $f(k)$. In general the question of the asymptotic behavior is a very difficult one as k approaches infinity along an arbitrary direction in the complex k plane. Of course the Riemann-Lebesgue lemma of Section 4.4 gives us a bound on this if $f(x)\in\mathscr{L}(-\infty,\infty)$ when k approaches infinity through *real* (positive or negative) values. For our present discussion we assume that outside of some sufficiently large circular region of radius K in the k plane $f_+(k)$ and $f_-(k)$ have convergent expansions in inverse powers of k as

$$f_+(k)=\frac{1}{\sqrt{2\pi}}\sum_{n=1}^{\infty}\frac{a_n}{(-ik)^n} \qquad |k|>K \tag{7.151}$$

$$f_-(k)=\frac{1}{\sqrt{2\pi}}\sum_{n=1}^{\infty}\frac{b_n}{(-ik)^n} \qquad |k|>K \tag{7.152}$$

Before we proceed let us point out that such inverse power series expansions often do not exist. In fact, even for such a simple function as $f(x)=1/(1+x^2)$, $f_+(k)$ and $f_-(k)$ will have an essential singularity at infinity (as may be seen by computing these Fourier transforms).

The structure we are assuming for the k plane is shown in Fig. 7.65. From Eq. 7.149 we have

$$f_+(x)=\frac{1}{\sqrt{2\pi}}\int_{-\infty+i\sigma}^{\infty+i\sigma} f_+(k)e^{-ikx}\,dk=\frac{1}{2\pi}\sum_{n=1}^{\infty} a_n\int_{-\infty+i\sigma}^{\infty+i\sigma}\frac{e^{-ikx}\,dk}{(-ik)^n}$$

where we take $\sigma>K$ so that the integrals converge. We can use residue calculus to evaluate these integrals as

$$\frac{1}{2\pi}\int_{-\infty+i\sigma}^{\infty+i\sigma}\frac{e^{-ikx}}{(-ik)^n}\,dk=\begin{cases}\dfrac{x^{n-1}}{(n-1)!} & x>0\\ 0 & x<0\end{cases}$$

so that we can write

$$f_+(x)=\sum_{n=1}^{\infty}\frac{a_n x^{n-1}}{(n-1)!}=\sum_{n=0}^{\infty}\frac{a_{n+1}x^n}{n!} \tag{7.153}$$

If Eq. 7.153 is a uniformly convergent power series with a finite radius of convergence about $z=0$ (and we have no guarantee in general that it is), then we can make the identification

$$a_n=f_+^{(n-1)}(0)$$

so that

$$f_+(k)=\frac{1}{\sqrt{2\pi}}\sum_{n=1}^{\infty}\frac{f_+^{(n-1)}(0)}{(-ik)^n}\qquad |k|>K \tag{7.154}$$

while by a similar argument we obtain for $f_-(k)$

$$f_-(k)=-\frac{1}{\sqrt{2\pi}}\sum_{n=1}^{\infty}\frac{f_-^{(n-1)}(0)}{(-ik)^n}\qquad |k|>K \tag{7.155}$$

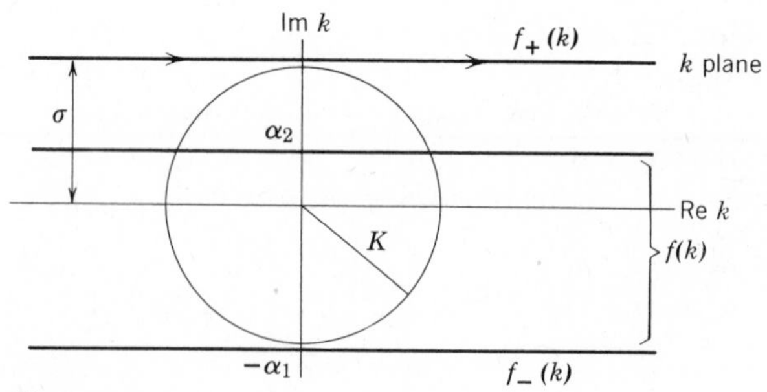

Figure 7.65

Finally we have from Eq. 7.150

$$f(k)=\frac{1}{\sqrt{2\pi}}\sum_{n=1}^{\infty}\frac{[f_+^{(n-1)}(0)-f_-^{(n-1)}(0)]}{(-ik)^n} \qquad |k|>K \qquad \textbf{(7.156)}$$

This shows that the asymptotic behavior of $f(k)$ in the k plane is determined by the behavior of $f(z)$ at the origin (and vice versa).

Example 7.20 Let $f(x)=e^{-|x|}$ so that

$$f_+(k)=\frac{1}{\sqrt{2\pi}}\int_0^{\infty} e^{-x}e^{ikx}\,dx=\frac{1}{\sqrt{2\pi}}\frac{(1+ik)}{(1+k^2)}$$

$$f_-(k)=\frac{1}{\sqrt{2\pi}}\int_{-\infty}^{0} e^{x}e^{ikx}\,dx=\frac{1}{\sqrt{2\pi}}\frac{(1-ik)}{(1+k^2)}$$

$$f(k)=f_+(k)+f_-(k)=\sqrt{\frac{2}{\pi}}\frac{1}{1+k^2}\xrightarrow[k\to\infty]{}\sqrt{\frac{2}{\pi}}\frac{1}{k^2}$$

We can also apply Eq. 7.156 with

$$f_+(x)=e^{-x} \qquad x\geq 0$$

$$f_-(x)=e^{x} \qquad x\leq 0$$

so that

$$f_+(0)=1=f_-(0)$$

$$f_+'(0)=-1=-f_-'(0)$$

which together imply

$$f(k)\xrightarrow[k\to\infty]{}\frac{1}{\sqrt{2\pi}}\frac{(-1-1)}{(-ik)^2}=\sqrt{\frac{2}{\pi}}\frac{1}{k^2}$$

as previously.

7.13 CLASSICAL FUNCTIONS

a. $F(a, b\,|c|\,z)$—The Hypergeometric Function

Many of the special functions of applied mathematics are particular cases of the *hypergeometric function* that is the solution to the differential equation

$$z(z-1)F''+[(a+b+1)z-c]F'+abF=0 \qquad \textbf{(7.157)}$$

In most of our treatment we will take a, b, and c to be real for convenience of notation although this is not necessary. We consider first only those solutions,

$F(a, b\,|c|\,z)$, which are holomorphic about $z=0$. Although we will discuss in detail constructing such solutions in Chapter 8, we will give here a simple development of the solution for reference later. We write

$$F(a, b\,|c|\,z)=\sum_{n=0}^{\infty} \alpha_n z^n \tag{7.158}$$

which assumes a solution holomorphic about $z=0$. Then we have

$$F'=\sum_{n=0}^{\infty} n\alpha_n z^{n-1}=\sum_{n=0}^{\infty} (n+1)\alpha_{n+1} z^n$$

$$F''=\sum_{n=0}^{\infty} n(n-1)\alpha_n z^{n-2}=\sum_{n=0}^{\infty} n(n+1)\alpha_{n+1} z^{n-1}=\sum_{n=0}^{\infty} (n+1)(n+2)\alpha_{n+2} z^n$$

Substituting these back into Eq. 7.157 we obtain

$$\sum_{n=0}^{\infty} [n(n-1)\alpha_n - n(n+1)\alpha_{n+1}+(1+a+b)n\alpha_n - c(n+1)\alpha_{n+1}+ab\alpha_n]z^n \equiv 0$$

Equating the coefficient of z^n to zero we have

$$\begin{aligned}(n+1)(n+c)\alpha_{n+1}&=[n(n-1)+n(1+a+b)+ab]\alpha_n\\ &=[n^2+(a+b)n+ab]\alpha_n\\ &=(n+a)(n+b)\alpha_n\end{aligned} \tag{7.159}$$

Let $\alpha_0 \equiv 1$ so that

$$\alpha_1=\frac{ab}{c}\,\alpha_0=\frac{ab}{c}$$

$$\alpha_{n+1}=\frac{(n+a)(n+b)}{(n+c)(n+1)}\,\alpha_n$$

However, since $\Gamma(z+1)=z\Gamma(z)$ and $\Gamma(n+1)=n!$, we easily see that

$$\alpha_n=\text{const.}\,\frac{\Gamma(a+n)\Gamma(b+n)}{\Gamma(c+n)n!}$$

The normalization condition $\alpha_0=1$ requires that

$$\text{const.}=\Gamma(c)/\Gamma(a)\Gamma(b)$$

or that

$$\alpha_n=\frac{\Gamma(c)}{\Gamma(a)\Gamma(b)}\,\frac{\Gamma(a+n)\Gamma(b+n)}{\Gamma(c+n)n!} \tag{7.160}$$

The solution given by Eq. 7.158 then becomes

$$\begin{aligned}F(a, b(c)z)&=1+\frac{ab}{c}\,z+\frac{a(a+1)b(b+1)}{2!\,c(c+1)}\,z^2+\cdots\\ &=\frac{\Gamma(c)}{\Gamma(a)\Gamma(b)}\sum_{n=0}^{\infty}\frac{\Gamma(a+n)\Gamma(b+n)}{\Gamma(c+n)n!}\,z^n\end{aligned} \tag{7.161}$$

Notice that this series does not converge if c is zero or a negative integer and that the series will terminate after a finite number of terms only if a or b is a negative integer.

In fact if $c=-N$ where N is zero or a positive integer then the recursion relation, Eq. 7.159, for $n=N$ becomes

$$(N+a)(N+b)\alpha_N=0$$

so that if neither a nor b is $-N$, then $\alpha_N=0$. This implies that

$$\alpha_n=0 \qquad n=0, 1, 2, \ldots, N$$

If we let $n=N+1+m$, then we obtain the recursion relation

$$\alpha_{N+2+m}=\frac{(N+m+1+a)(N+m+1+b)}{(N+1+m)(m+1)}\alpha_{N+1+m} \qquad m=0, 1, 2, \ldots \tag{7.162}$$

If we again choose $\alpha_{N+1}=1$ we obtain

$$\alpha_{N+1+m}=\frac{\Gamma(N+1)}{\Gamma(N+1+a)\Gamma(N+1+b)}\frac{\Gamma(N+1+m+a)\Gamma(N+1+m+b)}{\Gamma(N+1+m)m!} \tag{7.163}$$

so that

$$\begin{aligned} F(a, b\,|-N|\,z) &= z^{N+1}\sum_{m=0}^{\infty}\alpha_{N+1+m}z^m \\ &= \frac{N!\,z^{N+1}}{\Gamma(N+1+a)\Gamma(N+1+b)}\sum_{m=0}^{\infty}\frac{\Gamma(N+1+m+a)\Gamma(N+1+m+b)}{m!\,(N+m)!}z^m \end{aligned} \tag{7.164}$$

If we now return to Eq. 7.161 and apply the ratio test, we find that

$$\frac{\Gamma(a+n+1)\Gamma(b+n+1)\Gamma(c+n)n!}{\Gamma(c+n+1)\Gamma(a+n)\Gamma(b+n)(n+1)!}|z|=\frac{(a+n)(b+n)}{(c+n)(n+1)}|z|\xrightarrow[n\to\infty]{}|z|<1$$

which requires that

$$|z|<1$$

We see that $F(a, b\,|c|\,z)$ is holomorphic inside the region $|z|<1$. Of course there is one other linearly independent solution since we are dealing with a second order linear differential equation (cf. Section 8.2). In order to derive some functional relations just as with the gamma function for purposes of analytic continuation, we require a knowledge of this second solution about $z=0$. We have obtained $F(a, b\,|\,c\,|\,z)$ above by expanding F in the power series (7.158). and substituting this into the original differential equation and thus obtaining a recursion relation for the $\{\alpha_n\}$ in terms of α_0. In this case α_0 was chosen to be 1. The series (7.161) is valid for c not zero or a negative integer. As is often done we may attempt to find the second *linearly independent* solution by setting

$$F = z^\alpha f \qquad f = \sum_{n=0}^{\infty} \beta_n z^n$$

where we assume α is not a positive integer. That is, we assume f is holomorphic about $z=0$. Then we easily see that

$$F' = z^\alpha \left[\sum_{n=0}^{\infty} (\alpha+n)\beta_n z^{n-1} \right]$$

$$F'' = z^\alpha \left\{ \sum_{n=0}^{\infty} [\alpha(\alpha-1)+2\alpha n+n(n-1)]\beta_n z^{n-2} \right\}$$

If we substitute these into Eq. 7.157, we have

$$z^\alpha \left\{ \sum_{n=0}^{\infty} [\alpha(\alpha-1)+2\alpha n+n(n-1)+(a+b+1)(\alpha+n)+ab]\beta_n z^n - \sum_{n=0}^{\infty} [\alpha(\alpha-1)+2\alpha n+n(n-1)+c(\alpha+n)]\beta_n z^{n-1} \right\} \equiv 0$$

or

$$z^\alpha \left\{ \frac{1}{z}\alpha[(\alpha-1)+c]\beta_0 + \sum_{n=0}^{\infty} \gamma_n z^n \right\} \equiv 0$$

This is of the form $A(z)z^\alpha \equiv 0$ for all z in some region R. Since $z^\alpha \not\equiv 0$, then $A(z) \equiv 0$ by analyticity. This implies that

$$\alpha = 1-c \tag{7.165}$$

Again we see that c cannot be zero or a negative integer if we are to have a linearly independent solution. Therefore the second linearly independent solution is given as

$$F_2(z) = z^{1-c} f(z) \tag{7.166}$$

We can now compute the following derivatives,

$$F_2' = (1-c)z^{-c}f(z) + z^{1-c}f'(z)$$

$$F_2'' = -c(1-c)z^{-c-1}f(z) + 2(1-c)z^{-c}f'(z) + z^{1-c}f''(z)$$

and substitute these into Eq. 7.157 to find

$$z(z-1)f'' + [(a+b-2c+3)z + (c-2)]f' + (a-c+1)(b-c+1)f = 0$$

By comparison with Eq. 7.157 for $F(a, b \mid c \mid z)$ we see that if we take

$$a' = (b-c+1)$$

$$b' = (a-c+1)$$

$$c' = -(c-2)$$

then

$$f(z) = F(b-c+1, a-c+1 \mid 2-c \mid z)$$

Therefore finally the general solution to Eq. 7.157 about $z=0$ is just

$$AF(a, b\,|c|\,z)+Bz^{1-c}F(b-c+1, a-c+1\,|2-c|\,z) \tag{7.167}$$

where A and B are arbitrary constants. Again we must remember that c cannot be zero or a negative integer for this expression to be valid.

We now expand the nonsingular solution about $z=0$ in terms of the two linearly independent ones about $z=1$. First however we demonstrate that $z=+1$ is a singularity of $F(a, b\,|\,c\,|\,z)$. We see from Eq. 7.161 that the expansion coefficients of Eq. 7.158 are positive (or zero) for large enough n when a, b, and c are real. Let us assume that for an arbitrary $f(z)$ such that

$$f(z)=\sum_{n=0}^{\infty}\alpha_n z^n \tag{7.168}$$

$\alpha_n\geq 0$ all n, and that this series converges uniformly for $|z|<1$ and diverges for some $|z|=1$. Let us furthermore assume that $z=+1$ is *not* a singular point of $f(z)$. In that case there would exist a convergent Taylor expansion about the point, $x=1-\delta$, $0<\delta<1$, and the radius of convergence of this series would include the point, $z=+1$.

$$f(z)=\sum_{m=0}^{\infty}\frac{f^{(m)}(1-\delta)}{m!}(z-1+\delta)^m \tag{7.169}$$

Since $f(z)$ is holomorphic about $z=1-\delta$, we may differentiate Eq. 7.168 to evaluate the $f^{(m)}(1-\delta)$ for Eq. 7.169 to obtain

$$f(z)=\sum_{m=0}^{\infty}\frac{(z-1+\delta)^m}{m!}\sum_{n=0}^{\infty}\alpha_n n(n-1)\cdots(n-m+1)(1-\delta)^{n-m} \tag{7.170}$$

If we take $z=1+\delta$, then (7.170) is convergent by assumption and, since $\alpha_n\geq 0$ all n, this is a double series all of whose terms are positive (or zero). In Section A7.2 we prove that such a convergent series of nonnegative terms can be summed in any order to obtain a unique result. Using this result we have

$$\begin{aligned} f(z)&=\sum_{n=0}^{\infty}\alpha_n\sum_{m=0}^{\infty}\frac{n(n-1)\cdots(n-m+1)}{m!}(z-1+\delta)^m(1-\delta)^{n-m}\\ &=\sum_{n=0}^{\infty}\alpha_n\sum_{m=0}^{\infty}\frac{n!}{(n-m)!\,m!}(z-1+\delta)^m(1-\delta)^{n-m}\\ &=\sum_{n=0}^{\infty}\alpha_n(z-1+\delta+1-\delta)^n=\sum_{n=0}^{\infty}\alpha_n z^n \end{aligned} \tag{7.171}$$

where we have used the binomial expansion

$$(a+b)^p=\Gamma(p+1)\sum_{s=0}^{\infty}\frac{a^{p-s}b^s}{\Gamma(p+1-s)s!} \tag{7.172}$$

However Eq. 7.171 states that the original power series, Eq. 7.168, converges at

$z=1+\delta$, which is a contradiction since the radius of convergence was given as unity. Hence $z=+1$ is a singularity of $f(z)$.

Furthermore we can determine the nature of this singularity at $z=+1$ as follows. Suppose that

$$f(z)=\sum_{n=0}^{\infty} \alpha_n z^n$$

$$g(z)=\sum_{n=0}^{\infty} \beta_n z^n$$

are functions holomorphic within the unit circle, that both series diverge for $z=+1$, and that $\alpha_n \geq 0$, $\beta_n \geq 0$, $\forall n$. If

$$\alpha_n \xrightarrow[n\to\infty]{} \gamma\beta_n \tag{7.173}$$

where γ is some constant (independent of n), then

$$f(z) \xrightarrow[z\to+1]{} \gamma g(z)$$

in the sense that

$$\lim_{z\to+1} \frac{f(z)}{g(z)}=\gamma \tag{7.174}$$

By Eq. 7.173, for any ε, no matter how small, we can always find an N large enough so that

$$|\alpha_n-\gamma\beta_n|<\varepsilon\beta_n \qquad \forall n>N$$

For $0<x<1$ we can then write

$$\begin{aligned}
\left|\frac{f(x)}{g(x)}-\gamma\right| &= \frac{1}{|g(x)|}\left|\sum_{n=0}^{\infty} (\alpha_n-\gamma\beta_n)x^n\right| \\
&\leq \frac{1}{|g(x)|}\left\{\left|\sum_{n=0}^{N} (\alpha_n-\gamma\beta_n)x^n\right| + \left|\sum_{n=N+1}^{\infty} (\alpha_n-\gamma\beta_n)x^n\right|\right\} \\
&\leq \frac{1}{|g(x)|}\left\{\sum_{n=0}^{N} |\alpha_n-\gamma\beta_n| + \varepsilon \sum_{n=N+1}^{\infty} \beta_n x^n\right\} \\
&\leq \frac{1}{|g(x)|}\sum_{n=0}^{N} |\alpha_n-\gamma\beta_n| + \varepsilon
\end{aligned} \tag{7.175}$$

However since $g(x)$ diverges as $x \to 1^-$, we can always choose an x sufficiently close to $+1$ so that, for any fixed N, the first term on the r.h.s. of Eq. 7.175 can be made arbitrarily small (e.g., less than ε). Therefore we have

$$\left|\frac{f(x)}{g(x)}-\gamma\right|<2\varepsilon \tag{7.176}$$

for x close enough to $+1$.

Let us now apply these results to $F(a, b\,|c|\,z)$. By the binomial expansion we have

$$(1-z)^{-\alpha}=\sum_{n=0}^{\infty}\frac{\Gamma(\alpha+n)}{\Gamma(\alpha)n!}z^n \tag{7.177}$$

From Eq. 7.161 the expansion coefficients for $F(a, b\,|c|\,z)$ are

$$\alpha_n=\frac{\Gamma(c)}{\Gamma(a)\Gamma(b)}\frac{\Gamma(a+n)\Gamma(b+n)}{\Gamma(c+n)} \tag{7.178}$$

From Stirling's approximation, Eq. 7.133, we have

$$\begin{aligned}\Gamma(a+n)\xrightarrow[n\to\infty]{}&\sqrt{2\pi}\,(a+n)^{a+n-(1/2)}e^{-a-n}\\ \longrightarrow&\sqrt{2\pi}\,n^{a-(1/2)}e^{-a-n}(a+n)^n\\ =&\sqrt{2\pi}\,n^{a-(1/2)}e^{-a-n}n^n\left[\left(1+\frac{a}{n}\right)^{n/a}\right]^a\\ \longrightarrow&\sqrt{2\pi}\,a^{-1/2}e^{-a-n}n^ne^a\\ =&\sqrt{2\pi}\,n^{a-(1/2)}e^{-n}n^n\end{aligned}$$

so that

$$\frac{\Gamma(a+n)\Gamma(b+n)}{\Gamma(c+n)\Gamma(n+1)}\xrightarrow[n\to\infty]{}n^{a+b-c-1} \tag{7.179}$$

For $a+b-c>0$ the series $\sum_{n=0}^{\infty}\alpha_n$ will diverge as required for Eq. 7.176. We will treat the limiting case $a+b-c=0$ below. Similarly the coefficients of Eq. 7.177 have the asymptotic behavior

$$\frac{\Gamma(\alpha+n)}{\Gamma(\alpha)\Gamma(n+1)}\xrightarrow[n\to\infty]{}\frac{n^{\alpha-1}}{\Gamma(\alpha)} \tag{7.180}$$

Therefore, from Eqs. 7.173, 7.178–7.180, we seek a γ and an α such that

$$\frac{\Gamma(c)}{\Gamma(a)\Gamma(b)}n^{a+b-c}=\frac{\gamma n^{\alpha}}{\Gamma(\alpha)}$$

which requires

$$\alpha=a+b-c>0$$

$$\gamma=\frac{\Gamma(a+b-c)\Gamma(c)}{\Gamma(a)\Gamma(b)}$$

We finally obtain

$$F(a, b\,|c|\,z)\xrightarrow[z\to+1]{}\frac{\Gamma(c)\Gamma(a+b-c)}{\Gamma(a)\Gamma(b)}(1-z)^{c-a-b}\qquad a+b>c \tag{7.181}$$

Now if $a+b-c=0$, then Eq. 7.179 implies that

$$\alpha_n\xrightarrow[n\to\infty]{}\frac{\Gamma(c)}{\Gamma(a)\Gamma(b)}\frac{1}{n}$$

However an elementary power series is

$$\sum_{n=0}^{\infty} \frac{z^n}{n} = -\ln(1-z)$$

so that we may again apply Eq. 7.176 to conclude that

$$F(a, b\,|c|\,z) \xrightarrow[z\to+1]{} -\frac{\Gamma(c)}{\Gamma(a)\Gamma(b)} \ln(1-z) \qquad a+b=c \tag{7.182}$$

It is also easy to see directly from the differential equation for $F(a, b\,|c|\,z)$, Eq. 7.157, that when $a+b=c$ then $\ln(1-z)$ is approximately a solution (i.e., up to the most singular terms) near $z=+1$. We will discuss this point more in Section 8.2b.

If we return to Eq. 7.157 and make the change of variable

$$z' = 1-z$$

$$\frac{d}{dz} = \frac{-d}{dz'}$$

then the differential equation becomes

$$z'(z'-1)f'' + [(a+b+1)z' - (a+b+1-c)]f' + abf = 0 \tag{7.183}$$

If we compare this with Eq. 7.157 then with

$$a' = a$$

$$b' = b$$

$$c' = a+b+1-c$$

we see that a solution to Eq. 7.183 is just

$$f(z') = F(a, b\,|a+b-c+1|\,z') = F(a, b\,|a+b-c+1|\,1-z) \tag{7.184}$$

This is the regular solution about the point $z=1$. By comparison with the treatment for constructing the irregular solution about $z=0$, we see that the other linearly independent solution about $z=1$ is just

$$(1-z)^{c-a-b} F(c-b, c-a\,|c-a-b+1|\,1-z)$$

Since the solutions about $z=0$ and $z=1$ have a common domain of definition and are both solutions to the same differential equation, they must be expressible in terms of each other by the principle of analytic continuation (cf. Fig. 7.66). Therefore we can write

$$F(a, b\,|c|\,z) = \alpha F(a, b\,|a+b-c+1|\,1-z) + \beta F(c-b, c-a\,|c-a-b+1|\,1-z) \times (1-z)^{c-a-b}$$

where α and β are constants to be determined. If we consider each side of this relation as $z \to 1$, then since $F(\mu, \nu\,|\sigma|\,1-z) \to 1$, the dominant term on the

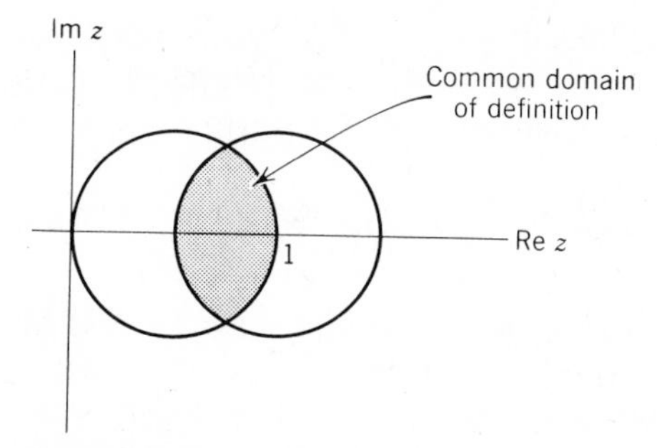

Figure 7.66

right is $\beta(1-z)^{c-a-b}$, $(a+b)>c$, while that on the left is

$$F(a, b\,|c|\,z) \xrightarrow[z\to 1]{} \frac{\Gamma(c)\Gamma(a+b-c)}{\Gamma(a)\Gamma(b)}(1-z)^{c-a-b}$$

so that

$$\beta = \frac{\Gamma(c)\Gamma(a+b-c)}{\Gamma(a)\Gamma(b)}$$

If we now consider the relation as $z \to 0$, then

$$1 \xrightarrow[z\to 0]{} \frac{(a+b-c)\Gamma(a+b-c)\Gamma(c-1)}{\Gamma(a)\Gamma(b)}\left[\alpha - \frac{\Gamma(c-a-b)\Gamma(c)}{\Gamma(c-a)\Gamma(c-b)}\right]z^{1-c} + \text{finite terms}$$

For $c>1$ the right side will become infinite as $z \to 0$ unless the coefficient of this term vanishes. Therefore in this case we obtain

$$\alpha = \frac{\Gamma(c-a-b)\Gamma(c)}{\Gamma(c-a)\Gamma(c-b)}$$

The joining relation sought is

$$F(a, b\,|c|\,z) = \frac{\Gamma(c)\Gamma(c-a-b)}{\Gamma(c-a)\Gamma(c-b)}F(a, b\,|a+b-c+1|\,1-z) + \frac{\Gamma(c)\Gamma(a+b-c)}{\Gamma(a)\Gamma(b)}(1-z)^{c-a-b}F(c-a, c-b\,|c-a-b+1|\,1-z) \tag{7.185}$$

which we have so far demonstrated for $(a+b)>c$, $c>1$.

We now consider constructing an integral representation for $F(a, b\,|c|\,z)$. We have seen earlier that the defining series is Eq. 7.161

$$F(a, b\,|c|\,z) = \frac{\Gamma(c)}{\Gamma(a)\Gamma(b)}\sum_{n=0}^{\infty}\frac{\Gamma(a+n)\Gamma(b+n)}{\Gamma(c+n)n!}z^n$$

Since

$$\Gamma(-t) \xrightarrow[t\to n]{} \frac{-(-1)^n}{(t-n)\Gamma(t+1)}$$

we can recognize $z^n/n!$ as part of the residue of a contour integral of $[-(-z)^t\Gamma(-t)]$ where the contour does not cross any of the singularities of $\Gamma(-t)$. Therefore we would expect an integral representation of the form

$$F(a, b\,|c|\,z)=\frac{\Gamma(c)}{\Gamma(a)\Gamma(b)}\oint\frac{\Gamma(a+t)\Gamma(b+t)}{2\pi i\Gamma(c+t)}\Gamma(-t)(e^{-\pi i}z)^t\,dt \qquad \textbf{(7.186)}$$

where C must avoid all the singularities of the integrand and must enclose all those (and *only* those) of $\Gamma(-t)$ (i.e., zero and the positive integers). This would suggest a contour from $-i\infty$ to $+i\infty$ bulging to the left of $t=0$ and closed in the right half t plane. However we must consider two important points. First the integrand must vanish on the large contour to the right and, second, we must not enclose any of the singularities of the rest of the integrand.

If neither a nor b is a negative integer or zero (and these are not very interesting cases since then the power series terminates after a finite number of terms and is, therefore, a finite-order polynomial in z), then the singularities of $\Gamma(a+t)\Gamma(b+t)$ are located at

$$t+a=\text{negative integer or zero}$$

$$t+b=\text{negative integer or zero}$$

We are also assuming that $a-b$ is not an integer since this would produce second-order poles in the integrand of Eq. 7.186 so that we would not recover the original power series given in Eq. 7.161. Furthermore c cannot be zero or a negative integer if the original power series for $F(a, b\,|c|\,z)$ is to be defined, as we have previously seen. We can then choose the contour as shown in Fig. 7.67 where the direction of the contour is taken as clockwise closed to the right. This contour remains valid if a and/or b are purely real *provided* neither is a negative integer or zero.

We must now consider neglecting the contribution from the large semicircular contour to the right. We first make use of the following relation for $\Gamma(z)$ that we have already derived (cf. Eq. 7.110)

$$\Gamma(z)\Gamma(1-z)\sin(\pi z)=\pi$$

With the aid of this relation and Stirling's approximation (cf. especially Eq. 7.179) the gamma functions in the integrand of Eq. 7.186 can then be expressed as follows,

$$\frac{\Gamma(a+t)\Gamma(b+t)}{\Gamma(c+t)\Gamma(t+1)}\xrightarrow[t\to\infty]{} t^{a+b-c-1}=R^{a+b-c-1}e^{i(a+b-c-1)\theta}$$

where $t=Re^{i\theta}$. Similarly we easily show that

$$\frac{1}{\sin(\pi t)}\xrightarrow[R\to\infty]{}\begin{cases}-2i\exp(i\pi R\cos\theta-\pi R\sin\theta) & 0<\theta<\pi\\ 2i\exp(i\pi R\cos\theta+\pi R\sin\theta) & 0>\theta>-\pi\end{cases}$$

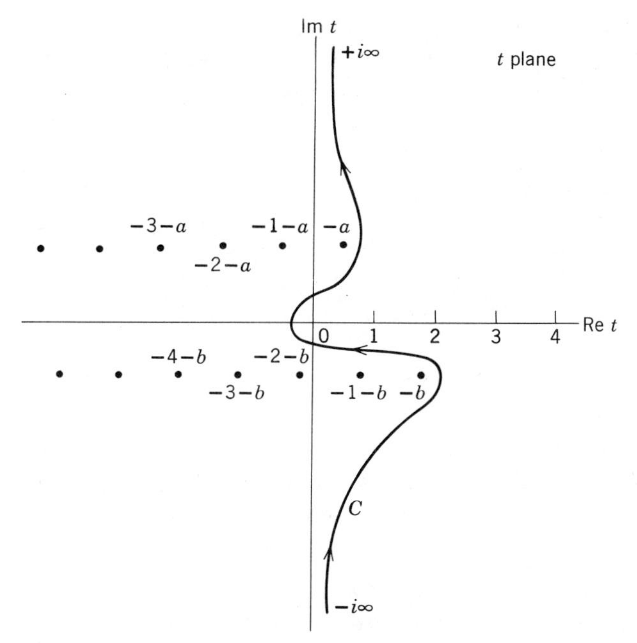

Figure 7.67

Also if $z=re^{i\phi}$ then we have

$$(e^{-\pi i}z)^t=[re^{i(\phi-\pi)}]^t=[e^{\ln r+i(\phi-\pi)}]^t=\exp\{R[(\ln r)\cos\theta+(\pi-\phi)\sin\theta]$$
$$+iR[(\phi-\pi)\cos\theta+(\ln r)\sin\theta]\}$$

Therefore the overall magnitude of the integrand of Eq. 7.186 is as $R\to\infty$,

$$\left|\frac{\Gamma(a+t)\Gamma(b+t)}{\Gamma(c+t)\Gamma(t+1)}\frac{(-z)^t}{\sin(\pi t)}\right|\xrightarrow[R\to\infty]{}2R^{a+b-c-1}e^{R\ln r\cos\theta}\begin{cases}e^{-R\phi\sin\theta} & 0<\theta<\pi/2\\ e^{R(2\pi-\phi)\sin\theta} & 0>\theta>-\pi/2\end{cases}\tag{7.187}$$

The one value of z ruled out is $\phi=0$ (or 2π) and $r=1$ [i.e., $z=+1$, a pole of $F(a,b\,|c|\,z)$]. The angles for θ have been reduced from $\pm\pi$ to $\pm\pi/2$ due to the $\cos\theta$ factor arising from $(e^{-\pi i}z)^t$. Therefore for the following ranges

$$0<|z|<1 \text{ (i.e., } 0<r<1 \qquad \ln r<0)$$

$$0\leq\phi\leq 2\pi \qquad -\frac{\pi}{2}<\theta<\frac{\pi}{2}$$

the integrand will vanish (exponentially) as $R\to\infty$ in the *right-half* t plane. Notice that the contour must approach $\pm i\infty$ to the *right* of the Im t axis. This establishes Eq. 7.186 where the path from $-i\infty$ to $+i\infty$ has been specified above.

We have seen that when the contour is closed to the right, then we require that $0<|z|<1$ for convergence, and we recover the original series representation for $F(a, b\,|c|\,z)$. However, if we consider closing the contour to the left then the integral again vanishes on the large semicircular portion of the contour provided

$$|z|>1 \text{ (i.e., } \ln r>0) \qquad \frac{\pi}{2}<\theta<\frac{3\pi}{2}$$

where we now obtain the residues of the poles of $\Gamma(a+t)$ and $\Gamma(b+t)$. As a technical point in obtaining the asymptotic behavior of the Γ-functions in the integrand in the left-hand t plane, we must first change all of the arguments to *negative* ones as

$$\Gamma(a+t)=\frac{1}{\Gamma(1-a-t)\sin[\pi(a+t)]}$$

before applying Stirling's approximation. The result obtained previously for the magnitude of the integrand remains valid for this new domain of z. We then find

$$\begin{aligned}\frac{\Gamma(a)\Gamma(b)}{\Gamma(c)}&F(a, b\,|c|\,z)\\&=\sum_{n=0}^{\infty}\frac{\Gamma(a+n)\Gamma(1-c+a+n)}{\Gamma(1+n)\Gamma(1+a-b+n)}\frac{\sin[\pi(c-a-n)]}{\cos(n\pi)\sin[\pi(b-a-n)]}(e^{-\pi i}z)^{-a-n}\\&\quad+\sum_{n=0}^{\infty}\frac{\Gamma(b+n)\Gamma(1-c+b+n)}{\Gamma(1+n)\Gamma(1-a+b+n)}\frac{\sin[\pi(c-b-n)]}{\cos(n\pi)\sin[\pi(a-b-n)]}(e^{-\pi i}z)^{-b-n}\\&=\frac{\Gamma(a)\Gamma(b-a)}{\Gamma(c-a)}(e^{-\pi i}z)^{-a}F(a, 1-c+a\,|1-b+a|\,1/z)\\&\quad+\frac{\Gamma(b)\Gamma(a-b)}{\Gamma(c-b)}(e^{-\pi i}z)^{-b}F(b, 1-c+b\,|1-a+b|\,1/z)\end{aligned} \tag{7.188}$$

If $[a-b]$ is an integer or zero, then one of these series diverges as we would expect from our previous comments on the validity of this integral representation.

These series are not really equal since they converge for $|z|<1$ and $|z|>1$, respectively. It is the *integral representation* with an appropriate contour that converges for (nearly) all values of z and hence is "the" function, $F(a, b\,|c|\,z)$. This representation allows analytic continuation from one region to another. We can see this as follows. If we take a z, $|z|<1$, but close to $z=+1$, then we can use Eqs. 7.186 and 7.185 (cf. Fig. 7.66) to connect $F(a, b\,|c|\,z)$ from a region with $|z|<1$ to a region with $|z|>1$. Now use Eq. 7.186 with the contour closed to the left to obtain Eq. 7.188. By the uniqueness theorem for analytic continuation and the common overlap region of Fig. 7.66, we see that the r.h.s. of (7.188) does represent $F(a, b\,|c|\,z)$ for all $|z|>1$.

We have treated only a few of the properties of the hypergeometric function and have excluded certain values of the parameters a, b, c from our considerations. There has been no attempt at completeness. What we have done illustrates the techniques we will use later and will serve as an introduction to further study for those so interested. Another integral representation for $F(a, b\,|c|\,z)$ is developed in Problem 7.52.

A special case of the hypergeometric function are the *Jacobi polynomials* defined as

$$P_n^{(\alpha,\beta)}(x)=\frac{\Gamma(1+\alpha+n)}{\Gamma(1+\alpha)n!}\,F\left(1+\alpha+\beta+n,\,-n\,|1+\alpha|\,\frac{1-x}{2}\right) \quad \textbf{(7.189)}$$

Many of the important orthogonal polynomials used in applied mathematics are special cases of the Jacobi polynomials. Some of the more important properties of these polynomials are developed in Section A7.3. We shall meet these polynomials again in Chapter 9 when we consider representations of the three-dimensional rotation group. As we will demonstrate in Section A7.3, the $P_n^{(\alpha,\beta)}(x)$, with $\alpha>-1$, $\beta>-1$, satisfy the differential equation (cf. Eq. 7.157)

$$(1-x^2)\frac{d^2}{dx^2}P_n^{(\alpha,\beta)}(x)+[\beta-\alpha-(\alpha+\beta+2)x]\frac{d}{dx}P_n^{(\alpha,\beta)}(x)$$
$$+n(n+\alpha+\beta+1)P_n^{(\alpha,\beta)}(x)=0 \quad \textbf{(7.190)}$$

the orthogonality relation

$$\int_{-1}^{1}(1-x)^\alpha(1+x)^\beta P_m^{(\alpha,\beta)}(x)P_n^{(\alpha,\beta)}(x)\,dx$$
$$=\frac{2^{\alpha+\beta+1}}{(2n+\alpha+\beta+1)}\frac{\Gamma(n+\alpha+1)\Gamma(n+\beta+1)}{\Gamma(n+1)\Gamma(n+\alpha+\beta+1)}\delta_{nm} \quad \textbf{(7.191)}$$

the symmetry property

$$P_n^{(\alpha,\beta)}(x)=(-1)^nP_n^{(\beta,\alpha)}(-x) \quad \textbf{(7.192)}$$

and the generating relation

$$P_n^{(\alpha,\beta)}(x)=\frac{(-1)^n}{2^n n!}(1-x)^{-\alpha}(1+x)^{-\beta}\frac{d^n}{dx^n}[(1-x)^{\alpha+n}(1+x)^{\beta+n}] \quad \textbf{(7.193)}$$

b. $P_n(z)$—The Legendre Functions

We have seen previously that the Legendre polynomials, $P_n(z)$ (cf. Section 4.3b), are solutions to the differential equation

$$(z^2-1)P_n''(z)+2zP_n'(z)-n(n+1)P_n(z)=0 \quad \textbf{(7.194)}$$

where we had taken n to be a positive integer or zero. For these cases we saw that (cf. Eq. 4.33)

$$P_n(z)=\frac{1}{2^n n!}\frac{d^n}{dz^n}(z^2-1)^n=\sum_{s=0}^{n}\frac{(-1)^{n-s}1\cdot3\cdot5\cdots(2s-1)}{(n-s)!\,(2s-n)!\,2^{n-s}}z^{2s-n}$$

From Cauchy's integral formula,

$$f^{(n)}(z)=\frac{n!}{2\pi i}\int_C \frac{f(t)\,dt}{(t-z)^{n+1}}$$

we can express $P_n(z)$ as

$$P_n(z)=\frac{1}{2\pi i}\int_C \frac{(t^2-1)^n\,dt}{2^n(t-z)^{n+1}} \tag{7.195}$$

which is an integral representation, where C is any closed contour about the point, $t=z$, in the t plane. If we substitute the integral representation into the original differential equation, Eq. 7.194, we find that

$$\begin{aligned}(z^2-1)P_n''(z)+2zP_n'(z)-n(n+1)P_n(z)&=-\frac{(n+1)}{2^n 2\pi i}\int_C \frac{d}{d}\left[\frac{(t^2-1)^{n+1}}{(t-z)^{n+2}}\right]dt\\&=-\frac{(n+1)}{2^n 2\pi i}\left[\frac{(t^2-1)^{n+1}}{(t-z)^{n+2}}\right]\Bigg|_{\substack{\text{end}\\\text{points}}}\end{aligned}$$

This will vanish as long as C is such that the integrand is single valued on the contour (i.e., the integrand returns to its original value after one circuit around the contour, C). For n a positive integer the integrand is single valued and the right hand side will therefore vanish. However if we let $n\to\alpha$, where α is arbitrary, then this quantity has branch points at

$$t=\pm 1 \qquad t=z \qquad t=\infty$$

We may keep the integrand single valued in the t plane by use of the contour shown in Fig. 7.68. That is, (cf. Example 7.14) the factor $(t-z)^{-\alpha}$ undergoes a phase change of $e^{-i2\pi\alpha}$ as we make one (counterclockwise) loop about $t=z$ while the factor $(t^2-1)^{\alpha}$ undergoes a phase change of $e^{i2\pi\alpha}$ as we make one (counterclockwise) loop about $t=+1$. Therefore the function $(t^2-1)^{\alpha}/(t-z)^{\alpha+1}$ does return to its initial value as we travel once around C. Since the integrand is single valued on C, $P_\alpha(z)$ will still satisfy the differential equation for arbitrary α. Of course there is now a branch cut in the z plane for $-\infty<x<-1$. Hence $P_\alpha(z)$ is single valued for positive integral values of α.

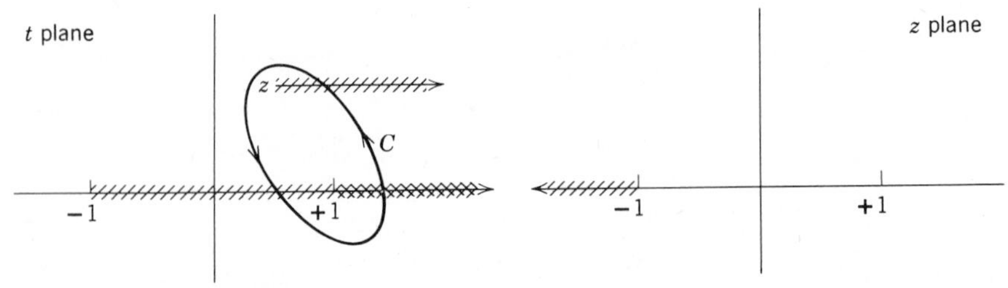

Figure 7.68

Since the differential equation

$$(z^2-1)P''_\alpha(z)+2zP'_\alpha(z)-\alpha(\alpha+1)P_\alpha(z)=0 \tag{7.196}$$

is invariant w.r.t. the substitution $\alpha \to -\alpha-1$, we know that there will exist some solution to Eq. 7.196, which will also possess this symmetry. We can see that $P_\alpha(z)$ is such a solution by comparing Eq. 7.196 with the hypergeometric equation, (7.157), and making the identifications

$$\begin{aligned} a&=-\alpha \\ b&=\alpha+1 \\ c&=1 \\ z &\to (\tfrac{1}{2})(1-z) \end{aligned}$$

so that

$$P_\alpha(z)=F\left(-\alpha,\, \alpha+1\, |1|\, \frac{1-z}{2}\right) \tag{7.197}$$

We easily check that the normalization is correct since from Eq. 7.195 for arbitrary α we have

$$P_\alpha(1)=\frac{1}{2\pi i}\int_C \frac{(t+1)^\alpha\, dt}{2^\alpha(t-1)}=1$$

which agrees with Eq. 7.197. Therefore we have

$$P_\alpha(z)=P_{-\alpha-1}(z) \tag{7.198}$$

It is also apparent from Eqs. 7.189 and 7.197 that the Jacobi polynomials with $\alpha=\beta=0$ are just the Legendre polynomials. From Eq. 7.182 we can see explicitly that $P_\alpha(z)$ has a branch point at $z=-1$ when α is not an integer since then

$$P_\alpha(z) \xrightarrow[z\to-1]{} \frac{\sin(\pi\alpha)}{\pi}\ln(1+z) \tag{7.199}$$

If we use the relation of Eq. 7.188 for $F(a, b, |c|\, z)$, we find

$$\begin{aligned} P_\alpha(z)=&\frac{\Gamma(2\alpha+1)}{\Gamma(\alpha+1)\Gamma(\alpha+1)}\left(\frac{z-1}{2}\right)^\alpha F\left(-\alpha, -\alpha\, |-2\alpha|\, \frac{2}{1-z}\right) \\ &+\frac{\Gamma(-2\alpha-1)}{\Gamma(-\alpha)\Gamma(-\alpha)}\left(\frac{z-1}{2}\right)^{-\alpha-1} F\left(\alpha+1, \alpha+1\, |2\alpha+2|\, \frac{2}{1-z}\right) \end{aligned} \tag{7.200}$$

Even if α is a positive integer this holds since then the second term vanishes because of the *two* factors of $\Gamma(-\alpha)$ in the denominator. Reference to the original series definition of $F(a, b, |c|\, z)$, Eq. 7.161, shows that the series terminates for α a positive integer or zero before the terms in the series can "blow up." This relation is useful for finding the asymptotic behavior of $P_\alpha(z)$,

Re $\alpha>0$, as $|z|\to\infty$. Since $F(a, b\,|c|\,0)=1$, we have

$$P_\alpha(z)\xrightarrow[|z|\to\infty]{}\frac{\Gamma(2\alpha+1)}{[\Gamma(\alpha+1)]^2}\frac{z^\alpha}{2^\alpha} \tag{7.201}$$

But since by Eq. 7.129

$$\Gamma(2\alpha+1)=\frac{2^{2\alpha}}{\sqrt{\pi}}\Gamma(\alpha+\tfrac{1}{2})\Gamma(\alpha+1)$$

we have

$$P_\alpha(z)\xrightarrow[|z|\to\infty]{}\frac{\Gamma(\alpha+\frac{1}{2})}{\sqrt{\pi}\,\Gamma(\alpha+1)}(2z)^\alpha \qquad \text{Re}\,\alpha>0 \tag{7.202}$$

As we have seen in Section A4.5, Eq. A4.47, the associated Legendre polynomials $P_n^m(z)$ are the solutions finite at $z=0$ of the differential equation

$$(1-z^2)\frac{d^2}{dz^2}P_n^m(z)-2z\frac{d}{dz}P_n^m(z)+n(n+1)P_n^m(z)-\frac{m^2}{1-z^2}P_n^m(z)=0 \tag{7.203}$$

where

$$P_n^m(z)\equiv(1-z^2)^{(1/2)m}\frac{d^m}{dz^m}P_n(z) \qquad m\geq 0 \qquad n\geq 0 \tag{7.204}$$

We can express these associated Legendre polynomials in terms of the hypergeometric function as follows. Since the residue of $\Gamma(z)$ at $z=-n$ is $(-1)^n/n!$, we see that for any integers m and n

$$\lim_{\varepsilon\to 0}\frac{\Gamma(-m+\varepsilon)}{\Gamma(-n+\varepsilon)}=(-1)^{n-m}\frac{n!}{m!} \tag{7.205}$$

If we use this relation and the defining series for $F(a, b\,|c|\,z)$ (cf. Eq. 7.161), we can write for $0\leq m\leq n$

$$F\left(m-n,\ m+n+1\,|m+1|\,\frac{1-z}{2}\right)=\frac{(n-m)!\,m!}{(n+m)!}\sum_{s=0}^{\infty}\frac{(-1)^s(n+m+s)!}{(m+s)!\,s!\,(n-m-s)!}\left(\frac{1-z}{2}\right)^s \tag{7.206}$$

If we use Eqs. 7.197, 7.204, and 7.206, we have

$$\begin{aligned}\frac{d^m}{dz^m}P_n(z)&=\frac{d^m}{dz^m}F\left(-n,\ n+1\,|1|\,\frac{1-z}{2}\right)\\&=\sum_{r=0}^{\infty}\frac{(-1)^r(n+r)!}{(n-r)!\,(r!)^2}\frac{r(r-1)\cdots(r-m+1)(-1)^m}{2^r}(1-z)^{r-m}\\&=\frac{(-1)^m}{2^m}\sum_{s=0}^{\infty}\frac{(-1)^s(n+m+s)!}{(n-m-s)!\,(m+s)!\,s!}\left(\frac{1-z}{2}\right)^s\\&=\frac{(n+m)!}{2^m m!\,(n-m)!}F\left(m-n,\ m+n+1\,|m+1|\,\frac{1-z}{2}\right)\end{aligned} \tag{7.207}$$

so that

$$P_n^m(z)=\frac{(n+m)!}{2^m m!\,(n-m)!}(1-z^2)^{m/2}F\left(m-n,\, m+n+1\,|m+1|\,\frac{1-z}{2}\right) \quad \textbf{(7.208)}$$

We now obtain a second linearly independent solution to Legendre's equation, Eq. 7.196. We begin by defining a function $Q_n(z)$, the *Legendre functions of the second kind,* for integer n as

$$Q_n(z)=\frac{1}{2}\int_{-1}^{1}\frac{P_n(t)\,dt}{(z-t)} \quad \textbf{(7.209)}$$

where this integral representation shows clearly that $Q_n(z)$ has a branch cut for $y=0$, $-1<x<1$ since, from Eq. 7.52,

$$\lim_{\varepsilon\to 0^+}[Q_n(x+i\varepsilon)-Q_n(x-i\varepsilon)]=-\pi i P_n(x) \quad \textbf{(7.210)}$$

If we use the generating expression

$$P_n(t)=\frac{1}{2^n n!}\frac{d^n}{dt^n}[(t^2-1)^n]$$

and integrate by parts n times, we obtain

$$Q_n(z)=\frac{1}{2^{n+1}n!}\int_{-1}^{1}\frac{dt}{(z-t)}\frac{d^n}{dt^n}[(t^2-1)^n]$$

$$=\frac{1}{2^{n+1}}\int_{-1}^{1}(1-t^2)^n(z-t)^{-n-1}\,dt$$

By direct calculation we see that

$$(z^2-1)\frac{d^2}{dz^2}Q_n(z)+2z\frac{d}{dz}Q_n(z)-n(n+1)Q_n(z)$$

$$=-\frac{(n+1)}{2^{n+1}}\int_{-1}^{1}\frac{(1-t^2)^n}{(z-t)^{n+1}}[(z-t)(n+1)(-2t)+(1-t^2)(n+2)]\,dt$$

$$=-\frac{(n+1)}{2^{n+1}}\int_{-1}^{1}\frac{d}{dt}\left[\frac{(1-t^2)^{n+1}}{(z-t)^{n+2}}\right]dt=a$$

Therefore $Q_n(z)$, at least for n integer, is also a solution to Legendre's equation. Since $Q_n(z)$ has a branch cut for $y=0$, $-1\le x\le 1$, while $P_n(z)$ does not have one here, $Q_n(z)$ is the second linearly independent solution to Legendre's equation.

It is fairly easy to see how to extend $Q_n(z)$ to arbitrary values of the index α, just as we did $P_n(z)$. Consider the expression (compare with Eq. 7.195)

$$Q_\alpha(z)=\frac{-1}{2^{\alpha+2}i\sin(\pi\alpha)}\oint_C\frac{(t^2-1)^\alpha\,dt}{(z-t)^{\alpha+1}} \quad \textbf{(7.211)}$$

where z cannot be on the real axis between -1 and $+1$ and where the contour, C, is shown in Fig. 7.69. If we do not let α be an integer or zero and keep z off the

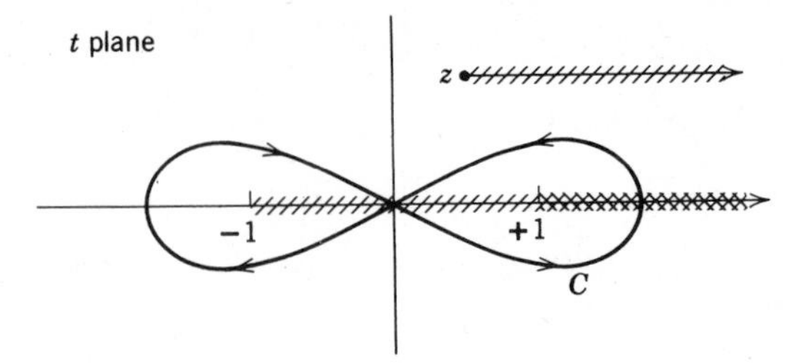

Figure 7.69

cut $-1\le t\le 1$, then just as for the $P_\alpha(z)$ we find that

$$(z^2-1)\frac{d^2}{dz^2}Q_\alpha(z)+2z\frac{d}{dz}Q_\alpha(z)-\alpha(\alpha+1)Q_\alpha(z)$$
$$=\frac{(\alpha+1)}{2^{\alpha+2}i\sin(\pi\alpha)}\oint_C\frac{d}{dt}\left[\frac{(1-t^2)^{\alpha+1}}{(z-t)^{\alpha+2}}\right]dt=0$$

where the last equality follows since the integrand returns to its original value after one circuit around C (cf. Example 7.14). If we restrict Re $\alpha>-1$, then we can collapse the contour of Fig. 7.69 into that shown in Fig. 7.70 with the indicated phases. The integral along C_1 can now be reduced to integrals just below and just above the line segment $0\le t\le +1$ and that along C_2 to similar integrals above and below $-1\le t\le 0$.

$$\oint_{C_1}\frac{(t^2-1)^\alpha\,dt}{(z-t)^{\alpha+1}}=\int_0^1\frac{e^{-\pi i\alpha}(1-t^2)^\alpha\,dt}{(z-t)^{\alpha+1}}+\int_1^0\frac{e^{i\pi\alpha}(1-t^2)^\alpha\,dt}{(z-t)^{\alpha+1}}$$
$$=-2i\sin(\pi\alpha)\int_0^1\frac{(1-t^2)^\alpha\,dt}{(z-t)^{\alpha+1}}$$
$$\oint_{C_2}\frac{(t^2-1)^\alpha\,dt}{(z-t)^{\alpha+1}}=\int_0^{-1}\frac{e^{\pi i\alpha}(1-t^2)^\alpha\,dt}{(z-t)^{\alpha+1}}+\int_{-1}^0\frac{e^{-\pi i\alpha}(1-t^2)^\alpha\,dt}{(z-t)^{\alpha+1}}$$
$$=-2i\sin(\pi\alpha)\int_{-1}^0\frac{(1-t^2)^\alpha\,dt}{(z-t)^{\alpha+1}}$$

Therefore for Re $\alpha>-1$ Eq. 7.211 reduces to

$$Q_\alpha(z)=\frac{1}{2^{\alpha+1}}\int_{-1}^1\frac{(1-t^2)^\alpha\,dt}{(z-t)^{\alpha+1}}\tag{7.212}$$

If we further restrict z to values such that $|z|>1$, then we can expand $(z-t)^{-\alpha-1}$ in a binomial series to obtain

$$Q_\alpha(z)=\frac{1}{(2z)^{\alpha+1}}\int_{-1}^1dt\sum_{s=0}^\infty\frac{\Gamma(\alpha+s+1)}{\Gamma(\alpha+1)s!}\left(\frac{t}{z}\right)^s(1-t^2)^\alpha$$
$$=\frac{1}{(2z)^{\alpha+1}}\sum_{m=0}^\infty\frac{\Gamma(\alpha+2m+1)}{\Gamma(\alpha+1)(2m)!}\frac{1}{z^{2m}}\int_{-1}^1t^{2m}(1-t^2)^\alpha\,dt$$

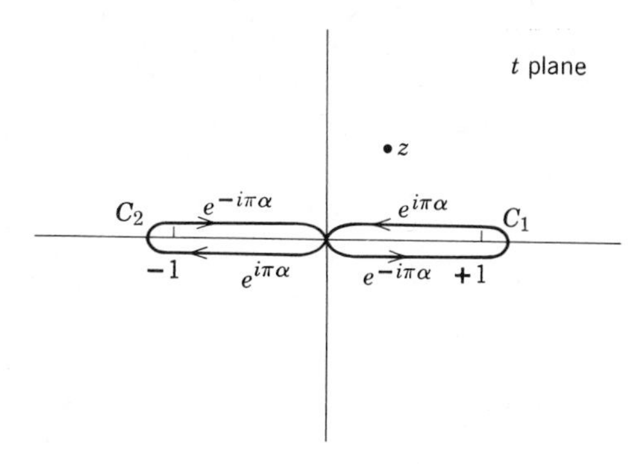

Figure 7.70

As we show in Section A7.3 (cf. Eq. A7.31),

$$B(\mu, \nu) \equiv \int_0^1 s^{\mu-1}(1-s)^{\nu-1}\, ds = \frac{\Gamma(\mu)\Gamma(\nu)}{\Gamma(\mu+\nu)}$$

so that, upon comparison with Eq. 7.161, we see that

$$Q_\alpha(z) = \frac{\sqrt{\pi}}{(2z)^{\alpha+1}} \frac{\Gamma(\alpha+1)}{\Gamma(\alpha+\frac{3}{2})} F\left(\frac{\alpha+2}{2}, \frac{\alpha+1}{2} \middle| \alpha+\frac{3}{2} \middle| \frac{1}{z^2}\right) \tag{7.213}$$

Of course since all the functions involved are analytic Eq. 7.213 is valid for all values of z and α. We can easily compute the first two Legendre functions of the second kind as (cf. Eq. 7.209)

$$Q_0(z) = \frac{1}{2}\int_{-1}^{1} \frac{dt}{z-t} = \tfrac{1}{2} \ln\left(\frac{z+1}{z-1}\right) \tag{7.214}$$

$$Q_1(z) = \frac{1}{2}\int_{-1}^{1} \frac{t\, dt}{z-t} = \tfrac{1}{2} z \ln\left(\frac{z+1}{z-1}\right) - 1 \tag{7.215}$$

In Section 4.3b we discussed the expansion of a continuous function $f(x)$ in terms of a series of Legendre polynomials $P_n(x)$ on the real line segment $-1 \le x \le 1$. In Section A7.4 we will prove that the expansion of a function $f(z)$, holomorphic in a region that includes the line segment $-1 \le x \le 1$, converges uniformly in the z plane within a unifocal ellipse (i.e., an ellipse with foci at $z = \pm 1$). This unifocal ellipse of convergence is the largest such ellipse that is free of singularities of $f(z)$. This illustrates an interesting and important feature of various expansions of holomorphic functions. We could expand a function holomorphic in a region including $-1 \le x \le 1$ in either a Taylor series or a Legendre series

$$f(z) = \sum_{n=0}^{\infty} \alpha_n z^n = \sum_{n=0}^{\infty} a_n P_n(z)$$

each of which has a different region of convergence (i.e., a circle and a unifocal ellipse, respectively). These two regions have some common overlap.

c. $J_n(z)$—The Bessel Function

The solutions to *Bessel's differential equation,*

$$z^2u''(z)+zu'(z)+(z^2-\nu^2)u(z)=0 \tag{7.216}$$

are extremely important in applied mathematics. Here ν is a real parameter (assumed nonnegative). We will discuss this equation in detail in Section 8.2b. For our present purposes we simply seek a solution to Eq. 7.216, which will be finite at $z=0$ and of the form

$$u(z)=z^\alpha f(z) \tag{7.217}$$

where we assume that $f(z)$ is holomorphic about $z=0$ with the expansion

$$f(z)=\sum_{s=0}^{\infty}\alpha_s z^s \tag{7.218}$$

If we substitute Eq. 7.217 into Eq. 7.216 and take the limit as z approaches zero, we find that

$$\alpha^2-\nu^2=0 \tag{7.219}$$

Since we want a $u(z)$ that will be finite at $z=0$, we take $\alpha=\nu$. If we now write

$$u(z)=z^\nu f(z) \tag{7.220}$$

we easily verify that $f(z)$ satisfies the differential equation

$$zf''(z)+(2\nu+1)f'(z)+zf(z)=0 \tag{7.221}$$

In Section A7.6 we express $f(z)$ in terms of the *confluent hypergeometric function*

$$F(a\,|c|\,z)\equiv\lim_{b\to 0}F(a,\,b+c\,|c|\,z/b)=\frac{\Gamma(c)}{\Gamma(a)}\sum_{n=0}^{\infty}\frac{\Gamma(a+n)}{\Gamma(c+n)n!}z^n \tag{7.222}$$

which satisfies the differential equation

$$zF''(a\,|c|\,z)+(c-z)F'(a\,|c|\,z)-aF(a\,|c|\,z)=0 \tag{7.223}$$

We can obtain the power series solution for $f(z)$ directly by substituting the expansion of Eq. 7.218 into Eq. 7.221 as

$$\sum_{s=1}^{\infty}[s(s+1)\alpha_{s+1}+(2\nu+1)(s+1)\alpha_{s+1}+\alpha_{s-1}]z^s+(2\nu+1)\alpha_1\equiv 0 \tag{7.224}$$

Since we have assumed ν is not negative, we obtain

$$\alpha_1=0 \tag{7.225}$$

$$\alpha_s=-\frac{\alpha_{s-2}}{s(s+2\nu)} \qquad s\geq 2 \tag{7.226}$$

which implies that $\alpha_{2s+1} \equiv 0$. If we let $s \to 2s$, then Eq. 7.226 becomes

$$\alpha_{2s+2} = \frac{-\alpha_{2s}}{4(s+1)(s+1+\nu)} \tag{7.227}$$

which has the obvious solution

$$\alpha_{2s} = \frac{\text{const.}\,(-1)^s}{\Gamma(s+1)\Gamma(s+\nu+1)2^{2s}} \tag{7.228}$$

If we make the standard choice for the normalization constant in Eq. 7.228, namely const. $= 2^{-\nu}$, then we have the series representation for the *Bessel function* of order ν,

$$J_\nu(z) = \left(\frac{z}{2}\right)^\nu \sum_{s=0}^{\infty} \frac{(-1)^s}{s!\,\Gamma(s+\nu+1)} \left(\frac{z}{2}\right)^{2s} \tag{7.229}$$

The ratio test shows that this series converges for all finite values of z (i.e., $|z| < \infty$).

In order to develop a representation that will allow us to investigate the asymptotic behavior of the solutions to Eq. 7.216 as $|z| \to \infty$, we will solve this differential equation by means of a general integral representation which will lend itself to a saddle-point asymptotic expansion. We define a differential operator L_z as

$$z^2u''(z) + zu'(z) + z^2u(z) - \nu^2u(z) = 0 \equiv L_z[u(z)] - \nu^2u(z) \tag{7.230}$$

and write

$$u(z) = \int_C e^{zf(t)} w(t)\, dt \tag{7.231}$$

where the contour, C, and the functions $f(t)$ and $w(t)$ must be found. The basic idea in solving linear differential equations by integral transforms is to find the simplest possible $f(t)$ and $w(t)$ that will make $u(z)$ satisfy the differential equation. The choices are not unique. In our case we have

$$\begin{aligned} L[u(z)] &\equiv \int_C [z^2f^2(t) + zf(t) + z^2]e^{zf(t)} w(t)\, dt \\ &= -\int_C \left\{\frac{d^2}{dt^2}[e^{zf(t)}]\right\} w(t)\, dt \end{aligned} \tag{7.232}$$

where we wish to determine an $f(t)$ to make this last equality identically true. This will be the case provided

$$z^2f^2(t) + zf(t) + z^2 + z\frac{d^2f}{dt^2} + z^2\left(\frac{df}{dt}\right)^2 \equiv 0$$

Since this must be identically true of all values of z, we must impose the condition

$$f''+f=0$$

so that (for example) $f(t)=\alpha \sin t$, and the further requirement

$$f'^2+f^2+1=0$$

which implies that

$$\alpha^2 \cos^2 t+\alpha^2 \sin^2 t=1$$

or

$$\alpha=\pm i$$

Acceptable choices for $f(t)$ are

$$f(t)=\pm i \sin t \tag{7.233}$$

Then Eq. 7.230 becomes

$$\int_C \left\{\frac{d^2}{dt^2}[e^{zf(t)}]+\nu^2 e^{zf(t)}\right\} w(t)\, dt=0$$

If we integrate by parts, we obtain

$$0=\frac{d}{dt} e^{zf(t)} w(t)\Big|_{\substack{\text{end}\\ \text{points}}} +\int_C \left\{-\frac{d}{dt} e^{zf(t)} \frac{dw}{dt}+\nu^2 e^{zf(t)} w(t)\right\} dt$$

$$=\left[\frac{d}{dt} e^{zf(t)} w(t)-e^{zf(t)} \frac{dw}{dt}\right]_{\substack{\text{end}\\ \text{points}}} +\int_C \left\{\frac{d^2 w}{dt^2}+\nu^2 w(t)\right\} e^{zf(t)}\, dt \tag{7.234}$$

We must later choose the contour, C, so as to make the boundary terms vanish. In that case we would have

$$w''+\nu^2 w=0$$

so that

$$w(t)=e^{\pm i\nu t}$$

or

$$u(z)=\int_C e^{\pm iz \sin t \pm i\nu t}\, dt \tag{7.235}$$

We must now enforce the end-point conditions

$$\left[\frac{d}{dt} e^{zf(t)} w(t)-e^{zf(t)} \frac{dw}{dt}\right]_{\substack{\text{end}\\ \text{points}}} \equiv 0$$

or

$$(z \cos t+\nu) e^{\pm iz \sin t \pm i\nu t}\Big|_{\substack{\text{end}\\ \text{points}}} =0$$

Since

$$\text{Re}\,[iz \sin t]=\text{Re}\,[\tfrac{1}{2}z(e^{it}-e^{-it})]=\text{Re}\left\{\frac{z}{2}[e^{i\,\text{Re}\,t} e^{-\text{Im}\,t}-e^{-i\,\text{Re}\,t} e^{\text{Im}\,t}]\right\}$$

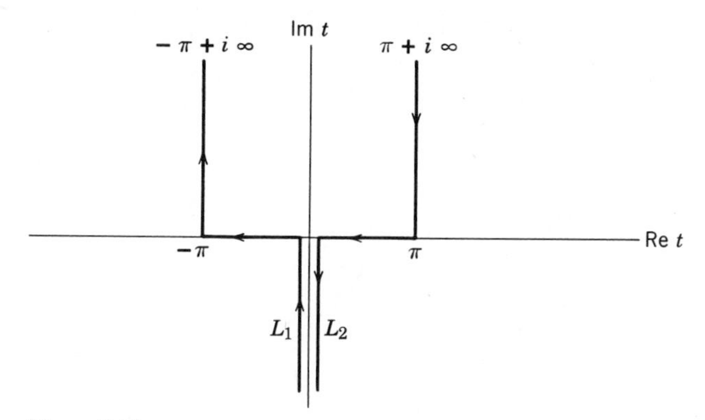

Figure 7.71

we see that Re $[iz \sin t] \to +\infty$ at the end points of both L_1 and L_2 in the t plane shown in Fig. 7.71 *if* Re $z > 0$. Therefore *a* choice of signs for (7.235) consistent with the boundary terms vanishing is

$$u(z) = \int_{L_1, L_2} e^{-iz \sin t + i\nu t}\, dt \tag{7.236}$$

We now define the *Hankel functions* of the first and second kind, respectively, as

$$H_\nu^{(1)}(z) = -\frac{1}{\pi} \int_{L_1} e^{-iz \sin t + i\nu t}\, dt \tag{7.237}$$

$$H_\nu^{(2)}(z) = -\frac{1}{\pi} \int_{L_2} e^{-iz \sin t + i\nu t}\, dt \tag{7.238}$$

These integral representations hold only for Re $z > 0$. If we define

$$\begin{aligned} g(z; t) &\equiv -iz \sin t + i\nu t \qquad & t &= \xi + i\eta \\ z &= x + iy & \nu &= a + ib \end{aligned} \tag{7.239}$$

then we have

$$\begin{aligned} \mathrm{Re}\,[g(z; t)] &= y \sin \xi \cosh \eta + x \cos \xi \sinh \eta - b\xi - a\eta \\ \mathrm{Im}\,[g(z; t)] &= -x \sin \xi \cosh \eta + y \cos \xi \sinh \eta + a\xi - b\eta \end{aligned} \tag{7.240}$$

If instead of 0 and $-\pi$ we take ξ_0 and $-\pi - \xi_0$ for L_1, then (as $\eta \to \pm\infty$) the representation is convergent along L_1' as long as (cf. Fig. 7.72)

$$y \sin \xi_0 - x \cos \xi_0 < 0$$

In the z plane the boundary curve for the convergence of this integral representation is

$$y \sin \xi_0 - x \cos \xi_0 = 0 \tag{7.241}$$

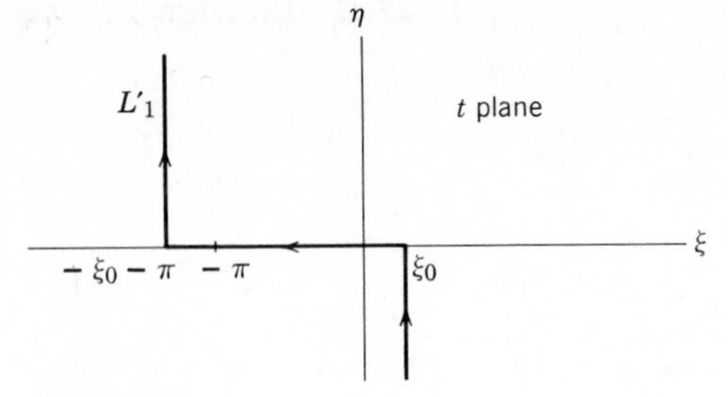

Figure 7.72

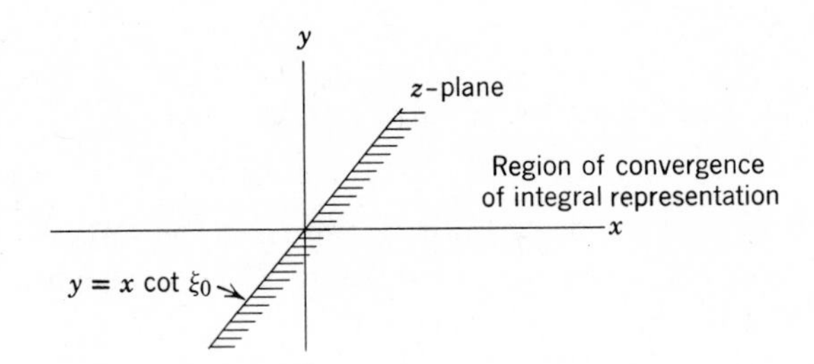

Figure 7.73

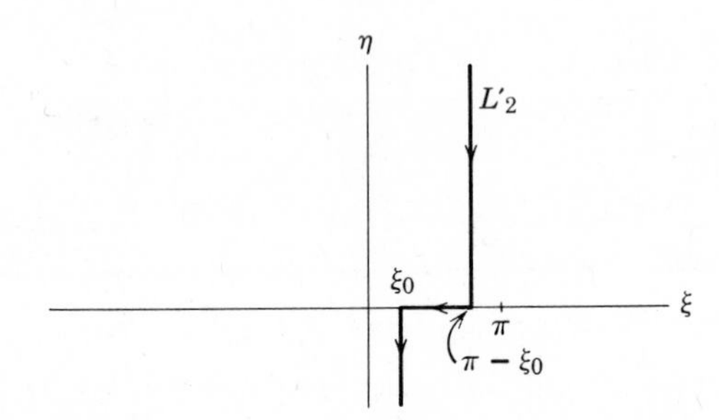

Figure 7.74

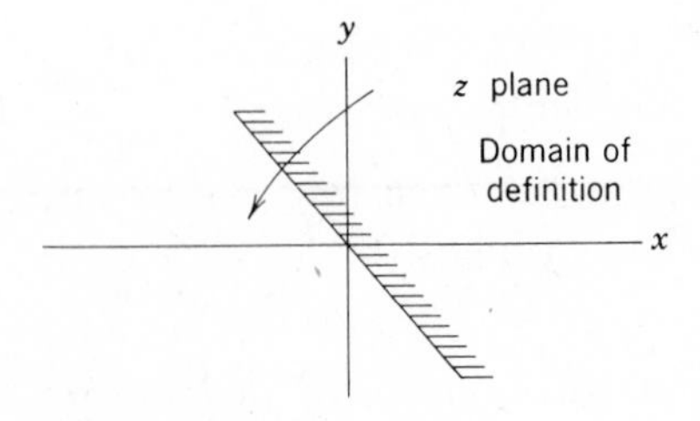

Figure 7.75

This curve demarcates a half-plane for the representation and partly overlaps that for $\operatorname{Re} z \equiv x > 0$ where both representations agree (cf. Fig. 7.73). The boundary line indicated in Fig. 7.73 rotates clockwise from the vertical position as ξ_0 increases from zero.

We can also vary the contour L_2 for $H_\nu^{(2)}(z)$ as shown in Fig. 7.74. If we check this for $\xi = \pi - \xi_0$ as $\eta \to +\infty$, we find from Eq. 7.240

$$\operatorname{Re}[g(z;t)] \underset{\substack{\eta \to +\infty \\ \xi = \pi - \xi_0}}{\longrightarrow} \frac{e^{\eta}}{2}[y \sin \xi_0 - x \cos \xi_0] \to -\infty$$

if $y \sin \xi_0 - x \cos \xi_0 < 0$, as previously for $H_\nu^{(1)}(z)$.

Therefore for both $H_\nu^{(1)}(z)$ and $H_\nu^{(2)}(z)$, as we vary ξ_0, $0 \to \xi_0 \to -2\pi$, we obtain representations that are valid for all finite values of $z = |z| e^{i\phi}$, $0 \leq \phi \leq 2\pi$. Although it may not be obvious at first sight, $H_\nu^{(1)}(z)$ and $H_\nu^{(2)}(z)$ have branch points at $z = 0$ whose order depends upon the value of ν. For suppose we let z describe a circle about $z = 0$ and then compare the values of $H_\nu^{(1)}(z)$ at $\phi = 0$ and at $\phi = 2\pi$. In the process we must vary ξ_0 so as to have at each step a well-defined integral representation. As $0 \to \phi \to 2\pi$ we must let $0 \to \xi_0 \to -2\pi$ [i.e., $\cos(\phi + \xi_0) > 0$] (cf. Fig. 7.75).

Let us begin by studying this for $H_\nu^{(1)}(z)$ in some detail (cf. Fig. 7.76). We see that

$$H_\nu^{(1)}(ze^{2\pi i}) = -\frac{1}{\pi}\int_{L_1} e^{-iz \sin t + i\nu t}\, dt = -H_\nu^{(2)}(z) - e^{-2\pi i\nu} H_\nu^{(2)}(z) - H_\nu^{(1)}(z)$$

where we have simply added and subtracted the dotted vertical segments of Fig. 7.76 to obtain the original contours L_1 and L_2 of Fig. 7.71 (after some translations by 2π) so that

$$H_\nu^{(1)}(ze^{2\pi i}) = -H_\nu^{(1)}(z) - (1 + e^{-2\pi i\nu}) H_\nu^{(2)}(z) \qquad \textbf{(7.242)}$$

As we will see shortly $H_\nu^{(1)}(z)$ and $H_\nu^{(2)}(z)$ are linearly independent. Hence $H_\nu^{(1)}(z)$ does not return to its original value so that $z = 0$ is a branch point.

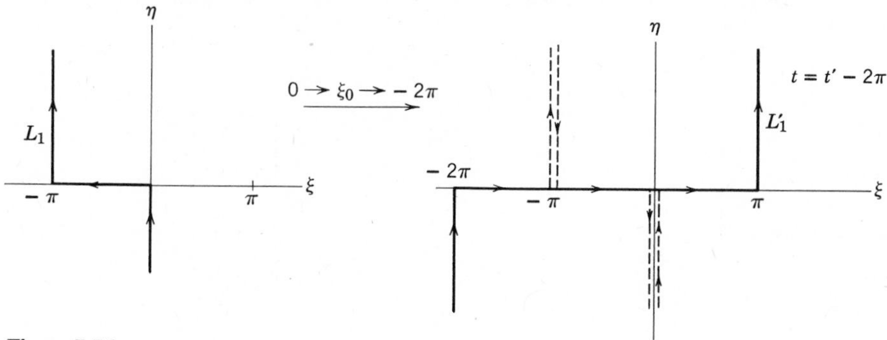

Figure 7.76

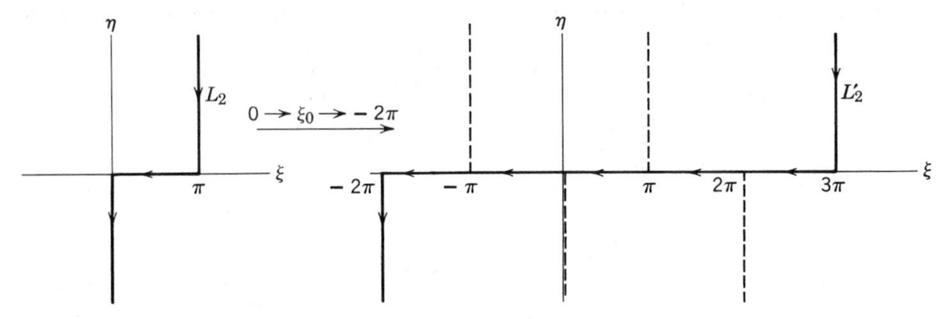

Figure 7.77

We can give a similar argument for $H_\nu^{(2)}(z)$ (cf. Fig. 7.77).

$$\begin{aligned}
H_\nu^{(2)}(ze^{2\pi i}) &= -\frac{1}{\pi}\int_{L_2'} e^{-iz\sin t + i\nu t}\, dt \\
&= e^{2\pi i\nu}H_\nu^{(2)}(z) + e^{2\pi i\nu}H_\nu^{(1)}(z) + H_\nu^{(2)}(z) + H_\nu^{(1)}(z) + e^{-2\pi i\nu}H_\nu^{(2)}(z) \\
&= (1+e^{2\pi i\nu})H_\nu^{(1)}(z) + (1+e^{-2\pi i\nu}+e^{2\pi i\nu})H_\nu^{(2)}(z) \qquad \textbf{(7.243)}
\end{aligned}$$

Finally if we consider the linear combination to be defined as

$$J_\nu(z) \equiv \tfrac{1}{2}[H_\nu^{(1)}(z) + H_\nu^{(2)}(z)] = -\frac{1}{2\pi}\int_L e^{-iz\sin t + i\nu t}\, dt \qquad \textbf{(7.244)}$$

we have from Eqs. (7.242) and (7.243)

$$J_\nu(ze^{2\pi i}) = \tfrac{1}{2}[H_\nu^{(1)}(ze^{2\pi i}) + H_\nu^{(2)}(ze^{2\pi i})] = e^{2\pi i\nu}J_\nu(z) \qquad \textbf{(7.245)}$$

We can also see this directly (cf. Fig. 7.78).

$$J_\nu(ze^{2\pi i}) = -\frac{1}{2\pi}\int_{L'} e^{-iz\sin t + i\nu t}\, dt = e^{2\pi i\nu}J_\nu(z)$$

This is what we might expect from the series representation of Eq. 7.229 although we have not yet shown Eqs. 7.229 and 7.244 to be equal. We will now

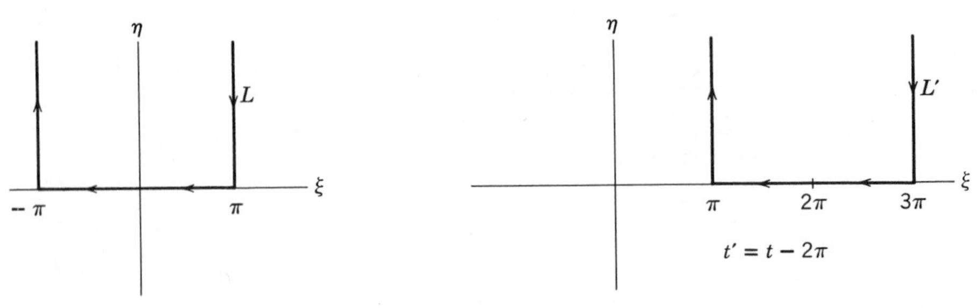

Figure 7.78

establish this equivalence. If we make the change of variable

$$\omega = \frac{z}{2} e^{-it} = \frac{z}{2} e^{\eta}(\cos \xi - i \sin \xi) \tag{7.246}$$

in the integral representation of Eq. 7.244, then as long as Re $z > 0$ we obtain

$$J_\nu(z) = \left(\frac{z}{2}\right)^\nu \frac{1}{2\pi i} \int_{\mathscr{L}} \exp\left(\omega - \frac{z^2}{4\omega}\right) \omega^{-(\nu+1)} \, d\omega \tag{7.247}$$

where the transformed contour may be rotated into $\mathscr{L}$ shown in Fig. 7.79. If we now expand $e^{-z^2/4\omega}$ in a Taylor series, we obtain

$$J_\nu(z) = \left(\frac{z}{2}\right)^\nu \sum_{s=0}^{\infty} \frac{(-1)^s}{s!} \left(\frac{z}{2}\right)^{2s} \left[\frac{1}{2\pi i} \int_{\mathscr{L}} e^{\omega} \omega^{-(\nu+1+s)} \, d\omega\right] \tag{7.248}$$

However the integral representation for $1/\Gamma(z)$, Eq. 7.114, becomes after the change of variable $t = e^{\pi i}\omega$

$$\frac{1}{\Gamma(z)} = \frac{1}{2\pi i} \int_{\mathscr{L}} e^{\omega} \omega^{-z} \, d\omega \tag{7.249}$$

We can then express Eq. 7.248 as

$$J_\nu(z) = \left(\frac{z}{2}\right)^\nu \sum_{s=0}^{\infty} \frac{(-1)^s}{s!\,\Gamma(\nu+1+s)} \left(\frac{z}{2}\right)^{2s}$$

which is just Eq. 7.229.

We can show that $H_\nu^{(1)}(z)$ and $H_\nu^{(2)}(z)$ are linearly independent as follows. If they were linearly dependent then there would have to exist constants $c_\nu^{(1)}$ and $c_\nu^{(2)}$ such that

$$c_\nu^{(1)} H_\nu^{(1)}(z) + c_\nu^{(2)} H_\nu^{(2)}(z) \equiv 0 \tag{7.250}$$

for *all* values of z. If we assume Im $z > 0$ and take $\xi_0 = -(\pi/2)$ in Fig. 7.72, we

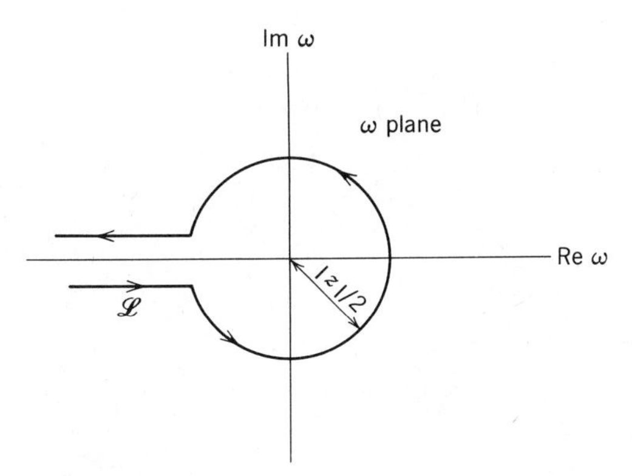

Figure 7.79

obtain a convergent representation of $H_\nu^{(1)}(z)$ from Eq. 7.237 as

$$H_\nu^{(1)}(z)=\frac{e^{-i\pi\nu/2}}{\pi i}\int_{-\infty}^{\infty} e^{iz\cosh\eta-\nu\eta}\,d\eta \tag{7.251}$$

so that

$$H_\nu^{(1)}(z)\xrightarrow[\substack{|z|\to\infty\\ \mathrm{Im}\,z>0}]{} 0 \tag{7.252}$$

Similarly if we take $\xi_0=\pi/2$ in Fig. 7.74 we obtain for Im $z<0$,

$$H_\nu^{(2)}(z)=-\frac{e^{-i\nu\pi/2}}{\pi i}\int_{-\infty}^{\infty} e^{-iz\cosh n-\nu\eta}\,d\eta \tag{7.253}$$

so that

$$H_\nu^{(2)}(z)\xrightarrow[\substack{|z|\to\infty\\ \mathrm{Im}\,z<0}]{} 0 \tag{7.254}$$

If we take $0<\xi_0\le\pi/2$ in Fig. 7.72 then we obtain a representation for $H_\nu^{(1)}(z)$ valid for Im $z<0$, as $|z|\to\infty$. Since the integrand of Eq. 7.237 is exponentially damped as $|z|\to\infty$, we can write

$$\begin{aligned}
H_\nu^{(1)}(y)&\xrightarrow[y\to-\infty]{}-\frac{1}{\pi}\int_{\xi_0}^{-\pi-\xi_0} e^{(y-ix)\sin\xi-b\xi+ia\xi}\,d\xi\\
&\longrightarrow-\frac{1}{\pi}\int_{0}^{-\pi} e^{(y-ix)\sin\xi-b\xi+ia\xi}\,d\xi\\
&=\frac{e^{b\pi/2}}{\pi}\int_{-\pi/2}^{\pi/2} e^{(-y+ix)\cos\xi}e^{b\xi}[-\sin(a\xi)-i\cos(a\xi)]\,d\xi\\
&=\frac{e^{b\pi/2}}{\pi}\int_{-\pi/2}^{\pi/2} e^{(-y)\cos\xi}e^{b\xi}[\sin(x\cos\xi-a\xi)-i\cos(x\cos\xi-a\xi)]\,d\xi
\end{aligned} \tag{7.255}$$

It is a simple matter to see that as long as

$$\frac{2}{\pi}|x|+|a|\le 1 \tag{7.256}$$

then the imaginary part of the integrand of Eq. 7.255 satisfies

$$e^{(-y)\cos\xi}e^{b\xi}\cos(x\cos\xi-a\xi)\ge 0$$

and there can be no cancellation in the integral so that, subject to Eq. 7.256,

$$H_\nu^{(1)}(z)\xrightarrow[\substack{|z|\to\infty\\ \mathrm{Im}\,z<0}]{} \infty \tag{7.257}$$

If we take $-(\pi/2)\le\xi_0<0$ in Fig. 7.74 a similar argument for $H_\nu^{(2)}(z)$ yields, again subject to Eq. 7.256,

$$H_\nu^{(2)}(z)\xrightarrow[\substack{|z|\to\infty\\ \mathrm{Im}\,z>0}]{} \infty \tag{7.258}$$

Therefore Eqs. 7.252, 7.254, 7.257, and 7.258 show that Eq. 7.250 cannot be satisfied for all values of z, at least for $|\text{Re } \nu| \leq 1$. Since we may use the same type argument as given in the beginning of Section 7.12a for the integral representation of $\Gamma(z)$ to show that the $H_\nu^{(1)}(z)$ and $H_\nu^{(2)}(z)$ as defined by Eqs. 7.237 and 7.238 are holomorphic functions of ν in the entire finite ν-plane, we could construct a proof based on analyticity in the variable ν to show that if Eq. 7.250 were satisfied for some region in the ν plane it would have to be satisfied everywhere, in particular for $|\text{Re } \nu| \leq 1$.

However it is fairly easy to extend the argument that $H_\nu^{(1)}(z)$ and $H_\nu^{(2)}(z)$ are linearly independent for all values of ν as follows. There is a simple relation between $H_\nu^{(1)}(z)$ and $H_{-\nu}^{(1)}(z)$ since by letting $t \to -t-\pi$, $L_1 \to -L_1$ of Eq. 7.237 we see that

$$\begin{aligned} H_{-\nu}^{(1)}(z) &= -\frac{1}{\pi}\int_{L_1} e^{-iz \sin t - i\nu t}\, dt \\ &= \frac{1}{\pi}\int_{-L_1} e^{-iz \sin t + i\nu t + i\nu\pi}\, dt = -\frac{e^{i\nu\pi}}{\pi}\int_{L_1} e^{-iz \sin t + i\nu t}\, dt \\ &= e^{i\nu\pi} H_\nu^{(1)}(z) \end{aligned} \qquad \textbf{(7.259)}$$

Similarly we can prove that

$$H_{-\nu}^{(2)}(z) = e^{-i\nu\pi} H_\nu^{(2)}(z) \qquad \textbf{(7.260)}$$

If we use Eqs. 7.244, 7.259, and 7.260, we can express $H_\nu^{(1)}(z)$ and $H_\nu^{(2)}(z)$ as follows, provided ν is not zero or an integer,

$$H_\nu^{(1)}(z) = \frac{-i}{\sin(\pi\nu)}[J_{-\nu}(z) - e^{-i\nu\pi} J_\nu(z)] \qquad \textbf{(7.261)}$$

$$H_\nu^{(2)}(z) = \frac{i}{\sin(\pi\nu)}[J_{-\nu}(z) - e^{i\nu\pi} J_\nu(z)] \qquad \textbf{(7.262)}$$

Since the series representation for $J_\nu(z)$, Eq. 7.229, remains valid for negative ν, we can see that

$$H_\nu^{(1)}(z) \xrightarrow[z\to 0]{} \frac{-i}{\sin(\pi\nu)}\left[\left(\frac{z}{2}\right)^{-\nu}\frac{1}{\Gamma(-\nu+1)} - e^{-i\nu\pi}\left(\frac{z}{2}\right)^{\nu}\frac{1}{\Gamma(\nu+1)}\right] \qquad \textbf{(7.263)}$$

$$H_\nu^{(2)}(z) \xrightarrow[z\to 0]{} \frac{i}{\sin(\pi\nu)}\left[\left(\frac{z}{2}\right)^{-\nu}\frac{1}{\Gamma(-\nu+1)} - e^{i\pi\nu}\left(\frac{z}{2}\right)^{\nu}\frac{1}{\Gamma(\nu+1)}\right] \qquad \textbf{(7.264)}$$

so that

$$H_\nu^{(1)}(z) + H_\nu^{(2)}(z) \xrightarrow[z\to 0]{} 0 \qquad \text{Re } \nu > 0 \qquad \textbf{(7.265)}$$

$$e^{i\nu\pi} H_\nu^{(1)}(z) + e^{-i\pi\nu} H_\nu^{(2)}(z) \xrightarrow[z\to 0]{} 0 \qquad \text{Re } \nu < 0 \qquad \textbf{(7.266)}$$

However neither set of coefficients $c_\nu^{(1)}$, $c_\nu^{(2)}$ given in Eqs. 7.265 or 7.266 can satisfy Eq. 7.250 for all values of ν since we have already shown that neither set will work for any ν such that $|\text{Re}\,\nu| \leq 1$. We must still treat the cast in which ν is zero or an integer.

The *Neumann function* is defined as

$$N_\nu(z) = \frac{1}{2i}[H_\nu^{(1)}(z) - H_\nu^{(2)}(z)] \tag{7.267}$$

Since the Jacobian of transformation between the pair of functions $J_\nu(z)$ (cf. Eq. 7.244), $N_\nu(z)$ and the pair $H_\nu^{(1)}(z)$, $H_\nu^{(2)}(z)$ is not zero, we see that $J_\nu(z)$ and $N_\nu(z)$ are linearly independent solutions to Bessel's equation (Eq. 7.216), at least for ν not zero or an integer. From Eqs. 7.261, 7.262, and 7.267 we can write

$$N_\nu(z) = \frac{1}{\sin(\pi\nu)}[\cos(\pi\nu)J_\nu(z) - J_{-\nu}(z)] \tag{7.268}$$

We see directly from Eq. 7.229 or Eqs. 7.244, 7.259, and 7.260 that for $\nu = n$, an integer or zero

$$J_{-n}(z) = (-1)^n J_n(z) \tag{7.269}$$

so that Eq. 7.268 is an indeterminant form as $\nu \to n$. If we use l'Hospital's rule we find

$$\begin{aligned}
N_n(z) \equiv \lim_{\nu\to n} N_\nu(z) &= \lim_{\nu\to n}\left[\frac{-\pi\sin(\nu\pi)J_\nu(z) + \cos(\nu\pi)\dfrac{\partial J_\nu(z)}{\partial\nu} - \dfrac{\partial J_{-\nu}}{\partial\nu}(z)}{\pi\cos(\pi\nu)}\right] \\
&= \lim_{\nu\to n}\frac{1}{\pi}\left[\frac{\partial J_\nu(z)}{\partial\nu} - (-1)^n\frac{\partial J_{-\nu}(z)}{\partial\nu}\right] \\
&= \frac{1}{\pi}[f(n;z) - (-1)^n f(-n;z)]
\end{aligned} \tag{7.270}$$

where

$$\begin{aligned}
f(\nu;z) \equiv \frac{\partial J_\nu(z)}{\partial\nu} &= \ln\left(\frac{z}{2}\right)\left(\frac{z}{2}\right)^\nu \sum_{s=0}^{\infty}\frac{(-1)^s}{s!\,\Gamma(\nu+1+s)}\left(\frac{z}{2}\right)^{2s} \\
&\quad - \left(\frac{z}{2}\right)^\nu \sum_{s=0}^{\infty}\frac{(-1)^s}{s!\,\Gamma(\nu+1+s)}\psi(\nu+1+s)\left(\frac{z}{2}\right)^{2s}
\end{aligned} \tag{7.271}$$

Here $\psi(z)$ is the logarithmic derivative of $\Gamma(z)$,

$$\psi(z) \equiv \frac{1}{\Gamma(z)}\frac{d}{dz}\Gamma(z) = \frac{d}{dz}[\ln\Gamma(z)] \tag{7.272}$$

If we recall that $\Gamma(z)$ has simple poles of residue $(-1)^n/n!$ at $z = -n$, then we see that $d[\Gamma(z)]^{-1}/dz$ is finite at $z=-n$ and we obtain

$$N_n(z)=\frac{1}{\pi}\ln\left(\frac{z}{2}\right)\left(\frac{z}{2}\right)^n\sum_{s=0}^{\infty}\frac{(-1)^s}{s!\,(n+s)!}\left(\frac{z}{2}\right)^{2s}-\frac{1}{\pi}\left(\frac{z}{2}\right)^n\sum_{s=0}^{\infty}\frac{(-1)^s}{s!\,(n+s)!}\psi(n+1+s)\left(\frac{z}{2}\right)^{2s}$$
$$+\frac{(-1)^n}{\pi}\left\{\ln\left(\frac{z}{2}\right)\left(\frac{z}{2}\right)^{-n}\sum_{s=0}^{\infty}\frac{(-1)^s}{s!\,\Gamma(-n+s+1)}\left(\frac{z}{2}\right)^{2s}\right.$$
$$\left.-\left(\frac{z}{2}\right)^{-n}\sum_{s=0}^{\infty}\frac{(-1)^s}{s!}\left[\frac{\psi(-n+1+s)}{\Gamma(-n+1+s)}\right]\left(\frac{z}{2}\right)^{2s}\right\}$$
$$=\frac{2}{\pi}\ln\left(\frac{z}{2}\right)J_n(z)-\frac{1}{\pi}\sum_{s=0}^{n-1}\frac{(n-s-1)!}{s!}\left(\frac{z}{2}\right)^{2s-n}$$
$$-\frac{1}{\pi}\left(\frac{z}{2}\right)^n\sum_{s=0}^{\infty}\frac{(-1)^s}{s!\,(n+s)!}[\psi(s+1)+\psi(n+s+1)]\left(\frac{z}{2}\right)^{2s} \quad \textbf{(7.273)}$$

As $z\to 0$ the $N_n(z)$ has the behavior

$$N_n(z)\xrightarrow[z\to 0]{}\begin{cases}\dfrac{2}{\pi}\ln\left(\dfrac{z}{2}\right) & n=0\\[2ex] \dfrac{-(n-1)!}{\pi}\left(\dfrac{2}{z}\right)^n & n\neq 0\end{cases} \quad \textbf{(7.274)}$$

so that $J_n(z)$ and $N_n(z)$ are linearly independent. Therefore $J_\nu(z)$ and $N_\nu(z)$ are linearly independent for all values of ν as are $H_\nu^{(1)}(z)$ and $H_\nu^{(2)}(z)$.

If we take $\nu=n$ in the integral representation for $J_\nu(z)$ (Eq. 7.244), then the contributions along the vertical parts of L cancel so that

$$J_n(z)=\frac{1}{2\pi}\int_{-\pi}^{\pi}e^{iz\sin t-int}\,dt=\frac{1}{\pi}\int_0^{\pi}\cos(z\sin t-nt)\,dt$$

From the fact that

$$\int_{-\pi}^{\pi}e^{i(n-m)t}\,dt=2\pi\delta_{nm}$$

we see that $J_n(z)$ is the nth Fourier coefficient in the expansion of $e^{iz\sin t}$,

$$e^{iz\sin t}=\sum_{n=-\infty}^{\infty}J_n(z)e^{int} \quad \textbf{(7.275)}$$

This is known as the *generating function* for $J_n(z)$. If we substitute

$$\zeta=e^{it} \qquad \frac{1}{\zeta}=e^{-it}$$

$$i\sin t=\frac{e^{it}-e^{it}}{2}=\frac{1}{2}\left(\zeta-\frac{1}{\zeta}\right)$$

then we obtain another form

$$e^{z[\zeta-(1/\zeta)]/2}=\sum_{n=-\infty}^{\infty}J_n(z)\zeta^n \quad \textbf{(7.276)}$$

We can also derive some useful recursion relations for the solutions to Bessel's equation. From Eqs. 7.237 and 7.238 we have for $j=1$ or 2

$$H^{(j)}_{\nu+1}(z)+H^{(j)}_{\nu-1}(z)=-\frac{1}{\pi}\int_{L_j} e^{-iz\sin t}\left[e^{i(\nu+1)t}+e^{i(\nu-1)t}\right]dt$$

$$=-\frac{2}{\pi}\int_{L_j} e^{-iz\sin t}e^{i\nu t}\cos t\,dt=-\frac{2}{\pi}\int_{L_j} e^{-iz\sin t}e^{i\nu t}\,d(\sin t)$$

$$=-\frac{2}{\pi}\left\{-\frac{1}{iz}e^{-iz\sin t}e^{i\nu t}\Big|_{\substack{\text{ends}\\ \text{of } L_j}}+\frac{\nu}{z}\int_{L_j} e^{-iz\sin t}e^{i\nu t}\,dt\right\}$$

$$=\frac{2\nu}{z}H^{(j)}_\nu(z)\qquad j=1,2 \tag{7.277}$$

and similarly

$$H^{(j)}_{\nu+1}(z)-H^{(j)}_{\nu-1}(z)=-\frac{1}{\pi}\int_{L_j} e^{-iz\sin t}\left[e^{i(\nu+1)t}-e^{i(\nu-1)t}\right]dt$$

$$=-\frac{2i}{\pi}\int_{L_j} e^{-iz\sin t}e^{i\nu t}\sin t\,dt=-2\frac{d}{dz}H^{(j)}_\nu(z)\qquad j=1,2 \tag{7.278}$$

It follows immediately that the $J_\nu(z)$ and $N_\nu(z)$ also satisfy these recursion relations since they are simply linear combinations of the $H^{(1)}_\nu(z)$ and the $H^{(2)}_\nu(z)$ (cf. Eqs. 7.244 and 7.267).

It is a simple matter to find the asymptotic behavior of the Hankel functions using Eqs. 7.115 and 7.128 once we make the identifications

$$f(t)=-i\sin t$$

$$w(t)=-\frac{e^{i\nu t}}{\pi}$$

so that the saddle point, t_0, is gotten as

$$\frac{df}{dt}=-i\cos t=0$$

or (cf. Fig. 7.71)

$$\begin{aligned} t_0&=-\frac{\pi}{2} &\quad \text{for}\quad & H^{(1)}_\nu(z)\\ t_0&=\frac{\pi}{2} &\quad \text{for}\quad & H^{(2)}_\nu(z)\end{aligned} \tag{7.279}$$

For $H^{(1)}_\nu(z)$ the path of steepest descent is (cf. Eq. 7.240)

$$\text{Im}\,(-iz\sin t)=\text{Im}\,(-iz\sin t_0)=x$$

or

$$-x \sin \xi \cosh \eta + y \cos \xi \sinh \eta = x \tag{7.280}$$

In the neighborhood of $t_0 = -\pi/2$, Eq. 7.280 reduces to

$$-\frac{x}{2}\left[\left(\xi+\frac{\pi}{2}\right)^2 - \eta^2\right] + y\left(\xi+\frac{\pi}{2}\right)\eta = 0$$

or with $(t-t_0) = \sigma e^{i\theta}$, $z = \rho e^{i\phi}$ to

$$\text{Im}\,[e^{i[(\pi/2)+\phi+2\theta]}] = \sin\left(\frac{\pi}{2}+\phi+2\theta\right) = 0$$

$$\text{Re}\,[-e^{i[(\pi/2)+\phi+2\theta]}] < 0$$

so that

$$\theta = \begin{cases} \frac{1}{2}\left(-\phi-\frac{\pi}{2}\right) \\ \frac{1}{2}\left(-\phi+\frac{3}{2}\pi\right) \end{cases}$$

A qualitative sketch of the path of steepest descent for $H_\nu^{(1)}(z)$ is shown in Fig. 7.80. Since

$$f[t=-(\pi/2)] = i = e^{i\pi/2}$$
$$f''[t=-(\pi/2)] = -i = e^{3\pi i/2}$$

we have from Eq. 7.128

$$H_\nu^{(1)}(z) \xrightarrow[|z|\to\infty]{} \sqrt{\frac{2}{\pi z}}\, e^{i[z-(\nu+(1/2))\pi/2]} \tag{7.281}$$

A similar argument for $H_\nu^{(2)}(z)$ produces

$$H_\nu^{(2)}(z) \xrightarrow[|z|\to\infty]{} \sqrt{\frac{2}{\pi z}}\, e^{-i[z-(\nu+1/2)\pi/2]} \tag{7.282}$$

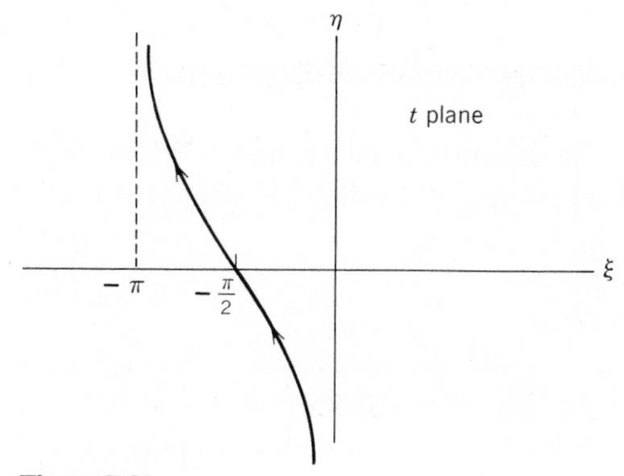

Figure 7.80

These two relations imply that

$$J_\nu(z) \xrightarrow[|z|\to\infty]{} \sqrt{\frac{2}{\pi z}} \cos\left[z-(\nu+\tfrac{1}{2})\frac{\pi}{2}\right] \tag{7.283}$$

and that

$$N_\nu(z) \xrightarrow[|z|\to\infty]{} \sqrt{\frac{2}{\pi z}} \sin\left[z-(\nu+\tfrac{1}{2})\frac{\pi}{2}\right] \tag{7.284}$$

The *spherical* Bessel, Neumann, and Hankel functions, which often arise in physical applications, are defined as

$$j_\nu(z) \equiv \sqrt{\frac{\pi}{2z}}\, J_{\nu+(1/2)}(z) \xrightarrow[|z|\to\infty]{} \frac{1}{z} \cos\left[z-\frac{\pi}{2}(\nu+1)\right] \tag{7.285}$$

$$n_\nu(z) \equiv \sqrt{\frac{\pi}{2z}}\, N_{\nu+(1/2)}(z) \xrightarrow[|z|\to\infty]{} \frac{1}{z} \sin\left[z-\frac{\pi}{2}(\nu+1)\right] \tag{7.286}$$

$$h_\nu^{(1)}(z) \equiv \sqrt{\frac{\pi}{2z}}\, H_{\nu+(1/2)}^{(1)}(z) \xrightarrow[|z|\to\infty]{} \frac{1}{z}\, e^{i[z-(\pi/2)(\nu+1)]} \tag{7.287}$$

$$h_\nu^{(2)}(z) = \sqrt{\frac{\pi}{2z}}\, H_{\nu+(1/2)}^{(2)}(z) \xrightarrow[|z|\to\infty]{} \frac{1}{z}\, e^{-i[z-(\pi/2)(\nu+1)]} \tag{7.288}$$

In particular, for $\nu=n$, we have (cf. Eqs. 7.110, 7.129, 7.268, and 7.229)

$$j_n(z) \xrightarrow[z\to 0]{} \frac{\sqrt{\pi}}{2}\left(\frac{z}{2}\right)^n \frac{1}{\Gamma(n+\frac{3}{2})} = \frac{2^{n+1}(n+1)!}{(2n+2)!}\, z^n \tag{7.289}$$

$$n_n(z) \xrightarrow[z\to 0]{} -\frac{\sqrt{\pi}}{2}\left(\frac{2}{z}\right)^{n+1} \frac{1}{\Gamma(-n+\frac{1}{2})\sin[\pi(n+\frac{1}{2})]} = -\frac{(2n)!}{2^n n!}\frac{1}{z^{n+1}} \tag{7.290}$$

A7.1 THE SOLUTION TO A DISPERSION-THEORY PROBLEM

As an application of the dispersion representation of Section 7.9d we will consider finding an analytic function $F(z)$ having the following properties.

i. $F^*(z^*) = F(z)$ **(A7.1)**
ii. $\operatorname{Im} F(x) = \rho(x)\,|F(x)|^2 \qquad x>0$ **(A7.2)**
with

$$\rho(x) = \frac{H(x-4)}{2\sqrt{x(x-4)}}$$

where H is the Heaviside step function

iii. $F(z)=F(4-z)$ **(A7.3)**

iv. $F(z)\xrightarrow[z\to 1]{}\dfrac{g}{z-1}$, where g is a real constant **(A7.4)**

v. $F(z)\xrightarrow[|z|\to\infty]{}\dfrac{g}{z-1}+\dfrac{g}{3-z}\equiv f(z)$ **(A7.5)**

If we assume that $F(z)$ has no other poles and cuts than those required by (ii), (iii), and (v), we see that there are simple poles at $z=1$ and $z=3$ and branch cuts along $4<x<+\infty$ [from (ii)] and along $-\infty<x<0$ [from (iii) and (ii)]. Since $f(z)$ has no branch points a new function $G(z)$ defined as

$$G(z)=\frac{f(z)}{F(z)}\xrightarrow[|z|\to\infty]{}1 \tag{A7.6}$$

will have the same cut structure as $F(z)$ but no poles at $z=1$ or $z=3$. It also follows that $G^*(z^*)=G(z)$ and that $G(z)=G(4-z)$. From Eq. A7.2 we see that for $x>0$

$$\text{Im}\left[\frac{1}{F(x)}\right]=\text{Im}\left[\frac{F^*(x)}{|F(x)|^2}\right]=-\frac{\text{Im } F(x)}{|F(x)|^2}=-\rho(x) \tag{A7.7}$$

so that

$$\text{Im } G(x)=-f(x)\rho(x) \tag{A7.8}$$

This new function $G(z)$ will have poles at the zeros of $F(z)$. First assume that $F(z)$ has no zeros. The z plane for $G(z)$ is shown in Fig. 7.81. If we apply Cauchy's integral theorem along the contour, C, we have

$$G(z)=\frac{1}{2\pi i}\int_C \frac{G(z')\,dz'}{(z'-z)}=1+\frac{1}{2\pi i}\int_4^\infty \frac{dx'}{(x'-z)}[G(x'+i\varepsilon)-G(x'-i\varepsilon)]$$

$$-\frac{1}{2\pi i}\int_0^{-\infty}\frac{dx'}{(x'-z)}[G(x'+i\varepsilon)-G(x'-i\varepsilon)]$$

$$=1+\frac{1}{\pi}\int_4^\infty \frac{dx'}{(x'-z)}\text{Im } G(x'+i\varepsilon)-\frac{1}{\pi}\int_4^\infty \frac{(-dx')}{(4-x'-z)}\text{Im } G(4-x'+i\varepsilon)$$

$$=1+\frac{1}{\pi}\int_4^\infty dx'\left[\frac{\text{Im } G(x'+i\varepsilon)}{(x'-z)}-\frac{\text{Im } G(x'-i\varepsilon)}{(x'+z-4)}\right]$$

$$=1+\frac{1}{\pi}\int_4^\infty dx'\,\text{Im } G(x'+i\varepsilon)\left[\frac{1}{(x'-z)}+\frac{1}{(x'+z-4)}\right]$$

$$=1-\frac{1}{2\pi}\int_4^\infty \frac{dx' f(x')}{\sqrt{x'(x'-4)}}\left[\frac{1}{x'-z}+\frac{1}{x'+z-4}\right] \tag{A7.9}$$

We must realize that we have to add to the r.h.s. of Eq. A7.9 any pole terms in $G(z)$ arising from the zeros of $F(z)$. If we use Eq. A7.5, which defines $f(z)$, and write

$$I(z)=\int_4^{\infty}\frac{dx'}{\sqrt{x'(x'-4)}}\left(\frac{1}{x'-1}+\frac{1}{3-x'}\right)\left(\frac{1}{x'-z}+\frac{1}{x'+z-4}\right) \tag{A7.10}$$

then we can evaluate this as

$$I(z)=\frac{1}{1-z}[J(1)-J(z)]+\frac{1}{3-z}[J(3)-J(z)]$$
$$+\frac{1}{z-3}[J(1)-J(4-z)]+\frac{1}{z-1}[J(3)-J(4-z)] \tag{A7.11}$$

where

$$J(z)=\int_4^{\infty}\frac{dx'}{\sqrt{x'(x'-4)}}\frac{1}{(x'-z)}=\frac{1}{\sqrt{-z(-z+4)}}\ln\left[\frac{z+2+\sqrt{-z(-z+4)}}{2}\right] \tag{A7.12}$$

We have chosen that branch of $J(z)$ such that $J(x)$ is real and positive whenever $x<4 (y=0)$; in fact,

$$J(x)=\frac{1}{\sqrt{x(-x+4)}}\tan^{-1}\left(\frac{\sqrt{x(-x+4)}}{-x+2}\right) \qquad 0<x<4 \qquad y=0$$

$$J(x)=\frac{1}{\sqrt{|x|\,(|x|+4)}}\ln\left[\frac{|x|+2+\sqrt{|x|\,(|x|+4)}}{2}\right] \qquad x<0 \qquad y=0$$

The final solution for $F(z)$ can now be written as

$$\frac{f(z)}{F(z)}=1-\frac{g}{2\pi}I(z)+\sum_j\frac{\text{Res}\,[G(z), z=z_j]}{(z-z_j)} \tag{A7.13}$$

where the $\{z_j\}$ are the zeros (assumed simple here) of $F(z)$. We see that the solution to the problem as posed in Eqs. A7.1–A7.5 is not complete until the zeros of $F(z)$ are specified.

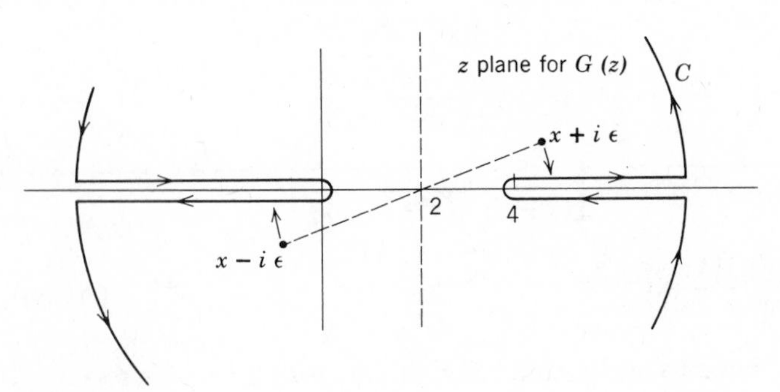

Figure 7.81

A7.2 A JUSTIFICATION FOR THE INTERCHANGE OF A DOUBLE INFINITE SUM WHEN $\alpha_{nm} \geq 0$, $\forall n, m$

We are given that the double sum

$$\sum_{n=0}^{\infty} \sum_{m=0}^{\infty} \alpha_{nm} = S \tag{A7.14}$$

is convergent when summed in the indicated order and that $\alpha_{nm} \geq 0$ for all n and m. This implies that the quantities $\{A_n\}$ defined as

$$A_n \equiv \sum_{m=0}^{\infty} \alpha_{nm} \tag{A7.15}$$

form a convergent series such that

$$\sum_{n=0}^{\infty} A_n = S \tag{A7.16}$$

However, $\alpha_{nm} \leq A_n$ for all n and m since $\alpha_{nm} \geq 0$ so that the quantities $\{A^{(m)}\}$

$$A^{(m)} \equiv \sum_{n=0}^{\infty} \alpha_{nm} \leq \sum_{n=0}^{\infty} A_n = S \tag{A7.17}$$

form a convergent sequence for all m. In fact we have

$$\sum_{m=0}^{M} A^{(m)} = \sum_{m=0}^{M} \sum_{n=0}^{\infty} \alpha_{nm} = \sum_{n=0}^{\infty} \sum_{m=0}^{M} \alpha_{nm} \leq \sum_{n=0}^{\infty} A_n = S \tag{A7.18}$$

so that $\sum_{m=0}^{M} A^{(m)}$ converges to a limit, for example, S', as $M \to \infty$.

$$S' = \sum_{m=0}^{\infty} A^{(m)} = \sum_{m=0}^{\infty} \sum_{n=0}^{\infty} \alpha_{nm} \leq S \tag{A7.19}$$

We can now repeat the same argument beginning with $\sum_{n=0}^{\infty} \alpha_{nm}$ in Eq. A7.15 and conclude that $S \leq S'$ so that, necessarily, $S' = S$ and

$$\sum_{n=0}^{\infty} \sum_{m=0}^{\infty} \alpha_{nm} = \sum_{m=0}^{\infty} \sum_{n=0}^{\infty} \alpha_{nm} \tag{A7.20}$$

A7.3 RECURSION AND ORTHOGONALITY RELATIONS FOR THE JACOBI POLYNOMIALS, $P_n^{(\alpha,\beta)}(x)$

We first establish Eqs. 7.190–7.193 of the text for the Jacobi polynomials as defined by Eq. 7.189. We begin with Leibritz' rule for the differentiation of a product of functions

$$(uv)^{(n)} = \sum_{s=0}^{n} \frac{n!}{s!\,(n-s)!}\, u^{(n-s)} v^{(s)} \tag{A7.21}$$

to obtain

$$\begin{aligned}
\frac{d^n}{dx^n}[(1-x)^{\alpha+n}(1+x)^{\beta+n}] &= \sum_{s=0}^{n} \frac{n!}{s!\,(n-s)!}[(1-x)^{\alpha+n}]^{(n-s)}[(1+x)^{\beta+n}]^{(s)} \\
&= \sum_{s=0}^{n} \frac{n!}{s!\,(n-s)!}(\alpha+n)(\alpha+n-1)\cdots(\alpha+s+1)(-1)^{n-s} \\
&\quad \times (1-x)^{\alpha+s}(\beta+n)(\beta+n-1)\cdots(\beta+n-s+1)(1+x)^{\beta+n-s} \\
&= (1-x)^{\alpha}(1+x)^{\beta}(-1)^n\Gamma(\alpha+n+1)\Gamma(\beta+n+1)n! \\
&\quad \times \sum_{s=0}^{n} \frac{(x-1)^s(1+x)^{n-s}}{s!\,(n-s)!\,\Gamma(\alpha+s+1)\Gamma(\beta+n-s+1)} \qquad \text{(A7.22)} \\
&= (-1)^n(1-x)^{\alpha}(1+x)^{\beta}\Gamma(\alpha+n+1)\Gamma(\beta+n+1)n! \\
&\quad \times \sum_{n=0}^{n} \frac{(x-1)^{n-r}(1+x)^r}{(n-r)!\,r!\,\Gamma(\alpha+n-r+1)\Gamma(\beta+r+1)} \qquad \text{(A7.23)}
\end{aligned}$$

This last sum is obviously a polynomial of degree n in x.

Let us now define $P_n^{(\alpha,\beta)}(x)$ as

$$P_n^{(\alpha,\beta)}(x) \equiv \frac{(-1)^n}{2^n n!}(1-x)^{-\alpha}(1+x)^{-\beta}\frac{d^n}{dx^n}[(1-x)^{\alpha+n}(1+x)^{\beta+n}] \tag{A7.24}$$

and establish that this is equivalent to Eq. 7.189. From Eqs. A7.22 and A7.23 it follows at once that $P_n^{(\alpha,\beta)}(x) = (-1)^n P_n^{(\beta,\alpha)}(-x)$. We observe that for $\alpha > -1$, $\beta > -1$, and $n \geq 1$

$$\begin{aligned}
&\int_{-1}^{1} x^m (1-x)^{\alpha}(1+x)^{\beta} P_n^{(\alpha,\beta)}(x)\,dx \\
&\quad = \frac{(-1)^n}{2^n n!}\int_{-1}^{1} x^m \frac{d^n}{dx^n}[(1-x)^{\alpha+n}(1+x)^{\beta+n}]\,dx = 0 \qquad m < n \qquad \text{(A7.25)}
\end{aligned}$$

since we may repeatedly integrate by parts as we did for the Legendre polynomials (cf. Section 4.3b). This implies that

$$\int_{-1}^{1} (1-x)^{\alpha}(1+x)^{\beta} P_n^{(\alpha,\beta)}(x) P_m^{(\alpha,\beta)}(x)\,dx = 0 \qquad m \neq n$$

or that

$$\int_{-1}^{1} (1-x)^{\alpha}(1+x)^{\beta} P_n^{(\alpha,\beta)}(x) P_m^{(\alpha,\beta)}(x)\,dx = c_n^{(\alpha,\beta)}\,\delta_{nm} \tag{A7.26}$$

where $c_n^{(\alpha,\beta)}$ must still be determined. If $m < n$, then we may integrate by parts to show that

$$\int_{-1}^{1} \frac{d}{dx}\left[(1-x)^{\alpha+1}(1+x)^{\beta+1}\frac{d}{dx}P_n^{(\alpha,\beta)}(x)\right]P_m^{(\alpha,\beta)}(x)\,dx$$

$$=-\int_{-1}^{1}\left[(1-x)^{\alpha+1}(1+x)^{\beta+1}\frac{d}{dx}P_n^{(\alpha,\beta)}(x)\right]\frac{d}{dx}P_m^{(\alpha,\beta)}(x)\,dx$$

$$=\int_{-1}^{1}P_n^{(\alpha,\beta)}(x)\frac{d}{dx}\left[(1-x)^{\alpha+1}(1+x)^{\beta+1}\frac{d}{dx}P_m^{(\alpha,\beta)}(x)\right]dx$$

$$=\int_{-1}^{1}(1-x)^{\alpha}(1+x)^{\beta}P_n^{(\alpha,\beta)}(x)\left\{[-(\alpha+1)(1+x)+(\beta+1)(1-x)]\frac{d}{dx}\right.$$

$$\left.\times P_m^{(\alpha,\beta)}(x)+(1-x^2)\frac{d^2}{dx^2}P_m^{(\alpha,\beta)}(x)\right\}dx$$

However every term in { } is a polynomial in x of degree m or lower (cf. Eq. A7.23) so that by Eq. A7.25

$$\int_{-1}^{1}(1-x)^{\alpha}(1+x)^{\beta}\left\{\frac{1}{(1-x)^{\alpha}(1+x)^{\beta}}\frac{d}{dx}\right.$$

$$\left.\times\left[(1-x)^{\alpha+1}(1+x)^{\beta+1}\frac{d}{dx}P_n^{(\alpha,\beta)}(x)\right]\right\}P_m^{(\alpha,\beta)}(x)\,dx=0 \qquad m<n \tag{A7.27}$$

Since it is clear that the term in { } of Eq. A7.27 is a polynomial of degree n in x, Eq. A7.27 states that (cf. Eq. A7.26)

$$\frac{1}{(1-x)^{\alpha}(1+x)^{\beta}}\frac{d}{dx}\left[(1-x)^{\alpha+1}(1+x)^{\beta+1}\frac{d}{dx}P_n^{(\alpha,\beta)}(x)\right]\equiv\sum_{m=0}^{n}a_m^{(\alpha,\beta)}P_m^{(\alpha,\beta)}(x)$$

can contain no $P_m^{(\alpha,\beta)}(x)$ for any $m<n$ so that

$$\frac{1}{(1-x)^{\alpha}(1+x)^{\beta}}\frac{d}{dx}\left[(1-x)^{\alpha+1}(1+x)^{\beta+1}\frac{d}{dx}P_n^{(\alpha,\beta)}(x)\right]=\text{const. }P_n^{(\alpha,\beta)}(x) \tag{A7.28}$$

We determine this constant by equating the coefficient of x^n on both sides of Eq. A7.28. If we denote this leading term by $k_n x^n$, then the x^n term on the l.h.s. of Eq. A7.28 is given as

$$\frac{1}{(1-x)^{\alpha}(1+x)^{\beta}}\frac{d}{dx}[(1-x)^{\alpha+1}(1+x)^{\beta+1}nk_n x^{n-1}]$$

$$=k_n[-(\alpha+1)(1+x)nx^{n-1}+(1-x)(\beta+1)nx^{n-1}+(1-x^2)n(n-1)x^{n-2}]$$

$$=-k_n n(\alpha+\beta+n+1)x^n+O(x^{n-1})=\text{const. }k_n x^n+O(x^{n-1})$$

so that

$$\text{const.}=-n(\alpha+\beta+n+1)$$

or

$$\frac{1}{(1-x)^\alpha(1+x)^\beta}\frac{d}{dx}\left[(1-x)^{\alpha+1}(1+x)^{\beta+1}\frac{d}{dx}P_n^{(\alpha,\beta)}(x)\right]+n(n+\alpha+\beta+1)P_n^{(\alpha,\beta)}(x)=0 \tag{A7.29}$$

which is equivalent to Eq. 7.190. It follows immediately from the hypergeometric equation, Eq. 7.157, and the definition of the $P_n^{(\alpha,\beta)}(x)$ of Eq. 7.189 that the polynomials of degree n given in Eq. 7.189 satisfy the differential equation (7.190). Since both polynomials of degree n defined by Eqs. 7.189 and A7.24 are solutions holomorphic about $x=1$ to the same second-order linear ordinary differential equation, these regular solutions must be proportional to each other (cf. Section 8.2b). However either from Eq. 7.189 or from Eqs. A7.22 and A7.24 we see that

$$P_n^{(\alpha,\beta)}(x=1)=\frac{\Gamma(n+\alpha+1)}{\Gamma(\alpha+1)n!} \tag{A7.30}$$

so that Eqs. 7.189 and A7.24 are equivalent definitions of the $P_n^{(\alpha,\beta)}(x)$.

We are now in a position to evaluate the normalization constant $c_n^{(\alpha,\beta)}$ of Eq. A7.26. We shall need the value of the integral

$$\begin{aligned}B(\mu,\nu)&\equiv\int_0^1 s^{\mu-1}(1-s)^{\nu-1}\,ds\\&=2\int_0^{\pi/2}(\sin\theta)^{2\mu-1}(\cos\theta)^{2\nu-1}\,d\theta\end{aligned}$$

From Eq. 7.104 we have

$$\begin{aligned}\Gamma(z)&=\int_0^\infty e^{-t}t^{z-1}\,dt\\&=2\int_0^\infty e^{-r^2}r^{2z-1}\,dr\end{aligned}$$

so that (cf. derivation of Eq. 7.110)

$$\begin{aligned}\Gamma(\alpha)\Gamma(\beta)&=4\int_0^\infty e^{-x^2}x^{2\alpha-1}\,dx\int_0^\infty e^{-y^2}y^{2\beta-1}\,dy\\&=4\int_0^\infty e^{-r^2}r^{2\alpha+2\beta-1}\,dr\int_0^{\pi/2}(\cos\theta)^{2\alpha-1}(\sin\theta)^{2\beta-1}\,d\theta\\&=\Gamma(\alpha+\beta)B(\alpha,\beta)\end{aligned}$$

or

$$\begin{aligned}B(\mu,\nu)&\equiv\int_0^1 s^{\mu-1}(1-s)^{\nu-1}\,ds\\&=\frac{\Gamma(\mu)\Gamma(\nu)}{\Gamma(\mu+\nu)}\end{aligned} \tag{A7.31}$$

If we now use Eq. A7.24 and integrate repeatedly by parts, we find that (for $\alpha>-1$, $\beta>-1$)

$$\int_{-1}^{1}(1-x)^{\alpha}(1+x)^{\beta}[P_n^{(\alpha,\beta)}(x)]^2\,dx=\frac{(-1)^n}{2^n n!}\int_{-1}^{1}\left\{\frac{d^n}{dx^n}[(1-x)^{n+\alpha}(1+x)^{n+\beta}]\right\}P_n^{(\alpha,\beta)}(x)\,dx$$
$$=\frac{1}{2^n n!}\int_{-1}^{1}(1-x)^{n+\alpha}(1+x)^{n+\beta}\frac{d^n}{dx^n}P_n^{(\alpha,\beta)}(x)\,dx \tag{A7.32}$$

If we use Eq. 7.189 for $P_n^{(\alpha,\beta)}(x)$ and Eq. 7.161 for $F(a,b|c|z)$, as well as the fact that (cf. Eq. 7.109)

$$\lim_{\varepsilon\to 0}\frac{\Gamma(-m+\varepsilon)}{\Gamma(-n+\varepsilon)}=(-1)^{n-m}\frac{n!}{m!}$$

we can write

$$P_n^{(\alpha,\beta)}(x)=\frac{\Gamma(1+\alpha+n)}{\Gamma(1+\alpha+\beta+n)}\sum_{s=0}^{n}\frac{(-1)^s\Gamma(1+\alpha+\beta+n+s)}{\Gamma(1+\alpha+s)s!\,(n-s)!}\left(\frac{1-x}{2}\right)^s \tag{A7.33}$$

so that the coefficient of x^n in $P_n^{(\alpha,\beta)}(x)$ is just

$$\frac{\Gamma(1+\alpha+\beta+2n)}{\Gamma(1+\alpha+\beta+n)n!\,2^n} \tag{A7.34}$$

If we use (A7.34) in Eq. A7.32 we obtain, with the aid of Eq. A7.31,

$$\int_{-1}^{1}(1-x)^{\alpha}(1+x)^{\beta}[P_n^{(\alpha,\beta)}(x)]^2\,dx$$
$$=\frac{1}{2^n n!}\frac{\Gamma(1+\alpha+\beta+2n)}{\Gamma(1+\alpha+b+n)2^n}\int_{-1}^{1}(1-x)^{n+\alpha}(1+x)^{n+\beta}\,dx$$
$$=\frac{1}{2^{2n}}\frac{\Gamma(1+\alpha+\beta+2n)}{\Gamma(1+\alpha+\beta+n)}\frac{2^{n+\alpha}2^{n+\beta+1}}{n!}\int_0^1 y^{n+\alpha}(1-y)^{n+\beta}\,dy$$
$$=\frac{\Gamma(1+\alpha+\beta+2n)2^{\alpha+\beta+1}}{\Gamma(1+\alpha+\beta+n)n!}\frac{\Gamma(n+\alpha+1)\Gamma(n+\beta+1)}{\Gamma(2n+\alpha+\beta+2)}$$
$$=\frac{2^{\alpha+\beta+1}\Gamma(n+\alpha+1)\Gamma(n+\beta+1)}{\Gamma(n+1)(2n+\alpha+\beta+1)\Gamma(n+\alpha+\beta+1)}\equiv c_n^{(\alpha,\beta)} \tag{A7.35}$$

which establishes Eq. 7.191.

We can obtain a useful recursion relation for the $P_n^{(\alpha,\beta)}(x)$ as follows. We first determine a constant a such that

$$P_n^{(\alpha,\beta)}(x)-axP_n^{(\alpha,\beta)}(x) \tag{A7.36}$$

is a polynomial if degree n in x. If Eq. A7.36 is to contain no x^{n+1} term, then from (A7.34) we see that a must be chosen as

$$a=\frac{(2n+\alpha+\beta+1)(2n+\alpha+\beta+2)}{2(n+1)(n+\alpha+\beta+1)} \tag{A7.37}$$

Therefore we may write

$$P_{n+1}^{(\alpha,\beta)}(x)-axP_n^{(\alpha,\beta)}(x)=\sum_{m=0}^{n} a_m^{(\alpha,\beta)}P_m^{(\alpha,\beta)}(x) \tag{A7.38}$$

so that by the orthogonality condition of Eq. A7.26 we have (cf. also Eq. A7.25)

$$\begin{aligned} c_m^{(\alpha,\beta)}a_m^{(\alpha,\beta)} &= \int_{-1}^{1}(1-x)^{\alpha}(1+x)^{\beta}[P_{n+1}^{(\alpha,\beta)}(x)-axP_n^{(\alpha,\beta)}(x)]P_m^{(\alpha,\beta)}(x)\,dx \\ &= -a\int_{-1}^{1}(1-x)^{\alpha}(1+x)^{\beta}P_n^{(\alpha,\beta)}(x)[xP_m^{(\alpha,\beta)}(x)]\,dx \\ &= 0, \qquad m>n-1 \end{aligned} \tag{A7.39}$$

or

$$P_{n+1}^{(\alpha,\beta)}(x)-axP_n^{(\alpha,\beta)}(x)=bP_n^{(\alpha,\beta)}(x)+cP_{n-1}^{(\alpha,\beta)}(x) \tag{A7.40}$$

where b and c must still be determined. If we use Eqs. A7.30 and 7.192 in Eq. A7.40 for $x=\pm1$, we obtain

$$\frac{\Gamma(n+\alpha+2)}{(n+1)!}-a\frac{\Gamma(n+\alpha+1)}{n!}=b\frac{\Gamma(n+\alpha+1)}{n!}+c\frac{\Gamma(n+\alpha)}{(n-1)!}$$

$$(-1)^{n+1}\frac{\Gamma(n+\beta+2)}{(n+1)!}+a(-1)^{n}\frac{\Gamma(n+\beta+1)}{n!}=\frac{b(-1)^{n}\Gamma(n+\beta+1)}{n!}+\frac{c(-1)^{n-1}\Gamma(n+\beta)}{(n-1)!}$$

or

$$\begin{aligned} \frac{(n+\alpha+1)(n+\alpha)}{(n+1)}-a(n+\alpha)&=b(n+\alpha)+cn \\ \frac{(n+\beta+1)(n+\beta)}{(n+1)}-a(n+\beta)&=-b(n+\beta)+cn \end{aligned} \tag{A7.41}$$

so that

$$b=\frac{(\alpha^2-\beta^2)(2n+\alpha+\beta+1)}{2(n+1)(n+\alpha+\beta+1)(2n+\alpha+\beta)} \tag{A7.42}$$

$$c=\frac{-(n+\alpha)(n+\beta)(2n+\alpha+\beta+2)}{(n+1)(n+\alpha+\beta+1)(2n+\alpha+\beta)} \tag{A7.43}$$

We have therefore obtained the recursion relation

$$\begin{aligned} 2(n+1)(n+\alpha+\beta+1)(2n+\alpha+\beta)P_n^{(\alpha,\beta)}(x) \\ =(2n+\alpha+\beta+1)[(2n+\alpha+\beta)(2n+\alpha+\beta+2)x+\alpha^2-\beta^2]P_n^{(\alpha,\beta)}(x) \\ -2(n+\alpha)(n+\beta)(2n+\alpha+\beta+2)P_{n-1}^{(\alpha,\beta)}(x) \end{aligned} \tag{A7.44}$$

If we use Eq. A7.33 we see that

$$\begin{aligned} \frac{d}{dx}P_n^{(\alpha,\beta)}(x) &= \frac{-\Gamma(1+\alpha+n)}{\Gamma(1+\alpha+\beta+n)}\sum_{s=0}^{n}\frac{(-1)^s\Gamma(1+\alpha+\beta+n+s)}{\Gamma(1+\alpha+s)(s-1)!\,(n-s)!\,2}\left(\frac{1-x}{2}\right)^{s-1} \\ &= \frac{1}{2}\frac{\Gamma(1+\alpha+n)}{\Gamma(1+\alpha+\beta+n)}\sum_{s=0}^{n-1}\frac{(-1)^s\Gamma(1+\alpha+1+\beta+1+n-1+s)}{\Gamma(1+\alpha+1+s)s!\,(n-1-s)!}\left(\frac{1-x}{2}\right)^{s-1} \\ &= \tfrac{1}{2}(n+\alpha+\beta+1)P_{n-1}^{(\alpha+1,\beta+1)}(x) \end{aligned} \tag{A7.45}$$

Again from Eq. A7.33 we can easily verify directly that

$$(2n+\alpha+\beta+2)\left(\frac{1-x}{2}\right)P_n^{(\alpha+1,\beta)}(x)=(n+\alpha+1)P_n^{(\alpha,\beta)}(x)-(n+1)P_{n+1}^{(\alpha,\beta)}(x) \tag{A7.46}$$

and by interchanging α and β and using Eq. 7.192 that

$$(2n+\alpha+\beta+2)\left(\frac{1+x}{2}\right)P_n^{(\alpha,\beta+1)}(x)=(n+\beta+1)P_n^{(\alpha,\beta)}(x)+(n+1)P_{n+1}^{(\alpha,\beta)}(x) \tag{A7.47}$$

If we use Eqs. A7.46 and A7.47 in Eq. A7.45 and then apply Eq. A7.44 we can show that

$$\begin{aligned}(2n+\alpha+\beta)(1-x^2)\frac{d}{dx}P_n^{(\alpha,\beta)}(x)\\ =n[(\alpha-\beta)-(2n+\alpha+\beta)x]P_n^{(\alpha,\beta)}(x)+2(n+\alpha)(n+\beta)P_n^{(\alpha,\beta)}(x)\end{aligned} \tag{A7.48}$$

A7.4 ELLIPSE OF CONVERGENCE OF THE LEGENDRE POLYNOMIAL EXPANSION

We will now show that the expansion of a function, holomorphic in a region that includes the line segment $-1\leq x\leq 1$, converges with a *unifocal* ellipse (i.e., one with foci at ± 1) in the z plane. This unifocal ellipse of convergence is the largest ellipse that is free of singularities of $f(z)$. We have already seen in Chapter 4 that for a continuous $f(x)$ on x $[-1, 1]$ we can write

$$f(x)=\sum_{n=0}^{\infty} a_n P_n(x) \qquad x\in[-1,1] \tag{A7.49}$$

$$a_n=\frac{(2n+1)}{2}\int_{-1}^{1} f(x)P_n(x)\,dx \tag{A7.50}$$

From the discontinuity equation (cf. Eq. 7.210) satisfied by $Q_n(z)$ across the cut, $-1\leq x\leq 1$, it is easily seen that the coefficients, $\{a_n\}$, of Eq. A7.50 can be expressed as

$$\begin{aligned}a_n&=\frac{(2n+1)}{2\pi i}\int_C f(t)Q_n(t)\,dt=\frac{-(2n+1)}{2\pi i}\int_{-1}^{1} f(t)[Q_n(t+i\varepsilon)-Q_n(t-i\varepsilon)]\,dt\\ &=\frac{(2n+1)}{2}\int_{-1}^{1} f(t)P_n(t)\,dt\end{aligned} \tag{A7.51}$$

where C is shown in Fig. 7.82 and $f(z)$ is assumed holomorphic in a region containing C. Therefore we can write

$$f(z)-\sum_{n=0}^{N} a_nP_n(z)=\frac{1}{2\pi i}\int_C f(t)\,dt\left\{\frac{1}{(t-z)}-\sum_{n=0}^{N}(2n+1)Q_n(t)P_n(z)\right\} \tag{A7.52}$$

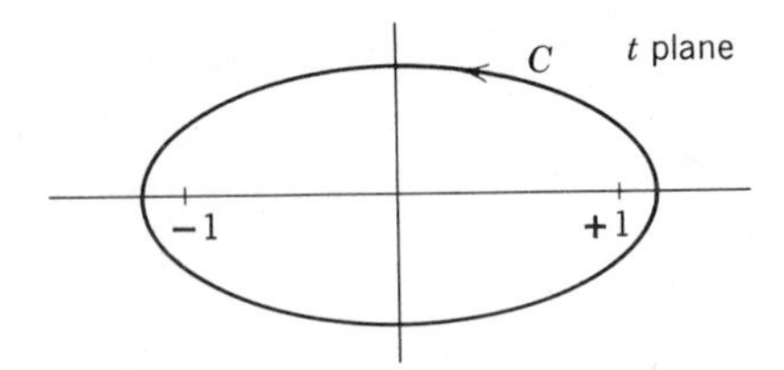

Figure 7.82

Directly from the integral representations of Eqs. 7.195 and 7.212 we obtain the recursion relations (also see Section A4.4, Eq. A4.34)

$$(2n+1)zP_n(z)-(n+1)P_{n+1}(z)-nP_{n-1}(z)=0 \tag{A7.53}$$

$$(2n+1)tQ_n(t)-(n+1)Q_{n+1}(t)-nQ_{n-1}(t)=0 \tag{A7.54}$$

If we multiply the first relation by $-Q_n(t)$ and the second by $P_n(z)$, add, and then sum, we find

$$\sum_{n=1}^{N}(2n+1)(t-z)Q_n(t)P_n(z)+(N+1)[Q_N(t)P_{N+1}(z)-P_N(z)Q_{N+1}(t)]$$
$$-P_1(z)Q_0(t)+Q_1(t)P_0(z)=0$$

With the aid of Eqs. 7.214 and 7.215 we obtain

$$\frac{1}{t-z}-\sum_{n=0}^{N}(2n+1)Q_n(t)P_n(z)=\frac{(N+1)}{(t-z)}[P_{N+1}(z)Q_N(z)-P_N(z)Q_{N+1}(t)] \tag{A7.55}$$

It is clear from Eq. A7.52 that we must now find in what region the norm of the r.h.s. of Eq. A7.55 can be made to vanish as $N\to\infty$.

If we take the integral representation for $P_n(z)$, Eq. 7.195, and take C to be a circle centered about $t=z$ and of radius $|z^2-1|^{1/2}$, then we may change variables as

$$t=z+(z^2-1)^{1/2}e^{i\varphi} \qquad -\pi\leq\varphi\leq\pi$$

to obtain

$$P_n(z)=\frac{1}{\pi}\int_0^{\pi}[z+\sqrt{z^2-1}\cos\varphi]^n\,d\varphi$$

Similarly, if we take the integral representation for $Q_n(z)$ and let

$$t=\frac{e^{\theta}\sqrt{z+1}-\sqrt{z-1}}{e^{\theta}\sqrt{z+1}+\sqrt{z-1}}$$

so that $-1\leq t\leq 1\to-\infty<\theta<\infty$, we obtain

$$Q_n(t)=\int_0^{\infty}[z+\sqrt{z^2-1}\cosh\theta]^{n-1}\,d\theta$$

Using these integral representations we have

$$P_{N+1}(z)Q_N(t)-P_N(z)Q_{N+1}(t)$$

$$=\frac{1}{\pi}\int_0^{\pi} d\varphi \int_0^{\infty} ds \frac{[z+\sqrt{z^2-1}\cos\varphi]^N}{[t+\sqrt{t^2-1}\cosh s]^{N+1}}\{z+\sqrt{z^2-1}\cos\varphi - [t+\sqrt{t^2-1}\cosh s]^{-1}\} \quad \textbf{(A7.56)}$$

We must show that this double integral vanishes as $N\to\infty$. It will be sufficient if we can show that

$$\left|\frac{z+\sqrt{z^2-1}\cos\varphi}{t+\sqrt{t^2-1}\cosh s}\right| \qquad \begin{matrix} 0\le\varphi\le\pi \\ 0\le s<\infty \end{matrix}$$

is always less than unity.

The equation for a unifocal ellipse with semimajor axis a and semiminor axis b is

$$\frac{x^2}{a^2}+\frac{y^2}{b^2}=1 \qquad a^2=b^2+1$$

If we define

$$a=\cosh\alpha \qquad b=\sinh\alpha$$

then we may write for any point $z=x+iy$ on this ellipse

$$z=\cosh(\alpha+i\theta)=\cosh\alpha\cos\theta+i\sinh\alpha\sin\theta$$

Similarly for t on another unifocal ellipse we may write

$$t=\cosh(\beta+i\chi)$$

Therefore we have the identity

$$\begin{aligned}|z+\sqrt{z^2-1}\cos\varphi|^2 &= |\cosh(\alpha+i\theta)+\sinh(\alpha+i\theta)\cos\varphi|^2\\ &=\{\cosh(\alpha+i\theta)\cosh(\alpha-i\theta)+[\cosh(\alpha+i\theta)\sinh(\alpha-i\theta)\\ &\quad+\cosh(\alpha-i\theta)\sinh(\alpha+i\theta)]\cos\varphi\\ &\quad+\sinh(\alpha+i\theta)\sinh(\alpha-i\theta)\cos^2\varphi\}\\ &=\tfrac{1}{2}\{[\cosh(2\alpha)+\cos(2\theta)]+[2\sinh(2\alpha)]\cos\varphi\\ &\quad+[\cosh(2\alpha)-\cos(2\theta)]\cos^2\varphi\}\end{aligned}$$

For real values of φ this has its maximum value when $\cos\varphi=1$ so that

$$\begin{aligned}|z+\sqrt{z^2-1}\cos\varphi| &\le |\cosh(\alpha+i\theta)+\sinh(\alpha+i\theta)|\\ &=|\cosh\alpha\cos\theta+i\sinh\alpha\sin\theta+\sinh\alpha\cos\theta+i\cosh\alpha\sin\theta|\\ &=|\cos\theta e^{\alpha}+i\sin\theta e^{\alpha}|=|e^{\alpha}e^{i\theta}|=e^{\alpha}\end{aligned}$$

Similarly we can show that

$$|t+\sqrt{t^2-1}\cosh s|^2 \geq e^{\beta}$$

from which we deduce

$$\left|\frac{z+\sqrt{z^2-1}\cos\varphi}{t+\sqrt{t^2-1}\cosh s}\right| \leq e^{(\alpha-\beta)}$$

for all φ and s on the integration ranges of Eq. A7.56. As long as $\beta > \alpha$ (i.e., t is on a larger unifocal ellipse than that on which z lies), we see that

$$\lim_{N\to\infty} |P_{N+1}(z)Q_N(t) - P_N(z)Q_{N+1}(t)| = 0$$

and that this limit is uniform in t for any fixed z. Finally, then, from Eq. A7.52 we see that the Legendre expansion, Eq. A7.49, converges for all z inside the largest possible unifocal ellipse enclosing the segment $-1 \leq x \leq 1$ and free of singularities of $f(z)$.

A7.5 PROOF OF THE CONVERGENCE OF A POLYNOMIAL EXPANSION BY MEANS OF CONFORMAL MAPPING

We now discuss another example of a complete set of orthogonal polynomials and use an elegant argument involving an elementary conformal transformation. The transformation

$$z = \frac{1}{2}\left(w + \frac{1}{w}\right) \tag{A7.57}$$

with the inverse

$$\begin{aligned} w &= z + \sqrt{z^2-1} \\ \frac{1}{w} &= z - \sqrt{z^2-1} \end{aligned} \tag{A7.58}$$

maps the entire z plane into the region exterior to the unit circle in the w plane as shown in Fig. 7.83. In fact it is easy to see that unifocal ellipses in the z plane are mapped into circles in the w plane as follows. Equation A7.57 can be written

$$w^2 - 2zw + 1 = 0$$

so that with

$$w = \sigma e^{i\theta}$$

$$z = \rho e^{i\varphi}$$

we find

$$z = x + iy = \rho(\cos\varphi + i\sin\varphi) = \frac{1}{2}\left[\left(\sigma + \frac{1}{\sigma}\right)\cos\theta + i\left(\sigma - \frac{1}{\sigma}\right)\sin\theta\right]$$

or

$$x=\frac{1}{2}\left(\sigma+\frac{1}{\sigma}\right)\cos\theta \tag{A7.59}$$

$$y=\frac{1}{2}\left(\sigma-\frac{1}{\sigma}\right)\sin\theta \tag{A7.60}$$

If we take circles (i.e., σ=const.) in the w plane, then we obtain unifocal ellipses of semimajor axis a and semiminor axis b given as

$$a=\frac{1}{2}\left(\sigma+\frac{1}{\sigma}\right)$$

$$b=\frac{1}{2}\left(\sigma-\frac{1}{\sigma}\right)$$

such that

$$a^2-b^2=\tfrac{1}{2}+\tfrac{1}{2}=1$$

However from elementary plane geometry we know that a^2-b^2 is always the distance from the origin to either focus.

Suppose that $f(z)$ is a real holomorphic function [i.e., $f^*(z^*)=f(z)$] and that z_0 lies on the boundary of the largest unifocal ellipse containing $-1\leq x\leq 1$ and free of singularities of $f(z)$ in the z plane with w_0, $|w_0|=R>1$, being the image of z_0 in the w plane. Let

$$g(w)\equiv f\left[\frac{1}{2}\left(w+\frac{1}{w}\right)\right]=f(z)$$

Since $g(w)$ is holomorphic in the annular region between $|w|=1$ and $|w|=R$, $g(w)$ will have a convergent Laurent expansion

$$g(w)=\sum_{m=-\infty}^{\infty}\alpha_m w^m \tag{A7.61}$$

$$\alpha_m=\frac{1}{2\pi i}\int_C \frac{g(w')\,dw'}{w'^{m+1}} \tag{A7.62}$$

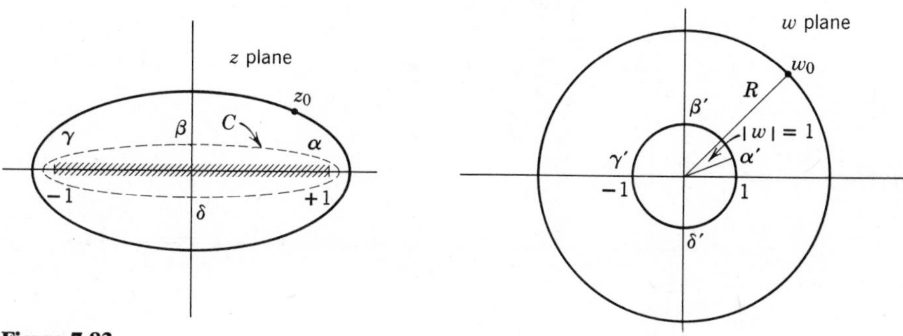

Figure 7.83

If we take the contour, C, to be the unit circle, then since here

$$w = e^{i\theta}$$

$$w^* = e^{-i\theta} = \frac{1}{w}$$

we have

$$g^*(w) = g(w^*) = g\left(\frac{1}{w}\right)$$

so that

$$\sum_{m=-\infty}^{\infty} \alpha_m^* w^{*m} = \sum_{m=-\infty}^{\infty} \alpha_m w^{*m} = \sum_{m=-\infty}^{\infty} \alpha_m w^{-m}$$

from which we conclude

$$\alpha_m^* = \alpha_m = \alpha_{-m} \qquad \forall m$$

This implies that $g(w) = g(1/w)$ everywhere, which allows us to write

$$g(w) = \alpha_0 + \sum_{m=1}^{\infty} \alpha_m \left(w^m + \frac{1}{w^m}\right) \tag{A7.63}$$

$$\alpha_m = \frac{1}{2\pi i}\int_C \frac{g(w)\,dw}{w^{m+1}} \qquad m = 0, 1, 2, \ldots$$

Since it is always possible to expand $[w + (1/w)]^m$ in a binomial series, we can invert this process and write

$$w^m + \frac{1}{w^m} = \sum_{s=0}^{m} \gamma_s^m \left[\frac{1}{2}\left(w + \frac{1}{w}\right)\right]^s \equiv \mathcal{P}_m(z) \equiv \mathcal{P}_m\left[\frac{1}{2}\left(w + \frac{1}{w}\right)\right]$$

where the sum contains only all even or all odd powers of s depending upon whether m is even or odd. If we now rewrite the Laurent expansion for $g(w)$ in terms of the variable, z, we obtain

$$f(z) = \alpha_0 + \sum_{m=1}^{\infty} \alpha_m \mathcal{P}_m(z) \tag{A7.64}$$

$$\alpha_m = \frac{1}{2\pi i}\oint \frac{g(w)\,dw}{w^{m+1}} = \frac{1}{2\pi i}\oint \frac{g(1/w)\,dw}{w^2}\,w^{m+1}$$

$$= \frac{1}{4\pi i}\oint g(w)\left[w^m + \frac{1}{w^m}\right]\frac{dw}{w}$$

$$= \frac{1}{4\pi i}\int_C f(z)\mathcal{P}_m(z)\frac{dz}{\sqrt{z^2-1}} = -\frac{1}{2\pi}\int_{-1}^{1} f(x)\mathcal{P}_m(x)\,\mathrm{Im}\left(\frac{1}{i\sqrt{1-x^2}}\right)dx$$

$$= \frac{1}{2\pi}\int_{-1}^{1} \frac{f(x)\mathcal{P}_m(x)\,dx}{\sqrt{1-x^2}} \tag{A7.65}$$

Furthermore these $\{P_m(z)\}$ satisfy a useful orthogonality relation,

$$\int_0^{2\pi} \mathcal{P}_m \mathcal{P}_n \frac{dw}{w} = \int_0^{2\pi} (e^{im\varphi} + e^{-im\varphi})(e^{in\varphi} + e^{-in\varphi}) i \, d\varphi$$

$$= \begin{cases} 4\pi i \, \delta_{mn} & m, n > 0 \\ 8\pi i & m = n = 0 \end{cases}$$

since m and n are positive integers or zero. But a simple calculation shows

$$\int_0^{2\pi} \mathcal{P}_m \mathcal{P}_n \frac{dw}{w} = \int_C \mathcal{P}_m(z) \mathcal{P}_n(z) \frac{dz}{\sqrt{z^2 - 1}}$$

$$= 2i \int_{-1}^{1} \mathcal{P}_m(x) \mathcal{P}_n(x) \operatorname{Im}\left(\frac{1}{i\sqrt{1 - x^2}}\right) dx$$

$$= 2i \int_{-1}^{1} \mathcal{P}_m(x) \mathcal{P}_n(x) \frac{dx}{\sqrt{1 - x^2}}$$

so that

$$\int_{-1}^{1} \mathcal{P}_m(x) \mathcal{P}_n(x) \frac{dx}{\sqrt{1 - x^2}} = \begin{cases} 2\pi \, \delta_{mn} & m, n > 0 \\ 4\pi & m = n = 0 \end{cases} \tag{A7.66}$$

Therefore the polynomials $\{\mathcal{P}_n(x)\}$ form a complete orthogonal set (at least for holomorphic functions), and the expansion (A7.64) converges uniformly within a unifocal ellipse.

If we compare Eq. A7.66 with Eq. 7.191 for the Jacobi polynomials with $\alpha = \beta = -\frac{1}{2}$, we would expect that

$$\mathcal{P}_n(x) \propto P_n^{[-(1/2), -(1/2)]}(x)$$

This is easily seen since from Eq. A7.25 for the $P_n^{[-(1/2), -(1/2)]}(x)$ we have

$$\int_{-1}^{1} \frac{dx}{\sqrt{1 - x^2}} x^m P_n^{[-(1/2), -(1/2)]}(x) = 0 \qquad \forall m < n$$

If we now expand the $P_n^{[-(1/2), -(1/2)]}(x)$ as

$$P^{[-(1/2), -(1/2)]}(x) = \sum_{m=0}^{n} \beta_m^n \mathcal{P}_m(x)$$

we can write

$$\langle \mathcal{P}_m \mid P_n^{[-(1/2), -(1/2)]} \rangle = 2\pi \beta_m^n$$

which is zero unless $n = m$ since $\mathcal{P}_m(x)$ is a polymonial of degree m in x. We conclude that

$$\mathcal{P}_n(x) = c(n) P_n^{[-(1/2), -(1/2)]}(x)$$

From the original definition of $\mathcal{P}_n(z)$ we see that $\mathcal{P}_n(1) = 2$ and from Eq. 7.189 we obtain

$$2=\frac{c(n)\Gamma(n+\frac{3}{2})}{\Gamma(\frac{3}{2})n!}=\frac{2c(n)\Gamma(n+\frac{3}{2})}{\sqrt{\pi}\,n!}$$

which finally yields

$$\mathcal{P}_n(x)=\frac{\sqrt{\pi}\,n!}{\Gamma(n+\frac{3}{2})}\,P_n^{[-(1/2),-(1/2)]}(x) \tag{A7.67}$$

In fact these $\mathcal{P}_n(x)$ are essentially the *Tchebichef polynomials.*

A7.6 CONFLUENT HYPERGEOMETRIC FUNCTION

If we make the change of variable $z \to z/b$ in the hypergeometric equation (Eq. 7.157), we find

$$bz\left(\frac{z}{b}-1\right)F''+\left[(a+b+1)\frac{z}{b}-c\right]bF'+abF=0$$

If we cancel the factor, b, common to all terms and take the limit as b approaches infinity, we obtain Eq. 7.223, the *confluent hypergeometric equation,*

$$zF''(a\,|c|\,z)+(c-z)F'(a\,|c|\,z)-aF(a\,|c|\,z)=0 \tag{A7.68}$$

Similarly Eq. 7.222 follows at once from Eq. 7.161 and Stirling's approximation, Eq. 7.133.

The substitution

$$f(z)=e^{-iz}g(z) \tag{A7.69}$$

in Eq. 7.221 yields for $g(z)$ the equation

$$zg''(z)+[(2\nu+1)-2iz]g'(z)-i(2\nu+1)g(z)=0 \tag{A7.70}$$

Comparison of Eqs. A7.68 and A7.70 allows us to make the identifications $c=2a=2\nu+1$, $z\to 2iz$ and to write $f(z)$ as

$$f(z)=\frac{e^{-iz}}{\Gamma(\nu+1)}\,F(\nu+\tfrac{1}{2}\,|2\nu+1|\,2iz) \tag{A7.71}$$

$$=\frac{e^{-iz}}{\Gamma(\nu+1)}\,\frac{\Gamma(2\nu+1)}{\Gamma(\nu+\frac{1}{2})}\sum_{s=0}^{\infty}\frac{\Gamma(\nu+\frac{1}{2}+s)}{\Gamma(2\nu+1+s)s!}(2iz)^s$$

$$=\sum_{s=0}^{\infty}\frac{(-1)^s}{s!\,\Gamma(\nu+s+1)}\left(\frac{z}{2}\right)^{2s} \tag{A7.72}$$

This last expansion is obtained by explicitly multiplying out the two infinite series for e^{-iz} and $F(\nu+\frac{1}{2}\,|2\nu+1|\,2iz)$ and showing that all odd powers of z cancel.

We may obtain an integral representation of the confluent hypergeometric function directly from its definition in Eq. 7.222 and the integral representation for $F(a, b\,|c|\,z)$ given in Table 7.1b (cf. Problem 7.52) as

$$
\begin{aligned}
F(a\,|c|\,z) &= \lim_{b\to\infty} F(a, b+c\,|c|\,z/b) \\
&= \lim_{b\to\infty} \frac{\Gamma(c)}{\Gamma(a)\Gamma(c-a)} \int_1^\infty (t-z/b)^{-b-c} t^b (t-1)^{c-a-1}\, dt \\
&= \lim_{b\to\infty} \frac{\Gamma(c)}{\Gamma(a)\Gamma(c-a)} \int_0^1 \left(1-\frac{zt}{b}\right)^{-b-c} t^{a-1}(1-t)^{c-a-1}\, dt \\
&= \lim_{b\to\infty} \frac{\Gamma(c)}{\Gamma(a)\Gamma(c-a)} \int_0^1 \left(1-\frac{zt}{b}\right)^{-(b/zt)(zt)-c} t^{a-1}(1-t)^{c-a-1}\, dt \\
&= \frac{\Gamma(c)}{\Gamma(a)\Gamma(c-a)} \int_0^1 e^{zt} t^{a-1}(1-t)^{c-a-1}\, dt \qquad \operatorname{Re} c > \operatorname{Re} a > 0
\end{aligned}
\tag{A7.73}
$$

since by definition

$$\lim_{h\to 0} (1+h)^{1/h} = e$$

We can also arrive at an integral representation for $F(a\,|c|\,z)$ by seeking a solution to Eq. A7.68, just as we did for the Bessel function in Section 7.13c, in the form

$$u(z) = \int_C e^{zf(t)} w(t)\, dt \tag{A7.74}$$

The same type manipulations that led to Eq. 7.235 show that acceptable choices for $f(t)$ and $w(t)$ are

$$
\begin{aligned}
f(t) &= t \\
w(t) &= At^{a-1}(t-1)^{c-a-1}
\end{aligned}
\tag{A7.75}
$$

where the contour C is shown in Fig. 7.84. With this contour $w(t)$ returns to its

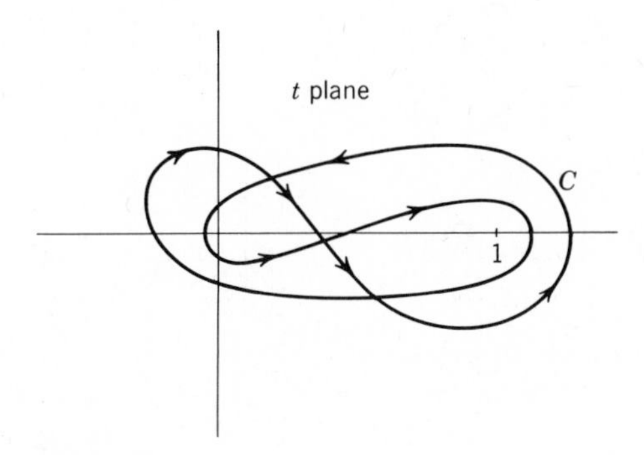

Figure 7.84

initial value after one complete loop around C so that the out-integrated terms (cf. Eq. 7.234) vanish. The representation

$$u(z)=A\int_C e^{zt}t^{a-1}(t-1)^{c-a-1}\,dt \quad \textbf{(A7.76)}$$

is valid for all values of a and c. However, if we now take Re $c>$Re $a>0$, we can deform the contour as shown in Fig. 7.85 to around $t=0$ and $t=1$ with the indicated phases for the integrand. We then obtain for $u(z)$

$$\begin{aligned}u(z)=A\Big\{&e^{\pi i(c-a-1)}\int_1^0 e^{zt}t^{a-1}(1-t)^{c-a-1}\,dt\\&+e^{2\pi i(a-1)}e^{\pi i(c-a-1)}\int_0^1 e^{zt}t^{a-1}(1-t)^{c-a-1}\,dt\\&+e^{2\pi i(a-1)}e^{-\pi i(c-a-1)}\int_1^0 e^{zt}t^{a-1}(1-t)^{c-a-1}\,dt\\&+e^{-\pi i(c-a-1)}\int_0^1 t^{a-1}(1-t)^{c-a-1}\,dt\Big\}\\=&\,4Ae^{\pi ia}\sin(\pi a)\sin[\pi(c-a)]\int_0^1 e^{zt}t^{a-1}(1-t)^{c-a-1}\,dt\end{aligned}$$

Since this $u(z)$ is regular at $z=0$ we may normalize it to $F(a\,|c|\,z=0)=1$ as

$$\begin{aligned}u(z=0)=1&=4Ae^{\pi ia}\sin(\pi a)\sin[\pi(c-a)]\int_0^1 t^{a-1}(1-t)^{c-a-1}\,dt\\&=4Ae^{\pi ia}\sin(\pi a)\sin[\pi(c-a)]\frac{\Gamma(a)\Gamma(c-a)}{\Gamma(c)}\\&=\frac{4Ae^{\pi ia}\pi^2}{\Gamma(1-a)\Gamma(a-c+1)\Gamma(c)}\end{aligned}$$

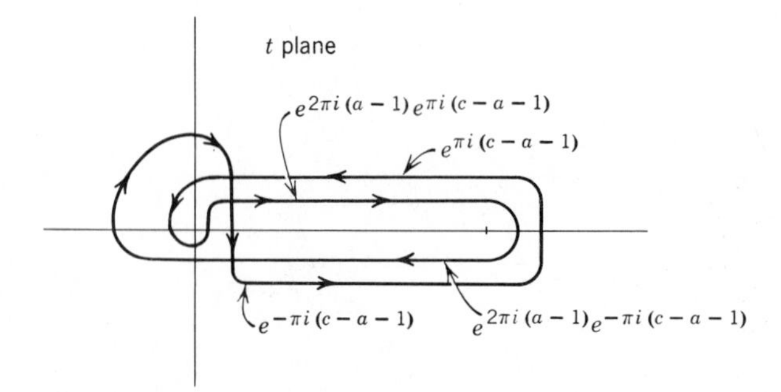

Figure 7.85

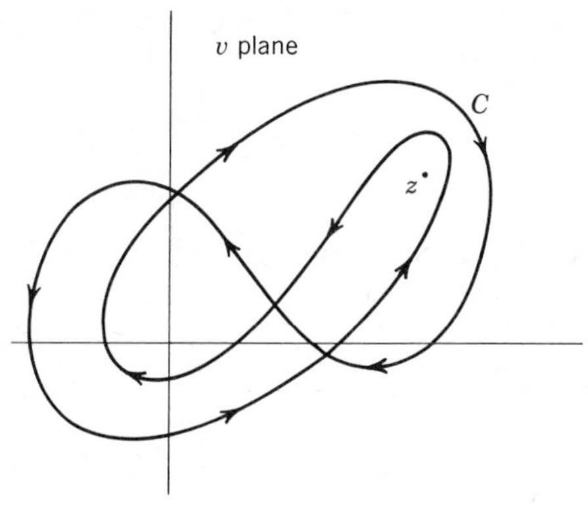

Figure 7.86

where we have used Eqs. A7.31 and 7.110. We finally have, then,

$$F(a\,|c|\,z)=\frac{e^{-\pi i a}}{4\pi^2}\Gamma(c)\Gamma(1-a)\Gamma(a-c+1)\int_C e^{zt}t^{a-1}(t-1)^{c-a-1}\,dt \qquad \textbf{(A7.77)}$$

We can use this integral representation to obtain the asymptotic behavior of $F(a\,|c|\,z)$ as $z\to\infty$. Let us first consider the case in which Re $z>0$ so that the dominant contribution to the integral in Eq. A7.77 comes from those portions of C just to the right of $t=1$ in Fig. 7.84. Let $t-1=-v/z$ so that

$$F(a\,|c|\,z)=\frac{-e^{-\pi i a}}{4\pi^2}\Gamma(c)\Gamma(1-a)\Gamma(a-c+1)z^{a-c}e^{z}\int_{C'} e^{-v}\left(1-\frac{v}{z}\right)^{a-1}(-v)^{c-a-1}\,dv$$

where C' is shown qualitatively in Fig. 7.86. Now the dominant contribution to the integral comes from those portions of C' near $v=0$. If we keep track of the phases of the integrand on C' (cf. Fig. 7.85), keep only those contributions from C' near $v=0$, and add on what is necessary to obtain C of Fig. 7.57, we obtain (cf. Eq. 7.113)

$$F(a\,|c|\,z)\xrightarrow[\text{Re } z>0]{|z|\to\infty}\frac{-e^{-\pi i a}\Gamma(c)\Gamma(1-a)\Gamma(a-c+1)}{4\pi^2}$$

$$\times z^{a-c}e^{z}[-1+e^{2\pi i(a-1)}]\int_C e^{-v}(e^{-\pi i}v)^{c-a-1}\,dv$$

$$=\frac{-(e^{\pi i a}-e^{-\pi i a})\Gamma(c)\Gamma(1-a)\Gamma(a-c+1)}{4\pi^2}z^{a-c}e^{z}2i\sin[\pi(c-a)]\Gamma(c-a)$$

$$=\frac{\Gamma(c)}{\Gamma(a)}z^{a-c}e^{z}$$

so that

$$F(a\,|c|\,z)\xrightarrow[z\to\infty]{}\frac{\Gamma(c)}{\Gamma(a)}z^{a-c}e^{z} \qquad \text{Re } z>0 \qquad \textbf{(A7.78)}$$

If Re $z<0$, then the dominant contribution comes from the neighborhood of $t=0$ in Fig. 7.84. We now set $t=-v/z$ and obtain by arguments similar to the above

$$F(a\,|c|\,z)=\frac{-\Gamma(c)\Gamma(1-a)\Gamma(a-c+1)}{4\pi^2}(-z)^{-a}\int_C e^{-v}(e^{-\pi i}v)^{a-1}\left(-\frac{v}{z}-1\right)^{c-a-1}dv$$

$$\xrightarrow[\substack{z\to\infty\\ \text{Re } z<0}]{}\frac{-\Gamma(c)\Gamma(1-a)\Gamma(a-c+1)[e^{\pi i(c-a)}-e^{-\pi i(c-a)}]}{\pi^2}(e^{-\pi i}z)^{-a}\int_C e^{-v}(e^{-\pi i}v)^{a-1}\,dv$$

$$=\frac{\Gamma(c)\Gamma(1-a)\Gamma(a-c+1)\sin[\pi(c-a)]}{\pi^2}(e^{-\pi i}z)^{-a}\sin(\pi a)\Gamma(a)$$

$$=\frac{\Gamma(c)}{\Gamma(c-a)}(e^{-\pi i}z)^{-a}$$

so that

$$F(a\,|c|\,z)\xrightarrow[z\to\infty]{}\frac{\Gamma(c)}{\Gamma(c-a)}(e^{-\pi i}z)^{-a}\qquad \text{Re } z<0 \tag{A7.79}$$

Finally consider the case in which Re $z=0$ so that $z=iy$. When $y>0$ the arguments given above should make it clear that the dominant contributions come from those portions of C just below $t=0$ and $t=1$. The net effect is that we pick up *both* contributions of Eq. A7.78 *and* of Eq. A7.79 so that

$$F(a\,|c|\,iy)\xrightarrow[y\to+\infty]{}\frac{\Gamma(c)}{\Gamma(a)}(iy)^{a-c}e^{iy}+\frac{\Gamma(c)}{\Gamma(c-a)}(e^{-\pi i}iy)^{-a} \tag{A7.80}$$

We can do an analogous calculation when $y<0$.

We gather all of these results together below.

$$F(a\,|c|\,z)\xrightarrow[z\to\infty]{}\begin{cases}\dfrac{\Gamma(c)}{\Gamma(a)}z^{a-c}e^{z} & -\dfrac{\pi}{2}<\phi<\dfrac{\pi}{2}\\[2ex] \dfrac{\Gamma(c)}{\Gamma(a)}(z)^{a-c}e^{z}+\dfrac{\Gamma(c)}{\Gamma(c-a)}(e^{-\pi i}z)^{-a} & \phi=\dfrac{\pi}{2}\\[2ex] \dfrac{\Gamma(c)}{\Gamma(c-a)}(e^{-\pi i}z)^{-a} & \dfrac{\pi}{2}<\phi<\dfrac{3\pi}{2}\\[2ex] \dfrac{\Gamma(c)}{\Gamma(a)}(e^{-2\pi i}z)^{a-c}e^{z}+\dfrac{\Gamma(c)}{\Gamma(c-a)}(e^{-\pi i}z)^{-a} & \phi=\dfrac{3\pi}{2}\end{cases} \tag{A7.81}$$

Notice that these asymptotic forms are consistent with $F(a\,|c|\,z)$ being a single-valued function on a large circle, as it must be since the series defining it (cf. Eq. 7.222) converges for all finite values of z. Further details on the asymptotic behavior of the confluent hypergeometric function can be found in Volume I, Section 6.13.1 of the *Bateman Manuscript Project.*

TABLE 7.1 Elementary Properties of Some Special Functions

a. Gamma Function

$$\Gamma(z)=\int_0^\infty e^{-t}t^{z-1}\,dt \qquad \text{Re } z>0$$

$$\Gamma(z+1)=z\Gamma(z)$$

$$\Gamma(n+1)=n!$$

$$\Gamma(\tfrac{1}{2})=\sqrt{\pi}$$

$$\text{Res}\,[\Gamma(z);\,z=-n]=\frac{(-1)^n}{n!}$$

$$\lim_{\varepsilon\to 0}\frac{\Gamma(-m+\varepsilon)}{\Gamma(-n+\varepsilon)}=\frac{(-1)^{m-n}n!}{m!}$$

$$\Gamma(z)\Gamma(1-z)=\pi\csc(\pi z)$$

$$\Gamma(2z+1)=\frac{2^{2z}\Gamma(z+\frac{1}{2})\Gamma(z+1)}{\Gamma(\frac{1}{2})}$$

$$\frac{1}{\Gamma(z)}=ze^{\gamma z}\prod_{n=1}^{\infty}\left[\left(1+\frac{z}{n}\right)e^{-z/n}\right]$$

$$\gamma\equiv\lim_{m\to\infty}\left(\sum_{n=1}^{m}\frac{1}{n}-\ln m\right)=0.5772157\cdots$$

$$\Gamma(z)\xrightarrow[|z|\to\infty]{}\sqrt{2\pi}\,z^{z-(1/2)}e^{-z}\left[1+\frac{1}{12z}+\frac{1}{288z^2}-\frac{1}{51840z^3}+O(z^{-4})\right] \qquad \text{Re } z>0$$

$$\frac{\Gamma(z+\alpha)\Gamma(z+\beta)}{\Gamma(z+\gamma)\Gamma(z+\delta)}\xrightarrow[|z|\to\infty]{}z^{\alpha+\beta-\gamma-\delta}$$

$$\psi(z)\equiv\frac{d}{dz}\ln[\Gamma(z)]=\frac{1}{\Gamma(z)}\frac{d\Gamma(z)}{dz}=-\gamma-\frac{1}{z}+\sum_{n=1}^{\infty}\left(\frac{1}{n}-\frac{1}{n+z}\right)$$

$$\psi(z+1)=\frac{1}{z}+\psi(z)$$

$$\psi(1)=-\gamma$$

$$\psi(n)=-\gamma+\sum_{m=1}^{n-1}\frac{1}{m} \qquad n\geq 2$$

$$\psi(z)\xrightarrow[z\to\infty]{}\ln z-\frac{1}{12z}-\frac{1}{12z^2}+\frac{1}{120z^4}+O(z^{-6})$$

$$B(\alpha,\beta)\equiv\int_0^1 s^{\alpha-1}(1-s)^{\beta-1}\,ds=\frac{\Gamma(\alpha)\Gamma(\beta)}{\Gamma(\alpha+\beta)}$$

b. Hypergeometric Functions

$$z(z-1)\frac{d^2}{dz^2}F(a,b\,|c|\,z)+[(a+b+1)z-c]\frac{d}{dz}F(a,b\,|c|\,z)+abF(a,b\,|c|\,z)=0$$

TABLE 7.1 (Cont.)

b. Hypergeometric Functions (Cont.)

$$F(a, b\,|c|\,z)=\frac{\Gamma(c)}{\Gamma(a)\Gamma(b)}\sum_{m=0}^{\infty}\frac{\Gamma(a+m)\Gamma(b+m)}{\Gamma(c+m)m!}z^m \qquad |z|<1 \qquad c\neq -n$$

$$F(a, b\,|-n|\,z)=\frac{n!\,z^{n+1}}{\Gamma(n+1+a)\Gamma(n+1+b)}\sum_{m=0}^{\infty}\frac{\Gamma(n+1+m+a)\Gamma(n+1+m+b)}{m!\,(m+n)!}z^m \qquad |z|<1$$

$$F_2(a, b\,|c|\,z)=z^{1-c}F(b-c+1, a-c+1\,|2-c|\,z) \qquad c\neq -n$$

$$F(a, b\,|c|\,z)\xrightarrow[z\to 1]{}\frac{\Gamma(c)\Gamma(a+b-c)}{\Gamma(a)\Gamma(b)}(1-z)^{c-a-b} \qquad \text{Re}\,(a+b)>\text{Re}\,c>0$$

$$F(a, b\,|a+b|\,z)\xrightarrow[z\to 1]{}-\frac{\Gamma(a+b)}{\Gamma(a)\Gamma(b)}\ln(1-z)$$

$$F(a, b\,|c|\,z)=\frac{\Gamma(c)\Gamma(c-a-b)}{\Gamma(c-a)\Gamma(c-b)}F(a, b\,|a+b-c+1|\,1-z)$$
$$+\frac{\Gamma(c)\Gamma(a+b-c)}{\Gamma(a)\Gamma(b)}(1-z)^{c-a-b}F(c-a, c-b\,|c-a-b+1|\,1-z) \qquad \text{Re}\,(a+b)>\text{Re}\,c>1$$

$$F(a, b\,|c|\,z)=(1-z)^{c-a-b}F(c-a, c-b\,|c|\,z)$$

$$\frac{\Gamma(a)\Gamma(b)}{\Gamma(c)}F(a, b\,|c|\,z)=\frac{\Gamma(a)\Gamma(b-a)}{\Gamma(c-a)}(e^{-\pi i}z)^{-a}F\left(a, 1-c+a\,|1-b+a|\,\frac{1}{z}\right)$$
$$+\frac{\Gamma(b)\Gamma(a-b)}{\Gamma(c-b)}(e^{-\pi i}z)^{-b}F\left(b, 1-c+b\,|1-a+b|\,\frac{1}{z}\right) \qquad a-b\neq\text{integer}$$

$$F(a, b\,|c|\,z)=\frac{\Gamma(c)}{\Gamma(b)\Gamma(c-b)}\int_1^{\infty}(t-z)^{-a}t^{a-c}(t-1)^{c-b-1}\,dt \qquad \text{Re}\,c>\text{Re}\,b>0$$

c. Jacobi Polynomials

$$(1-x^2)\frac{d^2}{dx^2}P_n^{(\alpha,\beta)}(x)+[\beta-\alpha-(\alpha+\beta+2)x]\frac{d}{dx}P_n^{(\alpha,\beta)}(x)+n(n+\alpha+\beta+1)P_n^{(\alpha,\beta)}(x)=0$$

$$P_n^{(\alpha,\beta)}(x)=\frac{\Gamma(1+\alpha+n)}{\Gamma(1+\alpha)n!}F\left(1+\alpha+\beta+n, -n\,|1+\alpha|\,\frac{1-x}{2}\right)$$
$$=\frac{(-1)^n}{2^n n!}(1-x)^{-\alpha}(1+x)^{-\beta}\frac{d^n}{dx^n}[(1-x)^{\alpha+n}(1+x)^{\beta+n}]$$

$$P_n^{(\alpha,\beta)}(x)=(-1)^n P_n^{(\beta,\alpha)}(-x)$$

$$\int_{-1}^{1}(1-x)^{\alpha}(1+x)^{\beta}P_m^{(\alpha,\beta)}(x)P_n^{(\alpha,\beta)}(x)\,dx$$
$$=\frac{2^{\alpha+\beta+1}}{(2n+\alpha+\beta+1)}\frac{\Gamma(n+\alpha+1)\Gamma(n+\beta+1)}{\Gamma(n+1)\Gamma(n+\alpha+\beta+1)}\delta_{nm} \qquad \alpha>-1 \qquad \beta>-1$$

$$2(n+1)(n+\alpha+\beta+1)(2n+\alpha+\beta)P_{n+1}^{(\alpha,\beta)}(x)=(2n+\alpha+\beta+1)[(2n+\alpha+\beta)(2n+\alpha+\beta+2)x+\alpha^2-\beta^2]P_n^{(\alpha,\beta)}(x)$$
$$-2(n+\alpha)(n+\beta)(2n+\alpha+\beta+2)P_{n-1}^{(\alpha,\beta)}(x)$$

$$\frac{d}{dx}P_n^{(\alpha,\beta)}(x)=\tfrac{1}{2}(n+\alpha+\beta+1)P_{n-1}^{(\alpha+1,\beta+1)}(x)$$

TABLE 7.1 (Cont.)

c. Jacobi Polynomials (Cont.)

$$(2n+\alpha+\beta)(1-x^2)\frac{d}{dx}P_n^{(\alpha,\beta)}(x)=[(\alpha-\beta)-(2n+\alpha+\beta)x]P_n^{(\alpha,\beta)}(x)+2(n+\alpha)(n+\beta)P_{n-1}^{(\alpha,\beta)}(x)$$

d. Legendre Functions (cf. also Table 4.1)

$$(z^2-1)\frac{d^2}{dz^2}P_\alpha(z)+2z\frac{d}{dz}P_\alpha(z)-\alpha(\alpha+1)P_\alpha(z)=0$$

$$P_\alpha(z)=F\left(-\alpha,\,\alpha+1\,|1|\,\frac{1-z}{2}\right)=\frac{1}{2\pi i}\oint_C\frac{(t^2-1)^\alpha\,dt}{2^\alpha(t-z)^{\alpha+1}}\qquad\text{(cf. Fig. 7.68)}$$

$$P_\alpha(z)\xrightarrow[z\to-1]{}\frac{\sin(\pi\alpha)}{\pi}\ln(1+z)\qquad\alpha\neq n$$

$$P_\alpha(z)=P_{-\alpha-1}(z)$$

$$P_\alpha(z)\xrightarrow[z\to\infty]{}\frac{\Gamma(a+\frac{1}{2})}{\sqrt{\pi}\,\Gamma(\alpha+1)}(2z)^\alpha\qquad\text{Re}\,\alpha>0$$

$$Q_\alpha(z)=\frac{-1}{2^{\alpha+2}i\sin(\pi\alpha)}\oint_C\frac{(t^2-1)^\alpha\,dt}{(z-t)^{\alpha+1}}\qquad\text{(cf. Fig. 7.69)}$$

$$=\frac{1}{2^{\alpha+1}}\int_{-1}^{1}\frac{(1-t^2)^\alpha dt}{(z-t)^{\alpha+1}}\qquad\text{Re}\,\alpha>-1$$

$$=\frac{\sqrt{\pi}}{(2z)^{\alpha+1}}\frac{\Gamma(\alpha+1)}{\Gamma(\alpha+\frac{3}{2})}F\left(\frac{\alpha+2}{2},\frac{\alpha+1}{2}\,|\alpha+\tfrac{3}{2}|\,\frac{1}{z^2}\right)$$

$$P_n(z)=\frac{1}{2^n n!}\frac{d^n}{dz^n}(z^2-1)$$

$$Q_n(z)=\frac{1}{2}\int_{-1}^{1}\frac{P_n(t)\,dt}{(z-t)}$$

$$\lim_{\varepsilon\to0^+}[Q_n(x+i\varepsilon)-Q_n(x-i\varepsilon)]=-\pi iP_n(x)\qquad -1<x<1$$

$$Q_0(z)=\frac{1}{2}\ln\left(\frac{z+1}{z-1}\right)$$

$$Q_1(z)=\frac{1}{2}z\ln\left(\frac{z+1}{z-1}\right)-1$$

$$(2n+1)zP_n(z)-(n+1)P_{n+1}(z)-nP_{n-1}(z)=0$$

$$(2n+1)zQ_n(z)-(n+1)Q_{n+1}(z)-nQ_{n-1}(z)=0$$

$$(1-z^2)\frac{dP_n(z)}{dz}=-nzP_n(z)+nP_{n-1}(z)$$

$$P_n^m(z)\equiv(1-z^2)^{(1/2)m}\frac{d^m}{dx^m}P_n(z)\qquad 0\le m\le n$$

$$=\frac{(n+m)!}{2^m m!\,(n-m)!}(1-z^2)^{(1/2)m}F\left(m-n,\,m+n+1\,|m+1|\,\frac{1-z}{2}\right)$$

TABLE 7.1 (Cont.)

e. Confluent Hypergeometric Function

$$z\frac{d^2}{dz^2}F(a\,|c|\,z)+(c-z)\frac{d}{dz}F(a\,|c|\,z)-aF(a\,|c|\,z)=0$$

$$F(a\,|c|\,z)=\lim_{b\to 0}F(a,b+c\,|\,c\,|\,z/b)=\frac{\Gamma(c)}{\Gamma(a)}\sum_{m=0}^{\infty}\frac{\Gamma(a+m)}{\Gamma(c+m)m!}z^m \qquad |z|<\infty \qquad c\neq -n$$

$$F(a\,|c|\,z)=\frac{\Gamma(c)}{\Gamma(a)\Gamma(c-a)}\int_0^1 e^{zt}t^{a-1}(1-t)^{c-a-1}\,dt \qquad \text{Re}\,c>\text{Re}\,a>0$$

$$F(a\,|c|\,z)\xrightarrow[z\to\infty]{}\begin{cases}\dfrac{\Gamma(c)}{\Gamma(a)}z^{a-c}e^z & \text{Re}\,z>0\\[2ex] \dfrac{\Gamma(c)}{\Gamma(c-a)}(-z)^{-a} & \text{Re}\,z<0\end{cases}$$

f. Bessel Functions

$$z^2\frac{d^2}{dz^2}J_\nu(z)+z\frac{d}{dz}J_\nu(z)+(z^2-\nu^2)J_\nu(z)=0$$

$$J_\nu(z)=\left(\frac{z}{2}\right)^\nu\sum_{m=0}^{\infty}\frac{(-1)^m}{m!\,\Gamma(m+\nu+1)}\left(\frac{z}{2}\right)^{2m}$$

$$=\left(\frac{z}{2}\right)^\nu\frac{e^{-iz}}{\Gamma(\nu+1)}F(\nu+\tfrac{1}{2}\,|2\nu+1|\,2iz)\xrightarrow[z\to 0]{}\frac{z^\nu}{2^\nu\Gamma(\nu+1)} \qquad \nu\geq 0$$

$$N_\nu(z)=\frac{1}{\sin(\pi\nu)}[\cos(\pi\nu)J_\nu(z)-J_{-\nu}(z)] \qquad \nu\neq n$$

$$N_n(z)=\frac{2}{\pi}\ln\left(\frac{z}{2}\right)J_n(z)-\frac{1}{\pi}\sum_{m=0}^{n-1}\frac{(n-m-1)!}{m!}\left(\frac{z}{2}\right)^{2m-n}$$
$$-\frac{1}{\pi}\left(\frac{z}{2}\right)^n\sum_{m=0}^{\infty}\frac{(-1)^m}{m!\,(m+n)!}[\psi(m+1)+\psi(n+m+1)]\left(\frac{z}{2}\right)^{2m}$$

$$N_n(z)\xrightarrow[z\to 0]{}\begin{cases}\dfrac{2}{\pi}\ln\left(\dfrac{z}{2}\right) & n=0\\[2ex] \dfrac{-(n-1)!}{\pi}\left(\dfrac{2}{z}\right)^n & n\geq 1\end{cases}$$

$$J_\nu(z)\xrightarrow[z\to\infty]{}\sqrt{\frac{2}{\pi z}}\cos\left[z-\left(\nu+\frac{1}{2}\right)\frac{\pi}{2}\right] \qquad \text{Re}\,z>0$$

$$N_\nu(z)\xrightarrow[z\to\infty]{}\sqrt{\frac{2}{\pi z}}\sin\left[z-\left(\nu+\frac{1}{2}\right)\frac{\pi}{2}\right] \qquad \text{Re}\,z>0$$

$$H_\nu^{(1)}(z)\equiv J_\nu(z)+iN_\nu(z)\xrightarrow[z\to\infty]{}\sqrt{\frac{2}{\pi z}}\,e^{i[z-(\nu+1/2)\pi/2]} \qquad \text{Re}\,z>0$$

TABLE 7.1 (Cont.)

$$H_\nu^{(2)}(z) \equiv J_\nu(z) - iN_\nu(z) \xrightarrow[z\to\infty]{} \sqrt{\frac{2}{\pi z}}\, e^{-i[z-(\nu+1/2)\pi/2]} \qquad \operatorname{Re} z > 0$$

$$e^{z[\zeta-(1/\zeta)]/2} = \sum_{n=-\infty}^{\infty} J_n(z)\zeta^n$$

$$H_{\nu+1}^{(j)}(z) + H_{\nu-1}^{(j)}(z) = \frac{2\nu}{z} H_\nu^{(j)}(z) \qquad j = 1, 2$$

$$H_{\nu+1}^{(j)}(z) - H_{\nu-1}^{(j)}(z) = -2\frac{d}{dz} H_\nu^{(j)}(z) \qquad j = 1, 2$$

$$H_{-\nu}^{(1)}(z) = e^{i\nu\pi} H_\nu^{(1)}(z)$$

$$H_{-\nu}^{(2)}(z) = e^{-i\nu\pi} H_\nu^{(2)}(z)$$

$$J_{-n}(z) = (-1)^n J_n(z)$$

$$N_{-n}(z) = (-1)^n N_n(z)$$

$$\frac{d}{dz} J_0(z) = -J_1(z)$$

$$j_\nu(z) \equiv \sqrt{\frac{\pi}{2z}}\, J_{\nu+1/2}(z) \xrightarrow[z\to\infty]{} \frac{1}{z}\cos\left[z - \frac{\pi}{2}(\nu+1)\right]$$

$$n_\nu(z) \equiv \sqrt{\frac{\pi}{2z}}\, N_{\nu+1/2}(z) \xrightarrow[z\to\infty]{} \frac{1}{z}\sin\left[z - \frac{\pi}{2}(\nu+1)\right]$$

$$j_n(z) \xrightarrow[z\to 0]{} \frac{2^{n+1}(n+1)!}{(2n+2)!} z^n$$

$$n_n(z) \xrightarrow[z\to 0]{} \frac{-(2n)!}{2^n n!} \frac{1}{z^{n+1}}$$

SUGGESTED REFERENCES

Some useful general references on complex variable theory are the following.

E. C. Titchmarsh, *The Theory of Functions.* This classic text contains many results that are difficult to find elsewhere. The proofs are usually elementary although sometimes tedious.

E. J. Townsend, *Functions of a Complex Variable.* This reference is clear, easy to read, and fairly complete with many examples to illustrate basic theorems. We have followed Townsend for many of the theorems in Sections 7.1–7.9.

E. T. Whittaker and G. N. Watson, *A Course of Modern Analysis.* Virtually everything needed for a background in classical applied mathematics is contained in this famous work. The style is not always easy to follow.

N. I. Muskhelishvili, *Singular Integral Equations.* This English translation of the Russian work is a fairly complete and elementary reference on singular integral equations.

R. Omnès, *On the Solution of Certain Singular Integral Equations of Quantum Field Theory*. The discussion of singular integral equations given in Section 7.10 follows closely the excellent presentation given in this paper.

The reader interested in the Dirichlet problem will find complete and varied discussions in these three books.

S. G. Mikhlin, *Integral Equations*. Chapter 4 applies Fredholm theory and complex variable theory to the solution of the Dirichlet problems in simply and multiply connected regions in two dimensions, as well as to the three-dimensional case.

Z. Nehari, *Conformal Mapping*. This is an exhaustive treatise on complex variable theory applied to harmonic functions and conformal mapping techniques. A proof of the Riemann mapping theorem is given.

F. G. Tricomi, *Integral Equations*. Section 2.7 presents a brief but very readable discussion of the Dirichlet problem using Fredholm theory.

The beautiful subject of asymptotic expansions, classical functions, and integral representations is well represented in these works.

A. Erdélyi, *Asymptotic Expansions*. This small lucid work contains many proofs and applications of extremely useful asymptotic expansions for classical functions.

A. Erdélyi (Ed.), *Higher Transcendental Functions* (2 Vols.), *Tables of Integral Transforms* (3 Vols.). These five volumes of the Bateman Manuscript Project contain an almost unbelievable wealth of material on classical functions. If you can't find what you're looking for here, or at least a reference to it, you're probably out of luck!

P. M. Morse and H. Feshbach, *Methods of Theoretical Physics* (Vol. I). Chapters 4 and 5 present a readable and useful discussion of several techniques of complex variable theory applied to the special functions of applied mathematics.

A. Sommerfeld, *Partial Differential Equations in Physics*. This book, by a master of the techniques of applied mathematics used in physical problems, is a pleasure to peruse and well worth the effort.

G. Szegö, *Orthogonal Polynomials*. This gives a rigorous and complete development of the expansion of functions in terms of general sets of orthogonal polynomials.

I. S. Gradshteyn and I. M. Ryzhik, *Table of Integrals, Series, and Products*. This is probably the best single volume reference for integrals and properties of functions, both elementary and advanced, met in applied mathematics.

PROBLEMS

7.1 Derive the following expansions and determine their radii of convergence:

a. $e^z = 1 + z + \frac{z^2}{2!} + \cdots$

b. $\sin z = z - \frac{z^3}{3!} + \frac{z^5}{5!} + \cdots$

c. $\cos z = 1 - \frac{z^2}{2!} + \frac{z^4}{4!} + \cdots$

That is, use the known properties of e^x, $\sin x$, $\cos x$, e^{iy}, where x and y are real variables, to show that e^z, $\sin z$, and $\cos z$ are holomorphic functions of z.

7.2 Show that

$$\frac{1}{1-z}=\sum_{n=0}^{\infty} z^n$$

What is the radius of convergence of this infinite series? Is this the complete domain of holomorphy of this $f(z)$? If not, what is? Why?

7.3 Prove Eq. 7.16.

7.4 Consider the function

$$f(z)=\frac{z^3}{(z-4i)(z+1)}$$

a. Expand in a Taylor series about $z=0$. What is the radius of convergence of this series as gotten directly from the coefficients (i.e., ratio test)? (*Hint*: First decompose by partial fractions and then use the result of Problem 7.2.)
b. Expand about $z=+2i$. What is the radius of convergence of this series?
c. Could you have predicted these radii from $f(z)$ directly?
d. What are the zeros and singularities of $f(z)$ and of what nature are they?
e. Sketch the z plane for $f(z)$ with the poles, zeros, and radii of convergence indicated.
f. What are the domains of meromorphy, holomorphy, and analyticity of $f(z)$?

7.5 Given the function

$$f(z)=\frac{z+3}{z(z^2-z-2)}$$

represent this function by an infinite series for values of z: (a) within the unit circle about the origin; (b) within the annular region between the concentric circles about the origin having respectively the radii 1 and 2; (c) exterior to the circle of radius 2.

7.6 Expand the function

$$f(z)=\frac{4z^2-4z+1}{z^3-z^4}$$

in a Laurent series about $z=0$. Locate the singularities of $f(z)$. What is the region of convergence of this expansion? Find the residues of this function at all of its poles. Save yourself some work by factoring and decomposing $f(z)$ into partial fractions *before* you attempt the expansion.

7.7 Can $|\sin z|$ ever be greater than unity? If so, at what points? How do the zeros of $\sin z$ compare with those of $\sin x$, where x is real? What solutions can the equation $\cos z=1$ have?

7.8 Consider the function

$$f(z)=\frac{e^z}{z^2+1}$$

Is $f(z)$ an analytic function? What is its region of existence? Find a region R in which the function is holomorphic. Locate the singular points and classify the singularities of the function at these points. Does the function have any zero points?

7.9 Let

$$f(z) = \frac{1}{(z-1)(z-3)}$$

If this function is expanded in a Taylor series about $z=2$, how large can the circle of convergence be? Expand the given function in powers of z (i.e., about $z_0=0$) and determine the circle of convergence.

7.10 Obtain the Laurent or Taylor series for the function

$$f(z) = \frac{1}{(z-1)(z-2)}$$

a. for $|z|<1$
b. for $|z-1|<1$
c. for $|z-2|<1$
(*Hint:* Use partial fractions.)
d. Now compute the value of the integral

$$\oint f(z)\,dz$$

around circles of radii $\frac{1}{2}$, $\frac{3}{2}$, and $\frac{5}{2}$, respectively, in the z plane.

7.11 Locate the zero points of $\sin 1/z$. What is the nature of this function at the limiting point of these zeros? From this conclusion what can be said of the singular points of $\csc z$? (*Hint*: Consider the point $z=\infty$ by the substitution $z=1/\zeta$ as $\zeta \to 0$.)

7.12 a. Use Liouville's theorem to prove the fundamental theorem of algebra; that is, that any polynomial of degree n,

$$P_n(z) \equiv a_0 + a_1 z + a_2 z^2 + \cdots + a_n z^n = \sum_{m=0}^{n} a_m z^m$$

has at least one root, $z=z_0$, such that

$$P_n(z_0) = 0$$

Be certain to establish that z_0 is a zero in the precise sense of that term.
b. Now prove that every polynomial of degree n has exactly n roots, where each root is counted a number of times equal to its multiplicity.

7.13 Given the function

$$f(z) = \frac{z^3+1}{z^3-2z^2+z-2}$$

determine the zero points and poles. Compute the residue at each of the latter.

7.14 Consider the integral

$$\int \frac{z^2\,dz}{z^3+1}$$

taken along a circle C about the origin as center. How large can the radius of C be taken and have the value of the integral zero? What would the integral be if the radius of C were taken $\frac{1}{2}$ unit larger?

7.15 By use of the theory of residues, evaluate the integral

$$\int_{-\infty}^{\infty} \frac{dx}{(x^2+2)^2}$$

7.16 Give an illustrative example of a single-valued analytic function having (a) no singular point, (b) no singular point in the finite region and a pole at infinity, (c) no singular point other than an essential singular point at infinity, (d) a finite number of poles in the finite region and an essential singular point at infinity, (e) an isolated essential singular point in the finite region.

7.17 Use the calculus of residues to evaluate the following integrals.

a. $\displaystyle\int_{-\infty}^{\infty} \frac{x\,dx}{(x^2+1)(x^2+2x+2)}$

b. $\displaystyle\int_{0}^{\pi} \frac{d\theta}{(a+\cos\theta)^2} \qquad a>1$

7.18 Evaluate the integral

$$\int_{0}^{2\pi} \frac{d\theta}{(\cos\theta+2)^2}$$

7.19 Use Cauchy's integral theorem to show that if $f(z)$ is a ratio of polynomials that vanishes sufficiently rapidly as $|z|\to\infty$, then the following infinite series may be summed as:

a. $\displaystyle\sum_{n=-\infty}^{\infty} f(n) = -\sum_{j} \text{Res}\,[\pi f(z_j)\cot(\pi z_j)]$

b. $\displaystyle\sum_{n=-\infty}^{\infty} (-1)^n f(n) = -\sum_{j} \text{Res}\,[\pi f(z_j)\csc(\pi z_j)]$

where the $\{z_j\}$ are the poles of $f(z)$. [*Hint*: Apply residue calculus to $\oint \pi f(z)\cot(\pi z)\,dz$ around a circle of radius R (centered at the origin) where R is not an integer. Then let $R\to\infty$.]

7.20 Given the multiple-valued function

$$w=(z-1)^{1/3}$$

locate the branch points, select a suitable system of branch cuts, and sketch the Riemann surface, showing how the various sheets are connected along the branch cuts.

7.21 Describe the Riemann surface for

$$w=\left(\frac{z-1}{z}\right)^{1/2}$$

7.22 Describe the Riemann surface for

$$w=\ln\left(\frac{z-1}{z+1}\right)$$

7.23 Show that

$$\int_0^{2\pi} e^{\cos\theta} \cos(n\theta - \sin\theta)\, d\theta = \frac{2\pi}{n!}$$

7.24 Show that

$$\int_0^{\infty} \frac{x^{-a}\, dx}{1+2x\cos\theta + x^2} = \frac{\pi}{\sin \pi a}\left(\frac{\sin a\theta}{\sin\theta}\right), \quad -1<a<1,\ -\pi<\theta<\pi$$

7.25 Evaluate the integral

$$\int_0^{\infty} \frac{\sqrt{x}\, dx}{(x^2+1)(x+2)^2}$$

7.26 Evaluate the integral

$$\int_1^{\infty} \frac{(x-1)^{1/3}\, dx}{1+x^2}$$

by choosing a suitable integration contour and applying Cauchy's integral formula directly. Justify dropping the contributions from any contours which you neglect. Do *not* simply quote the results of a formula derived from the text. Specify your phases clearly.

Evaluate the following integrals in Problems 7.27–7.30. State any necessary restrictions on the parameters α and a.

7.27 $\displaystyle\int_0^{\infty} \frac{x^{2\alpha-1}\, dx}{1+x^2} \qquad 0<\alpha<1$

7.28 $\displaystyle\int_0^{\infty} \frac{x^{a-1}}{1-x}\, dx$

7.29 $\displaystyle\int_0^{\infty} \frac{x^{a-1}}{x+e^{i\alpha}}\, dx$

7.30 $\displaystyle\int_0^{\infty} \frac{x^{a}}{(1+x^2)^2}\, dx$

7.31 Show that function z^{-2} represents the analytic continuation of the function

$$f(z) = \sum_{n=0}^{\infty} (n+1)(z+1)^n$$

beyond the region $|z+1|<1$.

7.32 If k is a real constant, show that the analytic continuation of

$$f(z) = \int_0^{\infty} e^{-zt} \sin kt\, dt$$

has simple poles at the points $z = \pm ik$.

7.33 Given the analytic expression

$$f(z) = \lim_{n\to\infty}\left(\frac{1}{1-z^n}\right)$$

show that $f(z)$ is an element of two distinct analytic functions according as we take $|z|<1$ or $|z|>1$.

7.34 If

$$f(z)=\sum_{n=0}^{\infty} z^{2^n}$$

show that $f(z)=f(z^2)+z$ and deduce from this that $|z|=1$ is a natural boundary of the function.

7.35 If n is a positive integer and Re $z>0$, integrate repeatedly by parts to show that

$$\int_0^n \left(1-\frac{t}{n}\right)^n t^{z-1}\, dt=\frac{n!\, n^z}{z(z+1)(z+2)\cdots(z+n)}$$

so that

$$\Gamma(z)=\lim_{n\to\infty}\left[\frac{n!\, n^z}{z(z+1)\cdots(z+n)}\right]=\lim_{n\to\infty}\left[\frac{e^{z\ln n}}{z\left(1+\frac{z}{1}\right)\left(1+\frac{z}{2}\right)\cdots\left(1+\frac{z}{n}\right)}\right]$$

$$=\lim_{n\to\infty}\left\{\frac{\exp\left[-z\left(1+\frac{1}{2}+\cdots+\frac{1}{n}-\ln n\right)\right]}{z(1+z)e^{-z}\left(1+\frac{z}{2}\right)e^{-1/2z}\cdots\left(1+\frac{z}{n}\right)e^{-z/n}}\right\}$$

Prove that

$$\lim_{n\to\infty}\left(\sum_{m=1}^{n}\frac{1}{m}-\ln n\right)=\gamma$$

where γ is a finite quantity (known as the Euler-Mascheroni constant), to arrive at

$$\frac{1}{\Gamma(z)}=ze^{\gamma z}\prod_{n=1}^{\infty}\left[\left(1+\frac{z}{n}\right)e^{-z/n}\right]$$

7.36 Let

$$f(z)=\sum_{n=0}^{\infty} a_n z^n \qquad |z|<1$$

where the series is given as being convergent for $|z|<1$, and

$$g(z)=\sum_{n=0}^{\infty}\frac{a_n}{n!}z^n$$

Show that the integral

$$F(z)=\int_0^{\infty} e^{-t}g(zt)\, dt$$

provides the continuation of $f(z)$ across any arc of the circle of convergence where $f(z)$ is holomorphic [i.e., it is given that $f(z)$ does have an analytic continuation across some arc of the unit circle]. Apply this to $f(z)=\sum_{n=0}^{\infty} z^n$.

7.37 A given function $f(z)$ is holomorphic in the strip $|\text{Im } z|<\alpha$ and in this strip has the property that $\lim_{z\to\infty} f(z)=0$. Show that

$$f(z)=f_+(z)-f_-(z)$$

where

$$f_+(z)=\frac{1}{2\pi i}\int_{-\infty-i\beta}^{\infty-i\beta}\frac{f(t)\,dt}{(t-z)}$$

$$f_-(z)=\frac{1}{2\pi i}\int_{-\infty+i\beta}^{\infty+i\beta}\frac{f(t)\,dt}{(t-z)}$$

and $\beta<\alpha$. Prove that f_- is holomorphic for Im $z<\beta$ and that f_+ is holomorphic for Im $z>-\beta$.

7.38 A function is represented by the following series within its radius of convergence.

$$f(z)=\sum_{n=0}^{\infty}\frac{1}{\Gamma(n+1)\Gamma(\frac{1}{2}-n)}\left(\frac{2n+1}{2n+3}\right)(-z)^n$$

Show that $z=1$ is a singularity of this function and find the nature of this singularity.

7.39 Consider the conformal mapping

$$w=e^z$$

What region in the z plane is mapped into the entire w plane? Into what curves are $x=$const. and $y=$const. mapped respectively? Prove that these mapped curves remain orthogonal trajectories in the w plane.

7.40 Prove that the general linear fractional transformation

$$w=\frac{\alpha z+\beta}{\gamma z+\delta}\qquad \alpha\delta-\beta\gamma\neq 0$$

where α, β, γ and δ are complex constants, can be considered geometrically as built up of these successive operations:

a. A translation.

b. An inversion in the unit circle.

c. A rotation and stretching followed by a translation. (*Hint*: Simply carry out these three transformations in succession.)

7.41 Given the two power series

$$g(t)=\sum_{n=1}^{\infty}\alpha_n(t-t_0)^n$$

$$t-t_0=\sum_{n=0}^{\infty}\frac{a_n}{(n+1)}g^{n+1}$$

substitute the second one into the first and equate powers of g^s to solve for a_0, a_1, and a_2 in terms of the α_n as

$$a_0=\frac{1}{\alpha_1}$$

$$a_1=-\frac{2}{\alpha_1^2}\left(\frac{\alpha_2}{\alpha_1}\right)$$

$$a_2=\frac{3}{\alpha_1^3}\left[2\left(\frac{\alpha_2}{\alpha_1}\right)^2-\left(\frac{\alpha_3}{\alpha_1}\right)\right]$$

Compute the Fourier transforms of the functions given in Problems 7.42–7.45. Take k to be a real variable in the transform, $f(k)$.

7.42 a. $f(x)=\dfrac{1}{x-ix_0}$

b. $f(x)=\dfrac{1}{(x-ix_0)(x+ix_1)}$ $\qquad \text{Re } x_0>0 \qquad \text{Re } x_1>0$

7.43 $f(x)=\tanh(k_0x)$
Simply take as given (cf. Section 8.1) the following representation of the Dirac delta function

$$\delta(k)=\frac{1}{2\pi}\int_{-\infty}^{\infty} e^{ikx}\,dx=\frac{1}{\pi}\int_0^{\infty}\cos(kx)\,dx$$

and use the results of Problem 7.19.

7.44 $f(x)=e^{-\lambda x}\ln|1-e^{-x}| \qquad -1<\text{Re }\lambda<0$

7.45 $f(x)=(\alpha+e^{-x})^{-1}e^{-\lambda x} \qquad 0<\text{Re }\lambda<1 \qquad -\pi<\arg\alpha<\pi$

7.46 a. Assume that a function $f(z)$ allows a Laurent expansion about the origin in an annular region including the unit circle so that

$$f(z)=\sum_{n=-\infty}^{\infty}\alpha_n z^n \qquad \alpha_n=\frac{1}{2\pi i}\oint\frac{f(\zeta)\,d\zeta}{\zeta^{n+1}}$$

Show that when z is on the unit circle this Laurent expansion (and its coefficients α_n) reduces to the Fourier series.
b. Therefore, if you are given a uniformly convergent Fourier series for $0\le\phi\le2\pi$, how would you recover the function in a larger region of the z plane? In particular what is the value of the function defined by

$$\sum_{n=0}^{\infty}\frac{\cos(n\phi)}{a^n} \qquad a>1 \qquad (a\text{—real constant})$$

at $z=0$? Is this extension unique?

7.47 You are given a function defined in the z plane on the unit circle as follows (cf. Fig. 7.87).

$$h(\theta)=\begin{cases}+1, & 0\le\theta<\alpha\\ -1, & \alpha<\theta<2\pi-\alpha\\ +1, & 2\pi-\alpha<\theta<2\pi\end{cases}$$

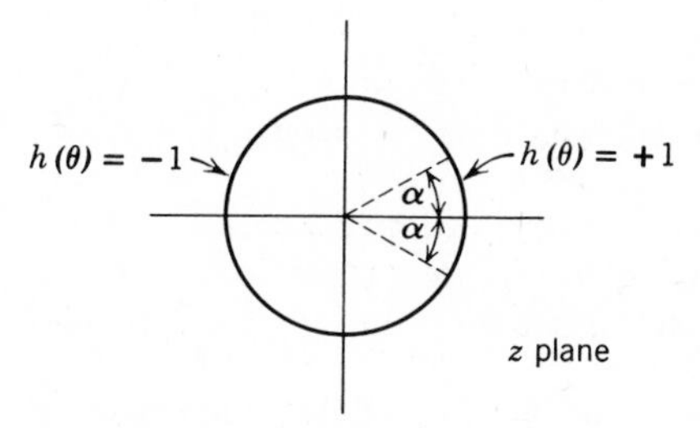

Figure 7.87

Construct an extension of this function, $f(z)$, which is holomorphic within the unit circle and such that $\text{Im}\, f(z=0)=0$. Then evaluate $f(z=0)$. Realize that

$$\ln(e^{2\pi i}e^{i\varphi}-z)=2\pi i+\ln(e^{i\varphi}-z) \qquad |z|<1$$

[*Hint*: Either use the Poisson integral formula, Eq. 7.75, or expand $h(\theta)$ in a Fourier series and use the results of Problem 7.46 plus the series

$$\ln(1-z)=-\sum_{n=1}^{\infty}\frac{z^n}{n}$$

to recover $f(z)$.] Use this result to construct a real function, $u(x, y)$, that is harmonic for $|z|<1$ and that reduces to $h(\theta)$ on the unit circle.

7.48 Use the method of steepest descent to find the asymptotic form of

$$F(x)\equiv\frac{x^{1/3}}{2\pi}\int_{-\infty}^{\infty}\exp\left[ix\left(\tfrac{1}{3}t^3+t\right)\right]dt\equiv\frac{x^{1/3}}{2\pi}\int_{-\infty}^{\infty}e^{xf(t)}\,dt$$

For simplicity assume that x is real and positive and let $x\to+\infty$. First prove that this integral exists by considering

$$\oint_C e^{xf(t)}\,dt$$

around the contour shown in Fig. 7.88 in the limit as R approaches infinity. Then proceed as follows.

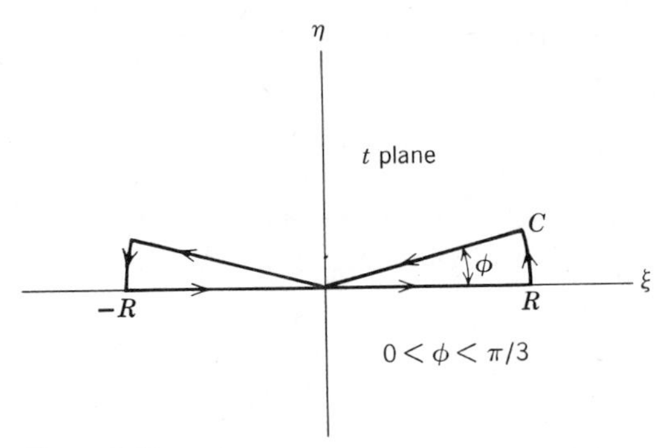

Figure 7.88

a. Locate the saddle point in the complex t plane ($t=\zeta+i\eta$).
b. Obtain the equation for the curves along which $\text{Im}\, f(t)=\text{const}$. (There are three of them in this case.)
c. Find out along which one of these curves $\text{Re}\, f(t)\to-\infty$ at *both* end points.
d. Verify that a saddle point lies on this curve and that the original contour $(-\infty\to t\to+\infty)$ can be continuously deformed into this path of steepest descent.
e. Make a Taylor expansion of $f(t)$ about the saddle point, t_0, [keeping terms of order $(t-t_0)^2$] and obtain the asymptotic form of $F(x)$.

7.49 Consider the function defined as

$$g(z) \equiv ie^{-i(\pi/2)z} \int_{(\pi/2)-i\infty}^{(\pi/2)+i\infty} e^{iz(e^{i\phi}+\phi)}\, d\phi$$

where the integration path is a vertical straight line. Find the asymptotic behavior of $g(z)$ as $|z| \to \infty$, Re $z > 0$. Why is this last restriction necessary? It is possible to relate this to $\Gamma(z)$.

7.50 Find the two leading terms in the asymptotic expansion of

$$F(z) = \int_{-\infty}^{\infty} e^{zf(t)}\, dt$$

where

$$f(t) = -\frac{t^4}{4} + \frac{t^3}{3} - \frac{t^2}{2} + t$$

In what region of the z plane will your expansion be valid?

7.51 Find an explicit nontrivial solution to

$$\phi(x) = \frac{1}{\pi} P \int_1^{\infty} \frac{\tan[\delta(x')]\phi(x')}{(x'-x)}\, dx'$$

where

$$\delta(x) = \frac{x}{1+x^2} \qquad x \in [1, \infty)$$

7.52 Consider the hypergeometric equation

$$z(z-1)u''(z) + [(a+b+1)z - c]u'(z) + abu(z) = 0 \equiv L_z[u]$$

and take a, b, and c to be real parameters with $c > b > 0$. Construct an integral representation for a solution as

$$u(z) = \int_C (z-t)^{\mu} w(t)\, dt$$

where the integration path, C, the index μ, and the function, $w(t)$, must be determined as follows.

a. Choose the index μ so that

$$L_z[u] \equiv \int_C L_z[(z-t)^{\mu}]w(t)\, dt = \int_C \mathscr{L}_t[(z-t)^{\mu}]w(t)\, dt$$

where

$$\mathscr{L}_t = \alpha(t)\frac{d^2}{dt^2} + \beta(t)\frac{d}{dt}$$

That is, you must determine μ, $\alpha(t)$, and $\beta(t)$ to make the last equality hold.

b. Now integrate by parts to obtain

$$L_z[u] = \{\text{out integrated terms}\}\big|_{\text{ends of } C} + \int_C (z-t)^{\mu} \mathscr{M}_t[w(t)]\, dt = 0$$

and set

$$\mathscr{M}_t[w(t)] = 0$$

to solve for (the simplest) nonzero $w(t)$.

c. Finally choose the integration path, C, so that the integral representation converges for all z as long as z is not a real number greater than unity and verify that the out-integrated terms do indeed vanish.

d. Now normalize your solution to $F(a, b|c| z)$ [i.e., such that $u(z=0)=1$].

7.53 Given the integral representation

$$I_n(z)=\frac{1}{2\pi}\int_{-\pi}^{\pi} e^{iz\cos t}e^{in(t-\pi/2)}\,dt$$

find the power series expansion for $I_n(z)$ when n is an integer. What is the relation between $I_n(z)$ and the Bessel function $J_n(z)$?

7.54 Prove that

$$\int_{-\infty+i\varepsilon}^{+\infty+i\varepsilon}\frac{dk\,\exp i[k\alpha-\sqrt{k^2-\gamma^2}\,\beta]}{\sqrt{k^2-\gamma^2}}=-2\pi i J_0[\gamma\sqrt{\alpha^2-\beta^2}]H(\beta-\alpha)$$

where $\alpha>0$, $\beta>0$, $\gamma>0$, $\varepsilon>0$. With proper deformation of the integration contour the following successive transformations are helpful:

$$-\beta=\sqrt{-\beta^2+\alpha^2}\sinh\theta \qquad k=\gamma\cos x$$

$$y=x+\frac{\pi}{2}$$

$$w=y+i\theta$$

Verify explicitly the stated result for $\gamma=0$.

8 Second-Order Linear Ordinary Differential Equations and Green's Functions

8.0 INTRODUCTION

Given the linear equation

$$Lu=f \tag{8.1}$$

we may either attempt to find the eigenvalues and eigenvectors of L as

$$Lu_j=\lambda_j u_j$$

and then expand f in terms of these $\{u_j\}$ if they are complete (cf. Section 3.0) or we may construct the inverse operator L^{-1} (cf. Sections 2.1c, 2.7, and 5.6) such that

$$LL^{-1}=I=L^{-1}L \tag{8.2}$$

and write the solution to Eq. 8.1 as

$$u=L^{-1}f \tag{8.3}$$

In nearly all of our previous work we have taken L to be a bounded linear operator. In this chapter L will be a linear differential operator and we will study the inverse of such an unbounded operator. Our aim will be not so much to develop specific techniques for solving linear differential equations, such as series expansions and applications to boundary-value problems, the latter being studied extensively in electrostatics, for example, but instead to study methods that are generally applicable to difficult problems which cannot be solved in closed form.

When L is a differential operator the inverse operator, L^{-1}, is an integral operator,

$$L^{-1}u(x)=\int g(x, t)u(t)\,dt \tag{8.4}$$

whose kernel, $g(x, t)$, is called a *Green's function.* As we often do for notational convenience, we have left the limits (a, b) off the integral in Eq. 8.4, although the integration is restricted to the same region as that for which the differential operator is defined. If we combine Eqs. 8.1–8.4 and do not worry about interchanging integrations and differentiations, we obtain

$$u(x)=LL^{-1}u(x)=L_x\int g(x, t)u(t)\,dt$$

$$=\int [L_x g(x, t)]u(t)\,dt \equiv \int \delta(x, t)u(t)\,dt \tag{8.5}$$

If Eq. 8.5 is to hold for all continuous $u(x)$, and if $\delta(x, t)$ is also to be continuous, then we can easily prove that $\delta(x, t)=0$, $x\neq t$, as follows. Since we can always redefine our origin, we take $u(x)=0$ except for $-\varepsilon<x<\varepsilon$ where it is continuous and positive. If we take $x=(\varepsilon+\alpha)$[or $x=-(\varepsilon+\alpha)$] for any $\alpha>0$, Eq. 8.5 becomes

$$0=\int_{-\varepsilon}^{\varepsilon} \delta(\varepsilon+\alpha, t)u(t)\,dt=2\varepsilon\,\delta(\varepsilon+\alpha, \tau)u(\tau) \qquad -\varepsilon<\tau<\varepsilon \tag{8.6}$$

where we have used the mean value theorem for integrals. Since α is arbitrary, while ε can be made as small as we please, we see that $\delta(x, t)=0$ except possibly when $x=t=0$. However by redefining our origin we can make x an arbitrary point in this argument so that

$$\delta(x, t)\equiv\delta(x-t)=0, \; x\neq t \tag{8.7}$$

If we now set $u(x)\equiv 1$ in Eq. 8.5, we obtain

$$\int \delta(x-t)\,dt=1 \tag{8.8}$$

But no function different from zero at one point only can make a contribution to an integral so that Eq. 8.8 is impossible for any true function $\delta(x-t)$.

We have arrived at this contradiction because we have not been mathematically precise in our derivation. We have interchanged a differentiation and an integration in Eq. 8.5. This is justified if the integrand is continuous and differentiable everywhere in the region of integration, which is surely not the case here. Nevertheless the concept of a δ "function" defined as

$$\int \delta(x)\phi(x)\,dx=\phi(0) \tag{8.9}$$

and of similar "functions" is useful in applied mathematics and has been justified mathematically by L. Schwartz with the theory of distributions (or ideal functions) as we will outline briefly.

However let us proceed formally for a short time. The most important property of the δ function is given in Eq. 8.9. In manipulations $\delta(x)$ is often treated algebraically, although we will show that any relations so obtained are to be understood as applying under an integral sign after multiplication by a continuous function vanishing at infinity. Thus we have

$$x\,\delta(x)=0$$

since

$$\int \phi(x)x\,\delta(x)\,dx=\phi(x)x|_{x=0}=0$$

Similarly we may formally verify the following identities.

i. $\delta(x)=\delta(-x)$

ii. $\delta'(x)=-\delta'(-x)$

iii. $x\,\delta(x)=0$

iv. $x\,\delta'(x)=-\delta(x)$

v. $\delta(ax)=|a|^{-1}\,\delta(x)$

vi. $\delta(x^2-a^2)=|2a|^{-1}[\delta(x-a)+\delta(x+a)]$

vii. $\int \delta(a-x)\,\delta(x-b)\,dx=\delta(a-b)$

viii. $f(x)\,\delta(x-a)=f(a)\,\delta(x-a)$

ix. $\delta[f(x)]=\dfrac{\delta(x-x_0)}{|f'(x_0)|} \qquad f(x_0)=0$

8.1 IDEAL FUNCTIONS

We now generalize our concept of a function in such a way that this new specification of a function will hold for all the continuous and piecewise continuous functions with which we are already familiar. However this new definition of a function will continue to be meaningful even when we can no longer specify the dependence of this function on an independent variable as $f(x)$.

a. Test Functions

In order to give our generalized definition of a function, we must first introduce *test* (or testing) *functions* which we denote by $\phi(x)$. A test function is a real function that

i. is continuous.
ii. has continuous derivatives of all orders.
iii. vanishes identically outside some finite interval.

An example of a test function is provided by

$$\phi(x)=\begin{cases}\exp[-1/(x-a)^2] & |x|\leq a \\ 0 & |x|\geq a\end{cases}$$

Notice that the space of test functions is not complete since the limit of a sequence of test functions need not be a test function.

b. Linear Functionals

One rigorous way to approach a generalized definition of a function is through *continuous linear functionals* of the test functions. Recall (cf. Section 1.8) that a functional $F(\phi)$ is *linear* provided

$$\begin{aligned} F(\phi_1+\phi_2)&=F(\phi_1)+F(\phi_2) \\ F(\lambda\phi)&=\lambda F(\phi) \end{aligned} \tag{8.10}$$

for all test functions $\phi_1(x)$ and $\phi_2(x)$ and for all λ. A linear functional is *continuous* if, whenever a sequence of test functions $\{\phi_n(x)\}$ is such that

$$\lim_{n\to\infty}\phi_n(x)=0$$

then

$$\lim_{n\to\infty}F(\phi_n)=0 \tag{8.11}$$

Let us first see how we could uniquely specify a continuous square-integrable function $f(x)$ by a continuous linear functional of the test functions. Since the space of continuous square-integrable functions has an inner product defined on it, we know (cf. Section 1.8, Th. 1.2) that a continuous linear functional on this space can always be represented by an inner product. We restrict our considerations to real $f(x)$, although this is not necessary. Define a continuous linear functional $F(\phi)$ of such an $f(x)$ as

$$F(\phi)\equiv\langle f\mid\phi\rangle=\int_{-\infty}^{\infty}f(x)\phi(x)\,dx \tag{8.12}$$

where we have indicated an infinite range of integration, although the range may also be finite. If the set of numbers $\{\langle f\mid\phi\rangle\}$ is given for *all* test functions $\phi(x)$, then $f(x)$ is uniquely specified in the sense that there is no other continuous square-integrable function that can reproduce the same set of numbers $\{\langle f\mid\phi\rangle\}$ for *all* $\phi(x)$. This is easily established by contradiction. Suppose that there were two functions, for example, $f_1(x)$ and $f_2(x)$, that gave the same set of values $\{\langle f \mid \phi\rangle\}$ for *all* $\phi(x)$, where both f_1 and f_2 are continuous. Then the function

$$g(x) \equiv f_1(x) - f_2(x)$$

is also continuous and

$$\langle g \mid \phi\rangle = 0$$

for *all* $\phi(x)$. Consider a $\phi(x)$ with finite *support* (region of nonvanishing values of ϕ) $a \le x \le b$. Since $g(x)$ is continuous here, we may subdivide $[a, b]$ into subintervals of positive, zero, or negative values of $g(x)$. Choose a set $\{\phi(x)\}$ of test functions successively positive on each subinterval and zero elsewhere. Then just as in the proof of Th. 6.1, we conclude that

$$\langle g \mid \phi\rangle = \int |g\phi|\, dx = 0$$

so that

$$g(x) \equiv 0 \qquad a \le x \le b$$

since $\{\phi(x) \not\equiv 0\}$ by construction. We may repeat this argument over each region in succession and conclude that $g(x) \equiv 0$ or that $f_1(x) \equiv f_2(x) \equiv f(x)$. It is easily seen that the argument can be extended to piecewise continuous or simply square-integrable functions, except that then we can only conclude that $g(x) = 0$ almost everywhere.

We can now extend our considerations to *ideal functions* (or *generalized functions* or *distributions*) that are defined to be continuous linear functionals of test functions. It is important to realize that these continuous linear functionals cannot necessarily be expressed as an inner product since we cannot always assign a finite norm to an ideal function. A simple but extremely important example of an ideal function is provided by the continuous linear functional

$$F(\phi) = \phi(0)$$

From the discussion of Section 8.0, in particular Eq. 8.9, we see that this defines the Dirac delta function that we have seen cannot be expressed in the form $\int_{-\infty}^{\infty} \delta(x)\phi(x)\, dx$. As a purely formal and notational convenience we often write $F(\phi)$ as an inner product (cf. Eq. 8.12) for any ideal function, but we must realize that the definition in terms of a continuous linear functional (or in terms of some equivalent definition; cf., e.g., Section 8.1d) must be used to derive any results rigorously.

c. Derivatives of Ideal Functions

Let us begin with a function $f(x)$ that has a continuous square-integrable derivative $f'(x)$. If we define a function $h(x) = f'(x)$, then just as in the previous section we can show that the set of numbers $\{\int_{-\infty}^{\infty} f'(x)\phi(x)\, dx\}$ for all test functions $\phi(x)$ uniquely specifies $f'(x)$. If we integrate by parts and recall that

$\phi(x)$ can have only finite support, then we obtain

$$\int_{-\infty}^{\infty} f'(x)\phi(x)\,dx = -\int_{-\infty}^{\infty} f(x)\phi'(x)\,dx$$

By use of Eq. 8.12 we may *define* the derivative of $f(x)$ by the continuous linear functional relation

$$F'(\phi) \equiv -F(\phi) \tag{8.13}$$

For a continuous differentiable $f(x)$ this is equivalent to the specification of $f(x)$ in terms of the usual limiting one at a point,

$$f'(x) \equiv \lim_{\Delta x \to 0} \left[\frac{f(x+\Delta x)-f(x)}{\Delta x}\right]$$

We now take Eq. 8.13 as the *definition* of the derivative of an ideal function. It is clear that ideal functions are infinitely differentiable with this definition.

Once again, *as a notational convenience* we often write

$$\int_{-\infty}^{\infty} f'(x)\phi(x)\,dx = -\int_{-\infty}^{\infty} f(x)\phi'(x)\,dx \tag{8.14}$$

for an ideal function although the precise definition is given by Eq. 8.13.

Example 8.1 The Dirac delta function has been defined as

$$F(\phi) \equiv \int_{-\infty}^{\infty} \delta(x)\phi(x)\,dx = \phi(0) \tag{8.15}$$

We can easily compute its derivative as follows.

$$F'(\phi) \equiv \int_{-\infty}^{\infty} \delta'(x)\phi(x)\,dx = -F(\phi') \equiv -\int_{-\infty}^{\infty} \delta(x)\phi'(x)\,dx = -\phi'(0)$$

In general we have

$$\int_{-\infty}^{\infty} \delta^{(n)}(x)\phi(x)\,dx = (-1)^n \left.\frac{d^n\phi(x)}{dx^n}\right|_{x=0} \tag{8.16}$$

Example 8.2 If we define the Heaviside unit (or step) function as

$$F(\phi) \equiv \int_{-\infty}^{\infty} H(x)\phi(x)\,dx \equiv \int_0^{\infty} \phi(x)\,dx \tag{8.17}$$

then we may evaluate its derivative as

$$F'(\phi) \equiv \int_{-\infty}^{\infty} H'(x)\phi(x)\,dx = -F(\phi') = -\int_{-\infty}^{\infty} H(x)\phi'(x)\,dx$$

$$= -\int_0^{\infty} \phi'(x)\,dx = -\phi(x)\Big|_0^{\infty} = \phi(0) \equiv \int_{-\infty}^{\infty} \delta(x)\phi(x)\,dx$$

Therefore *symbolically* we may write

$$H'(x) = \delta(x) \tag{8.18}$$

In fact $H(x)$ can be represented by a discontinuous function as (cf. Eq. 2.182)

$$H(x) \equiv \begin{cases} 1 & x > 0 \\ 0 & x < 0 \end{cases} \tag{8.19}$$

It is not defined at $x = 0$.

We have seen that ideal functions as defined above are infinitely differentiable (i.e., the derivative of an ideal function is again an ideal function). The product of an ideal function with an ordinary (e.g., continuous) function is again an ideal function. However the product of two ideal functions need not have any meaning [e.g., $\delta(x)\,\delta(x)$; cf. Problem 8.4].

d. Ideal Limits

Above we defined ideal functions by means of a linear functional

$$F(\phi) = \langle f \mid \phi \rangle$$

We can also define symbolic functions in terms of a *weak limit.* If $\langle (f_n - f_m) \mid \phi \rangle$ approaches zero as $n \to \infty$ and $m \to \infty$ uniformly for all test functions ϕ, then the sequence of functions $\{f_n\}$ is said to be *weakly convergent* (cf. Section A5.4). We are already familiar with strong convergence (i.e., if $\lim_{n\to\infty} \|f - f_n\| = 0$, then $\{f_n\} \to f$ strongly). Notice that strong convergence implies weak convergence but that the converse is not necessarily true (i.e., $|\langle f - f_n \mid \phi \rangle| \le \|f - f_m\|\,\|\phi\|$).

An ideal limit is defined as

$$f \equiv \lim_{n\to\infty} f_n \tag{8.20}$$

when $\{f_n\} \to f$ weakly. This characterizes an ideal function by the limit of the inner product,

$$\langle f \mid \phi \rangle = \lim_{n\to\infty} \langle f_n \mid \phi \rangle \tag{8.21}$$

This definition of an ideal function and that in terms of continuous linear functionals may be shown to be equivalent. From the definition (8.21) we see that the weak limit defines a linear functional. We need prove the converse, which we will do for a specific case only. Let

$$F_n(\phi) \equiv \int_{-\infty}^{\infty} \frac{n}{\sqrt{\pi}} e^{-n^2x^2} \phi(x)\,dx = \langle \delta_n \mid \phi \rangle$$

where

$$\delta_n(x) = \frac{n}{\sqrt{\pi}} e^{-n^2x^2} \qquad \ni \int_{-\infty}^{\infty} \delta_n(x)\, dx = 1 \qquad \forall n$$

Since $\phi(x)$ is a test function, we are guaranteed the existence of the bounds

$$\begin{aligned} |\phi(x)| &\leq M \\ |\phi'(x)| &\leq \bar{M} \end{aligned} \qquad \forall x$$

and we may apply the mean value theorem as

$$\phi(x) = \phi(0) + x\phi'(\xi) \qquad 0 \leq \xi \leq x$$

Therefore we may write

$$\begin{aligned}
\left| \frac{n}{\sqrt{\pi}} \int_{-\infty}^{\infty} e^{-n^2x^2} \phi(x)\, dx - \phi(0) \right| &= \left| \frac{n}{\sqrt{\pi}} \int_{-\infty}^{-R} e^{-n^2x^2} \phi(x)\, dx + \frac{n}{\sqrt{\pi}} \int_{R}^{\infty} e^{-n^2x^2} \phi(x)\, dx \right. \\
&\quad \left. + \frac{n}{\sqrt{\pi}} \int_{-R}^{R} e^{-n^2x^2} \phi(x)\, dx - \phi(0) \right| \\
&\leq 2M \frac{n}{\sqrt{\pi}} \int_{R}^{\infty} e^{-n^2x^2}\, dx + \left| \frac{n}{\sqrt{\pi}} \phi(0) \int_{-R}^{R} e^{-n^2x^2}\, dx - \phi(0) \right. \\
&\quad \left. + \frac{n}{\sqrt{\pi}} \int_{-R}^{R} x e^{-n^2x^2} \phi'[\xi(x)]\, dx \right| \\
&\leq 2M \frac{n}{\sqrt{\pi}} \int_{R}^{\infty} e^{-n^2x^2}\, dx + 2\bar{M} \frac{n}{\sqrt{\pi}} \int_{0}^{R} x e^{-n^2x^2}\, dx \\
&\quad + \left| \frac{n}{\sqrt{\pi}} \phi(0) \int_{-R}^{R} e^{-n^2x^2}\, dx - \frac{n}{\sqrt{\pi}} \phi(0) \int_{-\infty}^{\infty} e^{-n^2x^2}\, dx \right| \\
&= 2[M + \phi(0)] \frac{n}{\sqrt{\pi}} \int_{R}^{\infty} e^{-n^2x^2}\, dx + \frac{\bar{M}}{n\sqrt{\pi}} (1 - e^{-n^2R^2}) \\
&= \frac{2[M + \phi(0)]}{\sqrt{\pi}} \int_{nR}^{\infty} e^{-y^2}\, dy + \frac{\bar{M}}{\sqrt{\pi}} \frac{(1 - e^{-n^2R^2})}{n} \\
&\leq \frac{2[M + \phi(0)]}{\sqrt{\pi}} \int_{nR}^{\infty} y e^{-y^2}\, dy + \frac{\bar{M}}{\sqrt{\pi}} \frac{(1 - e^{-n^2R^2})}{n} \\
&= \frac{[M + \phi(0)]}{\sqrt{\pi}} e^{-n^2R^2} + \frac{\bar{M}}{\sqrt{\pi}} \frac{(1 - e^{-n^2R^2})}{n}
\end{aligned}$$

where we have taken $nR > 1$, which is always possible for any $R > 0$ once n is sufficiently large. This implies that

$$\phi(0) = \lim_{n\to\infty} \int_{-\infty}^{\infty} \frac{n}{\sqrt{\pi}} e^{-n^2x^2} \phi(x)\, dx = \lim_{n\to\infty} \langle \delta_n \mid \phi \rangle$$

so that $\lim_{n\to\infty} \langle \delta_n \mid \phi \rangle$ is indeed a continuous linear functional of the test functions $\phi(x)$. If we simply redefine our origin, we can write

$$\phi(x) = \lim_{n\to\infty} \int_{-\infty}^{\infty} \delta_n(x-\xi)\phi(\xi)\, d\xi$$

so that $\delta(x-\xi)$ can be represented as the limit of the sequence of continuous functions

$$\delta(x-\xi) = \lim_{n\to\infty} \frac{n}{\sqrt{\pi}}\, e^{-n^2(x-\xi)^2} \tag{8.22}$$

Of course this ideal limit has meaning only when weighted with a test function as

$$\langle \delta \mid \phi \rangle = \lim_{n\to\infty} \langle \delta_n \mid \phi \rangle$$

With Eq. 8.22 we can now easily prove the identities (i)–(ix) listed at the end of Section 8.0.

e. Derivatives of Discontinuous Functions

If a function has a discontinuity at a point, then the usual definition of a derivative in terms of a limit at that point does not yield a value for $f'(x)$. However we will see that the operational definition of the derivative of an ideal function, Eq. 8.13, can be used to identify this derivative with an ideal function. First consider a continuous function whose derivative has a discontinuity as shown in Fig. 8.1. Again $f'(x)$ as a limit at $x=a$ does not exist. However the relation

$$\int_{-\infty}^{\infty} f'(x)\phi(x)\, dx = -\int_{-\infty}^{\infty} f(x)\phi'(x)\, dx$$

will define $f'(x)$ almost everywhere (e.g., here, except at $x=a$). Since $f'(x)$ suffers only a *finite* discontinuity here, it can make no contribution to the integral.

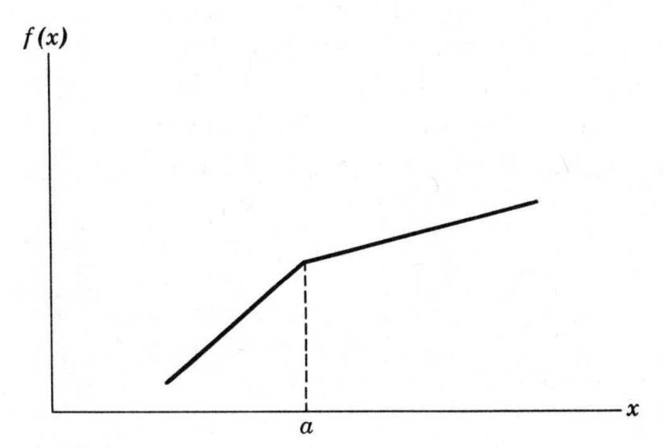

Figure 8.1

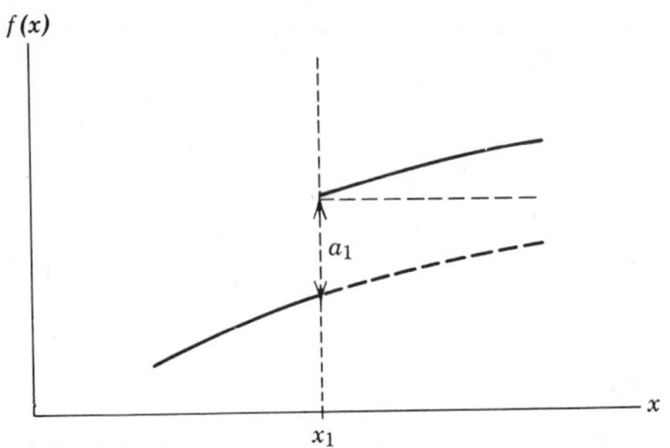

Figure 8.2

Next consider an $f(x)$ that undergoes a *finite* discontinuity a_1 at $x=x_1$ as in Fig. 8.2. We can construct a continuous function as

$$g(x)\equiv f(x)-a_1H(x-x_1)$$

and write

$$\int_{-\infty}^{\infty} f(x)\phi'(x)\,dx=\int_{-\infty}^{\infty} g(x)\phi'(x)+a_1\int_{-\infty}^{\infty} H(x-x_1)\phi'(x)\,dx$$

so that

$$=-\int_{-\infty}^{\infty} g'(x)\phi(x)\,dx-a_1\int_{-\infty}^{\infty} \delta(x-x_1)\phi(x)\,dx$$

$$\equiv-\int_{-\infty}^{\infty} f_i'(x)\phi(x)\,dx \tag{8.23}$$

Here again we have used the integral as a convenient notation for the linear functional. Therefore for our derivative that is an ideal function $f_i'(x)$ we write

$$f_i'(x)=f'(x)+a_1\,\delta(x-x_1) \tag{8.24}$$

since $f'(x)\equiv g'(x)$ whenever they are both defined. Of course we must realize that Eq. 8.24 is not a unique prescription for $f_i'(x)$ and that it is an ideal function. Finally, as with any relation between ideal functions, we must understand that Eq. 8.24 has meaning only after multiplied by a test function and integrated. In general, if $f(x)$ is a piecewise continuous function having finite discontinuities a_j at points x_j, $j=1, 2, 3, \ldots, n$, then we define the derivative of $f(x)$ in terms of an ideal function as

$$f_i'(x)=f'(x)+\sum_{j=1}^{n} a_j\,\delta(x-x_j) \tag{8.25}$$

One of the reasons we have taken the trouble to discuss the δ-function and other ideal functions is the following. We shall want solutions to the equation

$$L_x g(x,t)=\delta(x-t) \tag{8.26}$$

where L_x is a (second order) linear ordinary differential operator. If we find such a $g(x, t)$, then formally (for the moment) we can write

$$L_x \int g(x, t)f(t)\, dt = \int L_x g(x, t)f(t)\, dt = \int \delta(x-t)f(t)\, dt = f(x)$$

which implies that

$$u(x) \equiv \int g(x, t)f(t)\, dt$$

is a solution to the inhomogeneous differential equation

$$L_x u(x) = f(x)$$

Example 8.3 What we actually mean by all of this is the following in terms of sequences of continuous functions.

$$L_x g_n(x, t) = \delta_n(x-t)$$

We must now multiply everything by a test function and integrate,

$$\int dx\phi(x) \int L_x g_n(x, t)f(t)\, dt = \int dx\phi(x) \int \delta_n(x-t)f(t)\, dt$$

and then let $n \to \infty$,

$$\lim_{n\to\infty} \int dx\phi(x)\left[L_x \int g_n(x, t)f(t)\, dt\right] = \lim_{n\to\infty} \int dx\phi(x) \int \delta_n(x-t)f(t)\, dt$$

$$= \int dx\phi(x)f(x)$$

Therefore we obtain

$$u(x) \equiv \lim_{n\to\infty} \int g_n(x, t)f(t)\, dt$$

as a solution to

$$L_x u(x) = f(x)$$

at least almost everywhere.

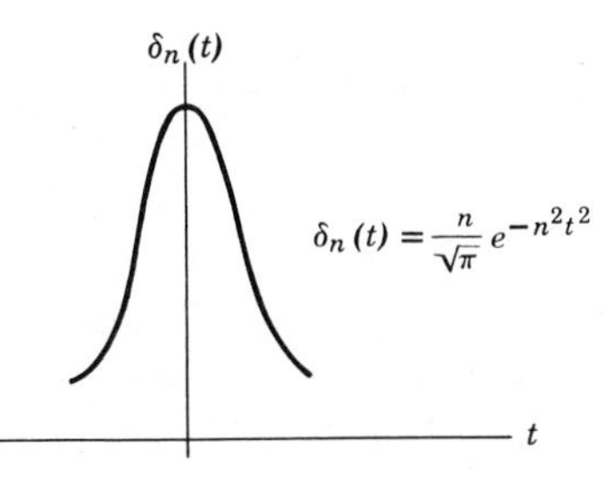

Figure 8.3

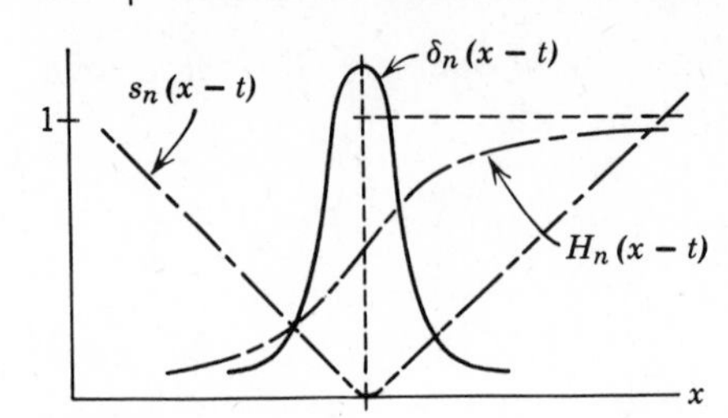

Figure 8.4

Consider the following sequence of operations (cf. Fig. 8.4).

$$\frac{d^2}{dx^2} g_n(x, t) = \delta_n(x-t)$$

$$\frac{d}{dx} g_n(x, t) = \tfrac{1}{2}[H_n(x-t) - H_n(t-x)]$$

$$g_n(x, t) = \tfrac{1}{2} s_n(x-t)$$

We can easily give explicit representations of $\delta_n(x)$, $H_n(x)$, and $s_n(x)$.

$$s_n(x) = \frac{1}{2n} \ln (2 + e^{2nx} + e^{-2nx}) \tag{8.27}$$

$$\frac{d}{dx}\left[\frac{s_n(x)}{2}\right] = \tanh (nx) = \tfrac{1}{2}[H_n(x) - H_n(-x)]$$

$$H_n(x) = \tfrac{1}{2}[\tanh (nx) + 1] = \frac{e^{nx}}{e^{nx} + e^{-nx}}$$

$$\frac{dH_n(x)}{dx} = \frac{2n}{(e^{nx} + e^{-nx})^2} \equiv \delta_n(x) \tag{8.28}$$

$$\int_{-\infty}^{\infty} \delta_n(x)\, dx \equiv 1 \qquad \forall n$$

$$\frac{d^2}{dx^2}[\tfrac{1}{2} s_n(x)] = \delta_n(x)$$

If we take the limit, $n \to \infty$, we conclude that

$$\frac{d^2}{dx^2} |x-t| = 2\, \delta(x-t) \tag{8.29}$$

Here we have been rather loose about interchanging limits, differentiations, and integrations. This can be rigorously justified for weak limits and for distributions (cf. Courant and Hilbert, Vol. II), although we will not attempt this here.

8.2 EXISTENCE AND UNIQUENESS THEOREMS FOR HOMOGENEOUS LINEAR SECOND-ORDER ORDINARY DIFFERENTIAL EQUATIONS

a. Ordinary Points

The form of differential operator we shall consider is

$$Lu = a(x)u''(x) + b(x)u'(x) + c(x)u(x) \tag{8.30}$$

where we shall assume that $a(x)$, $b(x)$, and $c(x)$ are continuous (or piecewise continuous) real functions of x on the finite *closed* interval $[a, b]$ and that $a(x)$ does not vanish anywhere on $[a, b]$. Any point at which $a(x) \neq 0$ is called an *ordinary point* of the differential equation

$$Lu = 0 \tag{8.31}$$

while any point x_0 such that $a(x_0) = 0$ is termed a *singular point* of the differential equation. We will duscuss singular points in Section 8.2b.

The simple substitutions

$$\begin{aligned} q(x) &= c(x) \\ p(x) &= \exp\left[\int \frac{b(x)}{a(x)}\, dx\right] \\ w^{-1}(x) &= a(x) \exp\left[-\int \frac{b(x)}{a(x)}\, dx\right] \end{aligned} \tag{8.32}$$

allow us to write the differential operator of Eq. 8.31 in the form

$$Lu = \frac{1}{w(x)} \frac{d}{dx}\left[p(x) \frac{du}{dx}\right] + q(x)u \tag{8.33}$$

where, by construction, $w(x)$ is of one sign on $[a, b]$ whenever $p(x)$ does not vanish on $[a, b]$. In fact we often take $w(x) = 1$ with no essential loss in generality. The form (Eq. 8.33) with $w(x) = 1$ is the canonical form of the *Sturm-Liouville operator.*

We now restate Th. 5.2 of Section 5.1b since it will be the starting point of our work in this chapter.

THEOREM 8.1

If for $x \in [a, b]$ the functions $a(x)$, $b(x)$, $c(x)$, and $f(x)$ are piecewise continuous and bounded while $a(x)$ is of one sign here, then for the differential equation

$$a(x)u''(x) + b(x)u'(x) + c(x)u(x) = f(x) \tag{8.34}$$

subject to the initial conditions

$$\begin{aligned} u(x_0) &= c_1 \\ u'(x_0) &= c_2 \qquad x_0 \in [a, b] \end{aligned} \tag{8.35}$$

there exists a unique solution such that $u(x)$ and $du(x)/dx$ are continuous for all $x \in [a, b]$.

We now make a few important observations. Recall (Section 5.1b) that Th. 8.1 was proved by establishing the equivalence of Eq. 8.34 with the Volterra integral equation

$$\phi(x)+\int_{x_0}^{x} K(x, y)\phi(y)\, dy=F(x) \tag{8.36}$$

where

$$K(x, y)=\frac{b(x)}{a(x)}+\frac{c(x)}{a(x)}(x-y)$$

$$F(x)=\frac{1}{a(x)}\{f(x)-c_2 b(x)-c(x)[c_1+c_2(x-x_0)]\} \tag{8.37}$$

with

$$u(x)=c_1+c_2(x-x_0)+\int_{x_0}^{x}(x-y)\phi(y)\, dy \tag{8.38}$$

Furthermore suppose we are given that $a(z)$, $b(z)$, $c(z)$, and $f(z)$ are *holomorphic* functions of z is a connected region R and that $a(z)\neq 0$ anywhere in R. Then both $K(z, z')$ and $F(z)$ will be holomorphic functions of z and z' within R. Let z_0 be a point interior to R and C be an ordinary curve *of finite length* from z_0 to z lying wholly within R. Just as in the proof of Th. 5.1 we can show (cf. Problem 8.1) that the solution to

$$\phi(z)+\int_{z_0}^{z} K(z, z')\phi'(z')\, dz'=F(z) \tag{8.39}$$

is given by the *uniformly convergent* series

$$\phi(z)=F(z)+\sum_{n=1}^{\infty}(-1)^n\int_{z_0}^{z} K_n(z, z')F(z')\, dz' \tag{8.40}$$

where

$$K_1(z, z')=K(z, z')$$

$$K_{n+1}(z, z')=\int_{z'}^{z} K(z, z'')K_n(z'', z')\, dz'' \qquad n\geq 1 \tag{8.41}$$

In extending the proof of Th. 5.1 to the complex plane we use

$$\left|\int_{z_0}^{z} f(z')g(z')\, dy\right|\leq\int_{z_0}^{z}|f(z')|\,|g(z')|\,|dz'|$$
$$\leq\left(\int_{z_0}^{z}|f(z')|^2\,|dz'|\right)^{1/2}\left(\int_{z_0}^{z}|g(z')|^2\,|dz'|\right)^{1/2} \tag{8.42}$$

This follows since a valid inner product is

$$\int_{z_0}^{z_1} f^*(z)g(z)\,|dz|$$

for which the Schwarz inequality yields

$$\left|\int_{z_0}^{z_1} f^*(z)g(z)\,|dz|\right| \le \left(\int_{z_0}^{z_1} |f(z)|^2\,|dz|\right)^{1/2}\left(\int_{z_0}^{z_1} |g(z)|^2\,|dz|\right)^{1/2} \tag{8.43}$$

Since Eq. 8.43 holds for *all* $f(z)$ and $g(z)$, it must hold for $|f(z)|$ and $|g(z)|$, from which Eq. 8.42 is obvious. Furthermore, if $K(z, z')$ is, for every value of z in R, a holomorphic function of z' in R and $f(z')$ is also holomorphic in R, then the function

$$\int_{z_0}^{z_1} K(z, z')f(z')\,dz'$$

is holomorphic for all z in R. Under these conditions Eq. 8.40 is a uniformly convergent series of holomorphic functions so that by Th. 7.6 the $\phi(z)$ of Eq. 8.39 is holomorphic everywhere within R. Finally $u(z)$ which is given as

$$u(z) = c_1 + c_2(z - z_0) + \int_{z_0}^{z} (z - z')\phi(z')\,dz' \tag{8.44}$$

and which satisfies

$$a(z)u''(z) + b(z)u'(z) + c(z)u(z) = f(z) \tag{8.45}$$

is holomorphic within R and must, therefore, have a uniformly convergent Taylor series about z_0. This gives us a very important result.

THEOREM 8.2

If $a(z)$, $b(z)$, $c(z)$, and $f(z)$ are holomorphic functions of z in some connected region R, and if $a(z)$ does not vanish anywhere within R, then the solution, $u(z)$, to the differential equation,

$$a(z)u''(z) + b(z)u'(z) + c(z)u(z) = f(z)$$

$$u(z_0) = c_1 \qquad u'(z_0) = c_2$$

exists, is unique, and is holomorphic within R. This $u(z)$ can be expanded in the uniformly convergent Taylor series

$$u(z) = \sum_{n=0}^{\infty} \alpha_n (z - z_0)^n \tag{8.46}$$

which converges for $|z - z_0| < r$ where r is the smallest distance from z_0 to the boundary of the region, R.

We now demonstrate that there exist two, and only two, linearly independent primitive solutions to the homogeneous equation

$$a(x)u''(x) + b(x)u'(x) + c(x)u(x) = 0 \tag{8.47}$$

A *primitive solution* of Eq. 8.47 is one with no boundary conditions specified. In this discussion we assume $x \in [a, b]$ although the same argument holds for the homogeneous version of Eq. 8.45 in a connected region of the z plane in which $a(z)$, $b(z)$, and $c(z)$ are holomorphic and in which $a(z)$ does not vanish anywhere.

We begin by considering the conditions under which three functions, $u_j(x)$, $j=1, 2, 3$, are linearly dependent. If we can find three constants, c_j, $j=1, 2, 3$, such that

$$\sum_{j=1}^{3} c_j u_j(x) = 0 \qquad \forall x \in [a, b] \tag{8.48}$$

then the $u_j(x)$ are by definition linearly dependent. If these functions are continuous and twice differentiable, then we have

$$\begin{aligned} &\sum_{j=1}^{3} c_j u_j(x) = 0 \\ &\sum_{j=1}^{3} c_j u_j'(x) = 0 \qquad \forall x \in [a, b] \\ &\sum_{j=1}^{3} c_j u_j''(x) = 0 \end{aligned} \tag{8.49}$$

If we consider Eqs. 8.49 as three simultaneous linear equations for the quantities $\{c_j\}$, then (cf. Th. 2.5) the necessary and sufficient condition for a nontrivial solution is the vanishing of the *Wronskian* $W(u_1, u_2, u_3)$,

$$W(u_1, u_2, u_3) \equiv \begin{vmatrix} u_1(x) & u_2(x) & u_3(x) \\ u_1'(x) & u_2'(x) & u_3'(x) \\ u_1''(x) & u_2''(x) & u_3''(x) \end{vmatrix} = 0 \qquad \forall x \in [a, b] \tag{8.50}$$

We must now prove that if Eq. 8.50 is satisfied then the $\{c_j\}$ are indeed *constants*. We already see that the Wronskian of a set of linearly independent differentiable functions cannot vanish identically.

Example 8.4 Before we give the proof, let us consider of the type situation we must be careful to cover in the proof. Let

$$A(\alpha) = \{a_{ij}(\alpha)\} = \begin{pmatrix} 1 & \alpha & \sqrt{\alpha^2 - 1} + \alpha \\ 0 & \alpha & \sqrt{\alpha^2 - 1} \\ 1 & -\alpha & -\sqrt{\alpha^2 - 1} + \alpha \end{pmatrix}$$

where α is an arbitrary parameter ($\alpha \neq 0$). We easily see that

$$\det A(\alpha) \equiv 0 \qquad \forall \alpha$$

The set of homogeneous equations

$$\sum_{j=1}^{3} a_{ij}(\alpha)x_j = 0 \qquad i = 1, 2, 3 \tag{8.51}$$

has the nontrivial solution

$$\mathbf{x}(\alpha) = \begin{pmatrix} 1 \\ \dfrac{\sqrt{\alpha^2-1}}{\alpha^2} \\ -\dfrac{1}{\alpha} \end{pmatrix}$$

The point is that the vanishing of det $A(\alpha)$ guarantees the existence of a solution to Eq. 8.51 but this solution $\mathbf{x}(\alpha)$ will generally depend on the parameter, α.

We must prove that a nontrivial solution for the $\{c_j\}$ of Eq. 8.49 exists that is *independent* of x. Let us begin by treating the case $W(u_1, u_2) = 0$,

$$\begin{vmatrix} u_1(x) & u_2(x) \\ u_1'(x) & u_2'(x) \end{vmatrix} = 0 = u_1^2 \frac{d}{dx}\left(\frac{u_2}{u_1}\right) = -u_2^2 \frac{d}{dx}\left(\frac{u_1}{u_2}\right) \qquad \forall x \in [a, b] \tag{8.52}$$

so that

$$u_2(x) = \text{const. } u_1(x) \tag{8.53}$$

provided $u_1(x) \neq 0$ [or $u_2(x) \neq 0$] anywhere on $[a, b]$. If $u_1(x)$ [or $u_2(x)$] does vanish at some point on $[a, b]$, we cannot necessarily conclude that Eq. 8.53 holds as the following example shows.

Example 8.5 For $x \in [-1, 1]$ let

$$u_1(x) = x^2$$

$$u_2(x) = \begin{cases} x^2 & x \leq 0 \\ 2x^2 & x \geq 0 \end{cases}$$

so that

$$W(u_1, u_2) = \begin{vmatrix} x^2 & x^2 \\ 2x & 2x \end{vmatrix} \equiv 0 \qquad x \leq 0$$

$$W(u_1, u_2) = \begin{vmatrix} x^2 & 2x^2 \\ 2x & 4x \end{vmatrix} \equiv 0 \qquad x \geq 0$$

which implies that $W(u_1, u_2) \equiv 0$, $x \in [-1, 1]$. However there are no nonzero *constants* c_1 and c_2 such that

$$c_1 u_1(x) + c_2 u_2(x) = 0 \qquad \forall x \in [-1, 1]$$

so that $u_1(x)$ and $u_2(x)$ are linearly independent. Therefore, not only must we require that $W(u_1, u_2)\equiv 0$ but also that at least one of the minors of the last row of $W(u_1, u_2)$ [i.e., in this case $u_1(x)$ or $u_2(x)$] not vanish *anywhere* on $[a, b]$ if we are to conclude that $u_1(x)$ and $u_2(x)$ *must* be linearly dependent everywhere on $[a, b]$.

Of course, if we knew that $u_1(z)$ and $u_2(z)$ were holomorphic in a region (rather than simply continuous), then when $W(u_1, u_2)=0$ and $u_1(z)=0$ [or $u_2(z)=0$] at some discrete set of points in R, we could still conclude $u_2(z)=$ const. $u_1(z)$ everywhere in R since we could apply Eq. 8.52 along all curves around these points.

We assume that none of the minors of the last row of Eq. 8.50 vanish identically so that no pair of the $u_j(x)$ can possibly be linearly dependent (since if any pair were, then certainly the set $\{u_j(x)\}$ would be linearly dependent already). We furthermore assume that one of these minors of the last row, say

$$\begin{vmatrix} u_2(x) & u_3(x) \\ u_2'(x) & u_3'(x) \end{vmatrix}$$

by choice, does not vanish anywhere on $x\in[a, b]$. If we expand Eq. 8.50 we obtain an expression of the form

$$\bar{a}(x)u_1''(x)+\bar{b}(x)u_1'(x)+\bar{c}(x)u_1(x)=0$$

where

$$\bar{a}(x)=\begin{vmatrix} u_2(x) & u_3(x) \\ u_2'(x) & u_3'(x) \end{vmatrix}$$

$$\bar{b}(x)=-\begin{vmatrix} u_2(x) & u_3(x) \\ u_2''(x) & u_3''(x) \end{vmatrix}$$

$$\bar{c}(x)=\begin{vmatrix} u_2'(x) & u_3'(x) \\ u_2''(x) & u_3''(x) \end{vmatrix}$$

If we replace $u_1(x)$ by $u_2(x)$ or by $u_3(x)$, we have as *identities*

$$W(u_2, u_2, u_3)\equiv 0\equiv W(u_3, u_2, u_3)$$

so that

$$\bar{a}(x)u_j''(x)+\bar{b}(x)u_j'(x)+\bar{c}(x)u_j(x)\equiv 0 \qquad \forall x\in[a, b] \qquad j=1, 2, 3 \qquad \textbf{(8.54)}$$

For any point $x_0\in[a, b]$ we have a nontrivial solution to Eq. 8.49 for a set of constants $\{c_j\}$. The function

$$u(x)\equiv\sum_{j=1}^{3} c_j u_j(x)$$

satisfies Eq. 8.54 everywhere on $[a, b]$ and is such that

$$u(x_0)=0=u'(x_0)$$

However by Th. 8.1 we see that $u(x)\equiv 0$, $\forall x\in[a, b]$, which establishes the existence of a set of *constants* $\{c_j\}$ satisfying Eq. 8.49. Equation 8.48 is identically satisfied so that the $\{u_j(x)\}$ are linearly dependent.

THEOREM 8.3

If three continuous, twice differentiable functions $u_j(x)$, $j=1, 2, 3$, are linearly dependent, then their Wronskian $W(u_1, u_2, u_3)$, vanishes identically. Conversely, if the Wronskian of three continuous, twice-differentiable functions vanishes identically, and if at least one of the minors of the last row does not vanish anywhere on $[a, b]$, then the three functions are linearly dependent here.

We have just proved Th. 8.3 for three functions, although the extension to n functions is straightforward.

Let us now return to our homogeneous differential equation, Eq. 8.47, and assume that there exist three linearly independent solutions. We can use Eq. 8.47 on $W(u_1, u_2, u_3)$ and our theorems on determinants (Sec. 2.1b) to show that

$$W(u_1, u_2, u_3)=\begin{vmatrix} u_1 & u_2 & u_3 \\ u_1' & u_2' & u_3' \\ -\frac{b}{a}u_1'-\frac{c}{a}u_1 & -\frac{b}{a}u_2'-\frac{c}{a}u_2 & -\frac{b}{a}u_3'-\frac{c}{a}u_3 \end{vmatrix}=0$$

We already have a nonsingular differential equation satisfied by the $\{u_j(x)\}$ so that we can drop the assumption about the minor of the last row of W in Th. 8.3 since the only purpose of this restriction was to provide the nonsingular differential equation of Eq. 8.54. This statement that $W(u_1, u_2, u_3)=0$ shows that there cannot exist three (or more) linearly independent primitive solutions to Eq. 8.47. We must now prove that we can always find a $u_1(x)$ and a $u_2(x)$ such that $W(u_1, u_2)\not\equiv 0$. If we write

$$W(u_1, u_2)=\begin{vmatrix} u_1(x) & u_2(x) \\ u_1'(x) & u_2'(x) \end{vmatrix}$$

and use Eq. 8.47, we find

$$\frac{dW}{dx}=u_1u_2''-u_1''u_2=u_1\left(-\frac{b}{a}u_2'-\frac{c}{a}u_2\right)-\left(-\frac{b}{a}u_1'-\frac{c}{a}u_1\right)u_2$$

$$=-\frac{b}{a}W$$

so that

$$W[u_1(x), u_2(x)]=W_0\exp\left(-\int_{x_0}^{x}\frac{b(x)}{a(x)}\,dx\right) \quad \textbf{(8.55)}$$

where

$$W_0\equiv W[u_1(x_0), u_2(x_0)]$$

Therefore we conclude that $W_0=0$ if and only if

$$W[u_1(x), u_2(x)]\equiv 0$$

However from Th. 8.1 we know that at any point $x_0\in[a, b]$ we can construct two solutions such that

$$u_1(x_0)=\alpha \qquad u_1'(x_0)=\beta$$

and

$$u_2(x_0)=\gamma \qquad u_2'(x_0)=\delta$$

Since α, β, γ, and δ are at our disposal, we need only choose them so that

$$\alpha\delta-\beta\gamma\equiv W_0\neq 0$$

This establishes the following important result, which we state for a region in the z plane. If we are given only continuity on a line segment $a\leq x\leq b$, instead of holomorphy in a region R, the theorem remains true if the phrase "continuous on $x\in[a, b]$" is substituted for "holomorphic in a region R" and "continuous and differentiable solutions on $x\in[a, b]$" for "holomorphic solutions in R."

THEOREM 8.4

If the functions $a(z)$, $b(z)$, and $c(z)$ are holomorphic in a region R and if $a(z)$ does not vanish anywhere in this region, then the differential equation

$$a(z)u''(z)+b(z)u'(z)+c(z)u(z)=0 \tag{8.56}$$

has two, and only two, linearly independent primitive holomorphic solutions in R.

We can also see now that since $u(z)$ has a uniformly convergent Taylor series expansion (cf. Th. 8.2), we may differentiate it, term by term, and substitute it into Eq. 8.56 to determine the expansion coefficients. In practice $a(z)$, $b(z)$, and $c(z)$ are usually low (i.e., first or second) degree polynomials. In any event though they have the Taylor series expansions

$$\begin{aligned} a(z)&=\sum_{n=0}^{\infty} a_n(z-z_0)^n \\ b(z)&=\sum_{n=0}^{\infty} b_n(z-z_0)^n \\ c(z)&=\sum_{n=0}^{\infty} c_n(z-z_0)^n \end{aligned} \tag{8.57}$$

while $u(z)$ has the expansion

$$u(z)=\sum_{n=0}^{\infty} \alpha_n(z-z_0)^n \tag{8.46}$$

If Eqs. 8.46 and 8.57 are substituted into Eq. 8.56 and the coefficients of $(z-z_0)^n$ equated to zero, we obtain recursion relations for the $\{\alpha_n\}$ in terms of the a_n, b_n, and c_n. It is only when $a(z)$, $b(z)$, and $c(z)$ are polynomials of finite degree that there is any hope of solving the recursion relations for the $\{\alpha_n\}$ in closed form. In particular the coefficient of $(z-z_0)^0$ is

$$2a_0\alpha_2+b_0\alpha_1+c_0\alpha_0=0 \tag{8.58}$$

Notice that $a_0\neq 0$ since we have assumed that $a(z_0)\neq 0$. We can assign α_0 and α_1 arbitrarily (e.g., $\alpha_0=1$, $\alpha_1=0$ and then $\alpha_0=0$, $\alpha_1=1$), which allows us to obtain the two linearly independent primitive solutions to Eq. 8.56.

Let us make an observation which we shall use later. If we obtain one solution, say $u_1(z)$, to Eq. 8.56, then we can use Eq. 8.55 as a first-order inhomogeneous differential equation to obtain $u_2(z)$. Since any first-order linear ordinary differential equation is easily reduced to quadrature, a second solution is seen to be

$$u_2(z)=Au_1(z)\int^{z}\frac{1}{u_1^2(z')}\exp\left[-\int_{z_0}^{z'}\frac{b(z'')}{a(z'')}\,dz''\right]dz'+Bu_1(z) \tag{8.59}$$

where A and B are arbitrary constants.

b. Singular Points

Any point z_0 at which $a(z_0)=0$, where $a(z)$ is still assumed holomorphic in R, is a *singular point* of Eq. 8.56. [Notice that we may assume that $b(z_0)$ and $c(z_0)$ are not both zero since, if they were, we could have divided out a common zero of some order from $a(z)$, $b(z)$, and $c(z)$.] Specifically, if the functions

$$(z-z_0)\frac{b(z)}{a(z)}$$

$$(z-z_0)^2\frac{c(z)}{a(z)}$$

are both holomorphic in the neighborhood of $z=z_0$, then z_0 is termed a *regular singular point* of the differential equation. If either or both of these functions cease to be holomorphic in the neighborhood of $z=z_0$, then z_0 is an *irregular singular point* of the differential equation. It is sufficient for our discussion to continue to assume that $a(z)$, $b(z)$, and $c(z)$ are holomorphic in R since if they had nonessential singularities in R (i.e., poles) we could always multiply Eq. 8.56 through by the appropriate terms of the form $(z-z_j)^m$ to remove these singularities. We will not consider the cases in which $a(z)$, $b(z)$, or $c(z)$ cease to be single valued in R or in which they contain essential singularities there. As we have mentioned previously, in many applications the coefficient functions $a(z)$, $b(z)$, and $c(z)$ are polynomials in z.

Since such a singular point z_0 gives rise to a singular integral equation (cf. Eq.

8.36), we do not know whether or not there exist any solutions to Eq. 8.56 regular about z_0. (We shall call a solution *regular* about z_0 if it is holomorphic in the neighborhood of z_0). We may attempt a solution of the form

$$u(z)=\sum_{n=0}^{\infty} \alpha_n(z-z_0)^{n+s} \tag{8.60}$$

where s is an index to be determined. If s is a positive integer or zero, then we have a regular solution. We begin by assuming that $a(z)$ has a *simple* zero at z_0 and substitute Eqs. 8.60 and 8.57 into Eq. 8.56 so that $a_0=0$, and $a_1\neq 0$ for this regular singular point. We find

$$\begin{aligned}(z-z_0)^{-1}s\alpha_0[(s-1)a_1+b_0]+\sum_{m=0}^{\infty}[s(s-1)\alpha_0 a_{m+2}+s\alpha_0 b_{m+1}\\ +s(s+1)\alpha_1 a_{m+1}](z-z_0)^m+\sum_{n=0}^{\infty}\sum_{m=0}^{\infty}[(n+2+s)(n+1+s)\alpha_{n+2}a_m\\ +(n+1+s)\alpha_{n+1}b_m+\alpha_n c_m](z-z_0)^{n+m}=0\end{aligned} \tag{8.61}$$

Since this must hold for all values of z in R we have

$$s\alpha_0[(s-1)a_1+b_0]=0 \tag{8.62}$$

$$\begin{aligned}s\alpha_0[(s-1)a_{n+2}+b_{n+1}]\\ +\sum_{m=0}^{n}\{(n-m+1+s)\alpha_{n-m+1}[(n-m+s)a_{m+1}+b_m]+\alpha_{n-m}c_m\}=0\end{aligned} \tag{8.63}$$

Equation 8.62 is the *indicial equation* and has the roots

$$s_1=0 \qquad s_2=1-\frac{b_0}{a_1} \tag{8.64}$$

provided $a_1\neq 0$. We will return to the case $a_1=0$ later. We also see from Eq. 8.62 that α_0 is arbitrary.

Since many of the special functions of applied mathematics are associated with second-order differential equations in which $a(z)$, $b(z)$, and $c(z)$ are second-degree polynomials, we write Eq. 8.63 for this case when $s=0$. The following can be used only when $s=0$ is an index.

$$\begin{aligned}(n+1)(na_1+b_0)\alpha_{n+1}+[n(n-1)a_2+nb_1+c_0]\alpha_n\\ +[(n-1)b_2+c_1]\alpha_{n-1}+c_2\alpha_{n-2}=0\end{aligned} \tag{8.65}$$

This equation holds for all integer $n\geq 0$ if we simply define $\alpha_{-2}=0=\alpha_{-1}$.

Although Eqs. 8.63–8.65, which follow from Eq. 8.56, are usually the simplest to use for actual calculation of the series solution, a general proof for the convergence of the series is more easily given if we rewrite Eq. 8.56 in the form

$$u''(z)+\beta(z)u'(z)+\gamma(z)u(z)=0 \tag{8.66}$$

We also take the regular singular point to be at $z_0=0$ without loss of generality. Since $z=0$ is to be a *regular* singular point, we assume that $z\beta(z)$ and $z^2\gamma(z)$ have Taylor expansions that converge in a region R.

$$z\beta(z)=\sum_{m=0}^{\infty}\beta_m z^m$$
$$z^2\gamma(z)=\sum_{m=0}^{\infty}\gamma_m z^m \qquad (8.67)$$
$$u(z)=\sum_{n=0}^{\infty}\alpha_n z^{n+s}$$

If, as previously, we substitute Eqs. 8.67 into Eq. 8.66 we obtain

$$\alpha_0[s(s-1)+s\beta_0+\gamma_0]=0 \qquad (8.68)$$

$$\alpha_n[(n+s)(n-1+s)+(n+s)\beta_0+\gamma_0] + \sum_{m=0}^{n-1}\alpha_m[(m+s)\beta_{n-m}+\gamma_{n-m}]=0 \qquad n=1,2,\ldots \qquad (8.69)$$

If we denote the roots of the indicial equation, Eq. 8.68, by s_1 and s_2 so that

$$(s-s_1)(s-s_2)=s(s-1)+s\beta_0+\gamma_0 \qquad s_1+s_2=1-\beta_0$$

then we may use Eq. 8.69 to obtain for $s=s_1$ (or $s=s_2$) and $n\geq|s_1-s_2|$

$$|\alpha_n|\, n(n-|s_1-s_2|)\leq|\alpha_n|\, n(n+s_1-s_2) \leq\sum_{m=0}^{n-1}|\alpha_m|\,[(m+|s_1|)\,|\beta_{n-m}|+|\gamma_{n-m}|] \qquad (8.70)$$

We know from Eq. 7.21 that the Taylor coefficients for $z\beta(z)$ and $z^2\gamma(z)$ satisfy the Cauchy inequalities

$$|\beta_m|\leq\frac{M}{r^m}$$
$$|\gamma_m|\leq\frac{M}{r^m} \qquad (8.71)$$

where r is the (smaller) radius of convergence of these two Taylor series so that

$$n(n-|s_1-s_2|)\,|\alpha_n|\leq M\sum_{m=0}^{n-1}|\alpha_m|\,(m+|s_1|+1)r^{-n+m}$$

Let us now define a set of real, nonnegative constants $\{\bar{\alpha}_n\}$ such that $|\alpha_n|\leq\bar{\alpha}_n$ and such that, again for $n\geq|s_1-s_2|$,

$$n(n-|s_1-s_2|)\bar{\alpha}_n=M\sum_{m=0}^{n-1}\bar{\alpha}_m(m+|s_1|+1)r^{-n+m} \qquad (8.72)$$

That is, if we choose $|\alpha_n|\leq\bar{\alpha}_n$ for all $n<|s_1-s_2|$, then it follows from induction that $\bar{\alpha}_n \geq |\alpha_n|$, $\forall n \geq |s_1 - s_2|$ since, from Eqs. 8.70, 8.71, and 8.72, we have

$$n(n-|s_1-s_2|)(\bar{\alpha}_n - |\alpha_n|) \geq M \sum_{m=0}^{n-1} (\bar{\alpha}_m - |\alpha_m|)(m+|s_1|+1)r^{-n+m} \geq 0$$

If we write Eq. 8.72 for n replaced by $(n-1)$ and subtract this result from Eq. 8.72, we find

$$n(n-|s_1-s_2|)\bar{\alpha}_n = \bar{\alpha}_{n-1} r^{-1}[M(n+|s_1-s_2|)+(n-1)(n-1-|s_1-s_2|)] \quad \textbf{(8.73)}$$

which implies that

$$\lim_{n\to\infty} \frac{\bar{\alpha}_n}{\bar{\alpha}_{n-1}} = \frac{1}{r}$$

Therefore by the comparison test on the series

$$\sum_{n=0}^{\infty} \alpha_n z^n \quad \textbf{(8.74)}$$

we see that Eq. 8.74 converges uniformly for all $|z| < r$. This establishes the following theorem.

THEOREM 8.5

If the differential equation

$$u''(z) + \beta(z)u'(z) + \gamma(z)u(z) = 0$$

is such that $(z-z_0)\beta(z)$ and $(z-z_0)^2\gamma(z)$ possess Taylor series expansions for $|z-z_0| < r$, so that z_0 is a regular singular point, then one primitive solution can be expressed as

$$u(z) = (z-z_0)^s \sum_{n=0}^{\infty} \alpha_n (z-z_0)^n$$

where s is either root of the indicial equation. This series is uniformly convergent for $|z-z_0| < r$ [i.e., $u(z)/(z-z_0)^s$ is holomorphic about z_0].

Since the proof just given holds both for $s = s_1$ and for $s = s_2$, it follows that if $s_1 - s_2$ is not zero or a positive integer, then this series solution technique will yield both linearly independent solutions. However for our purposes it will be sufficient to know that, once we have obtained one solution as outlined above, the other one can be gotten by use of Eq. 8.59.

Example 8.6 Before we go on to other general developments, it may prove instructive to apply these results to some of the differential equations we met in

Chapter 7. If we consider the hypergeometric equation, Eq. 7.157,

$$z(z-1)u''+[(a+b+1)z-c]u'+abu=0$$

and take $z_0=0$, then in terms of the notation of Eq. 8.56 we have

$$a(z)=-z+z^2$$
$$b(z)=-c+(a+b+1)z$$
$$c(z)=ab$$

so that Eqs. 8.64 and 8.65 become

$$s_1=0 \qquad s_2=1-c$$
$$(n+1)(n+c)\alpha_{n+1}=\alpha_n(n+a)(n+b)$$

We see from Th. 8.5 that since $s=0$ is a solution to the indicial equation, then one solution will always be regular in the neighborhood of z_0. If $c\neq -N$ (i.e., zero or a negative integer) we can use Eq. 8.59 and the behavior $u_1(z)\xrightarrow[z\to 0]{}$ const. to find the behavior of a second linearly independent solution as

$$\int \frac{b(z)}{a(z)}\,dz=\int \frac{[(a+b+1)z-c]\,dz}{z(z-1)}=\ln\left[(z-1)^{a+b+1-c}z^c\right]$$

$$\exp\left[-\int \frac{b(z)}{a(z)}\,dz\right]=\frac{(z-1)^{c-a-b-1}}{zc}$$

$$u_2(z)\xrightarrow[z\to 0]{}\text{const.}\int z\,dz\,z^{-c}\sim\begin{cases}\text{const. } z^{1-c} & c\neq 1\\ \text{const. } \ln z & c=1\end{cases}$$

This agrees with Eq. 7.166. If $c=-N$ the recursion relation (8.75) for α_n shows that $\alpha_n=0$, $0\leq n\leq N$ so that

$$u_1(z)\xrightarrow[z\to 0]{}\text{const. } z^{N+1}$$

and we may use Eq. 8.59 to deduce that

$$u_2(z)\xrightarrow[z\to 0]{}\text{const. } z^{N+1}\int_0^z z^{-2N-2}z^n\,dz\sim\text{const.}$$

About the point $z_0=1$ we have

$$a(z)=(z-1)+(z-1)^2$$
$$b(z)=(a+b+1-c)+(a+b+1)(z-1)$$
$$c(z)=ab$$

so that

$$s_1=0 \qquad s_2=c-a-b$$

An argument similar to that given above shows that there exists a solution such that $u_3(z) \xrightarrow[z\to 1]{}$ const. and another such that

$$u_4(z) \xrightarrow[z\to 1]{} \text{const.}\,(z-1)^{c-a-b}$$

as expected from Eq. 7.185 when Re $(a+b-c)>0$. Of course only two of these four solutions $\{u_j(x)\}$ can be linearly independent (cf. the joining relations of Eq. 7.185).

Example 8.7 Finally consider Bessel's equation, Eq. 7.216,

$$z^2u''(z)+zu'(z)+(z^2-\nu^2)u(z)=0$$

This satisfies the conditions of Th. 8.5. With $z_0=0$ we have

$$a(z)=z^2$$
$$b(z)=z$$
$$c(z)=-\nu^2+z^2$$

or

$$z\beta(z)=1$$
$$z^2\gamma(z)=-\nu^2+z^2$$

Since $\beta_0=1$, $\gamma_0=-\nu^2$, the indicial equation, Eq. 8.68, becomes

$$s^2-\nu^2=0$$

so that

$$s_1=\nu$$
$$s_2=-\nu$$

Let us take Re $\nu\geq 0$ since the case Re $\nu<0$ amounts to interchanging s_1 and s_2 and therefore to interchanging the two linearly independent solutions except when $\nu=0$, which we will treat separately. Equation 8.69 for $n=1$ and $s=s_1=\nu$ is

$$\alpha_1[(\nu+1)\nu+(\nu+1)-\nu^2]=\alpha_1(2\nu+1)=0$$

so that $\alpha_1=0$. For $n\geq 2$ Eq. 8.69 becomes

$$n(n+2\nu)\alpha_n+\alpha_{n-2}=0 \tag{8.76}$$

which requires that $\alpha_n=0$ for all odd n. If we let $n\to 2n$, we obtain

$$4n(n+\nu)\alpha_{2n}=-\alpha_{2(n-1)} \qquad n=1,2,3,\ldots$$

This leads to the solution of Eq. 7.229 considered previously. If ν is not zero or an integer, then a second linearly independent solution is obtained by the substitution $\nu\to-\nu$. In any event one solution is given by solving the recursion relation of Eq. 8.76 and has the behavior

$$u_1(z) \xrightarrow[z\to 0]{} \text{const.}\,z^{\nu}$$

that, with Eq. 8.59, leads to the behavior of the second linearly independent solution as

$$u_2(z) \xrightarrow[z\to 0]{} \begin{cases} \text{const. } z^{-\nu} & \nu \neq 0 \\ \text{const. } \ln z & \nu = 0 \end{cases}$$

We make one last general observation in this section. If the ratio $[b(z)/a(z)]$ has a pole of higher than the first order at $z = z_0$, (i.e., an *irregular* singular point) then we *expect* that the general solution to the differential equation will have an essential singularity at $z = z_0$, not simply a pole or branch point. This is easily seen from Eq. 8.59 since

$$\int \frac{b(z)}{a(z)}\, dz$$

will itself have a pole behavior so that

$$u_1(z) \int^z \frac{1}{u_1^2(z)} \exp\left[-\int_{z_0}^{z'} \frac{b(z'')}{a(z'')}\, dz''\right] dz'$$

will usually have an essential singularity.

8.3 STURM-LIOUVILLE PROBLEM FOR DISCRETE EIGENVALUES

We now show that the solutions to the differential equation

$$\frac{d}{dx}\left[p(x)\frac{du}{dx}\right] + g(x)u(x) = 0 \tag{8.77}$$

are oscillatory or nonoscillatory depending on the choices for $p(x)$ and $g(x)$. We assume the functions, $p(x)$ and $g(x)$, to be real, continuous, and bounded on the interval $a \le x \le b$ and $p(x) > 0$ here. Now let

$$p_1(x) \ge p_2(x) > 0 \qquad g_1(x) \le g_2(x) \qquad x \in [a, b]$$

$$\frac{d}{dx}\left[p_1(x)\frac{du}{dx}\right] + g_1(x)u = 0 \qquad \frac{d}{dx}\left[p_2(x)\frac{dv}{dx}\right] + g_2(x)v = 0$$

We also add the condition that $g_1(x)$ and $g_2(x)$ are not both identically zero simultaneously in any *finite* part of $[a, b]$.

Provided $v(x) \neq 0$ for any $x \in [x_1, x_2]$, we may verify the identity

$$\frac{d}{dx}\left\{\frac{u(x)}{v(x)}[p_1(x)u'(x)v(x) - p_2(x)u(x)v'(x)]\right\}$$

$$= [g_2(x) - g_1(x)]u^2(x) + [p_1(x) - p_2(x)]u'^2(x) + p_2(x)\frac{[u'(x)v(x) - u(x)v'(x)]^2}{v^2(x)}$$

Hence if $v(x)\neq 0$ for any $x\in[x_1, x_2]$ we may integrate this to obtain

$$\left[\frac{u}{v}(p_1u'v-p_2uv')\right]\Bigg|_{x_1}^{x_2}=\int_{x_1}^{x_2}[g_2(x)-g_1(x)]u^2\,dx$$
$$+\int_{x_1}^{x_2}[p_1(x)-p_2(x)]u'^2\,dx$$
$$+\int_{x_1}^{x_2}p_2(x)\frac{(u'v-uv')^2}{v^2}\,dx,\quad a\leq x_1\leq x\leq x_2\leq b \qquad \textbf{(8.78)}$$

If x_1 and x_2 are consecutive zeros of $u(x)$, and if $v(x)\neq 0$ for any $x\in[x_1, x_2]$, then the right side of Eq. 8.78 is positive while the left side vanishes. This contradiction implies that we were not justified in integrating the differential identity that led to Eq. 8.78 so that $v(x)$ has at least one zero on $x\in[x_1, x_2]$. We have assumed $g_1(x)$ and $g_2(x)$ are not both identically zero on a finite interval for the following reason. Suppose $p_1=p_2$ over some subinterval and $u'=0$ on the rest, so that $g_1\equiv 0$ here also. If u and v are proportional, the last integral in (8.78) vanishes. Finally, then, if $g_1=g_2$ over this range, the right side of Eq. 8.78 vanishes and there is no contradiction. This result also holds if $v(x)=0$ at x_1 and/or x_2. Suppose $v(x_1)=0$. Then the left side of Eq. 8.78 becomes

$$\lim_{x\to x_1}\left\{\frac{u}{v}[p_1u'v-p_2uv']\right\}=\lim_{x\to x_1}[(p_1-p_2)uu']=0$$

since the indeterminate form u/v equals u'/v' and neither u' nor v' at x_1 can be zero, since these satisfy a *homogeneous* second-order differential equation, and that would imply the identical vanishing of the solution (cf. Eqs. 8.36–8.38).

In any case then $v(x)$ must vanish at least once in the *open* interval $x_1<x<x_2$ [i.e., $x\in(x_1, x_2)$]. Therefore if $u(x_1)=0=v(x_1)$, $v(x)$ oscillates *more rapidly* than $u(x)$. This means that if $p(x)$ is *decreased* or left constant while $g(x)$ is *increased*, the solutions to this differential equation oscillate more rapidly (provided they oscillated to begin with). Now let

$$A\geq p(x)\geq c>0$$
$$B\geq g(x)\geq d \qquad x\in[a, b]$$

Then the solutions of

$$\frac{d}{dx}\left[p(x)\frac{du}{dx}\right]+g(x)u=0 \qquad \textbf{(8.79)}$$

do not oscillate more rapidly on $[a, b]$ than the solutions of

$$\frac{d}{dx}\left[c\frac{dv}{dx}\right]+Bv=0$$

or equivalently, of

$$\frac{d^2v}{dx^2}+\left(\frac{B}{c}\right)v=0 \tag{8.80}$$

i. If $B<0$, there is an exponential solution, $v(x)=\exp(\pm\sqrt{|B/c|}\,x)$, which has no zeros on $[a, b]$. The same is true for $B=0$. Therefore, if $g(x)\le 0$ throughout $[a, b]$, the solutions of Eq. 8.79 are nonoscillatory.

ii. If $B>0$, then the oscillatory solution $\sin(\sqrt{B/c}\,x)$ exists for Eq. 8.80. The distance between consecutive zeros is

$$\pi\sqrt{c/B}$$

Therefore if

$$\pi\sqrt{c/B}>(b-a)$$

$v(x)$ cannot have more than one zero and no oscillatory solutions to Eq. 8.80 exist. Consequently the solutions of Eq. 8.79 are nonoscillatory provided

$$\frac{B}{c}<\frac{\pi^2}{(b-a)^2}$$

Similarly the solutions of Eq. 8.79 oscillate at least as rapidly as those of

$$\frac{d}{dx}\left[A\frac{dv}{dx}\right]+dv=0$$

or, equivalently, of

$$\frac{d^2v}{dx^2}+\left(\frac{A}{d}\right)v=0 \tag{8.81}$$

If $d>0$, then the solutions of Eq. 8.81 are oscillatory (i.e., $\sin x$) and, therefore, so are those of Eq. 8.79. Hence a sufficient condition that the solutions of Eq. 8.79 have at least m zeros is that

$$m\pi\sqrt{A/d}\le(b-a)$$

or that

$$\frac{d}{A}\ge\frac{m^2\pi^2}{(b-a)^2}$$

In particular, a sufficient condition that Eq. 8.79 have one oscillatory solution on (a, b) is that

$$\frac{d}{A}\ge\frac{\pi^2}{(b-a)^2}$$

THEOREM 8.6

If $p(x)$ and $g(x)$ are real, continuous, and bounded on the closed interval $[a, b]$ as

$$A \geq p(x) \geq c > 0$$
$$B \geq g(x) \geq d$$

then the solutions to the differential equation

$$\frac{d}{dx}\left[p(x)\frac{du}{dx}\right] + g(x)u = 0$$

on $x \in [a, b]$ are nonoscillatory provided

$$\frac{B}{c} < \frac{\pi^2}{(b-a)^2}$$

and are oscillatory provided

$$\frac{d}{A} \geq \frac{\pi^2}{(b-a)^2}$$

We can now study the eigenvalue spectrum and eigenfunctions of the *Sturm-Liouville equation*

$$\frac{d}{dx}\left[p(x)\frac{du(x)}{dx}\right] + [q(x) + \lambda r(x)]u(x) = 0 \tag{8.82}$$

For example the functions $p(x)$, $q(x)$, and $r(x)$ sometimes result from the separation of variables in the partial differential equation

$$(\nabla^2 + k^2)\psi(\mathbf{r}) = 0$$

in some particular set of coordinates. We assume that $p(x)$, $q(x)$, and $r(x)$ are continuous and bounded and that $p(x) > 0$, $r(x) > 0$ on our finite range $[a, b]$. This will allow us to apply Th. 8.6. It is clear that for sufficiently large values of λ (i.e., real, positive)

$$\frac{[q(x) + \lambda r(x)]_{\text{min}}}{[p(x)]_{\text{max}}} \geq \frac{\pi^2}{(b-a)^2}$$

so that oscillatory solutions finally do exist. Furthermore, when λ decreases enough (possibly to negative values), the solutions cease to oscillate. Hence there will be some smallest eigenvalue, λ_0, such that the corresponding eigenfunction, $u_0(x)$, satisfies the two boundary conditions,

$$u_0(a) = 0 = u_0(b)$$

THEOREM 8.7

If $p(x)$, $q(x)$, and $r(x)$ are real, continuous, and bounded for $x \in [a, b]$ and if $p(x) > 0$, $r(x) > 0$ here, then there exists some least eigenvalue λ_0 such that

$$\frac{d}{dx}\left[p(x)\frac{d}{dx}u_0(x)\right] + [q(x) + \lambda_0 r(x)]u_0(x) = 0$$

where

$$u_0(a) = 0 = u_0(b)$$

In order to simplify the algebra of the following discussion we shall assume that our unmixed homogeneous boundary conditions are of the particularly simple form

$$u(a) = 0 = u(b) \tag{8.83}$$

although the results are valid for the more general unmixed homogeneous boundary conditions (cf. Problem 8.9)

$$\begin{aligned} \alpha_1 u(a) + \beta_1 u'(a) &= 0 \\ \alpha_2 u(b) + \beta_2 u'(b) &= 0 \end{aligned} \tag{8.84}$$

where α_j and β_j are real and such that $\alpha_1/\beta_1 < 0$, $\alpha_2/\beta_2 > 0$. If we begin with $u(a) = 0$, then from Th. 8.7 we know that for low enough λ there will be no zeros in the range $a < x \le b$. For some $\lambda = \lambda_0$ $u(b) = 0$. Then λ_0 is the lowest eigenvalue for this problem. We can always redefine $q(x)$ and $r(x)$ in the Sturm-Liouville equation, Eq. 8.82, so that $\lambda_0 = 0$ and all other eigenvalues are positive. It is also clear that we may assume the $\{u_n(x)\}$, subject to the boundary conditions (8.83) or (8.84), to be *real* functions for our proofs. All the $\{\lambda_n\}$ are real since, with homogeneous boundary conditions, the Sturm-Liouville operator is self-adjoint (cf. Sec. 8.4). (The proof using the inner product is the same as that given for hermitian matrices in Sec. 3.4.) Therefore we can order things so that $\lambda_{n+1} > \lambda_n$ and that $u_{n+1}(x)$ has more nodes (in fact exactly one more) than $u_n(x)$ on the range $[a, b]$.

We first show that $(\lambda_{n+1} - \lambda_n)$ is always finite (i.e., nonzero) if $(b - a)$ is finite. Let $\bar{\lambda}$ be such that $\lambda_n < \bar{\lambda} < \lambda_{n+1}$ and take $\bar{u}(x)$ to be the solution to Eq. 8.82 satisfying

$$\bar{u}(a) = 0 \qquad \bar{u}'(b) = 0$$

[For the boundary conditions of Eq. 8.84 we take $\alpha_1\bar{u}(a) + \beta_1\bar{u}'(a) = 0$ (cf. Problem 8.9).] This is always possible. Then since, as an identity, we have

$$\frac{d}{dx}\left[\bar{u}p\frac{du_n}{dx} - u_n p\frac{d\bar{u}}{dx}\right] = (\bar{\lambda} - \lambda_n)ru_n\bar{u}$$

we can write

$$p(b)\bar{u}(b)u_n'(b) = (\bar{\lambda} - \lambda_n)\int_a^b u_n(x)\bar{u}(x)r(x)\,dx \tag{8.85}$$

Note that $u_n'(b) \neq 0$ since it satisfies the differential equation and $u_n(b)=0$. Also $\bar{u}(b) \neq 0$ since it is, by assumption, not an eigenfunction. In fact the l.h.s. cannot be made arbitrarily small since then $x=b$ would be an accumulation point of zeros that would imply that $u_n \equiv 0$ by the existence theorem. That is, by the iterative solution (cf. Eqs. 8.36–8.38) $u(b)=0$ and $u'(b)=\varepsilon$ (an infinitesimal) would imply $u(x) \propto \varepsilon$, $\forall x$. But such a continuous function would be zero since, by the mean value theorem,

$$|f(x_0-\delta_1)-f(x_0+\delta_2)|=|f'(x_0)|\,(\delta_1+\delta_2)=\varepsilon(\delta_1+\delta_2)$$

which can be made arbitrarily small for a continuous range of values (i.e., of δ_1 and δ_2), which implies that $f(x)$ vanishes everywhere once $f(x_0)=0$. The integral on the r.h.s. of Eq. 8.85 cannot be infinite if $(b-a)$ is finite since all the quantities in the integrand are finite. Therefore we have

$$(\lambda_{n+1}-\lambda_n)>(\bar{\lambda}-\lambda_n)>0$$

so that $(\lambda_{n+1}-\lambda_n)$ is finite no matter how large n is.

If $(b-a)$ is finite, all the eigenvalues are separated by a finite distance and there can be no largest eigenvalue and this implies

$$\lim_{n\to\infty} \lambda_n=\infty \tag{8.86}$$

Hence the eigenvalues $\{\lambda_n\}$ form a denumerably infinite set. Also the corresponding eigenfunctions $\{u_n\}$ oscillate ever more rapidly as n increases. The eigenfunctions, $u_n(x)$, can never be degenerate (and linearly independent) in the one-dimensional case since, from Eq. 8.55, $W[u, v]=W_0 \exp[-\int b(x)/a(x)\,dx]$ so that if $u(a)=0=u(b)$ and $v(a)=0=v(b)$ then

$$W_0=W[u, v]|_{x=a}=\begin{vmatrix} 0 & 0 \\ u'(a) & v'(a) \end{vmatrix}=0$$

or

$$W[u, v]\equiv 0$$

The same result holds for the boundary conditions of Eq. 8.84.

We now give a proof that the eigenfunctions, $\{u_n(x)\}$, of the Sturm-Liouville equation form a complete set on the finite interval $(b-a)$. This result will allow us to expand any piecewise continuous function in the mean as

$$f(x)=\sum_n \alpha_n u_n(x)$$

$$\alpha_n=\int_a^b r(x)f(x)u_n^*(x)\,dx \tag{8.87}$$

We begin by obtaining the Sturm-Liouville equation,

$$Lu+\lambda ru=0$$

$$Lu=(pu')'+qu$$

as the solution to an extremum problem in the calculus of variations. Consider

$$I(u) \equiv \int_a^b (pu'u^{*\prime} - quu^*)\, dx \tag{8.88}$$

and make it an extremum subject to the constraint

$$J \equiv \int_a^b r\,|u|^2\, dx = 1 \tag{8.89}$$

We let

$$h[x, u(x), u^*(x), u'(x), u'^*(x)] = p\,|u'|^2 - q\,|u|^2 - \lambda r\,|u|^2$$

and use the Euler-Lagrange equation

$$\frac{\partial h}{\partial u^*} - \frac{d}{dx}\left(\frac{\partial h}{\partial u^{*\prime}}\right) = 0$$

$$\frac{\partial h}{\partial u^{*\prime}} = pu'$$

$$\frac{d}{dx}\left(\frac{\partial h}{\partial u^{*\prime}}\right) = (pu')'$$

$$\frac{\partial h}{\partial u^*} = -(q + \lambda r)u$$

$$(pu')' + (q + \lambda r)u = 0$$

We still take the boundary conditions to be

$$u(a) = 0 = u(b)$$

although the result is also valid for Eq. 8.84 (cf. Problem 8.9).

We vary u (*i.e.*, keep choosing new u's) until we obtain the lowest possible value for I. Let $\bar{u}$ be that function for which I is an absolute minimum. Since the Sturm-Liouville equation is the solution to this minimization problem, we know that

$$(p\bar{u}')' + (q + \bar{\lambda} r)\bar{u} = 0$$

Therefore we can write

$$\begin{aligned} I(\bar{u}) &= \int_a^b \left(p\left|\frac{d\bar{u}}{dx}\right|^2 - q\,|\bar{u}|^2\right) dx \\ &= \bar{u}^* p\frac{d\bar{u}}{dx}\Big|_a^b - \int_a^b \bar{u}^*\left[\frac{d}{dx}\left(p\frac{d\bar{u}}{dx}\right) + q\bar{u}\right] dx \\ &= \int_a^b r\bar{\lambda}\,|\bar{u}|^2\, dx = \bar{\lambda} \end{aligned}$$

where the outintegrated term vanishes by the boundary conditions. However, since this $\bar{\lambda}$ is the *smallest* possible value, we know that $\bar{\lambda}=\lambda_0$, the lowest eigenvalue, which we may take to be zero.

$$I(u_0)|_{\min}=\lambda_0=0$$

If we minimize $I(u)$ subject to the condition that this minimizing function be orthogonal to u_0, we obtain u_1 and so on.

Now if $f(x)$ is any once-differentiable function satisfying the boundary conditions, then $I(f)\geq 0$ since we have chosen $\lambda_0=0$. Furthermore, if $f_n(x)$ is such that

$$\int_a^b r(x)\,|f_n(x)|^2\,dx=1$$

$$\int_a^b f_n(x)r(x)u_m^*(x)\,dx=0 \qquad m=0,1,2,\ldots,n$$

and also satisfying the boundary conditions, then

$$I(f_n)\geq\lambda_{n+1}$$

as follows from the type argument leading to Eq. 3.41.

If the set $\{u_n(x)\}$ is to be complete, then we must be able to show that the mean square error between $f(x)$ and

$$\sum_{m=0}^{n}\alpha_m u_m(x)$$

with

$$\alpha_m\equiv\int_a^b u_m^*(x)f(x)r(x)\,dx \qquad \langle u_m\mid u_n\rangle=\delta_{mn}$$

can be made arbitrarily small as $n\to\infty$. That is

$$M_n\equiv\int_a^b\left|f(x)-\sum_{m=0}^{n}\alpha_m u_m(x)\right|^2 r(x)\,dx=\int_a^b|f(x)|^2\,r(x)\,dx-\sum_{m=0}^{n}|\alpha_m|^2$$

must be such that

$$\lim_{n\to\infty}M_n=0$$

If we take

$$f_n(x)\equiv\frac{1}{\sqrt{M_n}}\left[f(x)-\sum_{m=0}^{n}\alpha_m u_m(x)\right]$$

then we see by direct calculation that

$$\int_a^b f_n(x)u_m^*(x)r(x)\,dx=\begin{cases}0, & m\leq n\\ \alpha_m/\sqrt{M_n}, & m>n\end{cases}$$

and that

$$\int_a^b r(x)\,|f_n(x)|^2\,dx = 1$$

Therefore we obtain

$$\begin{aligned}
I(f_n) &= \int_a^b \left[p(x)\left|\frac{df_n}{dx}\right|^2 - q(x)\,|f_n|^2\right] dx \\
&= \frac{1}{M_n}\int_a^b \left[p\left|\frac{df}{dx}\right|^2 - q\,|f|^2\right] dx \\
&\quad - \frac{2\,\text{Re}}{M_n}\left\{\int_a^b \sum_{m=0}^{n} \alpha_m \left[p\frac{df}{dx}\frac{du_m^*}{dx} - qfu_m^*\right] dx\right\} \\
&\quad + \frac{1}{M_n}\sum_{s,m=0}^{n} \int_a^b \alpha_m\alpha_s^*\left[p\frac{du_m}{dx}\frac{du_s^*}{dx} - qu_m u_s^*\right] dx \\
&= \frac{1}{M_n}\left[I(f) - \sum_{m=0}^{n} |\alpha_m|^2\lambda_m\right] \geq \lambda_{n+1}
\end{aligned}$$

Since $\sum_{m=0}^{n} |\alpha_m|^2\lambda_m \geq 0$ because *all* $\lambda_m \geq 0$, we have

$$\frac{1}{M_n} I(f) \geq \lambda_{n+1}$$

or

$$M_n \leq \frac{I(f)}{\lambda_{n+1}}$$

However $I(f)$ is a positive (finite) quantity independent of n, so that as $n \to \infty$, $\lambda_{n+1} \to \infty$ and we have

$$\lim_{n\to\infty} M_n = 0$$

This establishes the completeness of these $\{u_n(x)\}$ for all twice-differentiable, square-integrable $f(x)$ satisfying the boundary conditions of these $\{u_n(x)\}$. We can easily extend this result to any piecewise continuous $f(x)$ by the type argument given at the end of Section 4.2. Of course we cannot make the final step to completeness for $\mathscr{L}_2(a, b)$ without Lebesgue integration theory.

THEOREM 8.8

The eigenfunctions, $\{u_n(x)\}$, of the Sturm-Liouville equation

$$\frac{d}{dx}\left[p(x)\frac{d}{dx}u_n(x)\right] + [q(x) + \lambda_n r(x)]u_n(x) = 0 \qquad x \in [a, b]$$

form a complete orthogonal set on range $[a, b]$ provided that $(b-a)$ is finite when $p(x)$, $q(x)$, and $r(x)$ are real continuous functions on $[a, b]$ and $p(x)>0$, $r(x)>0$ here and when the boundary conditions are $u(a)=0=u(b)$.

The case in which $p(x)$ vanishes at one or both of the end points of $[a, b]$ is often encountered in applications. The extension of Th. 8.8 to an important class of such differential equations is covered by the results of Problem 8.10. Also two special cases in which the range $[a, b]$ is infinite are treated in Section A8.3 and in Problem 8.10.

8.4 LINEAR DIFFERENTIAL OPERATORS†

a. Domain of a Linear Differential Operator

We take the form of our differential operator to be

$$L=a(x)\frac{d^2}{dx^2}+b(x)\frac{d}{dx}+c(x) \tag{8.90}$$

We shall assume that $a(x)$, $b(x)$, and $c(x)$ are *real* continuous functions and shall take the elements of S (our linear vector space) to be all complex-valued $u(x)\in\mathcal{L}_2[a, b]$ so that

$$\int_a^b |u(x)|^2\,dx<\infty$$

$$\langle u \mid v\rangle=\int_a^b u^*(x)v(x)\,dx \tag{8.91}$$

Sometimes we introduce a real nonnegative weight function $w(x)$ into our definition of inner product as

$$\langle u \mid v\rangle=\int_a^b w(x)u^*(x)v(x)\,dx \tag{8.92}$$

We must restrict L to operate on those u of S such that Lu exists and belongs to S.

However we must also specify some boundary conditions before the differential operator L is completely defined since we will want unique solutions to the problem

$$Lu=f$$

We express these boundary conditions as

$$\begin{aligned} B_1(u)&\equiv a_1u(a)+b_1u'(a)+c_1u(b)+d_1u'(b)=A\\ B_2(u)&\equiv a_2u(a)+b_2u'(a)+c_2u(b)+d_2u'(b)=B \end{aligned} \tag{8.93}$$

† Our discussion in Section 8.4 and in part of Section 8.6 is similar to that of the corresponding sections of Chapter 3 of Friedman's *Principles and Techniques of Applied Mathematics.*

where these conditions are assumed to be linearly independent, and the constants $a_1, a_2, \ldots, d_2$, A, and B are *real.* These include both homogeneous initial conditions and homogeneous boundary conditions.

Finally then the *domain* of L consists of those $u(x)$ such that $u''(x)$ is piecewise continuous, $B_1(u)=A$, $B_2(u)=B$, and $Lu \in S$. This domain is a *manifold* in S.

b. Adjoint and Hermitian Linear Differential Operators

As in Sec. 2.8 we define the adjoint of an operator by the relation

$$\langle v \mid Lu\rangle = \langle w \mid u\rangle \qquad w = L^{\dagger}v \tag{8.94}$$

where $L^{\dagger}$ is the adjoint of L. Here we must consider the boundary terms in the process of integration by parts. If, upon integrating by parts, the operators under the integral signs, L and $L^{\dagger}$, are identical, then L is said to be *formally self-adjoint.* If the two inner products are equal

$$\langle Lu \mid v\rangle = \langle u \mid Lv\rangle$$

and the boundary conditions on $u(x)$ and $v(x)$, which make the outintegrated terms vanish, are identical, then L is called *self-adjoint* or *hermitian.*

Example 8.8 The following simple case will illustrate these concepts.

$$L = x^2\frac{d^2}{dx^2} + 2x\frac{d}{dx} = \frac{d}{dx}\left(x^2\frac{d}{dx}\right) \qquad x \in [0, 1]$$

$$u(0) = 0 = u(1)$$

$$\begin{aligned}
\langle v \mid Lu\rangle &= \int_0^1 v^*(x)\frac{d}{dx}\left[x^2\frac{du(x)}{dx}\right]dx \\
&= v^*(x)x^2\frac{du}{dx}\bigg|_0^1 - \int_0^1 \frac{dv^*(x)}{dx}x^2\frac{du}{dx}\,dx \\
&= v^*(1)u'(1) + \int_0^1 \frac{d}{dx}\left[x^2\frac{dv^*}{dx}(x)\right]u(x)\,dx \\
&= v^*(1)u'(1) + \langle Lv \mid u\rangle
\end{aligned}$$

Therefore L is formally self-adjoint. If, furthermore, $v(0)=0=v(1)$, then L is self-adjoint. In other words, if the domain of L is that manifold of functions in S such that $u(0)=0=u(1)$, then L is self-adjoint or hermitian (in that manifold).

For our general second-order linear differential operator of Eq. 8.90 the *formal adjoint* is

$$L^{\dagger}v = \frac{d^2}{dx^2}(av) - \frac{d}{dx}(bv) + cv \tag{8.95}$$

since

$$\begin{aligned}
\langle L^{\dagger}v \mid u\rangle &= \int_a^b [(av^*)'' - (bv^*)' + cv^*]u\, dx \\
&= (av^*)'u\big|_a^b - \int_a^b (av^*)'u'\, dx - bv^*u\big|_a^b \\
&\quad + \int_a^b bv^*u'\, dx + \int_a^b cv^*u\, dx \\
&= (av^*)'u\big|_a^b - v^*au'\big|_a^b + \int_a^b v^*au''\, dx - bv^*u\big|_a^b \\
&\quad + \int_a^b v^*bu'\, dx + \int_a^b cv^*u\, dx \\
&= [(av^*)'u - v^*au' - bv^*u]\big|_a^b \\
&\quad + \int_a^b v^*(au'' + bu' + cu)\, dx \\
&= [(av^*)'u - v^*au' - bv^*u]\big|_a^b \\
&\quad + \langle v \mid Lu\rangle \equiv -P(v^*, u)\big|_a^b + \langle v \mid Lu\rangle
\end{aligned}$$

Therefore we have

$$\langle v \mid Lu\rangle - \langle L^{\dagger}v \mid u\rangle = P(v^*, u)\big|_a^b \tag{8.96}$$

where $P(v, u)$, defined as

$$P(v, u) \equiv avu' - u(av)' + bvu \tag{8.97}$$

is called the *bilinear concomitant* of v and u. It will be very important later when we construct Green's functions.

c. Self-Adjoint Second-Order Linear Differential Operators

We will see in Problem 8.18 that any linear second-order differential operator can be made formally self-adjoint since it can be written as

$$Lu = \frac{1}{w}(pu')' + qu \tag{8.98}$$

where $w(x)$ is the nonnegative weight function used in the inner product of Eq. 8.92.

We want to know when L is hermitian. If we use Eq. 8.96 and the inner product of Eq. 8.92, we obtain

$$\begin{aligned}
\langle v \mid Lu\rangle - \langle Lv \mid u\rangle &= P(v^*, u)\big|_a^b \\
&= [pv^*u' - upv^{*\prime} - up'v^* + p'v^*u]\big|_a^b \\
&= p(v^*u' - uv^{*\prime})\big|_a^b = \{p(x)W[v^*(x), u(x)]\}\big|_a^b
\end{aligned} \tag{8.99}$$

Therefore L will be self-adjoint if the concomitant P vanishes identically when u and v are in the same manifold. We now consider two special cases in which this is true. The most general conditions under which L will be self-adjoint subject to the homogeneous boundary conditions of Eq. 8.93 are derived in Section A8.2.

1. An *unmixed boundary* condition at $x=a$ is of the form

$$\alpha_1 u(a)+\beta_1 u'(a)=0$$

where again α_1 and β_1 are assumed real. If $u(x)$ satisfies unmixed boundary conditions at $x=a$ and at $x=b$, then $P(v^*, u)|_a^b=0$ since the relations

$$\begin{aligned} \alpha_1 u(a)+\beta_1 u'(a)&=0 \\ \alpha_2 u(b)+\beta_2 u'(b)&=0 \end{aligned} \tag{8.100}$$

imply that

$$\begin{aligned} P(v^*, u)|_a^b &= p(b)[v^*(b)u'(b)-v^{*\prime}(b)u(b)]-p(a)[v^*(a)u'(a)-v^{*\prime}(a)u(a)] \\ &= -p(b)\frac{1}{\beta_2}[\alpha_2 v^*(b)+\beta_2 v^{*\prime}(b)]u(b) \\ &\quad +p(a)\frac{1}{\beta_1}[\alpha_1 v^*(a)+\beta_1 v^{*\prime}(a)]u(a) \end{aligned}$$

Therefore, if u and v belong to the same manifold, $P(v^*, u)|_a^b=0$ and L is self-adjoint. Clearly the cases $\beta_1=0$ and/or $\beta_2=0$ yield the same results.

2. The boundary conditions are *periodic* if

$$u(a)=u(b) \qquad u'(a)=u'(b) \tag{8.101}$$

If we also have $p(a)=p(b)$, then

$$\begin{aligned} P(v^*, u)|_a^b &= p(a)[v^*(b)u'(b)-u(b)v^{*\prime}(b)-v^*(a)u'(a)+u(a)v^{*\prime}(a)] \\ &= p(a)\{u'(a)[v^*(b)-v^*(a)]+u(a)[v^{*\prime}(a)-v^{*\prime}(b)]\} \end{aligned}$$

so that L is again self-adjoint when operating on this manifold.

THEOREM 8.9

If L is formally self-adjoint and $u(x)$ and $v(x)$ [and therefore $u^*(x)$ and $v^*(x)$] are any primitive solutions to

$$Lu=0$$

then the concomitant $P(u, v)$ is a constant depending only on u and v. If, furthermore, $P(u, v)=0$ at any point x for which $p(x)\neq 0$, then $u(x)$ and $v(x)$ are linearly dependent.

PROOF

From Eq. 8.96 we see that

$$\langle v^* \mid Lu\rangle - \langle Lv^* \mid u\rangle \equiv \int_{x_1}^{x_2} (vLu - uLv)\, dx \equiv 0 = P[v(x), u(x)]\Big|_{x_1}^{x_2}$$

for all $x_1, x_2 \in [0, 1]$, so that

$$P(v, u)|_{x_1} = P(v, u)|_{x_2} = \text{const.}$$

If $P(v, u) = 0$ anywhere, it is identically zero everywhere on $[a, b]$ so that, from Eq. 8.99,

$$p(x)W[v(x), u(x)] = 0$$

If $p(x) \neq 0$, then $W(v, u) = 0$, which implies linear dependence here. Q.E.D.

8.5 GREEN'S FUNCTIONS

Let us begin a general introduction and orientation to the subject of Green's functions by classifying some of the types of boundary conditions that are often met in physical applications. Although we list these below for the three-dimensional case, these statements hold for two or more dimensions.

$$\begin{aligned} &\text{Dirichlet} && u(x, y, z)|_\Sigma = g(x, y, z) \\ &\text{Neumann} && \frac{\partial u}{\partial n}(x, y, z)\Big|_\Sigma = g(x, y, z) \\ &\text{Mixed} && \left[u(x, y, z) + h(x, y, z)\frac{\partial u}{\partial n}(x, y, z)\right]\Big|_\Sigma = g(x, y, z) \end{aligned} \qquad \textbf{(8.102)}$$

Here $g(x, y, z)$ and $h(x, y, z)$ are given functions defined on the closed surface Σ (or closed curve C in two dimensions) enclosing the region of definition of the problem, and n stands for the normal to this surface. In the one-dimensional case the analogues of these three cases are, respectively,

$$\begin{aligned} u(a) &= A & u(b) &= B \\ u'(a) &= A & u'(b) &= B \\ \alpha_1 u(a) + \beta_1 u'(a) &= A & \alpha_2 u(b) + \beta_2 u'(b) &= B \end{aligned}$$

If L is a linear differential operator, then the general problem we will study is to find a $u(x, y, z)$ such that

$$Lu = f(x, y, z)$$

where $f(x, y, z)$ is a given function defined in volume V enclosed by the surface Σ

and where $u(x, y, z)$ satisfies one of the boundary conditions on Σ listed in Eq. 8.102. Let us assume Dirichlet boundary conditions although the following discussion holds for any of the boundary conditions of Eq. 8.102. The general form of the problem is then

$$\begin{aligned} Lu &= f(x, y, z) \\ u|_\Sigma &= g(x, y, z) \end{aligned} \qquad \textbf{(8.103)}$$

We write

$$u(x, y, z) \equiv u_1(x, y, z) + u_2(x, y, z)$$

where u_1 is defined to satisfy

$$\begin{aligned} Lu_1 &= f \\ u_1|_\Sigma &= 0 \end{aligned} \qquad \textbf{(8.104)}$$

and u_2 to satisfy

$$\begin{aligned} Lu_2 &= 0 \\ u_2|_\Sigma &= g \end{aligned} \qquad \textbf{(8.105)}$$

(Recall that a special case of the problem posed in Eq. 8.105 is the Dirichlet problem treated in Section 7.11c.) We will show in Section 8.11 that once a solution to Eq. 8.104 has been found we can then obtain the solution to Eq. 8.105 and hence solve the general problem of Eq. 8.103. We will in fact construct a single function, the Green's function, which will allow us to write down the solution to Eq. 8.104.

In this Section we shall begin with the one-dimensional case,

$$\begin{aligned} Lu &= f \qquad \textbf{(8.106)} \\ B_1(u) &= A \\ B_2(u) &= B \end{aligned}$$

where L is given by Eq. 8.90 or Eq. 8.98 and $B_1(u)$ and $B_2(u)$ by Eq. 8.93. Again let $u = u_1 + u_2$ such that

$$Lu_1 = f \qquad B_1(u_1) = 0 = B_2(u_1) \qquad \textbf{(8.107)}$$

$$Lu_2 = 0 \qquad B_1(u_2) = A,\ B_2(u_2) = B \qquad \textbf{(8.108)}$$

We will see that once we have the two linearly independent primitive solutions to

$$Lv = 0$$

whose existence we discussed in Sec. 8.2, then we will be able to construct the solutions to Eqs. 8.107 and 8.108 and hence solve the general problem of Eq. 8.106. The final solution will be reduced to quadrature.

We will return to the three-dimensional case in Section 8.11.

a. Inverse of a Differential Operator

Example 8.9 We now consider a simple case of constructing the inverse of a differential operator. We can obtain a solution to

$$\frac{d^2u(x)}{dx^2}=f(x) \qquad x\in[0, 1]$$

as

$$u(x)=\int_0^1 g(x, t)f(t)\, dt$$

provided we can find a $g(x, t)$ such that

$$\frac{d^2}{dx^2}\, g(x, t)=\delta(x-t)$$

If we use Eq. 8.18 we have

$$\frac{d}{dx}\, g(x, t)=H(x-t)+\alpha(t)$$

and then

$$g(x, t)=(x-t)H(x-t)+x\alpha(t)+\beta(t)$$

where $\alpha(t)$ and $\beta(t)$ are arbitrary functions so far. Let us now impose some boundary conditions, say the periodic ones

$$u(0)=0=u(1)$$

Since $u(x)$ is given as

$$u(x)=\int_0^x (x-t)f(t)\, dt+\int_0^1 [x\alpha(t)+\beta(t)]f(t)\, dt$$

we have

$$u(0)\equiv\int_0^1 \beta(t)f(t)\, dt=0$$

Since this must hold for *all* $f(t)$, we conclude that $\beta(t)=0$. Similarly, since

$$u(1)=\int_0^1 (1-t)f(t)\, dt+\int_0^1 \alpha(t)f(t)\, dt=\int_0^1 [1-t+\alpha(t)]f(t)\, dt=0$$

we conclude that

$$\alpha(t)=t-1$$

so that the solution can be written as

$$u(x)=\int_0^x (x-t)f(t)\, dt+x\int_0^1 (t-1)f(t)\, dt$$

Notice that the Green's function,

$$g(x, t)=(x-t)H(x-t)+x(t-1) \qquad 0\leq x\leq 1 \qquad 0<t<1$$

satisfies the same boundary conditions as $u(x)$,

$$g(0, t)=0=g(1, t) \qquad \forall t\in(0, 1)$$

In fact for homogeneous boundary conditions we always construct the Green's function, $g(x, t)$ to satisfy the same boundary conditions as those prescribed for the solution to the differential equation.

b. An Existence Theorem

If we can solve

$$L_x g(x, t)=\delta(x-t) \tag{8.109}$$

then a solution to $Lu=f$ is given as

$$u(x)=\int_a^b g(x, t)f(t)\,dt \tag{8.110}$$

We consider the operator

$$L=\frac{d}{dx}\left[p(x)\frac{d}{dx}\right]+q(x)$$

We assume $p(x)$ continuous with a piecewise continuous derivative and $q(x)$ piecewise continuous on $[a, b]$ and $p(x)\neq 0$ anywhere here. First let $q(x)\equiv 0$ and denote the corresponding Green's function by $g_0(x, t)$. Then we have

$$\frac{d}{dx}\left[p(x)\frac{d}{dx}\,g_0(x, t)\right]=\delta(x-t)$$

We can integrate this as

$$p(x)\frac{d}{dx}\,g_0(x, t)=H(x-t)+\alpha(t)$$

so that

$$\frac{d}{dx}\,g_0(x, t)=\frac{1}{p(x)}\,H(x-t)+\frac{1}{p(x)}\,\alpha(t)$$

since $p(x)\neq 0$ here. Finally we obtain

$$g_0(x, t)=H(x-t)\int_t^x\frac{d\zeta}{p(\zeta)}+\alpha(t)\int_a^x\frac{d\zeta}{p(\zeta)}+\beta(t) \tag{8.111}$$

This $g_0(x, t)$ is a continuous function of x and $dg_0(x, t)/dx$ is continuous except at $x=t$ where it has a discontinuity of $1/p(t)$.

Now set $g(x, t)=g_0(x, t)+k(x, t)$ so that

$$L_x g \equiv L_x g_0 + L_x k = \delta(x-t) = \frac{d}{dx}\left[p\frac{dg_0}{dx}\right]+qg_0+\frac{d}{dx}\left[p\frac{dk}{dx}\right]+qk$$

or

$$L_x k = -q(x)g_0(x, t) \tag{8.112}$$

We now prove that $g(x, t)$ has the same general properties as $g_0(x, t)$. From Eq. 8.111 and Th. 8.1 we know that the solution to Eq. 8.112, $k(x, t)$, will be a continuous function of x for all t, where the c_1 and c_2 of Eq. 8.35 now become two arbitrary functions of t, say $\alpha(t)$ and $\beta(t)$. Furthermore since

$$p(x)\frac{d}{dx}k(x, t)\bigg|_{x_1}^{x_2} = -\int_{x_1}^{x_2} q(x')[g_0(x', t)+k(x', t)]\,dx'$$

we see that $p(x)\,dk(x, t)/dx$ is also a continuous function of x for all t. Finally, from Eq. 8.112 itself,

$$\frac{d}{dx}\left[p(x)\frac{d}{dx}k(x, t)\right] = -q(x)[g_0(x, t)+k(x, t)]$$

we see that $d/dx[p(x)\,dk(x, t)/dx]$ is piecewise continuous. Therefore the function

$$g(x, t) \equiv g_0(x, t)+k(x, t)$$

is continuous, its derivative, $dg(x, t)/dx$, has a discontinuity of $1/p(t)$ at $x=t$, and it satisfies Eq. 8.109.

THEOREM 8.10

If $p(x)$ is a positive, continuous, piecewise differentiable function on $x \in [a, b]$ while $q(x)$ is piecewise continouus there, then there exists a continuous function $g(x, t)$ such that

$$\frac{d}{dx}\left[p(x)\frac{d}{dx}g(x, t)\right]+q(x)g(x, t)=0$$

everywhere on $x \in [a, b]$ except at $x=t$ where its derivative satisfies the discontinuity relation

$$\frac{d}{dx}g(x, t)\bigg|_{x=t^+} - \frac{d}{dx}g(x, t)\bigg|_{x=t^-} = \frac{1}{p(t)}$$

This establishes the existence of a $g(x, t)$ such that Eq. 8.110 is a solution to Eq. 8.109. For homogeneous boundary conditions we must now make $g(x, t)$ satisfy identically the boundary conditions of the differential equation so that the solution, Eq. 8.110, will also.

8.6 VARIOUS BOUNDARY CONDITIONS

Since the functions $a(x)$, $b(x)$, and $c(x)$ of Eq. 8.90 and the constants $a_1, a_2, \ldots, d_2$, A, and B of Eq. 8.93 are assumed to be *real*, we may always take our solutions $u(x)$ to

$$Lu = f$$
$$B_1(u) = A$$
$$B_2(u) = B$$

to be real once we restrict f to be real. Throughout this section and in Section 8.8 we will assume that we have chosen real solutions to all differential equations considered. We take the differential operator, L, to be of the form

$$Lu = \frac{d}{dx}\left[p(x)\frac{d}{dx}u(x)\right] + q(x)u(x) \qquad x \in [a, b] \tag{8.113}$$

a. Unmixed Homogeneous Boundary Conditions

We begin by restricting our boundary conditions to be *unmixed* and homogeneous (cf. Eq. 8.100) so that

$$\begin{aligned} B_1(u) &\equiv \alpha_1 u(a) + \beta_1 u'(a) = 0 \\ B_2(u) &\equiv \alpha_2 u(b) + \beta_2 u'(b) = 0 \end{aligned} \tag{8.114}$$

We seek a solution to

$$\begin{aligned} Lu = f(x) \qquad x \in [a, b] \\ B_1(u) = 0 = B_2(u) \end{aligned} \tag{8.115}$$

With the two primitive linearly independent solutions $v_1(x)$ and $v_2(x)$ of

$$Lv = 0$$

we form suitable linear combinations $w_1(x)$ and $w_2(x)$ such that $w_1(x)$ satisfies

$$B_1(w_1) = 0$$

while $w_2(x)$ satisfies

$$B_2(w_2) = 0$$

Our goal is to construct a $g(x, t)$ such that

$$L_x g(x, t) = \delta(x - t)$$
$$B_1(g) = 0 = B_2(g)$$

We proceed by a very simple and direct construction argument based on the results of Th. 8.10. As a start let us write

$$g(x, t)=\begin{cases} w_1(x) & x<t \\ w_2(x) & x>t \end{cases}$$

which will satisfy the differential equation for $x \neq t$ and the proper boundary conditions. However this $g(x, t)$ is not continuous at $x=t$ nor does $dg(x, t)/dx$ satisfy the proper jump condition at $x=t$. We make it continuous as

$$g(x, t)=\begin{cases} w_1(x)w_2(t) & x \leq t \\ w_2(x)w_1(t) & x \geq t \end{cases}$$

We see by direct calculation that

$$g'(t^+, t)-g'(t^-, t) \equiv w_2'(t)w_1(t)-w_1'(t)w_3(t)=\frac{P(w_1, w_2)}{p(t)}$$

while it should be just $1/p(t)$. Therefore we finally obtain

$$\begin{aligned} g(x, t) &= \begin{cases} w_1(x)w_2(t)/P(w_1, w_2) & x \leq t \\ w_2(x)w_1(t)/P(w_1, w_2) & x \geq t \end{cases} \\ &= \frac{w_1(x)w_2(t)H(t-x)+w_2(x)w_1(t)H(x-t)}{P(w_1, w_2)} \end{aligned} \tag{8.116}$$

However from Th. 8.9 we know that

$$P(w_1, w_2)=p(t)W[w_1(t), w_2(t)]=\text{const.}$$

since L is self-adjoint. Of course this expression for $g(x, t)$ breaks down if $P(w_1, w_2)=0$. From Th. 8.9 we see that this will be true if and only if $w_1(x)$ and $w_2(x)$ are linearly dependent so that $W[w_1(t), w_2(t)] \equiv 0$. That is, $w_1(x)$ and $w_2(x)$ can satisfy the relation

$$c_1w_1(x)+c_2w_2(x)=0$$

for nonzero c_1 and c_2 if and only if

$$B_1(w_2)=0$$

and

$$B_2(w_1)=0$$

This would require that there exist a nontrivial function $w(x)$ such that

$$\begin{aligned} Lw &= 0 \\ B_1(w)=0 &= B_2(w) \end{aligned} \tag{8.117}$$

In this case $w(x)$ would be a nontrivial eigenfunction of L corresponding to the eigenvalue, $\lambda=0$. Therefore, if there is no nontrivial solution to Eq. 8.117, then the unique solution to the problem of Eq. 8.114 is

$$u(x)=\int_a^b g(x, t)f(t)\, dt \tag{8.118}$$

where $g(x, t)$ is given in Eq. 8.116.

We will return in Section 8.6c to the case in which a nontrivial solution to Eq. 8.117 exists.

b. Unmixed Inhomogeneous Boundary Conditions

We still restrict ourselves to the case in which the boundary conditions are unmixed (cf. Eq. 8.114) but now inhomogeneous,

$$\begin{aligned} B_1(u) &= A \\ B_2(u) &= B \end{aligned} \tag{8.119}$$

As indicated in Section 8.5 we obtain the solution to

$$\begin{aligned} Lu &= f \\ B_1(u) &= A \\ B_2(u) &= B \end{aligned} \tag{8.120}$$

by decomposing $u(x)$ as

$$u(x) = u_1(x) + u_2(x)$$

where

$$\begin{aligned} Lu_1 &= f \\ B_1(u_1) &= 0 = B_2(u_1) \end{aligned} \tag{8.121}$$

and

$$\begin{aligned} Lu_2 &= 0 \\ B_1(u_2) &= A \\ B_2(u_2) &= B \end{aligned} \tag{8.122}$$

In the previous section we have seen how to construct the solution to Eq. 8.121 provided there is no nontrivial solution to Eq. 8.117. If we again choose two solutions to

$$Lw_j = 0 \qquad j = 1, 2$$

such that

$$\begin{aligned} B_1(w_1) &= 0 \\ B_2(w_2) &= 0 \end{aligned}$$

and write

$$u_2(x) = \alpha w_1(x) + \beta w_2(x)$$

then we must determine α and β in order to satisfy

$$\begin{aligned} B_1(u_2) &\equiv \beta B_1(w_2) = A \\ B_2(u_2) &\equiv \alpha B_2(w_1) = B \end{aligned}$$

This will always be possible provided that neither $B_1(w_2)$ nor $B_2(w_1)$ is zero. Again neither of these can vanish if Eq. 8.117 has no nontrivial solution in which case

$$u_2(x) = \frac{Bw_1(x)}{[\alpha_2 w_1(b) + \beta_2 w_1'(b)]} + \frac{Aw_2(x)}{[\alpha_1 w_2(a) + \beta_1 w_2'(a)]} \tag{8.123}$$

c. Case of Nontrivial Eigenfunction with Unmixed Boundary Conditions

We now consider the simplest situation in which Eq. 8.117 has a nontrivial solution by assuming the boundary conditions are still unmixed. We shall prove that the problem

$$\begin{gathered} Lu = f \\ B_1(u) = 0 = B_2(u) \end{gathered} \tag{8.124}$$

has a (nonunique) solution when there exists a nontrivial $w(x)$ such that

$$\begin{gathered} Lw = 0 \\ B_1(w) = 0 = B_2(w) \end{gathered} \tag{8.125}$$

if and only if

$$\langle f \mid w \rangle \equiv \int_a^b f(x) w(x)\, dx = 0 \tag{8.126}$$

It is evident that Eq. 8.126 is a necessary condition for Eq. 8.124 to have a solution when Eq. 8.125 holds since the self-adjointness of L implies that

$$0 = \langle u \mid Lw \rangle - \langle Lu \mid w \rangle = -\langle f \mid w \rangle$$

We will establish that Eq. 8.126 is also sufficient by exhibiting a solution to Eq. 8.124 when Eqs. 8.125 and 8.126 are satisfied. Let $v(x)$ be any primitive solution to

$$Lv = 0$$

as long as $v(x)$ is linearly independent of $w(x)$. Theorem 8.4 guarantees the existence of such a solution.

We begin by constructing the solution to

$$\begin{gathered} Lu = f \\ u(a) = 0 = u'(a) \end{gathered} \tag{8.127}$$

A Green's function that satisfies these boundary conditions is

$$g(x, t) = \begin{cases} 0 & x < t \\ \alpha(t) w(x) + \beta(t) v(x) & x > t \end{cases}$$

The discontinuity condition on dg/dx implies that

$$\alpha(t)w'(t)+\beta(t)v'(t)=\frac{1}{p(t)}$$

while the continuity condition at $x=t$ requires

$$\alpha(t)w(t)+\beta(t)v(t)=0$$

so that

$$\alpha(t)=-\frac{\beta(t)v(t)}{w(t)}=\frac{v(t)}{P(v,\,w)}$$

$$\beta(t)=-\frac{w(t)}{P(v,\,w)}$$

Since $v(t)$ and $w(t)$ are linearly independent, $P(v, w)$ cannot vanish and we have

$$g(x,\,t)=\frac{1}{P(v,\,w)}[w(x)v(t)-v(x)w(t)]H(x-t)$$

The solution to Eq. 8.127 is just

$$u(x)\equiv\int_a^b g(x,\,t)f(t)\,dt=\frac{w(x)}{P}\int_a^x v(t)f(t)\,dt-\frac{v(x)}{P}\int_a^x w(t)f(t)\,dt \qquad \textbf{(8.128)}$$

However this $u(x)$ will also satisfy the boundary conditions of our original problem, Eq. 8.124, since, once $u(a)=0=u'(a)$, we have

$$B_1(u)\equiv\alpha_1 u(a)+\beta_1 u'(a)=0$$

We see that

$$u(b)=\frac{w(b)}{P}\int_a^b v(t)f(t)\,dt$$

since $\langle w\mid f\rangle=0$ by assumption and

$$u'(b)=\frac{w'(b)}{P}\int_a^b v(t)f(t)\,dt$$

for the same reason. Therefore, as claimed, we have

$$B_2(u)=\frac{B_2(w)}{P(v,\,w)}\int_a^b v(t)f(t)\,dt=0$$

d. The General Case

We begin with two linearly independent primitive solutions $v_1(x)$ and $v_2(x)$ to $Lv=0$ so that

$$W[v_1(x),\,v_2(x)]\equiv\frac{P(v_1,\,v_2)}{p(x)}\neq 0 \qquad \textbf{(8.129)}$$

We see (cf. Eq. 8.116) that the most general solution to the differential equation

$$Lu \equiv (pu')' + qu = f(x) \qquad x \in [a, b] \tag{8.130}$$

can be written

$$u(x) = \alpha v_1(x) + \beta v_2(x) + \frac{v_1(x)}{P(v_1, v_2)} \int_x^b v_2(t) f(t)\, dt + \frac{v_2(x)}{P(v_1, v_2)} \int_a^x v_1(t) f(x)\, dt \tag{8.131}$$

where no boundary conditions have yet been imposed. This can also be verified by substituting Eq. 8.131 into Eq. 8.130. We now require that

$$\begin{aligned} B_1(u) &= A \\ B_2(u) &= B \end{aligned} \tag{8.132}$$

where $B_1(u)$ and $B_2(u)$ are given in Eq. 8.93 subject to the requirement of self-adjointness (cf. Section A8.2, Eq. A8.16)

$$p(a)(d_1 c_2 - d_2 c_1) = p(b)(a_2 b_1 - a_1 b_2) \tag{8.133}$$

Since we assume $p(x) > 0$ for all $x \in [a, b]$, this self-adjointness of L implies that $P(v_1, v_2) = \text{const} \neq 0$. We must now choose α and β of Eq. 8.131 in such a way as to make Eq. 8.132 true for given values of A and B. The set of linear inhomogeneous algebraic equations for α and β is

$$\begin{aligned} \alpha B_1(v_1) + \beta B_1(v_2) &= A - [a_1 v_1(a) + b_1 v_1'(a)] \frac{\langle v_2 | f \rangle}{P(v_1, v_2)} - [c_1 v_2(b) + d_1 v_2'(b)] \frac{\langle v_1 | f \rangle}{P(v_1, v_2)} \\ \alpha B_2(v_1) + \beta B_2(v_2) &= B - [a_2 v_1(a) + b_2 v_1'(a)] \frac{\langle v_2 | f \rangle}{P(v_1, v_2)} - [c_2 v_2(b) + d_2 v_2'(b)] \frac{\langle v_1 | f \rangle}{P(v_1, v_2)} \end{aligned} \tag{8.134}$$

This system will have a unique solution for α and β provided

$$\begin{vmatrix} B_1(v_1) & B_1(v_2) \\ B_2(v_1) & B_2(v_2) \end{vmatrix} \neq 0$$

If this determinant does vanish, then from Section 2.1b we know that its columns are linearly dependent so that there must exist nonzero constants c_1 and c_2 such that

$$\begin{aligned} c_1 B_1(v_1) + c_2 B_1(v_2) &= 0 \\ c_1 B_2(v_1) + c_2 B_2(v_2) &= 0 \end{aligned}$$

in which case there would exist a nontrivial eigenfunction $w(x)$ such that

$$\begin{aligned} Lw &= 0 \\ B_1(w) &= 0 = B_2(w) \end{aligned} \tag{8.135}$$

Therefore, as we have previously seen, Eqs. 8.130 and 8.132 have a solution that is unique provided that there is no nontrivial solution to Eq. 8.135.

If Eq. 8.135 has just one nontrivial solution, for example, $v_1(x)$, such that

$$B_1(v_1)=0=B_2(v_1)$$

while $B_1(v_2)$ and $B_2(v_2)$ are not both zero, then the necessary and sufficient condition that Eq. 8.134 should have a solution is that

$$B_2(v_2)\left\{A-[a_1v_1(a)+b_1v_1'(a)]\frac{\langle v_2\,|\,f\rangle}{P(v_1,\,v_2)}-[c_1v_2(b)+d_1v_2'(b)]\frac{\langle v_1\,|\,f\rangle}{P(v_1,\,v_2)}\right\}$$
$$=B_1(v_2)\left\{B-[a_2v_1(a)+b_2v_1'(a)]\frac{\langle v_2\,|\,f\rangle}{P(v_1,\,v_2)}-[c_2v_2(b)+d_2v_2'(b)]\frac{\langle v_1\,|\,f\rangle}{P(v_1,\,v_2)}\right\} \quad \textbf{(8.136)}$$

By use of Eqs. 8.129, 8.133, and $B_1(v_1)=0=B_2(v_1)$ this may be reduced to

$$B_2(v_2)\left\{A-[c_1v_2(b)+d_1\,v_2'(b)]\frac{\langle v_1\,|\,f\rangle}{P(v_1,\,v_2)}\right\}$$
$$=B_1(v_2)\left\{B-[c_2v_2(b)+d_2\,v_2'(b)]\frac{\langle v_1\,|\,f\rangle}{P(v_1,\,v_2)}\right\} \quad \textbf{(8.137)}$$

With considerable algebraic labor Eq. 8.137 can be put into one of the following equivalent forms by use of the homogeneous boundary conditions satisfied by $v_1(x)$ and by $v_2(x)$. (The reader interested in these details should consult Section 11.31 of Ince's classic text.)

$$\begin{aligned}
(c_1B-c_2A)p(a)v_1'(a)+(a_1B-a_2A)p(b)v_1'(b)+(a_1c_2-c_1a_2)\langle v_1\,|\,f\rangle&=0\\
(d_1B-d_2A)p(a)v_1(a)+(b_1B-b_2A)p(b)v_1(b)+(d_1b_2-b_1d_2)\langle v_1\,|\,f\rangle&=0\\
(c_1B-c_2A)p(a)v_1(a)-(b_1B-b_2A)p(b)v_1'(b)+(c_1b_2-b_1c_2)\langle v_1\,|\,f\rangle&=0\\
(d_1B-d_2A)p(a)v_1'(a)-(a_1B-a_2A)p(b)v_1(b)+(a_1d_2-d_1a_2)\langle v_1\,|\,f\rangle&=0
\end{aligned} \quad \textbf{(8.138)}$$

That is, a necessary and sufficient condition that Eqs. 8.130 and 8.132 have a (nonunique) solution when Eq. 8.135 has just one linearly independent nontrivial solution is that one of the relations of Eq. 8.138 having a nonvanishing coefficient of $\langle v_1\,|\,f\rangle$ be satisfied.

If $B_1(v_2)=0=B_2(v_2)$ in addition to $B_1(v_1)=0=B_2(v_2)$, then the necessary and sufficient condition that Eqs. 8.130 and 8.132 have a (nonunique) solution is the vanishing of the right side of Eq. 8.134. This requirement can be put into the equivalent form

$$\begin{aligned}
(a_1B-a_2A)p(a)v_2(a)+(b_1B-b_2A)p(a)v_2'(a)+(a_1b_2-b_1a_2)\langle v_2\,|\,f\rangle&=0\\
(c_2B-c_1A)p(b)v_1(b)+(d_2A-d_1B)p(b)v_1'(a)+(c_1d_2-d_1c_2)\langle v_1\,|\,f\rangle&=0
\end{aligned} \quad \textbf{(8.139)}$$

In the special case of homogeneous self-adjoint boundary conditions (i.e.,

$A=B=0$ in Eq. 8.132) the necessary and sufficient condition simplifies considerably. Since L is self-adjoint it follows from $L(v_j)=0$, $B(v_j)=0$, $j=1$ and/or 2, that

$$0=\langle u \mid Lv_j\rangle-\langle Lu \mid v_j\rangle=-\langle f \mid v_j\rangle, \; j=1 \text{ and/or } 2$$

or

$$\langle v_j \mid f\rangle=0, \; j=1 \text{ and/or } 2 \tag{8.140}$$

so that Eq. 8.140 is a necessary condition for the existence of a (nonunique) solution to

$$\begin{gathered} L[u]=f \\ B_1(u)=0=B_2(u) \end{gathered} \tag{8.141}$$

when

$$\begin{gathered} L[v_j]=0 \\ B_1(v_j)=0=B_2(v_j) \qquad j=1 \text{ and/or } 2 \end{gathered} \tag{8.142}$$

Conversely, if $\langle v_1 \mid f\rangle=0$ when $B_1(v_1)=0=B_2(v_1)$ but $B_1(w_2)$ and $B_2(w_2)$ are not both zero, then Eq. 8.137 will be satisfied (with $A=B=0$). Also, if $B_1(v_1)=B_2(v_2)=0=B_1(v_2)=B_2(v_2)$, then the right side of Eq. 8.134 vanishes so that in either of these cases Eq. 8.140 is a sufficient condition for the system of Eqs. 8.141 and 8.142 to have a (nonunique) solution. Therefore Eq. 8.140 is the necessary and sufficient condition for the existence of a (nonunique) solution to the system (8.141)–(8.142). Of course this same result can be seen immediately from Eqs. 8.138 and 8.139.

THEOREM 8.11

With the requirements that $q(x)$ and $f(x)$ be piecewise continuous while $p(x)$ is differentiable and positive, the differential equation

$$\frac{d}{dx}\left[p(x)\frac{d}{dx}u(x)\right]+q(x)u(x)=f(x) \qquad x\in[a, b]$$

subject to the linearly independent boundary conditions

$$\begin{gathered} B_1(u)\equiv a_1u(a)+b_1u'(a)+c_1u(b)+d_1u'(b)=A \\ B_2(u)\equiv a_2u(a)+b_2u'(a)+c_2u(b)+d_2u'(b)=B \\ p(a)(d_1c_2-d_2c_1)=p(b)(a_2b_1-a_1b_2) \end{gathered}$$

possesses a solution that is unique (given by Eqs. 8.131 and 8.134) provided there exists no nontrivial solution to the eigenvalue problem

$$\begin{gathered} Lw=0 \\ B_1(w)=0=B_2(w) \end{gathered}$$

If there exists just one nontrivial solution to this eigenvalue problem, for example, $v_1(x)$, then a necessary and sufficient condition for a (nonunique) $u(x)$ to exist is that

$$B_2(v_2)\left\{A-[c_1v_2(b)+d_1v_2'(b)]\frac{\langle v_1 \mid f\rangle}{P(v_1, v_2)}\right\}=B_1(v_2)\left\{B-[c_2v_2(b)+d_2v_2'(b)]\frac{\langle v_1 \mid f\rangle}{P(v_1, v_2)}\right\}$$

or equivalently that one of Eqs. 8.138 be satisfied. If there are two linearly independent solutions, for example, $v_1(x)$ and $v_2(x)$, to the eigenvalue problem, then the necessary and sufficient condition for a (nonunique) $u(x)$ to exist is that the relations

$$AP(v_1, v_2)=[a_1v_1(a)+b_1v_1'(a)]\langle v_2 \mid f\rangle+[c_1v_2(b)+d_1v_2'(b)]\langle v_1 \mid f\rangle$$
$$BP(v_1, v_2)=[a_2v_1(a)+b_2v_1'(a)]\langle v_2 \mid f\rangle+[c_2v_2(b)+d_2v_2'(b)]\langle v_1 \mid f\rangle$$

or equivalently Eqs. 8.139, both be satisfied. In particular if $A=B=0$ so that the boundary conditions are homogeneous, then a (nonunique) $u(x)$ will exist if and only if

$$\langle v_j \mid f\rangle=0 \qquad j=1 \text{ and/or } 2$$

Notice that this result is very similar to Th. 2.14, which was proven for operators with closed range.

Now that we have established Th. 8.11, we see that as a practical matter one solves a problem of the type posed in Eqs. 8.130 and 8.132 by using any two linearly independent solutions $v_1(x)$ and $v_2(x)$ of $Lv=0$ in Eq. 8.131 and then determining the constants α and β from Eq. 8.134. The proofs and discussions of the special cases considered in Sections 8.6a–8.6c need no longer concern us.

As an aside for reference later we now point out an interesting and useful method of converting a linear differential equation subject to homogeneous boundary conditions into an integral equation that contains the homogeneous boundary conditions of the problem. If we have the differential equation

$$Lu+\lambda u=f \qquad x\in[a, b] \tag{8.143}$$

subject to some homogeneous boundary conditions, and if $g(x, t)$ satisfies these boundary conditions, and

$$L_x g(x, t)=\delta(x-t)$$

then Eq. 8.143 and its boundary conditions are completely equivalent to

$$u(x)=-\lambda\int_a^b g(x, t)u(t)\, dt+\int_a^b g(x, t)f(t)\, dt \tag{8.144}$$

8.7 EIGENFUNCTION EXPANSION OF THE GREEN'S FUNCTION

We now discuss a very useful representation of the Green's function in terms of eigenfunctions of the homogeneous equation

$$Lu_j + \lambda_j u_j = 0$$

We assume here that these eigenfunctions, subject to some set of homogeneous boundary conditions, form a complete orthonormal set (cf., for example, the Sturm-Liouville eigenfunctions of Section 8.3). The ştatement of completeness for this orthonormal set $\{u_j(x)\}$ is that in the mean any square-integrable function $h(x)$ can be expanded as

$$h(x) = \sum_j c_j u_j(x) = \sum_j \int u_j^*(t) h(t)\, dt\, u_j(x) = \int \delta(x-t) h(t)\, dt$$

so that completeness can be expressed as

$$\delta(x-t) = \sum_j u_j^*(t) u_j(x)$$

By the completeness of the $\{u_j(x)\}$ we can write

$$g(x, t) = \sum_j \alpha_j(t) u_j(x)$$

where this Green's function for Eq. 8.143 must satisfy

$$(L_x + \lambda) g(x, t) = \delta(x-t)$$

or

$$\sum_j \alpha_j(t)(-\lambda_j + \lambda) u_j(x) = \sum_j u_j^*(t) u_j(x)$$

so that

$$(\lambda - \lambda_j)\alpha_j(t) = u_j^*(t)$$

If λ is not an eigenvalue (i.e., not one of the $\{\lambda_j\}$), then we have

$$g(x, t) = \sum_j \frac{u_j^*(t) u_j(x)}{(\lambda - \lambda_j)} \tag{8.145}$$

If $\lambda = \lambda_m$, one of the eigenvalues, then Eq. 8.143 can have a solution if and only if

$$\int u_m^*(t) f(t)\, dt = 0$$

in which case

$$g_m(x, t) = \sum_{j \neq m} \frac{u_j^*(t) u_j(x)}{(\lambda_m - \lambda_j)}$$

8.8 A HEURISTIC DISCUSSION OF GREEN'S FUNCTIONS

In an attempt to simplify the understanding of Green's functions and their construction, we give the following heuristic discussion that may help unify the subject. Let us begin by *defining* a Green's function, $g(x, t)$, as a function such that:

i. $g(x, t)$ is a continuous function of x for all t and satisfies the *homogeneous* boundary conditions of the problem at $x=a$ and at $x=b$. [i.e., $a \leq x \leq b$, $a<t<b$].
ii. $g(x, t)$ has continuous first and second derivatives with respect to x except at $x=t$, where

$$g'(t^+, t)-g'(t^-, t)=\frac{1}{p(t)}$$

iii. $L_x g=0$, except at $x=t$.

We now show that

$$u(x)=\int_a^b g(x, t)f(t)\,dt \tag{8.146}$$

is a solution to the inhomogeneous differential equation

$$Lu \equiv (pu')'+qu=f(x)$$

From Eq. 8.146 we obtain

$$u'(x)=\int_a^b g'f\,dt$$

and

$$u''(x)=\int_a^{x-\delta} g''f\,dt+\int_{x-\delta}^{x+\delta} g''f\,dt+\int_{x+\delta}^{b} g''f\,dt$$

But in the limit that $\delta \to 0^+$ integration by parts yields for any differentiable $f(x)$

$$\int_{x-\delta}^{x+\delta} g''f\,dt=[g'(x, x+\delta)-g'(x, x-\delta)]f(x) \underset{\delta\to 0}{\longrightarrow} \frac{f(x)}{p(x)}$$

since $g(x, t)$ is symmetric for a self-adjoint operator (as we will prove below).

Therefore we have

$$pu''+p'u'+qu=\int_{\substack{a\\ x\neq t}}^{b} (pg''+p'g'+qg)f(t)\,dt+f(x)=f(x)$$

This is equivalent to stating that:

i. $g(x, t)$ is continuous and satisfies the boundary conditions.
ii. $g(x, t)$ has continuous first and second derivatives except at $x=t$ and is

such that

$$L_x g(x, t) = \delta(x - t)$$

Then we can write

$$\int_{t-\delta}^{t+\delta} L_x g(x, t)\, dx = \int_{t-\delta}^{t+\delta} \delta(x - t)\, dx = 1$$

so that

$$\int_{t-\delta}^{t+\delta} \left\{ \frac{d}{dx}\left[p(x) \frac{d}{dx} g(x, t) \right] + q(x) g(x, t) \right\} dx = 1$$

which implies

$$\frac{d}{dx} g(x, t) \bigg|_{t-\delta}^{t+\delta} = \frac{1}{p(t)}$$

Since L is self-adjoint and $g(x, t)$ satisfies homogeneous boundary conditions, we have

$$\begin{aligned} 0 &= \int_a^b [g(x, t) L_x g(x, s) - g(x, s) L_x g(x, t)]\, dx \\ &= \int_a^b [g(x, t)\, \delta(x - s) - g(x, s)\, \delta(x - t)]\, dx \\ &= g(s, t) - g(t, s) = 0 \end{aligned}$$

Therefore $g(x, t)$ is symmetric for a Sturm-Liouville operator satisfying (self-adjoint) homogeneous boundary conditions. Compare this result with Eq. 8.145 when L is self-adjoint so that the $\{\lambda_j\}$ are real and the $\{u_j(x)\}$ may be chosen real.

We can now use the continuity and jump conditions on $g(x, t)$ to construct the Green's function from the two linearly independent solutions of

$$Lw = 0$$

such that they satisfy the unmixed homogeneous boundary condition

$$B_1(w_1) = 0 = B_2(w_2)$$

As previously we easily find that (cf. Eq. 8.116)

$$g(x, t) = \begin{cases} \dfrac{w_2(t) w_1(x)}{p(t) W(w_1, w_2)} & a \leq x \leq t \\[2ex] \dfrac{w_1(t) w_2(x)}{p(t) W(w_1, w_2)} & t \leq x \leq b \end{cases}$$

if $W(w_1, w_2) \neq 0$.

8.9 ASYMPTOTIC BEHAVIOR OF THE SOLUTIONS TO A LINEAR DIFFERENTIAL EQUATION

We now consider a theorem that is extremely important in scattering theory and in other applications of asymptotic forms of solutions to differential equations. We show that if

$$\int^{\infty} |q(x)|\, dx < \infty \tag{8.147}$$

(e.g., that $|q(x)| < \text{const.}/x^{1+\varepsilon}$, $\varepsilon > 0$, as $x \to \infty$), then the differential equation

$$u'' + [k^2 + q(x)]u = 0 \tag{8.148}$$

with $q(x)$ bounded and piecewise continuous, has solutions such that

$$\begin{aligned} &\lim_{x\to\infty} |e^{-ikx}u_1(x) - 1| = 0 \\ &\lim_{x\to\infty} |e^{ikx}u_2(x) - 1| = 0 \end{aligned} \tag{8.149}$$

To prove this we write

$$w(x) = e^{-ikx}u(x) - 1 \tag{8.150}$$

and obtain

$$w''(x) + 2ikw'(x) + q(x)w(x) = -q(x) \tag{8.151}$$

subject to the boundary conditions

$$w(\infty) = 0 = w'(\infty) \tag{8.152}$$

The case $k = 0$, which is not covered by the theorem that we will prove, is often best handled by the substitution $y = 1/x$ (see Problem 8.20 for an example of this). Since Eqs. 8.151 and 8.152 constitute an inhomogeneous differential equation subject to homogeneous initial conditions, we may use the technique of Section 8.6c to construct the Green's function and obtain the equivalent integral equation

$$w(x) = \frac{1}{2ik} \int_x^{\infty} [1 - e^{-2ik(x-t)}] q(t) w(t)\, dt + f(x) \tag{8.153}$$

where for convenience we use $f(x)$ to stand for

$$f(x) = \frac{1}{2ik} \int_x^{\infty} [1 - e^{-2ik(x-t)}] q(t)\, dt \tag{8.154}$$

Just as in Section 2.7b we define an iterative solution as

$$w_{n+1}(x) = f(x) + \int_x^{\infty} \frac{[1 - e^{-2ik(x-t)}]}{2ik} q(t) w_n(t)\, dt \qquad n \geq 1$$

$$w_1(x) = f(x)$$

and we must establish that this sequence $\{w_n(x)\}$ actually does converge to the solution, $w(x)$, of Eq. 8.153.

Since we are considering the asymptotic behavior of the solutions to Eq. 8.148 for sufficiently large values of x, we may restrict our arguments to a region $x>R$ where $q(x)$ is bounded so that property (8.147) will insure that the $f(x)$ of Eq. 8.154 is continuous, bounded and such that $f(\infty)=0$. We see explicitly that

$$|w_2(x)|=\left|f(x)+\int_x^\infty \frac{[1-e^{-2ik(x-t)}]}{2ik}q(t)f(x)\,dt\right|$$

$$\leq|f(x)|+\int_x^\infty \frac{|q(t)|\,|f(t)|}{k}\,dt\leq f_m\left(1+\int_x^\infty \frac{|q(t)|}{k}\,dt\right)$$

where we have used f_m to denote the maximum value of $|f(x)|$ on the range of x under consideration. Let us now assume that $w_n(x)$ is bounded and then prove that $w_{n+1}(x)$ is also bounded. Since

$$F(x)\equiv\int_x^\infty \frac{|q(t)|}{k}\,dt\geq 0$$

is a monotonically decreasing function of x, we may write

$$\int_x^\infty \frac{|q(t)|}{k}F(t)\,dt\leq\int_x^\infty \frac{|q(t)|\,dt}{k}F(x)=[F(x)]^2$$

and we may take $F(x)<1$ for x sufficiently large. It is then a simple matter to establish by induction that

$$|w_n(x)|\leq f_m\sum_{s=0}^{n}[F(x)]^s\leq f_m\sum_{s=0}^{\infty}[F(x)]^s=\frac{f_m}{1-F(x)}<\infty \tag{8.155}$$

Although Eq. 8.155 does not give the sharpest bound possible on $w_n(x)$, it is sufficient for our purposes. Furthermore, if we define

$$v_n(x)\equiv w_n(x)-w_{n-1}(x)\qquad n\geq 2$$
$$v_1(x)\equiv f(x)$$

then we have

$$v_{n+1}(x)=\int_x^\infty \frac{[1-e^{-2ik(x-y)}]}{2ik}q(t)v_n(t)\,dt$$

from which we easily establish, again by induction, that

$$|v_n(x)|\leq f_m[F(x)]^{n-1}\qquad n\geq 2 \tag{8.156}$$

Since as an identity we may write

$$w_n(x)=\sum_{s=1}^{n}v_s(x)$$

we can use the Weierstrass M-test to conclude that the series

$$w_\infty(x) \equiv \sum_{s=1}^{\infty} v_s(s)$$

is uniformly convergent so that $w_\infty(x)$ exists. However it is clear that $w_\infty(x)$ equals $w(x)$, the solution to Eq. 8.153. This is the unique solution since the homogeneous version of Eq. 8.151 with the boundary conditions of Eq. 8.152 will have only zero as a solution. Therefore we have established the following theorem since we can apply exactly the same proof to the function $[e^{ikx}u(x)-1]$.

THEOREM 8.12

If

$$\int^{\infty} |q(x)|\, dx < \infty$$

then for any real k, $k \neq 0$, the differential equation

$$u''(x)+[k^2+q(x)]u(x)=0$$

has two linearly independent solutions, $u_1(x)$ and $u_2(x)$, such that

$$\lim_{x\to\infty} [e^{-ikx}u_1(x)-1]=0$$

$$\lim_{x\to\infty} [e^{ikx}u_2(x)-1]=0$$

As we will see in Problem 8.21, the equation

$$\frac{1}{w}(pu')'+qu=0 \tag{8.157}$$

has solutions that have the following asymptotic behavior for large positive values of x,

$$u(x) \xrightarrow[x\to+\infty]{} (wqp)^{-(1/4)} \exp\left[\pm i\int^{x}\left(\frac{wq}{p}\right)^{1/2} dx\right] \tag{8.158}$$

if

$$\int^{\infty} \left|g^{-1}\frac{d^2g}{dy^2}\right| dy < \infty \tag{8.159}$$

where

$$g=(wqp)^{1/4} \qquad y=\int^{x}(wq/p)^{1/2}\, dx \tag{8.160}$$

This assumes that $y \to \infty$ as $x \to \infty$. In fact if $p(x)$ grows no more rapidly asymptotically than *two* powers faster than $[w(x)q(x)]$, then this will be true.

Example 8.10 If we apply this to Bessel's equation,

$$u''(x)+\frac{1}{x}u'(x)+\left(1-\frac{\nu^2}{x^2}\right)u(z)=0$$

we can transform it into the proper form by setting

$$w(x)=x \qquad p(x)=x$$
$$q(x)=1-\frac{\nu^2}{x^2}$$

so that

$$Lu=\frac{1}{x}(xu')'+\left(1-\frac{\nu^2}{x^2}\right)u=0$$

We then obtain

$$[wqp]^{1/4}=(x^2-\nu^2)^{1/4}=g(x)$$
$$\frac{dg}{dx}=\frac{2x}{4(x^2-\nu^2)^{3/4}}=\frac{x}{2g^3}$$
$$y=\int^x\left(1-\frac{\nu^2}{x^2}\right)^{1/2}dx \qquad dy=\left(1-\frac{\nu^2}{x^2}\right)^{1/2}dx$$
$$\int^\infty\left|g^{-1}\frac{d^2g}{dy^2}\right|dy=\int^\infty\frac{(x^2+4\nu^2)x\,dx}{4(x^2-\nu^2)^{5/2}}<\infty \qquad x>\nu$$

Equation 8.158 implies that

$$u(x)\xrightarrow[x\to\infty]{}\frac{1}{(x^2-\nu^2)^{1/4}}\exp\left[\pm i\int^x\left(1-\frac{\nu^2}{x^2}\right)dx\right]$$
$$=\frac{1}{(x^2-\nu^2)^{1/4}}\exp\left\{\pm i\left[\sqrt{x^2-\nu^2}-\nu\tan^{-1}\left(\frac{\sqrt{x^2-\nu^2}}{\nu}\right)\right]\right\}$$
$$\xrightarrow[x\to\infty]{}\frac{1}{\sqrt{x}}e^{\pm i(x-\nu\pi/2)}$$

This differs only by a phase factor and a normalization from the asymptotic forms of $H^{(1)}(z)$ and $H^{(2)}(z)$ given previously in Eqs. 7.281 and 7.282.

8.10 THE CONTINUOUS SPECTRUM

So far we have considered mainly operators on a Hilbert space of square-integrable functions having representations as differential and integral operators with discrete spectra of eigenvalues. The corresponding eigenfunctions then belonged to $\mathcal{L}_2$. It often happens in physical applications that we must deal with operators having continuous spectra and with their eigenfunctions that are not square integrable. We will encounter these in our treatment of scattering problems with Green's functions in Section 8.11. Some operators may have a discrete spectrum only, some a continuous spectrum only, and others both a discrete and a continuous spectrum. In general, operators with continuous spectra are extremely difficult to handle in an elementary fashion with any degree of mathematical rigor.

We consider here the Sturm-Liouville equation with a continuous spectrum. In Section 8.3 we were able to prove that the spectrum of this equation was discrete *provided* the interval of definition, $[a, b]$, was *finite* while $p(x)$ and $r(x)$ were positive and bounded. If we have a Sturm-Liouville equation defined on an infinite interval, for example, $(-\infty, \infty)$, then the spectrum can become continuous and the eigenfunctions need no longer be square integrable. Consider the Sturm-Liouville equation

$$\frac{d^2}{dx^2}\psi(x)+k^2\psi(x)+q(x)\psi(x)=0 \qquad -\infty<x<\infty \tag{8.161}$$

Once we have studied this form we can extend the results to the general Sturm-Liouville operator by a change of variable (cf. Eqs. 8.158–8.160 and Problems 8.19 and 8.21). From Section 8.9 we know that if

$$\int^{\infty} |q(x)|\, dx<\infty$$

then the asymptotic forms of the solutions $\psi(x)$ are

$$\psi(x) \xrightarrow[x\to\infty]{} e^{\pm ikx}$$

Similarly if $\int^{-\infty} |q(x)|\, dx<\infty$, then we have the same result as $x \to -\infty$. We use $\psi(x)$ in this section, rather than $u(x)$, to denote nonsquare-integrable solutions. Of course only k such that $k^2<0$ can give square-integrable solutions. Therefore for real k these eigenfunctions are not normalizable in the usual sense and questions of completeness and of orthogonality must be treated carefully. We show below that the $\psi(k, x)$ are such that

$$\langle\psi(k)|\,\psi(k')\rangle \equiv \int \psi^*(k, x)\psi(k', x)\, dx=\delta(k-k')$$

so that these functions are orthogonal and normalized to a Dirac δ-function.

In order to simplify the discussion and the calculational details we write Eq.

8.161 in abstract operator notation as

$$(L-k^2)\psi(k)=V\psi(k) \tag{8.162}$$

where V is self-adjoint. We assume known the solutions to

$$\left(\frac{d^2}{dx^2}+k^2\right)\phi(k,x)=0$$

which we express in abstract form as

$$(L-k^2)\phi(k)=0 \tag{8.163}$$

If we use the notation of Section 2.5 (note change in sign convention in exponent here relative to Eq. 2.62), we can write these solutions as

$$\langle x \mid \phi(k)\rangle \equiv \phi(k,x)=\frac{1}{\sqrt{2\pi}}e^{ikx}$$

where

$$\langle \phi(k) \mid \phi(k')\rangle=\frac{1}{2\pi}\int_{-\infty}^{\infty} e^{i(-k+k')x}dx=\delta(k-k') \tag{8.164}$$

This is simply another statement of the Fourier integral theorem (Th. 4.9). We can see this in another way since we can write Eq. 8.22 as

$$\delta_n(x)=\frac{n}{\sqrt{\pi}}e^{-n^2x^2}=\frac{1}{2\pi}\int_{-\infty}^{\infty} e^{-(1/4)k^2/n^2}e^{ikx}\,dx$$

Since this represents a $\delta(x)$ in the sense that

$$\lim_{n\to\infty}\int_{-\infty}^{\infty}\delta_n(x)f(x)\,dx=f(0)$$

we have as a weak limit

$$\delta(x)=\lim_{n\to\infty}\delta_n(x)=\frac{1}{2\pi}\int_{-\infty}^{\infty} e^{ikx}\,dx$$

Since the $\{\phi(k)\}$ are complete (cf. Eqs. 8.164, 2.63, and 4.73), we may express any abstract vector $|f\rangle$ as

$$|f\rangle=\int dk\,|\phi(k)\rangle\langle\phi(k)\,|f\rangle$$

so that Eqs. 8.162 and 8.163 become, respectively,

$$\int\langle x|\,L\,|x'\rangle\langle x' \mid \psi(k)\rangle\,dx'-k^2\langle x \mid \psi(k)\rangle=\int\langle x|\,V\,|x'\rangle\langle x' \mid \psi(k)\rangle\,dx'$$

$$\int\langle x|\,L\,|x'\rangle\langle x' \mid \phi(k)\rangle\,dx'-k^2\langle x \mid \phi(k)\rangle=0$$

We obtain the representation of Eq. 8.161 with the identifications

$$\langle x|\, L\, |x'\rangle = -\delta(x-x')\,\frac{d^2}{dx^2}$$

$$\langle x|\, V\, |x'\rangle = \delta(x-x')q(x)$$

$$\langle x \mid \psi(k)\rangle \equiv \psi(k,x) \xrightarrow[x\to\infty]{} \phi(k,x)$$

where the asymptotic form of $\psi(k, x)$ follows from Th. 8.12.

We can invert Eq. 8.162 as

$$\psi(k) = \phi(k) + \lim_{\varepsilon\to 0^+} (L-k^2-i\varepsilon)^{-1} V\psi(k) \tag{8.165}$$

We easily verify that Eq. 8.165 is a solution to Eq. 8.162 by operating on both sides of Eq. 8.165 with $\lim\limits_{\varepsilon\to 0^+} (L-k^2-i\varepsilon) \equiv (L-k^2)$ to obtain

$$\lim_{\varepsilon\to 0^+} (L-k^2-i\varepsilon)\psi(k) = \lim_{\varepsilon\to 0^+} (L-k^2-i\varepsilon)\phi(k) + V\psi(k) = V\psi(k)$$

In fact, up to a sign, $(L-k^2=i\varepsilon)^{-1}$ is the Green's function for this problem. Later in this section and also in Section 8.11 we will consider specific examples and discuss the boundary, or asymptotic, conditions satisfied by this Green's function. We henceforth simply omit the symbol $\lim\limits_{\varepsilon\to 0^+}$ although this limit must always be understood as being taken. This limit, $\varepsilon \to 0^+$, is necessary in order to make the expression on the right of Eq. 8.165 well defined. We will discuss this in detail in a particular representation shortly.

The following relation will be needed.

$$\begin{aligned}
\langle \phi(k')\, |(L-k^2-i\varepsilon)^{-1} V|\, \psi(k)\rangle &= \int dk'' \left\langle \phi(k') \,\middle|\, \frac{1}{k''^2-k^2-i\varepsilon} \,\middle|\, \phi(k'') \right\rangle \langle \phi(k'') \mid V\psi(k)\rangle \\
&= \langle \phi(k')|\, (k'^2-k^2-i\varepsilon)^{-1} V\, |\psi(k)\rangle \\
&= \langle (k'^2-k^2+i\varepsilon)^{-1}\phi(k')\, |V|\, \psi(k)\rangle \\
&= \langle (L-k^2+i\varepsilon)^{-1}\phi(k')\, |V|\, \psi(k)\rangle
\end{aligned}$$

We have taken both k and k' to be real. For example, in the representation of Eq. 8.161, we would have explicitly

$$\langle \phi(k')|\, (L-k^2-i\varepsilon)^{-1}\, |V\psi(k)\rangle = \frac{1}{\sqrt{2\pi}} \int e^{-ik'x} \frac{1}{k'^2-k^2-i\varepsilon}\, V\psi(k,x)\, dx$$

$$\langle (L-k^2+i\varepsilon)^{-1}\phi(k') \mid V\psi(k)\rangle = \frac{1}{\sqrt{2\pi}} \int e^{-ik'x} \frac{1}{k'^2-k^2-i\varepsilon}\, V\psi(k,x)\, dx$$

We can also establish (directly from Eq. 8.165) the identities

$$\langle\phi(k) \mid V\psi(k')\rangle=\langle\psi(k) \mid V\psi(k')\rangle-\langle(L-k^2+i\varepsilon)^{-1}V\psi(k) \mid V\psi(k')\rangle$$
$$\langle V\psi(k) \mid \phi(k')\rangle=\langle V\psi(k) \mid \psi(k')\rangle-\langle V\psi(k) \mid (L-k'^2-i\varepsilon)^{-1}V\psi(k')\rangle$$

Therefore we may write

$$\begin{aligned}
\langle\psi(k) \mid \psi(k')\rangle&=\langle\phi(k) \mid \phi(k')\rangle+\langle\phi(k) \mid (L-k'^2-i\varepsilon)^{-1}V\psi(k')\rangle\\
&\quad+\langle(L-k^2+i\varepsilon)^{-1}V\psi(k) \mid \phi(k')\rangle\\
&\quad+\langle(L-k^2+i\varepsilon)^{-1}V\psi(k) \mid (L-k'^2-i\varepsilon)^{-1}V\psi(k')\rangle\\
&=\langle\phi(k) \mid \phi(k')\rangle+(k^2-k'^2-i\varepsilon)^{-1}\langle\phi(k) \mid V\psi(k')\rangle\\
&\quad+(k'^2-k^2-i\varepsilon)^{-1}\langle V\psi(k) \mid \phi(k')\rangle\\
&\quad+\langle(L-k^2+i\varepsilon)^{-1}V\psi(k) \mid (L-k'^2-i\varepsilon)^{-1}V\psi(k')\rangle\\
&=\langle\phi(k) \mid \phi(k')\rangle+(k^2-k'^2-i\varepsilon)^{-1}\{[\langle\psi(k) \mid V\psi(k')\rangle-\langle V\psi(k) \mid \psi(k')\rangle]\\
&\quad-\langle V\psi(k) | [(L-k^2-i\varepsilon)^{-1}-(L-k'^2-i\varepsilon)^{-1}] \mid V\psi(k')\rangle\}\\
&\quad+\langle(L-k^2+i\varepsilon)^{-1}V\psi(k) \mid (L-k'^2-i\varepsilon)^{-1}V\psi(k')\rangle\\
&=\langle\phi(k) \mid \phi(k')\rangle-\frac{(k^2-k'^2)}{(k^2-k'^2-i\varepsilon)}\\
&\quad\times\langle(L-k^2+i\varepsilon)^{-1}V\psi(k) \mid (L-k'^2-i\varepsilon)^{-1}V\psi(k')\rangle\\
&\quad+\langle(L-k^2+i\varepsilon)^{-1}V\psi(k) \mid (L-k'^2-i\varepsilon)^{-1}V\psi(k')\rangle\\
&=\langle\phi(k) \mid \phi(k')\rangle-\frac{i\varepsilon}{(k^2-k'^2-i\varepsilon)}\\
&\quad\times\langle(L-k^2+i\varepsilon)^{-1}V\psi(k) \mid (L-k'^2-i\varepsilon)^{-1}V\psi(k')\rangle
\end{aligned} \tag{8.166}$$

The important point is that $\psi(k, x)$, although not square integrable, is *bounded* while $q(x)$ is *absolutely* integrable so that $|V\psi(k)\rangle$ has finite norm as does $|(L-k^2-i\varepsilon)^{-1}V\psi(k)\rangle$ once $\|V\|<\infty$ (cf. Section A8.4). The second term on the right of Eq. 8.166 does vanish as $\varepsilon \to 0$ (since we are to take the limit $\varepsilon \to 0$ before we set $k^2=k'^2$).

Therefore, as claimed, we have the orthogonality relation

$$\langle\psi(k) \mid \psi(k')\rangle=\langle\phi(k) \mid \phi(k')\rangle=\delta(k-k')$$

We have *not* established the completeness of the $\{\psi(k)\}$. In fact if Eq. 8.161 has a discrete part to its eigenvalue spectrum as well, then the corresponding eigenfunctions $\{\psi_j\}$ must also be included with the $\{\psi(k)\}$ to form a complete set as

$$I=\sum_j |\psi_j\rangle\langle\psi_j|+\int dk\, |\psi(k)\rangle\langle\psi(k)|$$

We do not prove this here.

THEOREM 8.13

The solutions of the Sturm-Liouville equation

$$\left[\frac{d^2}{dx^2}+k^2+q(x)\right]\psi(k, x)=0 \qquad x\in(-\infty, \infty)$$

with k^2 real and positive and

$$\int^{\pm\infty} |q(x)|\, dx<\infty$$

satisfy the relation

$$\langle\psi(k) \mid \psi(k')\rangle=\delta(k-k')$$

We now look at this process of constructing the inverse of a differential operator with a continuous spectrum in a particular representation. Let us review briefly for reference the case of the discrete spectrum. In Section 8.7 we saw that when the $\{\lambda_j\}$ form a discrete spectrum and the $\{\varphi_j\}$ a complete set, then the inverse of

$$L\varphi_j=\lambda_j\varphi_j$$

is given as

$$(L-\alpha)^{-1}\varphi_j\equiv\psi_j=\sum_k \alpha_{jk}\varphi_k$$

which is defined only when operating on a state of finite norm. Since

$$(L-\alpha)(L-\alpha)^{-1}\varphi_j\equiv\varphi_j=(L-\alpha)\psi_j=\sum_k \alpha_{jk}(L-\alpha)\varphi_k=\sum_k \alpha_{jk}(\lambda_k-\alpha)\varphi_k$$

it follows that

$$\alpha_{jk}=\frac{\delta_{jk}}{\lambda_k-\alpha}$$

if $\alpha\neq\lambda_k$ for *any* k and

$$(L-\alpha)^{-1}\varphi_j=\frac{\varphi_j}{\lambda_j-\alpha}$$

so that

$$(L-\alpha)^{-1}f=(L-\alpha)^{-1}\sum_j c_j\varphi_j=\sum_j \frac{c_j\varphi_j}{(\lambda_j-\alpha)} \qquad c_j\equiv\langle f \mid \varphi_j\rangle$$

Let us consider the continuous spectrum and write Eq. 8.165 in the representation of Eq. 8.161 as

$$\langle x \mid \psi(k)\rangle = \langle x \mid \phi(k)\rangle + \lim_{\varepsilon\to 0}\int_{-\infty}^{\infty} dk' \langle x| (L - k^2 - i\varepsilon)^{-1} |\phi(k')\rangle\langle\phi(k')| V |\psi(k)\rangle$$

$$\psi(k, x) = \phi(k, x) + \lim_{\varepsilon\to 0}\int_{-\infty}^{\infty} \frac{dk'\phi(k', x)\langle\phi(k')| V |\psi(k)\rangle}{(k'^2 - k^2 - i\varepsilon)}$$

We have used the completeness of the $\{\phi(k)\}$ here. We can also write

$$\begin{aligned}\langle\phi(k')| V |\psi(k)\rangle &= \int dx'\, dx''\langle\phi(k') \mid x'\rangle\langle x'| V |x''\rangle\langle x'' \mid \psi(k)\rangle \\ &= \int dx'\, dx''\phi^*(k', x')q(x')\,\delta(x' - x'')\psi(k, x'') \\ &= \int dx'\phi^*(k', x')q(x')\psi(k, x')\end{aligned}$$

so that

$$\psi(k, x) = \phi(k, x) + \int dx' q(x')\left[\lim_{\varepsilon\to 0}\int_{-\infty}^{\infty} \frac{dk'\phi^*(k', x')\phi(k', x)}{(k'^2 - k^2 - i\varepsilon)}\right]\psi(k, x') \quad \textbf{(8.167)}$$

As we will do several times in Section 8.11, we may use contour integration to evaluate the k' integral in Eq. 8.167. Since $\phi(k, x) = 1/\sqrt{2\pi}\, e^{ikx}$, we can evaluate the inner integral in Eq. 8.167 as

$$\begin{aligned}\lim_{\varepsilon\to 0^+}\frac{1}{2\pi}\int_{-\infty}^{\infty}\frac{e^{-ik'x'}e^{ik'x}\,dk'}{(k'^2 - k^2 - i\varepsilon)} &= \lim_{\varepsilon\to 0^+}\frac{1}{2\pi}\int_{-\infty}^{\infty}\frac{e^{ik'(x-x')}\,dk'}{(k'^2 - k^2 - i\varepsilon)} \\ &= \frac{2\pi i}{2\pi}\left\{\frac{e^{ik(x-x')}}{2k}H(x - x') + \frac{e^{-ik(x-x')}}{2k}H(x' - x)\right\} \\ &= \frac{i}{2k}e^{ik|x-x'|}\end{aligned}$$

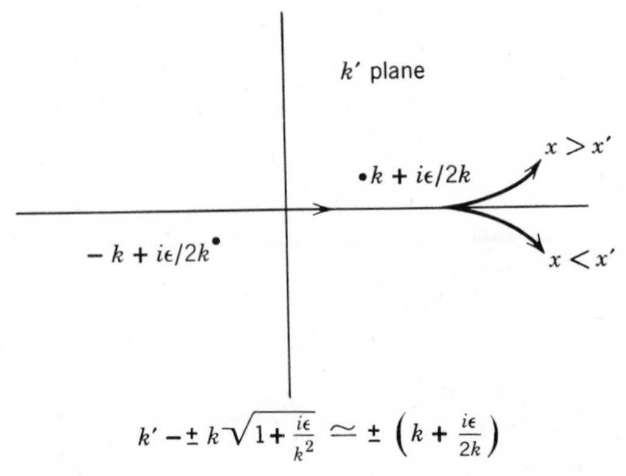

Figure 8.5

Finally we can write explicitly

$$\psi(k, x)=\phi(k, x)-\frac{1}{2ik}\int_{-\infty}^{\infty} dx' q(x') e^{ik|x-x'|}\psi(k, x') \tag{8.168}$$

Our Green's function,

$$-\langle x|\,(L-k^2-i\varepsilon)^{-1}\,|x'\rangle=\frac{1}{2ik}\,e^{ik|x-x'|}$$

$$\begin{aligned}
-\left(\frac{d^2}{dx^2}+k^2\right)\langle x|\,(L-k^2-i\varepsilon)^{-1}\,|x'\rangle &= \frac{1}{2ik}\left\{\frac{d}{dx}\left[\frac{d}{dx}\,e^{ik|x-x'|}\right]+k^2 e^{ik|x-x'|}\right\}\\
&= \frac{1}{2ik}\left\{\frac{d}{dx}\,[ike^{ik|x-x'|}\right.\\
&\quad \left.\times(H(x-x')-H(x'-x))]+k^2 e^{ik|x-x'|}\right\}\\
&= \frac{1}{2ik}\,\{-k^2 e^{ik|x-x'|}[H(x-x')-H(x'-x)]^2\\
&\quad +2ike^{ik|x-x'|}\,\delta(x-x')+k^2 e^{ik|x-x'|}\}\\
&= \delta(x-x')
\end{aligned}$$

satisfies the asymptotic conditions

$$-\langle x|\,(L-k^2-i\varepsilon)^{-1}\,|x'\rangle \to \begin{cases} \dfrac{1}{2ik}\,e^{ikx}e^{-ikx'} & x\to+\infty \quad x' \text{ fixed}\\[2mm] \dfrac{1}{2ik}\,e^{-ikx}e^{ikx'} & x\to-\infty \quad x' \text{ fixed}\end{cases}$$

If we had taken $(L-k^2+i\varepsilon)^{-1}$ in Eqs. 8.165 or 8.167 we would have obtained

$$-\langle x|\,(L-k^2+i\varepsilon)^{-1}\,|x'\rangle=-\frac{1}{2ik}\,e^{-ik|x-x'|}\to \begin{cases} -\dfrac{1}{2ik}\,e^{-ikx}e^{ikx'} & x\to+\infty \quad x' \text{ fixed}\\[2mm] -\dfrac{1}{2ik}\,e^{ikx}e^{-ikx'} & x\to-\infty \quad x' \text{ fixed}\end{cases}$$

If we had taken $\varepsilon=0$ in Eqs. 8.165 or 8.167, then we would have had to take the principal value as

$$\psi(k)=\phi(k)+P(L-k^2)^{-1}V\psi(k)$$

with the Green's function (cf. Eq. 7.52)

$$\begin{aligned}
-\langle x|\,P(L-k^2)^{-1}\,|x'\rangle &= -\tfrac{1}{2}\{\langle x|\,(L-k^2-i\varepsilon)^{-1}\,|x'\rangle+\langle x|\,(L-k^2+i\varepsilon)^{-1}\,|x'\rangle\}\\
&= \frac{1}{4ik}\,[e^{ik|x-x'|}-e^{-ik|x-x'|}]\\
&= \frac{1}{2k}\sin\,(k\,|x-x'|)
\end{aligned}$$

That is, our choice of Green's function now implies a particular *asymptotic behavior* of the solution $\psi(k, x)$ as $x \to \pm\infty$. Previously, in the case of a finite domain (or for square-integrable solutions), we had made the Green's function satisfy prescribed boundary conditions at the end points of the interval $[a, b]$.

8.11 PHYSICAL APPLICATIONS OF GREEN'S FUNCTIONS

We begin by classifying a general second-order linear partial differential equation of the form

$$\sum_{i,j=1}^{n} a_{ij}(x_1, \ldots, x_n) \frac{\partial^2}{\partial x_i \partial x_j} \psi(x_1, \ldots, x_n) + f\left(\psi; x_1, \ldots, x_n; \frac{\partial \psi}{\partial x_1}, \ldots, \frac{\partial \psi}{\partial x_n}\right) = 0$$

where the x_i, $i = 1, 2, \ldots, n$, are the independent variables, the $a_{ij}(x_1, \ldots, x_n)$, i, $j = 1, 2, \ldots, n$, are real continuously differentiable functions of the x_i, f is real continuous and differentiable in all of its arguments, and $\psi(x_1, \ldots, x_n)$ is the solution to this partial differential equation. We define the a_{ij} so that $a_{ij} = a_{ji}$. Such differential equations are classified according to certain properties of the *principal part* defined as

$$\sum_{i,j=1}^{n} a_{ij} \frac{\partial^2 \psi}{\partial x_i \, \partial x_j}$$

Let us change variables according to the real invertible transformations

$$\xi_i = \xi_i(x_1, \ldots, x_n) \qquad i = 1, 2, \ldots, n$$

so that

$$\frac{\partial \psi}{\partial x_j} = \sum_{k=1}^{n} \frac{\partial \xi_k}{\partial x_j} \frac{\partial \psi}{\partial \xi_k} \qquad \frac{\partial^2 \psi}{\partial x_i \, \partial x_j} = \sum_{k,l=1}^{n} \frac{\partial \xi_k}{\partial x_j} \frac{\partial \xi_l}{\partial x_i} \frac{\partial^2 \psi}{\partial \xi_k \, \partial \xi_l} + \cdots$$

where we have kept only second derivative terms of ψ in the second equation. Then the principal part of the transformed partial differential equation is

$$\sum_{i,j=1}^{n} \alpha_{ij}(\xi_1, \ldots, \xi_n) \frac{\partial^2 \psi}{\partial \xi_i \, \partial \xi_j}$$

where

$$\alpha_{ij} = \alpha_{ji} = \sum_{k,l=1}^{n} a_{kl} \frac{\partial \xi_i}{\partial x_l} \frac{\partial \xi_j}{\partial x_k} = \sum_{k,l=1}^{n} \left(\frac{\partial \xi_i}{\partial x_k}\right) a_{kl} \left(\frac{\partial \xi_j}{\partial x_l}\right) \equiv \sum_{k,l=1}^{n} u_{ik} a_{kl} u_{jl}$$

$$= \sum_{k,l=1}^{n} U_{ik} a_{kl} U_{lj}^{\mathrm{T}}$$

Since the transformations $\xi_i = \xi_i(x_1, \ldots, x_n)$ have been assumed invertible, we know that both $U = \{\partial\xi_i/\partial x_k\}$ and $U^{-1} = \{\partial x_i/\partial\xi_k\}$ exist so that $\det U \neq 0$. If we define the real symmetric matrices (which necessarily have all of their eigenvalues real)

$$A = \{a_{ij}\} = A^T$$
$$B = \{\alpha_{ij}\} = UAU^T = B^T$$

then we may argue that both A and B have the same number of positive, zero, and negative eigenvalues as follows. Consider any real symmetric matrix $A = \{a_{ij}\}$ and the corresponding eigenvalue problem

$$A\varphi_j = \lambda_j\varphi_j \qquad j = 1, 2, \ldots, n$$

From Th. 3.4 we know that the real orthogonal matrix having the φ_j for its columns

$$S \equiv \{\varphi_j\}$$

will diagonalize A as

$$S^T A S = \{\delta_{ij}\lambda_j\}$$

For convenience let us order the $\{\lambda_j\}$ so that $\lambda_j < 0$, $1 \le j \le k$, $\lambda_j = 0$, $k+1 \le j \le l$, $\lambda_j > 0$, $l+1 \le j \le n$. In analogy with the proof of Th. 3.6 we define a matrix R as

$$R = (|\lambda_1|^{-1/2}\varphi_1 \cdots |\lambda_k|^{-1/2}\varphi_k; \varphi_{k+1} \cdots \varphi_l; \lambda_{l+1}^{-1/2}\varphi_{l+1} \cdots \lambda_n^{-1/2}\varphi_n)$$

so that

$$R^T R = \begin{pmatrix} |\lambda_1|^{-1} & & & & & & & & \\ & \ddots & & & & & & \mathbf{0} & \\ & & |\lambda_k|^{-1} & & & & & & \\ & & & 1 & & & & & \\ & & & & \ddots & & & & \\ & & & & & 1 & & & \\ & \mathbf{0} & & & & & \lambda_{l+1} & & \\ & & & & & & & \ddots & \\ & & & & & & & & \lambda_n \end{pmatrix}$$

which shows that det $R \neq 0$ and hence that R is not singular. Direct calculation shows that

$$R^{T}AR = \begin{pmatrix} -1 & & & & & & & \\ & \ddots & & & & & & \\ & & -1 & & & & \mathbf{0} & \\ & & & 0 & & & & \\ & & & & \ddots & & & \\ & & & & & 0 & & \\ & & \mathbf{0} & & & & 1 & \\ & & & & & & & \ddots \\ & & & & & & & & 1 \end{pmatrix}$$

That is, A may be brought into a diagonal form in which the first k elements are -1, the next $(l-k)$ are 0, and the last $(n-l)$ are $+1$. Since this is true for *any* real symmetric matrix, it follows that the number of negative, zero, and positive eigenvalues is left invariant under any nonsingular transformations of the form UAU^{T}.

This fact allows us to classify a general second-order partial differential equation by its principal part as follows. If at any point $(x_1^0, x_2^0, \ldots, x_n^0)$ the eigenvalues of the matrix $A=\{a_{ij}(x_1^0, \ldots, x_n^0)\}$:

i. Are all positive or all negative, the differential equation is termed *elliptic*.
ii. Are all of one sign, except for one that has the opposite sign, the differential equation is termed *hyperbolic*.
iii. Have some zeros, the differential equation is termed *parabolic*.

This classification is not exhaustive but is sufficient for our work.

We see that all second-order linear ordinary differential equations in one variable are elliptic. The scalar *Helmholtz equation*,

$$(\nabla^2+k^2)\psi(\mathbf{r}) \equiv \left(\frac{\partial^2}{\partial x^2}+\frac{\partial^2}{\partial y^2}+\frac{\partial^2}{\partial z^2}+k^2\right)\psi(\mathbf{r})=\rho(\mathbf{r})$$

and the time-independent *Schrödinger equation*,

$$(\nabla^2+k^2)\psi(\mathbf{r})=q(\mathbf{r})\psi(\mathbf{r})$$

are also elliptic partial differential equations. The scalar *wave equation*,

$$\left(\nabla^2-\frac{1}{c^2}\frac{\partial^2}{\partial t^2}\right)\psi(\mathbf{r}, t)=\rho(\mathbf{r}, t)$$

is a hyperbolic partial differential equation. The *diffusion equation,*

$$\left(\nabla^2 - a^2 \frac{\partial}{\partial t}\right)\psi(\mathbf{r}, t) = \rho(\mathbf{r}, t)$$

is a parabolic partial differential equation.

We will construct the Green's functions for these partial differential operators. In general the Green's function for an elliptic linear differential operator is a continuous function while that for a hyperbolic linear differential operator is an ideal function. We will not prove this in general but simply verify it for the cases just listed.

a. An Example

As a simple introduction to the general approach we employ in this section, let us consider the quantum-mechanical scattering in one dimension of a beam of particles from a potential barrier as depicted in Fig. 8.6. The relevant equation for a steady-state problem of this type is the one-dimensional time-independent Schrödinger equation

$$\left[-\frac{\hbar^2}{2m}\frac{d^2}{dx^2} + V(x)\right]\psi(k, x) = E\psi(k, x)$$

where the energy, E, is positive, $\hbar$ is Planck's constant divided by 2π, m is the mass of the particle being scattered, and $V(x)$ is a real function. With the substitutions

$$q(x) = \frac{2m}{\hbar^2} V(x) \qquad k^2 = \frac{2m}{\hbar^2} E > 0$$

we obtain

$$\frac{d^2}{dx^2}\psi(k, x) + k^2\psi(k, x) = q(x)\psi(k, x)$$

We assume that $\int^{\pm\infty} |q(x)|\, dx < \infty$.

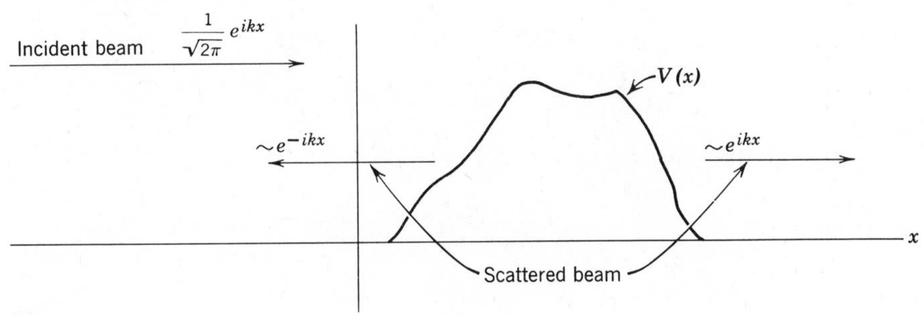

Figure 8.6

In such a quantum-mechanical treatment of this problem we represent a steady beam of particles incident from the left of Fig. 8.6 by the plane wave

$$\frac{1}{\sqrt{2\pi}} e^{ikx}$$

while the reflected and transmitted beams are represented, respectively, by the plane waves

$$\frac{1}{\sqrt{2\pi}} e^{-ikx} \quad \text{and} \quad \frac{1}{\sqrt{2\pi}} e^{ikx}$$

We will take as the two linearly independent solutions to the homogeneous equation

$$\left(\frac{d^2}{dx^2}+k^2\right)\phi(k, x)=0$$

the functions

$$\phi_1(k, x)=\frac{1}{\sqrt{2\pi}} e^{-ikx}$$

$$\phi_2(k, x)=\frac{1}{\sqrt{2\pi}} e^{ikx}$$

This problem is now that posed in Eq. 8.161 and can be reduced to the integral equation of Eq. 8.168. However let us construct the Green's function directly as follows. We know that the solution to our problem is a superposition of a solution to the homogeneous equation plus one to the inhomogeneous equation

$$\psi(k, x)=\phi(k, x)+\int_{-\infty}^{\infty} g(x, x')q(x')\psi(k, x')\, dx' \tag{8.169}$$

As we have discussed in Section 8.10, our boundary conditions are prescribed in the form of asymptotic conditions on $\psi(k, x)$. Since we are sending in a steady beam of particles from the left in Fig. 8.6, we want only these progressive plane waves if $q(x)\equiv 0$ (*i.e., no scattering*) so that $\phi(k, x)=\phi_2(k, x)=(1/\sqrt{2\pi})e^{ikx}$. Furthermore the second term to the right of Eq. 8.169 represents the effect of the scattering potential so that we demand for the Green's function

$$g(x, x')\begin{cases} \xrightarrow[x\to-\infty]{} \sim e^{-ikx} \propto \phi_1(x) & x<x' \\ \xrightarrow[x\to+\infty]{} \sim e^{ikx} \propto \phi_2(x) & x>x' \end{cases}$$

The same arguments that led to Eq. 8.116 allow us to write

$$g(x, x')=\begin{cases} \dfrac{\phi_1(x)\phi_2(x')}{P(\phi_1, \phi_2)} & x\leq x' \\[2ex] \dfrac{\phi_1(x')\phi_2(x)}{P(\phi_1, \phi_2)} & x\geq x' \end{cases}$$

We easily see that

$$P(\phi_1, \phi_2) = W(\phi_1, \phi_2) = \frac{ik}{\pi}$$

which gives

$$g(x, x') = \begin{cases} \dfrac{e^{-ik(x-x')}}{2ik} & x \leq x' \\ \dfrac{e^{-ik(x-x')}}{2ik} & x \geq x' \end{cases}$$

or

$$g(x, x') = \frac{e^{ik|x-x'|}}{2ik}$$

Therefore Eq. 8.169 becomes

$$\psi(k, x) = \frac{e^{ikx}}{\sqrt{2\pi}} + \int_{-\infty}^{\infty} \frac{e^{ik|x-x'|}}{2ik} q(x')\psi(k, x')\, dx' \tag{8.170}$$

This differs from Eq. 8.168 by a sign in the second term because of the difference in sign associated with $q(x)$ in Eq. 8.161 and in the Schrödinger equation of this section.

Before we finish our discussion of this example let us construct the Green's function $g(x, x')$ by another method that is similar to that used in Section 8.10 and that we will use extensively later in the present section. As we saw in Section 8.10 we may represent $\delta(x-x')$ as

$$\delta(x-x') = \frac{1}{2\pi} \int_{-\infty}^{\infty} e^{ik'(x-x')}\, dk'$$

We seek a $g(x, x')$ such that

$$\left(\frac{d^2}{dx^2} + k^2\right) g(x, x') = \delta(x-x') = \frac{1}{2\pi} \int_{-\infty}^{\infty} e^{ik'(x-x')}\, dk'$$

If we operate on both sides of this equation with $[(d^2/dx^2) + k^2]^{-1}$ as we did in Section 8.10, we obtain

$$g(x, x') = -\frac{1}{2\pi} \int_{-\infty}^{\infty} \frac{e^{ik'(x-x')}}{(k'^2 - k^2)}\, dk'$$

where the integration path is shown in Fig. 8.7 (compare this with Fig. 8.5). Therefore we must close the contour in the upper- or lower-half k' plane of Fig. 8.7 according as $x > x'$ or $x < x'$, respectively. This yields, with the contour shown in Fig. 8.7,

$$g(x, x') = \frac{e^{ik|x-x'|}}{2ik}$$

as before. If we had chosen a contour below $k' = -k$ and above $k' = k$, we would

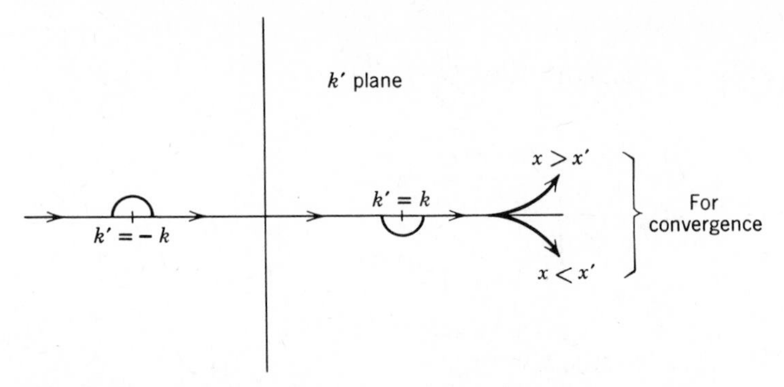

Figure 8.7

have obtained

$$\tilde{g}(x, x') = -\frac{e^{-ik|x-x'|}}{2ik}$$

This would correspond to outgoing plane waves as $x \to -\infty$ and incoming ones as $x \to +\infty$. As we have stated the problem (i.e., waves incident from the *left*), $g(x, x')$ fits our physical boundary conditions. Here we see that the boundary conditions are contained in the choice of the contour in the integral representation of $g(x, x')$. We have seen similar circumstances before with integral representations.

Let us return to Eq. 8.170 and examine explicitly the asymptotic behavior of $\psi(k, x)$ as $x \to \pm\infty$. We can rewrite Eq. 8.170 as

$$\psi(k, x) = \frac{e^{ikx}}{\sqrt{2\pi}} + \frac{e^{ikx}}{2ik}\int_{-\infty}^{x} e^{-ikx'} q(x')\psi(k, x')\, dx' + \frac{e^{-ikx}}{2ik}\int_{x}^{\infty} e^{ikx'} q(x')\psi(k, x')\, dx'$$

If we first consider the case $x \to +\infty$, then we have

$$\left|\int_{x}^{\infty} e^{ikx'} q(x')\psi(k, x')\, dx'\right| \le \int_{x}^{\infty} |q(x')||\psi(k, x')|\, dx' \le \psi_{\max}\int_{x}^{\infty} |q(x')|\, dx' \xrightarrow[x\to+\infty]{} 0$$

where we have used the fact that $|\psi(k, x)|$ is bounded (cf. Th. 8.12). A similar argument shows that

$$\left|\int_{-\infty}^{x} e^{-ikx'} q(x')\psi(k, x')\, dx'\right| \xrightarrow[x\to-\infty]{} 0$$

Therefore we have

$$\psi(k, x) \xrightarrow[x\to-\infty]{} \frac{e^{ikx}}{\sqrt{2\pi}} + \frac{e^{-ikx}}{2ik}\int_{-\infty}^{\infty} e^{ikx'} q(x')\psi(k, x')\, dx'$$

$$\psi(k, x) \xrightarrow[x\to+\infty]{} \frac{e^{ikx}}{\sqrt{2\pi}} + \frac{e^{ikx}}{2ik}\int_{-\infty}^{\infty} e^{-ikx'} q(x')\psi(k, x')\, dx'$$

We define the reflection and transmission coefficients $R(k)$ and $T(k)$, respectively, as

$$R(k) \equiv \frac{\pi}{ik} \int_{-\infty}^{\infty} \frac{e^{ikx'}}{\sqrt{2\pi}} q(x') \psi(k, x')\, dx' = \frac{\pi}{ik} \langle \phi_1(k) |\, q\, | \psi(k) \rangle$$

$$T(k) \equiv 1 + \frac{\pi}{ik} \int_{-\infty}^{\infty} \frac{e^{-ikx'}}{\sqrt{2\pi}} q(x') \psi(k, x')\, dx' = 1 + \frac{\pi}{ik} \langle \phi_2(k) |\, q\, | \psi(k) \rangle$$

so that

$$\psi(k, x) \xrightarrow[x \to -\infty]{} \phi_2(k, x) + R(k) \phi_1(k, x)$$

$$\psi(k, x) \xrightarrow[x \to +\infty]{} T(k) \phi_2(k, x)$$

Notice that in order to calculate these coefficients $R(k)$ and $T(k)$ we need to know the exact solutions $\psi(k, x)$ of the scattering problem. Just as we did in the proof of Th. 8.12 we can define a convergent sequence of functions $\{\psi_n(k, x)\}$ as

$$\psi_{n+1}(k, x) = \frac{e^{ikx}}{\sqrt{2\pi}} + \int_{-\infty}^{\infty} \frac{e^{ik|x-x'|}}{2ik} q(x') \psi_n(k, x')\, dx'$$

$$\psi_1(k, x) \equiv \frac{e^{ikx}}{\sqrt{2\pi}} = \phi_2(k, x)$$

We can then use this iterative scheme for $\psi(k, x)$ to define the convergent sequences

$$R_n(k) = \frac{\pi}{ik} \langle \phi_1(k) |\, q\, | \psi_n(k) \rangle$$

$$T_n(k) = 1 + \frac{\pi}{ik} \langle \phi_2(k) |\, q\, | \psi_n(k) \rangle$$

such that

$$\lim_{n \to \infty} R_n(k) = R(k)$$

$$\lim_{n \to \infty} T_n(k) = T(k)$$

b. The Scalar Helmholtz Equation

In the remainder of this chapter we will discuss Green's functions for some of the important partial differential equations of applied mathematics. There will be less rigor here than in the preceding sections of this chapter since the discussion of a Green's function for a general linear partial differential equation is an extremely difficult mathematical problem. For some partial differential operators the "Green's function" is no longer an ordinary function, as it was for the case of an ordinary linear differential operator which we have studied, but is itself a distribution. The reader interested in a rigorous presentation for the existence and nature of Green's functions for partial differential operators can consult the second volume of Courant and Hilbert.

We begin with the *scalar Helmholtz equation*

$$(\nabla^2+k^2)\psi(\mathbf{r})=\rho(\mathbf{r})$$

where as usual

$$\nabla^2=\frac{\partial^2}{\partial x^2}+\frac{\partial^2}{\partial y^2}+\frac{\partial^2}{\partial z^2}$$

and $\rho(\mathbf{r})$ is a given source function, or inhomogeneous term. We seek a solution $\psi(\mathbf{r})$ in a region of space V bounded by a closed surface Σ. Boundary conditions on $\psi(\mathbf{r})$ or on $\partial\psi(\mathbf{r})/\partial n$ are specified everywhere on the enclosing surface Σ. We often assume Dirichlet boundary conditions although all of the arguments given hold for Neumann or mixed ones as well (cf. Section 8.5, Eq. 8.102). We now prove that when the Green's function, $G(\mathbf{r}, \mathbf{r}')$, defined by

$$(\nabla^2+k^2)G(\mathbf{r}, \mathbf{r}')=\delta(\mathbf{r}-\mathbf{r}')$$

is subject to any *homogeneous* boundary condition of the types just mentioned, then it satisfies the important symmetry, or reciprocity, relation

$$G(\mathbf{r}, \mathbf{r}')=G(\mathbf{r}', \mathbf{r})$$

Since we obtain similar results for the Green's functions of other partial differential equations by similar arguments, we now give the proof for this case in some detail. If we multiply the differential equation

$$(\nabla^2+k^2)G(\mathbf{r}, \mathbf{r}')=\delta(\mathbf{r}-\mathbf{r}')$$

by $G(\mathbf{r}, \mathbf{r}'')$ and

$$(\nabla^2+k^2)G(\mathbf{r}, \mathbf{r}'')=\delta(\mathbf{r}-\mathbf{r}'')$$

by $G(\mathbf{r}, \mathbf{r}')$, subtract the results and integrate over the volume, V, we obtain

$$\begin{aligned}G(\mathbf{r}', \mathbf{r}'')-G(\mathbf{r}'', \mathbf{r}')&=\int_V [G(\mathbf{r}, \mathbf{r}'')\,\nabla^2 G(\mathbf{r}, \mathbf{r}')-G(\mathbf{r}, \mathbf{r}')\,\nabla^2 G(\mathbf{r}, \mathbf{r}'')]\,d\mathbf{r}\\&=\int_V \nabla\cdot[G(\mathbf{r}, \mathbf{r}'')\,\nabla G(\mathbf{r}, \mathbf{r}')-G(\mathbf{r}, \mathbf{r}')\,\nabla G(\mathbf{r}, \mathbf{r}'')]\,d\mathbf{r}\\&=\int_\Sigma [G(\mathbf{r}, \mathbf{r}'')\,\nabla G(\mathbf{r}, \mathbf{r}')-G(\mathbf{r}, \mathbf{r}')\,\nabla G(\mathbf{r}, \mathbf{r}'')]\cdot d\mathbf{S}=0\end{aligned}$$

where the last equality follows from the homogeneous boundary conditions satisfied by the Green's function on Σ. We have also used Gauss' theorem,

$$\int_V dv\,\nabla\cdot\mathbf{a}=\int_\Sigma \mathbf{a}\cdot d\mathbf{S}$$

If we are willing to accept the existence of such a Green's function (which we have not proved here although we will explicitly exhibit some for particular cases later), then it is a simple matter to write down the solution to the scalar Helmholtz equation. We multiply

$$(\nabla'^2+k^2)\psi(\mathbf{r}')=\rho(\mathbf{r}')$$

by $G(\mathbf{r}, \mathbf{r}')$ and, after using the reciprocity relation on $G(\mathbf{r}, \mathbf{r}')$, we multiply

$$(\nabla'^2+k^2)G(\mathbf{r}, \mathbf{r}')=\delta(\mathbf{r}-\mathbf{r}')$$

by $\psi(\mathbf{r}')$, subtract these results, integrate over V, use Gauss' theorem, and obtain

$$\psi(\mathbf{r})=\int_V G(\mathbf{r}, \mathbf{r}')\rho(\mathbf{r}')\,d\mathbf{r}'+\int_\Sigma [\psi(\mathbf{r}')\,\nabla' G(\mathbf{r}, \mathbf{r}')-G(\mathbf{r}, \mathbf{r}')\,\nabla'\psi(\mathbf{r}')]\cdot d\mathbf{S}'$$

Again $G(\mathbf{r}, \mathbf{r}')$ is to satisfy homogeneous boundary conditions on Σ, although we have made no requirement that $\psi(\mathbf{r})$ necessarily satisfy *homogeneous* boundary conditions on Σ. Once $\psi(\mathbf{r})$ and $\partial\psi/\partial n$ are specified on Σ, $\psi(\mathbf{r})$ is completely determined everywhere within V and on Σ. Suppose, for example, that $G|_\Sigma=0$ while $\psi|_\Sigma\equiv\psi_\Sigma$, a specified function on Σ. If we now let $\mathbf{r}\to\mathbf{r}_0$ from *within* Σ, where $\mathbf{r}_0$ locates a point *on* Σ, then we obtain

$$\psi(\mathbf{r}_0)\equiv\psi_\Sigma=\int_\Sigma \psi_\Sigma(\mathbf{r}_0')\frac{\partial G(\mathbf{r}_0, \mathbf{r}_0')}{\partial n_0'}\,dS_0'$$

which states that $\partial G(\mathbf{r}_0, \mathbf{r}_0')/\partial n_0'$ must behave like a $\delta(\mathbf{r}_0-\mathbf{r}_0')$ (i.e., like a surface delta function).

Let us summarize this as follows. The solution to the problem

$$(\nabla^2+k^2)\psi(\mathbf{r})=\rho(\mathbf{r})$$
$$\psi|_\Sigma=\psi_\Sigma(\mathbf{r}_0) \qquad \mathbf{r}_0\in\Sigma$$

can be written as

$$\psi(\mathbf{r})=\psi_1(\mathbf{r})+\psi_2(\mathbf{r})$$

where

$$(\nabla^2+k^2)\psi_1(\mathbf{r})=\rho(\mathbf{r})$$
$$\psi_1|_\Sigma=0$$
$$\psi_1(\mathbf{r})=\int_V G(\mathbf{r}, \mathbf{r}')\rho(\mathbf{r}')\,d\mathbf{r}'$$
$$(\nabla^2+k^2)G(\mathbf{r}, \mathbf{r}')=\delta(\mathbf{r}-\mathbf{r}')$$
$$G(\mathbf{r}, \mathbf{r}_0')=0 \qquad \mathbf{r}_0'\in\Sigma$$
$$(\nabla^2+k^2)\psi_2(\mathbf{r})=0$$
$$\psi_2|_\Sigma=\psi_\Sigma(\mathbf{r}_0)$$
$$\psi_2(\mathbf{r})=\int_\Sigma \psi_\Sigma(\mathbf{r}_0')\frac{\partial}{\partial n_0'}G(\mathbf{r}, \mathbf{r}_0')\,dS_0'$$

That is, $\hat{\mathbf{n}}_0' \cdot \nabla' G(\mathbf{r}, \mathbf{r}_0')$ generates the solution to the homogeneous differential equation satisfying inhomogeneous boundary conditions on Σ (cf. the discussion of Section 8.5). The same result holds for Neumann or mixed boundary conditions on Σ.

Example 8.11 It is instructive to apply this approach to the one-dimensional case studied in Sections 8.6a and 8.6b. For illustration let us consider the problem

$$Lu(x) = f(x) \qquad x \in [a, b]$$

$$u(a) = A \qquad u(b) = B$$

and take the Green's function of Eq. 8.116,

$$g(x, t) = \frac{w_1(x) w_2(t) H(t-x) + w_2(x) w_1(t) H(x-t)}{P(w_1, w_2)}$$

$$w_1(a) = 0 = w_2(b)$$

so that

$$g(a, t) = 0 = g(b, t) \qquad a < t < b$$

By the same manipulations we used on the scalar Helmholtz equation we see that

$$\begin{aligned} u(x) &= \int_a^b g(x, t) f(t)\, dt + \int_a^b \frac{d}{dt}\left[u(t)p(t)\frac{\partial}{\partial t} g(x, t) - g(x, t)p(t)\frac{du(t)}{dt}\right] dt \\ &= \int_a^b g(x, t) f(t)\, dt + u(b)p(b)\frac{\partial}{\partial t} g(x, t)\bigg|_{t=b} - u(a)p(a)\frac{\partial}{\partial t} g(x, t)\bigg|_{t=a} \\ &\quad - u'(b)p(b)g(x, b) + u'(a)p(a)g(x, a) \\ &\equiv u_1(x) + u_2(x) \end{aligned}$$

Explicitly we have

$$\frac{\partial}{\partial t} g(x, t) = \frac{w_1(x) w_2'(t) H(t-x) + w_2(x) w_1'(t) H(x-t)}{P(w_1, w_2)}$$

so that for the present case

$$u_2(x) = \frac{Bp(b) w_1(x) w_2'(b)}{P(w_1, w_2)} - \frac{Ap(a) w_2(x) w_1'(a)}{P(w_1, w_2)}$$

where

$$Lu_2(x) = 0$$

$$u_2(a) = A \qquad u_2(b) = B$$

If we apply this same argument to the general *unmixed* inhomogeneous boundary conditions (cf. Eqs. 8.114 and 8.119), we obtain Eq. 8.123.

---

c. The Schrödinger Equation

For the time-independent Schrödinger equation in three dimensions,

$$\left[-\frac{\hbar^2}{2m}\nabla^2+V(\mathbf{r})-E\right]\psi(\mathbf{r})=0$$

we can define

$$q(\mathbf{r})=\frac{2m}{\hbar^2}V(\mathbf{r})$$

$$k^2=\frac{2m}{\hbar^2}E$$

so that this equation becomes

$$(\nabla^2+k^2)\psi(\mathbf{r})=q(\mathbf{r})\psi(\mathbf{r}) \qquad \textbf{(8.171)}$$

We convert this into an integral equation just as we did the one-dimensional problem in Section 8.11a by constructing the Green's function for the corresponding scalar Helmholtz equation in free space (i.e., in an infinite spatial domain) as the solution to

$$(\nabla^2+k^2)G(\mathbf{r},\mathbf{r}')=\delta(\mathbf{r}-\mathbf{r}')$$

The solutions of the corresponding homogeneous problem,

$$(\nabla^2+k^2)\phi(\mathbf{k},\mathbf{r})=0$$

are

$$\phi(\mathbf{k},\mathbf{r})=\frac{1}{(2\pi)^{3/2}}e^{i\mathbf{k}\cdot\mathbf{r}}$$

These are normalized to $\delta(\mathbf{k}-\mathbf{k}')$ as

$$\langle\phi(\mathbf{k}')\mid\phi(\mathbf{k})\rangle=\frac{1}{(2\pi)^3}\int d\mathbf{r}\, e^{i(\mathbf{k}-\mathbf{k}')\cdot\mathbf{r}}=\delta(\mathbf{k}-\mathbf{k}')$$

This is easily seen since $\delta(\mathbf{r}-\mathbf{r}')$ must have the property

$$1=\int_{\substack{\text{all}\\ \text{space}}}\delta(\mathbf{r}-\mathbf{r}')\,d\mathbf{r}'=\int_{-\infty}^{\infty}dx'\int_{-\infty}^{\infty}dy'\int_{-\infty}^{\infty}dz'\,\delta(\mathbf{r}-\mathbf{r}')$$

so that

$$\delta(\mathbf{r}-\mathbf{r}')=\delta(x-x')\,\delta(y-y')\,\delta(z-z')$$

However since we know that

$$\delta(x-x')=\frac{1}{2\pi}\int_{-\infty}^{\infty}e^{ik'(x-x')}\,dk'$$

it follows that

$$\delta(\mathbf{r}-\mathbf{r}')=\frac{1}{(2\pi)^3}\int_{-\infty}^{\infty} dk_x'\int_{-\infty}^{\infty} dk_y'\int_{-\infty}^{\infty} dk_z' e^{i[k_x'(x-x')+k_y'(y-y')+k_z'(z-z')]}$$

$$=\frac{1}{(2\pi)^3}\int d\mathbf{k}' e^{i\mathbf{k}'\cdot(\mathbf{r}-\mathbf{r}')}$$

This allows us to pose the Green's function equation as

$$(\nabla^2+k^2)G(\mathbf{r},\mathbf{r}')=\frac{1}{(2\pi)^3}\int d\mathbf{k}' e^{i\mathbf{k}'\cdot(\mathbf{r}-\mathbf{r}')}$$

If we now operate on both sides of this equation with $(\nabla^2+k^2)^{-1}$, we find an integral representation for $G(\mathbf{r},\mathbf{r}')$

$$G(\mathbf{r},\mathbf{r}')=-\frac{1}{(2\pi)^3}\int \frac{e^{i\mathbf{k}'\cdot(\mathbf{r}-\mathbf{r}')}\,d\mathbf{k}'}{(k'^2-k^2)} \tag{8.172}$$

where we must still specify the integration contour. In order to evaluate this integral we set $\boldsymbol{\rho}=(\mathbf{r}-\mathbf{r}')$ and take $\boldsymbol{\rho}$ to be aligned along the k_z' axis and define $w=\cos\theta$ so that

$$G(\mathbf{r},\mathbf{r}')=\frac{-1}{(2\pi)^3}\int \frac{e^{i\mathbf{k}'\cdot\boldsymbol{\rho}}\,d\mathbf{k}'}{(k'^2-k^2)}=\frac{-1}{(2\pi)^3}\int_0^{\infty} k'^2\,dk'\int_{-1}^{1} dw\int_0^{2\pi}\frac{d\varphi e^{ik'\rho w}}{(k'^2-k^2)}$$

$$=\frac{-1}{(2\pi)^2}\int_0^{\infty}\frac{k'^2\,dk'}{(k'^2-k^2)}\frac{(e^{ik'\rho}-e^{-ik'\rho})}{ik'\rho}=\frac{-1}{2i\rho(2\pi)^2}\int_{-\infty}^{\infty}\frac{k'\,dk'}{(k'^2-k^2)}(e^{ik'\rho}-e^{-ik'\rho})$$

$$=\frac{-1}{(2\pi)^2 i\rho}\int_{-\infty}^{\infty}\frac{k'\,dk'}{(k'^2-k^2)}e^{ik'\rho}$$

We again have the usual contour in the k' plane shown in Fig. 8.8. We must close the contour in the upper-half k' plane to insure convergence. The final result is

$$G(\mathbf{r},\mathbf{r}')=-\frac{2\pi i}{(2\pi)^2 i\rho}\frac{k}{2k}e^{ik\rho}=-\frac{e^{ik\rho}}{4\pi\rho}$$

$$=-\frac{e^{ik|\mathbf{r}-\mathbf{r}'|}}{4\pi|\mathbf{r}-\mathbf{r}'|} \tag{8.173}$$

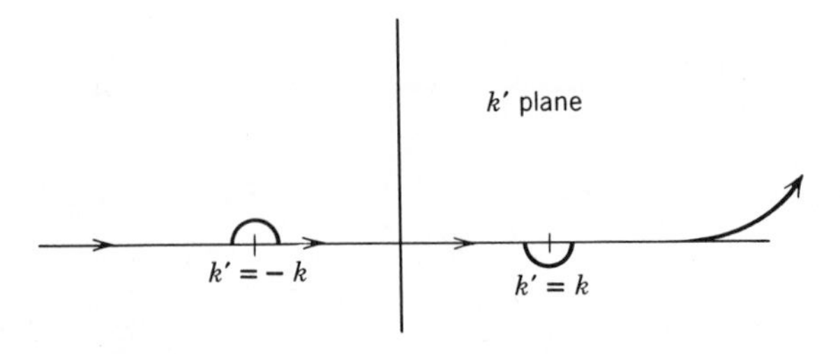

Figure 8.8

These boundary conditions are such that we have outgoing spherical waves as $|\mathbf{r}| \to \infty$ since

$$G(\mathbf{r}, \mathbf{r}') \xrightarrow[|\mathbf{r}|\to\infty]{} -\frac{e^{ik|\mathbf{r}|}}{4\pi|\mathbf{r}|}$$

Therefore, just as in the case of the one-dimensional scattering problem, we can represent the scattering solution to the Schrödinger equation for plane waves incident from the left (i.e., moving in the direction of increasing x along the x axis) as

$$\psi(\mathbf{r}) = \frac{e^{ikx}}{(2\pi)^{3/2}} - \frac{1}{4\pi}\int \frac{e^{ik|\mathbf{r}-\mathbf{r}'|}}{|\mathbf{r}-\mathbf{r}'|} q(\mathbf{r}')\psi(\mathbf{r}')\, d\mathbf{r}'$$

d. The Scalar Wave Equation

We now study the Green's function for the *scalar wave equation*

$$\left(\nabla^2 - \frac{1}{c^2}\frac{\partial^2}{\partial t^2}\right)\psi(\mathbf{r}, t) = \rho(\mathbf{r}, t) \qquad \textbf{(8.174)}$$

The Green's function is the solution for a point source emitting at a time $t = t'$. Therefore $G(\mathbf{r}, \mathbf{r}'; t, t')$ must satisfy

$$\left(\nabla^2 - \frac{1}{c^2}\frac{\partial^2}{\partial t^2}\right)G(\mathbf{r}, \mathbf{r}'; t, t') = \delta(\mathbf{r}-\mathbf{r}')\,\delta(t-t') \qquad \textbf{(8.175)}$$

subject to the initial conditions

$$\begin{aligned} G(\mathbf{r}, \mathbf{r}'; t, t') &= 0 \\ \frac{\partial}{\partial t} G(\mathbf{r}, \mathbf{r}'; t, t') &= 0 \qquad t < t' \end{aligned} \qquad \textbf{(8.176)}$$

and to homogeneous boundary conditions on some closed spatial surface Σ. The conditions in Eq. 8.176 are the so-called causality conditions that appear physically reasonable, at least macroscopically.

In the case of the scalar Helmholtz equation

$$(\nabla^2 + k^2)\psi(\mathbf{r}) = \rho(\mathbf{r}) \qquad \textbf{(8.177)}$$

we saw that the Green's function satisfied the symmetry relation $G(\mathbf{r}, \mathbf{r}') = G(\mathbf{r}', \mathbf{r})$. This symmetry relation followed from the self-adjoint character of the operator. Formally the Helmholtz equation (8.177) can be obtained from the scalar wave equation, (8.174), if $\rho(\mathbf{r}, t) = f(\mathbf{r})$ only and if we substitute $\psi(r, t) \to e^{\pm ict}\varphi(\mathbf{r})$. In fact, however, they are really very different types of differential equations, as we will see, since one is hyperbolic and the other is elliptic.

In the present case the reciprocity relation for the Green's function of Eq.

8.175 cannot simply read

$$G(\mathbf{r}, \mathbf{r}'; t, t') = G(\mathbf{r}', \mathbf{r}; t', t) \tag{8.177}$$

since if $t > t'$ the right side is zero by causality, Eq. 8.176. Instead, the time must be reversed so that

$$G(\mathbf{r}, \mathbf{r}'; t, t') = G(\mathbf{r}', \mathbf{r}; -t', -t) \tag{8.178}$$

We prove this as follows using the same type manipulations as in the corresponding proof for the scalar Helmholtz equation.

$$G_2 \equiv G(\mathbf{r}, \mathbf{r}''; -t, -t'') \to \nabla^2 G(\mathbf{r}, \mathbf{r}'; t, t') - \frac{1}{c^2}\frac{\partial^2}{\partial t^2} G(\mathbf{r}, \mathbf{r}'; t, t')$$

$$= \delta(\mathbf{r}-\mathbf{r}')\,\delta(t-t')$$

$$G_1 \equiv G(\mathbf{r}, \mathbf{r}'; t, t') \to \nabla^2 G(\mathbf{r}, \mathbf{r}''; -t, -t'') - \frac{1}{c^2}\frac{\partial^2}{\partial t^2} G(\mathbf{r}, \mathbf{r}''; -t, -t'')$$

$$= \delta(\mathbf{r}-\mathbf{r}'')\,\delta(t-t'')$$

If we carry out the indicated multiplications, subtract, and integrate over all space and time, we have

$$\int_{-\infty}^{\infty} dt \int dv \left\{ [G_2 \nabla^2 G_1 - G_1 \nabla^2 G_2] - \frac{1}{c^2}\left[G_2 \frac{\partial^2 G_1}{\partial t^2} - G_1 \frac{\partial^2 G_2}{\partial t^2} \right] \right\}$$
$$= G(\mathbf{r}', \mathbf{r}''; -t', -t'') - G(\mathbf{r}'', \mathbf{r}'; t'', t') \tag{8.179}$$

We again use the identity

$$\nabla \cdot [G_2 \nabla G_1 - G_1 \nabla G_2] = G_2 \nabla^2 G_1 - G_1 \nabla^2 G_2$$

and Gauss' theorem so that the first term in the integral of Eq. 8.179 becomes

$$\int_{-\infty}^{\infty} dt \int_{\Sigma} d\mathbf{S} \cdot [G_2 \nabla G_1 - G_1 \nabla G_2] = 0$$

since both G_1 and G_2 satisfy the same homogeneous boundary conditions on Σ. Finally the second term on the left side of Eq. 8.179 becomes

$$\int dv \left\{ G(\mathbf{r}, \mathbf{r}''; -t, -t'') \frac{\partial}{\partial t} G(\mathbf{r}, \mathbf{r}'; t, t') - G(\mathbf{r}, \mathbf{r}'; t, t') \frac{\partial}{\partial t} G(\mathbf{r}, \mathbf{r}''; -t, -t'') \right\} \Bigg|_{-\infty}^{\infty}$$

which vanishes by the causality conditions, Eq. 8.176. Therefore, as claimed, the reciprocity relation Eq. 8.178 frollows from Eq. 8.179 with the use of Eq. 8.176.

Notice that in the present case G depends on the shape of the surface for a finite boundary. Once we have constructed G for a given geometry, then we obtain the solution to Eq. 8.174 as follows. As previously perform the multiplications indicated by

$$G(\mathbf{r}, \mathbf{r}'; t, t') \to \nabla'^2\psi(\mathbf{r}', t) - \frac{1}{c^2}\frac{\partial^2}{\partial t'^2}\psi(\mathbf{r}', t') = \rho(\mathbf{r}', t')$$

$$\psi(\mathbf{r}', t') \to \left(\nabla'^2 - \frac{1}{c^2}\frac{\partial^2}{\partial t'^2}\right)G(\mathbf{r}, \mathbf{r}'; t, t') = \delta(\mathbf{r}-\mathbf{r}')\,\delta(t-t')$$

subtract these equations, integrate over the primed variables, and find that

$$\int_{-\infty}^{\infty} dt' \int dv' \left\{G\,\nabla'^2\psi - \psi\,\nabla'^2 G - \frac{1}{c^2}\left[G\frac{\partial^2\psi}{\partial t'^2} - \psi\frac{\partial^2 G}{\partial t'^2}\right]\right\}$$

$$= \int_{-\infty}^{\infty} dt' \int dv' G(\mathbf{r}, \mathbf{r}'; t, t')\rho(\mathbf{r}', t') - \psi(\mathbf{r}, t)$$

Therefore, after the same type use of Gauss' theorem as above, we obtain

$$\psi(\mathbf{r}, t) = \int_{-\infty}^{\infty} dt' \int dv' G(\mathbf{r}, \mathbf{r}'; t, t')\rho(\mathbf{r}', t')$$

$$-\int_{-\infty}^{\infty} dt' \int_{\Sigma} d\mathbf{S}' \cdot [G\,\nabla'\psi - \psi\,\nabla' G] - \frac{1}{c^2}\int dv'\left[G\frac{\partial\psi}{\partial t'}\bigg|_{-\infty} - \psi\frac{\partial G}{\partial t'}\bigg|_{-\infty}\right] \qquad \textbf{(8.180)}$$

where $t_0 = -\infty$ could just as well be replaced by any initial time t_0 when the problem began. The right side of Eq. 8.180 contains only G, ρ, and the values of ψ, $\partial\psi/\partial t$ specified initially and ψ, $\nabla\psi$ specified on the closed boundary Σ. It therefore represents a complete solution to the problem.

We now consider the Green's function for an infinite spatial region so that boundary effects are neglected. We construct the solution to Eq. 8.175 using the spectral decomposition for the δ functions.

$$G(\mathbf{r}, \mathbf{r}'; t, t') = \left(\nabla^2 - \frac{1}{c^2}\frac{\partial^2}{\partial t^2}\right)^{-1} \delta(\mathbf{r}-\mathbf{r}')\,\delta(t-t')$$

$$\delta(t-t') = \frac{1}{2\pi}\int_{-\infty}^{\infty} e^{i\omega(t-t')}\,d\omega$$

$$\delta(\mathbf{r}-\mathbf{r}') = \frac{1}{(2\pi)^3}\int e^{i\mathbf{k}\cdot(\mathbf{r}-\mathbf{r}')}\,d\mathbf{k}$$

$$G(\mathbf{r}, \mathbf{r}'; t, t') = (2\pi)^{-4}\left(\nabla^2 - \frac{1}{c^2}\frac{\partial^2}{\partial t^2}\right)^{-1} \int e^{i\mathbf{k}'(\mathbf{r}-\mathbf{r}')}\,d\mathbf{k} \int_{-\infty}^{\infty} e^{i\omega(t-t')}\,d\omega$$

$$= \frac{-1}{(2\pi)^4}\int \frac{e^{i\mathbf{k}\cdot(\mathbf{r}-\mathbf{r}')}e^{i\omega(t-t')}\,d\mathbf{k}\,d\omega}{k^2 - \omega^2/c^2}$$

$$= \frac{-1}{(2\pi)^4}\int_{-\infty}^{\infty} d\omega \int_0^{\infty} k^2\,dk \int_{-1}^{1} dw \int_0^{2\pi} \frac{d\varphi e^{ik\rho w}e^{i\omega(t-t')}}{k^2 - \omega^2/c^2}$$

$$= \frac{-1}{(2\pi)^3}\int_{-\infty}^{\infty} d\omega \int_0^{\infty} \frac{k^2\,dk(e^{ik\rho} - e^{-ik\rho})e^{i\omega(t-t')}}{(k^2 - \omega^2/c^2)ik\rho}$$

$$=\frac{-1}{2(2\pi)^3}\int_{-\infty}^{\infty} d\omega \int_{-\infty}^{\infty} \frac{k\,dk(e^{ik\rho}-e^{-ik\rho})}{i\rho(k^2-\omega^2/c^2)}\,e^{i\omega(t-t')}$$

$$=\frac{-1}{(2\pi)^3 2i\rho}\int_{-\infty}^{\infty} d\omega\left[\int_{-\infty}^{\infty}\frac{k\,dke^{ik\rho}}{(k-\omega/c)(k+\omega/c)}\right.$$

$$\left.-\int_{-\infty}^{\infty}\frac{k\,dke^{-ik\rho}}{(k-\omega/c)(k+\omega/c)}\right]e^{i\omega(t-t')}$$

$$=\frac{-1}{(2\pi)^3 i\rho}\int_{-\infty}^{\infty} d\omega e^{i\omega(t-t')}\int_{-\infty}^{\infty}\frac{k\,dke^{ik\rho}}{(k-\omega/c)(k+\omega/c)}$$

$$=\frac{-2\pi i}{(2\pi)^3 i\rho}\int_{-\infty}^{\infty} d\omega e^{i\omega(t-t')}\,\frac{e^{-i(\omega/c)\rho}}{2}$$

$$=\frac{-1}{4\pi\rho}\frac{1}{2\pi}\int_{-\infty}^{\infty} e^{-i\omega(\rho/c-\tau)}\,d\omega=\frac{-1}{4\pi}\frac{\delta(\rho/c-\tau)}{\rho} \qquad \textbf{(8.181)}$$

Here we have defined

$$\rho=|\mathbf{r}-\mathbf{r}'| \qquad \tau=t-t'$$

Notice that the choice of contour in the k plane (cf. Fig. 8.9) reflects the causality condition. As we have seen before, the deformation of contour about $k=\pm\omega/c$ has replaced $\lim_{\varepsilon\to 0^+}$ in

$$\left(\nabla^2-\frac{1}{c^2}\frac{\partial^2}{\partial t^2}-i\varepsilon\right)^{-1}$$

Therefore, if in the infinite domain the initial conditions on the field are $\psi=0=\partial\psi/\partial t$ at $t=-\infty$, Eq. 8.180 becomes

$$\psi(\mathbf{r},t)=-\frac{1}{4\pi}\int_{-\infty}^{\infty} dt'\int d\mathbf{r}'\,\frac{\delta[(|\mathbf{r}-\mathbf{r}'|/c)-(t-t')]}{|\mathbf{r}-\mathbf{r}'|}\,\rho(\mathbf{r}',t')$$

$$=-\frac{1}{4\pi}\int\frac{d\mathbf{r}'}{|\mathbf{r}-\mathbf{r}'|}\,\rho\left[\mathbf{r}',\left(t-\frac{|\mathbf{r}-\mathbf{r}'|}{c}\right)\right] \qquad \textbf{(8.182)}$$

Finally we can find a spectral representation of the Green's function

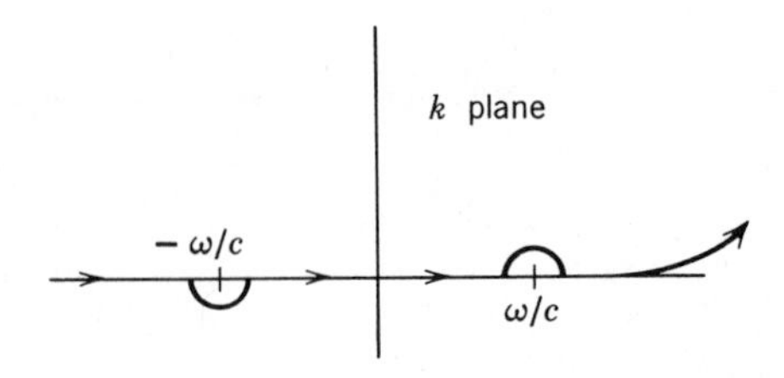

Figure 8.9

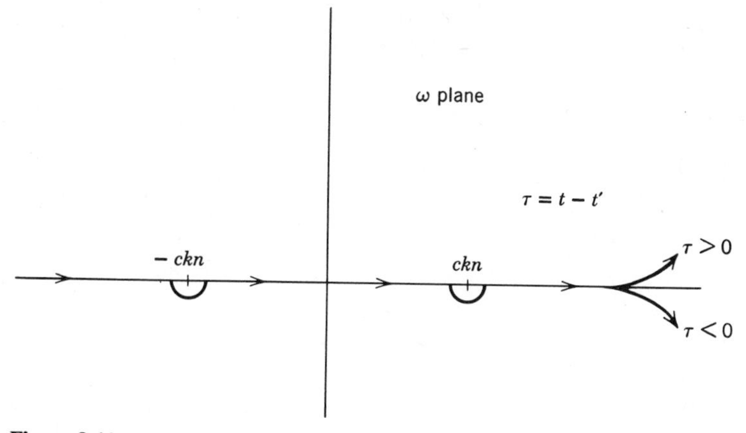

Figure 8.10

$G(\mathbf{r}, \mathbf{r}'; t, t')$ for a finite spatial domain. We write

$$\left(\nabla^2 - \frac{1}{c^2}\frac{\partial^2}{\partial t^2}\right)G(\mathbf{r}, \mathbf{r}'; t, t') = \delta(\mathbf{r}-\mathbf{r}')\,\delta(t-t') = \sum_n u_n^*(\mathbf{r}')u_n(\mathbf{r})\,\delta(t-t')$$

where $\{u_n(\mathbf{r})\}$ is a complete set of solutions to

$$(\nabla^2 + k_n^2)u_n(\mathbf{r}) = 0$$

satisfying the homogeneous boundary conditions specified on the closed spatial surface, Σ. As before we calculate G to be

$$\begin{aligned}
G(\mathbf{r}, \mathbf{r}'; t, t') &= \left(\nabla^2 - \frac{1}{c^2}\frac{\partial^2}{\partial t^2}\right)^{-1} \sum_n u_n^*(\mathbf{r}')u_n(\mathbf{r}) \frac{1}{2\pi}\int_{-\infty}^{\infty} e^{i\omega(t-t')}\,d\omega \\
&= \frac{1}{2\pi}\sum_n u_n^*(\mathbf{r}')u_n(\mathbf{r}) \int_{-\infty}^{\infty} \frac{e^{i\omega(t-t')}\,d\omega}{\omega^2/c^2 - k_n^2} \\
&= \frac{c^2}{2\pi}\sum_n u_n^*(\mathbf{r}')u_n(\mathbf{r}) \int_{-\infty}^{\infty} \frac{e^{i\omega(t-t')}\,d\omega}{(\omega + ck_n)(\omega - ck_n)}
\end{aligned}$$

If $(t-t')>0$ we must close the contour upward for convergence while if $(t-t')<0$, then downward. The choice of contour is again dictated by causality.

$$\begin{aligned}
G(\mathbf{r}, \mathbf{r}'; t, t') &= \frac{c^2}{2\pi}\sum_n u_n^*(\mathbf{r}')u_n(\mathbf{r})H(t-t')2\pi i\left[\frac{e^{ick_n(t-t')}}{2ck_n} - \frac{e^{-ick_n(t-t')}}{2ck_n}\right] \\
&= -c\sum_n u_n^*(\mathbf{r}')u_n(\mathbf{r})\frac{\sin[ck_n(t-t')]}{k_n}H(t-t') \qquad \textbf{(8.183)}
\end{aligned}$$

This obviously satisfies both causality and the boundary conditions.

It is worth stressing again that the Green's function depends only on the geometry and the initial and boundary conditions so that once it has been constructed it can be used to solve *any* physical problem for the specified configuration.

e. The Diffusion or Heat Equation

The *diffusion equation*

$$\left(\nabla^2 - a^2 \frac{\partial}{\partial t}\right)\psi(\mathbf{r}, t) = \rho(\mathbf{r}, t) \tag{8.184}$$

unlike the scalar wave equation, is not invariant under time reversal (i.e., it differentiates between past and future; $t \to -t$). As in the case of the scalar wave equation, we demand that

$$G(\mathbf{r}, \mathbf{r}'; t, t') = 0 \qquad t < t' \tag{8.185}$$

Also since

$$\left(\nabla^2 - a^2 \frac{\partial}{\partial t}\right) G(\mathbf{r}, \mathbf{r}'; t, t') = \delta(\mathbf{r} - \mathbf{r}')\, \delta(t - t')$$

it follows that

$$\left(\nabla'^2 + a^2 \frac{\partial}{\partial t'}\right) G(\mathbf{r}, \mathbf{r}'; t, t') = \delta(\mathbf{r} - \mathbf{r}')\, \delta(t - t')$$

once we use the reciprocity relation

$$G(\mathbf{r}, \mathbf{r}'; t, t') = G(\mathbf{r}', \mathbf{r}; -t', -t)$$

which may be proved as it was for the scalar wave equation. By manipulations similar to those that led to Eq. 8.180 we find

$$\begin{aligned}\psi(\mathbf{r}, t) = &\int_0^\infty dt' \int dv'\, G(\mathbf{r}, \mathbf{r}'; t, t') \rho(\mathbf{r}, t') \\ &+ \int_0^\infty dt' \int_\Sigma d\mathbf{S} \cdot [G(\mathbf{r}, \mathbf{r}'; t, t')\, \nabla' \psi(\mathbf{r}', t') - \psi(\mathbf{r}', t')\, \nabla' G(\mathbf{r}, \mathbf{r}'; t, t')] \\ &+ a^2 \int dv' [\psi(\mathbf{r}', t') G(\mathbf{r}, \mathbf{r}'; t, t']\Big|_{t'=0}\end{aligned} \tag{8.186}$$

We have here chosen to begin the problem at $t = 0$ rather than at $t = -\infty$.

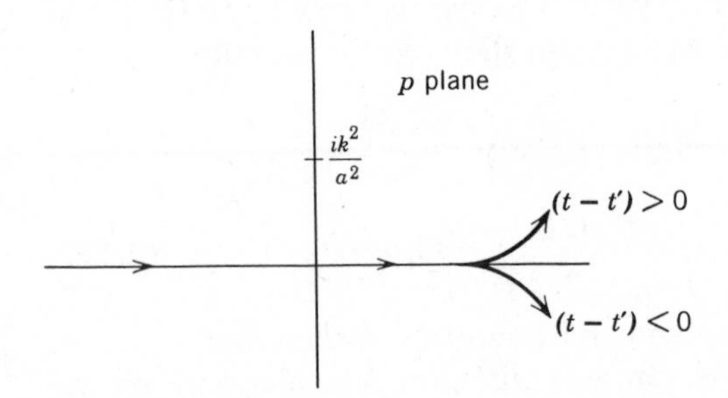

Figure 8.11

As previously we may construct the Green's function for the infinite spatial domain by use of the spectral representation for the δ functions.

$$\left(\nabla^2 - a^2 \frac{\partial}{\partial t}\right) G(\mathbf{r}, \mathbf{r}'; t, t') = \delta(\mathbf{r}-\mathbf{r}')\, \delta(t-t') \qquad \mathbf{(8.187)}$$

$$G(\mathbf{r}, \mathbf{r}'; t, t') = \left(\nabla^2 - a^2 \frac{\partial}{\partial t}\right)^{-1} \frac{1}{(2\pi)^4} \int d\mathbf{k} e^{i\mathbf{k}\cdot\boldsymbol{\rho}} \int_{-\infty}^{\infty} e^{ip(t-t')}\, dp$$

$$= \frac{-1}{(2\pi)^4} \int d\mathbf{k} \int_{-\infty}^{\infty} \frac{dp e^{i\mathbf{k}\cdot\boldsymbol{\rho}} e^{ip(t-t')}}{k^2 + a^2 pi}$$

$$= \frac{-1}{ia^2(2\pi)^4} \int d\mathbf{k} \int_{-\infty}^{\infty} \frac{dp e^{i\mathbf{k}\cdot\boldsymbol{\rho}} e^{ip(t-t')}}{p - ik^2/a^2}$$

$$= \frac{-1}{ia^2(2\pi)^3} \int_0^{\infty} k\, dk \int_{-\infty}^{\infty} \frac{dp[e^{ik\rho} - e^{-ik\rho}] e^{ip(t-t')}}{i\rho(p - ik^2/a^2)}$$

$$= \frac{1}{a^2(2\pi)^3} \int_{-\infty}^{\infty} \frac{k\, dk}{\rho} e^{ik\rho} \int_{-\infty}^{\infty} \frac{dp e^{ip(t-t')}}{p - ik^2/a^2}$$

$$= \frac{2\pi i}{a^2(2\pi)^3} \int_{-\infty}^{\infty} k\, dk e^{ik\rho} e^{-(k^2/a^2)(t-t')} H(t-t')$$

$$= \frac{-1}{a^2 4\pi^2 \rho} \int_{-\infty}^{\infty} k\, dk \sin(k\rho) e^{-(k^2/a^2)(t-t')} H(t-t')$$

$$= \frac{2}{a^2 4\pi^2 \rho} \frac{\partial}{\partial \rho} \int_0^{\infty} dk \cos(k\rho) e^{-(k^2/a^2)(t-t')} H(t-t')$$

$$= \frac{2}{a^2 4\pi^2 \rho} \frac{\partial}{\partial \rho} \left[\frac{\sqrt{\pi} a}{2\sqrt{t-t'}} e^{-(\rho^2/4)[a^2/(t-t')]}\right] H^{(t-t')}$$

$$= \frac{-1}{4\pi} \frac{a}{2\sqrt{\pi}\,(t-t')^{3/2}} \exp\left[\frac{-a^2 |\mathbf{r}-\mathbf{r}'|^2}{4(t-t')}\right] H(t-t') \qquad \mathbf{(8.188)}$$

Again we can compute G for a finite spatial domain by using a spectral representation in terms of the eigenfunctions of the scalar Helmholtz equation,

$$(\nabla^2 + k_n^2) u_n(\mathbf{r}) = 0$$

for $\delta(\mathbf{r}, \mathbf{r}')$.

$$\left(\nabla^2 - a^2 \frac{\partial}{\partial t}\right) G(\mathbf{r}, \mathbf{r}'; t, t') = \sum_n u_n^*(\mathbf{r}') u_n(\mathbf{r})\, \delta(t-t')$$

The result is

$$G(\mathbf{r}, \mathbf{r}'; t, t') = -\frac{H(t-t')}{a^2} \sum_n \exp\left[-\frac{k_n^2}{a^2}(t-t')\right] u_n^*(\mathbf{r}') u_n(\mathbf{r}) \qquad \mathbf{(8.189)}$$

Example 8.12 As a simple example consider the one-dimensional homogeneous diffusion equation

$$\left(\frac{\partial^2}{\partial x^2}-a^2\frac{\partial}{\partial t}\right)u(x,t)=0 \qquad 0\le x\le l$$

subject to the conditions

$$\left.\begin{array}{ll}\text{i.} & u(0,t)=0,\\ \text{ii.} & u(l,t)=0,\end{array}\right\} \qquad \text{boundary or "surface" terms}$$

$$\text{iii.}\quad u(x,0)=f(x) \qquad \text{initial condition}$$

We can first solve this directly as an eigenvalue problem. Let

$$u(x,t)=X(x)T(t)$$

so that

$$X''(x)T(t)-a^2X(x)T'(t)=0$$

$$\frac{X''(x)}{X(x)}\equiv a^2\frac{T'(t)}{T(t)}=-k^2 \qquad \text{a constant}$$

We then have

$$X''(x)+k^2X(x)=0 \qquad T'(t)+(k/a)^2T(t)=0$$

$$X(x)=\alpha\sin kx+\beta\cos kx$$

with

$$X(0)=0$$

which requires $\beta=0$ and

$$X(l)\equiv\alpha\sin kl=0$$

which requires $k_n=n\pi/l$, $n=1, 2, \ldots$. Also we see that

$$\frac{dT}{T}=-\left(\frac{k_n}{a}\right)^2 dt$$

or that $T=A\exp[(-k_n/a)^2t]$. We can finally write for the solution to the original problem

$$u_n(x,t)=\sin(n\pi x/l)e^{-(k_n/a)^2t}$$

or

$$u(x,t)=\sum_{n=1}^{\infty}\alpha_n u_n(x,t)$$

The initial condition

$$u(x,0)\equiv f(x)=\sum_n \alpha_n\sin(n\pi x/l) \qquad 0\le x\le l$$

requires that

$$\alpha_n=\frac{2}{l}\int_0^l f(x)\sin\left(\frac{n\pi x}{l}\right)dx$$

However we can also solve this by a Green's function technique by use of Eq. 8.186. In the present case $\rho(x, t) \equiv 0$ and we have homogeneous boundary conditions on the "surface" so that

$$u(x, t) = -a^2 \int_0^l G(x, x'; t, 0) u(x', 0)\, dx'$$

Again we must solve for an orthonormal set $\{u_n(x)\}$ such that

$$u_n'' + k_n^2 u_n = 0$$

and subject to the boundary conditions. The result is easily seen to be

$$u_n(x) = \sqrt{\frac{2}{l}} \sin\left(\frac{n\pi x}{l}\right) \qquad k_n = \frac{n\pi}{l}$$

so that

$$G(x, x'; t, t') = -\frac{H(t-t')}{a^2} \frac{2}{l} \sum_{n=1}^{\infty} e^{-(k_n^2/a^2)(t-t')} \sin\left(\frac{n\pi x'}{l}\right) \sin\left(\frac{n\pi x}{l}\right)$$

Therefore the solution is again given as

$$u(x, t) = H(t) \frac{2}{l} \sum_{n=1}^{\infty} \left[\int_0^l f(x') \sin\left(\frac{n\pi x'}{l}\right) dx'\right] \sin\left(\frac{n\pi x}{l}\right) \exp\left(-\frac{k_n^2}{a^2} t\right)$$

A8.1 PROOF OF THE COMPLETENESS OF THE STURM-LIOUVILLE EIGENFUNCTIONS VIA THE HILBERT-SCHMIDT THEOREM

We will give an alternative proof of the completeness of the eigenfunctions of the Sturm-Liouville equation

$$\frac{d}{dx}\left[p(x) \frac{d}{dx} u_n(x)\right] + [q(x) + \lambda_n r(x)] u_n(x) = 0 \qquad x \in [a, b] \tag{A8.1}$$

subject to the unmixed homogeneous boundary conditions

$$B_1(u_n) = 0 = B_2(u_n) \tag{A8.2}$$

where the $B_j(u_n)$ are given in either Eq. 8.83 or Eq. 8.84. We know from Section 8.3 that the $\{\lambda_n\}$ are denumerably infinite with an accumulation point at infinity.

We assume for this proof that $\lambda = 0$ is *not* an eigenvalue. If it were we could simply redefine $q(x)$, $r(x)$, and λ so that it would not be. Let $w_1(x)$ and $w_2(x)$ be the linearly independent solutions to

$$Lw_j \equiv \frac{d}{dx}\left[p(x) \frac{d}{dx} w_j(x)\right] + q(x) w_j(x) = 0 \qquad j = 1, 2 \tag{A8.3}$$

such that

$$B_1(w_1) = 0 = B_2(w_2) \tag{A8.4}$$

These $w_1(x)$ and $w_2(x)$ must be linearly independent since $\lambda=0$ is assumed not to be an eigenvalue. A Green's function, $g(x, t)$, such that

$$\begin{aligned} L_x g(x, t) &= \delta(x-t) \\ B_1[g(x, t)] &= 0 = B_2[g(x, t)] \end{aligned} \tag{A8.5}$$

is given as (cf. Eq. 8.116)

$$g(x, t) = \frac{w_1(x)w_2(t)H(t-x) + w_2(x)w_1(t)H(x-t)}{P(w_1, w_2)} = g(t, x) \tag{A8.6}$$

where $P(w_1, w_2) = \text{const.}$ since L is self-adjoint (cf. Th. 8.9).

If we use Eq. 8.144 we can rewrite the eigenvalue problem of Eqs. A8.1 and A8.2 as

$$u_n(x) = -\lambda_n \int_a^b g(x, t) r(t) u_n(t)\, dt \tag{A8.7}$$

where

$$B_1(u_n) = 0 = B_2(u_n)$$

As in Section 8.3 we assume that $r(x) > 0$. Since the $\{u_n(x)\}$ are orthonormal w.r.t. the weight function $r(x)$ as (cf. Eq. 8.92)

$$\langle u_n \mid u_m \rangle \equiv \int_a^b u_n^*(x) u_m(x) r(x)\, dx = \delta_{nm} \tag{A8.8}$$

then if we define

$$\begin{aligned} \varphi_n(x) &\equiv \sqrt{r(x)}\; u_n(x) \\ K(x, t) &\equiv \sqrt{r(x)}\; g(x, t) \sqrt{r(t)} = K(t, x) \end{aligned} \tag{A8.9}$$

we obtain

$$\varphi_n(x) = -\lambda_n \int_a^b K(x, t) \varphi_n(t)\, dt$$

with

$$\langle \varphi_n \mid \varphi_m \rangle \equiv \int_a^b \varphi_n^*(x) \varphi_m(x)\, dx = \delta_{nm} \tag{A8.10}$$

We also see at once that

$$\begin{aligned} \|K\|^2 &\equiv \int_a^b dx \int_a^b dt\, |K(x, t)|^2 < \infty \\ k^2(x) &\equiv \int_a^b dt\, |K(x, t)|^2 < C \end{aligned} \tag{A8.11}$$

so that the Hilbert-Schmidt theorem applies. This tells us that the $\{\varphi_n(x)\}$ span the range of K. From the result of Section A2.4 we need simply establish that the null space of K contains only the zero vector to prove that the $\{\varphi_n(x)\}$ is a complete set. Let $f(x)$ be *any* function in the null space of K so that

$$\int_a^b K(x, t)f(t)\,dt \equiv 0 \qquad \forall x \in [a, b]$$

This implies that (cf. Eq. A8.9)

$$\int_a^b g(x, t)\sqrt{r(t)}\, f(t)\,dt \equiv 0 \tag{A8.12}$$

or

$$L_x \int_a^b g(x, t)\sqrt{r(t)}\, f(t)\,dt = \sqrt{r(x)}\, f(x) \equiv 0 \tag{A8.13}$$

This requires that $f(x) \equiv 0$, which establishes the completeness of the $\{u_n(x)\}$.

A8.2 GENERAL CONDITION FOR A SECOND-ORDER LINEAR ORDINARY DIFFERENTIAL OPERATOR TO BE SELF-ADJOINT

We now prove that the Sturm-Liouville operator

$$Lu \equiv \frac{d}{dx}\left[p(x)\frac{du}{dx}(x)\right] + q(x)u(x) \qquad x \in [a, b] \tag{A8.14}$$

subject to the general homogeneous boundary conditions

$$B_1(u) = 0 = B_2(u) \tag{A8.15}$$

where $B_1(u)$ and $B_2(u)$ are given in Eqs. 8.93, is self-adjoint if and only if

$$p(a)(d_1c_2 - d_2c_1) = p(b)(a_2b_1 - a_1b_2) \tag{A8.16}$$

We will, without loss of generality, assume that all solutions to the differential equation are real. From Eq. 8.99 we see that the concomitant is given as

$$\begin{aligned}\langle v \mid Lu\rangle - \langle Lu \mid v\rangle &\equiv P(v, u)|_a^b \\ &= p(b)[v(b)u'(b) - u(b)v'(b)] - p(a)[v(a)u'(a) - u(a)v'(a)]\end{aligned} \tag{A8.17}$$

The vanishing of Eq. A8.17 is the necessary and sufficient condition that L be self-adjoint.

Since the boundary conditions, $B_1(u)$ and $B_2(u)$, must be linearly independent, not all of the determinants

$$\begin{matrix} a_1b_2 - b_1a_2 & a_1c_2 - a_2c_1 & a_1d_2 - a_2d_1 \\ b_1c_2 - b_2c_1 & b_1d_2 - b_2d_1 & c_1d_2 - c_2d_1 \end{matrix} \tag{A8.18}$$

can be zero. This follows from the conditions for the linear independence of two vectors, for example,

$$f_1=\sum_{j=1}^{4} \alpha_j x_j \qquad f_2=\sum_{j=1}^{4} \beta_j x_j$$

If these were to be linearly dependent, then there would exist nonzero c_1 and c_2 such that

$$c_1 f_1 + c_2 f_2 = 0$$

or

$$\sum_{j=1}^{4} (c_1\alpha_j + c_2\beta_j)x_j = 0$$

which implies

$$c_1\alpha_j + c_2\beta_j = 0 \qquad j=1, 2, 3, 4$$

This is possible for nonzero c_1 and c_2 if and only if

$$\frac{\alpha_j}{\beta_j} = \text{const.} \qquad j=1, 2, 3, 4$$

which would require the vanishing of the six determinants in Eq. A8.18.

Let us assume that

$$a_1 b_2 - a_2 b_1 \neq 0 \tag{A8.19}$$

The argument we are about to give can equally well be started by assuming that *any* one of the six determinants in Eq. A8.18 is nonzero. We now choose two more linearly independent boundary conditions, say

$$\begin{aligned} B_3(u) &= u(b) \\ B_4(u) &= u'(b) \end{aligned} \tag{A8.20}$$

All four of these $B_j(u)$, $j=1, 2, 3, 4$, are linearly independent since the determinant of the entire system is

$$\Delta \equiv \begin{vmatrix} a_1 & b_1 & c_1 & d_1 \\ a_2 & b_2 & c_2 & d_2 \\ 0 & 0 & 1 & 0 \\ 0 & 0 & 0 & 1 \end{vmatrix} = a_1 b_2 - a_2 b_1 \neq 0 \tag{A8.21}$$

We next introduce four boundary conditions on $v(x)$, say $\bar{B}_j(v)$, $j=1, 2, 3, 4$, such that

$$P(v, u)\Big|_a^b = \sum_{j=1}^{4} B_j(u)\bar{B}_j(v) \tag{A8.22}$$

so that

$$\begin{aligned} p(b)[v(b)u'(b) - u(b)v'(b)] - p(a)[v(a)u'(a) - u(a)v'(a)] \\ = [a_1 u(a) + b_1 u'(a) + c_1 u(b) + d_1 u'(b)]\bar{B}_1(v) \\ + [a_2 u(a) + b_2 u'(a) + c_2 u(b) + d_2 u'(b)]\bar{B}_2(v) \\ + u(b)\bar{B}_3(v) + u'(b)\bar{B}_4(v) \end{aligned} \tag{A8.23}$$

This yields a system of equations for $\bar{B}_j(v)$ as

$$a_1\bar{B}_1(v)+a_2\bar{B}_2(v)=p(a)v'(a)$$
$$b_1\bar{B}_1(v)+b_2\bar{B}_2(v)=-p(a)v(a)$$
$$c_1\bar{B}_1(v)+c_2\bar{B}_2(v)+\bar{B}_3(v)=-p(b)v'(b)$$
$$d_1\bar{B}_1(v)+d_2\bar{B}_2(v)+\bar{B}_4(v)=p(b)v(b)$$

Since $\Delta\neq 0$ is given, we obtain from Cramer's rule

$$\begin{aligned}
\bar{B}_1(v)&=\frac{p(a)}{\Delta}[a_2v(a)+b_2v'(a)]\\
\bar{B}_2(v)&=\frac{-p(a)}{\Delta}[a_1v(a)+b_1v'(a)]\\
\bar{B}_3(v)&=p(b)v'(b)+\frac{1}{\Delta}[(b_2c_1-b_1c_2)p(a)v'(a)+(a_2c_1-a_1c_2)p(a)v(a)]\\
\bar{B}_4(v)&=-p(b)v(b)+\frac{1}{\Delta}[(b_2d_1-d_2b_1)p(a)v'(a)+(a_2d_1-d_2a_1)p(a)v(a)]
\end{aligned} \tag{A8.24}$$

If we now require that $P(v, u)|_a^b=0$, then, since $B_1(u)=0=B_2(u)$, we obtain

$$\bar{B}_3(v)=0=\bar{B}_4(v) \tag{A8.25}$$

This defines the manifold of the adjoint. If L is to be self-adjoint, then this manifold must be the same as that originally defined by $B_1(u)=0=B_2(u)$ so that

$$\begin{aligned}
\bar{B}_3(v)&=\alpha_1B_1(v)+\alpha_2B_2(v)\\
\bar{B}_4(v)&=\beta_1B_1(v)+\beta_2B_2(v)
\end{aligned} \tag{A8.26}$$

If we equate the coefficients of $v(a)$, $v'(a)$, $v(b)$, and $v'(b)$ in the first equation above, we obtain

$$p(b)=\alpha_1\, d_1+\alpha_2\, d_2$$
$$\frac{(b_2c_1-b_1c_2)}{\Delta}\,p(a)=\alpha_1b_1+\alpha_2b_2$$
$$\frac{(a_2c_1-a_1c_2)}{\Delta}\,p(a)=\alpha_1a_1+\alpha_2a_2$$
$$0=\alpha_1c_1+\alpha_2c_2$$

From the second and third equations we find

$$\alpha_1=-\frac{p(a)}{\Delta}\,c_2$$
$$\alpha_1=\frac{p(a)}{\Delta}\,c_1$$

so that the fourth equation is identically satisfied while the first requires

$$p(b)=\frac{p(a)}{\Delta}(-c_2d_1+d_2c_1)$$

or

$$p(b)(a_2b_1-a_1b_2)=p(a)(c_2d_1-c_1d_2)$$

as claimed.

A8.3 AN ORTHONORMAL BASIS FOR $\mathscr{L}_2(-\infty, \infty)$

We now establish that the functions

$$\varphi_n(x)=\frac{1}{\sqrt{2^n n!\sqrt{\pi}}}e^{-x^2/2}H_n(x) \qquad n=0, 1, 2, \ldots \tag{A8.27}$$

where the $H_n(x)$ are the *Hermite polynomials* (cf. Table 8.1e and Problem 8.12), form a complete set for all once-differentiable square-integrable $f(x)$. Once we have this result we easily extend it to all piecewise continuous $f(x)$ by the type argument given at the end of Section 4.2. As always we cannot make the final step to completeness for $\mathscr{L}_2(-\infty, \infty)$ without some knowledge of the Lebesgue theory of integration.

We begin with the differential equation

$$\frac{d^2}{dx^2}y(x)+(\lambda-x^2)y(x)=0 \qquad x\in(-\infty, \infty) \tag{A8.28}$$

subject to the boundary conditions

$$y(-\infty)=0=y(+\infty) \tag{A8.29}$$

We must first establish that the eigenvalue spectrum for λ is discrete. Let

$$y(x)=e^{-x^2/2}u(x) \tag{A8.30}$$

so that $u(x)$ satisfies the equation

$$\frac{d^2}{dx^2}u(x)-2x\frac{d}{dx}u(x)+(\lambda-1)u(x)=0 \tag{A8.31}$$

If we compare this with the confluent hypergeometric equation of Eq. A7.68,

$$zF''(a\,|c|\,z)+(c-z)F'(a\,|c|\,z)-aF(a\,|c|\,z)=0$$

we see that functions

$$F\left(-\frac{\lambda-1}{4}\,\middle|\,\frac{1}{2}\,\middle|\,x^2\right) \qquad \text{and} \qquad xF\left(-\frac{\lambda-1}{4}+\frac{1}{2}\,\middle|\,\frac{3}{2}\,\middle|\,x^2\right)$$

are two linearly independent solutions to Eq. A8.31 so that

$$u(x)=AF\left(-\frac{\lambda-1}{4}\,\middle|\,\frac{1}{2}\,\middle|\,x^2\right)+BxF\left(-\frac{\lambda-1}{4}+\frac{1}{2}\,\middle|\,\frac{3}{2}\,\middle|\,x^2\right) \tag{A8.32}$$

From Eq. A7.78 we see that as $x\to\pm\infty$ the $u(x)$ of Eq. A8.32 will grow as e^{x^2} so that Eq. A8.29 cannot be satisfied unless one of these confluent hypergeometric functions terminates (i.e., is a polynomial). This requires that

$$-\frac{(\lambda-1)}{4}=-N \qquad N=0,1,2,\ldots \tag{A8.33}$$

or that

$$-\frac{(\lambda-1)}{4}+\frac{1}{2}=-N \qquad N=0,1,2,\ldots$$

This yields the eigenvalues $\lambda_n=2n+1$, where $n=0, 1, 2, \ldots$, so that Eq. A8.31 becomes

$$\frac{d^2}{dx^2}u_n(x)-2x\frac{d}{dx}u_n(x)+2nu_n(x)=0 \tag{A8.34}$$

As stated in Table 8.1e the solutions are the Hermite polynomials. This implies that the $\{\varphi_n(x)\}$ of Eq. A8.27 are the only solutions to Eqs. A8.28 and A8.29 and that the eigenvalue spectrum is denumerably infinite with $\lambda_{\min}=0$ and such that $\lim_{n\to\infty}\lambda_n=\infty$. We can now apply the variational completeness proof given for Th. 8.8 in Section 8.3 or the proof via the Hilbert-Schmidt theorem in Section A8.1.

A8.4 EXPLICIT PROOF THAT $\langle\psi(k)\,|\,\psi(k')\rangle=\langle\phi(k)\,|\,\phi(k')\rangle$ FOR THE CONTINUOUS SPECTRUM

We begin with Eq. 8.168 for $\psi(k, x)$ in a particular representation,

$$\psi(k,x)=\phi(k,x)-\frac{1}{2ik}\int_{-\infty}^{\infty}dx'q(x')e^{ik|x-x'|}\psi(k,x')$$

It is easy to see that

$$\begin{aligned}\langle x\,|(L-k^2-i\varepsilon)^{-1}|\,x'\rangle&=\int_{-\infty}^{\infty}dk'\langle x|\,(k'^2-k^2-i\varepsilon)^{-1}\,|\phi(k')\rangle\langle\phi(k')\,|\,x'\rangle\\&=\int_{-\infty}^{\infty}\frac{dk'\langle x\,|\,\phi(k')\rangle\langle\phi(k')\,|\,x'\rangle}{(k'^2-k^2-i\varepsilon)}=\frac{1}{2\pi}\int_{-\infty}^{\infty}\frac{dk'e^{ik'x}e^{-ik'x'}}{(k'^2-k^2-i\varepsilon)}\\&=\frac{i}{2k}e^{ik|x-x'|}\end{aligned} \tag{A8.35}$$

$$\begin{aligned}\langle x\,|(L-k^2-i\varepsilon)^{-1}|\,V\psi(k)\rangle&=\int dx'\,dx''\langle x|\,(L-k^2-i\varepsilon)^{-1}\,|x'\rangle\langle x'|\,V\,|x''\rangle\langle x''\,|\,\psi(k)\rangle\\&=\int dx'\frac{1}{2\pi}\int dk''\frac{e^{ik''x}e^{-ik''x'}}{(k''^2-k^2-i\varepsilon)}q(x')\psi(k,x')\end{aligned} \tag{A8.36}$$

TABLE 8.1 Orthogonal Polynomial Solutions to the Sturm-Liouville Equation

---

a. Jacobi Polynomials

$$(1-x^2)\frac{d^2}{dx^2}P_n^{(\alpha,\beta)}(x)+[\beta-\alpha-(\alpha+\beta+2)x]\frac{d}{dx}P_n^{(\alpha,\beta)}(x)+n(n+\alpha+\beta+1)P_n^{(\alpha,\beta)}(x)=0$$

$$P_n^{(\alpha,\beta)}(x)=\frac{(-1)^n}{2^n n!}(1-x)^{-\alpha}(1+x)^{-\beta}\frac{d^n}{dx^n}[(1-x)^{\alpha+n}(1+x)^{\beta+n}]$$

$$=\frac{\Gamma(1+\alpha+n)}{\Gamma(1+\alpha)n!}F\left(1+\alpha+\beta+n,\,-n\,|1+\alpha|\,\frac{1-x}{2}\right)$$

$$\int_{-1}^{1}(1-x)^{\alpha}(1+x)^{\beta}P_n^{(\alpha,\beta)}(x)P_m^{(\alpha,\beta)}(x)\,dx$$

$$=\frac{2^{\alpha+\beta+1}}{(2n+\alpha+\beta+1)}\,\frac{\Gamma(n+\alpha+1)\Gamma(n+\beta+1)}{n!\,\Gamma(n+\alpha+\beta+1)}\,\delta_{nm}\qquad \alpha>-1\qquad \beta>-1$$

b. Gegenbauer Polynomials

$$(1-x^2)\frac{d^2}{dx^2}P_n^{(\lambda)}(x)-2(\lambda+1)x\frac{d}{dx}P_n^{(\lambda)}(x)+n(n+2\lambda+1)P_n^{(\lambda)}(x)=0$$

$$P_n^{(\lambda)}(x)=\frac{\Gamma(\lambda+1)\Gamma(n+2\lambda+1)}{\Gamma(2\lambda+1)\Gamma(n+\lambda+1)}P_n^{(\lambda,\lambda)}(x)$$

$$=\frac{(-1)^n}{n!\,2^n}\frac{\Gamma(\lambda+1)\Gamma(n+2\lambda+1)}{\Gamma(n+\lambda+1)\Gamma(2\lambda+1)}\frac{1}{(1-x^2)^{\lambda}}\frac{d^n}{dx^n}[(1-x^2)^{n+\lambda}]$$

$$=\frac{\Gamma(n+2\lambda+1)}{n!\,\Gamma(2\lambda+1)}F\left(n+2\lambda+1,\,-n\,|\lambda+1|\,\frac{1-x}{2}\right)$$

$$\int_{-1}^{1}(1-x^2)^{\lambda}P_n^{(\lambda)}(x)P_m^{(\lambda)}(x)\,dx=\frac{2^{2\lambda+1}\Gamma(n+2\lambda+1)}{(2n+2\lambda+1)n!}\left[\frac{\Gamma(\lambda+1)}{\Gamma(2\lambda+1)}\right]^2\delta_{nm}\qquad \lambda>-1$$

c. Tchebycheff Polynomials

$$(1-x^2)\frac{d^2}{dx^2}T_n(x)-x\frac{d}{dx}T_n(x)+n^2T_n(x)=0$$

$$T_n(x)=\frac{2^{2n}\cdot(n!)^2}{(2n)!}P_n^{(-1/2,-1/2)}(x)$$

$$=F\left(n,\,-n\,\middle|\,\frac{1}{2}\,\middle|\,\frac{1-x}{2}\right)=\cos(n\theta)\qquad x=\cos\theta$$

$$\int_{-1}^{1}T_n(x)T_m(x)(1-x^2)^{-1/2}\,dx=\frac{\pi}{2}\delta_{nm}$$

d. Laguerre Polynomials

$$x\frac{d^2}{dx^2}L_n^{\alpha}(x)+[(\alpha+1)-x]\frac{d}{dx}L_n^{\alpha}(x)+nL_n^{\alpha}(x)=0$$

TABLE 8.1 (Contd.)

$$L_n^\alpha(x) = \frac{\Gamma(\alpha+n+1)}{n!}\frac{e^x}{x^\alpha}\frac{d^n}{dx^n}[x^{\alpha+n}e^{-x}]$$

$$= \frac{[\Gamma(\alpha+n+1)]^2}{n!\,\Gamma(\alpha+1)}F(-n\,|1+\alpha|\,x)$$

$$\int_0^\infty x^\alpha e^{-x}L_n^\alpha(x)L_m^\alpha(x)\,dx = \frac{[\Gamma(\alpha+n+1)]^3}{n!}\delta_{nm}$$

e. Hermite Polynomials

$$\frac{d^2}{dx^2}H_n(x) - 2x\frac{d}{dx}H_n(x) + 2nH_n(x) = 0$$

$$H_n(x) = (-1)^n e^{x^2}\frac{d^n}{dx^n}(e^{-x^2})$$

$$= \begin{cases} (-1)^{(1/2)n}\dfrac{n!}{\Gamma\left(\dfrac{n}{2}+1\right)}F(-\tfrac{1}{2}n\,|\tfrac{1}{2}|\,x^2) & n = 0, 2, 4, \ldots \\[2ex] 2(-1)^{(1/2)(n-1)}\dfrac{n!}{\Gamma\left(\dfrac{1}{2}n+\dfrac{1}{2}\right)}xF(-\tfrac{1}{2}n+\tfrac{1}{2}\,|\tfrac{3}{2}|\,x^2) & n = 1, 3, 5, \ldots \end{cases}$$

$$\int_{-\infty}^\infty e^{-x^2}H_n(x)H_m(x)\,dx = 2^n n!\,\sqrt{\pi}\,\delta_{nm}$$

so that we may write

$$\langle (L-k^2+i\varepsilon)^{-1}V\psi(k)\,|\,(L-k'^2-i\varepsilon)^{-1}V\psi(k')\rangle \equiv \langle\mathcal{M}\rangle$$

$$= \int dx\int dx'\int dx''\int dk''\int dk'''\,\frac{1}{4\pi}\frac{e^{-ik''x}e^{ik''x'}}{(k''^2-k^2+i\varepsilon)}q(x')\psi^*(k,x')$$

$$\times\frac{e^{ik'''x}e^{-ik'''x''}}{(k'''^2-k'^2-i\varepsilon)}q(x'')\psi(k',x'')$$

$$= \int dx'\int dx''\int dk''\,\frac{1}{2\pi}\frac{e^{ik''x'}}{(k''^2-k^2+i\varepsilon)}q(x')\psi^*(k,x')\frac{e^{-ik''x''}}{(k''^2-k'^2-i\varepsilon)}q(x'')\psi(k',x'') \tag{A8.37}$$

Let us define

$$f(k,x') \equiv q(x')\psi(k,x')$$

$$F(k,k'') \equiv \frac{1}{\sqrt{2\pi}}\int_{-\infty}^\infty dx'e^{ik''x'}f(k,x') \xrightarrow[k''\to\infty]{} O\left(\frac{1}{k''}\right) \tag{A8.38}$$

where the last bound follows from the Riemann-Lebesgue lemma (Th. 4.8) which applies with suitable asymptotic restrictions on $q(x)$. Therefore

TABLE 8.2 The Sturm-Liouville Differential Equation Subject to Self-Adjoint Boundary Conditions

---

$$Lu \equiv \frac{1}{w(x)}\frac{d}{dx}\left[p(x)\frac{d}{dx}u(x)\right]+q(x)u(x)=f(x) \qquad x\in[a,b]$$

$$B_1(u)\equiv a_1u(a)+b_1u'(a)+c_1u(b)+d_1u'(b)=A$$

$$B_2(u)\equiv a_2u(a)+b_2u'(a)+c_2u(b)+d_2u'(b)=B$$

$$p(a)(d_1c_2-d_2c_1)=p(b)(a_2b_1-a_1b_2)$$

$$P[v_1(x),v_2(x)]\equiv p(x)W[v_1(x),v_2(x)]=p(x)[v_1(x)v_2'(x)-v_2(x)v_1'(x)]$$

$$\equiv P(v_1,v_2) \qquad \text{a constant}$$

$$Lv_j(x)=0 \qquad v_j^*(x)=v_j(x) \qquad j=1,2$$

$$W[v_1(x),v_2(x)]\neq 0$$

$$u(x)=\alpha v_1(x)+\beta v_1(x)+\frac{v_1(x)}{P(v_1,v_2)}\int_x^b v_2(t)f(t)w(t)\,dt+\frac{v_2(x)}{P(v_1,v_2)}\int_a^x v_1(t)f(t)w(t)\,dt$$

$$\alpha B_1(v_1)+\beta B_1(v_2)=A-[a_1v(a)+b_1v_1'(a)]\frac{\langle v_2\,|\,f\rangle}{P(v_1,v_2)}-[c_1v_2(b)+d_1v_2'(b)]\frac{\langle v_1\,|\,f\rangle}{P(v_1,v_2)}$$

$$\alpha B_2(v_1)+\beta B_2(v_2)=B-[a_2v_1(a)+b_2v_1'(a)]\frac{\langle v_2\,|\,f\rangle}{P(v_1,v_2)}-[c_2v_2(b)+d_2v_2'(b)]\frac{\langle v_1\,|\,f\rangle}{P(v_1,v_2)}$$

$$\langle v_j\,|\,f\rangle\equiv\int_a^b v_j(x)f(x)w(x)\,dx \qquad j=1,2$$

a. If there is no nontrivial solution $w(x)$ such that

$$L[w]=0$$

$$B_1(w)=0=B_2(w)$$

then α and β are uniquely determined so that a unique solution to the original problem does exist.

b. i. If a nontrivial $v_1(x)$ exists such that

$$L[v_1]=0$$

$$B_1(v_1)=0=B_2(v_1)$$

while not both $B_1(v_2)$ and $B_2(v_2)$ are zero, then the necessary and sufficient condition for the existence of a (nonunique) solution to the original problem is

$$B_2(v_2)\left\{A-[c_1v_2(b)+d_1v_2'(b)]\frac{\langle v_1\,|\,f\rangle}{P(v_1,v_2)}\right\}=B_1(v_2)\left\{B-[c_2v_2(b)+d_2v_2'(b)]\frac{\langle v_1\,|\,f\rangle}{P(v_1,v_2)}\right\}$$

Also see Eq. 8.138.

ii. If $B_1(v_2)=0=B_2(v_2)$ as well, then the necessary and sufficient condition for the existence of a (nonunique) solution to the original problem is

$$AP(v_1,v_2)=[a_1v_1(a)+b_1v_1'(a)]\langle v_2\,|\,f\rangle+[c_1v_2(b)+d_1v_2'(b)]\langle v_1\,|\,f\rangle$$

$$BP(v_1,v_2)=[a_2v_1(a)+b_2v_1'(a)]\langle v_2\,|\,f\rangle+[c_2v_2(b)+d_2v_2'(b)]\langle v_1\,|\,f\rangle$$

Also see Eq. 8.139.

iii. In the case of homogeneous boundary conditions (i.e., $A=B=0$), the necessary and sufficient condition for the existence of a (nonunique) solution to the original problem is

$$\langle v_1\,|\,f\rangle=0$$

when $B_1(v_1)=0=B_2(v_1)$, but $B_1(v_2)$ and $B_2(v_2)$ are not both zero, and

$$\langle v_j\,|\,f\rangle=0 \qquad j=1,2$$

when $B_1(v_j)=0=B_2(v_j)$, $j=1,2$.

$$\langle\mathcal{M}\rangle=\int\frac{dk''F(k,k'')F^*(k',k'')}{(k''^2-k^2+i\varepsilon)(k''^2-k'^2-i\varepsilon)}$$

$$=\frac{1}{(k^2-k'^2)}\left\{P\int\frac{dk''F(k,k'')F^*(k',k'')}{(k''^2-k^2)}-\pi iF(k,k)F^*(k',k)\right.$$

$$\left.-P\int\frac{dk''F(k,k'')F^*(k',k'')}{(k''^2-k'^2)}+\pi iF(k,k')F^*(k',k')\right\} \qquad \textbf{(A8.39)}$$

where we have decomposed the original denominator by partial fractions and then used Eq. 7.52. The principal-value integrals exist because of the asymptotic condition stated in Eq. A8.38. Therefore the second term in Eq. 8.166 is finite when $k'^2\neq k^2$ and vanishes when $\varepsilon\to 0^+$.

TABLE 8.3 Green's Functions in the Infinite Spatial Domain for Some Partial Differential Operators

a. Scalar Helmholtz Equation

$$(\nabla^2+k^2)G(\mathbf{r},\mathbf{r}')=\delta(\mathbf{r}-\mathbf{r}')$$

$$G(\mathbf{r},\mathbf{r}')=-\frac{e^{ik|\mathbf{r}-\mathbf{r}'|}}{4\pi\,|\mathbf{r}-\mathbf{r}'|}$$

b. Scalar Wave Equation

$$\left(\nabla^2-\frac{1}{c^2}\frac{\partial^2}{\partial t^2}\right)G(\mathbf{r},\mathbf{r}';t,t')=\delta(\mathbf{r}-\mathbf{r}')\,\delta(t-t')$$

$$G(\mathbf{r},\mathbf{r}';t,t')=-\frac{1}{4\pi\,|\mathbf{r}-\mathbf{r}'|}\,\delta\!\left(\frac{|\mathbf{r}-\mathbf{r}'|}{c}-(t-t')\right)$$

c. Diffusion or Heat Equation

$$\left(\nabla^2-a^2\frac{\partial}{\partial t}\right)G(\mathbf{r},\mathbf{r}';t,t')=\delta(\mathbf{r}-\mathbf{r}')\,\delta(t-t')$$

$$G(\mathbf{r},\mathbf{r}';t,t')=-\frac{1}{4\pi}\frac{a}{2\sqrt{\pi}\,(t-t')^{3/2}}\exp\left[-\frac{a^2\,|\mathbf{r}-\mathbf{r}'|^2}{4(t-t')}\right]H(t-t')$$

d. Dissipative Wave Equation

$$\left(\nabla^2-a^2\frac{\partial}{\partial t}-\frac{1}{c^2}\frac{\partial^2}{\partial t^2}\right)G(\mathbf{r},\mathbf{r}';t,t')=\delta(\mathbf{r}-\mathbf{r}')\,\delta(t-t')$$

$$G(\mathbf{r},\mathbf{r}';t,t')=-\frac{c\exp[-\frac{1}{2}a^2c^2(t-t')]}{4\pi\,|\mathbf{r}-\mathbf{r}'|}\left\{\delta[c(t-t')-|\mathbf{r}-\mathbf{r}'|]\right.$$

$$\left.+\frac{a^2c\,|\mathbf{r}-\mathbf{r}'|}{2\sqrt{|\mathbf{r}-\mathbf{r}'|^2-c^2(t-t')^2}}\,J_1\!\left[\frac{a^2c}{2}\sqrt{|\mathbf{r}-\mathbf{r}'|^2-c^2(t-t')^2}\right]H[c(t-t')-|\mathbf{r}-\mathbf{r}'|]\right\}H(t-t')$$

SUGGESTED REFERENCES

F. W. Byron and R. W. Fuller, *Mathematics of Classical and Quantum Physics* (Vol. I). Section 5.10 contains a clear and complete discussion of the Sturm-Liouville differential equation.

R. Courant and D. Hilbert, *Methods of Mathematical Physics* (Vol. II). This advanced text deals with a rigorous treatment of partial differential equations and their Green's functions. The Appendix to Chapter VI contains a very readable introduction to ideal functions.

J. W. Dettman, *Mathematical Methods in Physics and Engineering*. Section 4.2 gives a detailed variational proof of the completeness of the solutions to the Sturm-Liouville equation for all of the commonly applied boundary conditions.

B. Friedman, *Principles and Techniques of Applied Mathematics.* Chapter 3 gives a complete discussion of Green's functions for second-order linear differential operators under various boundary conditions.

P. M. Morse and H. Feshbach, *Methods of Theoretical Physics* (Vol. I). Section 5.2 presents a fairly thorough, semirigorous treatment of series solutions to second-order linear ordinary differential equations. Chapter 7 gives a useful physical presentation of the Green's functions for the important partial differential equations of physics and engineering. It gives much more detail than our brief discussion.

E. L. Ince, *Ordinary Differential Equations.* This is *the* classic text on ordinary differential equations, in both the real and complex domains. Open the book anyplace, read, and you'll learn something interesting and useful. The proofs are complete and given in great detail.

E. C. Titchmarsh, *Eigenfunction Expansions* (Part I). Chapters II and III present a rigorous and detailed representation-dependent proof of the completeness of the improper eigenfunctions of the Sturm-Liouville equation.

PROBLEMS

8.1 Complete the proof of Th. 8.2, especially the uniform convergence of the series for the $\phi(z)$ of Eq. 8.40.

8.2 Prove that

$$x^m \, \delta^{(n)}(x) = 0$$

if $m \geq n+1$.

8.3 Prove that if $f(x)$ is a continuous function with a continuous derivative while $h(x)$ is an ideal function then

$$[f(x)h(x)]' = f(x)h'(x) + f'(x)h(x)$$

As a special case show explicitly that

$$\frac{d}{dx}[H(x)f(x)] = f'(x)H(x) + f(0)\,\delta(x)$$

8.4 A distribution (or ideal function) $f_i(x)$ can be represented as an ideal limit of a sequence of continuous functions $\{f_n(x)\}$ as

$$F(\phi)=\lim_{n\to\infty}\int_{-\infty}^{\infty} f_n(x)\phi(x)\,dx$$

However the product of two distributions is not necessarily a distribution. Use the sequences of ideal functions given in Eqs. 8.22 and 8.28 for $\delta_n(x)$ and $H_n(x)$ to show that:

a. $\delta(x)\delta(x)$ is not defined as a distribution
b. $\delta(x)H(x)=\frac{1}{2}\delta(x)$

Hint: Evaluate

$$\lim_{n\to\infty}\int_{-\infty}^{\infty} \delta_n(x)\,\delta_n(x)\phi(x)\,dx$$

and

$$\lim_{n\to\infty}\int_{-\infty}^{\infty} H_n(x)\,\delta_n(x)\phi(x)\,dx$$

where $\phi(x)$ is a test function.

8.5 Show that a solution to

$$u''(x)=f(x)$$

is

$$u(x)=\frac{1}{2}\int |x-t|\,f(t)\,dt$$

8.6 Evaluate the integral

$$I(a,b)=\int_{-1}^{1} \delta(x^2-a^2)H(x-b)\,dx$$

where a and b are real parameters such that $a\neq\pm1$, $a\neq b$.

8.7 Given the differential equation

$$y''+xy'+y=0$$

assume the solutions are regular about $x=0$ so that they can be expanded in power series as

$$y(x)=\sum_{n=0}^{\infty} \alpha_n x^n$$

Obtain a recursion relation for the expansion coefficients $\{\alpha_n\}$. Then generate an even solution $y_1(x)=y_1(-x)$ by setting $\alpha_0=1$, $\alpha_1=0$ and next an odd solution $y_2(x)=-y_2(-x)$ by setting $\alpha_0=0$, $\alpha_1=1$. Prove that these series converge and find their radii of convergence. How do you know that $y_1(x)$ and $y_2(x)$ are linearly independent?

8.8 Consider the differential equation

$$xy''-2y'+y=0$$

and develop two linearly independent solutions about $x=0$. Once you have developed a regular solution $y_1(x)$ about $x = 0$, generate a second linearly independent one $y_2(x)$ by use of Eq. 8.59. How do $y_1(x)$ and $y_2(x)$ behave in the neighborhood of $x=0$? Do you see any paradox here since $y=\text{const.}$ is not a solution to the differential equation? How do you resolve this? Prove whether the point $x=\infty$ is a regular or an irregular singular point.

8.9 Show that under the general unmixed boundary conditions of Eq. 8.84 the eigenvalues of the Sturm-Liouville equation are still such that $\lim_{n\to\infty} \lambda_n = \infty$. Then carry through the proof of Th. 8.8 for these general boundary conditions. *Hint:* Instead of using Eq. 8.88 to define $I(u)$ let

$$I(u) = \int_a^b [p\,|u'|^2 - q\,|u|^2]\,dx - \frac{\alpha_1}{\beta_1}|u(a)|^2\,p(a) + \frac{\alpha_2}{\beta_2}|u(b)|^2\,p(b)$$

8.10 Consider the Sturm-Liouville problem in the form

$$\frac{1}{w(x)}\frac{d}{dx}\left[w(x)a(x)\frac{du}{dx}\right] + [c(x)-\lambda]u(x) = 0 \qquad x\in[a,b]$$

where

$$a(x) = a_0 + a_1x + a_2x^2$$

$$\frac{1}{w(x)}\frac{d}{dx}[w(x)a(x)] = b_0 + b_1x \qquad c(x) = c_0$$

all constants being real. You are given that $[w(x)a(x)]$ vanishes at the end points $x=a$ and $x=b$, but not at points interior to (a, b).

a. Show that for the eigenvalues $\{\lambda_n\}$ given as

$$\lambda_n = n(n-1)a_2 + nb_1 + c_0 \qquad n = 0, 1, 2, \ldots$$

there always exist nontrivial polynomial solutions $u_n(x)$

$$u_n(x) = \sum_{m=0}^{n} \alpha_m^{(n)} x^m$$

b. Show that these polynomials have the property that

$$\langle u_m \mid u_n\rangle \equiv \int_a^b u_m^*(x)u_n(x)w(x)\,dx = 0 \qquad m\neq n$$

c. Using an argument similar to that given in Section A7.3 for the Jacobi polynomials, show that the expression

$$\frac{1}{w(x)}\frac{d^n}{dx^n}[a^n(x)w(x)]$$

is a polynomial of degree n such that

$$\int_a^b x^r\left\{\frac{1}{w(x)}\frac{d^n}{dx^n}[a^n(x)w(x)]\right\}w(x)\,dx = 0 \qquad r<n$$

and deduce from this that

$$u_n(x) = c_n \frac{1}{w(x)} \frac{d^n}{dx^n} [a^n(x) w(x)]$$

where c_n is a normalization constant. It is useful to establish first that

$$\frac{d^m}{dx^m} [a^n(x) w(x)] = p_m(x) w(x) a^{n-m}(x) \qquad m \leq n$$

where $p_m(x)$ is a polynomial of degree m.

d. Show that for any continuous n-fold differentiable function $f(x)$

$$\int_a^b f(x) u_n(x) w(x)\, dx = (-1)^n c_n \int_a^b f^{(n)}(x) a^n(x) w(x)\, dx$$

so that

$$\int_a^b u_n^2(x) w(x)\, dx = (-1)^n n!\, c_n \alpha_n^{(n)} \int_a^b a^n(x) w(x)\, dx$$

e. If $w(x) \xrightarrow[x\to\infty]{} Ax^\alpha$, then show that

$$\alpha_n^{(n)} = c_n a_2^n \frac{\Gamma(2n+\alpha+1)}{\Gamma(n+\alpha+1)}$$

f. If $w(x) \xrightarrow[x\to\infty]{} Bx^s e^{\beta x}$, then show that $a_2 = 0$ and that

$$\alpha_n^{(n)} = c_n b_1^n$$

g. If $w(x) \xrightarrow[x\to\infty]{} Ce^{\gamma x^2}$, then show that $a_2 = a_1 = b_0 = 0$ and that

$$\alpha_n^{(n)} = c_n b_1^n$$

8.11 Consider the differential equation for the Jacobi polynomials for $x \in [-1, 1]$

$$(1-x^2)y'' + [\beta - \alpha - (\alpha+\beta+2)x]y' + n(n+\alpha+\beta+1)y = 0 \qquad \alpha > -1 \qquad \beta > -1$$

and n a positive integer or zero. Construct explicitly the solution that is regular about $x = +1$. Normalize these so that they agree with Eq. 7.189. Also find the behavior of the irregular solution in the neighborhood of $x = +1$.

8.12 Use the results of Problem 8.10 to construct *orthonormal* polynomials in the following cases where

$$a(x)u''(x) + b(x)u'(x) + c(x)u(x) = 0$$

a. $a(x) = (1-x^2)$, $b(x) = -2(\lambda+1)x$, $c(x) = n(n+2\lambda+1)$,
$w(x) = (1-x^2)^\lambda$, $x \in [-1, 1]$ (Gegenbauer polynomials)

b. $a(x) = (1-x^2)$, $b(x) = -x$, $c(x) = n^2$
$w(x) = (1-x^2)^{-1/2}$, $x \in [-1, 1]$ (Tschebycheff polynomials)

c. $a(x)=x$, $b(x)=(\alpha+1)-x$, $c(x)=n$
$w(x)=x^{\alpha}e^{-x}$, $x\in[0,\infty)$ (Laguerre polynomials)

d. $a(x)=1$, $b(x)=-2x$, $c(x)=2n$
$w(x)=e^{-x^2}$, $x\in(-\infty,\infty)$ (Hermite polynomials)

8.13 Consider the differential equation

$$\frac{d}{dx}\left[p(x)\frac{du}{dx}\right]=f(x) \qquad x\in[0,1]$$

where $p(x)$ and $f(x)$ are continuous and differentiable while $p(x)>0$, subject to the boundary conditions

$$u(0)=\alpha \qquad u(1)=\beta$$

Write the solution for arbitrary $f(x)$. Does such a $u(x)$ always exist? Is it unique? Explain your answers. Construct the solution for

$$p(x)=(1+x)^2 \qquad f(x)=2x \qquad \alpha=\beta=0$$

8.14 If

$$u''-\frac{2}{x^2}u=f(x) \qquad x\in[0,1]$$

where $f(x)$ is square integrable and

$$u(0)=0 \qquad u(1)=1$$

construct the solution to this problem.

8.15 Solve the problem

$$\frac{1}{x}u''-\frac{1}{x^2}u'=f(x) \qquad x\in[0,1]$$

$$u(0)+u'(1)=1$$

$$u'(0)-u(1)=0$$

Discuss the existence and uniqueness of the solution.

8.16 Obtain the solution to

$$x^2u''+2xu'-2u=f(x) \qquad x\in[0,1] \qquad u(0)=0 \qquad u'(1)=1$$

8.17 Show that the differential operator

$$Lu\equiv\frac{1}{w(x)}\frac{d}{dx}\left[p(x)\frac{du(x)}{dx}\right]+q(x)u(x) \qquad x\in[0,1]$$

is formally self-adjoint with respect to the inner product

$$\langle u, v\rangle=\int_0^1 w(x)u^*(x)v(x)\,dx$$

Here $w(x)$ is a real positive weight function and $p(x)$ and $q(x)$ are real functions.

8.18 Prove that the substitutions

$$\frac{1}{w(x)}=a(x)\exp\left[-\int\frac{b(x)}{a(x)}\,dx\right] \qquad p(x)=\exp\left[\int\frac{b(x)}{a(x)}\,dx\right] \qquad q(x)=c(x)$$

will transform the differential operator

$$Lu\equiv a(x)u''(x)+b(x)u'(x)+c(x)u(x)$$

into a formally self-adjoint operator.

8.19 Prove that if

$$u''+b(x)u'+c(x)u=0$$

and if $v(x)$ is defined as

$$u(x)=v(x)\exp\left[-\frac{1}{2}\int b(x)\,dx\right]$$

then the differential equation for $v(x)$ has no first-derivative term.

8.20 Find the asymptotic behavior of the hypergeometric function, $u(x)$, which satisfies

$$x(x-1)u''+[(a+b+1)x-c]u'+abu=0$$

Compare your result with Eq. 7.188. What are the restrictions on a, b, and c?

8.21 Consider the standard Sturm-Liouville equation

$$\frac{1}{w(x)}\frac{d}{dx}\left[p(x)\frac{du(x)}{dx}\right]+q(x)u(x)=0$$

Show that if we define

$$v(y)\equiv gu$$

where

$$g\equiv(wpq)^{1/4}$$

and change the independent variables as

$$y=\int^{x}(qw/p)^{1/2}\,dx$$

then $v(y)$ satisfies the equation

$$\frac{d^2v(y)}{dy^2}+[1-Q(y)]v(y)=0$$

where

$$Q(y)=\frac{1}{g(y)}\frac{d^2g(y)}{dy^2}$$

Now prove Eqs. 8.158 and 8.159. Are any asymptotic restrictions necessary on $p(x)$, $q(x)$, and $w(x)$?

8.22 Find the asymptotic behavior of the two linearly independent solutions to

$$u''+(a+bx^2)u=0$$

8.23 Consider the differential equation

$$\left[\left(1+\frac{2\alpha}{x}\right)\frac{d^2}{dx^2}+\left(1+\frac{\alpha}{x}\right)\left(\frac{2}{x}\frac{d}{dx}-\frac{n(n+1)}{x^2}\right)+1-\frac{2\beta}{x}+\frac{\gamma}{x^2}\right]y(x)=0$$

where α, β, and γ are real positive numbers, n is a positive integer or zero, and $x\in[0,\infty)$. Locate the singular points of this differential equation on its range of definition and find the behavior of the regular and irregular solutions about $x=0$. Then calculate the asymptotic behavior of the solutions as $x\to\infty$. In the special case in which $\alpha=\gamma=0$ find the asymptotic form of the solution that is regular at $x=0$. (*Hint:* When $\alpha=\gamma=0$ set $y(x)=x^n e^{ix} f(x)$ and use Eq. A7.81.)

8.24 Supply the steps necessary to obtain Eq. 8.186.

8.25 Let $\{\chi_\alpha\}$ be the eigenfunctions of an operator H_0 with eigenvalues $\{E_\alpha\}$, where α stands for some complete set of labels (discrete and/or continuous). You are given that the $\{\chi_\alpha\}$ are orthonormal and complete

$$\langle\chi_\alpha\mid\chi_\beta\rangle=\delta_{\alpha\beta}\qquad \sum_\gamma|\chi_\gamma\rangle\langle\chi_\gamma|=I$$

Let V be an operator such that

$$\|V\chi_\alpha\|<\infty\qquad \forall\alpha$$

and define

$$H=H_0+V$$

Let $\{\psi_\alpha\}$ be the eigenfunctions of H with eigenvalues $\{E_\alpha\}$,

$$H\psi_\alpha=E_\alpha\psi_\alpha$$

Show that the matrix element

$$T_{\beta\alpha}\equiv\langle\chi_\beta|\,V\,|\psi_\alpha\rangle$$

satisfies the "integral equation"

$$T_{\beta\alpha}=\langle x_\beta|\,V\,|\chi_\alpha\rangle-\sum_\gamma\frac{\langle\chi_\beta|\,V\,|\chi_\gamma\rangle T_{\gamma\alpha}}{(E_\gamma-E_\alpha-i\varepsilon)}$$

8.26 Consider the scalar wave equation in one spatial dimension,

$$\left(\frac{\partial^2}{\partial x^2}-\frac{1}{c^2}\frac{\partial^2}{\partial t^2}\right)\psi(x,t)=\rho(x,t)$$

a. Write the differential equation satisfied by the Green's function, $G(x, x'; t, t')$.
b. State the causality condition for the Green's function.
c. Assume that $x\in[a,b]$ (a finite interval) and that the boundary conditions are

$$G(a,x';t,t')=0=G(b,x';t,t')$$
$$\psi(a,t)=c_1\qquad \psi(b,t)=c_2$$

Prove the reciprocity relation

$$G(x,x';t,t')=G(x',x;-t',-t)$$

d. Obtain an expression for $\psi(x, t)$ in terms of the source function $\rho(x, t)$, the Green's function, and the initial and boundary values of ψ.

e. Finally obtain explicitly the Green's function for the infinite spatial domain, $x \in (-\infty, \infty)$, being careful to relate any integration contour to the causality condition to be satisfied by $G(x, x'; t, t')$.

8.27 Show that the Green's function for

$$\left(\nabla^2 - a^2\frac{\partial}{\partial t} - \frac{1}{c^2}\frac{\partial^2}{\partial t^2}\right)\psi(\mathbf{r}, t) = \rho(\mathbf{r}. t)$$

in the infinite domain is given as

$$G(\mathbf{r}, \mathbf{r}'; t, t') = \frac{-c \exp\left(-\frac{1}{2}a^2c^2\tau\right)}{4\pi\rho}\left\{\delta(c\tau - \rho) + \frac{a^2c\rho}{2\sqrt{\rho^2 - c^2\tau^2}}J_1\left(\frac{a^2c}{2}\sqrt{\rho^2 - c^2\tau^2}\right)H(c\tau - \rho)\right\}H(\tau)$$

where

$$\boldsymbol{\rho} = \mathbf{r} - \mathbf{r}' \qquad \tau = t - t'$$

(*Hint:* Use the result of problem 7.54.)

8.28 For the scalar wave equation in the infinite spatial domain take the source function to be

$$\rho(\mathbf{r}, t) = q_0\, \delta(\mathbf{r} - \mathbf{v}_0 t)$$

where $\mathbf{v}_0$ is a constant vector. Assume that $\psi = 0 = \partial\psi/\partial t$ at $t = -\infty$ and *explicitly* evaluate the integral for $\psi(\mathbf{r}, t)$.

9 Group Theory

9.0 INTRODUCTION

We will now study a subject that has proven itself to be an enormously useful tool for applied mathematics in recent times. Certainly a consideration of groups does not follow compellingly from the topics we have treated previously. There may appear to be a certain break in the continuity of the material presented, but this seems unavoidable at the level of this text. The feeling that elementary group theory is "difficult" arises largely because many physical scientists receive practically no exposure to the subject during their early training. This chapter comes at the end of this text since several properties of special functions developed previously are required for constructing explicit representations of the groups treated. Many examples are presented so that the student can see concretely what the various definitions and theorems really mean.

Groups are usually classified as finite or infinite (either discrete or continuous) according to the number of elements they contain. The study of certain finite groups (i.e., crystal symmetry ones) is important in solid state physics. Some knowledge of continuous groups, in particular of Lie groups, is essential for an understanding of quantum mechanics, for example.

In the next four sections of this chapter we present a brief introduction to the basics of abstract group theory for finite groups following as closely as possible the notation of Hamermesh, since that text has become a standard reference on group theory for physical scientists, although the arguments are essentially those of Wigner. The remainder of the chapter develops some elementary results for Lie groups, after the fashion of Racah and of Wigner. Particular attention is given to the three-dimensional rotation group.

This material on group theory, relative to the entire field of group theory, is far more elementary than the previous material in this text is to the field of classical analysis.

9.1 DEFINITIONS AND ELEMENTARY THEOREMS

a. Definition of an Abstract Group and Examples

As we have already stated in Section 1.1a, an *abstract group* G is a collection of distinct elements, $\{g_j\}$, for which a law of composition or multiplication "·" is given such that the product of any two elements is well-defined and satisfies (for *all* elements of G).

i. $g_i \in G,\ g_j \in G \Rightarrow g_k \equiv g_i \cdot g_j \in G,\ \forall g_i,\ g_j \in G$—*closure*.
ii. $g_i \cdot (g_j \cdot g_k) = (g_i \cdot g_j) \cdot g_k$—*associative* operation.
iii. $\exists e \in G \ni g_i \cdot e = g_i = e \cdot g_i,\ \forall g_i \in G$—a unique *identity* element.
iv. $g_i \in G \Rightarrow \exists g_i^{-1} \in G \ni g_i^{-1} g_i = e = g_i \cdot g_i^{-1}$—a unique *inverse* for every element $g_1 \in G$.

That the left and right identity elements are equal and unique and that the left and right inverses are equal and unique can be proven from the group axioms themselves and need not be assumed separately (cf. Problem 9.1).

If, furthermore, the operation is *commutative*,

v. $g_i \cdot g_j = g_j \cdot g_i,\ \forall g_i,\ g_j \in G$

then G is a *commutative* or *abelian group*.

A group having a finite number of distinct elements—g—is said to be a (finite) group of *order* g.

Aside from the obvious examples of rotations and of integers, we list the following.

Example 9.1 The group of spatial reflections is a group of order 2. Its elements are the identity, I, and the reflection through the origin, P; $P^2 = I$.

Example 9.2 The group of permutations of n objects is of order $n!$ This *symmetric group of degree n*, S_n, is an extremely important discrete group. The permutations themselves are the group elements. If $n = 3$ we have for the various permutations or group elements

$$123 \rightarrow \begin{cases} 123 \\ 132 \\ 213 \\ 231 \\ 312 \\ 321 \end{cases}$$

so that these form a group. Explicitly the group multiplication is, for instance,

$$(132)(213)=(231)$$

Example 9.3 Consider the three-element group with elements a, b, e where, as usual, e denotes the identity element. We can easily deduce the results of group multiplication as follows.

i.
$$a \cdot b = e \tag{9.1}$$

This follows since there are, by assumption, three and only three elements. Notice that if we try

$$a \cdot b = a$$

then the existence of an inverse implies that

$$b = e$$

which is a contradiction. Similarly the choice

$$a \cdot b = b$$

implies that

$$a = e$$

which is again a contradiction.

ii. $$a \cdot a = \begin{cases} a \Rightarrow a = e \Rightarrow \text{contradiction} \\ b \\ e \Rightarrow (a \cdot a) \cdot b = b = a \cdot (a \cdot b) = a \cdot e = a \Rightarrow a = b \end{cases}$$

$$\Rightarrow \text{contradiction.}$$

$$\therefore \quad a \cdot a = b \tag{9.2}$$

iii. Similarly to (ii) we find that

$$b \cdot b = a \tag{9.3}$$

iv. Also to avoid a contradiction we must choose

$$b \cdot a = e \tag{9.4}$$

As often proves convenient, we display the results of Eqs. 9.1–9.4 in a *group multiplication table.*

Table 9.1

	e	a	b
e	e	a	b
a	a	b	e
b	b	e	a

We see that this three-element group is abelian and that we may write the group elements as $\{e, a, a^2\}$ with $a^3 = e$. This is a simple example of a *cyclic group* of order n in which all of the elements can be expressed as $e, a, a^2, \ldots, a^{n-1}$, such that $a^n = e$.

A group having an uncountable infinity of elements, each of which depends on one or more continuous parameters, is called a *continuous group.* A simple example would be $g(\alpha) = e^{i\alpha}$ where α is a real parameter.

Two groups G and G' are termed *isomorphic* (which we denote as $G \approx G'$) provided their elements can be put into a one-to-one correspondence with each other and provided this correspondence is preserved under the respective laws of multiplication for the two groups. Symbolically we can state the requirements for an isomorphism between G and G' as

i. $g_i \in G \Leftrightarrow \exists g_i' \in G'$ (*uniquely*)
ii. $g_i \cdot g_j = g_k \Leftrightarrow g_i' \cdot g_j' = g_k'$ (for all members of G and G')

It is important to realize that the laws of multiplication may be different for each group. Two groups that are isomorphic have the same structure so that they both correspond to the same abstract group.

Example 9.4 The group of spatial reflections is isomorphic to the group of permutations of two objects as

$$\left.\begin{matrix} 1 \rightarrow 1 \\ 2 \rightarrow 2 \end{matrix}\right\} I \qquad \left.\begin{matrix} 1 \rightarrow 2 \\ 2 \rightarrow 1 \end{matrix}\right\} P$$

These are two realizations of the abstract group of order 2.

Example 9.5 The symmetries of the plane equilateral triangle (i.e., the rotations about various symmetry axes that leave the triangle configuration invariant; cf. Problems 9.18 and 9.19) from a group isomorphic to S_3.

Example 9.6 The three-element cyclic group $\{e, a, b\}$ of Example 9.3 is isomorphic to the subgroup of S_3 consisting of the cyclic permutations

$$123 \rightarrow 312 \rightarrow 231 \rightarrow 123$$

Let H be a subset of the elements of G (which we denote by $H \subset G$). If H is a group (with the same law of multiplication as G), then H is a *subgroup* of G. Every group has two *trivial* (or *improper*) subgroups:

i. $H = e$—the identity element only
ii. $H = G$—the entire group

We are usually interested in finding all of the *proper* (i.e., not improper) or *nontrivial* subgroups of a group G. A subset of elements of G that possesses closure and all inverses is necessarily a subgroup.

b. Cayley's Theorem

The symmetric group, S_n, is a remarkable group in that every group G of order n is isomorphic to a subgroup of S_n. This result, often referred to as *Cayley's theorem*, shows the extraordinary importance of the symmetric group in a discussion of finite groups. The structure of *any* finite group is contained in that of S_n or in one of its subgroups. We simply mention this important result, but we will not prove the theorem since that would require an extensive development of the properties of S_n. We will not require use of Cayley's theorem in our work. Example 9.6 is a very simple special case of this theorem.

c. Lagrange's Theorem

We will now prove a theorem that will be of considerable use in analyzing the subgroups of a finite group.

THEOREM 9.1 (Lagrange's theorem)

Order m of a subgroup H of a finite group G is an integral divisor of order g of the group.

PROOF

If subgroup H is the entire group, G, then $m = g$ and the theorem is trivially true. Let the elements of H be $e, h_2, h_3, \ldots, h_m$. If $a \in G$ but $a \notin H$, form the elements

$$ae = a, ah_2, ah_3, \ldots, ah_m.$$

which we will denote compactly as aH. We point out two important properties of these elements aH:

i. *All* of the ah_j, $j = 1, 2, \ldots, m$, are distinct since, if we had $ah_i = ah_j$, $i \neq j$, this would imply that

$$h_i = h_j$$

which is a contradiction.

ii. None of the ah_j is contained in H since, if one were, then we would have $ah_i = h_j$ so that

$$a = h_j h_i^{-1}$$

which would make $a \in H$, again a contradiction.

Therefore aH is a set of m *distinct* elements, none of which is contained in H. Since this is assumed to be a finite group, we can proceed and finally exhaust G after k such steps as

$$G = H + a_2 H + a_3 H + \cdots + a_k H \qquad \text{Q.E.D.} \quad \textbf{(9.5)}$$

Here k is termed the *index* of the subgroup H of G.

Example 9.7 Consider the permutation group, S_3, and let H be the cyclic permutations. That is, H consists of the elements

$$\begin{cases} 123 \\ 312 \\ 231 \end{cases}$$

while those elements of G not in H are

$$\begin{cases} 132 \\ 213 \\ 321 \end{cases}$$

Let $a_2 = 132$ so that

$$a_2 H = \begin{cases} 132 \\ 321 \\ 213 \end{cases}$$

We see that

$$G \equiv S_3 = H + a_2 H$$

so that H has index 2 in S_3.

---

d. Cosets, Conjugate Classes, and Invariant Subgroups

Of course, in Th. 9.1, we could just as well have decomposed G as

$$G = H + Hb_2 + Hb_3 + \cdots + Hb_k \tag{9.6}$$

In Eq. 9.5 we have decomposed G by *left cosets* and in Eq. 9.6 by *right cosets.* Given a group G we can *always* take any element a and form a cyclic subgroup as e, a, $a^2, \ldots, a^{m-1}$, until $a^m = e$. By Lagrange's theorem, if order n of G is a prime number, then G is necessarily cyclic.

If there exists an element $u \in G$ such that $a \in G$ and $b \in G$ are related as

$$uau^{-1} = b$$

then b is *conjugate* to element a (often written $a \sim b$). We easily verify that

i. a is a conjugate to a.
ii. If a is conjugate to b, then b is conjugate to a.
iii. If a is conjugate to b and b is conjugate to c, then a is conjugate to c.

This implies that we can take any finite group G and separate it into sets of elements that are conjugate to each other (*conjugate classes*).

We define a *class* of G as the set of conjugates of an element $a \in G$. The element a belongs to this set. The class conjugate to a and that conjugate to b are the same provided a and b are conjugate. If a and b are not conjugate, these classes have *no* elements in common. Since each element of G belongs to a class, we can decompose G into classes. Any element of G that commutes with all elements of G forms a class by itself. The identity is such an element, as are the elements of an abelian group, each of which is in a class by itself.

Example 9.8 Consider the rotations through an angle φ about two different axes fixed in space and passing through a common origin (cf. Problems 9.23 and

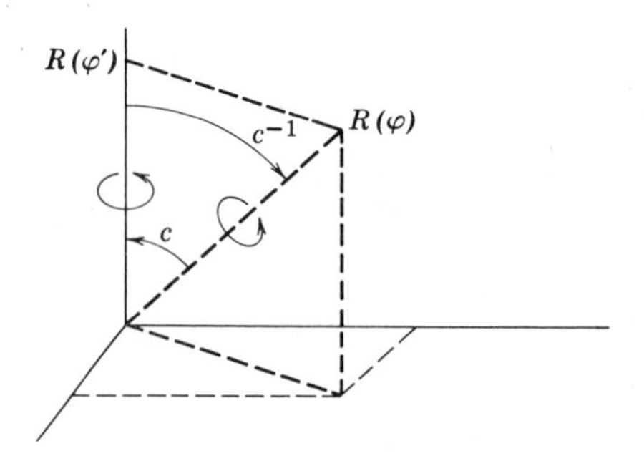

Figure 9.1

9.24). As indicated in Fig. 9.1 we have the relation

$$R'(\varphi)=cR(\varphi)c^{-1} \qquad 0\leq\varphi<\pi$$

so that these rotations form a class of the three-dimensional spatial rotations.

Let us take any subgroup H of G and any $a\in G$ and form the set of elements aHa^{-1}. This set of elements is a *conjugate subgroup* of H in G. If it happens that for *all* $a\in G$,

$$aHa^{-1}=H \tag{9.7}$$

then H is an *invariant* subgroup of G. Equation 9.7 states that all of the left and right cosets of H are identical. The elements of H, as a collection, commute with all the elements of G if H is an invariant subgroup. The necessary and sufficient condition that H be an invariant subgroup of G is that it contain the elements of G in *complete classes,* which means that H contains either all or none of the elements of each class of G. We can see this as follows. Take some $a\in G$ and $u_j\in G$ so that the class conjugate to a is the set of elements (not necessarily all distinct)

$$g_j=u_jau_j^{-1},\ j=1,2,\ldots,g \text{ (the order of } G)$$

If one g_j, for example, g_1, is in an invariant subgroup H, then we have

$$u_1^{-1}g_1u_1=a\in H$$

which implies that all these g_j belong to H. Conversely, it now becomes obvious that if one element of a class is not in H, then none of the elements of that class can be in H. Every group G always possesses the trivial (i.e., improper) invariant subgroups $H=G$ and $H=e$. A group with no proper (i.e., nontrivial) invariant subgroups is *simple.* If none of the (proper) invariant subgroups of a group is abelian, then the group is *semisimple.*

Since for any invariant subgroup H and any elements $a\in G$, $b\in G$, we have with respect to coset multiplication,

$$(aH)(bH)=a(Hb)H=a(bH)H=ab(HH)=abH$$
$$H(aH)=(Ha)H=(aH)H=a(HH)=aH$$
$$(a^{-1}H)(aH)=a^{-1}(Ha)H=a^{-1}a(HH)=H$$

we see that H and its cosets, taken as group elements, form a group with respect to this multiplication. This is the *factor group,* often written G/H, of the invariant subgroup of H in G. Its order is the index of H in G.

We now consider some examples in quite a bit of detail in order to clarify these concepts.

Example 9.9 A subgroup of the symmetric group S_n is the *alternating group* A_n consisting of all *even* permutations of n objects. A_n is of order $(1/2)n!$. Recall from Section 2.1b that a permutation is even or odd according to whether it takes an even or odd number of transpositions to bring the permutation $ijk \cdots$ into the standard arrangement $1, 2, 3, \ldots, n$, a transposition being the interchange of the positions of two elements in a permutation. Consider the case of A_3, which has $(1/2)6! = 3$ elements, where

$$\begin{array}{lll} e = (123) & e^{-1} = e & aa = b \\ a = (231) & a^{-1} = b & bb = a \\ b = (312) & b^{-1} = a & ab = e = ba \end{array}$$

which establishes that A_3 is a group. This group has already been met in Example 9.6. It is clear that A_3 has no nontrivial subgroup so that it must be simple.

Example 9.10 Take $G = S_3$ and denote the $3! = 6$ elements as follows:

$$\begin{array}{ll} e = 123\text{—identity element} & e^{-1} = 123 = e \\ a = 312 & a^{-1} = 231 = b \\ b = 231 & b^{-1} = 312 = a \\ c = 132 & c^{-1} = 132 = c \\ d = 321 & d^{-1} = 321 = d \\ f = 213 & f^{-1} = 213 = f \end{array}$$

$$H = \begin{cases} e = 123 \\ a = 312 \\ b = 231 \end{cases}$$

$$cH = \begin{cases} 132 \\ 321 \\ 213 \end{cases} \quad Hc = \begin{cases} 132 \\ 213 \\ 321 \end{cases} \quad \therefore \quad cH = Hc$$

$$dH = \begin{cases} 321 \\ 213 \\ 132 \end{cases} \quad Hd = \begin{cases} 321 \\ 132 \\ 213 \end{cases} \quad \therefore \quad dH = Hd$$

$$fH = \begin{cases} 213 \\ 132 \\ 321 \end{cases} \quad Hf = \begin{cases} 213 \\ 321 \\ 132 \end{cases} \quad \therefore \quad fH = Hf$$

Since H is a nontrivial invariant subgroup of S_3, we see that S_3 is not simple. Also, since $ab = ba = e$, we see that H is abelian so that S_3 is not semisimple.

Furthermore the decomposition by cosets is, for example,

$$S_3 = H + cH = H + Hc$$

In fact, in this case we have

$$Hc = Hd = Hf$$

Since the index of H in S_3 is two, H *must* be invariant (cf. Problem 9.10). We now find the conjugate classes,

$$e \sim e \text{ only}$$

$$\begin{aligned}
eae^{-1} &= ea \quad = a && = a && = a \\
aaa^{-1} &= ae \quad = a && = a && = a \\
bab^{-1} &= ba^2 = (231)(231) && = (312) && = a \\
cac^{-1} &= cac = (132)(213) && = (231) && = b \\
dad^{-1} &= dad = (321)(132) && = (231) && = b \\
faf^{-1} &= faf = (213)(321) && = (231) && = b
\end{aligned}$$

Therefore $\{a, b\}$ is the class conjugate to a so that $a \sim b$.

$$\begin{aligned}
ece^{-1} &= cc \quad = c && = c && = c \\
aca^{-1} &= acb = (312)(213) && = (321) && = d \\
bcb^{-1} &= bca = (231)(321) && = (213) && = f \\
ccc^{-1} &= ce \quad = c && = c && = c \\
dcd^{-1} &= dcd = (321)(312) && = (213) && = f \\
fcf^{-1} &= fcf = (213)(231) && = (321) && = d
\end{aligned}$$

Therefore $\{c, d, f\}$ is the class conjugate to c so that $d \sim c$, $f \sim c$, $d \sim f$. We can decompose S_3 by classes as

$$S_3 = e + \{a, b\} + \{c, d, f\}$$

Finally then S_3 contains *three* conjugate classes.

If we now take $H = \{e, c\}$, which is an abelian subgroup, we find the following results.

$$aH = \begin{cases} a = a \\ 213 = f \end{cases} \qquad Ha = \begin{cases} a = a \\ 321 = d \end{cases} \qquad \therefore \quad aH \neq Ha$$

$$bH = \begin{cases} b = b \\ 321 = d \end{cases} \qquad Hb = \begin{cases} b = b \\ 213 = f \end{cases} \qquad \therefore \quad bH \neq Hb$$

$$dH = \begin{cases} d = d \\ 231 = b \end{cases} \qquad Hd = \begin{cases} d = d \\ 312 = a \end{cases} \qquad \therefore \quad dH \neq Hd$$

$$fH = \begin{cases} f = f \\ 312 = a \end{cases} \qquad Hf = \begin{cases} f = f \\ 231 = b \end{cases} \qquad \therefore \quad fH \neq Hf$$

This implies that $H=\{e, c\}$ is not an invariant subgroup of S_3. This allows us to decompose S_3 as

$$S_3=\{e, c\}+\{a, f\}+\{b, d\}$$
$$=H+aH+bH$$

If we apply Lagrange's theorem to S_3, we see that the orders of the *possible* subgroups are 1, 2, 3, and 6. From the results of this example we have for these subgroups of orders 1, 2, 3, and 6, respectively

1—e—trivial

$$2\text{—}\begin{cases}\{e, c\}\\ \{e, d\}\\ \{e, f\}\end{cases}$$

3— $\{e, a, b\}$—abelian and invariant

6— S_3—trivial

Example 9.11 Consider the possible groups of order four with elements $\{e, a, b, c\}$, all assumed distinct. Lagrange's theorem implies that the only possible subgroups are those of orders 1, 2, and 4. We leave it to the student to verify some of the following statements.

i. If $a^2=b$, the *only possible* choices are

$$e=a^4$$
$$a=a$$
$$b=a^2$$
$$c=a^3$$

ii. If $a^2=c$, we *must* have

$$e=a^4$$
$$a=a$$
$$b=a^3$$
$$c=a^2$$

iii. If $a^2=e$, $c^2=a$, then we must have

$$e=b^4$$
$$a=b^2$$
$$b=b$$
$$c=b^3$$

It is clear that cases (i)–(iii) are cyclic (and therefore abelian) groups, that they contain only the two trivial subgroups, and that they are isomorphic to each other. The group multiplication table for (i) is given below.

TABLE 9.2

	e	a	b	c
e	e	a	b	c
a	a	b	c	e
b	b	c	e	a
c	c	e	a	b

iv. If we take $a^2=e$, $c^2=e$, then the group multiplication table is the following.

TABLE 9.3

	e	a	b	c
e	e	a	b	c
a	a	e	c	b
b	b	c	e	a
c	c	b	a	e

This group is also abelian, but contains the nontrivial subgroup $\{e, a\}$ [or $\{e, c\}$] and is, therefore, not isomorphic to any of the above.

These four cases, i–iv, exhaust the possible structures of groups of order four, all of which are abelian. In case iv the factor group of $H=\{e, a\}$ in G has elements $H=\{e, a\}$ and $bH=\{b, c\}$. Since these groups are abelian, they all contain *four* classes. We will return to this example after we have discussed representations of groups in Sections 9.2–9.4.

Example 9.12 Let $G=S_4$. This is already quite a large group since it contains $4!=24$ elements. By Lagrange's theorem it can contain subgroups of orders 1, 2, 3, 4, 6, 8, 12, and 24. From our previous discussions we know that the subgroups of orders 1, 2, 3, and 4 are abelian. In order to decide whether or not any of the subgroups is an invariant subgroup, we can save ourselves a lot of tedious calculation if we realize that any invariant subgroup must contain elements by complete classes. It is left to the student to show that there are just five classes in S_4. We list these

	No. of Elements
C_1: (1234)	1
C_2: (2143), (3412), (4321)	3
C_3: (2134), (3214), (4231), (1324), (1432), (1243)	6
C_4: (2341), (2413), (3421), (3142), (4312), (4123)	6
C_5: (2314), (3124), (2431), (3241), (4132), (4213), (1342), (1423)	8

Since any subgroup must contain the identity element (i.e., C_1 above), we see at once that the subgroups of orders 2, 3, 6, and 8 cannot be invariant subgroups. Therefore the only *possible* nontrivial invariant subgroups are

$$H_4 = C_1 + C_2 \tag{9.8}$$

$$H_{12} = C_1 + C_2 + C_5 = H_4 + C_5 \tag{9.9}$$

We easily see that H_4 is a group and that *every* element h_j of H_4 is such that $h_j^2 = e$, so that $h_j = h_j^{-1}$ for this subgroup and

$$h_k = h_i h_j = h_k^{-1} = (h_i h_j)^{-1} = h_j^{-1} h_i^{-1} = h_j h_i$$

so that the subgroup H_4 is an abelian invariant subgroup. This implies that S_4 is neither simple nor semisimple. We leave it to the reader to show that H_{12} is also a subgroup (and hence invariant).

---

e. Homomorphism

Two groups G and G' are *homomorphic* provided that the elements of G can be put into a correspondence with those of G' (possibly in a many-to-one fashion) and provided that this correspondence is preserved under the respective laws of multiplication of the two groups.

9.2 LINEAR REPRESENTATIONS OF GROUPS

Just as in Section 2.5 we saw how to represent a bounded linear operator by a matrix in a linear vector space, so we now seek specific representations of the elements of abstract groups. We restrict ourselves to representations in terms of linear operators (usually finite-dimensional matrices but sometimes linear differential operators as well).

If a set of operators in a vector space satisfies the group postulates, then the set forms a group. Specifically, if we can establish a homomorphism between the elements of a group G and the group of operators, $\{D(G)\}$, acting on a linear vector space, then the $\{D(G)\}$ form a *representation* of G in this space. If this space is of dimension n, then we have an n-dimensional representation of G. We will use Wigner's notation $D(R)$ for the representation of the element R of G. When the representation is given in terms of matrices, we use $D_{ij}(R)$ for the matrix elements. The elements of these representations must have the following properties.

$$\begin{aligned} D(RS) &= D(R)D(S) \\ D(R^{-1}) &= [D(R)]^{-1} \\ D(E) &= I \end{aligned} \tag{9.10}$$

Here E is the identity element and the law of multiplication for matrix representations is ordinary matrix multiplication. When we have several different representations, each of dimensionality n_μ, we write these as $D_{ij}^{(\mu)}(R)$.

When the correspondence between the elements of G and those of $D(G)$ is an isomorphism, the representation is *faithful*. In this case the order of G is the order of $D(G)$. In accordance with the definition of similar matrices, we say that two representations $D'(G)$ and $D(G)$ are *equivalent* provided that for *all* R there exists *one* C such that

$$D'(R) = CD(R)C^{-1} \tag{9.11}$$

We can obviously construct an invariant of a matrix representation through its trace

$$\chi(R) \equiv \sum_j D_{jj}(R) \tag{9.12}$$

where $\chi(R)$ is the *character* of R in the representation D. Equation 9.11 implies that equivalent representations have the same set of characters, $\chi^{(\mu)}(R)$. Conjugate elements $S = URU^{-1}$ have identical characters since

$$\chi(S) = TrD(S) = Tr[D(U)D(R)D^{-1}(U)] = TrD(R) = \chi(R)$$

All elements in a given class have the same character (which is simply a number). In the μth representation G has a set of characters $\chi_j^{(\mu)}$, $j = 1, 2, \ldots, k$ where k is the number of classes in G. We often find it convenient to treat χ^μ as a vector in a space of dimension k.

Let us now consider the construction of representations. Suppose that we are given the effect of a group of transformations $\{R\}$ some space (e.g., spatial rotations) as

$$\mathbf{x} \to R\mathbf{x} \tag{9.13}$$

We can associate with transformation R a linear operator O_R defined by its action on an arbitrary function $\psi(\mathbf{x})$ as

$$O_R\psi(\mathbf{x}) \equiv \psi(R\mathbf{x}) \tag{9.14}$$

That is, we replace arguments x_i of $\psi(\mathbf{x}) \equiv \psi(x_1, x_2, \ldots, x_n)$ by the values $\sum_{j=1}^{n} r_{ij}x_j$. We prove that these operators $\{O_R\}$ have the group properties of the $\{R\}$ by considering the successive transformations

$$\mathbf{x} \to R\mathbf{x} \to S(R\mathbf{x}) \equiv T\mathbf{x} \tag{9.15}$$

According to Eq. 9.14 we have

$$O_S[O_R\psi(\mathbf{x})] = O_S\psi(R\mathbf{x}) = \psi(SR\mathbf{x})$$

and also, from Eqs. 9.14 and 9.15,

$$O_T\psi(\mathbf{x}) = \psi(T\mathbf{x}) = \psi(SR\mathbf{x})$$

so that

$$O_T\psi(\mathbf{x}) = O_S O_R\psi(\mathbf{x})$$

or

$$O_T \equiv O_{SR} = O_S O_R \tag{9.16}$$

when $T = SR$. This shows that the correspondence, $\{R\} \to \{O_R\}$, is preserved under multiplication. Similarly if

$$\mathbf{x} \to R\mathbf{x} \to R^{-1}(R\mathbf{x}) \equiv \mathbf{x}$$

then

$$O_R^{-1}[O_R\psi(\mathbf{x})] = O_{R^{-1}}\psi(R\mathbf{x}) = \psi(\mathbf{x})$$

so that

$$O_{R^{-1}}O_R = I$$

or

$$O_{R^{-1}} = [O_R]^{-1} \tag{9.17}$$

This means that if $R \to O_R$, then $R^{-1} \to O_{R^{-1}} = [O_R]^{-1}$. The existence of the identity, $O_I = I$, and the associativity property are obvious. This establishes a homomorphism between the group transformation elements, $\{R\}$, and the set of linear operators, $\{O_R\}$, which establishes that we do have a representation of the group.

If we can find a set of linearly independent functions $\{\psi_i(\mathbf{x})\}$ such that the $\{O_R\psi_i(\mathbf{x})\}$ can be expressed in terms of the original set as

$$O_R\psi_i(\mathbf{x}) \equiv \psi_i(R\mathbf{x}) = \sum_{j=1}^{n} D_{ji}(R)\psi_j(\mathbf{x}) \qquad i = 1, 2, \ldots, n \tag{9.18}$$

then these $n \times n$ matrices, $\{D(R\ \}$, form a representation of the group since, from Eqs. 9.18 and 9.16,

$$O_S O_R\psi_i(\mathbf{x}) = \sum_{j=1}^{n} D_{ji}(R)O_S\psi_j(\mathbf{x}) = \sum_{j,k=1}^{n} D_{ji}(R)D_{kj}(S)\psi_k(\mathbf{x})$$

$$= \sum_{j=1}^{n} [D(S)D(R)]_{ji}\psi_j(\mathbf{x}) = O_{SR}\psi_i(\mathbf{x}) = \sum_{j=1}^{n} [D(SR)]_{ji}\psi_j(\mathbf{x})$$

so that

$$D(S)D(R) = D(SR) \tag{9.19}$$

Example 9.13 Let us take as our group the rotations R_1, R_2, R_3, R_4 in the x-y plane (i.e., about the z axis) through angles of 0°, 90°, 180°, and 270°, respectively. Under a counterclockwise rotation through an angle φ we have

$$x \to x\cos\varphi + y\sin\varphi$$

$$y \to -x\sin\varphi + y\cos\varphi$$

Let $\psi(x, y)$ be an arbitrary function and set

$$\psi_1 = \psi(x, y)$$

$$\psi_2 = \psi(y, -x)$$

$$\psi_3 = \psi(-x, -y)$$

$$\psi_4 = \psi(-y, x)$$

As always we have the representation for the identity element as

$$D(R_1) = I = \begin{pmatrix} 1 & 0 & 0 & 0 \\ 0 & 1 & 0 & 0 \\ 0 & 0 & 1 & 0 \\ 0 & 0 & 0 & 1 \end{pmatrix}$$

Applying Eq. 9.18 we have

$$O_{R_2}\psi_1 = \psi(y, -x) = \psi_2 = \sum_{j=1}^{4} D_{j1}(R_2)\psi_j$$

$$O_{R_2}\psi_2 = \psi(-x, -y) = \psi_3 = \sum_{j=1}^{4} D_{j2}(R_2)\psi_j$$

$$O_{R_2}\psi_3 = \psi(-y, x) = \psi_4 = \sum_{j=1}^{4} D_{j3}(R_2)\psi_j$$

$$O_{R_2}\psi_4 = \psi(x, y) = \psi_1 = \sum_{j=1}^{4} D_{j4}(R_2)\psi_j$$

Since $\psi(x, y)$ is arbitrary, we can conclude that

$$D(R_2) = \begin{pmatrix} 0 & 0 & 0 & 1 \\ 1 & 0 & 0 & 0 \\ 0 & 1 & 0 & 0 \\ 0 & 0 & 1 & 0 \end{pmatrix}$$

We can similarly apply Eq. 9.18 for R_3 and R_4 to find

$$D(R_3) = \begin{pmatrix} 0 & 0 & 1 & 0 \\ 0 & 0 & 0 & 1 \\ 1 & 0 & 0 & 0 \\ 0 & 1 & 0 & 0 \end{pmatrix} \qquad D(R_4) = \begin{pmatrix} 0 & 1 & 0 & 0 \\ 0 & 0 & 1 & 0 \\ 0 & 0 & 0 & 1 \\ 1 & 0 & 0 & 0 \end{pmatrix}$$

This is just the four-element cyclic group of Example 9.11 (cf. Table 9.2).

9.3 UNITARY REPRESENTATIONS

When the operators forming the representation of a group G are unitary, we have a *unitary representation.*

THEOREM 9.2

Every matrix representation $\{D(G)\}$ of a *finite* group G is equivalent to a unitary representation.

PROOF

We begin by defining an hermitian matrix M as

$$M \equiv \sum_{R \in G} D(R) D^{\dagger}(R) = M^{\dagger}$$

From Section 3.3 we know that there exists a unitary matrix U that will diagonalize M as

$$D_M = UMU^{\dagger} = \sum_{R \in G} UD(R)U^{\dagger}UD^{\dagger}(R)U^{\dagger} \equiv \sum_{R \in G} \bar{D}(R)\bar{D}^{\dagger}(R) \qquad \textbf{(9.20)}$$

where

$$\bar{D}(R) \equiv UD(R)U^{\dagger}$$

If we write $D_M = \{d_i \, \delta_{ij}\}$, then we have

$$d_i = \sum_R \sum_k \bar{D}_{ik}(R) \bar{D}^*_{ik}(R) = \sum_R \sum_k |\bar{D}_{ik}(R)|^2 > 0$$

That is, all of the diagonal matrix elements of D_M are real and positive. [Notice that if any of the $d_i = 0$, then $\bar{D}_{ik}(R) = 0$, $\forall k$ and R, so that $\det \bar{D}(R) = 0$, $\forall R$, which would contradict the assumption that the $\{D(R)\}$ form a representation.] This means that there exist real diagonal matrices $D_M^{1/2}$ and $D_M^{-1/2}$. This allows us to rewrite Eq. 9.20 as

$$\sum_R D_M^{-1/2} \bar{D}(R) \bar{D}^{\dagger}(R) D_M^{-1/2} = I$$

Let us define a representation $\{D'(R)\}$ equivalent to the $\{D(R)\}$ as

$$D'(R) \equiv D_M^{-1/2} UD(R)U^{\dagger} D_M^{1/2} = D_M^{-1/2} \bar{D}(R) D_M^{1/2}$$

We can then easily see that $D'(R)$ is unitary as follows.

$$\begin{aligned} D'(R)D'^{\dagger}(R) &= D_M^{-1/2} \bar{D}(R) D_M \bar{D}^{\dagger}(R) D_M^{-1/2} \\ &= D_M^{-1/2} \sum_S \bar{D}(R) \bar{D}(S) \bar{D}^{\dagger}(S) \bar{D}^{\dagger}(R) D_M^{-1/2} \\ &= D_M^{-1/2} \sum_S \bar{D}(RS) \bar{D}^{\dagger}(RS) D_M^{-1/2} \\ &= D_M^{-1/2} \sum_R \bar{D}(R) \bar{D}^{\dagger}(R) D_M^{-1/2} = I \end{aligned}$$

We have used an obvious property of summation over all the elements of the group that, for any function of the group elements, $F(R)$, we have the identity

$$\sum_{\forall S \in G} F(RS) = \sum_{\forall S \in G} F(S)$$

for any $R \in G$. That is, because of the closure property both sums run over exactly the same elements, only possibly in different order. Q.E.D.

9.4 IRREDUCIBLE REPRESENTATIONS

We have previously studied the reducibility of matrix representations of operators in terms of invariant subspaces (cf. Section 3.1). For our present purposes we say that a representation is reducible provided every element of the representation can be brought simultaneously into the block-diagonal form

$$D(R) = \begin{pmatrix} D^{(1)}(R) & 0 \\ 0 & D^{(2)}(R) \end{pmatrix} \tag{9.21}$$

by a *single* similarity transformation (i.e., for *all* $R \in G$). This means that our representation space decomposes into a *direct sum* of two invariant subspaces, one of dimension m and the other of dimension $(n-m)$. We write this as

$$D = D^{(1)} \oplus D^{(2)}$$

or simply as

$$D = D^{(1)} + D^{(2)}$$

We are often interested in finding the *irreducible* (i.e., not reducible) *representations* of a group G. Thus we write for an arbitrary representation D

$$D = D^{(1)} + D^{(2)} + \cdots + D^{(N)} \tag{9.22}$$

where each $D^{(\nu)}$ is an irreducible representation. Among these $D^{(\nu)}$ may be several that are equivalent to each other (of the same dimension naturally). We may use the same symbol for equivalent representations.

$$D = a_1 D^{(1)} + \cdots + a_r D^{(\nu)} = \sum_\nu a_\nu D^{(\nu)} \tag{9.23}$$

Some $D^{(\nu)}$ may have *same* n_μ and be inequivalent. Here the $\{a_\nu\}$ are positive integers indicating the number of times a given irreducible representation occurs in the reduction of D.

a. Schur's Lemma

We now formulate general criteria for the reducibility of representations.

THEOREM 9.3 (Schur's lemma)

If $\{D(R)\}$, of dimension n, and $\{D'(R)\}$, of dimension m, are irreducible representations of a group G and A is a matrix such that

$$D(R)A = AD'(R) \tag{9.24}$$

for *all* $R \in G$, then:

i. If $n \neq m$, then $A = 0$.
ii. If $n = m$, then either $A = 0$ or $\det A \neq 0$, in which latter case $D(R)$ and $D'(R)$ are equivalent.
iii. If $D(R) = D'(R)$, $\forall R \in G$, then $A = aI$, where a is a constant. Furthermore that $A = aI$ be the only matrix such that $[D(R), A] = 0$, $\forall R \in G$, is the necessary and sufficient condition for $D(R)$ to be irreducible.

PROOF

As a result of Th. 9.2 we need only consider unitary representations. We begin by proving (iii) first. We are given

$$D(R)A = AD(R) \qquad \forall R \in G$$

or

$$[D(R), A] = 0$$

$$D(R)D^{\dagger}(R) = I$$

so that

$$A^{\dagger}D^{\dagger}(R) = D^{\dagger}(R)A^{\dagger}$$

or

$$D(R)A^{\dagger} = A^{\dagger}D(R)$$

which is

$$[D(R), A^{\dagger}] = 0$$

We can always decompose A in terms of two hermitian matrices as

$$A = A_{+} + iA_{-}$$

where

$$A_{+} = \frac{1}{2}(A + A^{\dagger}) = A_{+}^{\dagger}$$

$$A_{-} = \frac{1}{2i}(A - A^{\dagger}) = A_{-}^{\dagger}$$

from which it follows that

$$[D(R), A_{+}] = 0 = [D(R), A_{-}]$$

Therefore it is sufficient to consider hermitian matrices that commute with the $\{D(R)\}$. Let us denote this matrix by M so that we now have

$$[D(R), M] = 0 \qquad \forall R \in G$$

$$D(R)D^{\dagger}(R) = I$$

$$M = M^{\dagger}$$

Since M is hermitian it may be diagonalized by a unitary matrix U as

$$D_M = UMU^\dagger$$

so that with $\bar{D}(R) \equiv UD(R)U^\dagger$ we have

$$\bar{D}(R)D_M = D_M D(R)$$

If we take $D_M = \{d_i\, \delta_{ij}\}$, then in component form this last equation becomes

$$\bar{D}_{ij}(R)\, d_j = d_i \bar{D}_{ij}(R)$$

or

$$\bar{D}_{ij}(R)(d_i - d_j) = 0 \qquad \forall i, jR$$

For convenience we will order the eigenvalues, $\{d_j\}$, such that $d_1 \leq d_2 \leq \cdots \leq d_n$. If D_M is not a constant multiple of I (i.e., if $d_1 \neq d_j$, $\forall i, j$), then $\bar{D}_{ij}(R)$, $\forall R \in G$, is block diagonal and the original $D(R)$ is reducible, contrary to our assumption. Therefore we have

$$A_+ = a_+ I$$

$$A_- = a_- I$$

or

$$A = (a_+ + ia_-)I \equiv aI$$

as claimed. Notice that this proof of (iii) has also established that if $D(R)$ is reducible as, say,

$$\bar{D}(R) = UD(R)U^\dagger = \begin{array}{c} \\ m \\ \\ \end{array}\!\!\begin{pmatrix} \overset{m}{D^{(1)}(R)} & 0 \\ 0 & \underset{(n-m)}{D^{(2)}(R)} \end{pmatrix}\!(n-m)$$

then there always exists a matrix A, aside from $A = aI$, such that

$$[\bar{D}(R), A] = 0 \qquad \forall R \in G$$

For example we may take

$$A = \begin{array}{c} \\ m \\ \\ \end{array}\!\!\begin{pmatrix} \overset{m}{a_1 I} & \\ & \underset{(n-m)}{a_2 I} \end{pmatrix}\!(n-m)$$

so that with $A' = U^\dagger A U \neq aI$ we have

$$\begin{aligned} [D(R), A'] &= D(R)U^\dagger AU - U^\dagger AUD(R) \\ &= U^\dagger\{UD(R)U^\dagger A - AUD(R)U^\dagger\}U = U^\dagger[\bar{D}(R), A]U = 0 \end{aligned}$$

Therefore, if there exists no other matrix aside from a constant multiple of the identity that commutes with *every* $D(R)$ of a representation, then this representation is irreducible.

We now proceed to cases (i) and (ii), again restricting our attention to unitary irreducible representations $D(R)$ and $D'(R)$. From the adjoint of Eq. 9.24 and the unitary property of the representations we obtain

$$A^\dagger D^\dagger(R) = D'^\dagger(R) A^\dagger$$

or

$$A^\dagger D(R^{-1}) = D'(R^{-1}) A^\dagger$$

From this last equation plus Eq. 9.24 for $R \to R^{-1}$, that is,

$$D(R^{-1})A = AD'(R^{-1})$$

we see that

$$D(R^{-1})AA^\dagger = AA^\dagger D(R^{-1}) \qquad \forall R \in G$$

However in our proof of (iii) we have just shown that this requires

$$AA^\dagger = aI$$

where a is some constant. The proof of (ii) is now immediate. If $n = m$, then A is a square matrix so that

$$\det(AA^\dagger) = |\det A|^2 = a$$

If $a \neq 0$, then $\det A \neq 0$ and A^{-1} exists so that

$$D'(R) = A^{-1} D(R) A$$

and the two representations would be equivalent. If $a = 0$, then

$$AA^\dagger = 0$$

or

$$\sum_j |a_{ij}|^2 = 0 \qquad \forall i$$

which requires that

$$a_{ij} = 0 \qquad \forall i, j$$

This means $A = 0$ so that the proof of (ii) is complete.

Finally, if $n \neq m$, then we may assume $m < n$, so that A is an $n \times m$ rectangular matrix such that

$$AA^\dagger = aI$$

where, of course, $AA^\dagger$ is an $n \times n$ matrix. Just as in our discussion of the Gram determinant in Section 2.3a, we can imbed A in an $n \times n$ square matrix B which has all zeros in the last $(n-m)$ columns so that

$$AA^\dagger = BB^\dagger = aI$$

or

$$a = |\det B|^2 = 0$$

since B has at least one column of zeros. This again implies that $A = 0$. Q.E.D.

b. Completeness

We can now derive some important orthogonality and completeness relations for irreducible representations. Construct the matrix

$$A = \sum_{S \in G} D(S)XD(S^{-1})$$

where the sum runs over *all* the group elements and X is an arbitrary matrix and the $D(R)$ is an irreducible representation.

$$D(R)A = \sum_S D(R)D(S)XD(S^{-1}) = \sum_S D(R)D(S)XD(S^{-1})D(R^{-1})D(R)$$

$$= \left[\sum_S D(RS)XD([RS]^{-1})\right]D(R) \equiv \sum_S D(S)XD(S^{-1})D(R) = AD(R)$$

Therefore by Th. 9.3 (iii) we conclude that

$$A = aI$$

The value of a depends on the choice for X. Let X have all elements zero except one such that $X_{lm} = 1$ and let $a = a_{lm}$ so that

$$\sum_S D_{il}(S)D_{mj}(S^{-1}) = a_{lm}\,\delta_{ij}$$

If we set $i = j$ and sum over i, we find

$$\sum_S \sum_i D_{il}(S)D_{mi}(S^{-1}) = na_{lm} = \sum_S D_{ml}(SS^{-1})$$

$$= \sum_S D_{ml}(E) = \sum_S \delta_{ml} = g\,\delta_{ml}$$

or

$$a_{lm} = \frac{g}{n}\,\delta_{ml}$$

This important result shows that

$$\sum_S D_{il}(S)D_{mj}(S^{-1}) = \frac{g}{n}\,\delta_{lm}\,\delta_{ij} = \sum_S D_{il}(S)D^*_{jm}(S) \tag{9.25}$$

where the last form holds for unitary representations.

We can also construct an A satisfying Th. 9.3(i) as follows. If we are given two inequivalent irreducible representations, $D(R)$ and $D'(R)$ (of dimensions n and m, respectively), of G, we define an $m \times n$ matrix A as

$$A = \sum_{S \in G} D'(S)XD(S^{-1})$$

where X is an arbitrary $m \times n$ matrix.

$$D'(R)A = \sum_S D'(R)D'(S)XD(S^{-1})$$

$$= \sum_S D'(RS)XD([RS]^{-1})D(R) = AD(R)$$

According to Schur's lemma we must have

$$\sum_S D'(S)XD(S^{-1})=0 \qquad \forall X$$

If we again choose X as before, this becomes

$$\sum_S D_{il}(S)D_{mj}(S^{-1})=0 \tag{9.26}$$

We can combine Eqs. 9.25 and 9.26 into the single statement

$$\sum_R D_{il}^{(\mu)}(R)D_{mj}^{(\nu)}(R^{-1})=\frac{g}{n_\mu}\,\delta_{\mu\nu}\,\delta_{ij}\,\delta_{lm}=\sum_R D_{il}^{(\mu)}(R)D_{jm}^{(\nu)*}(R) \tag{9.27}$$

where again the last form holds for unitary representations.

We see that each irreducible representation $D^{(\mu)}$ provides us with n_μ^2 orthogonal vectors. If we consider a space of dimension g (for a *finite* group), then since $\{D_{ij}^{(\mu)}(R)\}$ is a vector in a space of dimension g (i.e., the "index" R labels the components of these vectors),

$$\sum_{\mu=1}^{N} n_\mu^2 \le g \tag{9.28}$$

so that the number, N, of inequivalent irreducible representations for a finite group is finite.

From Eqs. 9.27 and 9.12 we see that

$$\sum_R \chi^{(\mu)}(R)\chi^{(\nu)}(R^{-1})=g\,\delta_{\mu\nu}=\sum_R \chi^{(\mu)}(R)\chi^{(\nu)*}(R)$$

$$=\sum_{j=1}^{k} g_j\chi_j^{(\mu)}\chi_j^{(\nu)*} \qquad \mu, \nu=1, 2, \ldots, N \tag{9.29}$$

where k is the number of classes in G and g_j is the number of elements of G in jth class. Since $X_j^{(\mu)}$ is a vector in a k-dimensional space, we obtain the important result

$$N \le k \tag{9.30}$$

We now prove that the equality signs hold in Eqs. 9.28 and 9.30. Without loss of generality we assume unitary irreducible representations in the discussion. From Eq. 9.23

$$D(R)=\sum_\nu a_\nu D^{(\nu)}(R)$$

we have

$$\chi_j=\sum_\nu a_\nu \chi_j^{(\nu)} \tag{9.31}$$

for R in the class K_j of G, there being g_j elements in class K_j. Equation 9.29 implies that

$$\sum_j \chi_j^{(\mu)*}\chi_j g_j=\sum_\nu a_\nu \sum_j g_j\chi_j^{(\mu)*}\chi_j^{(\nu)}=\sum_\nu a_\nu g\,\delta_{\mu\nu}=g a_\mu$$

or

$$a_\mu=\frac{1}{g}\sum_j g_j\chi_j^{(\mu)*}\chi_j \tag{9.32}$$

This gives the number of times a particular irreducible representation occurs in $D(R)$. From Eqs. 9.31 and 9.32 we conclude that

$$\sum_j \chi_j^* \chi_j g_j = \sum_{\mu,\nu} a_\mu a_\nu \chi_j^{*(\mu)} \chi^{(\nu)} = \sum_{\mu,\nu} a_\mu a_\nu g \, \delta_{\mu\nu} = g \sum_\mu a_\mu^2 \tag{9.33}$$

If we choose an irreducible representation, then

$$a_\mu = \delta_{\mu\nu}$$

so that

$$\sum_j g_j \, |\chi_j|^2 = g \tag{9.34}$$

Equation 9.34 is a criterion for irreducibility.

In order to prove that the equality actually holds in Eq. 9.28, we introduce the *regular representation.* Suppose we have a group G of order g with elements $R_1, R_2, \ldots, R_g$. If we consider these as coordinates (or basis vectors) in a space of dimension g, then if we take any element, for example, S_i, and successively multiply each S_j by it, we simply rearrange the $\{S_j\}$ among themselves,

$$S_i S_j = S_{i_j}$$

where $i_j = \{1, 2, \ldots, g\}$ *in some order.* That is, one works out the group multiplication table for G to obtain the correspondence between j and i_j. We now choose the $D(S_i)$ to be a $g \times g$ matrix whose (i_j, j) element is one and all of whose other elements in the i_jth row and jth column are zero. Of course $D(e) \equiv I$. In general we write

$$D_{jl}(S_i) = \delta_{j;i_l} \tag{9.35}$$

Obviously det $D(S_i) \neq 0$ so that $D^{-1}(S_i) \equiv D(S_i^{-1})$ exists. We must now check that such a prescription actually does provide a representation. This is easily done by direct calculation. If $S_a S_j = S_{a_j}$ we write

$$D_{jl}(S_a) = \delta_{j;a_l}$$

and if $S_b S_j = S_{b_j}$ we write

$$D_{jl}(S_b) = \delta_{j;b_l}$$

then for

$$S_{a_b} S_j = S_{(a_b)_j}$$

we write

$$D_{jl}(S_{a_b}) = \delta_{j;(a_b)_l}$$

This implies that

$$[D(S_a)D(S_b)]_{jl} = \sum_{k=1}^{g} D_{jk}(S_a) D_{kl}(S_b) = \sum_{k=1}^{g} \delta_{j;a_k} \, \delta_{k;b_l}$$

$$= \delta_{j;a_{(b_l)}}$$

But since

$$S_{a_{(b_l)}} = S_a S_{b_l} = S_a S_b S_l$$

and

$$S_{(a_b)_l} = S_{a_b} S_l = S_a S_b S_l$$

we see that

$$\delta_{j;a(b_l)} \equiv \delta_{j;(a_b)_l}$$

or

$$[D(S_a)D(S_b)]_{jl} = D_{jl}(S_{ab})$$

Therefore we have an *isomorphism* and hence a *faithful* representation. Furthermore the regular representation obtained in this manner is always unitary (in fact, orthogonal, since all of the matrices are by construction real) because

$$\sum_l D_{jl}(S_i)[D^T(S_i)]_{lk} = \sum_l \delta_{j;i_l}\, \delta_{k;i_l} = \delta_{jk}$$

or

$$D(R)D^{\dagger}(R) = I \qquad \forall R \in G$$

Example 9.14 At this point a simple example will do more than any amount of abstract discussion to clarify the concept of the regular representation. Consider the four-element *cyclic* group (e, a, b, c) [cf. Example 9.11(i)] where $b = a^2$, $c = a^3$, $a^4 = e$ (cf. Table 9.2). Let us make the correspondence

$$e \rightarrow \begin{pmatrix} 1 \\ 0 \\ 0 \\ 0 \end{pmatrix} \qquad a \rightarrow \begin{pmatrix} 0 \\ 1 \\ 0 \\ 0 \end{pmatrix} \qquad b \rightarrow \begin{pmatrix} 0 \\ 0 \\ 1 \\ 0 \end{pmatrix} \qquad c \rightarrow \begin{pmatrix} 0 \\ 0 \\ 0 \\ 1 \end{pmatrix}$$

Since

$$ae = a = a$$

$$aa = a^2 = b$$

$$ab = a^3 = c$$

$$ac = a^4 = e$$

we see that

$$D(a) = \begin{pmatrix} 0 & 0 & 0 & 1 \\ 1 & 0 & 0 & 0 \\ 0 & 1 & 0 & 0 \\ 0 & 0 & 1 & 0 \end{pmatrix}$$

We can also make the correspondence required for Eq. 9.35 since

$$S_2 S_j = S_{2_j}$$

implies that

$$2_1 = 2$$

$$2_2 = 3$$

$$2_3 = 4$$

$$2_4 = 1$$

so that

$$D_{jl}(a)=\delta_{j;2_l}$$

becomes

$$D_{14}(a)=\delta_{1;2_4}=\delta_{11}=1$$
$$D_{21}(a)=\delta_{2;2_1}=\delta_{22}=1$$
$$D_{32}(a)=\delta_{3;3_2}=\delta_{33}=1$$
$$D_{43}(a)=\delta_{4;2_3}=\delta_{44}=1$$

which yields the same $D(a)$ as obtained previously. Similarly, since

$$\begin{array}{ll} be=a^2e=b & ce=a^3=c \\ ba=a^3\;=c & ca=a^4=e \\ bb=a^4\;=e & cb=a^5=a \\ bc=a^5\;=a & cc=a^6=b \end{array}$$

we see that

$$D(b)=\begin{pmatrix} 0 & 0 & 1 & 0 \\ 0 & 0 & 0 & 1 \\ 1 & 0 & 0 & 0 \\ 0 & 1 & 0 & 0 \end{pmatrix} \qquad D(c)=\begin{pmatrix} 0 & 1 & 0 & 0 \\ 0 & 0 & 1 & 0 \\ 0 & 0 & 0 & 1 \\ 1 & 0 & 0 & 0 \end{pmatrix}$$

From Eq. 9.35 we see that all of the diagonal elements of these regular representations are zero except for the $D(e)=I$ since if $D_{jj}(S_i)=S_{j;i_j}=1$, then $j=i_j$ so that $S_iS_j=S_{i_j}=S_i$. This requires that $S_i=e$, as claimed. Therefore in this representation the character is such that

$$\chi(r)=\begin{cases} 0 & R\neq e \\ g & R=e \end{cases}$$

If we now express the regular representation in terms of inequivalent irreducible representations, then by Eq. 9.31 we have

$$\chi_j=\sum_\nu a_\nu \chi_j^{(\nu)}$$

where j runs over the classes. Since the identity forms a class by itself,

$$\chi_1=g \tag{9.36}$$

while for the νth irreducible representation

$$\chi_1^{(\nu)}=n_\nu \tag{9.37}$$

for the class of the identity again. Eqs. 9.31, 9.36, and 9.37 reduce to

$$g=\sum_\nu a_\nu n_\nu \tag{9.38}$$

However from Eq. 9.32 we can evaluate a_μ as

$$a_\mu \equiv \frac{1}{g} \sum_j g_j \chi_j^{(\mu)*} \chi_j = \frac{1}{g} g n_\mu = n_\mu = \chi_1^{(\mu)} \qquad \textbf{(9.39)}$$

which states that the number of times (i.e., a_ν) an (inequivalent) irreducible representation is contained in the regular representation is equal to the degree (or dimension) of that irreducible representation.

Example 9.15 Let us verify Eq. 9.39 in the example of the four-element cyclic group (cf. Example 9.14). This is very easily done since matrices $D(e)$, $D(a)$, $D(b)=D^2(a)$, and $D(c)=D^3(a)$ form a set of mutually commuting (since the group is abelian) unitary matrices so that there exists a single unitary matrix U that will simultaneously diagonalize all of the $D(S_i)$. Furthermore we can write down the diagonal forms of the $D(S_i)$, say $\bar{D}(S_i)=\{\delta_{kl}\lambda_l^i\}$, as

$$\bar{D}(S_i)=U^+D(S_i)U$$

since for this group

$$\det D(a) = -1 = \prod_{k=1}^{4} \lambda_k^a$$

$$\det D(b) = \ \ 1 = \prod_{k=1}^{4} \lambda_k^b$$

$$\det D(c) = -1 = \prod_{k=1}^{4} \lambda_k^c$$

$$\det D(e) = \ \ 1 = \prod_{k=1}^{4} \lambda_k^e$$

while

$$\lambda_k^b = (\lambda_k^a)^2 \qquad \lambda_k^c = (\lambda_k^a)^3 \qquad (\lambda_k^a)^4 = 1 \qquad k=1, 2, 3, 4$$

However from Example 9.14 we see explicitly that

$$\det |D(a) - \lambda I| = \lambda^4 - 1 = 0$$

which implies that the $\{\lambda_k^a\}$ are the fourth roots of unity,

$$\lambda_k^a = \{i, -1, -i, 1\}$$
$$\lambda_k^b = \{-1, 1, -1, 1\}$$
$$\lambda_k^c = \{-i, -1, i, 1\}$$
$$\lambda_k^e = \{1, 1, 1, 1\}$$

or that

$$D(a)=\begin{pmatrix} i & 0 & 0 & 0 \\ 0 & -1 & 0 & 0 \\ 0 & 0 & -i & 0 \\ 0 & 0 & 0 & 1 \end{pmatrix}$$

$$D(b)=\begin{pmatrix} -1 & 0 & 0 & 0 \\ 0 & 1 & 0 & 0 \\ 0 & 0 & -1 & 0 \\ 0 & 0 & 0 & 1 \end{pmatrix}$$

$$D(c)=\begin{pmatrix} -i & 0 & 0 & 0 \\ 0 & -1 & 0 & 0 \\ 0 & 0 & i & 0 \\ 0 & 0 & 0 & 1 \end{pmatrix}$$

The interested student can verify that the unitary transformation that accomplishes the diagonalization is

$$U=\frac{1}{2}\begin{pmatrix} 1 & 1 & 1 & 1 \\ -i & -1 & i & 1 \\ -1 & 1 & -1 & 1 \\ i & -1 & -i & 1 \end{pmatrix}$$

Naturally all of these representations are completely reducible in terms of the four inequivalent one-dimensional irreducible representations since this group is abelian. As we expect from Eq. 9.39, each irreducible representation is contained in the regular representation exactly once.

Finally we obtain our desired result by combining Eqs. 9.38 and 9.39 to yield

$$g=\sum_{\nu=1}^{N} n_{\nu}^{2} \tag{9.40}$$

which is a strengthened form of Eq. 9.28. The result of Eq. 9.40 is perfectly general since the regular representation was used in the derivation but does not appear in the final result.

If we now combine Eqs. 9.27 and 9.40, then since for fixed μ (i.e., n_μ) there are n_μ^2 of the $D_{ij}^{(\mu)}(R)$ and since $g=\sum_{\nu=1}^{N} n_\nu^2$, we see that the $D_{ij}^{(\mu)}(R)$ $(R \in G)$ form a *complete set* of g vectors in a g-dimensional space. These vectors have components (when orthonormalized by Eq. 9.27)

$$\sqrt{\frac{n_\nu}{g}}\, D_{ij}^{(\nu)}(R_1) \qquad \sqrt{\frac{n_\nu}{g}}\, D_{ij}^{(\nu)}(R_2) \qquad \cdots \qquad \sqrt{\frac{n_\nu}{g}}\, D_{ij}^{(\nu)}(R_g) \tag{9.41}$$

Now any time we have a complete set of orthonormal vectors,

$$\langle \psi_i \mid \psi_j \rangle = \delta_{ij} \qquad \psi_j = \{\xi_i^{(j)}\}$$

we can always find a unitary transformation that will map them onto the orthonormal set $\varphi_j = \{\delta_{ij}\}$ since $\psi_j = U\varphi_j$ requires

$$\xi_i^{(j)} = u_{ik}\varphi_k^{(j)} = u_{ij}$$

so that $\langle \varphi_i \mid \varphi_j \rangle = \delta_{ij}$ becomes

$$UU^{\dagger} = I$$

with $U^{-1} = U^{\dagger}$. But in the basis $\{\varphi_j\}$ not only is

$$\langle \varphi_i \mid \varphi_j \rangle = \delta_{ij}$$

but also

$$\sum_j \varphi_l^{(j)} \varphi_{l'}^{(j)} = \delta_{ll'} \tag{9.42}$$

so that

$$\sum_j \xi_l^{*(j)} \xi_{l'}^{(j)} = \sum_j u_{lj}^{*} u_{l'j} = (U^{\dagger} U)_{ll'} = \delta_{ll'} \tag{9.43}$$

Therefore we may state the completeness of the $D_{ij}^{(\nu)}(R)$ as

$$\sum_{\nu=1}^{N} \sum_{i,j=1}^{n_\nu} \sqrt{\frac{n_\nu}{g}}\, D_{ij}^{(\nu)}(R) \sqrt{\frac{n_\nu}{g}}\, D_{ij}^{(\nu)*}(R') = \delta_{RR'} \tag{9.44}$$

Let us now sum Eq. 9.44 over two classes, K_l and K_m, such that $R \in K_l$ and $R' \in K_m$. The r.h.s. vanishes unless $K_l = K_m$ because otherwise these classes have *no* elements in common. If $K_l = K_m$, then the right side is g_l. We define a matrix $M_l^{(\nu)}$ as

$$M_l^{(\nu)} \equiv \sum_{R \in K_l} D^{(\nu)}(R) \tag{9.45}$$

For any $R' \in G$ we have

$$\begin{aligned} D^{(\nu)}(R') M_l^{(\nu)} D^{(\nu)}(R')^{-1} &= \sum_{R \in K_l} D^{(\nu)}(R') D^{(\nu)}(R) D^{(\nu)}(R^{-1}) \\ &= \sum_{R \in K_l} D^{(\nu)}(R'RR'^{-1}) = \sum_{R \in K_l} D^{(\nu)}(R) = M_l^{(\nu)} \end{aligned}$$

so that

$$D^{(\nu)}(R') M_l^{(\nu)} = M_l^{(\nu)} D^{(\nu)}(R') \qquad \forall R' \in G \tag{9.46}$$

Schur's lemma [Th. 9.3(iii)] requires that

$$M_l^{(\nu)} = a_l^{(\nu)} I \tag{9.47}$$

Since

$$\begin{aligned} Tr(M_l^{(\nu)}) = a_l^{(\nu)} n_\nu &= \sum_{R \in K_l} Tr[D^{(\nu)}(R)] \\ &= \sum_{R \in K_l} \chi^{(\nu)}(R) = g_l \chi_l^{(\nu)} \end{aligned}$$

we deduce that

$$M_l^{(\nu)} = \frac{g_l}{n_\nu} \chi_l^{(\nu)} I \tag{9.48}$$

If we now use Eqs. 9.44 and 9.45, we find that

$$\sum_{\nu=1}^{N} \sum_{i,j=1}^{n_\nu} n_\nu [M_l^{(\nu)}]_{ij} [M_m^{(\nu)*}]_{ij} = \sum_{\nu,i,j} n_\nu \left[\sum_{R \in K_l} D^{(\nu)}(R) \right]_{ij} \left[\sum_{R' \in K_m} D^{*(\nu)}(R') \right]_{ij}$$

$$= \delta_{lm} \sum_{R \in K_l} g = \delta_{lm} g g_l \tag{9.49}$$

Finally, from Eq. 9.48, we obtain

$$\sum_{\nu=1}^{N} \sum_{i,j=1}^{n_\nu} n_\nu \frac{g_l}{n_\nu} \chi_l^{(\nu)} \frac{g_m}{n_\nu} \chi_m^{(\nu)*} = \delta_{ij}$$

or

$$\sum_{\nu=1}^{N} \chi_l^{(\nu)} \chi_m^{(\nu)*} = \frac{g}{g_m} \delta_{lm} \qquad l, m = 1, 2, \ldots, k \tag{9.50}$$

Therefore we have generated k orthogonal vectors in a space of dimension N so that $k \leq N$. However from the conclusion of Eq. 9.30, that is, $N \leq k$, we conclude that

$$k = N \tag{9.51}$$

This very important result states that the number of inequivalent irreducible representations equals the number of classes in the group.

We summarize the important basic results (for unitary representations of finite groups) below.

1. $k = N$
2. $\sum_{\nu=1}^{N} n_\nu^2 = g$
3. $\sum_{\nu=1}^{N} \sum_{i=1}^{n_\nu} \sum_{j=1}^{n_\nu} \sqrt{\frac{n_\nu}{g}} D_{ij}^{(\nu)}(R) \sqrt{\frac{n_\nu}{g}} D_{ij}^{(\nu)*}(R') = \delta_{RR'}$
4. $\sum_{R \in G} \sqrt{\frac{n_\mu}{g}} D_{lk}^{(\mu)*}(R) \sqrt{\frac{n_\nu}{g}} D_{ij}^{(\nu)}(R) = \delta_{il}\, \delta_{kj}\, \delta_{\mu\nu}$
5. $\sum_{\nu=1}^{N} \sqrt{\frac{g_i}{g}} \chi_i^{(\nu)} \sqrt{\frac{g_j}{g}} \chi_j^{(\nu)*} = \delta_{ij}$
6. $\sum_{i=1}^{k} \sqrt{\frac{g_i}{g}} \chi_i^{(\mu)*} \sqrt{\frac{g_i}{g}} \chi_i^{(\nu)} = \delta_{\mu\nu}$

Notice that the sums over i in (5) and (6) are equivalent to the sum over the group elements, $R \in G$. We have used the following notations:

k—number of classes.
N—number of (inequivalent) irreducible representations.

g—order of the group G.
g_i—number of elements in the ith class.
n_ν—dimensionality of the νth irreducible representation.

THEOREM 9.4 (Burnside's theorem)

The necessary and sufficient condition that an n-dimensional representation $D(R)$ be an irreducible representation of a group is that there exist n^2 linearly independent matrices in set $\{D(R)\}$.

PROOF

We prove the sufficiency by contradiction. That is, assume that there are n^2 linearly independent matrices in the set and that $D(R)$ is reducible so that it can be decomposed as

$$D^{(n)}(R)=D^{(n_1)}(R)+D^{(n_2)}(R) \qquad \forall R\in G$$

where $n=n_1+n_2$. However a space of dimension n has an obvious set of n^2 linearly independent basis matrices, $\{e_{ij}\}$, where $e_{ij}=\{\delta_{ir}\ \delta_{js}\}$ [i.e., e_{ij} has a 1 in the (i, j) position only and zeros everywhere else]. Similarly the space of dimension n_1 has n_1^2 of these basis matrices and that of dimension n_2 just n_2^2 of them. Since the $D^{(n)}(R)$ have been assumed reducible, we could express all of these $D^{(n)}(R)$ in terms of (*at most*) $n_1^2+n_2^2$ linearly independent matrices. But

$$n^2=(n_1+n_2)^2=n_1^2+n_2^2+2n_1n_2>n_1^2+n_2^2$$

so that no set of n^2 matrices in the set, $\{D^{(n)}(R)\}$, could have been linearly independent. This contradiction implies that the $D(R)$ must be irreducible.

We establish the stated condition as being necessary by explicitly exhibiting a set of n^2 linearly independent matrices that can be formed from the $D(R)$. Let a set of n^2 matrices, $\{M_{kl}\}$, be defined as

$$M_{kl}\equiv\sum_{R\in G} D_{kl}^*(R)D(R) \qquad k, l=1, 2, \ldots, n$$

As usual we can take the irreducible representation to be unitary. If we use the orthogonality relation of Eq. 9.27 and the obvious fact that by definition

$$D(R)=\sum_{i,j=1}^{n} D_{ij}(R)e_{ij}$$

we see that

$$M_{kl}=\sum_R\sum_{i,j} D_{kl}^*(R)D_{ij}(R)e_{ij}=\frac{g}{n}\sum_{i,j}\delta_{ik}\ \delta_{jl}e_{ij}=\frac{g}{n}\,e_{kl}$$

This demonstrates that the set of n^2 matrices $\{M_{kl}\}$ is linearly independent. Since $n\times n$ matrices form a linear vector space w.r.t. addition and since no more than n^2 of these matrices can be linearly independent (cf. Section 1.4), we see that n^2 of the $\{D(R)\}$ must be linearly independent [for the linearly independent set, $\{M_{kl}\}$, is expressible in terms of the $\{D(R)\}$]. Q.E.D.

An interesting implication of Th. 9.4 (or of Eq. 9.39) is that the regular representation is always reducible unless $g=1$ (*i.e.*, except for a group with just one element).

Example 9.16 We now apply these results on representations to the four-element *noncyclic* group that we discussed in Example 9.11 (iv) and whose multiplication table is given in Table 9.3. Since this is an abelian group we have for its representatives $D(R)D(S)=D(S)D(R)$, $\forall R$ and $S\in G$. Since all of these can be chosen to be unitary and since they all commute, all of the $D(R)$ can be completely diagonalized *simultaneously* by a single unitary transformation. This means that *all* of the representations of this group can be reduced to one-dimensional ones. Since there are four classes in this group, there are just four inequivalent irreducible representations, each of dimension one. For these four inequivalent representations, $\nu=1, 2, 3, 4$, we can satisfy Table 9.3 as follows.

TABLE 9.4

ν \ i	e	a	b	c
1	1	1	1	1
2	1	−1	1	−1
3	1	1	−1	−1
4	1	−1	−1	1

An acceptable four-dimensional (regular) representation is given below.

$$D(e)=\begin{pmatrix}1&0&0&0\\0&1&0&0\\0&0&1&0\\0&0&0&1\end{pmatrix}\qquad D(a)=\begin{pmatrix}0&1&0&0\\1&0&0&0\\0&0&0&1\\0&0&1&0\end{pmatrix}$$

$$D(b)=\begin{pmatrix}0&0&1&0\\0&0&0&1\\1&0&0&0\\0&1&0&0\end{pmatrix}\qquad D(c)=\begin{pmatrix}0&0&0&1\\0&0&1&0\\0&1&0&0\\1&0&0&0\end{pmatrix}$$

Furthermore the real orthogonal matrix

$$S=\frac{1}{2}\begin{pmatrix}1 & 1 & 1 & 1\\ 1 & -1 & 1 & -1\\ 1 & 1 & -1 & -1\\ 1 & -1 & -1 & 1\end{pmatrix}$$

will simultaneously bring all the elements of this representation into the following diagonal forms.

$$\bar{D}(e)=\begin{pmatrix}1 & 0 & 0 & 0\\ 0 & 1 & 0 & 0\\ 0 & 0 & 1 & 0\\ 0 & 0 & 0 & 1\end{pmatrix} \qquad \bar{D}(a)=\begin{pmatrix}1 & 0 & 0 & 0\\ 0 & -1 & 0 & 0\\ 0 & 0 & 1 & 0\\ 0 & 0 & 0 & -1\end{pmatrix}$$

$$\bar{D}(b)=\begin{pmatrix}1 & 0 & 0 & 0\\ 0 & 1 & 0 & 0\\ 0 & 0 & -1 & 0\\ 0 & 0 & 0 & -1\end{pmatrix} \qquad \bar{D}(c)=\begin{pmatrix}1 & 0 & 0 & 0\\ 0 & -1 & 0 & 0\\ 0 & 0 & -1 & 0\\ 0 & 0 & 0 & 1\end{pmatrix}$$

Notice that this regular representation is different (inequivalent) from that for the four-element cyclic group of Example 9.15 (as it must be).

9.5 DEFINITIONS OF CONTINUOUS GROUPS AND OF LIE GROUPS

Let us begin with a simple example of a continuous group. Take the elements $g(a)$ to depend on a real continuous parameter as

$$g(a)=e^{ia} \tag{9.52}$$

If we use the property that $e^{2\pi ni}=1$, where n is any integer or zero, we can restrict the range of a to be $0\leq a<2\pi$. It is clear that these elements satisfy all of the group properties of Section 9.1. We say that a group is an *r-parameter continuous group* when all of its elements can be labeled by r *real* continuously varying parameters a_λ, $\lambda=1, 2, \ldots, r$. We denote these group elements as $g(a_1, a_2, \ldots, a_r)\equiv g(a)$. We deal only with those continuous groups whose elements can be labeled with a finite number of such real parameters. The essential requirement we make is that a "small" change in the group parameter will produce but a "small" change in the group element.

Since a group must possess an identity element, there must exist some set of parameters $a^\circ\equiv(a_1^\circ, a_2^\circ, \ldots, a_r^\circ)$ such that $g(a^\circ)$ behaves as the identity or equivalently that

$$g(a^\circ)g(a)=g(a)g(a^\circ)=g(a) \tag{9.53}$$

As a notational convenience we often write $a^\circ = 0$ so that the identity corresponds to $a_\lambda = 0$, $\lambda = 1, 2, \ldots, r$, although this need not be the case. The existence of an inverse requires that for any parameter a there exists an $\bar{a}$ such that

$$g(a)g(\bar{a}) = g(\bar{a})g(a) = g(0)$$

or that

$$g(\bar{a}) = [g(a)]^{-1} \tag{9.54}$$

The closure property requires that for any parameters a and b there must exist a c such that

$$g(c) = g(a)g(b) \tag{9.55}$$

where $g(c)$ is also a member of the group. These real parameters c_λ, $\lambda = 1, 2, \ldots, r$, must be real functions of the real parameters, a and b,

$$c_\lambda = \phi_\lambda (a_1, a_2, \ldots, a_r; b_1, b_2, \ldots, b_r) \qquad \lambda = 1, 2, \ldots, r \tag{9.56}$$

or

$$c = \phi(a; b)$$

We say that we have an r-parameter *Lie group* when the functions of Eq. 9.56 can be expanded in a uniformly convergent Taylor series of all of their arguments.

Similarly we define an *r-parameter Lie group of transformations* as

$$x_i' = f_i(x_1, x_2, \ldots, x_n; a_1, a_2, \ldots, a_r) \qquad i = 1, 2, \ldots, n \tag{9.57}$$

or

$$x' = f(x; a)$$

Since these are to form a group, there must exist an inverse, or a set of parameters $\bar{a}$, so that

$$x'' = f(x'; \bar{a}) = f[f(x; a); \bar{a}] = x \tag{9.58}$$

In other words, the transformations (9.57) must be invertible for x in terms of x' so that the Jacobian of transformation must be nonsingular,

$$\left| \frac{\partial f}{\partial x} \right| \neq 0$$

The closure property requires that

$$\begin{aligned} x_i'' &= f_i(x'; b) = f_i[f_1(x; a), \ldots, f_n(x; a); b] \\ &= f_i(x; c) = f_i[x; \phi(a; b)] \end{aligned} \tag{9.59}$$

or

$$f[f(x, a); b] = f[x; \phi(a; b)] \tag{9.60}$$

Again for this Lie group we assume that the $\phi(a; b)$ are infinitely differentiable functions of their arguments. As we will see in Section 9.7, the requirement of Eq. 9.60 is an extremely stringent one.

9.6 EXAMPLES OF LIE GROUPS

a. Orthogonal Group in n Dimensions, $O(n)$

We begin with the two-dimensional case. These are the transformations that leave x^2+y^2 invariant in a real two-dimensional space so that

$$\begin{aligned} x' &= Ox \\ OO^{\mathrm{T}} &= I \end{aligned} \tag{9.61}$$

where O is a real matrix. The four parameters of these matrices are subject to three algebraic relations so that we have a one-parameter group. If we allow rotations only, but no spatial reflections, then we also have

$$\det O = +1$$

This is denoted as $O^+(2)$ and can be characterized by the matrix

$$\begin{pmatrix} \cos\phi & \sin\phi \\ -\sin\phi & \cos\phi \end{pmatrix} \tag{9.62}$$

For the three-dimensional case O is a real 3×3 matrix such that

$$\begin{aligned} x' &= Ox \\ OO^{\mathrm{T}} &= I \end{aligned} \tag{9.63}$$

We have six conditions on nine parameters, so that we have a three-parameter group. If we again rule out spatial reflections, we obtain $O^+(3)$, the rotation group in three dimensions. We will study this in detail later.

Finally for the n-dimensional case we require that the real $n\times n$ matrices O satisfy

$$OO^{\mathrm{T}} = I \tag{9.64}$$

This leaves $[n(n-1)]/2$ independent parameters.

b. Unitary Group in n Dimensions, $U(n)$

In two dimensions we require that the 2×2 complex matrices U satisfy

$$UU^{\dagger} = I \tag{9.65}$$

This leaves $\langle x \mid x\rangle$ invariant in a two-dimensional complex space. Since we obtain four relations among eight real parameters, we have a group with four real parameters.

For the n-dimensional complex space we have

$$UU^{\dagger} = I \tag{9.66}$$

This group depends upon n^2 real parameters.

c. Special (or Unimodular) Unitary Group in *n* Dimensions, *SU*(*n*)

In two dimensions we have

$$UU^{\dagger}=I \qquad \det U=+1 \tag{9.67}$$

so that $\det U=e^{i\varphi}$. This group has *three* real parameters. We will see that this group is homomorphic to $O^{+}(3)$.

For n dimensions we still require Eq. 9.67. This has n^2-1 independent real parameters.

d. Complex Orthogonal Group in Four Dimensions, *M*(4)

On the complex 4×4 matrices, M, we impose the restriction

$$MM^{T}=I \tag{9.68}$$

This group depends on *twelve* real parameters $[32-2(10)=12]$. That subgroup for which $\det M=+1$ is denoted by $M^{+}(4)$. If we write the elements of M as $\{\alpha_{ij}\}$, then another important subgroup of $M(4)$ can be characterized as

$$\begin{aligned} &\alpha_{ij},\ \textit{real} && i,j=1,2,3 \\ &\alpha_{i4},\ \alpha_{4i},\ \textit{imaginary} && i=1,2,3 \\ &\alpha_{44},\ \textit{real} && \end{aligned} \tag{9.69}$$

This is the *homogeneous Lorentz group, L*. It has *six* real parameters [i.e., 16 real parameters $-$ 10 constraints]. A subgroup of this is the *proper Lorentz transformations, L_p*, characterized by

$$\det M=+1 \qquad \alpha_{44}\geq+1 \tag{9.70}$$

e. Complex Unimodular Group in Two Dimensions, *C*(2)

The 2×2 complex matrices, C, must satisfy

$$\det C=+1 \tag{9.71}$$

This group has *six* real parameters.

We note here the following homomorphisms:

$$O^{+}(3)\sim SU(2) \qquad M^{+}(4)\sim C(2)\times C'(2)$$
$$O^{+}(4)\sim SU(2)\times SU'(2) \qquad L_P\sim C(2)$$

The importance of these homomorphisms is that once we find the irreducible representations of $SU(2)$ and of $C(2)$, we can build up those for the other groups above.

9.7 INFINITESIMAL GENERATORS AND GROUP PARAMETERS†

Suppose we are given a Lie group of transformations, Eq. 9.57. We can reach a neighboring "point" $x_i'+dx_i'$ near original "point" x_i' through a small variation da_λ of parameters a_λ as

$$x_i'+dx_i'=f_i(x_1, \dots, x_n; a_1+da_1, \dots, a_r+da_r) \qquad i=1, 2, \dots, n \tag{9.72}$$

or we may take

$$x_i'\equiv f_i(x_1', \dots, x_n'; 0, \dots, 0) \tag{9.73}$$

which is the identity transformation and take a set of parameter variations δa_λ such that

$$x_i'+dx_i'=f_i(x_1', \dots, x_n'; \delta a_1, \dots, \delta a_r) \qquad i=1, 2, \dots, n \tag{9.74}$$

so that

$$dx_i'=\sum_{\lambda=1}^{r} \frac{\partial f_i}{\partial a_\lambda}\bigg|_{a=0} \delta a_\lambda \equiv \sum_{\lambda=1}^{r} u_{i\lambda}(x')\, \delta a_\lambda \tag{9.75}$$

From Eqs. 9.60, 9.72, and 9.74 we see that

$$a_\lambda+da_\lambda=\phi_\lambda(a_1, \dots, a_r; \delta a_1, \dots, \delta a_r) \qquad \lambda=1, 2, \dots, r \tag{9.76}$$

so that

$$da_\lambda=\sum_{\mu=1}^{r} \frac{\partial \phi_\lambda}{\partial b_\mu}(a_1, \dots, a_r; b_1, \dots, b_r)\bigg|_{b=0} \delta a_\mu \equiv \sum_{\mu=1}^{r} \Gamma_{\lambda\mu}(a)\, \delta a_\mu \tag{9.77}$$

where

$$\Gamma_{\lambda\mu}(0)=\delta_{\lambda\mu} \tag{9.78}$$

Since by the Lie group property we must be able to solve for the $\{\delta a_\lambda\}$ in terms of the $\{da_\lambda\}$, we can write

$$\delta a_\kappa=\sum_{\lambda=1}^{r} \Delta_{\kappa\lambda}(a)\, da_\lambda \qquad \kappa=1, 2, \dots, r \tag{9.79}$$

where

$$\Delta_{\kappa\lambda}(0)=\delta_{\kappa\lambda} \tag{9.80}$$

Of course, if we substitute Eq. 9.79 into Eq. 9.77, we must obtain an identity that requires

$$\sum_{\mu=1}^{r} \Gamma_{\lambda\mu}(a)\, \Delta_{\mu\nu}(a)=\delta_{\lambda\nu} \qquad \lambda, \nu=1, 2, \dots, r \qquad \forall a \tag{9.81}$$

† The basic ideas and arguments for much of Sections 9.7 and 9.8 come from Section 1.1 of Giulio Racah's famous lectures *Group Theory and Spectroscopy* delivered at Princeton in 1951. We have supplied many of the details of the discussion for the reader.

In this and the following we assume that all of the parameters, $\{a_\lambda\}$, are *essential;* that is, no two transformations $x_i'=f(x_i; a)$, Eq. 9.57, with different values of the parameters are the same for all values of x. This set of r parameters is the *minimum* set required to specify the transformations uniquely and completely. Notice that none of the $\{u_{i\lambda}(x)\}$ defined in Eq. 9.75 can be identically zero for all values of x if all of the parameters $\{a_\lambda\}$ are essential. For suppose that just one of the transformations $x_i'=f_i(x; a)$, for example, x_1', did not depend on one of the parameters, for example, a_1, for any value of x. By the group property of the transformation we could choose a set of values of $\{a_\lambda\}$ so that $x_1' \to x_2'$ which, by assumption, would now also depend on a_1.

We will now prove that if all of the parameters, $\{a_\lambda\}$, are essential, then the $\{u_{i\lambda}(x)\}$ defined in Eq. 9.75 are linearly independent; that is, that there exists no relation of the form

$$\sum_{\lambda=1}^{r} c_\lambda u_{i\lambda}(x)=0 \qquad i=1, 2, \ldots, n \qquad \forall x$$

unless $c_\lambda=0$, for all $\lambda=1, 2, \ldots, r$. This result will be needed later. From Eqs. 9.75 and 9.78 we see that

$$dx_i'=\sum_{\lambda,\mu=1}^{r} u_{i\lambda}(x')\, \Delta_{\lambda\mu}(a)\, da_\mu$$

so that

$$\frac{\partial x_i'}{\partial a_\lambda}=\sum_{\kappa=1}^{r} u_{i\kappa}(x')\, \Delta_{\kappa\lambda}(a) \tag{9.82}$$

Let us begin with a simple case of one variable x in which there are just two parameters, assumed to be *nonessential.* Then there would exist values δa_1 and δa_2 for all values of x such that

$$f(x; a_1, a_2)\equiv f(x; a_1+\delta a_1, a_2+\delta a_2)$$

so that to first order in the $\{\delta a_\lambda\}$,

$$u_1(x)\, \delta a_1+u_2(x)\, \delta a_2=0$$

Since a_1 and a_2 are nonessential, there must exist a single essential parameter, for example α, such that

$$a_1=a_1(\alpha) \qquad a_2=a_2(\alpha)$$

which implies that

$$\lim_{\delta\alpha\to 0} \frac{\delta a_1}{\delta\alpha}=c_1$$

$$\lim_{\delta\alpha\to 0} \frac{\delta a_2}{\delta\alpha}=c_2$$

or that

$$c_1 u_1(x)+c_2 u_2(x)=0$$

Conversely, assume that $u_1(x)$ and $u_2(x)$ are linearly dependent in which case

$$dx = u_1(x)\,\delta a_1 + u_2(x)\,\delta a_2 = u_1(x)\left(\delta a_1 - \frac{c_1}{c_2}\,\delta a_2\right) \equiv u_1(x)\,\delta\alpha$$

That is, infinitely many values of δa_1 and δa_2 will produce the same dx for all values of x. Therefore, at least for this simple case, we see that the $\{a_\lambda\}$ are nonessential if and only if the $\{u_{i\lambda}(x)\}$ are linearly dependent or, what is equivalent, that the $\{a_\lambda\}$ are essential if and only if the $\{u_{i\lambda}(x)\}$ are linearly independent.

Let us now consider the general case in which

$$x_i' = f_i(x_1, \ldots, x_n; a_1, \ldots, a_r) \qquad i = 1, 2, \ldots, n$$

If not all of the $\{a_\lambda\}$, $\lambda = 1, 2, \ldots, r$, are essential, then there must exist some smaller set of parameters, for example, $\{\alpha_\mu\}$, $\mu = 1, 2, \ldots, m$, $m < r$, such that

$$x_i' = f_i(x_1, \ldots, x_n; a_1, \ldots, a_r) = F_i(x_1, \ldots, x_n; \alpha_1, \ldots, \alpha_m) \qquad i = 1, 2, \ldots, n$$

where

$$\alpha_\mu = \alpha_\mu(a_1, \ldots, a_r) \qquad \mu = 1, 2, \ldots, m \tag{9.83}$$

If we now define

$$\alpha_\mu = \alpha_\mu(a_1, \ldots, a_r) \equiv 0 \qquad \mu = m+1, \ldots, r$$

then the $r \times r$ determinant with elements $\partial\alpha_\mu/\partial a_\lambda$ (i.e., the Jacobian) will vanish identically so that the rows (or columns) of this determinant must be related as (cf. Section 2.1b)

$$\sum_{\lambda=1}^{r} c_\lambda(a)\,\frac{\partial\alpha_\mu}{\partial a_\lambda} = 0 \qquad \mu = 1, 2, \ldots, m \tag{9.84}$$

Since we assume that all functions can be expanded in Taylor series of their arguments, it follows for any function of the $\{\alpha_\mu\}$, for example, $g(\alpha_1, \ldots, \alpha_m)$, that Eq. 9.84 also applies as

$$\sum_{\lambda=1}^{r} c_\lambda(a)\,\frac{\partial g(\alpha_1, \ldots, \alpha_m)}{\partial a_\lambda} = \sum_{\lambda=1}^{r}\sum_{\mu=1}^{m} c_\lambda(a)\,\frac{\partial\alpha_\mu}{\partial a_\lambda}\,\frac{\partial g}{\partial\alpha_\mu} = 0$$

Therefore we have for the $F_i(x; \alpha)$

$$0 = \sum_{\lambda=1}^{r} c_\lambda(a)\,\frac{\partial F_i}{\partial a_\lambda} \equiv \sum_{\lambda=1}^{r} c_\lambda(a)\,\frac{\partial f_i(x; a)}{\partial a_\lambda} \tag{9.85}$$

so that when $a_\lambda = 0$, $\lambda = 1, 2, \ldots, r$, we obtain the desired statement of linear dependence

$$\sum_{\lambda=1}^{r} c_\lambda u_{i\lambda}(x) = 0 \qquad i = 1, 2, \ldots, n \tag{9.86}$$

Conversely, if we assume that Eq. 9.86 holds so that not all of the $\{c_\lambda\}$ are zero (assume by choice that $c_1 \neq 0$ for the following argument), then we may write

$$dx_i \equiv \sum_{\lambda=1}^{r} u_{i\lambda}(x)\, \delta a_\lambda = \sum_{\lambda=2}^{r} u_{i\lambda}(x)\, \delta a_\lambda - \left[\frac{1}{c_1} \sum_{\lambda=2}^{r} c_\lambda u_{i\lambda}(x)\right] \delta a_1$$

$$= \sum_{\lambda=2}^{r} u_{i\lambda}(x)\left(\delta a_\lambda - \frac{1}{c_1}\, \delta a_1\right) \equiv \sum_{\lambda=2}^{r} u_{i\lambda}(x)\, \delta \alpha_\lambda \tag{9.87}$$

which states that not all of the $\{a_\lambda\}$ are essential. We therefore see that the $\{u_{i\lambda}(x)\}$ are linearly independent if and only if all of the $\{a_\lambda\}$ are essential.

We now consider the change of a function $F(x)$ produced by the infinitesimal transformations of Eq. 9.72. Use of Eq. 9.75 shows that

$$dF \equiv \sum_{i=1}^{n} \frac{\partial F}{\partial x_i}\, dx_i = \sum_{i=1}^{n} \frac{\partial F}{\partial x_i} \sum_{\lambda=1}^{r} u_{i\lambda}(x)\, \delta a_\lambda$$

$$= \sum_{\lambda=1}^{r} \delta a_\lambda \left(\sum_{i=1}^{n} u_{i\lambda}(x) \frac{\partial}{\partial x_i}\right) F \equiv \sum_{\lambda=1}^{r} \delta a_\lambda X_\lambda F \tag{9.88}$$

where we have defined the *infinitesimal generators*, X_λ, of the group as

$$X_\lambda \equiv \sum_{i=1}^{n} u_{i\lambda}(x) \frac{\partial}{\partial x_i} \qquad \lambda = 1, 2, \ldots, r \tag{9.89}$$

It is clear that the operator

$$I + \sum_{\lambda} X_\lambda\, \delta a_\lambda \tag{9.90}$$

differs only slightly from the identity.

Example 9.17 As a simple case consider the rotation group in two dimensions. We rotate the coordinate system counterclockwise here so that

$$x' = x \cos\phi + y \sin\phi \simeq x + y\, \delta\phi$$
$$y' = -x \sin\phi + y \cos\phi \simeq -x\, \delta\phi + y$$

from which we obtain

$$X = -x \frac{\partial}{\partial y} + y \frac{\partial}{\partial x}$$

and

$$dx_i = X x_i\, \delta\phi \qquad i = 1, 2$$

9.8 STRUCTURE CONSTANTS

We now prove the very important result that for a Lie group of transformations the commutators of the infinitesimal generators can be expressed as linear combinations of the generators themselves.

THEOREM 9.5

The infinitesimal generators, $\{X_\lambda\}$, of any Lie group of transformations satisfy the relations

$$[X_\rho, X_\sigma] = c^\kappa_{\rho\sigma} X_\kappa \qquad \rho, \sigma = 1, 2, \ldots, r \tag{9.91}$$

where the *structure constants* $c^\kappa_{\rho\sigma}$ satisfy the identities

$$c^\kappa_{\rho\sigma} = -c^\kappa_{\sigma\rho} \tag{9.92}$$

$$c^\mu_{\rho\sigma} c^\nu_{\mu\tau} + c^\mu_{\sigma\tau} c^\nu_{\mu\rho} + c^\mu_{\tau\rho} c^\nu_{\mu\sigma} = 0 \qquad \rho, \sigma, \nu, \tau = 1, 2, \ldots, r \tag{9.93}$$

PROOF

In this proof and the following discussion all repeated indices are summed over unless otherwise stated. For a Lie group of transformations

$$x_i = f_i(x_1^\circ, \ldots, x_n^\circ; a_1, \ldots, a_\nu)$$

where $x^\circ = \{x_i^\circ\}$ is some reference point or configuration in the coordinate space, we have from Eq. 9.82, on relabeling the coordinates,

$$\frac{\partial x_i}{\partial a_\lambda} = u_{i\kappa}(x)\, \Delta_{\kappa\lambda}(a) \qquad i = 1, 2, \ldots, n \qquad \lambda = 1, 2, \ldots, r \tag{9.94}$$

Equation 9.94 describes how the coordinates (or variables), $\{x_i\}$, change with the parameters, $\{a_\lambda\}$. We treat the $\{a_\lambda\}$ as the set of independent variables [not the $\{x_i\}$]. From our continuity requirements on this Lie group of transformations we have

$$\frac{\partial^2 x_i}{\partial a_\lambda\, \partial a_\mu} = \frac{\partial^2 x_i}{\partial a_\mu\, \partial a_\lambda} \tag{9.95}$$

It is important to realize that in Eqs. 9.94 and 9.95 we are computing the *total* variation produced by a variation in the parameters $\{a_\lambda\}$ [i.e., including the variation induced by the change in the $\{x_i\}$ as these parameters $\{a_\lambda\}$ vary]. We also ask for the *explicit* variation of certain quantities as the set, $\{a_\lambda\}$, varies [while the $\{x_i\}$ remain unchanged]. This latter is the more conventional use of the partial derivative symbol, $\partial/\partial a_\lambda$. Combining Eqs. 9.94 and 9.95 and the identity

$$\frac{\partial u_{i\kappa}}{\partial a_\lambda} = \frac{\partial u_{i\kappa}}{\partial x_j} \frac{\partial x_j}{\partial a_\lambda} = \frac{\partial u_{i\kappa}}{\partial x_j} u_{j\nu}\, \Delta_{\nu\lambda}$$

we obtain

$$u_{i\kappa}\left(\frac{\partial \Delta_{\kappa\mu}}{\partial a_\lambda} - \frac{\partial \Delta_{\kappa\lambda}}{\partial a_\mu}\right) + \left(u_{j\nu} \frac{\partial u_{i\kappa}}{\partial x_j} - u_{j\kappa} \frac{\partial u_{i\nu}}{\partial x_j}\right) \Delta_{\kappa\mu}\, \Delta_{\nu\lambda} = 0 \tag{9.96}$$

If we use the orthogonality relation (9.81) and define

$$c^\kappa_{\tau\sigma}(a) \equiv \left(\frac{\partial \Delta_{\kappa\mu}}{\partial a_\lambda} - \frac{\partial \Delta_{\kappa\lambda}}{\partial a_\mu}\right) \Gamma_{\mu\tau} \Gamma_{\lambda\sigma} \tag{9.97}$$

we find

$$\left(u_{j\tau}\frac{\partial u_{i\sigma}}{\partial x_j}-u_{j\sigma}\frac{\partial u_{i\tau}}{\partial x_j}\right)=c^{\kappa}_{\tau\sigma}(a)u_{i\kappa}(x) \tag{9.98}$$

If we differentiate Eq. 9.98 w.r.t. a_ρ *explicitly* [i.e., realizing that the $\{u_{i\kappa}(x)\}$ do not depend upon the $\{a_\lambda\}$ explicitly], we see that

$$\frac{\partial c^{\kappa}_{\tau\sigma}}{\partial a_\rho}u_{i\kappa}(x)=0 \tag{9.99}$$

From the linear independence of the $\{u_{i\kappa}\}$ we conclude that

$$\frac{\partial c^{\kappa}_{\tau\sigma}}{\partial a_\rho}=0 \qquad \kappa,\tau,\sigma,\rho=1,2,\ldots,r \tag{9.100}$$

which states that the $c^{\kappa}_{\tau\sigma}$ are indeed constants that depend on the group only and not on its parameters $\{a_\lambda\}$. Therefore from Eq. 9.89 we find that

$$\begin{aligned}[X_\rho,X_\sigma]\equiv X_\rho X_\sigma-X_\sigma X_\rho&=u_{i\rho}\frac{\partial}{\partial x_i}\left(u_{j\sigma}\frac{\partial}{\partial x_j}\right)-u_{j\sigma}\frac{\partial}{\partial x_j}\left(u_{i\rho}\frac{\partial}{\partial x_i}\right)\\&=\left[u_{i\rho}\frac{\partial u_{j\sigma}}{\partial x_i}-u_{i\sigma}\frac{\partial u_{j\rho}}{\partial x_i}\right]\frac{\partial}{\partial x_j}=c^{\kappa}_{\rho\sigma}u_{j\kappa}\frac{\partial}{\partial x_j}=c^{\kappa}_{\rho\sigma}X_\kappa\end{aligned} \tag{9.101}$$

The $c^{\kappa}_{\rho\sigma}$ are the *structure constants* of the Lie group and from their definition in (9.97) they obviously satisfy Eq. 9.92,

$$c^{\kappa}_{\rho\sigma}=-c^{\kappa}_{\sigma\rho}$$

Furthermore, from the Jacobi identity for any operators A, B, C,

$$[A,[B,C]]+[B,[C,A]]+[C,[A,B]]=0$$

these structure constants are seen to satisfy Eq. 9.93. Q.E.D.

Lie proved the remarkable result that given the structure constants satisfying Eqs. 9.92 and 9.93 we can finally construct the transformation functions, $f_i(x;a)$. (The interested reader can find a proof of this in Eisenhart's book.) That is, Eq. 9.91 is a *test* for a Lie group. Therefore, once we know the structure constants, we know everything about the Lie group. We now give a nonrigorous argument of this for the representations.

Suppose we have an r-parameter Lie group

$$G=\{g_a\}$$

and a representation in a linear space in terms of $n\times n$ matrices, $\{D(a)\}$. Denote the basis vectors of the representation space by ψ,

$$\psi=(\psi_1,\psi_2,\ldots,\psi_n)$$

To any transformation $g_a \in G$ there will correspond a transformation

$$\psi \rightarrow \psi' = D(a)\psi$$

or in component form

$$\psi_i \rightarrow \psi_i' = D_{ij}(a)\psi_j \tag{9.102}$$

Notice that our convention on the indices of $D(a)$ for this discussion is just the opposite of that in Eq. 9.18. Because of the continuity in a we can write

$$\begin{aligned} \psi_i \rightarrow \psi_i' &= \left[\delta_{ij} + \left(\frac{\partial D_{ij}}{\partial a_1}\right)_{a=0} a_1 + \cdots + \left(\frac{\partial D_{ij}}{\partial a_\nu}\right)_{a=0} a_\nu + O(a^2)\right]\psi_j \\ &= (\delta_{ij} + a_\lambda X_\lambda^{ij})\psi_j \end{aligned} \tag{9.103}$$

to first order in the a's, where the set of r matrices $\{X_\lambda\}$ is defined by

$$X_\lambda \equiv \left.\left(\frac{\partial D(a)}{\partial a_\lambda}\right)\right|_{a=0} \qquad \lambda = 1, 2, \ldots, r \tag{9.104}$$

The $n \times n$ matrices, $\{X_\lambda\}$, are *constant* matrices. The transformation

$$\psi \rightarrow \psi' = (I + a_\lambda X_\lambda)\psi$$

corresponds to an infinitesimal transformation g_a with $|a_\lambda| \ll 1, \forall\lambda$. Since a Lie group is determined by its structure constants, we see that the $\{X_\lambda\}$ of Eq. 9.104 must satisfy Eq. 9.91. Because of the closure and continuity properties of Eqs. 9.55 and 9.56 we can obtain any element of the group for finite parameters $a = (a_1, a_2, \ldots, a_r)$ as

$$g_a = \lim_{k\rightarrow\infty} (g_{a/k})^k \tag{9.105}$$

Equation 9.105, which we have not established here with any rigor, states the essential property of Lie groups; namely, that any element with finite parameters can be built up continuously from those elements in the neighborhood of the identity element. Once the forms of the infinitesimal transformations have been specified for a Lie group, the general form for finite transformations is also determined. This is a tremendous advantage since it is often relatively easy to write down the form of an infinitisimal transformation as we will see in examples.

From Eq. 9.105 we might expect that the representations for finite transformations can be obtained as

$$D(a) = \lim_{k\rightarrow\infty}\left(I + \frac{a_\lambda}{k} X_\lambda\right)^k = \exp(a_\lambda X_\lambda) \tag{9.106}$$

so that

$$\psi \rightarrow \psi' = e^{a_\lambda X_\lambda}\psi$$

In fact for any matrix M we have

$$\left(I + \frac{1}{k}M\right)^k = k! \sum_{s=0}^{k} \frac{1}{s!\,(k-s)!}\left(\frac{1}{k}M\right)^s = \sum_{s=0}^{k} \frac{1}{s!} M^s \left[\frac{k!}{(k-s)!\,k^s}\right]$$

while for any s Stirling's approximation gives (cf. also Section A2.2, Eqs. A2.10 and A2.11)

$$\frac{k!}{(k-s)!\,k^s}=\frac{\Gamma(k+1)}{\Gamma(k+1-s)k^s}\xrightarrow[k\to\infty]{}\frac{k^s}{k^s}=1$$

so that

$$\lim_{k\to\infty}\left(I+\frac{1}{k}M\right)^k=\sum_{s=0}^{\infty}\frac{M^s}{s!}=e^M$$

Of course if M can be diagonalized (e.g., say M is normal), then the proof is particularly simple (cf. Section 3.7). Therefore, once an $n\times n$ matrix representation of the infinitesimal operators (or generators), $\{X_\lambda\}$, has been found, Eq. 9.106 allows us to write down at once an $n\times n$ representation for the finite transformations. Since these matrix representatives of the $\{X_\lambda\}$ satisfy Eq. 9.91, we have a simple method of computing the structure constants of the group.

Example 9.18 Consider the rotation group in three dimensions, $O(3)$. This is to leave invariant the form $\langle x\mid x\rangle=x^2+y^2+z^2=\langle x'\mid x'\rangle$ as it must since

$$x'=Ox$$
$$OO^T=I$$

By considering successive rotations about the x, y, z axes, we can easily see that O for infinitesimal transformations can be written as (i.e., a *counterclockwise* rotation of the coordinate system)

$$O=\begin{pmatrix}1 & \delta\alpha_3 & -\delta\alpha_2\\ -\delta\alpha_3 & 1 & \delta\alpha_1\\ \delta\alpha_2 & -\delta\alpha_1 & 1\end{pmatrix}$$
$$OO^T=I+O(\delta\alpha^2)$$

In fact this group is $O^+(3)$ as represented here since $\det O=+1$. All transformations continuously connected with the identity must have $\det O=+1$. Since

$$\begin{aligned}x'&=x+y\,\delta\alpha_3-z\,\delta\alpha_2\\ y'&=-x\,\delta\alpha_3+y+z\,\delta\alpha_1\\ z'&=x\,\delta\alpha_2-y\,\delta\alpha_1+z\end{aligned}\tag{9.107}$$

we have

$$\begin{aligned}\delta x&=\qquad\quad y\,\delta\alpha_3-z\,\delta\alpha_2\\ \delta y&=-x\,\delta\alpha_3\qquad\quad+z\,\delta\alpha_1\\ \delta z&=x\,\delta\alpha_2-y\,\delta\alpha_1\end{aligned}$$

so that

$$\delta x_i=\sum_{k=1}^{3}\delta\alpha_k X_k x_i$$

and

$$X_1 = z\frac{\partial}{\partial y} - y\frac{\partial}{\partial z}$$

$$X_2 = -z\frac{\partial}{\partial x} + x\frac{\partial}{\partial z}$$

$$X_3 = y\frac{\partial}{\partial x} - x\frac{\partial}{\partial y}$$

We easily verify that

$$\begin{aligned} [X_1, X_2] &= X_3 \\ [X_2, X_3] &= X_1 \\ [X_3, X_1] &= X_2 \end{aligned} \tag{9.108}$$

We also have a simple representation of these $\{X_\rho\}$ in terms of 3×3 matrices as

$$X_1 = \begin{pmatrix} 0 & 0 & 0 \\ 0 & 0 & 1 \\ 0 & -1 & 0 \end{pmatrix} \qquad X_2 = \begin{pmatrix} 0 & 0 & -1 \\ 0 & 0 & 0 \\ 1 & 0 & 0 \end{pmatrix} \qquad X_3 = \begin{pmatrix} 0 & 1 & 0 \\ -1 & 0 & 0 \\ 0 & 0 & 0 \end{pmatrix}$$

with

$$\delta\mathbf{x} = \sum_{k=1}^{3} \delta\alpha_k X_k \mathbf{x}$$

$$\mathbf{x} = \begin{pmatrix} x \\ y \\ z \end{pmatrix}$$

Notice that these matrix elements can be written as

$$X_{jk}^{(i)} = \varepsilon_{ijk}$$

Example 9.19 Consider the rotation group in four dimensions in a real space, $O(4)$. From Section 9.6a we know that this group of transformations leaves invariant the quantity

$$x^2 + y^2 + z^2 + t^2 = \text{const.}$$

and that there are

$$\frac{4(3)}{2} = 6$$

parameters and, therefore, six infinitesimal generators that can be represented as

$$X_1 = z\frac{\partial}{\partial y} - y\frac{\partial}{\partial z} \qquad X_2 = x\frac{\partial}{\partial z} - z\frac{\partial}{\partial x} \qquad X_3 = y\frac{\partial}{\partial x} - x\frac{\partial}{\partial y}$$

$$X_4 = x\frac{\partial}{\partial t} - t\frac{\partial}{\partial x} \qquad X_5 = y\frac{\partial}{\partial t} - t\frac{\partial}{\partial y} \qquad X_6 = z\frac{\partial}{\partial t} - t\frac{\partial}{\partial z}$$

It is obvious that there are 15 commutation relations. However, if we define a new set of generators,

$$Y_j = \frac{X_j + X_{j+3}}{2} \qquad Z_j = \frac{X_j - X_{j+3}}{2} \qquad j = 1, 2, 3$$

we can summarize their 15 commutation relations as

$$[Y_i, Y_j] = \varepsilon_{ijk} Y_k$$
$$[Z_i, Z_j] = \varepsilon_{ijk} Z_k$$
$$[Y_i, Z_j] = 0 \qquad \forall i, j = 1, 2, 3$$

Therefore $O(4)$ is isomorphic to the direct product of two $O(3)$ groups as claimed in Section 9.6e.

9.9 CASIMIR OPERATORS AND THE RANK OF A GROUP

The program for finding representations of a Lie group should now be clear. One takes the basic commutation relations

$$[X_\rho, X_\sigma] = c^\kappa_{\rho\sigma} X_\kappa$$

where the structure constants $c^\kappa_{\rho\sigma}$ are known (say from the representation space defining the group), and finds matrix representatives of the generators $\{X_\lambda\}$ in an n-dimensional space. The matrix representation of the $\{X_\lambda\}$ in the defining space [e.g., a real two-dimensional space for $O(2)$] is termed the *fundamental representation.* One then uses Eq. 9.106 to write the general representation as

$$D(a) = \exp(a_\lambda X_\lambda)$$

In order to label the irreducible representations we must find a quantity (or quantities) that remains constant within a given representation (e.g., a candidate might be the dimension of the representation for finite dimensional matrices). Suppose we manage to find a set of operators, $\{C_i\}$, each member of which commutes with *all* of the group generators X_λ,

$$[X_\lambda, C_i] = 0 \qquad \textbf{(9.109)}$$

so that

$$[D(a), C_i] = 0 \qquad \textbf{(9.110)}$$

for all members of a given irreducible representation. According to Th. 9.3 Eq. 9.110 implies that the operators, $\{C_i\}$, are multiples of the identity operator

$$C_i = c_i I \qquad \textbf{(9.111)}$$

where the constants, $\{c_i\}$, characterize the irreducible representation; that is, the

$\{c_i\}$ may vary from one irreducible representation to another, but they remain fixed for all members of a given irreducible representation. This allows us to use these $\{c_i\}$ as labels for the irreducible representations. Such operators C_i are the *Casimir operators* of the group. In general it is a very difficult problem to find all of the independent Casimir operators for an arbitrary Lie group. We will see below that it is rather simple for the rotation group, $O^+(3)$.

Example 9.20 We list here the Casimir operators for two familiar Lie groups. For $O(3)$ the Casimir operator is

$$C = \sum_{\lambda=1}^{3} X_\lambda^2 \tag{9.112}$$

while for $O(4)$ we have two Casimir operators,

$$C_1 = \sum_{\lambda=1}^{3} Y_\lambda^2 \qquad C_2 = \sum_{\lambda=1}^{3} Z_\lambda^2$$

These Casimir operators are polynomials in the generators, X_λ. They can easily be seen to satisfy Eq. 9.109.

We define the *rank* of a group as the maximum number of mutually commuting linearly independent operators formed from linear combinations of the group generators, $\{X_\lambda\}$. That is, for some fixed operator A,

$$A = \sum_\mu \alpha_\mu X_\mu \tag{9.113}$$

we seek all of the solutions $\{X\}$ to

$$[A, X] = 0$$

where

$$X = \sum_\nu x_\nu X_\nu$$

We then vary A to, for example, A' and evaluate $[A', X]$ for all $\{X\}$ that were solutions to $[A, X] = 0$ and keep only those X for which $[A', X] = 0$ also. We continue this process until we obtain all of the independent mutually commuting linear operators for the group. This number is the rank of the group.

Example 9.21 Consider $O(3)$ and let

$$A = \sum_{\mu=1}^{3} \alpha_\mu X_\mu$$

$$X = \sum_{\mu=1}^{3} x_\nu X_\nu$$

so that

$$O=[A, X]=\sum_{\mu,\nu=1}^{3} \alpha_\mu x_\nu [X_\mu, X_\nu]=\sum_{\mu,\nu,\rho} \alpha_\mu x_\nu \varepsilon_{\mu\nu\rho} X_\rho$$

Since the $\{X_\rho\}$ are linearly independent, we have

$$\sum_{\mu,\nu} \alpha_\mu \varepsilon_{\mu\nu\rho} x_\nu = 0 \qquad \rho=1, 2, 3$$

or, equivalently, the system of homogeneous linear equations

$$\begin{aligned} -\alpha_2 x_1 + \alpha_1 x_2 \qquad\qquad &= 0 \\ \alpha_3 x_1 \qquad\qquad - \alpha_1 x_3 &= 0 \\ -\alpha_3 x_2 + \alpha_2 x_3 &= 0 \end{aligned}$$

Since the determinant of the coefficients of the $\{x_\nu\}$ vanishes identically, there exists a nontrivial solution for the $\{x_\nu\}$. In fact there is just one linearly independent nontrivial solution (cf. Section 2.2b) given, up to an arbitrary overall multiplicative constant, by

$$x_\nu = \alpha_\nu \qquad \nu=1, 2, 3$$

so that $X=A$ which states that $O(3)$ is a rank *one* Lie group.

We can easily show that the Casimir operator of Eq. 9.112 is the only (independent) Casimir operator for $O^+(3)$. Since the infinitesimal generators $\{X_i\}$ satisfy the commutation relations

$$[X_i, X_j]=\varepsilon_{ijk} X_k \tag{9.114}$$

and since $(I+\delta\alpha_j X_j)$ generates the infinitesimal rotations, we see that the $\{X_i\}$ of Eq. 9.108 transform as

$$\begin{aligned} \delta X_i &\equiv (I+\delta\alpha_j X_j) X_i (I-\delta\alpha_k X_k) - X_i \\ &\simeq \delta\alpha_j (X_j X_i - X_i X_j) = \delta\alpha_j \varepsilon_{jik} X_k \end{aligned} \tag{9.115}$$

Comparison of Eqs. 9.115 and 9.107 shows that the $\{X_i\}$ transform as the components of a vector, $\mathbf{X}=(X_1, X_2, X_3)$. Since a Casimir operator is to satisfy

$$[C, X_i]=0$$

C must be a scalar operator w.r.t. $O^+(3)$. There are only two possible scalars we can build

$$C_1 = \delta_{ij} X_i X_j = X_i X_i$$

$$C_2 = \varepsilon_{ijk} X_i X_j X_k$$

However since

$$C_2 = \tfrac{1}{2}(\varepsilon_{ijk}X_iX_j - \varepsilon_{ijk}X_jX_i)X_k = \tfrac{1}{2}\varepsilon_{ijk}[X_i, X_j]X_k$$
$$= \tfrac{1}{2}\varepsilon_{ijk}\varepsilon_{ijl}X_lX_k = \delta_{kl}X_lX_k = C_1$$

we see that C_1 and C_2 are not independent. Furthermore since

$$[X^2, X_k] = \sum_{j=1}^{3} (X_jX_jX_k - X_jX_kX_j + X_jX_kX_j - X_kX_jX_j)$$
$$= \sum_{j=1}^{3} \{X_j[X_j, X_k] + [X_j, X_k]X_j\} = \sum_{j=1}^{3} (X_j\varepsilon_{jkl}X_l + \varepsilon_{jkl}X_lX_j)$$
$$= \sum_{j=1}^{3} \varepsilon_{jkl}(X_jX_l + X_lX_j) = 0$$

it follows that $C = X^2$ is indeed a Casimir operator as claimed.

Therefore for $O^+(3)$ the largest set of mutually commuting independent operators is X^2 and any *one* of the X_i. Conventionally this *complete* set of operators is taken to be X^2 and X_3. Notice also from Eq. 9.108 that the 3×3 matrices $\{X_j\}$ are such that the *hermitian* matrices

$$J_j = iX_j \qquad j = 1, 2, 3 \tag{9.116}$$

satisfy

$$[J_i, J_j] = i\varepsilon_{ijk}J_k \tag{9.117}$$

so that

$$\delta\mathbf{x} = i\sum_{j=1}^{3} \delta\alpha_j J_j\mathbf{x} \tag{9.118}$$

9.10 HOMOMORPHISM BETWEEN THE PROPER ROTATION GROUP $O^+(3)$ AND $SU(2)$

We begin by studying the group, $SU(2)$, defined as those 2×2 complex unitary matrices such that

$$U = \begin{pmatrix} a & b \\ c & d \end{pmatrix} \qquad UU^\dagger = I \qquad \det U = +1 \tag{9.119}$$

The infinitesimal transformations have the form

$$U = \begin{pmatrix} 1+\delta a & \delta b \\ \delta c & 1+\delta d \end{pmatrix} \tag{9.120}$$

so that

$$UU^\dagger \simeq \begin{pmatrix} 1+\delta a^* + \delta a & \delta c^* + \delta b \\ \delta c + \delta b^* & 1 + \delta d + \delta d^* \end{pmatrix} = \begin{pmatrix} 1 & 0 \\ 0 & 1 \end{pmatrix}$$
$$\det U \simeq 1 + \delta a + \delta d = 1$$

or

$$\begin{aligned}
&\delta a^* + \delta a = 0 \Rightarrow \delta a^* = -\delta a \Rightarrow \delta a = \frac{i}{2}\,\delta\alpha_3 \\
&\delta c^* + \delta b = 0 \Rightarrow \delta b = -\delta c^* \Rightarrow \delta b = \tfrac{1}{2}\delta\alpha_2 + \frac{i}{2}\,\delta\alpha_1 \\
&\delta d + \delta d^* = 0 \\
&\delta a + \delta d = 0 \Rightarrow \delta d = -\delta a
\end{aligned} \tag{9.121}$$

where the last set of equalities in the first two equations are definitions of the parameters α_1, α_2, and α_3. Therefore the form for these infinitesimal transformations can be written as

$$\begin{aligned}
U &= \begin{pmatrix} 1+\frac{i}{2}\,\delta\alpha_3 & \frac{1}{2}\delta\alpha_2+\frac{i}{2}\,\delta\alpha_1 \\ -\frac{1}{2}\delta\alpha_2+\frac{i}{2}\,\delta\alpha_1 & 1-\frac{i}{2}\,\delta\alpha_3 \end{pmatrix} \\
&= I+\frac{i}{2}\,\delta\alpha_3\begin{pmatrix}1 & 0\\ 0 & -1\end{pmatrix}+\frac{i}{2}\,\delta\alpha_2\begin{pmatrix}0 & -i\\ i & 0\end{pmatrix}+\frac{i}{2}\,\delta\alpha_1\begin{pmatrix}0 & 1\\ 1 & 0\end{pmatrix} \\
&= I+i\sum_{j=1}^{3}\delta\alpha_j\tfrac{1}{2}\sigma_j
\end{aligned} \tag{9.122}$$

where the $\{\sigma_j\}$ are the *Pauli spin matrices*. Notice that I and $\boldsymbol{\sigma}$ form a complete set of 2×2 hermitian matrices. We can easily check that

$$\sigma_i\sigma_j = \delta_{ij} + i\varepsilon_{ijk}\sigma_k$$

so that

$$[\tfrac{1}{2}\sigma_i, \tfrac{1}{2}\sigma_j] = i\varepsilon_{ijk}(\tfrac{1}{2}\sigma_k) \tag{9.123}$$

which states that the matrices $(1/2)\sigma_j$ have the same commutation relations as the J_j of Eq. 9.116 for $O^+(3)$. The group, $SU(2)$, is a *rank one* group. Also since both $O^+(3)$ and $SU(2)$ are Lie groups continuously connected with the identity, we might *expect* them to be at least homomorphic.

THEOREM 9.6

The groups, $SU(2)$ and $O^+(3)$, are homomorphic. To every element of $O^+(3)$ there correspond two elements of $SU(2)$.

PROOF

Let M be the traceless hermitian matrix

$$M \equiv \mathbf{x}\cdot\boldsymbol{\sigma} = \sum_{j=1}^{3} x_j\sigma_j = \begin{pmatrix} z & x+iy \\ x-iy & -z \end{pmatrix}$$

so that

$$\det M = -(x^2+y^2+z^2)$$

Under the action of a similarity transformation of a U of $SU(2)$, we obtain

$$M' = UMU^\dagger$$

However since

$$\text{Tr } M' = \text{Tr } M$$

$$\det M' = \det (UMU^\dagger) = \det (UU^\dagger) \det M = \det M$$

we have

$$M' = \mathbf{x}' \cdot \boldsymbol{\sigma} = \begin{pmatrix} z' & x'+iy' \\ x'-iy' & -z' \end{pmatrix}$$

$$\text{Tr } M' = 0$$

so that

$$(x'^2+y'^2+z'^2) = (x^2+y^2+z^2)$$

which implies that $\langle x \mid x \rangle$ is left invariant under these transformations. For an infinitesimal transformation we can write $U \simeq I + i \sum_{j=1}^{3} \delta\beta_j \sigma_j$ so that

$$\begin{aligned}
\delta M = \sum_{j=1}^{3} \delta x_j \sigma_j &= M' - M = UMU^+ - M \\
&\simeq \left(I + i \sum_l \delta\beta_l \sigma_l\right)\left(\sum_j x_j \sigma_j\right)\left(I - i \sum_k \delta\beta_k \sigma_k\right) - M \\
&= \left(I + i \sum_l \delta\beta_l \sigma_l\right)\left(\sum_j x_j \sigma_j - i \sum_{j,k} x_j \, \delta\beta_k \sigma_j \sigma_k\right) - M \\
&\simeq \sum_j x_j \sigma_j - i \sum_{j,k} x_j \, \delta\beta_k \sigma_j \sigma_k + i \sum_{l,j} \delta\beta_l x_j \sigma_l \sigma_j - M \\
&= -i \sum_{j,k} x_j \, \delta\beta_k (\sigma_j \sigma_k - \sigma_k \sigma_j) = -i \sum_{j,k} x_j \, \delta\beta_k [\sigma_j, \sigma_k] \\
&= 2 \sum_{j,k,l} x_j \, \delta\beta_k \varepsilon_{jkl} \sigma_l = 2 \sum_{j,k,l} x_l \, \delta\beta_k \varepsilon_{lkj} \sigma_j
\end{aligned}$$

or

$$\delta x_j = 2 \sum_{k,l} x_l \, \delta\beta_k \varepsilon_{lkj} \tag{9.124}$$

If we compare this with Eq. 9.107, we find that

$$\begin{aligned}
\delta x_1 &= 2\delta\beta_3 x_2 - 2\delta\beta_2 x_3 \equiv y \, \delta\alpha_3 - z \, \delta\alpha_2 \\
\delta x_2 &= -2\delta\beta_3 x_1 + 2\delta\beta_1 x_3 \equiv -x \, \delta\alpha_3 + z \, \delta\alpha_1 \\
\delta x_3 &= 2\delta\beta_2 x_1 - 2\delta\beta_1 x_2 \equiv x \, \delta\alpha_2 - y \, \delta\alpha_1
\end{aligned}$$

or, more compactly,

$$\delta\alpha_j = 2\delta\beta_j \tag{9.125}$$

which agrees with our convention in Eq. 9.122. We can build up the finite representations easily since these are Lie groups. We already know that for a rotation φ about the z axis the element of $O^+(3)$ is

$$R(\varphi)=\begin{pmatrix} \cos\varphi & \sin\varphi & 0 \\ -\sin\varphi & \cos\varphi & 0 \\ 0 & 0 & 1 \end{pmatrix} \qquad 0\le\varphi<2\pi$$

so that a corresponding element of $SU(2)$ is

$$U(\varphi)=\exp\left[i\frac{\varphi}{2}\sigma_3\right]=\begin{pmatrix} e^{i\varphi/2} & 0 \\ 0 & e^{-i\varphi/2} \end{pmatrix} \qquad 0\le\varphi<4\pi$$

However since

$$R(\varphi+2\pi)=R(\varphi)$$

while

$$U(\varphi+2\pi)=\begin{pmatrix} -e^{i\varphi/2} & 0 \\ 0 & -e^{-i\varphi/2} \end{pmatrix}=-U(\varphi)$$

we see that, as claimed,

$$\begin{matrix} +U(\varphi) \\ -U(\varphi) \end{matrix} \rightarrow R(\varphi) \tag{9.126}$$

so that the homomorphism is two to one. Q.E.D.

In fact, by the methods of Section 3.7, one easily verifies the useful relation

$$U_j(\alpha)\equiv e^{i(\alpha/2)\sigma_j}=I\cos\left(\frac{\alpha}{2}\right)+i\sigma_j\sin\left(\frac{\alpha}{2}\right) \qquad j=1,2,3 \tag{9.127}$$

More generally (cf. Problem 3.23) we have

$$\exp\left(i\frac{\varphi}{2}\boldsymbol{\sigma}\cdot\hat{\mathbf{n}}\right)=\cos\left(\frac{\varphi}{2}\right)+i\sin\left(\frac{\varphi}{2}\right)\boldsymbol{\sigma}\cdot\hat{\mathbf{n}}$$

However we know that in terms of the Euler angles, α, β, γ, for successive counterclockwise rotations of the coordinates about the z, y', and z'' axes, respectively (cf. Fig. 9.2; our conventions here are the same as those of Edmonds

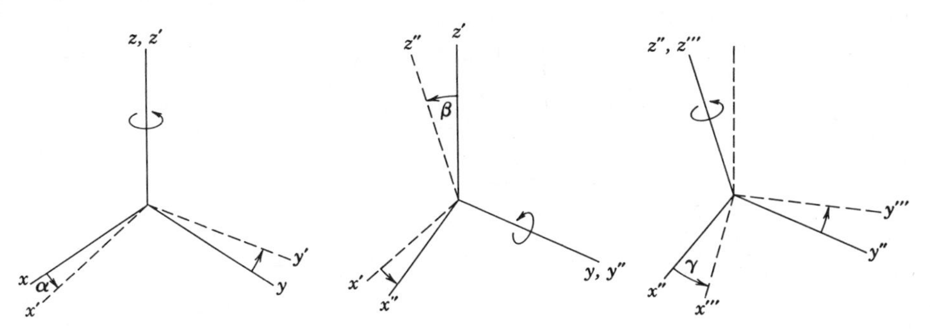

Figure 9.2

and of Rose, but different from those of Goldstein), we can represent an arbitrary element of $O^{+}(3)$ as

$$R(\alpha\beta\gamma)=R_{z''}(\gamma)R_{y'}(\beta)R_{z}(\alpha) \tag{9.128}$$

We have already seen in Eq. 9.127 that

$$R_z(\gamma) \leftrightarrow U_z(\gamma)=\begin{pmatrix} e^{i\gamma/2} & 0 \\ 0 & e^{-i\gamma/2} \end{pmatrix}=\cos\left(\frac{\gamma}{2}\right)+i\sigma_3 \sin\left(\frac{\gamma}{2}\right)$$

$$R_y(\beta) \leftrightarrow U_y(\beta)=\cos\left(\frac{\beta}{2}\right)+i\sigma_2 \sin\left(\frac{\beta}{2}\right)=\begin{pmatrix} \cos\left(\frac{\beta}{2}\right) & \sin\left(\frac{\beta}{2}\right) \\ -\sin\left(\frac{\beta}{2}\right) & \cos\left(\frac{\beta}{2}\right) \end{pmatrix}$$

$$R_z(\alpha) \leftrightarrow U_z(\alpha)=\begin{pmatrix} e^{i\alpha/2} & 0 \\ 0 & e^{-i\alpha/2} \end{pmatrix}=\cos\left(\frac{\alpha}{2}\right)+i\sigma_3 \sin\left(\frac{\alpha}{2}\right)$$

so that

$$R(\alpha\beta\gamma) \leftrightarrow U(\alpha\beta\gamma)=\begin{pmatrix} e^{i(\gamma+\alpha)/2}\cos\left(\frac{\beta}{2}\right) & e^{i(\gamma-\alpha)/2}\sin\left(\frac{\beta}{2}\right) \\ -e^{-i(\gamma-\alpha)/2}\sin\left(\frac{\beta}{2}\right) & e^{-i(\gamma+\alpha)/2}\cos\left(\frac{\beta}{2}\right) \end{pmatrix} \tag{9.129}$$

9.11 IRREDUCIBLE REPRESENTATIONS OF *SU*(2)

We must now construct the irreducible representations of $SU(2)$. We then have irreducible representations of $O^{+}(3)$ as we shall see.

a. Spinor Representations

We begin with the fundamental representations of $SU(2)$ given in Eq. 9.119. This can be written as

$$U=\begin{pmatrix} a & b \\ -b^* & a^* \end{pmatrix} \tag{9.130}$$

subject to the constraint

$$|a|^2+|b|^2=1 \tag{9.131}$$

Suppose we take as a set of basis vectors for this space

$$\hat{\mathbf{x}}_1=\begin{pmatrix} 1 \\ 0 \end{pmatrix} \qquad \hat{\mathbf{x}}_2=\begin{pmatrix} 0 \\ 1 \end{pmatrix} \tag{9.132}$$

so that

$$\hat{\mathbf{x}}_1' \equiv U\hat{\mathbf{x}}_1=\begin{pmatrix} a \\ -b^* \end{pmatrix}=a\hat{\mathbf{x}}_1-b^*\hat{\mathbf{x}}_2$$

$$\hat{\mathbf{x}}_2' \equiv U\hat{\mathbf{x}}_2=\begin{pmatrix} b \\ a^* \end{pmatrix}=b\hat{\mathbf{x}}_1+a^*\hat{\mathbf{x}}_2$$

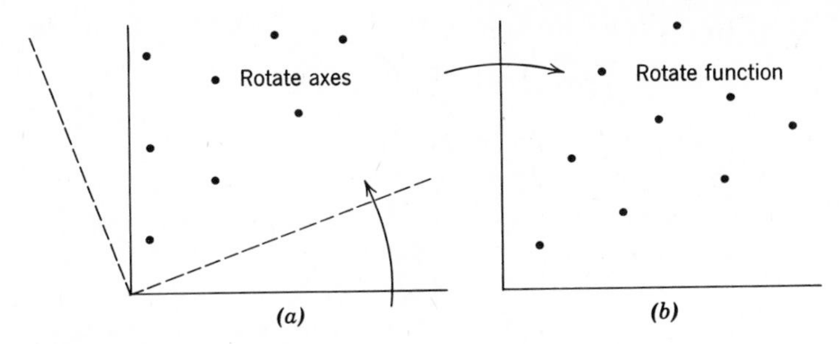

Figure 9.3

or

$$\hat{\mathbf{x}}_i' = \sum_{j=1}^{2} U_{ji}\hat{\mathbf{x}}_j \equiv \sum_{j=1}^{2} (U^T)_{ij}\hat{\mathbf{x}}_j \tag{9.133}$$

which agrees with the convention we adopted for representations in Eq. 9.18. However, if we write a two-component vector in this complex space as $\begin{pmatrix} u \\ v \end{pmatrix}$ (a *spinor*), then the components, u and v, transform as

$$\begin{aligned} u' &= au + bv = U_{11}u + U_{12}v \\ v' &= -b^*u + a^*v = U_{21}u + U_{22}v \end{aligned} \tag{9.134}$$

so that u and v transform according to U while $\hat{\mathbf{x}}_1$ and $\hat{\mathbf{x}}_2$ transform according to U^T (cf. Section 2.6a). That is, the components of a vector and the basis functions transform according to matrices that are the transposes of each other (here $b \to -b^*$). So far we have considered rotations of the coordinate axes in constructing U. According to our convention of Eq. 9.18 we have for the basis functions $\{\psi_i(\mathbf{r})\}$

$$\psi_i(R\mathbf{r}) \equiv O_R\psi_i(\mathbf{r}) = \sum_j D_{ji}(R)\psi_j(\mathbf{r})$$

Suppose we have a function $f(\mathbf{r})$ that assigns numbers to every point in space (i.e., the map of the function). We can compare the effect of rotating the frame with that of rotating the map as depicted in Fig. 9.3. We see that applying R to the coordinate axes produces the same relative effect as applying R^{-1} to the (map of the) function. Therefore, once we obtain representations for $U(R)$, the rotation matrices, $D(R)$, that rotate the basis functions will be given as

$$D(R) = U(R^{-1}) \tag{9.135}$$

Before we construct representations in higher dimensional spaces, we make an important observation. The real orthogonal matrix

$$S = \begin{pmatrix} 0 & 1 \\ -1 & 0 \end{pmatrix}$$

is such that for the U of Eq. 9.130

$$SUS^{-1} = U^* \tag{9.136}$$

so that U and U^* are equivalent matrices. This implies that conjugate representations of $SU(2)$ are always equivalent so that once we have U, then U^* does not provide an inequivalent representation. For arbitrary Lie groups conjugate representations are not always equivalent [cf. the group $C(2)$ of Section 9.6e and problem 9.28].

In order to find irreducible representations in an $(n+1)$-dimensional space, we need a set of $(n+1)$ *linearly independent* basis functions. Clearly the polynomials

$$u^n,\ u^{n-1}v,\ u^{n-2}v^2, \ldots,\ uv^{n-1},\ v^n$$

form a suitable basis. To agree with normal conventions, we set $n=2j$, $j=0, 1/2, 1, 3/2, 2, \ldots$, and define the polynomials

$$f^j_m(u, v) \equiv \frac{u^{j+m}v^{j-m}}{\sqrt{(j+m)!\,(j-m)!}} \qquad m=j, j-1, \ldots, -j \tag{9.137}$$

For a fixed value of j these $(2j+1)$ polynomials are linearly independent. Therefore we can find a $(2j+1)\times(2j+1)$ representative of an element of U in this space as (cf. Eq. 9.134)

$$\begin{aligned} Uf^j_m(u, v) \equiv f^j_m(u', v') &= \sum_{m'=-j}^{j} U_{mm'} f^j_{m'}(u, v) \\ &= \frac{(au+bv)^{j+m}(-b^*u+a^*v)^{j-m}}{\sqrt{(j+m)!\,(j-m)!}} \end{aligned} \tag{9.138}$$

We need only use the binomial theorem for the expansions

$$(au+bv)^{j+m} = \sum_{k=0}^{j+m} \frac{(j+m)!}{k!\,(j+m-k)!}\, a^{j+m-k}u^{j+m-k}b^k v^k$$

$$(-b^*u+a^*v)^{j-m} = \sum_{l=0}^{j-m} (-1)^{j-m-l} \frac{(j-m)!}{l!\,(j-m-l)!}\, b^{*(j-n-l)}u^{j-m-l}a^{*l}v^l$$

to obtain

$$Uf^j_m(u, v) = \sum_{k=0}^{j+m}\sum_{l=0}^{j-m} \frac{(-1)^{j-m-l}\sqrt{(j+m)!\,(j-m)!}}{k!\,l!\,(j+m-k)!\,(j-m-l)!}\, a^{j+m-k}a^{*l}b^k b^{*(j-n-l)}u^{2j-k-l}v^{k+l}$$

If we let $j-k-l=m'$, then

$$Uf^j_m(u, v) = \sum_{m'=-j}^{j} U^j_{mm'} f^j_{m'}(u, v)$$

where

$$\begin{aligned} U^j_{mm'} = \sum_{k=0}^{j+m} & \frac{(-1)^{m'-m+k}\sqrt{(j+m)!\,(j-m)!\,(j+m')!\,(j-m')!}}{k!\,(j-m'-k)!\,(j+m-k)!\,(m'-m+k)!} \\ & \times a^{j+m-k}a^{*(j-m'-k)}b^k b^{*(m'-m+k)} \end{aligned} \tag{9.139}$$

We can see that the matrix defined by Eq. 9.139 is unitary as follows. Since $|u|^2+|v|^2$ is an invariant

$$|u'|^2+|v'|^2=|u|^2+|v|^2$$

we have

$$\begin{aligned}
\sum_{m=-j}^{j} f_m^{*j}(u', v')f_m^j(u', v') &= \sum_{m=-j}^{j} \frac{[|au+bv|^2]^{j+m}[|-b^*u+a^*v|^2]^{j-m}}{(j+m)!\,(j-m)!} \\
&= \frac{1}{(2j)!}(2j)! \sum_{s=0}^{2j} \frac{[|au+bv|^2]^s[|-b^*u+a^*v|^2]^{2j-s}}{s!\,(2j-s)!} \\
&= \frac{[|u'|^2+|v'|^2]^{2j}}{(2j)!} = \frac{[|u|^2+|v|^2]^{2j}}{(2j)!} \\
&= \sum_{m=-j}^{j} f_m^{*j}(u, v)f_m^j(u, v)
\end{aligned}$$

which, with Eq. 9.138, states that U is unitary. These also form a representation since

$$\begin{aligned}
f_m^j(u'', v'') &= \sum_{m'} U_{mm'}^j(a', b')f_m^j(u', v') \\
&= \sum_{m',m''} U_{mm'}^j(a', b')U_{m'm''}(a, b)f_{m''}^j(u, v) \\
&\equiv \sum_{m''} U_{mm''}^j(a'', b'')f_{m''}^j(u, v)
\end{aligned}$$

where

$$\begin{pmatrix} u' \\ v' \end{pmatrix} = U(a, b)\begin{pmatrix} u \\ v \end{pmatrix}$$

$$\begin{pmatrix} u'' \\ v'' \end{pmatrix} = U(a', b')\begin{pmatrix} u' \\ v' \end{pmatrix} = U(a', b')U(a, b)\begin{pmatrix} u \\ v \end{pmatrix} \equiv U(a'', b'')\begin{pmatrix} u \\ v \end{pmatrix}$$

the last line following from the group property of these 2×2 matrices U defined in Eq. 9.130.

We will return to the question of the irreducibility of the representation of Eq. 9.139 in Section 9.14b.

b. The Rotation Matrices

From the parameter identifications of Eqs. 9.129 and 9.130, we can use Eqs. 9.135 and 9.139 to obtain the matrix elements of the conventionally defined *rotation matrices* $D(R)$ as

$$D_{mm'}^j(R)=[U(R^{-1})]_{mm'}^j=[U^\dagger(R)]_{mm'}^j=U_{m'm}^{j*}(R)$$

or

$$D_{mm'}^j(\alpha\beta\gamma)=U_{m'm}^{j*}(\alpha\beta\gamma)\equiv e^{-im\alpha}d_{mm'}^j(\beta)e^{-im'\gamma} \tag{9.140}$$

where the real functions $d^j_{mm'}(\beta)$ are given as

$$d^j_{mm'}(\beta) \equiv \sum_{k=0}^{j+m'} \frac{(-1)^k \sqrt{(j+m')!\,(j-m')!\,(j+m)!\,(j-m)!}}{k!\,(j-m-k)!\,(j+m'-k)!\,(m-m'+k)!} \times \left(\cos\frac{\beta}{2}\right)^{2j+m'-m-2k} \left(-\sin\frac{\beta}{2}\right)^{m-m'+2k} \quad \textbf{(9.141)}$$

In terms of the Jacobi polynomials of Eq. 7.189 these can be expressed as

$$d^j_{mm'}(\beta) = \sqrt{\frac{(j+m')!\,(j-m')!}{(j+m)!\,(j-m)!}} \left(\cos\frac{\beta}{2}\right)^{-m-m'} \left(-\sin\frac{\beta}{2}\right)^{m-m'} P^{(m-m',-m-m')}_{j+m'}(\cos\beta) \quad \textbf{(9.142)}$$

We can also verify the identities

$$d^j_{m'm}(\beta) = d^j_{mm'}(-\beta) = (-1)^{m'-m}\, d^j_{-m'-m}(\beta) = (-1)^{m'-m}\, d^j_{mm'}(\beta)$$

We can use Eq. 9.141 to prove that the $D^j(\alpha\beta\gamma)$ are irreducible representations. According to Th. 9.3 [which we shall see how to extend to $SU(2)$ and $O^+(3)$ in Section 9.14a] it will be sufficient to show that any matrix that commutes with $D^j(\alpha\beta\gamma)$ is a multiple of the unit matrix. Assume there is a matrix A independent of α, β, and γ (since A must be the *same* for all the D^j), such that

$$(AD^j)_{mm'} = (D^j A)_{mm'} \qquad \forall \alpha, \beta, \gamma$$

From Eq. 9.140 this becomes

$$\sum_k A_{mk} e^{-ik\alpha}\, d^j_{km'}(\beta) e^{-im'\gamma} = \sum_k e^{-im\alpha}\, d^j_{mk}(\beta) e^{ik\gamma} A_{km'} \quad \textbf{(9.143)}$$

From Eqs. 9.141 or 9.142 we see that

$$d^j_{km'}(0) = \delta_{km'}$$

so that for $\beta = \gamma = 0$ Eq. 9.143 reduces to

$$A_{mm'} e^{-im'\alpha} = e^{-im\alpha} A_{mm'}$$

so that $A_{mm'}$ must be diagonal. Equation 9.143 with $\alpha = \gamma = 0$ becomes

$$A_{mm}\, d^j_{mm'}(\beta) = d^j_{mm'}(\beta) A_{m'm'}$$

If we set $m' = j$ then, from Eqs. 9.129, 9.130, 9.139 and 9.140, we see that

$$d^j_{mj}(\beta) = (-1)^j \sqrt{\frac{(2j)!}{(j+m)!\,(j-m)!}} \left(\cos\frac{\beta}{2}\right)^{j+m} \left(\sin\frac{\beta}{2}\right)^{j-m}$$

which is not identically zero as a function of β. As a result the relation

$$A_{mm}\, d^j_{mj}(\beta) = d^j_{mj}(\beta) A_{jj}$$

implies that

$$A_{mm} = A_{jj} \qquad \forall m$$

so that A is a multiple of the identity matrix, which means that the $\{D^j(\alpha\beta\gamma)\}$ form irreducible representations.

c. Representations in the Space of Spherical Harmonics

We now have representations of any dimension, $n=1, 2, 3, \dots,$ in terms of this spinor basis. For instance for $j=1$ we have explicitly

$$D^1_{m'm}(\alpha\beta\gamma)=\begin{matrix} \\ m'=1 \\ \\ m'=0 \\ \\ m'=-1 \end{matrix}\overset{\begin{matrix} m=1 & \qquad\qquad m=0 & \qquad\qquad m=-1 \end{matrix}}{\begin{pmatrix} e^{-i\alpha}\dfrac{1+\cos\beta}{2}e^{-i\gamma} & -e^{-i\alpha}\dfrac{\sin\beta}{\sqrt{2}} & e^{-i\alpha}\dfrac{1-\cos\beta}{2}e^{i\gamma} \\ \dfrac{\sin\beta}{\sqrt{2}}e^{-i\gamma} & \cos\beta & -\dfrac{\sin\beta}{\sqrt{2}}e^{i\gamma} \\ e^{i\alpha}\dfrac{1-\cos\beta}{2}e^{-i\gamma} & \dfrac{e^{i\alpha}\sin\beta}{\sqrt{2}} & e^{i\alpha}\dfrac{1+\cos\beta}{2}e^{i\gamma} \end{pmatrix}} \tag{9.144}$$

This is the same as $R(\alpha\beta\gamma)$ (cf. Eq. 9.216) up to a similarity transformation. We have previously studied the spherical harmonics (cf. Eq. 4.47),

$$Y_l^m(\theta, \varphi)\equiv(-1)^{(m+|m|)/2}\sqrt{\frac{2l+1}{4\pi}\frac{(l-|m|)!}{(l+|m|)!}}\,P_l^{|m|}(\cos\theta)e^{im\varphi}$$

$$m=-l, -l+1, \dots, l \qquad l\geq 0$$

$$Y_l^{m*}(\theta, \varphi)=(-1)^m Y_l^{-m}(\theta, \varphi)$$

which are a complete set of functions over the unit sphere. In particular for $l=1$ these functions are

$$Y_1^1(\theta, \varphi)=-\sqrt{\frac{3}{8\pi}}\sin\theta e^{i\varphi}$$

$$Y_1^0(\theta, \varphi)=\sqrt{\frac{3}{4\pi}}\cos\theta$$

$$Y_1^{-1}(\theta, \varphi)=\sqrt{\frac{3}{8\pi}}\sin\theta e^{-i\varphi}$$

For the rotation of a unit vector of angles (θ, φ) through the Euler angles, α, β, γ, we have

$$\mathbf{r}'=R(\alpha\beta\gamma)\mathbf{r}$$

or in component form,

$$\begin{pmatrix} \sin\theta'\,\cos\varphi' \\ \sin\theta'\,\sin\varphi' \\ \cos\theta' \end{pmatrix}=R(\alpha\beta\gamma)\begin{pmatrix} \sin\theta\cos\varphi \\ \sin\theta\sin\varphi \\ \cos\theta \end{pmatrix}$$

which implies that

$$\begin{aligned}\cos\theta'&=\sin\beta\cos\alpha\sin\theta\cos\varphi+\sin\beta\sin\alpha\sin\theta\sin\varphi+\cos\beta\cos\theta \\ &=\sin\beta\sin\theta\cos(\varphi-\alpha)+\cos\beta\cos\theta\end{aligned}$$

But this is identical with the result

$$Y_1^0(\theta', \varphi') = \sum_{m'=-1}^{1} D^1_{m'0}(\alpha\beta\gamma) Y_1^{m'}(\theta, \varphi)$$

or

$$\sqrt{\frac{3}{4\pi}} \cos\theta' = e^{-i\alpha} \frac{\sin\beta}{\sqrt{2}} \sqrt{\frac{3}{8\pi}} \sin\theta e^{i\varphi} + \cos\beta \sqrt{\frac{3}{4\pi}} \cos\theta + e^{i\alpha} \frac{\sin\beta}{\sqrt{2}} \sqrt{\frac{3}{8\pi}} \sin\theta e^{-i\varphi}$$

$$= \sqrt{\frac{3}{4\pi}} [\sin\beta \sin\theta \cos(\varphi - \alpha) + \cos\beta \cos\theta]$$

In fact we easily verify the relation

$$Y_1^m(\theta', \varphi') = \sum_{m'=-1}^{1} D^1_{m'm}(\alpha\beta\gamma) Y_1^{m'}(\theta, \varphi)$$

In the next section we shall prove that, in general,

$$O_R Y_l^m(\theta, \varphi) \equiv Y_l^m(\theta', \varphi') = \sum_{m'=-l}^{l} D^l_{m'm}(\alpha\beta\gamma) Y_l^{m'}(\theta, \varphi) \tag{9.145}$$

This will give us a representation of the spatial rotations in the space spanned by the spherical harmonics. In this manner we obtain only those representations for spaces of dimension $2l+1$ (i.e., *odd* dimensions) (cf. Section 9.14b).

9.12 ALGEBRA OF THE ANGULAR MOMENTUM OPERATORS

We have seen previously in Eqs. 9.116 and 9.117 that the three hermitian *angular momentum operators*, $\{J_i\}$, which are the generators of the spatial rotations, satisfy the commutation relations

$$[J_i, J_j] = i\varepsilon_{ijk} J_k \tag{9.146}$$

and that

$$J^2 \equiv \sum_{j=1}^{3} J_j^2$$

is such that

$$[J^2, J_j] = 0 \tag{9.147}$$

so that we can, for example, simultaneously diagonalize both J^2 and J_3. We now present a general discussion of the eigenvalues and eigenvectors of J^2 and of J_3 and of the transformation properties of these eigenvectors under spatial rotations.

a. Spectra of J^2 and of J_3

Equations 9.146 and 9.147 state that there exist simultaneous eigenvectors of J^2 and of J_3, which we shall denote by ψ_{jm}. Since J^2 and J_3 are hermitian, we also know that these ψ_{jm} will form a complete set of basis vectors for our representation space, at least for finite-dimensional spaces. (Recall that $n=2j+1$.) If we define

$$J^{\pm} \equiv J_1 \pm iJ_2, \ni (J^+)^\dagger = J^- \tag{9.148}$$

we find that

$$[J^2, J^{\pm}] = 0$$
$$[J_3, J^{\pm}] = \pm J^{\pm}$$
$$[J^+, J^-] = 2J_3$$

We choose a set of simultaneous basis states as

$$J^2 \psi_{jm} = n_j \psi_{jm} \tag{9.149}$$

$$J_3 \psi_{jm} = m\psi_{jm} \tag{9.150}$$

Since the diagonal matrix elements of the square of a hermitian matrix are nonnegative (i.e., we could go to a diagonal basis) and since

$$J_2^2 + J_1^2 \equiv J^2 - J_3^2$$
$$(J_1^2 + J_2^2)\psi_{jm} = (n_j - m^2)\psi_{jm}$$

we conclude that

$$n_j - m^2 \geq 0 \tag{9.151}$$

where n_j and m are real. We now define new vectors $\phi_{jm}^{\pm}$

$$\phi_{jm}^{\pm} \equiv J^{\pm}\psi_{jm} \tag{9.152}$$

so that

$$J^2\phi_{jm}^{\pm} = J^2 J^{\pm}\psi_{jm} = J^{\pm}J^2\psi_{jm} = n_j J^{\pm}\psi_{jm} = n_j \phi_{jm}^{\pm}$$
$$J_3\phi_{jm}^{\pm} = J_3 J^{\pm}\psi_{jm} = (\pm J^{\pm} + J^{\pm}J_3)\psi_{jm} = J^{\pm}(m \pm 1)\psi_{jm} = (m \pm 1)\phi_{jm}^{\pm}$$

This states that the $\phi_{jm}^{\pm}$ are also eigenvectors of J^2 and of J_3 but ϕ_{jm}^{+} corresponds to the eigenvalue $(m+1)$ of J_3. Hence J^+ acts as a raising operator. However for a given j (and hence n_j) there must be some *largest* value of m since

$$-\sqrt{n_j} \leq m \leq +\sqrt{n_j}$$

from Eq. 9.151. Let this largest value be m_2.

Therefore, if we apply J^+ to ψ_{jm_2} it must yield zero or else we would have an eigenvector corresponding to m_2+1, contrary to our assumption; hence,

$$J^+\psi_{jm_2} = 0$$

Similarly, if m_1 is the smallest eigenvalue of J_3, then

$$J^-\psi_{jm_1} = 0$$

Since as an identity

$$J^2 \equiv J^{\mp}J^{\pm} + J_3(J_3 \pm 1)$$

we can see that

$$J^-J^+\psi_{jm_2}=[n_j-m_2(m_2+1)]\psi_{jm_2}=0 \tag{9.153}$$

and

$$J^+J^-\psi_{jm_1}=[n_j+m_1(m_1-1)]\psi_{jm_1}=0$$

which together imply that

$$m_2(m_2+1)=m_1(m_1-1)$$

or

$$(m_2+m_1)(m_2-m_1+1)=0 \tag{9.154}$$

But $m_2 \geq m_1$, so we conclude from Eq. 9.154 that

$$m_1=-m_2 \tag{9.155}$$

Since successive eigenvalues of J_3 for $\phi^{\pm}_{jm}$ differ by unity, m_2-m_1 is a nonnegative integer. If we set

$$m_2-m_1 \equiv 2j \qquad j=0, \tfrac{1}{2}, 1, \tfrac{3}{2}, 2, \ldots$$

then we have for the largest and smallest values of m, respectively,

$$\begin{aligned} m_2&=j \\ m_1&=-j \end{aligned} \tag{9.156}$$

so that

$$m=j, j-1, \ldots, -j$$

which states that there are $(2j+1)$ values of m for each j. From Eqs. 9.153 and 9.156 we have

$$n_j=j(j+1) \tag{9.157}$$

This allows us to span a space of dimension $(2j+1)$.

We can summarize these results as

$$J^2\psi_{jm}=j(j+1)\psi_{jm} \tag{9.158}$$

$$J_3\psi_{jm}=m\psi_{jm} \tag{9.159}$$

$$J^{\pm}\psi_{jm} \equiv \phi^{\pm}_{jm}=N^{\pm}\psi_{j,m\pm1} \tag{9.160}$$

It is relatively simple to evaluate the normalization factor, $N^{\pm}$, as follows.

$$\begin{aligned} \langle J^{\pm}\psi_{jm}, J^{\pm}\psi_{jm}\rangle &= |N^{\pm}|^2\langle\psi_{j,m\pm1} \mid \psi_{j,m\pm1}\rangle = |N^{\pm}|^2 \\ &= \langle\psi_{jm}, J^{\mp}J^{\pm}\psi_{jm}\rangle = \langle\psi_{jm}, [J^2-J_3(J_3\pm1)]\psi_{jm}\rangle \\ &= j(j+1)-m(m\pm1)=(j\mp m)(j\pm m+1) \end{aligned}$$

If we choose the arbitrary phase factor to be $+1$, then we obtain the conventional expression

$$N^{\pm}=[(j\mp m)(j\pm m+1)]^{1/2} \tag{9.161}$$

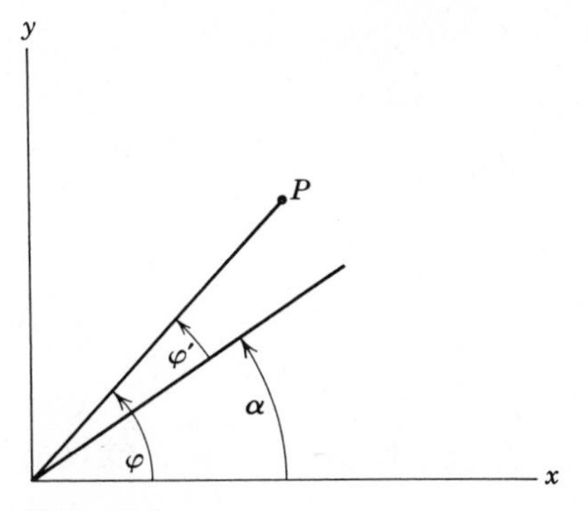

Figure 9.4

b. Rotation of Angular Momentum Eigenfunctions

The operator that generates a finite rotation through an angle θ about an axis specified by the unit vector $\hat{\mathbf{n}}$ is, for a (map of a) function,

$$R = e^{-i\theta\hat{\mathbf{n}}\cdot\mathbf{J}}$$

where

$$\hat{\mathbf{n}}\cdot\mathbf{J} \equiv n_1 J_1 + n_2 J_2 + n_3 J_3 \tag{9.162}$$

This is easily seen from Fig. 9.4 if we take $\hat{\mathbf{n}}$ to be along the z axis. In terms of the commutation relations of Eq. 9.146 and the representation of the X_j [recall that $J_j = iX_j$ from Eq. 9.116] for $O^+(3)$ in Section 9.8 in terms of differential operators, we can express the J_j in spherical coordinates as

$$J_1 = i\left(\sin\varphi \frac{\partial}{\partial\theta} + \cot\theta\cos\varphi\frac{\partial}{\partial\varphi}\right)$$

$$J_2 = i\left(-\cos\varphi\frac{\partial}{\partial\theta} + \cot\theta\sin\varphi\frac{\partial}{\partial\varphi}\right)$$

$$J_3 = -i\frac{\partial}{\partial\varphi} \tag{9.163}$$

Consider the function $f(\varphi) = e^{im\varphi}$ and rotate the coordinate system through an angle α as indicated, but leave the map (i.e., values) of $f(\varphi)$ fixed. Referred to the new frame, the field value at point P will be

$$\begin{aligned} f(\varphi') &= e^{im(\varphi-\alpha)} = e^{-im\alpha}e^{im\varphi} \\ &= e^{-\alpha(\partial/\partial\varphi)}e^{im\varphi} = e^{-i\alpha[-i(\partial/\partial\varphi)]}e^{im\varphi} = e^{-i\alpha J_3}f(\varphi) \end{aligned} \tag{9.164}$$

(Here we are following the convention of Rose, which is different from that of Edmonds.) In fact, for an arbitrary scalar function $f(x, y, z)$ we have

$$f(x', y', z') = Rf(x, y, z) \tag{9.165}$$

For an infinitesimal rotation about the z axis (cf. Eq. 9.62) we have

$$x' = x + y\,d\theta$$
$$y' = -x\,d\theta + y$$
$$z' = z$$

so that

$$\begin{aligned} f(x', y', z') &= f(x + y\,d\theta, -x\,d\theta + y, z) \\ &= f(x, y, z) + d\theta\left(y\frac{\partial}{\partial x} - x\frac{\partial}{\partial y}\right)f(x, y, z) \\ &= (1 - id\theta J_3)f(x, y, z) \simeq e^{-id\theta J_3}f(x, y, z) \equiv Rf(x, y, z) \end{aligned} \qquad \textbf{(9.166)}$$

Since

$$[J^2, e^{-i\theta\hat{\mathbf{n}}\cdot\mathbf{J}}] = 0$$

we see that the vector $R\psi_{jm}$ will still be an eigenvector of J^2 with eigenvalue $j(j+1)$ but not necessarily an eigenvector of J_3.

According to our convention of Eq. 9.18, we can generate representations of the rotations as

$$R\psi_{jm} \equiv e^{-i\theta\hat{\mathbf{n}}\cdot\mathbf{J}}\psi_{jm} = \sum_{m'} \mathcal{D}^j_{m'm}(R)\psi_{jm'} \qquad \textbf{(9.167)}$$

We use the symbol $\mathcal{D}^j_{m'm}(\alpha\beta\gamma)$ here since we do not yet know logically that they are the same as the $D^j_{mm'}(\alpha\beta\gamma)$ of Eq. 9.140. In terms of the Euler angles introduced in Section 9.10, Fig. 9.2, we have, from Eq. 9.128,

$$R(\alpha\beta\gamma) = e^{-i\gamma J_z''}e^{-i\beta J_y'}e^{-i\alpha J_z} \qquad \textbf{(9.168)}$$

However since $e^{i\alpha J_z}$ is a unitary operator, we can transform from one coordinate frame to another as

$$\begin{aligned} e^{-i\beta J_y'} &= e^{-i\alpha J_z}e^{-i\beta J_y}e^{i\alpha J_z} \\ e^{-i\gamma J_z''} &= e^{-i\beta J_y'}e^{-i\gamma J_z'}e^{i\beta J_y'} \\ &= e^{-i\alpha J_z}e^{-i\beta J_y}e^{i\alpha J_z}e^{-i\alpha J_z}e^{-i\gamma J_z}e^{i\alpha J_z}e^{-i\alpha J_z}e^{i\beta J_y}e^{i\alpha J_z} \end{aligned}$$

This allows us to write Eq. 9.168 as

$$R(\alpha\beta\gamma) = e^{-i\alpha J_z}e^{-\beta J_y}e^{-i\gamma J_z} \qquad \textbf{(9.169)}$$

where all the operators are now referred to the original coordinate system (i.e., before rotation). Since the $\psi_{jm} = |jm\rangle$ form a complete orthonormal set for our representation space, we can combine Eqs. 9.167 and 9.169 to obtain

$$R\,|jm\rangle \equiv \sum_{m'} D^j_{m'm}(\alpha\beta\gamma)\,|jm'\rangle = \sum_{m'} |jm'\rangle\langle jm'|\,R(\alpha\beta\gamma)\,|jm\rangle$$

so that

$$\begin{aligned} D^j_{m'm}(\alpha\beta\gamma) &= \langle jm'|\,R(\alpha\beta\gamma)\,|jm\rangle = \langle jm'|e^{-i\alpha J_z}e^{-i\beta J_y}e^{-i\gamma J_z}\,|jm\rangle \\ &= e^{-i\alpha m'}\langle jm'|\,e^{-i\beta J_y}\,|jm\rangle e^{-i\gamma m} \end{aligned} \qquad \textbf{(9.170)}$$

It only remains to be shown that $\langle jm'|\, e^{-i\beta J_y}\, |jm\rangle$ of Eq. 9.170 is identical to the $d^j_{m'm}(\beta)$ of Eq. 9.141. Since we can build up finite rotations from the infinitesimal ones, we shall demonstrate this equality only for infinitesimal β.

$$\begin{aligned}
\langle jm'|\, e^{-i\beta J_y}\, |jm\rangle &\simeq \delta_{m'm} - i\beta\langle jm'|\, J_y\, |jm\rangle \\
&= \delta_{m'm} - \tfrac{1}{2}\beta\langle jm'|\,(J^+ - J^-)\,|jm\rangle \\
&= \delta_{m'm} - \frac{\beta}{2}\Big[\sqrt{(j-m)(j+m+1)}\langle jm' \mid j, m+1\rangle - \sqrt{(j+m)(j-m+1)} \\
&\qquad\qquad \times \langle jm' \mid j, m-1\rangle\Big] \\
&= \delta_{m'm} - \frac{\beta}{2}\Big[\sqrt{(j-m)(j+m+1)}\ \delta_{m',m+1} - \sqrt{(j+m)(j-m+1)}\ \delta_{m',m-1}\Big]
\end{aligned} \tag{9.171}$$

Here we have used Eqs. 9.148 and 9.161. If we return to Eq. 9.141 and keep only the terms with $m'-m+2k=0$ and $m'-m+2k=1$, then because of the factor $(m-m'+k)!$ in the denominator of the sum the only terms to survive are

$$d^j_{m'm}(\beta) \simeq \delta_{m'm} - \frac{\beta}{2}\Big[\sqrt{(j-m)(j+m+1)}\ \delta_{m',m+1} - \sqrt{(j-m)(j+m+1)}\ \delta_{m',m-1}\Big]$$

so that we have established the relation

$$\langle jm'|\, e^{-i\beta J_y}\, |jm\rangle = d^j_{m'm}(\beta) \tag{9.172}$$

which establishes that $\mathscr{D}^j_{m'm}(\alpha\beta\gamma) \equiv D^j_{m'm}(\alpha\beta\gamma)$ or that Eq. 9.167 can be written as

$$e^{-i\theta\hat{\mathbf{n}}\cdot\mathbf{J}}\psi_{jm} = \sum_{m'} D^j_{m'm}(\alpha\beta\gamma)\psi_{jm'} \tag{9.173}$$

Example 9.22 Let us take $j=\frac{1}{2}$, where $J_y = \frac{1}{2}\sigma_y$ and $\sigma_y = \sigma_2$ is one of the Pauli matrices defined in Eq. 9.122. From Eq. 9.127 we have

$$e^{-i(\beta/2)\sigma_y} = I\cos\left(\frac{\beta}{2}\right) - i\sigma_y \sin\left(\frac{\beta}{2}\right)$$

so that

$$\begin{aligned}
\langle \tfrac{1}{2}|\, \sigma_y\, |\tfrac{1}{2}\rangle &= 0 \\
\langle \tfrac{1}{2}|\, \sigma_y\, |-\tfrac{1}{2}\rangle &= -i \\
\langle -\tfrac{1}{2}|\, \sigma_y\, |\tfrac{1}{2}\rangle &= i \\
\langle -\tfrac{1}{2}|\, \sigma_y\, |-\tfrac{1}{2}\rangle &= 0
\end{aligned}$$

which implies that

$$d^{1/2}_{m'm}(\beta) = \begin{array}{c} \\ m'=\frac{1}{2} \\ m'=-\frac{1}{2} \end{array}\begin{array}{c} \begin{array}{cc} m=\frac{1}{2} & \quad m=-\frac{1}{2} \end{array} \\ \begin{pmatrix} \cos\left(\frac{\beta}{2}\right) & -\sin\left(\frac{\beta}{2}\right) \\ \sin\left(\frac{\beta}{2}\right) & \cos\left(\frac{\beta}{2}\right) \end{pmatrix} \end{array}$$

or

$$D^{1/2}_{m'm}(\alpha\beta\gamma)=\begin{pmatrix} e^{-i(\alpha/2)}\cos\left(\frac{\beta}{2}\right)e^{-i(\gamma/2)} & -e^{-i(\alpha/2)}\sin\left(\frac{\beta}{2}\right)e^{i(\gamma/2)} \\ e^{i(\alpha/2)}\sin\left(\frac{\beta}{2}\right)e^{-i(\gamma/2)} & e^{i(\gamma/2)}\cos\left(\frac{\beta}{2}\right)e^{i(\gamma/2)} \end{pmatrix} \tag{9.174}$$

which agrees with Eqs. 9.129 and 9.140.

It is now a simple matter to find the effect produced on the spherical harmonics by the rotation operator. From Eqs. 9.163 we easily verify that

$$J^2 \equiv J_1^2+J_2^2+J_3^2=-\left[\frac{1}{\sin\theta}\frac{\partial}{\partial\theta}\left(\sin\theta\frac{\partial}{\partial\theta}\right)+\frac{1}{\sin^2\theta}\frac{\partial^2}{\partial\varphi^2}\right] \tag{9.175}$$

$$J_3=-i\frac{\partial}{\partial\varphi}$$

It is a straightforward exercise to use the definition of the $Y_l^m(\theta,\varphi)$ in Eq. 4.47 and the differential equation satisfied by the associated Legendre polynomials, $P_l^m(\cos\theta)$ given in Eq. 7.203, to prove that

$$\begin{aligned} J^2 Y_l^m(\theta,\varphi)&=l(l+1)Y_l^m(\theta,\varphi)\\ J_3 Y_l^m(\theta,\varphi)&=mY_l^m(\theta,\varphi)\end{aligned} \tag{9.176}$$

Since the $\{Y_l^m(\theta,\varphi)\}$ form an orthonormal set of eigenfunctions, we can use Eq. 9.173 to write at once

$$RY_l^m(\theta,\varphi)\equiv Y_l^m(\theta',\varphi')=\sum_{m'=-l}^{l} D^l_{m'm}(\alpha\beta\gamma)Y_l^{m'}(\theta,\varphi)$$

as claimed in Eq. 9.145. This relation also allows us to obtain an important addition theorem for spherical harmonics (cf. problem 9.25),

$$P_l(\cos\gamma)=\frac{4\pi}{(2l+1)}\sum_{m=-l}^{l} Y_l^{*m}(\theta,\varphi)Y_l^m(\theta',\varphi') \tag{9.177}$$

where $\cos\gamma=\hat{\mathbf{n}}\cdot\hat{\mathbf{n}}'$, with $\hat{\mathbf{n}}$ and $\hat{\mathbf{n}}'$ being unit vectors specified by the angles (θ,φ) and (θ',φ'), respectively, in spherical coordinates.

Furthermore use of Eqs. 9.142, 7.189, 7.208, 4.47 and the relation

$$F(a,b\,|c|\,z)=(1-z)^{c-a-b}F(c-a,c-b\,|c|\,z)$$

which may be verified from Eqs. 7.157 and 7.161, allows us to show from Eq. 9.140 that

$$\begin{aligned} D^l_{m0}(\alpha\beta 0)&=e^{-im\alpha}d^l_{m0}(\beta)\\ &=\frac{e^{-im\alpha}l!\,(-1)^m}{\sqrt{(l+m)!\,(l-m)!}}\left(\tan\frac{\beta}{2}\right)^m P_l^{(m,-m)}(\cos\beta)\\ &=\frac{e^{-im\alpha}(-1)^m}{m!}\left(\tan\frac{\beta}{2}\right)^m\sqrt{\frac{(l+m)!}{(l-m)!}}\,F\left(l+1,-l\,|m+1|\,\frac{1-\cos\beta}{2}\right)\end{aligned}$$

$$=\frac{e^{-im\alpha}(-1)^m}{m!}\left(\tan\frac{\beta}{2}\right)^m\sqrt{\frac{(l+m)!}{(l-m)!}}\left(\frac{1+\cos\beta}{2}\right)^m$$

$$\times F\left(m-l,\, m+l+1\,|m+1|\,\frac{1-\cos\beta}{2}\right)$$

$$=e^{-im\alpha}\frac{(-1)^m}{m!}\left(\sin\frac{\beta}{2}\cos\frac{\beta}{2}\right)^m\sqrt{\frac{(l+m)!}{(l-m)!}}$$

$$\times F\left(m-l,\, m+l+1\,|m+1|\,\frac{1-\cos\beta}{2}\right)$$

$$=e^{-im\alpha}(-1)^m\sqrt{\frac{(l-m)!}{(l+m)!}}\left[\frac{(l+m)!}{2^m(l-m)!}\right](\sin\beta)^m$$

$$\times F\left(m-l,\, m+l+1\,|m+1|\,\frac{1-\cos\beta}{2}\right)$$

$$=e^{-im\alpha}(-1)^m\sqrt{\frac{(l-m)!}{(l+m)!}}\,P_l^m(\cos\beta)=\sqrt{\frac{4\pi}{2l+1}}\,Y_l^{m*}(\beta,\alpha) \tag{9.178}$$

Use of Eqs. 4.47, 9.145 (with $m=0$), and 9.178 will yield Eq. 9.177.

9.13 COUPLING OF TWO ANGULAR MOMENTA

a. Product Basis and the Coupled Representation

We will now discuss what is often referred to as the coupling of two angular momenta. Mathematically we are simply seeking representations of the rotation operator in a direct product space. Consider two separate sets of angular momentum operators,

$$\begin{aligned}
&[J_i^{(1)}, J_j^{(1)}]=i\varepsilon_{ijk}J_k^{(1)} &\qquad &[J_i^{(2)}, J_j^{(2)}]=i\varepsilon_{ijk}J_k^{(2)}\\
&J^{(1)^2}\,|j_1, m_1\rangle=j_1(j_1+1)\,|j_1, m_1\rangle &\qquad &J^{(2)^2}\,|j_2, m_2\rangle=j_2(j_2+1)\,|j_2, m_2\rangle\\
&J_3^{(1)}\,|j_1, m_1\rangle=m_1\,|j_1, m_1\rangle &\qquad &J_3^{(2)}\,|j_2, m_2\rangle=m_2\,|j_2, m_2\rangle
\end{aligned} \tag{9.179}$$

Construct a new operator as

$$J_i\equiv J_i^{(1)}+J_i^{(2)} \qquad i=1, 2, 3 \tag{9.180}$$

in the direct product space. Actually, rather than the standard notation of Eq. 9.180, a more precise expression would be

$$J_i\equiv J_i^{(1)}\otimes I^{(2)}+I^{(1)}\otimes J_i^{(2)} \qquad i=1, 2, 3$$

Our basis states are then the product basis ones,

$$|j_1m_1\rangle\,|j_2m_2\rangle$$

or

$$|j_1m_1\rangle\otimes|j_2m_2\rangle$$

They span a space of dimension $(2j_1+1)(2j_2+1)$. Since $J_i^{(1)}$ and $J_i^{(2)}$ operate in different spaces, they commute

$$[J_i^{(1)}, j_j^{(2)}]=0 \qquad \forall i, j$$

from which we easily prove that

$$[J_i, J_j]=i\varepsilon_{ijk}J_k \tag{9.181}$$

so that the $\{J_i\}$ are also angular momentum operators such that

$$\begin{aligned} J^2\,|JM\rangle &= J(J+1)\,|J, M\rangle \\ J_3\,|JM\rangle &= M\,|JM\rangle \\ J^{\pm}\,|JM\rangle &= N^{\pm}\,|J, M\pm 1\rangle \end{aligned} \tag{9.182}$$

Since we are in a space of dimension $(2j_1+1)(2j_2+1)$, we must finally obtain *exactly* these many orthonormal states $|JM\rangle$. If we observe that

$$J^2 \equiv J^{(1)^2}+J^{(2)^2}+J^{(1)+}J^{(2)-}+J^{(1)-}J^{(2)+}+2J_3^{(1)}J_3^{(2)}$$

we see that

$$\begin{aligned} J^2\,|j_1j_1\rangle\,|j_2j_2\rangle &= [j_1(j_1+1)+j_2(j_2+1)+2j_1j_2]|j_1j_1\rangle\,|j_2j_2\rangle \\ &= (j_1+j_2)(j_1+j_2+1)\,|j_1j_1\rangle\,|j_2j_2\rangle \\ J_3\,|j_1j_1\rangle\,|j_2j_2\rangle &= (j_1+j_2)\,|j_1j_1\rangle\,|j_2j_2\rangle \end{aligned}$$

so that the state

$$|JJ\rangle \equiv |j_1j_1\rangle\,|j_2j_2\rangle \qquad J=j_1+j_2 \tag{9.183}$$

is a normalized eigenstate of J^2 and of J_3. A more complete notation for the vector of Eq. 9.183 would be $|J, J; j_1, j_2\rangle$. If we apply J^- to the state of Eq. 9.183, we find that

$$\begin{aligned} J^-\,|JJ\rangle &= [(J+J)(J-J+1)]^{1/2}\,|J, J-1\rangle = \sqrt{2J}\,|J, J-1\rangle \\ &= (J^{(1)-}+J^{(2)-})\,|j_1j_1\rangle\,|j_2j_2\rangle \\ &= [\sqrt{2j_1}\,|j_1, j_1-1\rangle\,|j_2j_2\rangle+\sqrt{2j_2}\,|j_1j_1\rangle\,|j_2, j_2-1\rangle] \end{aligned} \tag{9.184}$$

Here $|J, J-1\rangle$ is guaranteed to be an eigenstate of J^2 and of J_3 (with $M=J-1$). We can repeatedly apply J^- until we reach the state $|J, -J\rangle$. In this fashion we generate $2(j_1+j_2)+1$ orthonormal eigenstates.

Now consider the state

$$|J, J-1\rangle = \frac{1}{\sqrt{2J}}[\sqrt{2j_1}\,|j_1, j_1-1\rangle\,|j_2j_2\rangle+\sqrt{2j_2}\,|j_1j_1\rangle\,|j_2, j_2-1\rangle] \qquad J=j_1+j_2 \tag{9.185}$$

We can construct a state orthogonal to this and an eigenvector of J_3 with $M=J-1=j_1+j_2-1$ as

$$|\alpha, \beta\rangle \equiv a\,|j_1, j_1-1\rangle\,|j_2j_2\rangle+b\,|j_1j_1\rangle\,|j_2, j_2-1\rangle \tag{9.186}$$

where a and b may be chosen real such that

$$a^2+b^2=1$$

to insure proper normalization and

$$\sqrt{2j_1}\,a+\sqrt{2j_2}\,b=0$$

to make $|\alpha, \beta\rangle$ orthgonal to $|J, J-1\rangle$. We find for a and b

$$a=\pm\sqrt{\frac{j_2}{j_1+j_2}}$$

$$b=\mp\sqrt{\frac{j_1}{j_1+j_2}}$$

However by direct calculation we see that

$$\begin{aligned}
J^2\,|\alpha, \beta\rangle = &\{a[j_1(j_1+1)+j_2(j_2+1)]\,|j_1, j_1-1\rangle\,|j_2 j_2\rangle \\
&+a[(j_1-j_1+1)(j_1+j_1-1+1)(j_2+j_2)]^{1/2}\,|j_1 j_1\rangle\,|j_2, j_2-1\rangle \\
&+a[2(j_1-1)(j_2)]\,|j_1, j_1-1\rangle\,|j_2 j_2\rangle+b[j_1(j_1+1)+j_2(j_2+1)]\,|j_1 j_1\rangle\,|j_2, j_2-1\rangle \\
&+b[(2j_1)(2j_2)]^{1/2}\,|j_1, j_1-1\rangle\,|j_2 j_2\rangle+b[2j_1(j_2-1)]\,|j_1 j_1\rangle\,|j_2, j_2-1\rangle\} \\
=&\{a[j_1(j_1+1)+j_2(j_2+1)+2(j_1-1)j_2]+2\sqrt{j_1 j_2}\,b\}\,|j_1, j_1-1\rangle\,|j_2 j_2\rangle \\
&+\{a2\sqrt{j_1 j_2}+b[j_1(j_1+1)+j_2(j_2+1)+2j_1(j_2-1)]\}\,|j_1 j_1\rangle\,|j_2, j_2-1\rangle \\
=&\,a[j_1(j_1+1)+j_2(j_2+1)+2(j_1-1)j_2-2j_1]\,|j_1, j_1-1\rangle\,|j_2 j_2\rangle \\
&+b[j_1(j_1+1)+j_2(j_2+1)+2j_1(j_2-1)-2j_2]\,|j_1 j_1\rangle\,|j_2, j_2-1\rangle
\end{aligned}$$

so that finally

$$J^2\,|\alpha, \beta\rangle=(j_1+j_2-1)(j_1+j_2)\,|\alpha, \beta\rangle$$

which implies that we may make the choice

$$|\alpha, \beta\rangle=|j_1+j_2-1,\, j_1+j_2-1\rangle \tag{9.187}$$

We choose the phase by imposing the conventional (Condon-Shortley) requirement

$$\langle j_1+j_2,\, j_1+j_2-1|\, J_3^{(1)}\,|j_1+j_2-1,\, j_1+j_2-1\rangle\equiv\frac{1}{\sqrt{2J}}[\sqrt{2j_1}\,(j_1-1)a+\sqrt{2j_2}\,j_1 b]>0$$

so that

$$j_1[\sqrt{2j_1}\,a+\sqrt{2j_2}\,b]-\sqrt{2j_1}\,a\equiv-\sqrt{2j_1}\,a>0$$

or

$$a=-\sqrt{\frac{j_2}{j_1+j_2}}\qquad b=\sqrt{\frac{j_1}{j_1+j_2}} \tag{9.188}$$

That is, in general we require that all nondiagonal matrix elements of $J_3^{(1)}$, $\langle J, M|\, J_3^{(1)}\, |J', M\rangle$, are real and nonnegative. By applying J^- to $|j_1+j_2-1, j_1+j_2-1\rangle$ repeatedly we generate $2(j_1+j_2-1)+1=2(j_1+j_2)-1$ more orthonormal eigenfunctions, none of which are zero (cf. discussion of Section 9.12a).

So far we have taken the vector $|J, J; j_1, j_2\rangle$, $J=(j_1+j_2)$, of Eq. 9.183 and applied the lowering operator, J^-, to generate $2(j_1+j_2)+1$ orthogonal vectors. Next we took the vector, $|J, J-1; j_1, j_2\rangle$, $J=(j_1+j_2)$ (cf. Eq. 9.185), which is a linear combination of $|j_1, j_1-1\rangle\, |j_2, j_2\rangle$ and $|j_1, j_1\rangle\, |j_2, j_2-1\rangle$ and constructed another vector $|J, J; j_1, j_2\rangle$, $J=(j_1+j_2-1)$, orthogonal to it (cf. Eq. 9.187). With J^- we obtain $2(j_1+j_2-1)+1$ more orthogonal vectors. The vectors $|J, J-2; j_1, j_2\rangle$, $J=(j_1+j_2)$, and $|J, J-1; j_1, j_2\rangle$, $J=(j_1+j_2-1)$, are each linear combinations of the vectors $|j_1, j_1-2\rangle\, |j_2, j_2\rangle$, $|j_1, j_1-1\rangle\, |j_2, j_2-1\rangle$, and $|j_1, j_1\rangle\, |j_2, j_2-2\rangle$. Just as we did for Eq. 9.186 we can now construct a third linear combination of these three vectors that will be orthogonal to both $|J, J-2; j_1, j_2\rangle$, $J=(j_1+j_2)$, and to $|J, J-1; j_1, j_2\rangle$, $J=(j_1+j_2-1)$. We again normalize this vector to unity and impose the Condon-Shortley phase convention, choosing all constants to be real. A straightforward calculation shows that this produces a vector $|J, J; j_1, j_2\rangle$ with $J=(j_1+j_2-2)$ so that repeated application of J^- will yield $2(j_1+j_2-2)+1$ more orthogonal states. We continue this process of constructing new orthogonal vectors [i.e., next go to the vector $|J, J-1; j_1, j_2\rangle$, $J=(j_1+j_2-2)$, and proceed as previously to construct a new vector $|J, J; j_1, j_2\rangle$, $J=(j_1+j_2-3)$] until we arrive at two [if $j_1-j_2=(2n+1)/2$] or at one (if $j_1-j_2=n$) vectors. Therefore in all we generate

$$\sum_{J=|j_1-j_2|}^{j_1+j_2} (2J+1)=(2j_1+1)(2j_2+1)$$

orthonormal eigenfunctions of J^2 and J_3 and hence span the product space. The transformation from the product basis, $|j_1m_1\rangle\, |j_2m_2\rangle$ to the coupled representation, $|JM; j_1j_2\rangle$, is accomplished by the *Clebsch-Gordan coefficients* as

$$|JM; j_1j_2\rangle = \sum_{m_1m_2} \langle j_1m_1; j_2m_2\,|\,JM; j_1j_2\rangle\, |j_1m_1; j_2m_2\rangle \tag{9.189}$$

where $M=m_1+m_2$.

We see that these Clebsch-Gordan coefficients, $\langle j_1m_1; j_2m_2\,|\,JM; j_1j_2\rangle$, can be chosen to be the elements of a real orthogonal matrix as follows. We began with the state $|JJ\rangle=|j_1j_1; j_2j_2\rangle$ and have applied J^- to generate $2(j_1+j_2)+1$ orthonormal states all of whose coefficients are real by the choice of phase made in Eq. 9.161. With the choice of phase in Eq. 9.188 we insure real coefficients for the next $2(j_1+j_2)-1$ states and so forth. Compactly we can express this as

$$\Phi_\alpha = U_{\alpha;ij}\phi_{ij} \tag{9.190}$$

where $\alpha=(JM; j_1j_2)$, $i=(j_1m_1)$, $j=(j_2m_2)$. Since we go from one orthonormal basis to another, we have

$$U^\dagger U = I$$

while the reality condition,

$$U = U^*$$

implies that

$$UU^T = I \tag{9.191}$$

as claimed.

Example 9.23 It may be instructive to apply these general ideas to a specific, simple, useful case. Let $j_1 = 1$ and $j_2 = \frac{1}{2}$. We begin with the highest state in the representation for $J = \frac{3}{2}$,

$$|\tfrac{3}{2}, \tfrac{3}{2}\rangle = |1, 1\rangle\, |\tfrac{1}{2}, \tfrac{1}{2}\rangle \tag{9.192}$$

where on the left side of Eq. 9.192 we have used an abbreviated notation $|\frac{3}{2}, \frac{3}{2}\rangle$ rather than $|\frac{3}{2}, \frac{3}{2}; 1, \frac{1}{2}\rangle$. If we let

$$t^2\, |1, m\rangle = 2\, |1, m\rangle$$

$$t_3\, |1, m\rangle = m\, |1, m\rangle$$

$$\tau^2\, |\tfrac{1}{2}, \mu\rangle = \tfrac{3}{4}\, |\tfrac{1}{2}, \mu\rangle$$

$$\tau_3\, |\tfrac{1}{2}, \mu\rangle = \mu\, |\tfrac{1}{2}, \mu\rangle$$

and define $\mathbf{T} = \mathbf{t} + \boldsymbol{\tau}$,

$$T^2 = (\mathbf{t} + \boldsymbol{\tau}) \cdot (\mathbf{t} + \boldsymbol{\tau}) = t^2 + \tau^2 + t^+\tau^- + t^-\tau^+ + 2t_3\tau_3$$

then we readily verify, as we have already proved in general, that

$$T^2\, |\tfrac{3}{2}, \tfrac{3}{2}\rangle = [2 + \tfrac{3}{4} + 2(1)(\tfrac{1}{2})]\, |\tfrac{3}{2}, \tfrac{3}{2}\rangle = (\tfrac{3}{2})(\tfrac{3}{2} + 1)\, |\tfrac{3}{2}, \tfrac{3}{2}\rangle$$

so that $T = \frac{3}{2}$, as claimed. Also we have

$$T_3\, |\tfrac{3}{2}, \tfrac{3}{2}\rangle = (1 + \tfrac{1}{2})\, |\tfrac{3}{2}, \tfrac{3}{2}\rangle = \tfrac{3}{2}\, |\tfrac{3}{2}, \tfrac{3}{2}\rangle$$

so that again we obtain the expected result

$$T_3 = \tfrac{3}{2}$$

We construct the rest of the states corresponding to $T = \frac{3}{2}$ in a similar fashion.

$$T^-\, |\tfrac{3}{2}, \tfrac{3}{2}\rangle = \sqrt{3}\, |\tfrac{3}{2}, \tfrac{1}{2}\rangle = \sqrt{2}\, |1, 0\rangle\, |\tfrac{1}{2}, \tfrac{1}{2}\rangle + |1, 1\rangle\, |\tfrac{1}{2}, -\tfrac{1}{2}\rangle$$

$$\therefore\quad |\tfrac{3}{2}, \tfrac{1}{2}\rangle = \sqrt{\tfrac{2}{3}}\, |1, 0\rangle\, |\tfrac{1}{2}, \tfrac{1}{2}\rangle + \frac{1}{\sqrt{3}}\, |1, 1\rangle\, |\tfrac{1}{2}, -\tfrac{1}{2}\rangle$$

$$T^-\, |\tfrac{3}{2}, \tfrac{1}{2}\rangle = 2\, |\tfrac{3}{2}, -\tfrac{1}{2}\rangle = \frac{1}{\sqrt{3}}\, 2\, |1, -1\rangle\, |\tfrac{1}{2}, \tfrac{1}{2}\rangle + \frac{2}{\sqrt{3}}\, \sqrt{2}\, |1, 0\rangle\, |\tfrac{1}{2}, -\tfrac{1}{2}\rangle$$

$$\therefore\quad |\tfrac{3}{2}, -\tfrac{1}{2}\rangle = \frac{1}{\sqrt{3}}\, |1, -1\rangle\, |\tfrac{1}{2}, \tfrac{1}{2}\rangle + \sqrt{\tfrac{2}{3}}\, |1, 0\rangle\, |\tfrac{1}{2}, -\tfrac{1}{2}\rangle$$

$$T^-\, |\tfrac{3}{2}, -\tfrac{1}{2}\rangle = \sqrt{3}\, |\tfrac{3}{2}, -\tfrac{3}{2}\rangle = \sqrt{3}\, |1, -1\rangle\, |\tfrac{1}{2}, -\tfrac{1}{2}\rangle$$

$$\therefore\quad |\tfrac{3}{2}, -\tfrac{3}{2}\rangle = |1, -1\rangle\, |\tfrac{1}{2}, -\tfrac{1}{2}\rangle$$

We now construct a state $|\frac{1}{2}, \frac{1}{2}\rangle$ orthogonal to the $|\frac{3}{2}, \frac{1}{2}\rangle$ one as

$$|\tfrac{1}{2}, \tfrac{1}{2}\rangle = a\,|1, 1\rangle\,|\tfrac{1}{2}, -\tfrac{1}{2}\rangle + b\,|1, 0\rangle\,|\tfrac{1}{2}, \tfrac{1}{2}\rangle$$

with a and b real such that

$$a^2 + b^2 = 1$$

The requirement

$$\langle\tfrac{3}{2}, \tfrac{1}{2}|\, t_3\, |\tfrac{1}{2}, \tfrac{1}{2}\rangle > 0$$

fixes a and b such that

$$|\tfrac{1}{2}, \tfrac{1}{2}\rangle = \sqrt{\tfrac{2}{3}}\,|1, 1\rangle\,|\tfrac{1}{2}, -\tfrac{1}{2}\rangle - \frac{1}{\sqrt{3}}\,|1, 0\rangle\,|\tfrac{1}{2}, \tfrac{1}{2}\rangle$$

$$T^{-}|\tfrac{1}{2}, \tfrac{1}{2}\rangle = |\tfrac{1}{2}, -\tfrac{1}{2}\rangle = \frac{1}{\sqrt{3}}\,|1, 0\rangle\,|\tfrac{1}{2}, -\tfrac{1}{2}\rangle - \sqrt{\tfrac{2}{3}}\,|1, -1\rangle\,|\tfrac{1}{2}, \tfrac{1}{2}\rangle$$

Therefore the product representation space of $3 \times 2 = 6$ dimensions decomposes into the direct sum of a four-dimensional one ($T = \frac{3}{2}$) and a two dimensional one ($T = \frac{1}{2}$) or, symbolically,

$$D' \otimes D^{1/2} = D^{3/2} \oplus D^{1/2} \quad \textbf{(9.193)}$$

---

b. Clebsch-Gordan Theorem and Selection Rules

We now derive a generalization of Eq. 9.193 for the coupling of any two angular momenta j_1 and j_2. We begin by applying the rotation operator

$$R(\alpha\beta\gamma) = e^{-i\theta\hat{\mathbf{n}}\cdot\mathbf{J}} = e^{-\theta\hat{\mathbf{n}}\cdot\mathbf{J}^{(1)}}e^{-i\theta\hat{\mathbf{n}}\cdot\mathbf{J}^{(2)}}$$

to both sides of Eq. 9.189. If we use Eqs. 9.173 and 9.189 on this result, we obtain

$$\sum_{m_1', m_2'} D^{j_1}_{m_1' m_1}(\alpha\beta\gamma) D^{j_2}_{m_2' m_2}(\alpha\beta\gamma)\, |j_1 m_1'\rangle\, |j_2 m_2'\rangle$$

$$= \sum_{J,M} \langle JM; j_1 j_2 \,|\, j_1 m_1; j_2 m_2\rangle \sum_{M'} D^J_{M'M}(\alpha\beta\gamma)\, |JM'; j_1 j_2\rangle$$

$$= \sum_{J,M'} \langle JM; j_1 j_2 \,|\, j_1 m_1; j_2 m_2\rangle D^J_{M'M}(\alpha\beta\gamma)$$

$$\times \sum_{m_1', m_2'} \langle j_1 m_1'; j_2 m_2' \,|\, JM'; j_1 j_2\rangle\, |j_1 m_1'; j_2 m_2'\rangle$$

so that

$$D^{j}_{m_1' m_1}(\alpha\beta\gamma) D^{j_2}_{m_2' m_2}(\alpha\beta\gamma) = \sum_J \langle JM; j_1 j_2 \,|\, j_1 m_1; j_2 m_2\rangle$$

$$\times D^J_{M'M}(\alpha\beta\gamma) \langle j_1 m_1'; j_2 m_2' \,|\, JM'; j_1 j_2\rangle \quad \textbf{(9.194)}$$

where $M' = m_1' + m_2'$, $M = m_1 + m_2$. This is equivalent to a statement of the

Clebsch-Gordan theorem for $SU(2)$ or for $O^+(3)$,

$$D^{j_1} \otimes D^{j_2} = D^{j_1+j_2} \otimes D^{j_1+j_2-1} \otimes \cdots \oplus D^{|j_1-j_2|} \equiv \sum_{s=0}^{j_1+j_2-|j_1-j_2|} \oplus D^{j_1+j_2-s} \quad \textbf{(9.195)}$$

We can also derive some simple but important selection rules, the generalization of which we will give in Section 9.15. Suppose we are given an operator T that commutes with all of the angular momentum operators (i.e., it is *rotationally invariant*),

$$[T, J_j] = 0 \qquad j = 1, 2, 3 \quad \textbf{(9.196)}$$

From Eqs. 9.148 and 9.196 we easily see that

$$[T, J^2] = 0 = [T, J^{\pm}] \quad \textbf{(9.197)}$$

Equations 9.158–9.161 imply

$$0 \equiv \langle jm | [T, J^2] | j'm' \rangle = [j'(j'+1) - j(j+1)] \langle jm | T | j'm' \rangle$$

so that since $j \geq 0$ and $j' \geq 0$

$$\langle jm | T | j'm' \rangle = \delta_{jj'} \langle jm | T | jm' \rangle \quad \textbf{(9.198)}$$

Also we have

$$0 \equiv \langle jm | [T, J_3] | jm' \rangle = (m - m') \langle jm | T | jm' \rangle$$

so that

$$\langle jm | T | j'm' \rangle = \delta_{jj'} \, \delta_{mm'} \langle jm | T | jm \rangle \quad \textbf{(9.199)}$$

Finally we find that

$$0 \equiv \langle j, m+1 | [T, J^+] | j, m \rangle$$

$$= \sqrt{(j-m)(j+m+1)} \langle j, m+1 | T | j, m+1 \rangle - \sqrt{(j+m+1)(j-m)} \langle jm | T | jm \rangle \quad \textbf{(9.200)}$$

As long as $j \neq m$ we conclude that

$$\langle j, m+1 | T | j, m+1 \rangle = \langle jm | T | jm \rangle \quad \textbf{(9.201)}$$

If $j = m$, we obtain the desired result from $[T, J^-] = 0$. Equation 9.201 states that, for a fixed j, $\langle jm | T | jm \rangle$ is independent of m. Combining Eqs. 9.199 and 9.201 we obtain

$$\langle jm | T | j'm' \rangle = \delta_{jj'} \, \delta_{mm'} \langle j \| T \| j \rangle \equiv \delta_{jj'} \, \delta_{mm'} T^j \quad \textbf{(9.202)}$$

9.14 INTEGRATION OF ROTATION GROUP PARAMETERS

The discussion we now give of invariant integration over the parameters of a continuous group is applicable to a large class of continuous groups, not just to the rotation group. The main result of the section, Eq. 9.220, can also be obtained directly by more elementary means through the representation of the rotation matrices in terms of the Jacobi polynomials, as indicated below. Therefore the reader who is interested only in this orthogonality relation for the rotation group may pass over the subject of invariant integration.

For discrete groups we defined a sum over the group elements as

$$\sum_{R\in G} F(R)$$

for a function $F(R)$ of the group elements. In Section 9.3 we saw that this sum had the property

$$\sum_{S\in G} F(RS) = \sum_{S\in G} F(S)$$

which was essential for our derivations of the unitarity and the orthogonality of the irreducible representations of these groups. We now consider defining for continuous groups the analogue of such a sum. For continuous groups each element R is labeled by the r parameters of the group, $R = R(a_1, a_2, \ldots, a_r)$. If in our parameter space we let $g(R)$ denote the density of group elements in the neighborhood of R and if $F(R)$ is a function depending only on element R, then we define the *Hurwitz integral* as

$$\sum_R F(R) \to \int F(R)\, dR = \int F(R) g(R)\, da_1 \ldots da_r$$
$$= \int F[R(a_1, \ldots, a_r)] g[R(a_1, \ldots, a_r)]\, da_1 \cdots da_r \qquad \textbf{(9.203)}$$

where the integral runs over all the values of the parameters in the space. It should be clear that groups for which the "volume" integral in parameter space is positive and finite

$$0 < \int dR = \int g[R(a_1, \ldots, a_r)]\, da_1 \cdots da_r < \infty$$

are relatively easy to work with since we can then readily extend the unitarity and orthogonality relations for the irreducible representations, as well as Schur's lemma (Th. 9.3), to such continuous groups provided we can show that

$$\int F(SR)\, dR = \int F(R)\, dR \qquad \forall S \in G \qquad \textbf{(9.204)}$$

More explicitly, if $S = S(a_1', \ldots, a_r')$, then we have

$$R(a_1'', \ldots, a_r'') = S(a')R(a)$$

where $a_\lambda'' = \phi_\lambda(a'; a)$ (cf. Eq. 9.56). This means that

$$\int F[S(a_1', \ldots, a_r')R(a_1, \ldots, a_r)] g[R(a_1, \ldots, a_r)]\, da_1 \cdots da_r$$
$$= \int F[R(a_1'', \ldots, a_r'')] g[R(a_1, \ldots, a_r)]\, da_1 \cdots da_r$$
$$= \int F[R(a_1'', \ldots, a_r'')] g\{R[a_1(a', a''), \ldots, a_r(a', a'')]\} \left|\frac{\partial a}{\partial a''}\right| da_1'' \cdots da_r''$$

which we hope to make equal to

$$\int F[R(a_1'', \ldots, a_r'')]g[R(a_1'', \ldots, a_r'')]\, da_1'' \cdots da_r''$$

Here $|\partial a/\partial a''|$ is simply the Jacobian of transformation. This equality will hold provided

$$g\{R[a_1(a', a''), \ldots, a_r(a', a'')]\}\, |\partial a/\partial a''| \equiv g[R(a_1'', \ldots, a_r'')] \qquad \forall S, R \in G$$

If we define

$$\rho(a) \equiv g[R(a_1, \ldots, a_r)]$$

and take our reference element to be the identity element [i.e., $R(a=0)=I$], then we can write

$$\rho(a) \left| \left. \frac{\partial \phi(\alpha; a)}{\partial \alpha} \right|_{\alpha=0} \right| = \rho(0) \qquad \textbf{(9.205)}$$

where $\rho(0)$ is an arbitrary normalization constant and where the Jacobian in Eq. 9.205 is given explicitly as

$$\left| \left. \frac{\partial \phi(\alpha; a)}{\partial \alpha} \right|_{\alpha=0} \right| = \begin{vmatrix} \left. \frac{\partial \phi_1(\alpha; a)}{\partial \alpha_1} \right|_{\alpha=0} & \left. \frac{\partial \phi_2(\alpha; a)}{\partial \alpha_1} \right|_{\alpha=0} & \cdots & \left. \frac{\partial \phi_r(\alpha; a)}{\partial \alpha_1} \right|_{\alpha=0} \\ \left. \frac{\partial \phi_1(\alpha; a)}{\partial \alpha_2} \right|_{\alpha=0} & \left. \frac{\partial \phi_2(\alpha; a)}{\partial \alpha_2} \right|_{\alpha=0} & \cdots & \left. \frac{\partial \phi_r(\alpha; a)}{\partial \alpha_2} \right|_{\alpha=0} \\ \vdots & \vdots & \vdots & \vdots \\ \left. \frac{\partial \phi_1(\alpha; a)}{\partial \alpha_r} \right|_{\alpha=0} & \left. \frac{\partial \phi_2(\alpha; a)}{\partial \alpha_r} \right|_{\alpha=0} & \cdots & \left. \frac{\partial \phi_r(\alpha; a)}{\partial \alpha_r} \right|_{\alpha=0} \end{vmatrix} \qquad \textbf{(9.206)}$$

a. Orthogonality of the Rotation Matrices

Once we realize that such a density function exists, we can often find $\rho(a)$ by elementary means. Using this $\rho(a)$ we can then repeat the proofs of Ths. 9.2 and 9.3 to deduce that

$$\int D_{il}^{(\mu)}(R) D_{mj}^{*(\nu)}(R)\, dR = \frac{\delta_{\mu\nu}\, \delta_{im}\, \delta_{lj}}{n_\nu} \int dR \qquad \textbf{(9.207)}$$

and for the characters that

$$\int \chi^{(\mu)}(R) \chi^{*(\nu)}(R)\, dR = \frac{\delta_{\mu\nu}}{n_\mu} \int dR \qquad \textbf{(9.208)}$$

Let us apply this to find the characters of the three-dimensional rotation group $O^+(3)$. We know (cf. Problem 9.23) that any spatial rotation is simply equivalent to a rotation about an axis fixed in space. Let this rotation angle be ϕ, $0 \leq \phi < \pi$. We may therefore use this angle (and the two angles specifying the axis about which ϕ is measured) to label the *classes* of the elements of the rotation group. From Eq. 9.140 we know that such a rotation is given as

$$\begin{pmatrix} e^{-il\phi} & 0 & 0 & \cdots & 0 \\ 0 & e^{-i(l-1)\phi} & 0 & \cdots & 0 \\ 0 & 0 & e^{-i(l-2)\phi} & \cdots & 0 \\ \cdot & \cdot & \cdot & \cdot & \cdot \\ \cdot & \cdot & \cdot & \cdot & \cdot \\ \cdot & \cdot & \cdot & \cdot & \cdot \\ 0 & 0 & 0 & \cdots & e^{il\phi} \end{pmatrix} = D^l(\phi, 0, 0) \tag{9.209}$$

Therefore the character of this class is

$$\chi^l(\phi) = e^{-il\phi} \sum_{k=0}^{2l} (e^{i\phi})^k = e^{-il\phi} \frac{e^{i(2l+1)\phi} - 1}{e^{i\phi} - 1} = \frac{\sin[(l+\frac{1}{2})\phi]}{\sin(\phi/2)} \tag{9.210}$$

Since $l \geq 0$ and $l' \geq 0$, we want, in analogy with Eq. 9.29, to have

$$\int_0^\pi \frac{\sin[(l+\frac{1}{2})\phi]\sin[(l'+\frac{1}{2})\phi]}{\sin^2(\phi/2)} g(\phi)\, d\phi$$

$$= \int_0^\pi \{\cos[(l-l')\phi] - \cos[(l+l'+1)\phi]\}\left[\frac{g(\phi)\, d\phi}{2\sin^2(\phi/2)}\right] \equiv \delta_{ll'} \int_0^\pi g(\phi)\, d\phi \tag{9.211}$$

We see that a *consistent* choice is

$$g(\phi) \propto \sin^2(\phi/2) \tag{9.212}$$

Equation 9.212 can also be obtained directly from Eq. 9.206 (cf. Section A9.1).

It is important to realize here that this ϕ is *not* in general one of the Euler angles used in Fig. 9.2 to label the $D^l(\alpha\beta\gamma)$. Therefore, in terms of the group parameters α, β, γ used there, this is not the proper density function. Of course we could transform from the $(\theta', \varphi', \phi)$ of Fig. 9.5 to the Euler angles, (α, β, γ).

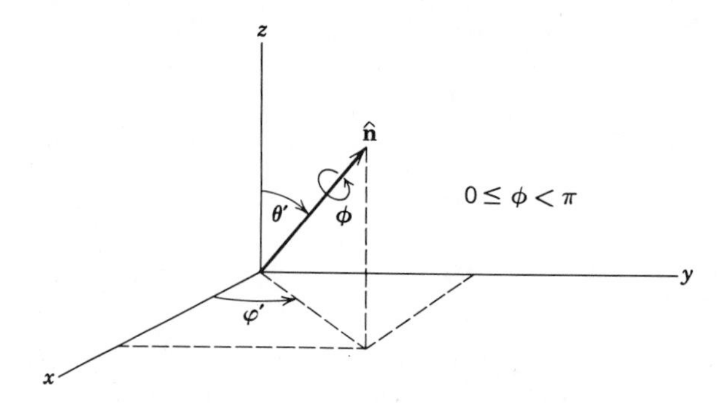

Figure 9.5

We then find that

$$g(\alpha\beta\gamma)=\frac{1}{8\pi^2}\sin\beta \tag{9.213}$$

so that

$$\frac{1}{8\pi^2}\int_0^{2\pi} d\alpha\int_0^{\pi}\sin\beta\, d\beta\int_0^{2\pi} d\gamma\, D^{*l}_{m_1'm_1}(\alpha\beta\gamma)D^{l'}_{m_2'm_2}(\alpha\beta\gamma)=\frac{\delta_{ll'}\,\delta_{m_1m_2}\,\delta_{m_1'm_2'}}{(2l+1)} \tag{9.214}$$

as we will now prove.

The angle of rotation, ϕ, about the fixed axis is given in terms of the Euler angles as

$$\cos\left(\frac{\phi}{2}\right)=\cos\left(\frac{\alpha+\gamma}{2}\right)\cos\left(\frac{\beta}{2}\right) \tag{9.215}$$

(cf. Problem 2.10). The rotation matrix in terms of the Euler angles is given as

$$R(\alpha\beta\gamma)=\begin{pmatrix}\cos\alpha\cos\beta\cos\gamma-\sin\alpha\sin\gamma & \sin\alpha\cos\beta\cos\gamma+\cos\alpha\sin\gamma & -\sin\beta\cos\gamma\\ -\cos\alpha\cos\beta\sin\gamma-\sin\alpha\cos\gamma & -\sin\alpha\cos\beta\sin\gamma+\cos\alpha\cos\gamma & \sin\beta\sin\gamma\\ \cos\alpha\sin\beta & \sin\alpha\sin\beta & \cos\beta\end{pmatrix} \tag{9.216}$$

The eigenvalues of this matrix are 1, $e^{\pm i\phi}$. We want the eigenvector corresponding to eigenvalue +1; that is

$$R(\alpha\beta\gamma)\chi(\theta',\varphi')=\chi(\theta',\varphi')=\begin{pmatrix}\sin\theta'\cos\varphi'\\ \sin\theta'\sin\varphi'\\ \cos\theta'\end{pmatrix}\equiv\begin{pmatrix}x_1\\ x_2\\ x_3\end{pmatrix} \tag{9.217}$$

We must find a $g(\alpha\beta\gamma)$ by transforming $\sin^2(\phi/2)$ such that

$$\int D^{l'}_{m_1m_1'}(\alpha\beta\gamma)D^{*l}_{m_2m_2'}(\alpha\beta\gamma)g(\alpha\beta\gamma)\, d\alpha\, d\beta\, d\gamma=\frac{\delta_{ll'}\,\delta_{m_1m_2}\,\delta_{m_1'm_2'}}{(2l+1)} \tag{9.218}$$

However, from the form of $D^{l}_{mm'}[\alpha\beta\gamma)$ given in Eq. 9.140, we see that these are orthogonal with respect to the m's if $g(\alpha\beta\gamma)$ is independent of α and γ. Therefore in computing $g(\beta)$ we shall take α and γ to be infinitesimal and keep only first-order terms in these variables.

Now if $\alpha=\gamma=0$, then $\phi=\beta$ and we would have a rotation about the y axis so that the solution to Eq. 9.217 would be $\theta'=\pi/2$, $\varphi'=\pi/2$. If we now take $\theta'=\pi/2+\varepsilon_1$, $\varphi'=\pi/2+\varepsilon_2$, where ε_1 and ε_2 are infinitesimal quantities, then

$$\chi(\theta',\varphi')\simeq\begin{pmatrix}-\varepsilon_2\\ 1\\ -\varepsilon_1\end{pmatrix}$$

so that, to first order, the eigenvalue problem of Eq. 9.217 reduces to

$$\sin\beta\varepsilon_1+(1-\cos\beta)\varepsilon_2=-\alpha\cos\beta-\gamma$$
$$(1-\cos\beta)\varepsilon_1-\sin\beta\varepsilon_2=-\alpha\sin\beta$$

the solution to which is

$$\varepsilon_1=\frac{-(\gamma+\alpha)}{2}\cot\left(\frac{\beta}{2}\right)$$
$$\varepsilon_2=\frac{-\gamma+\alpha}{2} \tag{9.219}$$

again to first order in α and γ. We easily find, from Eqs. 9.217 and 9.219, that

$$\frac{\partial\phi}{\partial\alpha}=0=\frac{\partial\phi}{\partial\gamma}\qquad\frac{\partial\phi}{\partial\beta}=1$$

$$\frac{\partial\varphi'}{\partial\alpha}=-\frac{\partial\varphi'}{\partial\gamma}=\frac{1}{2}\qquad\frac{\partial\varphi'}{\partial\beta}=0$$

$$\frac{\partial\theta'}{\partial\alpha}=\frac{\partial\theta'}{\partial\gamma}=-\frac{1}{2}\cot\left(\frac{\beta}{2}\right)$$

$$\frac{\partial\theta'}{\partial\beta}=\frac{(\gamma+\alpha)}{4\sin^2\left(\frac{\beta}{2}\right)}$$

so that

$$\left|\frac{\partial(\phi,\varphi',\theta')}{\partial(\alpha,\beta,\gamma)}\right|=\frac{1}{2}\cot\left(\frac{\beta}{2}\right)$$

which finally yields

$$\sin^2\left(\frac{\phi}{2}\right)d\phi\,d\varphi'\,d\theta'\equiv\sin^2\left(\frac{\beta}{2}\right)\left|\frac{\partial(\phi,\varphi',\theta')}{\partial(\alpha,\beta,\gamma)}\right|d\alpha\,d\beta\,d\gamma=\frac{1}{4}\sin\beta\,d\alpha\,d\beta\,d\gamma$$

Since we want the density function normalized as

$$\int_0^{2\pi}d\alpha\int_0^{\pi}d\beta\int_0^{2\pi}d\gamma\,g(\beta)=1$$

we have

$$g(\alpha\beta\gamma)\equiv g(\beta)=\frac{1}{8\pi^2}\sin\beta$$

as claimed in Eq. 9.213. A more careful derivation of this result is given in Section A9.2.

We can also obtain the result of Eq. 9.214 for arbitrary j (i.e., either integer or half odd integer) from Eqs. 9.140 and 9.142 plus the orthogonality relation for the Jacobi polynomials given in Eq. 7.191

$$\frac{1}{8\pi^2}\int_0^{2\pi} d\alpha \int_0^{\pi} d\beta \sin\beta \int_0^{2\pi} d\gamma D^{*j}_{m_1m_1'}(\alpha\beta\gamma)D^{j'}_{m_2m_2'}(\alpha\beta\gamma)$$

$$=\frac{1}{8\pi^2}\int_0^{2\pi} d\alpha \int_0^{\pi} d\beta \sin\beta \int_0^{2\pi} d\gamma e^{im_1\alpha} d^j_{m_1m_1'}(\beta)e^{im_1'\gamma}e^{-im_2\alpha}\, d^{j'}_{m_2m_2'}(\beta)e^{-im_2'\gamma}$$

$$=\tfrac{1}{2}\delta_{m_1m_2}\,\delta_{m_1'm_2'}\left[\frac{(j+m_1')!\,(j-m_1')!\,(j'+m_1')!\,(j'-m_1')!}{(j+m_1)!\,(j-m_1)!\,(j'+m_1)!\,(j'-m_1)!}\right]^{1/2}\int_{-1}^{1} d(\cos\beta)\left(\cos^2\frac{\beta}{2}\right)^{-m_1-m_1'}$$

$$\times\left(\sin^2\frac{\beta}{2}\right)^{m_1-m_1'} P^{(m_1-m_1',-m_1-m_1')}_{j+m_1'}(\cos\beta)P^{(m_1-m_1',-m_1-m_1')}_{j'+m_1'}(\cos\beta)$$

$$=\frac{\delta_{m_1m_2}\,\delta_{m_1'm_2'}\,\delta_{jj'}}{(2j+1)} \tag{9.220}$$

b. Completeness of the Rotation Matrices

It is clear that the general proof of the irreducibility of the $D^j(\alpha\beta\gamma)$ given in Section 9.11b shows that the $D^l(\alpha\beta\gamma)$ (l integer) are irreducible representations of $O^+(3)$. That there are no other continuous single-valued irreducible representations of $O^+(3)$ can be shown as follows. We have already seen in Eq. 9.211 that the characters of these irreducible representations, $\chi^l(\phi)$, are orthogonal. If there existed some other irreducible representation with a character $\chi(\phi)$, then this character would have to satisfy

$$\int_0^{\pi} d\phi(1-\cos\phi)\chi^l(\phi)\chi(\phi)=0 \qquad l=0, 1, 2, \ldots \tag{9.221}$$

However from Eqs. 9.210 and 9.221 we have

$$0=\int_0^{\pi} d\phi(1-\cos\phi)[\chi^{l+1}(\phi)-\chi^l(\phi)]\chi(\phi)$$

$$=2\int_0^{\pi} d\phi[(1-\cos\phi)\chi(\phi)]\cos(l\phi) \qquad l=1, 2, \ldots$$

Therefore with Eq. 9.221 for $l=0$ this would require that

$$\int_0^{\pi} d\phi[(1-\cos\phi)\chi(\phi)]\cos(l\phi)=0 \qquad l=0, 1, 2, \ldots \tag{9.222}$$

It is easy to see that $\{\cos l\phi\}$, $l=0, 1, \ldots$, $\phi\in[0, \pi]$ is a complete set so that Eq. 9.222 implies that $\chi(\phi)\equiv 0$. If we set $\phi=\frac{1}{2}(\theta+\pi)$, then Eq. 9.222 is of the form

$$0=\int_{-\pi}^{\pi} g(\phi)\cos\left[\frac{l}{2}(\theta+\pi)\right]d\theta=\begin{cases}(-1)^n\displaystyle\int_{-\pi}^{\pi} g(\theta)\cos(n\theta)\,d\theta & l=2n\\ -(-1)^n\displaystyle\int_{-\pi}^{\pi} g(\theta)\sin[(n+\tfrac{1}{2})\theta]\,d\theta & l=2n+1\end{cases}$$

We can subtract the expressions for $l=2n+1$ and $l=2(n+1)+1$ to obtain

$$0=\int_{-\pi}^{\pi}[g(\theta)\cos(\tfrac{1}{2}\theta)]\sin[(n+1)\theta]\,d\theta=0 \qquad n=0,1,2,\ldots$$

The expression for $l=2n$, that is,

$$\int_{-\pi}^{\pi} g(\theta)\cos(n\theta)\,d\theta=0 \qquad n=0,1,2,\ldots$$

implies that $g(\theta)$ must be an odd function of θ for $\theta\in[-\pi,\pi]$ (cf. Section 4.3a). Since $h(\theta)\equiv g(\theta)\cos(\frac{1}{2}\theta)$ must then also be an odd function, we conclude that $h(\theta)\equiv 0$, $\theta\in[-\pi,\pi]$, so that $g(\theta)\equiv 0$, $\theta\in[-\pi,\pi]$, as claimed.

Our conclusion that $\chi(\phi)\equiv 0$ for all ϕ implies that there are no other continuous irreducible representations of $O^+(3)$ since the character of the class of the identity cannot be zero (i.e., this character would just be the trace of the unit matrix).

It should be evident that this same argument can be used to prove that the $\{D^j(\alpha\beta\gamma)\}$ are all of the irreducible representations of $SU(2)$ where j is either integer or half odd integer. It is important to realize that the parameter ϕ can now vary over the range $0\leq\phi<2\pi$ [rather than $0\leq\phi<\pi$ as for $0^+(3)$].

9.15 TENSOR OPERATORS AND THE WIGNER-ECKART THEOREM

We find the law of transformation for an operator Q as follows. If

$$\psi'=O_R\psi \tag{9.223}$$

then we require that

$$Q'\psi'=(Q\psi)'=Q'O_R\psi=O_R(Q\psi)$$

so that

$$Q'O_R=O_RQ$$

or

$$Q'=O_RQ\,O_R^{-1} \tag{9.224}$$

An operator $\mathbf{V}$ with components V_i, $i=1,2,3$, is a vector operator provided it transforms as a vector; that is,

$$\mathbf{V}'=O_R\mathbf{V}O_R^{-1}=R\mathbf{V} \tag{9.225}$$

In component form Eq. 9.225 becomes

$$V_i'=\sum_j (R)_{ij}V_j \tag{9.226}$$

It is often more convenient to use spherical components defined as

$$\begin{aligned}\tilde{V}_1&\equiv-\frac{V_1+iV_2}{\sqrt{2}}\equiv-\frac{V_+}{\sqrt{2}}\\ \tilde{V}_0&=V_3\\ \tilde{V}_{-1}&\equiv\frac{V_1-iV_2}{\sqrt{2}}\equiv\frac{V_-}{\sqrt{2}}\end{aligned} \tag{9.227}$$

so that

$$\mathbf{V}\cdot\mathbf{W}=\sum_{m=-1}^{1}(-1)^m\tilde{V}_{-m}\tilde{W}_m \tag{9.228}$$

These quantities transform as $j=1$ eigenfunctions; that is, (cf. Eqs. 9.144, 9.216, and 9.173),

$$\tilde{V}'_m\equiv O_R V_m O_R^{-1}=\sum_{m'} D^1_{m'm}(R)\tilde{V}_{m'} \tag{9.229}$$

In general an *irreducible spherical tensor* of rank μ is defined as an operator with $2\mu+1$ components that transforms as

$$O_R T_m^{(\mu)} O_R^{-1}=\sum_{m'} D^{\mu}_{m'm}(R) T_{m'}^{(\mu)} \tag{9.230}$$

For $\mu=2$ this is different from a Cartesian tensor of rank two, which would have nine (i.e., 3^2) components and would be reducible since

$$T_{ij}\to D^1\otimes D^1=D^0+D^1+D^2$$

That is, this Cartesian tensor reduces to irreducible tensors of ranks 5, 3, and 1 (i.e., symmetric traceless, antisymmetric, and the trace).

Suppose we now want matrix elements of such an irreducible tensor of rank μ between states $|jm\rangle$ and $|j'm'\rangle$. From Eqs. 9.173 and 9.194, 9.223, and 9.230 we deduce that

$$\begin{aligned}\langle j'm'|\,T_k^{(\mu)}|jm\rangle &\equiv\langle O_R j'm',\,O_R T_k^{(\mu)} O_R^{-1} O_R jm\rangle\\ &=\sum_{\substack{m'',m''',\\k'}} D^{*j'}_{m''m'}(R) D^{\mu}_{k'k}(R) D^{j}_{m'''m}(R)\langle j'm''|\,T_{k'}^{(\mu)}\,|jm'''\rangle\\ &=\sum_{\substack{m'',m''',\\k',J}} D^{*j'}_{m''m'}(R)\langle \mu k',\,jm'''\,|JM';\,\mu j\rangle\langle JM;\,\mu j|\,\mu k;\,jm\rangle\\ &\quad\times D^{J}_{M'M}(R)\langle j'm''\,|T_{k'}^{(\mu)}\,|jm'''\rangle\end{aligned}$$

where $M'=m'''+k'$, $M=m+k$. If we now integrate over the group parameters and use the orthonormality relation of Eq. 9.220, we obtain

$$\langle j'm'|\,T_k^{(\mu)}|jm\rangle=\sum_{\substack{m'',m'''\\k',J}}\langle \mu k';\,jm'''\,|JM';\,\mu j\rangle\langle JM;\,\mu j\,|\,\mu k;\,jm\rangle \times\frac{\delta_{Jj'}\,\delta_{M'm''}\,\delta_{m'M}}{(2j'+1)}\langle j'm''|\,T_k^{(\mu)}|jm'''\rangle$$

or

$$\begin{aligned}\langle j'm'|\,T_k^{(\mu)}|jm\rangle &=\langle j',\,m+k;\,\mu j\,|\,\mu k;\,jm\rangle\,\delta_{m';m+k}\sum_{\substack{m'',m'''\\k'}}\\ &\quad\times\langle \mu k';\,jm'''\,|\,j',\,k'+m'';\,\mu j\rangle\,\delta_{m'''+k';m''}\langle j'm''\,|\,T_{k'}^{(\mu)}\,|\,jm'''\rangle\end{aligned} \tag{9.231}$$

Notice that the sum in Eq. 9.231, which is defined as the *reduced matrix element* $\langle j'\| T^{(\mu)} \|j\rangle$, is independent of m', k, and m and depends only on μ, j, and j'. We therefore have obtained the *Wigner-Eckart theorem*

$$\langle j'm'| T_k^{(\mu)} |jm\rangle = \langle j', m+k; \mu j \mid \mu k; jm\rangle\, \delta_{m';m+k} \langle j'\| T^{(\mu)} \|j\rangle \tag{9.232}$$

This is a generalization of Eq. 9.202, which was the expression for a spherical tensor of rank 0 (i.e., a scalar operator).

A9.1 A CALCULATION OF THE $O^+(3)$ INVARIANT INTEGRATION DENSITY FUNCTION IN TERMS OF THE CLASS PARAMETER

We begin by parameterizing the spatial rotations by the Cartesian components of the vector $\mathbf{w} = \phi\hat{\mathbf{n}}$ where $\hat{\mathbf{n}}$ is the unit vector left invariant by a rotation $R(\alpha\beta\gamma)$ (cf. Problem 9.23)

$$R(\alpha\beta\gamma)\hat{\mathbf{n}} = \hat{\mathbf{n}}$$

and ϕ is the angle of rotation about this axis $(0 \le \phi < \pi)$. If we take

$$R(0\phi 0) = \begin{pmatrix} \cos\phi & 0 & -\sin\phi \\ 0 & 1 & 0 \\ \sin\phi & 0 & \cos\phi \end{pmatrix} \tag{A9.1}$$

then

$$\hat{\mathbf{n}}_0 = \begin{pmatrix} 0 \\ 1 \\ 0 \end{pmatrix}$$

$$R(0\phi 0)\hat{\mathbf{n}}_0 = \hat{\mathbf{n}}_0$$

We now perform an infinitesimal rotation [cf. Eq. 9.216]

$$R(\varepsilon_1\varepsilon_2\varepsilon_3) = \begin{pmatrix} 1 & \varepsilon_3 & -\varepsilon_2 \\ -\varepsilon_3 & 1 & \varepsilon_1 \\ \varepsilon_2 & -\varepsilon_1 & 1 \end{pmatrix}$$

so that

$$R(\alpha'\beta'\gamma') = R(0\phi 0)R(\varepsilon_1\varepsilon_2\varepsilon_3) \tag{A9.2}$$

and seek an $\hat{\mathbf{n}}'$ such that

$$R(\alpha'\beta'\gamma')\hat{\mathbf{n}}' = \hat{\mathbf{n}}' \tag{A9.3}$$

By direct calculation we find to first order in the infinitesimals

$$R(\alpha'\beta'\gamma') = \begin{pmatrix} \cos\phi - \varepsilon_2\sin\phi & (\varepsilon_1+\varepsilon_3)\cos\phi & -\sin\phi - \varepsilon_2\cos\phi \\ -\varepsilon_1 - \varepsilon_3 & 1 & 0 \\ \sin\phi + \varepsilon_2\cos\phi & (\varepsilon_1+\varepsilon_3)\sin\phi & \cos\phi - \varepsilon_2\sin\phi \end{pmatrix} \tag{A9.4}$$

Since the eigenvalues of $R(\alpha'\beta'\gamma')$ are 1, $e^{\pm i\phi'}$, we see that $\mathrm{Tr}\, R(\alpha'\beta'\gamma') = 1+2\cos\phi'$ so that

$$\phi' = \phi + \varepsilon_2 \tag{A9.5}$$

If we set

$$\hat{\mathbf{n}}' = \begin{pmatrix} \delta_1 \\ 1+\delta_2 \\ \delta_3 \end{pmatrix} \tag{A9.6}$$

where the $\{\delta_j\}$ are infinitesimals, then Eq. A9.3 becomes, again to first order,

$$\begin{pmatrix} \delta_1 \cos\phi + (\varepsilon_1+\varepsilon_3)\cos\phi - \delta_3 \sin\phi \\ 1+\delta_2 \\ \delta_1 \sin\phi + (\varepsilon_1+\varepsilon_3)\sin\phi + \delta_3\cos\phi \end{pmatrix} = \begin{pmatrix} \delta_1 \\ 1+\delta_2 \\ \delta_3 \end{pmatrix} \tag{A9.7}$$

Since $|\hat{\mathbf{n}}'| = 1$ we see that

$$1+2\delta_2 = 1$$

or

$$\delta_2 = 0 \tag{A9.8}$$

If we solve Eq. A9.7 for δ_1 and δ_3 we obtain

$$\begin{aligned} \delta_1 &= \tfrac{1}{2}\varepsilon_1 \cot\left(\frac{\phi}{2}\right) - \tfrac{1}{2}\varepsilon_3 \\ \delta_3 &= \tfrac{1}{2}\varepsilon_1 + \tfrac{1}{2}\varepsilon_3 \cot\left(\frac{\phi}{2}\right) \end{aligned} \tag{A9.9}$$

This allows us to write

$$\mathbf{w} \equiv \phi'\hat{\mathbf{n}}' = \begin{pmatrix} \dfrac{\phi}{2}\varepsilon_1 \cot\left(\dfrac{\phi}{2}\right) - \dfrac{\phi}{2}\varepsilon_3 \\ \phi + \varepsilon_2 \\ \dfrac{\phi}{2}\varepsilon_1 + \dfrac{\phi}{2}\varepsilon_3 \cot\left(\dfrac{\phi}{2}\right) \end{pmatrix}$$

and then to compute the Jacobian of transformation

$$\left|\frac{\partial \mathbf{w}}{\partial \boldsymbol{\varepsilon}}\right| = \begin{vmatrix} \dfrac{\phi}{2}\cot\left(\dfrac{\phi}{2}\right) & 0 & -\dfrac{\phi}{2} \\ 0 & 1 & 0 \\ \dfrac{\phi}{2} & 0 & \dfrac{\phi}{2}\cot\left(\dfrac{\phi}{2}\right) \end{vmatrix}$$

$$= \frac{\phi^2}{4}\left[\cot^2\left(\frac{\phi}{2}\right) + 1\right] = \frac{\phi^2}{2(1-\cos\phi)} \equiv \left|\frac{\partial \mathbf{w}}{\partial \hat{\mathbf{n}}}\right| \tag{A9.11}$$

From Eqs. 9.205 and 9.206 we see that

$$\rho(\mathbf{w}) \propto \frac{2(1-\cos\phi)}{\phi^2} \tag{A9.12}$$

If we express $\mathbf{w}$ in Cartesian coordinates as

$$\mathbf{w} \equiv \phi\hat{\mathbf{n}} = (w_x, w_y, w_z)$$

and integrate over the group parameters in a spherical basis in $\hat{\mathbf{n}}$-space, we obtain

$$\int \rho(\mathbf{w})\, d\mathbf{w} = \int d\Omega_{\hat{\mathbf{n}}} \int_0^\pi \phi^2\, d\phi\, \frac{2(1-\cos\phi)}{\phi^2} = 8\pi \int_0^\pi (1-\cos\phi)\, d\phi = 8\pi^2$$

so that the invariant integration density function such that

$$\int_0^\pi g(\phi)\, d\phi = 1$$

is

$$g(\phi) = \frac{1}{\pi}(1-\cos\phi) = \frac{2}{\pi}\sin^2\left(\frac{\phi}{2}\right) \tag{A9.13}$$

A9.2 A DIRECT CALCULATION OF THE INVARIANT INTEGRATION DENSITY FUNCTION FOR $O^+(3)$

We now calculate the density function of Eq. 9.213 directly from Eqs. 9.205 and 9.206. We take as our reference element $(0, \pi/2, 0)$ since as we will see $(0, 0, 0)$ proves inconvenient because both sides of Eq. 9.205 vanish here.

$$R\left(0, \frac{\pi}{2}, 0\right) = \begin{pmatrix} 0 & 0 & -1 \\ 0 & 1 & 0 \\ 1 & 0 & 0 \end{pmatrix} \tag{A9.14}$$

We have for infinitesimal α_1, β_1, and γ_1

$$R\left(\alpha_1, \frac{\pi}{2}+\beta_1, \gamma_1\right) = \begin{pmatrix} -\beta_1 & \gamma_1 & -1 \\ -\alpha_1 & 1 & \gamma_1 \\ 1 & \alpha_1 & -\beta_1 \end{pmatrix} \tag{A9.15}$$

so that $R(\alpha_1, (\pi/2)+\beta_1, \gamma_1)$ followed by a *finite* $R(\alpha\beta\gamma)$ (cf. Eq. 9.216) is

$$R(\alpha'\beta'\gamma') = R(\alpha\beta\gamma) R\left(\alpha, \frac{\pi}{2}+\beta_1, \gamma_1\right) \tag{A9.16}$$

where

$$\begin{aligned} \alpha' &= \alpha'(\alpha_1, \beta_1, \gamma_1; \alpha, \beta, \gamma) \equiv \alpha_0 + \varepsilon_1 \\ \beta' &= \beta'(\alpha_1, \beta_1, \gamma_1; \alpha, \beta, \gamma) \equiv \beta_0 + \varepsilon_2 \\ \gamma' &= \gamma'(\alpha_1, \beta_1, \gamma_1; \alpha, \beta, \gamma) \equiv \gamma_0 + \varepsilon_3 \end{aligned} \tag{A9.17}$$

According to Eq. 9.206 we must eventually compute the Jacobian of transformation from α', β', γ' to α_1, β_1, γ_1.

If we use A9.14, 9.216, and A9.17 we find that

$$\begin{aligned}
&\cos\alpha_0 \sin\beta_0 = -\cos\beta \\
&\sin\alpha_0 \sin\beta_0 = \sin\alpha \sin\beta \\
&\cos\beta_0 = -\cos\alpha \sin\beta \\
&\sin\beta_0 \sin\gamma_0 = \cos\alpha \cos\beta \sin\gamma + \sin\alpha \sin\gamma \\
&\sin\beta_0 \cos\gamma_0 = \cos\alpha \cos\beta \cos\gamma - \sin\alpha \sin\gamma \\
&-\cos\alpha_0 \cos\beta_0 \sin\gamma_0 - \sin\alpha_0 \cos\gamma_0 = \sin\beta \sin\gamma
\end{aligned} \tag{A9.18}$$

If we next perform the matrix multiplication indicated in Eq. A9.16 and equate matrix elements, we obtain, after some algebra,

$$\begin{aligned}
&\varepsilon_2 \sin\beta_0 = \beta_1 \cos\beta - \gamma_1 \sin\alpha \sin\beta \\
&\varepsilon_1 \sin\alpha_0 \sin\beta_0 = \varepsilon_2 \cos\alpha_0 \cos\beta_0 + \beta_1 \cos\alpha \sin\beta + \alpha_1 \sin\alpha \sin\beta \\
&\varepsilon_3 \sin\beta_0 \sin\gamma_0 = \varepsilon_2 \cos\beta_0 \cos\gamma_0 + \gamma_1(\sin\alpha \cos\beta \cos\gamma + \cos\alpha \sin\gamma) \\
&\qquad + \beta_1 \sin\beta \cos\gamma \;\; \cos\gamma
\end{aligned} \tag{A9.19}$$

so that

$$\begin{aligned}
\varepsilon_1 &= \varepsilon_1(\alpha_1, \beta_1, \gamma_1) \\
\varepsilon_2 &= \varepsilon_2(\beta_1, \gamma_1) \\
\varepsilon_3 &= \varepsilon_3(\beta_1, \gamma_1)
\end{aligned} \tag{A9.20}$$

The Jacobian therefore simplifies to

$$\left|\frac{\partial(\alpha', \beta', \gamma')}{\partial(\alpha_1, \beta_1, \gamma_1)}\right| = \frac{\partial \varepsilon_1}{\partial \alpha_1} \begin{vmatrix} \dfrac{\partial \varepsilon_2}{\partial \beta_1} & \dfrac{\partial \varepsilon_3}{\partial \beta_1} \\ \dfrac{\partial \varepsilon_2}{\partial \gamma_1} & \dfrac{\partial \varepsilon_3}{\partial \gamma_1} \end{vmatrix}$$

From Eqs. A9.18 and A9.19 we find that

$$\frac{\partial \varepsilon_1}{\partial \alpha_1} = \frac{\sin\alpha \sin\beta}{\sin\alpha_0 \sin\beta_0} = 1$$

$$\frac{\partial \varepsilon_2}{\partial \beta_1} = \frac{\cos\beta}{\sin\beta_0}$$

$$\frac{\partial \varepsilon_2}{\partial \gamma_1} = -\frac{\sin\alpha \sin\beta}{\sin\beta_0}$$

$$\frac{\partial \varepsilon_3}{\partial \beta_1} = \frac{1}{\sin\beta_0 \sin\gamma_0}\left(\frac{\cos\beta}{\sin\beta_0} \cos\beta_0 \cos\gamma_0 + \sin\beta \cos\gamma\right)$$

$$\frac{\partial \varepsilon_3}{\partial \gamma_1} = \frac{1}{\sin\beta_0 \sin\gamma_0}\left(-\frac{\sin\alpha \sin\beta}{\sin\beta_0} \cos\beta_0 \cos\gamma_0 + \sin\alpha \cos\beta \cos\gamma + \cos\alpha \sin\gamma\right)$$

so that, from Eq. A9.18,

$$\left|\frac{\partial(\alpha', \beta', \gamma')}{\partial(\alpha_1, \beta_1, \gamma_1)}\right| = \frac{\sin\alpha \sin\beta}{\sin\alpha_0 \sin^3\beta_0 \sin\gamma_0}(\sin\alpha\cos\gamma + \cos\alpha\sin\gamma\cos\beta)$$

$$= \frac{1}{\sin\beta_0} \tag{A9.21}$$

From Eqs. 9.205 and 9.206 we finally obtain

$$g(\alpha\beta\gamma) \propto \sin\beta \tag{A9.22}$$

SUGGESTED REFERENCES

The following general references on group theory will provide a useful introduction to and background on the subject.

L. P. Eisenhart, *Continuous Groups of Transformations.* This is a very mathematical and rigorous discussion of continuous groups of transformations. Section 15 of Chapter I presents a proof of Lie's theorem, mentioned in Section 9.8 of Chapter 9, that the infinitesimal form of a Lie transformation determines the finite form.

M. Hamermesh, *Group Theory and Its Applications to Physical Problems.* This is one of the best and most readable texts on abstract group theory and specific applications to the symmetric, rotation, and Lorentz groups.

G. Racah, *Group Theory and Spectroscopy.* This famous set of lectures, originally delivered in 1951 and long out of print, has finally become available in Band 37 of *Ergebnisse der exakten Naturwissenschaften.* Lecture 1 contains an extremely clear and beautiful presentation of the basic theory of Lie groups.

M. Tinkham, *Group Theory and Quantum Mechanics.* This provides an elementary treatment of the applications of group theory to quantum mechanics.

E. P. Wigner, *Group Theory.* This is a classic, seminal text on the rotation group and its application to quantum mechanics.

The two references listed below are very useful and thorough works on the detailed applications and calculations involved in applying the rotation group to quantum mechanical systems.

A. R. Edmonds, *Angular Momentum in Quantum Mechanics.*

M E. Rose, *Elementary Theory of Angular Momentum.*

J. D. Talman, *Special Functions, A Group Theoretic Approach.* This book, based on lectures by E. P. Wigner, presents an introduction to the theory of Lie groups and then approaches the topic of the classical functions of applied mathematics via the representations of these groups.

PROBLEMS

9.1 Prove that the left and right identity elements are equal and unique. Then show that the left and right inverses are equal and unique using the axioms listed in Section 9.1.

9.2 Consider group S_4 (i.e., the group of permutations of four objects). Show that the cyclic permutations form a subgroup H of this group. Decompose S_4 into left cosets and then into right cosets. Is H an invariant subgroup?

9.3 Consider elements $H_4 = C_1 + C_2$ of Example 9.12. Verify that C_1 and C_2 are indeed classes. Prove that this invariant subgroup is isomorphic to the abstract group of Example 9.11(iv) (cf. Table 9.3). Show that the factor group S_4/H_4 is isomorphic to S_3.

9.4 Prove that all irreducible representations of a finite abelian group are one-dimensional. Then take the 2×2 matrix representation of rotations in two dimensions and reduce it into a direct sum.

9.5 Consider the three element group.

a. How many irreducible representations are there?

b. List all of the characters for each irreducible representation.

c. List the distinct irreducible representations.

9.6 Prove that for a finite group two representations (reducible or not) are equivalent if and only if all of the characters of one group are identical to all of the corresponding characters of the other group.

9.7 Write down the regular representation of the group of order five and reduce it fully in terms of the irreducible representations of the group.

9.8 a. Give a convincing argument that all groups of order *less than six* must be abelian.

b. Use the results of Example 9.10 to find the number of inequivalent irreducible representations of S_3. What are the dimensions of the irreducible representations?

c. Find all of these inequivalent irreducible representations explicitly in terms of matrices. You should not attack this problem blindly. Take a few minutes and consider the following facts.

i. All of the irreducible representations of an abelian group (or subgroup) are one dimensional.

ii. Given any irreducible representation of the full group, those corresponding to an abelian subgroup can be simultaneously diagonalized (why?).

d. Which, if any, of these irreducible representations is faithful?

e. Write down a representation equivalent to the regular representation of S_3.

9.9 Consider the group built up from the unit matrix and the Pauli spin matrices. Take the elements of this group to be $\pm I$, $\pm i\sigma_1$, $\pm i\sigma_2$, and $\pm i\sigma_3$ where

$$\sigma_1 = \begin{pmatrix} 0 & 1 \\ 1 & 0 \end{pmatrix} \qquad \sigma_2 = \begin{pmatrix} 0 & -i \\ i & 0 \end{pmatrix} \qquad \sigma_3 = \begin{pmatrix} 1 & 0 \\ 0 & -1 \end{pmatrix}$$

What is the group multiplication table? Find all the subgroups and classes. Are there any invariant subgroups? How many inequivalent irreducible representations are there? What dimensions do they have?

9.10 Consider an arbitrary group G with a subgroup H. Show that if H of index two then H is invariant.

9.11 Find the subgroup of S_5 to which the cyclic group of order five is isomorphic. Demonstrate the isomorphism explicitly by performing all the relevant multiplications of the elements of S_5.

9.12 Let G and G' be two groups. Show that in a homomorphism of G onto G', the elements of G that are mapped into the identity of G' form an invariant subgroup, H, of G.

9.13 Show that group G' of the previous problem is isomorphic to factor group G/H where H is as defined in the previous problem.

9.14 By direct calculation show that

$$\nabla^2 = \frac{\partial^2}{\partial x^2} + \frac{\partial^2}{\partial y^2} + \frac{\partial^2}{\partial z^2}$$

is invariant under *any* spatial rotation.

9.15 Show that any 2×2 matrix that commutes with the three Pauli matrices must be a multiple of the unit matrix.

9.16 Show that the most general 2×2 unitary matrix whose eigenvalues are 1 and -1 can be written as

$$U = \begin{pmatrix} \cos\theta & e^{i\phi}\sin\theta \\ e^{-i\phi}\sin\theta & -\cos\theta \end{pmatrix}$$

9.17 Let:

$$I = \begin{pmatrix} 1 & 0 \\ 0 & 1 \end{pmatrix} \quad A = \begin{pmatrix} 0 & 1 \\ -1 & 0 \end{pmatrix} \quad B = \begin{pmatrix} -1 & 0 \\ 0 & -1 \end{pmatrix} \quad C = \begin{pmatrix} 0 & -1 \\ 1 & 0 \end{pmatrix}$$

Show that these matrices are reducible. Reduce them.

9.18 Consider three identical objects (labeled α, β, γ) at the vertices of an equilateral triangle. This configuration is left invariant under successive rotations of 120° about the z axis and by ones of 180° about the axes labeled 1, 2, 3 in the x-y plane (cf. Fig. 9.6). Let

$$G_1 = R_z(0) = \begin{pmatrix} 1 & 0 \\ 0 & 1 \end{pmatrix} \qquad G_2 = R_z(2\pi/3) = \begin{pmatrix} -1/2 & -\sqrt{3}/2 \\ \sqrt{3}/2 & -1/2 \end{pmatrix}$$

$$G_3 = R_z(4\pi/3) = \begin{pmatrix} -1/2 & \sqrt{3}/2 \\ -\sqrt{3}/2 & -1/2 \end{pmatrix} \qquad G_4 = R_1(\pi) = \begin{pmatrix} -1 & 0 \\ 0 & 1 \end{pmatrix}$$

$$G_5 = R_2(\pi) = \begin{pmatrix} 1/2 & -\sqrt{3}/2 \\ -\sqrt{3}/2 & -1/2 \end{pmatrix} \qquad G_6 = R_3(\pi) = \begin{pmatrix} 1/2 & \sqrt{3}/2 \\ \sqrt{3}/2 & -1/2 \end{pmatrix}$$

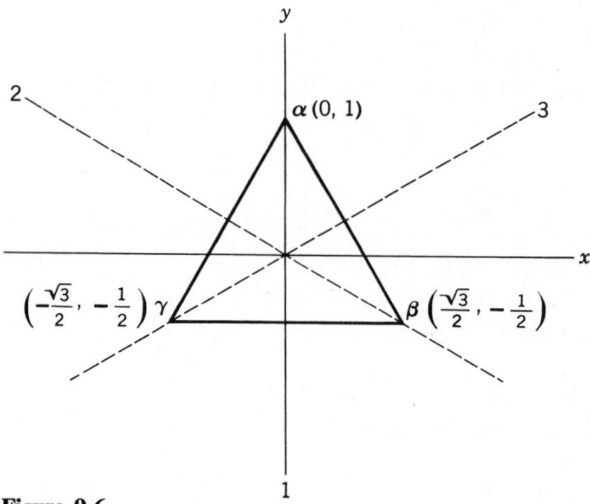

Figure 9.6

where

$$R_z(\theta) \equiv \begin{pmatrix} \cos\theta & -\sin\theta \\ \sin\theta & \cos\theta \end{pmatrix}$$

since we are rotating the objects (or, equivalently, the vectors locating them) counterclockwise about the various axes (cf. Section 2.6a, Eq. 2.86).

a. Construct the group multiplication table for this group (known as D_3, the *dihedral group* with a threefold axis of symmetry).
b. Prove that the representation given is irreducible.
c. Prove that (I, G_2, G_3) is a subgroup. Is it invariant?
d. Prove that (I, G_4), (I, G_5), and (I, G_6) all form two-element subgroups but that none of these is an invariant subgroup of D_3.

9.19 Group D_3 of problem 9.18 may be considered as a permutation group of three objects. That is, G_2 rotates β into α, α into γ, and γ into β. In three dimensions this is

$$\begin{pmatrix} 0 & 1 & 0 \\ 0 & 0 & 1 \\ 1 & 0 & 0 \end{pmatrix}\begin{pmatrix} \alpha \\ \beta \\ \gamma \end{pmatrix} = \begin{pmatrix} \beta \\ \gamma \\ \alpha \end{pmatrix}$$

a. Develop 3×3 representations for the rest of D_3.
b. Reduce these 3×3 representations to the 2×2 ones of problem 9.18 (cf. Problem 9.8).

9.20 Show that an $n\times n$ real orthogonal matrix has $n(n-1)/2$ *independent* parameters.

9.21 Show that an $n\times n$ special unitary matrix has n^2-1 *independent* (real) parameters.

9.22 The unimodular (or *special linear*) group of Section 9.6e, $C(2)$, consists of all 2×2 matrices (with complex elements) having determinant $+1$. Show that such matrices form a group.

9.23 Prove Euler's theorem on the rotation of a rigid body; that is, that any pure spatial rotation (or successive series of pure spatial rotations) is equivalent to a single rotation about some axis fixed in space.

9.24 Prove whether or not $R_z(\varphi)$ is an invariant subgroup of $O^+(3)$. Also show that the statement of Example 9.8 about the classes of the rotation group is correct. (*Hint:* Use Eq. 9.127 and prove that $(\boldsymbol{\alpha}\cdot\boldsymbol{\sigma})(\boldsymbol{\beta}\cdot\boldsymbol{\sigma}) = (\boldsymbol{\alpha}\cdot\boldsymbol{\beta}) + i(\boldsymbol{\alpha}\times\boldsymbol{\beta})\cdot\boldsymbol{\sigma}$ where $\boldsymbol{\alpha}$ and $\boldsymbol{\beta}$ are any two vectors.)

9.25 Take $D^l(\alpha\beta\gamma)$ to be unitary and show that

$$\sum_{m=-l}^{l} Y_l^{m*}(\theta, \varphi) Y_l^m(\theta', \varphi')$$

is a scalar under spatial rotations. Use this result to obtain the addition theorem for spherical harmonies

$$P_l(\cos\gamma) = \frac{4\pi}{(2l+1)} \sum_{m=-l}^{l} Y_l^{m*}(\theta, \varphi) Y_l^m(\theta', \varphi')$$

where γ is the angle between two vectors with directions (θ, φ) and (θ', φ'). Compare this with the result of Eq. 9.145 with $m=0$ (cf. Eq. 4.47) and Eq. 9.178.

9.26 Prove that the general $SU(2)$ element can be written as (cf. Problem 9.16)

$$U(\xi, \eta, \zeta) = \begin{pmatrix} e^{i\xi}\cos\eta & e^{i\zeta}\sin\eta \\ -e^{-i\zeta}\sin\eta & e^{-i\xi}\cos\eta \end{pmatrix}$$

This is constructed by the following counterclockwise rotations:

i. A rotation of ϕ about the z axis.
ii. A rotation of θ about the new x' axis.
iii. A rotation of ψ about the new z'' axis.

Use Eq. 9.127 to show that

$$\phi = \xi - \zeta + \frac{\pi}{2} = \alpha + \frac{\pi}{2}$$

$$\theta = 2\eta = \beta$$

$$\psi = \xi + \zeta - \frac{\pi}{2} = \gamma - \frac{\pi}{2}$$

where the parameters α, β, γ are those of Eq. 9.129.

9.27 Compute the direct product $D^{1/2}(\alpha\beta\gamma) \otimes D^{1/2}(\alpha\beta\gamma)$ explicitly and then find a constant 4×4 unitary matrix S that will reduce this direct product to the block-diagonal form $D^0(\alpha\beta\gamma) \oplus D^1(\alpha\beta\gamma)$.

9.28 Prove that conjugate representations of the group $C(2)$ of Section 9.6e are inequivalent.

Appendix I

Elementary Real Analysis

In this appendix we list some results familiar from elementary real analysis (or advanced calculus). Complete proof or full discussion will not always be given since the purpose of the appendix is to facilitate reference to and review of facts the reader is assumed to have seen previously. The texts by Buck, Kaplan, and Lass cited in the bibliography contain complete treatments of these subjects.

A function is said to be *piecewise continuous* on an interval provided it is continuous on that interval with the exception of a finite number of points at which the function undergoes finite discontinuities.

A point of a set is a *limit point* of the set provided every neighborhood of this point contains infinitely many points of the set.

Theorem I.1 (Bolzano-Weierstrass theorem). Every bounded infinite set has at least one limit point.

Theorem I.2. Every bounded monotone sequence converges.

Theorem I.3. If $f(x)$ is continuous on a closed finite interval $[a, b]$, then it is uniformly continuous there.

A function $f(x, y)$ is said to be *continuous* in a region R if for every $(x_0, y_0) \in R$ and for any given $\varepsilon > 0$ it is possible to find a $\delta > 0$ so that whenever

$$0 < (x - x_0)^2 + (y - y_0)^2 < \delta$$

it follows that

$$|f(x, y) - f(x_0, y_0)| < \varepsilon$$

In Section 7.12b we needed to know the nature of the extremal points of a real function of two real variables, $f(x, y)$. That is, we are given that at (x_0, y_0)

$$\left.\frac{\partial f}{\partial x}\right|_{(x_0, y_0)} = 0 = \left.\frac{\partial f}{\partial y}\right|_{(x_0, y_0)}$$

From Taylor's theorem we know that

$$\Delta f \equiv f(x_0+\Delta x, y_0+\Delta y) - f(x_0, y_0) \simeq \alpha_{11}(\Delta x)^2 + (\alpha_{12}+\alpha_{21})\,\Delta x \Delta y + \alpha_{22}(\Delta y)^2$$

where

$$\alpha_{11} = \left.\frac{\partial^2 f}{\partial x^2}\right|_{(x_0,y_0)} \qquad \alpha_{12} = \alpha_{21} = \left.\frac{\partial^2 f}{\partial x\,\partial y}\right|_{(x_0,y_0)} \qquad \alpha_{22} = \left.\frac{\partial^2 f}{\partial y^2}\right|_{(x_0,y_0)}$$

The point (x_0, y_0) will be a relative maximum provided $\Delta f < 0$ for all Δx, Δy in the neighborhood of (x_0, y_0) and a relative minimum provided $\Delta f > 0$ for all Δx, Δy. This means that we must examine the real quadratic form

$$\langle A \rangle = \sum_{i,j=1}^{2} \alpha_{ij} \xi_i \xi_j \qquad \xi_1^2 + \xi_2^2 = 1$$

Since $A = \{\alpha_{ij}\}$ is a real self-adjoint matrix, we can express $\langle A \rangle$ in diagonal form as (cf. Section 3.3)

$$\langle A \rangle = \lambda_1 \eta_1^2 + \lambda_2 \eta_2^2 = (\lambda_1 - \lambda_2)\eta_1^2 + \lambda_2 \qquad \eta_1^2 + \eta_2^2 = 1 \tag{I.1}$$

with

$$\lambda_1 = \frac{\alpha_{11} + \alpha_{22} + \sqrt{(\alpha_{11} - \alpha_{22})^2 + 4(\alpha_{12})^2}}{2}$$

$$\lambda_2 = \frac{\alpha_{11} + \alpha_{22} - \sqrt{(\alpha_{11} - \alpha_{22})^2 + 4(\alpha_{12})^2}}{2}$$

Since Eq. I.1 shows that $\langle A \rangle$ is a linear function of η_1^2 and since

$$\lambda_1 - \lambda_2 = \sqrt{(\alpha_{11} - \alpha_{22})^2 + 4(\alpha_{12})^2} \geq 0$$

we see that $\langle A \rangle$ attains its maximum value when $\eta_1^2 = 1$,

$$\langle A \rangle_{\text{Max}} = \lambda_1$$

and its minimum value when $\eta_1^2 = 0$,

$$\langle A \rangle_{\text{Min}} = \lambda_2$$

The condition for a relative maximum then becomes

$$\lambda_1 < 0$$

or

$$\alpha_{11} + \alpha_{22} < 0$$

$$(\alpha_{11} + \alpha_{22})^2 > (\alpha_{11} - \alpha_{22})^2 + 4(\alpha_{12})^2$$

The conditions that (x_0, y_0) be a relative maximum for $f(x, y)$ are then

$$\alpha_{11} + \alpha_{22} < 0$$

$$(\alpha_{12})^2 - \alpha_{11}\alpha_{22} < 0$$

Similarly the condition for a relative minimum is

$$\lambda_2 > 0$$

or

$$\alpha_{11} + \alpha_{22} > 0$$

$$(\alpha_{12})^2 - \alpha_{11}\alpha_{22} < 0$$

If $\lambda_1>0$ when $\lambda_2\leq 0$, or $\lambda_1\geq 0$ when $\lambda_2<0$, then $\langle A\rangle$ changes sign as $0\rightarrow\eta_1^2\rightarrow 1$ and (x_0, y_0) is a saddle point or minimax. Either of these conditions implies that

$$(\alpha_{12})^2-\alpha_{11}\alpha_{22}>0$$

is true when (x_0, y_0) is a saddle point. Finally if $\lambda_1=\lambda_2=0$, or equivalently

$$(\alpha_{12})^2-\alpha_{11}\alpha_{22}=0$$

then the nature of the critical point (x_0, y_0) is undetermined and we must examine higher order terms in the Taylor expansion of $f(x, y)$. We state this useful result as a theorem.

Theorem I.4. If $f(x, y)$ is continuous with continuous first and second partial derivatives in a region containing the point (x_0, y_0) for which

$$\left.\frac{\partial f}{\partial x}\right|_{(x_0,y_0)}=0=\left.\frac{\partial f}{\partial y}\right|_{(x_0,y_0)}$$

and if

$$\alpha_{11}\equiv\left.\frac{\partial^2 f}{\partial x^2}\right|_{(x_0,y_0)}\qquad \alpha_{12}\equiv\left.\frac{\partial^2 f}{\partial x\,\partial y}\right|_{(x_0,y_0)}\qquad \alpha_{22}\equiv\left.\frac{\partial^2 f}{\partial y^2}\right|_{(x_0,y_0)}$$

then:

i. (x_0, y_0) is a relative maximum when $(\alpha_{12})^2-\alpha_{11}\alpha_{22}<0$ and $\alpha_{11}+\alpha_{22}<0$.
ii. (x_0, y_0) is a relative minimum when $(\alpha_{12})^2-\alpha_{11}\alpha_{22}<0$ and $\alpha_{11}+\alpha_{22}>0$.
iii. (x_0, y_0) is a saddle point or minimax when $(\alpha_{12})^2-\alpha_{11}\alpha_{22}>0$.
iv. (x_0, y_0) remains of undetermined nature when $(\alpha_{12})^2-\alpha_{11}\alpha_{22}=0$.

Theorem I.5a. (First mean value theorem for integrals): If $f(x)$ and $g(x)$ are continuous and bounded in $[a, b]$ and if $g(x)>0$, $\forall x\in[a, b]$, then

$$\int_a^b f(x)g(x)\,dx=f(\bar{x})\int_a^b g(x)\,dx \qquad a\leq\bar{x}\leq b$$

PROOF. We can easily obtain this result from the mean value theorem for a continuous differentiable function $F(x)$,

$$F(b)-F(a)=F'(\bar{x})(b-a) \qquad a\leq\bar{x}\leq b$$

Now let

$$dy=g(x)\,dx$$
$$h(y)=f[x(y)]$$
$$\beta=y(b)$$
$$\alpha=y(a)$$
$$F(y)=\int_\alpha^y h(y')\,dy'$$

so that

$$F(\beta)-F(\alpha)\equiv\int_a^b f(x)g(x)\,dx=\left.\frac{dF}{dy}\right|_{\bar{y}}(\beta-\alpha)=\left.\left(\frac{dF}{dx}\frac{dx}{dy}\right)\right|_{\bar{y}}(\beta-\alpha)$$

$$=f(\bar{x})(\beta-\alpha)=f(\bar{x})\int_\alpha^\beta dy=f(\bar{x})\int_a^b g(x)\,dx \qquad \text{Q.E.D.}$$

Theorem I.5b. (Second mean value theorem for integrals): If $f(x)$ and $k(x)$ are integrable on $x\in[a, b]$ and if $k(x)$ is a positive nondecreasing function there, then there exists a ξ, $a \le \xi \le b$, such that

$$\int_a^b f(x)k(x)\,dx = k(b)\int_\xi^b f(x)\,dx$$

PROOF. We give a proof only for those functions $f(x)$ and $k(x)$ that are bounded and piecewise continuous (having at most a finite number of finite discontinuities). If $f(x)$ is of one sign on $[a, b]$ (say positive), then the result is obvious since

$$F(\xi) \equiv k(b)\int_\xi^b f(x)\,dx$$

is a positive continuous increasing function of ξ, $\xi \in [a, b]$, beginning at zero when $\xi = b$ and increasing to $k(b)\int_a^b f(x)\,dx$ when $\xi = a$. However we easily see that

$$\int_a^b k(x)f(x)\,dx \le k(b)\int_a^b f(x)\,dx$$

and

$$\int_a^b k(x)f(x)\,dx \ge k(a)\int_a^b f(x)\,dx \ge 0$$

so that

$$0 \le \int_a^b k(x)f(x)\,dx \le k(b)\int_a^b f(x)\,dx$$

or

$$F(\xi)_{\min} \le \int_a^b k(x)f(x)\,dx \le F(\xi)_{\max}$$

It then follows that

$$\int_a^b k(x)f(x)\,dx = \int_\xi^b f(x)\,dx$$

for some ξ, $a \le \xi \le b$.

Now suppose that $f(x)$ changes sign just once on $a \le x \le b$, say at x_0, and assume $f(x) \ge 0$, $x_0 \le x \le b$ [if $f(x) \le 0$, then make the following argument for $-f(x)$]. We then have

$$\int_a^b f(x)\,dx = -\int_a^{x_0} |f(x)|\,dx + \int_{x_0}^b |f(x)|\,dx$$

Also in this case,

$$\text{Max}\left[k(b)\int_\xi^b f(x)\,dx\right] = k(b)\int_{x_0}^b |f(x)|\,dx \qquad a \le \xi \le b$$

and

$$\text{Min}\left[k(b)\int_\xi^b f(x)\,dx\right] = \begin{cases} k(b)\int_a^b f(x)\,dx & \text{if this is negative} \\ 0 & \text{otherwise} \end{cases}$$

But since

$$\begin{aligned}\int_a^b k(x)f(x)\,dx &= -\int_a^{x_0} k(x)\,|f(x)|\,dx + \int_{x_0}^b k(x)\,|f(x)|\,dx \\ &\le -k(a)\int_a^{x_0} |f(x)|\,dx + k(b)\int_{x_0}^b |f(x)|\,dx \\ &\le k(b)\int_{x_0}^b |f(x)|\,dx \equiv \text{Max}\left[k(b)\int_\xi^b f(x)\,dx\right]\end{aligned}$$

and

$$\int_a^b k(x)\,f(x)\,dx = -\int_a^{x_0} k(x)\,|f(x)|\,dx + \int_{x_0}^b k(x)\,|f(x)|\,dx$$

$$\geq -k(x_0^-)\int_a^{x_0} |f(x)|\,dx + k(x_0^+)\int_{x_0}^b |f(x)|\,dx$$

$$= k(x_0^+)\left[\int_{x_0}^b |f(x)|\,dx - \int_a^{x_0} |f(x)|\,dx\right]$$

$$+[k(x_0^+)-k(x_0^-)]\int_a^{x_0} |f(x)|\,dx$$

$$\geq k(x_0^+)\int_a^b f(x)\,dx \begin{cases} \geq 0 & \text{if } \int_a^b f(x)\,dx \geq 0 \\ \geq k(b)\int_a^b f(x)\,dx & \text{if } \int_a^b f(x)\,dx < 0 \end{cases}$$

we see that

$$\text{Min}\left[k(b)\int_\xi^b f(x)\,dx\right] \leq \int_a^b k(x)f(x)\,dx \leq \text{Max}\left[k(b)\int_\xi^b f(x)\,dx\right]$$

so that

$$\int_a^b k(x)f(x)\,dx = k(b)\int_\xi^b f(x)\,dx, \qquad \text{some } \xi, \qquad a \leq \xi \leq b$$

The extension to the general case when $f(x)$ changes sign an arbitrary (but finite) number of times on $[a, b]$ is now straightforward. Q.E.D.

Theorem I.6 (Mean value theorem for two variables). If $f(x, y)$ is continuous, having continuous first partial derivatives, in a region R, then for any points $(x+\Delta x, y+\Delta y)$ and (x, y) in R there exists a point $(\bar{x}, \bar{y})$, $x \leq \bar{x} \leq x+\Delta x$, $y \leq \bar{y} \leq y+\Delta y$, such that

$$f(x+\Delta x, y+\Delta y) - f(x, y) = \left.\frac{\partial f}{\partial x}\right|_{(\bar{x},\bar{y})} \Delta x + \left.\frac{\partial f}{\partial y}\right|_{(\bar{x},\bar{y})} \Delta y$$

PROOF. Let α be a real variable, $0 \leq \alpha \leq 1$, and define

$$F(\alpha) = f(x+\alpha\,\Delta x, y+\alpha\,\Delta y)$$

Apply the mean value theorem to $F(\alpha)$ as

$$F(1)-F(0) \equiv f(x+\Delta x, y+\Delta y) - f(x, y) = \left.\frac{dF}{d\alpha}\right|_{\bar{\alpha}} (1-0)$$

$$= \left.\frac{\partial}{\partial x}\right|_{\bar{\alpha}} \Delta x + \left.\frac{\partial f}{\partial y}\right|_{\bar{\alpha}} \Delta y = \left.\frac{\partial f}{\partial x}\right|_{(\bar{x},\bar{y})} \Delta x + \left.\frac{\partial f}{\partial y}\right|_{(\bar{x},\bar{y})} \Delta y$$

where

$$(\bar{x}, \bar{y}) = (x+\bar{\alpha}\,\Delta x, y+\bar{\alpha}\,\Delta y) \qquad 0 \leq \alpha \leq 1$$

Q.E.D.

Theorem I.7. The alternating series

$$\sum_{n=0}^{\infty} (-1)^n a_n \qquad a_n \geq 0$$

converges if $a_{n+1} \leq a_n$, $\forall n$, and if $\lim_{n\to\infty} a_n = 0$.

PROOF. If N is odd we have

$$S_N = \sum_{n=0}^{N} (-1)^n a_n = (a_0 - a_1) + (a_2 - a_3) + \cdots + (a_{N-1} - a_N)$$

$$= a_0 - (a_1 - a_2) - (a_3 - a_4) - \cdots - a_N$$

so that

$$0 \leq S_N \leq a_0$$

and S_N increases (or remains) constant as N increases through odd integers. Since

$$S_{N+1} = S_N + a_{N+1}$$

we see that

$$\lim_{N\to\infty} S_N = S$$

by Th. I.2. Q.E.D.

Theorem I.8 (Weierstrass M-test). If $\{u_n(x)\}$ are defined for all $x \in [a, b]$ and if

$$|u_n(x)| \leq \alpha_n \qquad \forall n \qquad \text{and} \qquad \forall x \in [a, b]$$

and if

$$\sum_{n=0}^{\infty} \alpha_n < \infty$$

then

$$\sum_{n=0}^{\infty} u_n(x)$$

is absolutely and uniformly convergent for $x \in [a, b]$.

PROOF. The conditions stated clearly insure absolute convergence since

$$\left|\sum_{n=0}^{\infty} u_n(x)\right| \leq \sum_{n=0}^{\infty} |u_n(x)| \leq \sum_{n=0}^{\infty} \alpha_n < \infty$$

Furthermore the series converges uniformly since the remainder after N terms, $R_N(x)$, is just

$$|R_N(x)| = \left| \sum_{n=N+1}^{\infty} u_n(x) \right| \leq \sum_{n=N+1}^{\infty} |u_n(x)| \leq \sum_{n=N+1}^{\infty} \alpha_n$$

Since $\sum_{n=0}^{\infty} \alpha_n$ is given as being convergent, then

$$\sum_{n=N+1}^{\infty} \alpha_n < \varepsilon$$

where ε may be chosen arbitrarily small for N sufficiently large. Therefore we have the *uniform* bound

$$|R_N(x)| < \varepsilon \qquad \forall x \in [a, b]$$

for N sufficiently large. Q.E.D.

Theorem I.9. If $f(x)$ is defined and continuous for $x \geq R$, decreasing as x increases such that $\lim_{x\to\infty} f(x)=0$, then if $f(n)=a_n$ the series $\sum_{n=0}^{\infty} a_n$ converges or diverges according as

$$\int_R^\infty f(x)\,dx$$

converges or diverges.

PROOF. The conditions stated imply that $\{a_n\}$ are all positive decreasing for $n>N\geq R$ so that

$$0\leq S_M(N)\equiv \sum_{n=N+1}^{N+M} a_n = \sum_{n=N+1}^{N+M} f(n) \leq \int_N^{N+M} f(x)\,dx \leq \int_N^\infty f(x)\,dx < \infty$$

Since $S_M(N)$ is a positive, bounded, increasing function of M, Th. I.2 implies it has a finite limit. Similarly, we see that

$$\bar{S}_M(N)\equiv \sum_{n=N}^{N+M} a_n = \sum_{n=N}^{N+M} f(n) \geq \int_N^{N+M} f(x)\,dx$$

If the integral diverges as $M\to\infty$ so does the series. Q.E.D.

Theorem I.10. A uniformly convergent series of continuous functions defines a continuous function.

PROOF. Define

$$f(x)=\sum_{n=0}^{\infty} u_n(x) \qquad x\in[a, b]$$

and choose an $n\geq N$ sufficiently large so that the partial sum S_n satisfies

$$|f(x)-S_n(x)|<\varepsilon \qquad \forall x\in[a, b]$$

Since $S_N(x)$, being a *finite* sum of continuous functions, is itself a continuous function, we may choose an $x_0\in[a, b]$ and a δ such that

$$|S_N(x)-S_N(x_0)|<\varepsilon \qquad |x-x_0|<\delta$$

Since the series is given as being uniformly convergent, we have

$$\begin{aligned}|f(x)-f(x_0)| &= |f(x)-S_N(x)+S_N(x)-S_N(x_0)+S_N(x_0)-f(x_0)| \\ &\leq |f(x)-S_N(x)|+|S_N(x)-S_N(x_0)| \\ &\quad +|f(x_0)-S_N(x_0)|<3\varepsilon \qquad |x-x_0|<\delta\end{aligned}$$

Q.E.D.

Notice that Th. I.10 equivalently states that the following limits may be interchanged for a uniformly convergent sum of continuous functions.

$$\lim_{x\to x_0}\left[\lim_{N\to\infty}\sum_{n=0}^{N} u_n(x)\right]=\lim_{N\to\infty}\left[\lim_{x\to x_0}\sum_{n=0}^{N} u_n(x)\right]$$

By the same type argument used in the proof of Th. I.10 one can establish the following two theorems.

Theorem I.11. A uniformly convergent series of continuous functions can be integrated term by term

$$\int_a^b \sum_{n=0}^{\infty} u_n(x)\, dx = \sum_{n=0}^{\infty} \int_a^b u_n(x)\, dx$$

Theorem I.12. If the functions $\{u_n(x)\}$ have continuous derivatives and if the series

$$f(x) = \sum_{n=0}^{\infty} u_n(x)$$

is convergent, then

$$\frac{df}{dx} = \sum_{n=0}^{\infty} u_n'(x)$$

provided this last series is uniformly convergent.

Furthermore, the proofs of Ths. I.8 and I.10–I.12, which are stated for infinite series, can be given for improper integrals to yield the following results.

Theorem I.13. If $g(t)$ is continuous for $t \in [c, \infty)$ and if $f(x, t)$ is continuous for $t \in [c, \infty)$ whenever $x \in [a, b]$ and if

$$|f(x, t)| \leq g(t) \qquad \forall x \in [a, b]$$

then if

$$\int_c^{\infty} g(t)\, dt < \infty$$

it follows that

$$\int_c^{\infty} f(x, t)\, dt$$

is uniformly and absolutely convergent for all $x \in [a, b]$.

Theorem I.14. If $f(x, t)$ is continuous for $x \in [a, b]$, $t \in [c, \infty)$ and if

$$\int_c^{\infty} f(x, t)\, dt$$

is uniformly convergent for $x \in [a, b]$, then

$$F(x) = \int_c^{\infty} f(x, t)\, dt$$

is continuous for $x \in [a, b]$.

Theorem I.15. If $f(x, t)$ and $\partial f/\partial x$ is continuous for $x \in [a, b], t \in [c, \infty)$ and if

$$F(x) \equiv \int_c^{\infty} f(x, t)\, dt$$

converges while

$$\int_c^{\infty} \frac{\partial f(x, t)}{\partial x}\, dt$$

converges uniformly, then

$$\frac{dF(x)}{dx}=\int_c^\infty \frac{\partial f(x,t)}{\partial t}\,dt \qquad x\in[a,b]$$

Theorem I.16. If $u(x, y)$ and $v(x, y)$ are differentiable functions in a region R and functionally dependent in R, then

$$\frac{\partial(u,v)}{\partial(x,y)}\equiv\begin{vmatrix}\frac{\partial u}{\partial x} & \frac{\partial u}{\partial y}\\ \frac{\partial v}{\partial x} & \frac{\partial v}{\partial y}\end{vmatrix}=0$$

everywhere in R. Conversely, if $\partial(u, v)/\partial(x, y)\equiv 0$ in R and $\nabla u\neq 0$, $\nabla v\neq 0$, then u and v are functionally dependent in R.

PROOF. If u and v are functionally dependent, then

$$F[u(x,y), v(x,y)]\equiv 0$$

so that

$$\frac{\partial F}{\partial u}\frac{\partial u}{\partial x}+\frac{\partial F}{\partial v}\frac{\partial v}{\partial x}=0=\frac{\partial F}{\partial u}\frac{\partial u}{\partial y}+\frac{\partial F}{\partial v}\frac{\partial v}{\partial y}$$

Since $\partial F/\partial u$ and $\partial F/\partial v$ are not zero, we must require for consistency of the equations (cf. Section 2.2b) that

$$\frac{\partial(u,v)}{\partial(x,y)}=0$$

Conversely, since $\partial(u, v)/\partial(x, y)=0$ is equivalent to

$$\nabla u\times\nabla v=0$$

we see that the curves

$$u(x,y)=\text{const.}$$

$$v(x,y)=\text{const.} \tag{I.2}$$

coincide. If we consider this mapping from the (x, y) to the (u, v) plane, we see that Eqs. I.2 map the curves into a single point in the (u, v) plane. Since $\nabla u\neq 0$ and $\nabla v\neq 0$, then if we move along a curve in the (x, y) plane normal to the curves of Eq. I.2, u and v must change producing a curve in the (u, v) plane, for example,

$$u=h(v)$$

so that

$$0=u(x,y)-h[v(x,y)]\equiv F[u(x,y), v(x,y)] \qquad \text{Q.E.D.}$$

Theorem I.17. The transformations

$$u=f(x,y) \tag{I.3}$$

$$v=g(x,y) \tag{I.4}$$

are invertible for $x=x(u, v)$, $y=y(u, v)$ in a region R provided $\partial(u, v)/\partial(x, y)\neq 0$ anywhere in R.

PROOF. Since $\partial(u, v)/\partial(x, y) \neq 0$ it follows that $\partial u/\partial x$ and $\partial u/\partial y$ cannot both be zero. Assume that $\partial u/\partial y \neq 0$. This allows us to solve Eq. I.3 for y as

$$y = \phi(u, x)$$

so that

$$u - f[x, \phi(u, x)] = 0$$

and

$$\frac{\partial \phi}{\partial x} = -\frac{\partial f/\partial x}{\partial f/\partial \phi} = -\frac{\partial f/\partial x}{\partial f/\partial y}$$

Equation I.4 becomes

$$v - g[x, \phi(u, x)] = 0 \equiv h(u, v, x)$$

so that

$$x = \psi(u, v)$$

provided $\partial h/\partial x \neq 0$. But we have identically

$$\frac{\partial h}{\partial x} = -\frac{\partial g}{\partial x} - \frac{\partial g}{\partial \phi}\frac{\partial \phi}{\partial x} = -\frac{\partial g}{\partial x} + \frac{\partial g}{\partial \phi}\left(\frac{\partial f}{\partial x} \Big/ \frac{\partial f}{\partial y}\right)$$

$$= \frac{-\left[\dfrac{\partial g}{\partial x}\dfrac{\partial f}{\partial y} - \dfrac{\partial g}{\partial y}\dfrac{\partial f}{\partial x}\right]}{(\partial f/\partial y)} = \frac{-[\partial(u, v)/\partial(x, y)]}{(\partial f/\partial y)} \neq 0$$

where the last step follows from our assumption on $\partial(u, v)/\partial(x, y)$. Finally then we can write

$$x = \psi(u, v) \equiv x(u, v)$$

$$y = \phi(u, x) = \phi[u, \psi(u, v)] \equiv y(u, v) \qquad \text{Q.E.D.}$$

Appendix II

Lebesgue Integration and Functional Analysis

In this appendix we attempt to give the reader some idea what is meant by the expressions "set of measure zero," "almost everywhere," and "Lebesgue square integrable." We also list without any proofs several important results from Lebesgue theory and from functional analysis which we have referred to in the text, especially in Chapter 5.

Since we have not assumed any background in Lebesgue integration theory on the part of the student and since we cannot develop that background here, our discussion will not be rigorous or complete. In fact it will be a mathematician's nightmare of imprecision and vagueness. Nevertheless the student having no familiarity with measure theory may find it helpful. In any event the theorems stated will provide a convenient reference. The student who wishes to study these topics should consult the texts by Goffman, Graves, Riesz and Nagy, Schmeidler, Schwartz, and Taylor listed in the bibliography.

We begin with the concept of the *measure* of a set E of points. The *exterior Lebesgue measure* of a set E is the greatest lower bound of the sum of the lengths of a set of intervals covering E for all such coverings. This measure is often denoted by $m_e(E)$. When the exterior measure of a set E is zero we say E is a *set of measure zero.* Also the phrase "*almost everywhere*" is often used to mean "except on a set of measure zero." It follows that every subset of a set of measure zero also has measure zero, as does the sum of any denumerable number of sets of measure zero. That is, a set of points E has measure zero provided the set can be covered by a denumerable sequence of intervals whose total length can be made arbitrarily small. For example, all of the rational numbers contained on $[0, 1]$ form a set of measure zero while the entire (nondenumerably infinite) set of points on $[0, 1]$ do not have measure zero.

In defining the *Riemann integral* one writes

$$\int_a^b f(x)\,dx \equiv \lim_{\substack{n\to\infty \\ \max \Delta_j x \to 0}} \sum_{j=1}^{n} f(\bar{x}_j)\,\Delta_j x \qquad \textbf{(II.1)}$$

where $\{\Delta_j x\}=\{x_j - x_{j-1}\}$ is a subdivision of the interval $[a, b]$ and $x_{j-1} \leq \bar{x}_j \leq x_j$ and $\sum_{j=1}^{n} \Delta_j x = (b-a)$. The integral is defined provided that the limit of this sum exists for *all* possible subdivisions and any choice of the $\{\bar{x}_j\}$.

To define the *Lebesgue integral* one introduces the concept of a *measurable function.* A function $f(x)$, $x \in [a, b]$, is said to be measurable if the set of points is measurable for which

$$m \leq f(x) < M \qquad m < M$$

It is possible to show that such measurable functions are the limit almost everywhere of a sequence of step functions. If $\{\varphi_n(x)\}$ is a sequence of step functions bounded almost everywhere and if this sequence $\{\varphi_n(x)\}$ converges almost everywhere to $f(x)$, then the Lebesgue integral is defined as

$$\int_a^b f(x)\,dx \equiv \lim_{n\to\infty} \int_a^b \varphi_n(x)\,dx$$

and $f(x)$ is called summable or integrable in the Lebesgue sense. A function different from zero only on a set of measure zero can make no contribution to such an integral. The Riemann integral is a special case of the Lebesgue integral. Of course the Lebesgue integral often exists when the Riemann integral does not.

A function $f(x)$ is *absolutely continuous* provided

$$\lim_{m_e(E')=0} \sum_j [f(b_j) - f(a_j)] = 0$$

where

$$E' \equiv \sum_j (b_j - a_j)$$

and $\{(a_j, b_j)\}$ is *any* finite or infinite set of nonoverlapping subintervals of $[a, b]$. If $f(x)$ satisfies a *Lipschitz condition*

$$|f(x_1) - f(x_2)| \leq C\,|x_1 - x_2| \qquad \forall x_1, x_2 \in [a, b]$$

then it will be absolutely continuous.

The space $\mathscr{L}_2[a, b]$ consists of all functions $f(x)$ such that

$$\int_a^b |f(x)|^2\,dx < \infty$$

where the integral is defined in the Lebesgue sense.

We now list several important theorems.

Theorem II.1 (Riesz-Fischer theorem). A sequence of functions $\{f_n\} \in \mathscr{L}_2$ converges to an $f \equiv \lim_{n\to\infty} f_n \in \mathscr{L}_2$ if and only if

$$\|f_n - f_m\| \to 0$$

as $m, n \to \infty$.

Theorem II.2 (Fubini's theorem). If $f(x, y)$ is summable, then the order of integration may be interchanged as

$$\int_a^b \left[\int_c^d f(x, y)\,dy\right] = \int_c^d \left[\int_a^b f(x, y)\,dx\right] dy \equiv \int_a^b \int_c^d f(x, y)\,dx\,dy$$

Theorem II.3 (Beppo-Levi's theorem). The infinite series

$$\sum_{n=0}^{\infty} u_n(x)$$

of summable functions $\{u_n(x)\}$ converges almost everywhere to a summable function for $x \in [a, b]$ and can be integrated term by term provided

$$\sum_{n=0}^{\infty} \int_a^b |u_n(x)|\, dx < \infty$$

Theorem II.4. If $f(x)$ is summable, then

$$f(x) = \frac{d}{dx} \int_a^x f(\xi)\, d\xi$$

almost everywhere.

Theorem II.5. If $f(x)$ is absolutely continuous, then

$$\int_a^x f'(\xi)\, d\xi = f(x) - f(a)$$

Theorem II.6. If $f(x)$ is summable for $x \in [a, b]$ and $g(y)$ is absolutely continuous for $y \in [c, d]$, then

$$\int_a^b f(x)\, dx = \int_c^d f[g(y)]g'(y)\, dy$$

where

$$a = g(c) \qquad b = g(d)$$

Theorem II.7. The Hermite functions,

$$\varphi_n(x) = \frac{(-1)^n}{\sqrt{2^n n!}\, \pi^{1/4}}\, e^{x^2/2} \frac{d^n(e^{-x^2})}{dx^n} \qquad n = 0, 1, 2, \ldots$$

form a complete orthonormal set of functions for $\mathcal{L}_2(-\infty, \infty)$.

Notice that Th. II.7 shows that $\mathcal{L}_2(-\infty, \infty)$ (and therefore any $\mathcal{L}_2[a, b]$) is a separable Hilbert space (cf. Section 1.5) and also that any $f(x) \in \mathcal{L}_2$ can be approximated arbitrarily closely in the norm by a continuous function

$$g_N(x) = \sum_{n=0}^{N} \langle \varphi_n \mid f \rangle \varphi_n(x)$$

since

$$\|f\| = \|f - g_N + g_N\| \leq \|f - g_N\| + \|g_N\|$$

or

$$\big|\, \|f\| - \|g_N\|\, \big| \leq \|f - g_N\| < \varepsilon$$

where the last inequality is simply a statement of the completeness of the $\{\varphi_n\}$.

SYMBOLS AND NOTATIONS

$\Rightarrow$	**implies**
$\Leftarrow$	**is implied by**
$\Leftrightarrow$	**if and only if**
$\exists$	**there exist(s)**
$\forall$	**for all**
$\nexists$	**there does not exist**
$\ni$	**such that**
$\equiv$	**is defined as, or, is identically equal to**
$\perp$	**orthogonal to**
$\therefore$	**therefore**
$\in$	**belongs to**
$\notin$	**does not belong to**
$[a, b]$	**a closed interval from *a* to *b* including the end points**
(a, b)	**an open interval from *a* to *b* not including the end points**
w.r.t.	**with respect to**
l.h.s.	**left hand side (of an equation)**
r.h.s.	**right hand side (of an equation)**
Q.E.D.	**quod erat demonstrandum (indicates the end of a proof)**
$\subset$	**is contained in; is a subset of**
$\{\ \}$	**a sequence or set**
$\{a_{ij}\}$	**the elements of a matrix A**
$\mathcal{R}(L)$	**the range of the operator *L***

Bibliography

GENERAL WORKS DEALING WITH APPLIED MATHEMATICS

Arfken, George, *Mathematical Methods for Physicists* (2nd Ed.) (Academic Press, 1970).

Byron, Frederick W. and Robert W. Fuller, *Mathematics of Classical and Quantum Physics* (2 Vols.) (Addison-Wesley, 1969).

Courant, R. and D. Hilbert, *Methods of Mathematical Physics* (2 Vols.) (Wiley-Interscience, 1953).

Dettman, John W., *Mathematical Methods in Physics and Engineering* (2nd Ed.) (McGraw-Hill, 1969).

Friedman, Bernard, *Principles and Techniques of Applied Mathematics*, (Wiley, 1956).

Jackson, J. D., *Mathematics of Quantum Mechanics* (W. A. Benjamin, 1962).

Lass, Harry, *Elements of Pure and Applied Mathematics* (McGraw-Hill, 1957).

Mathews, Jon and R. L. Walker, *Mathematical Methods of Physics* (W. A. Benjamin, 1969).

Morse, Philip M. and Herman Feshbach, *Methods of Theoretical Physics* (2 Vols.) (McGraw-Hill, 1953).

Schwartz, Laurent, *Mathematics for the Physical Sciences* (Hermann, 1966).

ANALYSIS AND LEBESGUE THEORY OF INTEGRATION

Buck, R. Creighton, *Advanced Calculus* (McGraw-Hill, 1956).

Goffman, Casper, *Real Functions* (Rinehart and Co., 1953).

Graves, Lawrence M., *The Theory of Functions of Real Variables* (McGraw-Hill, 1956).

Kaplan, Wilfred, *Advanced Calculus* (Addison-Wesley, 1952).

Lass, Harry, *Vector and Tensor Analysis* (McGraw-Hill, 1950).

Phillips, H. B., *Vector Analysis* (Wiley, 1933).

LINEAR SPACES AND FUNCTIONAL ANALYSIS

Beals, Richard, *Topics in Operator Theory* (University of Chicago Press, 1971).

Birkhoff, Garrett and Saunders MacLane, *A Survey of Modern Algebra* (The MacMillan Co., 1953).

Halmos, Paul R., *Finite-Dimensional Vector Spaces* (D. Van Nostrand, 1958).

Halmos, Paul R., *Introduction to Hilbert Space* (Chelsea Publishing Co., 1951).

Marcus, Marvin and Henryk Minc, *A Survey of Matrix Theory and Inequalities* (Allyn and Bacon, 1964).

Murdock, D. C., *Linear Algebra for Undergraduates* (Wiley, 1957).

Riesz, Frigyes and Bela Sz.-Nagy, *Functional Analysis* (2nd Ed.) (Frederick Ungar Publishing Co., 1955).

Schmeidler, Werner, *Linear Operators in Hilbert Space* (Academic Press, 1965).

Smirnov, V. I., *Linear Algebra and Group Theory* (Dover Publications, 1970).

Taylor, Angus E., *Introduction to Functional Analysis* (Wiley, 1958).

FOURIER SERIES AND INTEGRAL TRANSFORMS

Byerly, William E., *An Elementary Treatise on Fourier Series* (Dover Publications, 1959).

Churchill, Ruel V., *Fourier Series and Boundary Value Problems* (McGraw-Hill, 1941).

Hobson, E. W., *Theory of Spherical and Ellipsoidal Harmonics* (Cambridge University Press, 1931).

Sneddon, Ian N., *Fourier Transforms* (McGraw-Hill, 1951).

Titchmarsh, E. C., *Introduction to the Theory of Fourier Integrals* (Oxford University Press, 1937).

DIFFERENTIAL AND INTEGRAL EQUATIONS

Ince, E. L., *Ordinary Differential Equations* (Dover Publications, 1956).

Lovitt, W. V., *Linear Integral Equations* (McGraw-Hill, 1924).

Mikhlin, S. G., *Integral Equations* (The MacMillan Co., 1964).

Muskhelishvili, N. I., *Singular Integral Equations* (P. Noordhoff N. V., 1953).

Omnès, R., *Il Nuovo Cimento 8*, 316 (1958).

Titchmarsh, E. C., *Eigenfunction Expansions* (2 Vols.), (Oxford University Press, 1962).

Tricomi, F. G., *Integral Equations* (Wiley-Interscience, 1957).

CALCULUS OF VARIATIONS

Bliss, G. A., *Calculus of Variations* (Open Court, 1925).

Courant, R., and Robbins, H., "Plateau's Problem," in *The World of Mathematics* (Ed. J. R. Newman), pp. 901–909 (Simon and Schuster, 1956).

Sagan, H., *Introduction to the Calculus of Variations* (McGraw-Hill, 1969).

Weinstock, Robert, *Calculus of Variations* (McGraw-Hill, 1952).

Yourgrau, Wolfgang and Stanley Mandelstam, *Variational Principles in Dynamics and Quantum Theory* (Isaac Pitman and Sons, 1968).

COMPLEX VARIABLES AND SPECIAL FUNCTIONS

Boas, Ralph P., *Entire Functions* (Academic Press, 1954).

Churchill, Ruel V., *Introduction to Complex Variables* (McGraw-Hill, 1948).

Erdelyi, A., *Asymptotic Expansions* (Dover Publications, 1956).

Erdelyi, A. (Ed.), *Higher Transcendental Functions* (2 Vols.), *Tables of Integral Transforms* (3 Vols.) (McGraw-Hill, 1954).

Gradshteyn, I. S., and Ryzhik, I. M., *Table of Integrals, Series, and Products* (Academic Press, 1965).

Nehari, Z., *Conformal Mapping* (McGraw-Hill, 1952).

Sommerfeld, Arnold, *Partial Differential Equations in Physics* (Academic Press, 1949).

Szegö, G., *Orthogonal Polynomials* (American Mathematical Society, 1959).

Talman, James D., *Special Functions, A Group Theoretic Approach* (W. A. Benjamin, 1968).

Titchmarsh, E. C., *The Theory of Functions* (Oxford University Press, 1939).

Townsend, E. J., *Functions of a Complex Variable* (Henry Holt and Co., 1915).

Whittaker, E. T. and G. N. Watson, *A Course of Modern Analysis* (Cambridge University Press, 1927).

GROUP THEORY

Boerner, Hermann, *Representations of Groups* (North-Holland Publishing Co., 1963).

Cohn, P. M., *Lie Groups* (Cambridge University Press, 1957).

Edmonds, A. R., *Angular Momentum in Quantum Mechanics* (Princeton University Press, 1957).

Eisenhart, L. P., *Continuous Groups of Transformations* (Dover Publications, 1961).

Hamermesh, Morton, *Group Theory and Its Applications to Physical Problems* (Addison-Wesley, 1962).

Heine, Volker, *Group Theory in Quantum Mechanics* (Pergamon Press, 1960).

Lipkin, Harry J., *Lie Groups for Pedestrians* (North-Holland Publishing Co., 1966).

Pontrjagin, L., *Topological Groups* (Princeton University Press, 1939).

Racah, G., "Group Theory and Spectroscopy," in *Ergebnisse der exakten Naturwissenschaften* (37 Band) (Springer-Verlag, 1965).

Rose, M. E., *Elementary Theory of Angular Momentum* (Wiley, 1957).

Ruhl, W., *The Lorentz Group and Harmonic Analysis* (W. A. Benjamin, 1970).

Tinkham, Michael, *Group Theory and Quantum Mechanics* (McGraw-Hill, 1964).

Wigner, Eugene P., *Group Theory* (Academic Press, 1959).

PHYSICAL APPLICATIONS

Goldstein, Herbert, *Classical Mechanics* (Addison-Wesley, 1950).

Mandl, F., *Quantum Mechanics* (2nd Ed.) (Butterworths, 1957).

Messiah, Albert, *Quantum Mechanics* (2 Vols.) (North-Holland Publishing Co., 1966).

Roman, Paul, *Theory of Elementary Particles* (2nd Ed.) (North-Holland Publishing Co., 1961).

Index